SENSORY
COMMUNICATION

SENSORY COMMUNICATION

CONTRIBUTIONS TO THE SYMPOSIUM ON
PRINCIPLES OF SENSORY COMMUNICATION
July 19 – August 1, 1959, Endicott House, M.I.T.

WALTER A. ROSENBLITH, *Editor*

 THE M.I.T. PRESS
MASSACHUSETTS INSTITUTE OF TECHNOLOGY
CAMBRIDGE, MASSACHUSETTS

The symposium reported herein was partially supported by the Air Force Office of Scientific Research under contract AF 49 (638)-421, monitored by the Information Sciences Division, Directorate of Mathematical Sciences. AFOSR 796.

ISBN 978-0-262-51842-0

Library of Congress Catalog No. 61-8798

PRINTED IN THE UNITED STATES OF AMERICA

Preface

The postwar period has seen a surfeit of interdisciplinary symposia. These symposia were a manifestation of several not unrelated facts: the scientific universe was expanding at a rate that was hard to grasp or assess, and the frontiers between the sciences seemed to be moving almost as much as science itself. Communication between disciplines had become more and more sketchy, and even within a given discipline communication had often become problematical. At the same time many of the traditional fields among the life and behavioral sciences found themselves profoundly affected by the technological advances that had their origin in the physical sciences. Experts in these technologies acquired a taste for the challenges of these less structured fields, but they sometimes lacked perspective and respect for the toughness of the problems that they proposed to tackle.

It was with these difficulties that the numerous symposia attempted to grapple. Like other scientific meetings these interdisciplinary symposia were unequal in quality. At the outset the freshness of the confrontation, the unorthodox approaches, promised a great deal. But as time wore on it became clear that while the symposia were productive of suggestive ideas they were no substitute for the workaday interaction of experimentation and theory making. Responsible workers in the behavioral and life sciences became increasingly squeamish about the one-day symposium in which mathematicians, physicists, and engineers vented frequently the belief that the intelligent application of some rather elementary notions from mathematics and physics should yield spectacular results in the solution of a variety of thorny problems. The brain, perception, learning, thinking, that is to say, all topics that in Warren Weaver's phrase are characterized by "organized complexity," were the explicit target of these optimistic predictions.

The biological scientists who had been trained in the more traditional approaches adopted a retaliatory stance in which they were only too often satisfied to point to the truly appalling complexities of their experimental data. The two parties separated, leaving behind numerous volumes that were represented to be high-fidelity transcriptions of the tape recordings of these encounters. Relatively few among the symposiasts got beyond the hors d'oeuvres and stayed, as it were, for dinner.

And yet these intellectual skirmishes made their imprint upon the kind of questions that laboratory and computer scientists began to ask and on the results they obtained. Here were at least partial payments on the perhaps overgenerous promissory notes.

Throughout the nineteenth century sensory problems had proved attractive to a variety of scientists who had also worked in the physical sciences— Young, Ohm, Helmholtz, Fechner, Mach, to name just a few. The twentieth century perpetuated and modified this tradition. Electronic and communications engineers concerned with the telephone and with radio, television, and photography discovered their stake in the understanding of man's sensory performance. The engineers were largely responsible for pointing out how fruitful the application of theoretical concepts à la Wiener-Shannon might be in an analysis of human information handling. The advent of large computers and of complex weapons systems put a premium on our ability to couple man to these machines that seem so capable of simulating and amplifying many aspects of man's intelligent behavior.

Thus there arose, albeit in a changed context, a renewed concern with perception. Attempts to program computers to recognize certain patterns, to translate from one language into another, and more generally to solve certain classes of logical problems led to an emphasis upon man's cognitive functions, to a reconsideration of the role that sensory processes play in this type of symbolic behavior. But the formulation that was couched in terms of the processing of sensory information enabled the scientists of the post-World War II period to transcend (perhaps only to bypass or even sidestep) the traditional dichotomy between sensation and perception.

It thus appeared worth while to bring together a group of scientists and engineers who had themselves done research on problems of sensory communication and who were willing to listen to neurophysiologists expound up-to-date neurophysiology, or psychophysicists talk about contemporary psychophysics, instead of being satisfied with their own version of the other man's science.

These considerations led late in 1957 to the not entirely original idea of a multidisciplinary international symposium on principles of sensory communication. Before too long the Office of Scientific Research of the U. S. Air Force expressed interest in sponsoring such a meeting, provided that the collaboration of a representative group of scientists could be assured. In the fall of 1958 an international organizing committee was formed consisting of F. Bremer, T. H. Bullock, A. Fessard, R. Galambos, W. D. Neff, C. Pfaffmann, S. S. Stevens, and W. A. Rosenblith.

Thanks to the active support of this committee we were able to assemble a group whose broad-ranging competence in the area of sensory communication was unquestionable. We regretted, however, that a variety of circumstances forced a few scientists to decline our invitation and left us with a somewhat less "balanced representation" than we had hoped for.

If communication on sensory communication was to take place, the number of symposium participants had to be limited. The number of full-fledged participants who could be accommodated was further limited by our conviction that appropriate living and dining arrangements are requisite for the effective interchange of ideas. We were grateful, therefore, to our guests (practically all of whom were from the Boston area) who were willing to attend single sessions only during the first week, and who nevertheless participated so fruitfully in the deliberations of the working groups during the second week. We appreciated also the helpful participation of H. O. Wooster, L. Butsch, H. O. Parrack, H. Savely, and J. Steele, who attended the meetings as representatives of the sponsor, the U. S. Air Force, whose generous, nondirective support it is a pleasure to acknowledge.

The symposium was held during the last two weeks of July 1959 at M.I.T.'s Endicott House in Dedham, a Boston suburb. Sunday night Professor Jerome B. Wiesner, Director of the Research Laboratory of Electronics and Chairman of the Steering Committee of its Center for Communication Sciences, opened the proceedings by extending the Institute's official welcome. Thereafter, the participants and guests at this inaugural session listened with evident pleasure to E. G. Boring, Edgar Pierce Professor of Psychology Emeritus at Harvard University, who applied his wisdom, insight, and wit to a review of the history of afferent communication.

After this auspicious send-off, the more formal sessions started on Monday morning and continued throughout the first week, every morning and every evening after dinner. The afternoons were left open for laboratory visits,* informal discussions, walks in the park, or dips in the Endicott House pool. The chapters in this volume correspond by and large to the quasi-formal presentations of the first week. After a week-end break most of the participants reassembled in several working and discussion groups. There was no attempt to record these exchanges of ideas and data on tape, and we can only hope that the section on comments and the final versions of chapters in this volume reflect some of the many cogent points that were raised, often in an informal manner.

Of the several pleasant social occasions that accompanied the symposium one deserves special mention: a reception and tea at Brandegee House, tendered to the participants by the American Academy of Arts and Sciences. Hudson Hoagland and Ralph Burhoe acted as hosts on behalf of the Academy Council and the Committee on Informal Gatherings. Hoagland's warm words of welcome evoked a delightfully spirited response from W. A. H. Rushton, F.R.S.

Rushton said, "It was during the American Revolution that this Academy was founded, and I suppose the chief factors were the activity of great Boston characters and the disappearance of the King of England. You never know what circumstances will found an Academy, and with our Royal Society (upon which you said your Academy was modeled) the conditions were just the other way. For with the death of Oliver Cromwell the King

* It is a pleasure to thank E. H. Land, F. A. Webster, D. R. Griffin, K. D. Roeder, J. Y. Lettvin, H. R. Maturana, S. W. Kuffler, D. H. Hubel, and T. N. Wiesel, who arranged these laboratory visits and demonstrations, which were much appreciated by the symposium participants.

of England reappeared and it was the Puritans who went—to Boston as I suppose.

"The descendants of these Boston emigrees—John Adams and the rest—were the men who gave you your noble and liberal Charter, parts of which we should like to take as an ideal for the matter and manner of our meetings. For in matter we too wish 'to promote and encourage medical discoveries; mathematical disquisitions; philosophical inquiries and experiments,' and in manner I hope that our meetings 'may tend to advance the interest, honour, dignity and happiness of free, independent and virtuous people.' "

A given scientific gathering owes a large share of its success to a great number of people, many of whom are nonscientists. Acknowledgment of this debt would gain in meaning were it accompanied by a functional flow diagram of auxiliary activities. How otherwise can one do justice to the efforts of those who struggle with the ever-present inadequacies of slide projectors, who act as chauffeurs and guides, who perform the multifarious duties that make intellectual discussions at an international scale possible? This meeting had its oversize and overtime share of such efforts. No one showed more dedication in planning and operation than my assistant, Mrs. Aurice Albert. Her devotion to the success of the symposium fully merited the high tribute the participants paid her in the final session.

The atmosphere of the symposium owed much to the good management of Mrs. Gertrude Winquist, who presides over Endicott House. Most of the participants lived in these comfortable and congenial surroundings, and all of them socialized there repeatedly.

The staff of the Research Laboratory of Electronics, under the direction of R. A. Sayers, provided many services with customary efficiency and dispatch. My colleagues from the Laboratory's Communications Biophysics group unselfishly gave of their time in a variety of ways; for example, they provided several demonstrations of the processing of neuroelectric data by means of high-speed electronic computers. My colleague, Dr. Eda Berger Vidale, contributed in many ways to the production of this volume: in particular, she spent untold hours in the compilation of the subject index.

Finally, I should like to express my personal gratitude to Geraldine Stone. She has been much more than an assistant to the editor of this volume or an intermediary between the authors and the editor. Her knowledgeable, persistent, sensitive, and indefatigable efforts are visible on almost every page of this book. Miss Stone collaborated with Miss Constance D. Boyd of the M.I.T. Press, who contributed her measure of serious and patient labors. No editor could hope for more capable and dedicated associates.

WALTER A. ROSENBLITH

Cambridge, Massachusetts
May, 1961

Participants

Fred Attneave
Department of Psychology
University of Oregon

H. B. Barlow
Physiological Laboratory
Cambridge

Lloyd M. Beidler
Division of Physiology
Florida State University

M. A. Bouman
Institute for Perception RVO-TNO
Soesterberg, The Netherlands

Robert M. Boynton
Department of Psychology
University of Rochester

Mary A. B. Brazier
Brain Research Institute
University of California at Los Angeles

F. Bremer
Laboratoire de Pathologie Générale
Université de Bruxelles

Theodore H. Bullock
University of California at Los Angeles

P. Buser
Centre de Physiologie nerveuse du C.N.R.S.
Faculté des Sciences, Paris

Colin Cherry
Electrical Engineering Department
Imperial College, London

Hallowell Davis
Central Institute for the Deaf
St. Louis, Missouri

Gösta Ekman
Psychological Laboratory
University of Stockholm

Peter Elias
Department of Electrical Engineering
Massachusetts Institute of Technology

A. Fessard
Collège de France, Paris

Robert Galambos
Department of Neurophysiology
Walter Reed Army Medical Center

Frank A. Geldard
University of Virginia

Timothy H. Goldsmith
Department of Zoology, Yale University

Raúl Hernández-Peón
Brain Research Unit
Medical Center of Mexico City

Richard Jung
Abteilung für klinische Neurophysiologie
Universität Freiburg/Br., Germany

Yasuji Katsuki
Department of Physiology
Tokyo Medical and Dental University

Wolf D. Keidel
Physiologisches Institut
Universität Erlangen, Germany

Sven Landgren
Department of Physiology
Veterinärhögskolan, Stockholm

Affiliations given are current in 1960–61.

J. C. R. Licklider
Bolt Beranek and Newman Inc.
Cambridge, Massachusetts

Donald B. Lindsley
Department of Psychology
University of California at Los Angeles

D. M. MacKay
Department of Communication
University College
Keele, Staffordshire, England

Vernon B. Mountcastle
Department of Physiology
The Johns Hopkins University
School of Medicine

William D. Neff
Laboratory of Physiological Psychology
University of Chicago

Carl Pfaffmann
Department of Psychology, Brown University

Irwin Pollack
Operational Applications Laboratory
Air Force Cambridge Research Center
Bedford, Massachusetts

Floyd Ratliff
The Rockefeller Institute, New York

Werner Reichardt
Forschungsgruppe Kybernetik
Max-Planck-Institut für Biologie
Tübingen, Germany

Kenneth D. Roeder
Department of Biology, Tufts University

Jerzy E. Rose
Laboratory of Neurophysiology
Medical School, University of Wisconsin

Walter A. Rosenblith
Center for Communication Sciences
Research Laboratory of Electronics
Massachusetts Institute of Technology

Burton S. Rosner
West Haven Veterans Administration Hospital
and Yale University School of Medicine

W. A. H. Rushton
Trinity College, Cambridge

Ulf Söderberg
Nobel Institute for Neurophysiology
Karolinska Institutet, Stockholm

S. S. Stevens
Harvard University

Hessel de Vries *
Natuurkundig Laboratorium
Groningen, The Netherlands

Patrick D. Wall
Department of Biology and
Center for Communication Sciences
Massachusetts Institute of Technology

Clinton N. Woolsey
Laboratory of Neurophysiology
Medical School, University of Wisconsin

Yngve Zotterman
Department of Physiology
Veterinärhögskolan, Stockholm

* Hl. de Vries died in December 1959.

Contents

SENSORY
COMMUNICATION

1

S. S. STEVENS

Harvard University

The Psychophysics of Sensory Function

An inquiry into the nature of the sensory process begins properly with psychophysics, the hundred-year-old discipline concerned with the responses that organisms make to the energies of the environment. We live in a restless world of energetic forces, some of which affect us and some of which, like radio waves, impinge upon us and pass unnoticed because we have no sense organs able to transduce them. But we see lights, hear sounds, taste substances, and smell vapors, and it is these elementary facts of psychophysics that stir our interest in the anatomy and physiology of the mechanisms that make sensation possible. An orderly and systematic account of sensory communication must include a delineation of *what* is perceived as well as an explanation of *how* perception is accomplished. In this sense, psychophysics defines the challenge: it tells what the organism can do and it asks those who are inspired by such mysteries to try, with scalpel, electrode, and test tube, to advance our understanding of how such wonders are performed.

It must be confessed at the outset that psychophysics has often failed to do its part of the job with distinction. Its task is not easy. For one thing, long-standing prejudices, derived in great measure from a chronic dualistic metaphysics, have triggered a variety of stubborn objections whenever it has been proposed that sensation may be amenable to orderly and quantitative investigation. You cannot, the objectors complain, measure the inner, private, subjective strength of a sensation. Perhaps not, in the sense the objectors have in mind, but in a different and very useful sense the strength of a sensation can, as we shall see, be fruitfully quantified. We must forgo arguments about the private life of the mind and ask sensible objective questions about the input-output relations of sensory transducers as these relations are disclosed in the behavior of experimental organisms, whether men or animals.

1

Another difficulty is that psychophysics had an unfortunate child-hood. Although Plateau in the 1850's made a half-hearted attempt to suggest the proper form of the function relating apparent sensory intensity to stimulus intensity, he was shouted down by Fechner, who saddled the infant discipline with the erroneous "law" that bears his name (see Stevens, 1957*b*). Perhaps the hardest task before us is to clear the scientific bench top of the century-old dogma that sensation intensity grows as the logarithm of stimulus intensity (Fechner's law). The relation is not a logarithmic function at all. On more than a score of sensory continua it has now been shown that apparent, or subjective, magnitude grows as a *power function* of stimulus intensity, and the exponents of the power function have been found to range from about 0.33 for brightness to about 3.5 for electric shock (60 cps) applied to the fingers. There seems to exist, in other words, a simple and pervasive psychophysical law, a law that was once conjectured by Plateau and later abandoned by him, a law that is congenial not only to the mounting empirical evidence, but also to certain reasonable principles of theory construction (Luce, 1959). There will be more to say about the power law, but first a few words about Fechner.

The misconception began when Fechner, in 1850, espoused the view that error itself provides a unit of measurement. He called it the just noticeable difference (jnd). Under most circumstances the jnd is a statistical concept, a measure of the dispersion or variability of a discriminatory response, in short, a measure of error. In deriving his logarithmic law, Fechner made the erroneous assumption that error is constant all up and down the psychological scale. Although he was willing to assume that the stimulus level error is relative, that is, $\Delta\phi = k\phi$ (Weber's law), he assumed that at the psychological level $\Delta\psi$ equals a constant. From these two assumptions he derived the relation $\psi = k \log \phi$, and thereby caused much mischief.

It is curious indeed that Fechner, a physicist, should have assumed that error, or variability of judgment, is constant all up and down the psychological continuum. Most variables do not behave that way. On the continua with which a physicist most often deals, error is usually not constant, but tends to vary with magnitude. It is percentage error that typically stays constant: precision can generally be stated as one part in so many.

Suppose Fechner had taken this as his model, not only for the stimulus jnd $\Delta\phi$, but also for the subjective jnd $\Delta\psi$. He then could have written

$$\Delta\psi/\psi = k\Delta\phi/\phi$$

from which it would follow that the psychological magnitude ψ is a power function of the physical magnitude ϕ. But he fought off this suggestion when it was first made (by Brentano), and with Fechner's temporary victory psychophysics entered upon a period of futility during which there seemed to be no more interesting work to do than measure the jnd. And the logarithmic law became "an idol of the den".

So much for the past. Since the 1930's psychophysics has been staging a comeback. New interest in the age-old problem of sensory response has been kindled by the invention of procedures for assessing the over-all input-output operating characteristics of the intact sensory system. These methods show that sensory response grows according to a power law. So rarely does it happen in the study of behavior that a simple relation can be shown to hold under many diverse kinds of stimulation, that the widespread invariance of the power law becomes a matter of large significance.

Measurement

The problem of the laws that govern the reactions of sentient organisms is intimately bound up with the problem of measurement. Since the theory of measurement was thoroughly explored in another symposium (Stevens, 1959b), it need not divert us here. It may be helpful, however, to refer to Table 1, which attempts a systematic classification of scales of measurement in a compact form (Stevens, 1946a, 1951). The four scales listed, *nominal, ordinal, interval,* and *ratio,* are those most commonly used in the business of science, and all of them get involved in research on sensory communication.

The nominal scale, the most general of the lot, is not always thought of as a form of measurement, mainly because names or letters, rather than numbers, are most often employed to designate the categories or classes used in nominal scaling. Yet this ubiquitous and important form of measurement goes on constantly, for it includes the process of identifying and classifying. Mostly we take only a casual interest in such problems, but our interest has a way of turning into animated curiosity when it becomes a question of doing the detective work necessary to pin the proper labels on the functional parts of the central nervous system. The identification of the "areas" associated with this or that sensory process constitutes a lively exercise in nominal scaling. And needless to say, much of our scientific effort in this field never goes beyond the essential and basic nominal level. The ordinary

Table 1.   A Classification of Scales of Measurement

Measurement is the assignment of numbers to objects or events according to rule. The rules and the resulting kinds of scales are tabulated below. The basic operations needed to create a given scale are all those listed in the second column, down to and including the operation listed opposite the scale. The third column gives the mathematical transformations that leave the scale form invariant. Any number x on a scale can be replaced by another number x' where x' is the function of x listed in column 2. The fourth column lists, cumulatively downward, examples of statistics that show invariance under the transformations of column 3 (the mode, however, is invariant only for discrete variables).

Scale	Basic empirical operations	Mathematical group-structure	Permissible statistics (invariantive)	Typical examples
Nominal	Determination of equality	Permutation group $x' = f(x)$ where $f(x)$ means any one-to-one substitution	Number of cases Mode "Information" measures Contingency correlation	"Numbering" of football players Assignment of type or model numbers to classes
Ordinal	Determination of greater or less	Isotonic group $x' = f(x)$ where $f(x)$ means any increasing monotonic function	Median Percentiles Order correlation (type 0: interpreted as a test of order)	Hardness of minerals Grades of leather, lumber, wool, and so forth Intelligence-test raw scores
Interval	Determination of the equality of intervals or of differences	Linear or affine group $x' = ax + b$ $a > 0$	Mean Standard deviation Order correlation (type I: interpreted as r) Product moment (r)	Temperature (Fahrenheit and Celsius) Position on a line Calendar time Potential energy Intelligence-test "standard scores" (?)
Ratio	Determination of the equality of ratios	Similarity group $x' = cx$ $c > 0$	Geometric mean Harmonic mean Per cent variation	Length, density, numerosity, time intervals, work, and so forth Temperature (Kelvin) Loudness (sones) Brightness (brils)

determination of thresholds, which involves the categorization of stimuli into classes (for example, seen and not seen), is another important instance of nominal scaling (Stevens, 1958).

The key to the nature of the four kinds of scales lies in a powerful but simple principle: the concept of invariance. When we have carried out a series of empirical operations, balancings, comparisons, orderings, and so on, we assign a set of numbers to reflect the outcome of the operations. This is the essence of measurement. But what kind of measurement have we achieved? That depends on the answer to the decisive question: in what ways can the scale numbers be transformed without loss of empirical information? As shown in Table 1, each of the scales has its group of permissible transformations.

The ratio scale, the scale of greatest interest, allows only multiplication by a constant, as when we change from inches to centimeters. No more general transformation is allowed. If an arbitrary constant were to be added to the measured diameters of a set of nerve fibers, for example, the resulting numbers would tell us less than we knew before. We would have lost some valuable information, namely, our knowledge of the *ratios* among the fiber diameters. Which fiber is twice as thick as some given fiber would now, with the altered scale values, be impossible to tell. In general, therefore, the more restricted are the admissible transformations, the more the scale is able to tell us.

As regards the measurement of sensation, the schema of Table 1 suggests that our aspiration should be to measure, where possible, on a ratio scale. This would call for assigning numbers to sensory magnitudes in such a way that anything more drastic than multiplication by a constant would result in a loss of information. Several variations on such a procedure have been elaborated (Stevens, 1958), but before we consider the resulting scales, certain distinctions need to be made. (Some nominal scaling needs to be done!)

Sensory Qualities

An obvious thing about sensations is that they differ in both kind and amount. Sweet is different from sour, but both may vary from weak to strong. The sensory qualities get named and classified (nominal scaling), but we try to measure the subjective intensities on higher-order scales.

The distinctive quality aroused by a given sensory excitation presents a baffling problem for which no plausible explanation is yet available. Why does a sound differ from a taste in the way it does? This qualitative aspect of the sensory world confronts us with a baffling succession of discontinuous leaps as we go from one sense modality to another, and no-one seems to know why. On the other hand, it is at precisely this level that the anatomist and the neurophysiologist join the game and perform some of their most effective work in tracing pathways for the various modalities, and even for some of the separate qualities within a modality. Clearly these problems of topography need to be clarified before an understanding of the sensory mechanisms can be anchored in the soup and substance of neural process. Connections by themselves may not explain it all, but connections are there, and it seems improbable that they count for nothing.

Two Kinds of Continua

Psychophysics progresses beyond the elementary task of naming sensory qualities as soon as it becomes concerned with sensations that appear to lie on a continuum of some sort. A continuum seems clearly to be involved when sensations vary in strength or intensity, but certain other attributes of sensory response seem also to form continua in the ordinary sense of the term.

It would greatly simplify the mission of psychophysics if all the sensory continua obeyed the same rules, and did so in an invariant fashion. It turns out, however, that a basic distinction needs to be made between two kinds of continua, prothetic and metathetic. Loudness, for example, is prothetic; pitch is metathetic. An important difference between the psychophysical functions governing pitch and loudness is this: the jnd for pitch represents a constant distance on the scale of subjective pitch, measured in *mels*, whereas the jnd for loudness represents an increasing distance on the subjective scale of loudness, measured in *sones* (Stevens and Volkmann, 1940). In other words, provided they are measured in subjective units, the jnd for pitch is constant, but the jnd for loudness grows rapidly larger as loudness is increased. The uniformity of sensitivity or resolving power on the pitch continuum and the nonuniformity on the loudness continuum entail several other functional differences between pitch and loudness. These are discussed elsewhere (Stevens, 1957*b*).

The prothetic continua (loudness, brightness, and subjective intensity in general) seem to be concerned with *how much*. The metathetic continua (pitch, apparent azimuth, apparent inclination) have to do with *what kind* or *where* (position). Corresponding to these two functional classes, there seem to be two basic physiological mechanisms. Sensory discrimination can be mediated by either of two processes, the one *additive*, the other *substitutive* (Stevens, 1946*b*). We detect, for example, an increase in loudness when excitation is added to excitation already present. We detect a change in pitch when new excitation is substituted for excitation that has been removed. Or, to consider another modality, we can tell when a light pressure changes to a strong pressure at a given point on the skin (addition of excitation), and we can also tell when a stimulus is moved from one to another location (substitution of excitation). Whether all perceptual continua that behave in the prothetic manner are mediated by additive physiological processes is not certain, of course, but in at least some instances it seems evident that the existence of two basic kinds of physiological mechanisms is reflected in the behavior of the psychological scales and functions which we construct from subjective measurements in the sensory domain.

Most of what follows is concerned with prothetic continua, for they seem the more interesting and well behaved. It should be noted, however, that the pitch continuum provides an example of a rather exciting attempt to match up and thereby "explain" several psycho-physical functions by means of a physiological substratum. Position of maximal excitation on the basilar membrane appears to relate in a straightforward linear manner to several sensory functions, including the mel scale of subjective pitch, the jnd, and the so-called critical bandwidth (Békésy and Rosenblith, 1951; Zwicker, Flottorp, and Stevens, 1957).

Three Kinds of Sensory Measures

Three separate classes of sensory scales are distinguished (and sometimes confused) in psychophysics:

1. *Discriminability scales.* These are constructed in the tradition of Fechner, or his modern counterpart, Thurstone. Some measure of jnd, variability, confusion, or resolving power is employed as a unit, and a scale is constructed by counting off such units.

2. *Category scales (partition scales).* These are constructed by one or another variation on the procedure that Plateau invented when he required observers to partition a segment of a continuum into equal-appearing intervals. (Plateau had eight artists paint a gray that seemed to lie halfway between black and white.) Bisection is

one partitioning procedure; asking a listener to assign a series of tones to *n* equally spaced categories is another.

3. *Magnitude scales.* These are ratio scales of apparent magnitude, constructed by one or another of four principal methods, of which "fractionation" is perhaps the best known and "magnitude estimation" the most useful (see Stevens, 1956*b*, 1959*b*). Under the method of magnitude estimation the observer simply estimates the apparent strength or intensity of his subjective impressions relative to a stand-

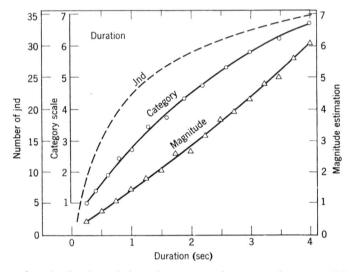

Fig. 1. Three kinds of psychological measures of apparent duration. *Triangles:* mean magnitude estimations by 12 observers who judged the apparent durations of white noises. *Circles:* mean category judgments by 16 observers on a scale from 1 to 7. The two end stimuli (0.25 and 4.0 sec) were presented at the outset to indicate the range, and each observer twice judged each duration on a 7-point scale. *Dashed curve:* discriminability scale obtained by counting off jnd.

ard, or modulus, set either by himself or by the experimenter. The power functions obtained by this procedure can, and indeed should, be validated by direct cross-modality matches—a procedure that does not require the observer to make numerical estimations.

An important difference between prothetic and metathetic continua is this: on metathetic continua all three kinds of scales tend to be linearly related one to another; on prothetic continua the three kinds of scales are nonlinearly related (Stevens and Galanter, 1957; Stevens, 1959*c*). Typical examples of the relations among the three kinds of scales on prothetic continua are shown in Fig. 1, for apparent duration (of a noise), and in Fig. 2, for apparent intensity of vibration applied to the fingertip. On all prothetic continua the magnitude scale is a power function, the discriminability (jnd) scale approximates a logarithmic function, and the category scale assumes a form intermediate between the other two. Over the different sense modalities, these relations among the three scales are strikingly invariant; they constitute one of the really stable aspects of psychophysics.

Of the three kinds of measures shown in Figs. 1 and 2, the one that seems most directly related to the over-all input-output function of a

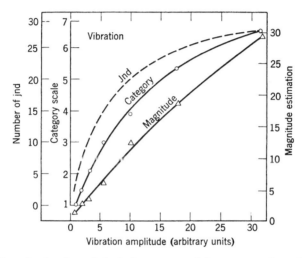

Fig. 2. Three kinds of psychological measures of the apparent intensity of a 60-cycle vibration applied to the finger tip. Procedures were essentially similar to those for Fig. 1. For details, see Stevens (1959d).

sensory system is the magnitude scale. The scale obtained by counting off jnd is really only that. At most it tells us how resolving power varies with stimulus magnitude. The category scale is at best only an interval scale on which the zero point is arbitrary. It is not a ratio scale. But since it is nonlinearly related to the ratio scale of apparent magnitude, the category scale turns out, in fact, to be not even a good interval scale. The reasons for the curvature of the category scale have been discussed elsewhere (Stevens and Galanter, 1957), but, roughly speaking, it is as though the observer, when he tries to partition a continuum into equal intervals, finds himself biased by the fact that a given difference at the low end of the scale is more noticeable or impressive than the same difference at the high end of the scale.

This asymmetry is not present on metathetic continua, and therefore the category scale is not systematically curved.

Operating Characteristics

Sense organs serve as the transducers that convert the energies of the environment into neural form. Like any transducer, each sense organ has its dynamic operating characteristic, defined by the input-output relation. It is only recently that much attention has been paid to the dynamics of sensory function—the manner in which the sensory system responds to variations in input intensity. Future efforts in this direction promise interesting rewards, however, for the form of the over-all dynamic process is now becoming more fully understood.

Conceivably, of course, all sense organs could have the same operating characteristic. All sensations would then grow at the same rate with increasing stimulus intensity. That this is far from true can be readily verified by a simple comparison. Note, for example, what happens when the luminance of a spot of light is doubled. Then note what happens when a 60-cycle current passing through the fingers is doubled. Doubling the luminance of a spot of light in a dark field has surprisingly little effect on its apparent brightness. As estimated by the typical (median) observer, the apparent increase is only about 25 per cent. But doubling the current through the fingers makes the sensation of shock seem about ten times as strong. The dynamic operating characteristics of these two sensory systems are clearly and dramatically different.

Closer investigation reveals, however, that both brightness and shock have a fundamental feature in common. In both instances the psychological magnitude ψ is related to the physical magnitude ϕ by

$$\psi = k\phi^n$$

The exponent n has the value 0.33 for brightness and 3.5 for shock. The value of k depends merely on one's choice of units. As will be shown below, the physical measure used to express ϕ needs to take account of threshold.

The power function has the convenient feature that in log-log coordinates it plots as a straight line whose slope is equal to the value of the exponent. Figure 3 illustrates this fact and shows how the slow growth of brightness contrasts with the rapid growth of electric shock. Also included for comparison is the function obtained by

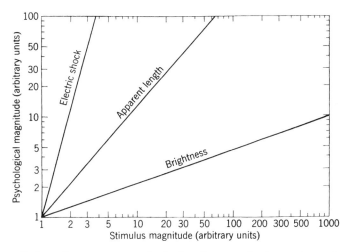

Fig. 3. Scales of apparent magnitude for three prothetic continua plotted in log-log coordinates. The slope of the line corresponds to the exponent of the power function governing the growth of the psychological magnitude.

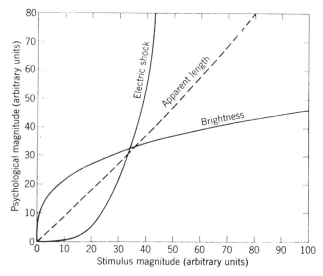

Fig. 4. In linear coordinates the subjective magnitude functions are concave upward or downward depending on whether the power-function exponent is greater or less than 1.0.

asking observers to make magnitude estimations of the apparent length of various lines. Here, as we should expect, the slope (exponent) of the function is not very different from 1.0. This is another way of saying that to most people a length of 100 centimeters looks about twice as long as a length of 50 centimeters.

The same three functions shown in Fig. 3 are plotted in linear coordinates in Fig. 4. The function for apparent length is almost a straight line (exponent about 1.1), but electric shock grows as an accelerating function and brightness as a decelerating function.

Exponents

The number of prothetic continua on which the psychophysical power law has been shown to hold to at least a first-order approximation now exceeds two dozen. In the author's experience, there appears to be no exception. (Hence the temerity of calling it a *law*.)

Table 2 lists the exponents of the power functions for some of the continua explored thus far. Although this table extends and revises the list presented earlier (Stevens, 1957*b*), it must still be regarded as tentative and incomplete, for there is virtually no limit to the number of different combinations of sense organs and stimuli that are waiting to be studied.

All the exponents in Table 2 were determined by the method of magnitude estimation. Many of them have been confirmed in other laboratories and by other methods, such as fractionation. Many of them, as we shall see, have also been validated by cross-modality intercomparisons. Nevertheless, it must be understood that the exact value of an exponent is difficult to determine with precision, and some of those listed in Table 2 must be regarded as a first approximation only. In all cases, of course, the exponent represents an average value and is not necessarily appropriate to a particular individual. At least ten observers were used to determine each of the exponents in Table 2, although some exponents (for example, loudness, brightness, lifted weights) have been determined in several laboratories and on large numbers of observers.

The particular version of the method of magnitude estimation used in our most recent experiments—the one arrived at after some years of trial and error—is extremely simple. In an experiment on loudness, for example, the procedure may be as follows. The experimenter presents a "standard" sound of moderate intensity and tells the observer to consider its loudness to have the value of "10." The experimenter then presents in irregular order a series of intensities above and below the standard and instructs the observer to assign to each stimulus a number proportional to the apparent loudness. In other words, the question is: if the standard is 10, what is each of the other stimuli? The observer is told to use any numbers that seem appropriate, fractions, decimals, or whole numbers, and to judge each stimulus as *he* hears it. The standard is usually presented only at the

Table 2. Representative Exponents of the Power Functions Relating Psychological Magnitude to Stimulus Magnitude on Prothetic Continua

Continuum	Exponent	Stimulus conditions
Loudness	0.6	Binaural
Loudness	0.54	Monaural
Brightness	0.33	5° target—dark-adapted eye
Brightness	0.5	Point source—dark-adapted eye
Lightness	1.2	Reflectance of gray papers
Smell	0.55	Coffee odor
Smell	0.6	Heptane
Taste	0.8	Saccharine
Taste	1.3	Sucrose
Taste	1.3	Salt
Temperature	1.0	Cold—on arm
Temperature	1.6	Warmth—on arm
Vibration	0.95	60 cps—on finger
Vibration	0.6	250 cps—on finger
Duration	1.1	White-noise stimulus
Repetition rate	1.0	Light, sound, touch, and shocks
Finger span	1.3	Thickness of wood blocks
Pressure on palm	1.1	Static force on skin
Heaviness	1.45	Lifted weights
Force of handgrip	1.7	Precision hand dynamometer
Autophonic level	1.1	Sound pressure of vocalization
Electric shock	3.5	60 cps, through fingers

beginning of the series, although a stimulus having the same intensity as the standard may appear as one of the stimuli to be judged along with the others. With a series of six to ten stimuli, each stimulus is usually presented twice, but the order of the stimuli is made different for each observer. In the averaging of the data from a group of observers it is usual to compute the geometric means of the estimates, although sometimes the median provides a more representative measure. Since the distributions of responses are usually skewed, the arithmetic mean is seldom an appropriate statistic.

Cross-Modality Comparisons

Few scientists fail to sense an uneasy concern about the foregoing procedure, which seems to rely merely on the observer's expression of opinion, and which seems also to depend on his having a moderately sophisticated understanding of the number system. This is a proper concern, because naïveté about numbers, and especially about the concept of proportion, certainly impedes the ability of some observers to perform well in this kind of experiment. The matching of numbers to sensation intensity is not something that a person does with fine precision, or that he feels great certainty about, even though the typical graduate student can usually manage a consistent set of estimates.

The interesting question, however, is not whether we are uneasy about the procedure, but whether the experiments on magnitude estimation can predict other empirical consequences that can be put to test. In particular, can we confirm the power law without asking observers to make any numerical estimations at all? If so, can we proceed to verify the relations among the exponents listed in Table 2? An affirmative answer to these questions is suggested by the results of a method in which the observer equates the apparent strengths of the sensations produced in two different modalities. By means of such cross-modality matches, made at various levels of stimulus intensity, an "equal-sensation function" can be mapped out, and its form can be compared with the form predicted by the magnitude scales for the two modalities involved.

If, given an appropriate choice of units, two modalities are governed by the equations

$$\psi_1 = \phi_1{}^m$$

and

$$\psi_2 = \phi_2{}^n$$

and if the subjective values ψ_1 and ψ_2 are equated by cross-modality matches at various levels, then the resulting equal-sensation function will have the form

$$\phi_1{}^m = \phi_2{}^n$$

In terms of logarithms

$$\log \phi_1 = n/m \log \phi_2$$

In other words, in log-log coordinates the equal-sensation function should be a straight line whose slope is given by the ratio of the two exponents.

The experimental question is whether observers can make cross-modality matches, and whether their matches can, in fact, be predicted from the ratio scales of apparent magnitude determined independently by magnitude estimation. The ability of observers to make the simple judgment of apparent equality has been well established in other contexts. Heterochromatic photometry and the mapping of equal-loudness contours provide two well known examples of procedures that involve the judgment of apparent equality of sensory intensity—a judgment made in the presence of an obvious qualitative difference. It is but a small step to extend these procedures to cross-modality equations. As a matter of fact, some cross-modality equations seem less difficult than some equations within a single modality.

In principle, of course, cross-modality matches can be made between every sensory continuum and every other one. Since this potential enterprise involves heroic numbers of experiments, only certain illustrative tests have been completed. They are sufficient, however, to demonstrate the general validity of the ratio scales of subjective magnitude. A few of these cross-modality experiments will be described.

Loudness versus Vibration

Two stimuli that are relatively easy to equate for apparent strength are sound and mechanical vibration. The sound employed was a band of noise of moderately low frequency, and the vibration was a single frequency (60 cps) delivered to the end of the middle finger (Stevens, 1959a).

The matching of the apparent intensities of sound and vibration was carried out in two complementary experiments. In one experiment the level of the sound was adjusted to match the vibration; in the other the level of the vibration was adjusted to match the sound. The sound and vibration were presented simultaneously. (In many experiments the stimuli have been presented successively for one reason or another.) Each of ten observers made two adjustments at each level in each experiment.

The results are shown in Fig. 5. The circles represent the means of the decibel levels to which the sound was adjusted, and the squares represent the means of the decibel levels to which the vibration was adjusted. The coordinate scales are in decibels relative to the approximate thresholds of the two kinds of stimuli.

The interesting point to note is that the slope of the line in Fig. 5 is 0.6, which is close to the slope that is called for by the ratio of the exponents of the two magnitude functions. It is also apparent that the relation is essentially linear, which is consistent with the fact that, over the ranges of the stimuli involved, both loudness and vibration are governed by power functions.

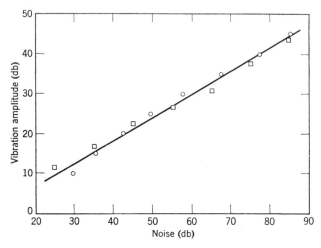

Fig. 5. An equal-sensation function relating 60-cycle vibration on the finger tip to the intensity of a band of noise. The observers adjusted the loudness to match the vibration (circles) and the vibration to match the loudness (squares). The stimulus values are measured in terms of logarithmic scales (decibels).

The departure of some of the points from the straight line in Fig. 5 is in large measure due to the interesting fact that, depending on which stimulus is adjusted, the slope turns out to be slightly different. The situation is analogous to the two regression lines in a correlation plot. This "regression" or "centering tendency" is common, if not universal, in matching procedures, and it points up the desirability of a balanced design in which each stimulus is made to serve as both the standard and the variable (see Stevens, 1955*b*). The matching of loudness and vibration turned out to be surprisingly easy. Some of the observers, who happened to have served in loudness matching experiments, expressed the opinion that matching loudness to vibration seemed easier than matching the loudnesses of two tones of widely different pitch or quality (cf. Stevens, 1956*a*). The consistency of the judgments seemed to bear this out.

Other Comparisons

Cross-modality matches similar to those between vibration and loudness have been made for other pairs of continua, notably vibration versus electric shock, and electric shock versus loudness (S. S. Stevens, 1959a). This "round robin" of cross-modality comparisons completes an interesting circle in a process of validation, for it turns out that all the matches are consistent with the predictions derived from the ratio scales determined by magnitude estimation. Furthermore, from the equal-sensation functions determined for two of the three pairs of continua, the function for the third pair can be predicted, and this prediction is verified to within a good approximation.

Another procedure by which the operating characteristics of sensory systems can be compared is cross-modality *ratio matching*. The observer is asked to make the apparent ratio between one pair of stimuli match the apparent ratio between some other pair. In particular, he may adjust stimulus D so that he achieves the relations: A is to B as C is to D.

This procedure is exemplified by an experiment conducted by J. C. Stevens in which the observer adjusted a pair of loudnesses to match a ratio defined by a pair of brightnesses (see S. S. Stevens, 1957b). Figure 6 shows the median settings made by 15 observers, each of whom twice adjusted the loudness of the second of two noises until the apparent ratio between the noises equaled the apparent ratio between two luminous targets seen against a dark surround. The tendency of the points in Fig. 6 to fall near the 45-degree diagonal means that whatever ratio (number of decibels) the experimenter set between the lights, the observer set approximately the same ratio between the sounds. The largest ratio used (40 db) represents a stimulus ratio of 10,000 to 1.

The outcome shown in Fig. 6 is what would be expected if both brightness and loudness were governed by a power law and if the two exponents were approximately the same size. The consensus of many experiments on loudness (S. S. Stevens, 1955b) shows that the exponent is approximately 0.3 when the stimulus is measured in terms of sound energy (0.6 when measured in terms of sound pressure). The exponent for brightness is approximately 0.33. Thus, in terms of the energy delivered to the sense organs, the exponents for loudness and brightness are approximately the same size. The similarity of these exponents is illustrated by Fig. 7, which exhibits the results of two of the many experiments that have been performed to determine ratio scales of subjective magnitude for loudness and brightness.

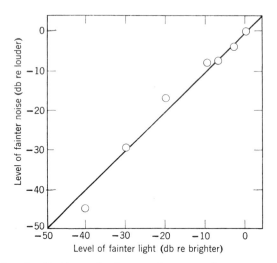

Fig. 6. Results of adjusting a loudness ratio to match an apparent brightness ratio defined by a pair of luminous circles. One of the circles was made dimmer than the other by the amount shown on the abscissa. The observer produced white noises by pressing one or the other of two keys, and he adjusted the level of one noise (ordinate) to make the loudness ratio seem equal to the brightness ratio. The brighter light was about 99 db re 10^{-10} lambert, and the louder noise was about 92 db re 0.0002 dyne per square centimeter.

It should be pointed out that the use of decibel scales simplifies the stimulus specification for vision and audition, and facilitates comparisons between their sensory dynamics. The foregoing examples demonstrate some of these advantages. Although the application of decibel measures to visual stimuli is not yet common practice, there is much to recommend it (see S. S. Stevens, 1955a). As a matter of fact, except for the rigidities of professional custom, the application of the decibel notation to the measurement of light presents less difficulty than its application to sound, because the decibel is defined in terms of energy flow

$$N_{db} = 10 \log \frac{E_1}{E_0}$$

and it is only by a kind of bastardized extension that the decibel gets used with measures of sound pressure. The energy in a sound wave

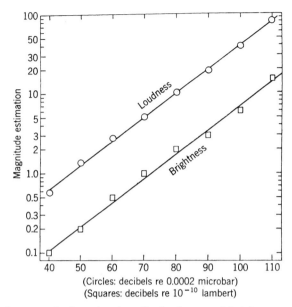

(Circles: decibels re 0.0002 microbar)
(Squares: decibels re 10^{-10} lambert)

Fig. 7. Median magnitude estimations for loudness and brightness.

For the loudness of a 1000-cycle tone, each of 32 observers made 2 estimates at each level. Since no standard modulus was designated, each observer chose his own, and the resulting numerical estimates were transformed to a common modulus at the 80-db level.

For brightness each of 28 dark-adapted observers made 2 estimates of each stimulus level. The target subtended an angle of about 5 degrees and was illuminated for about 3 sec. Once at the beginning of each session the observer was shown a stimulus of 70 db (14 observers) or 80 db (14 observers) and told to call it "10." The estimates were transformed to a common modulus at 70 db.

is proportional to the square of the sound pressure, but only under very special conditions. With light, on the other hand, we are concerned only with energy measures, relative or absolute, and there is no need to become entangled in measures that are nonlinearly related to energy.

Force of Handgrip

Like any other sensation, the subjective impression of muscle tension can be measured on a ratio scale of psychological magnitude. By squeezing a precision dynamometer (Fig. 8), an observer can produce a sensation of apparent force and at the same time activate a dial that indicates the actual force exerted. Two pertinent questions pose themselves: (1) How does the feeling of apparent force relate to the physical force exerted? (2) What happens when observers try to report the apparent magnitude of other kinds of sensations by squeezing the dynamometer instead of by making numerical estimations? In other words, we face a problem in scaling and a problem in cross-modality matching.

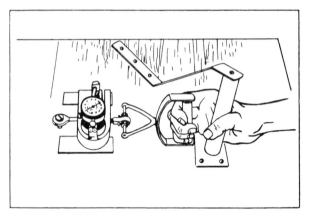

Fig. 8. One of the hand dynamometers used. It consisted of a pair of handles, one of which was connected through a ball joint to a calibrated Dillon force gauge.

The scaling problem was attacked by J. C. Stevens and Mack (1959), who found that the apparent force of handgrip grows as the 1.7 power of the physical force applied. They used several methods, but principally magnitude estimation and magnitude production. The method of magnitude production, which has not been mentioned before in this chapter, proved to be an easy and convenient procedure for scaling the continuum of apparent muscular force. Instead of producing stimuli and asking observers to judge their magnitudes (as in magnitude estimation), the experimenter named numbers in an irregular order, and the observer exerted forces that seemed to him proportional to the numbers. With magnitude estimation the observer squeezed until the experimenter signaled that a sufficient force had been achieved, and then the observer estimated its apparent magnitude. The results of both procedures approximated power functions, but there was a tendency for magnitude production to give a slightly higher exponent than magnitude estimation.

Figures 9 and 10 show the results for the individual observers in two different experiments, each of which employed the two different methods. The two experiments involved different kinds of dynamometers: the one shown in Fig. 8, which had a stiff, noncompliant force gauge, and a more compliant dynamometer whose handle moved through about 1.5 inches for a pull of 40 pounds. The two dyna-mometers gave similar results. It is clear from Figs. 9 and 10 that the results for each observer approximate a power function. It is also

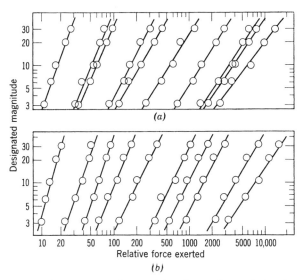

Fig. 9. Functions for apparent force of handgrip obtained by the method of magnitude production. Each curve is for a single observer, and the position of the curve on the abscissa is arbitrary. The experimenter designated the numerical values (ordinate), and the observer produced the appropriate squeezes (ab-scissa). (*a*) Medians of 7 squeezes by each observer using the dynamometer shown in Fig. 8. (*b*) Medians of 10 squeezes by each observer using the more compliant dynamometer.

clear that the slope may vary from one observer to another. It is not clear, however, that this variation in slope (exponent) means any-thing more than that people differ in what they understand by relative magnitudes. Whether the action of one man's sensory transducers is different from another's cannot be told with certainty from a single experiment of this sort. It is possible, on the other hand, that a battery of cross-modality comparisons might well provide definitive evidence of abnormal sensory function. The use of such a battery for the detection of recruitment in hard-of-hearing patients has been proposed elsewhere (Stevens, 1959*c*).

Handgrip versus Nine Other Continua

Although a subjective scale for force of handgrip proves interesting in its own right, the convenient ability of handgrip to serve as an indicator of other subjective magnitudes has led to even more exciting results.   Instead of asking observers to emit numbers in response to stimuli, we can ask them to emit squeezes of appropriate sizes.   In this manner, observers have matched apparent force to apparent sensory intensity on nine different continua and have produced the

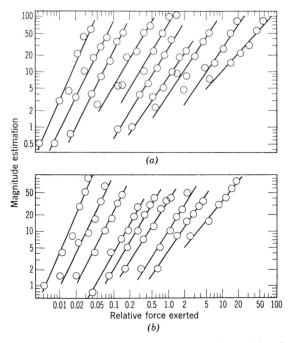

Fig. 10.   Functions for apparent force of handgrip obtained by the method of magnitude estimation.   Each curve is for a single observer, and the position of the curve on the abscissa is arbitrary.   (*a*) Medians of 6 estimates by each observer using the dynamometer shown in Fig. 8.   The forces estimated were 4, 10, 15, 22, 30, and 40 pounds.   (*b*) Medians of 10 estimates by each observer using the more compliant dynamometer.   The forces estimated were 5, 11, 17, 23, 29, and 35 pounds.

results shown in Fig. 11 (J. C. Stevens, Mack, and S. S. Stevens, 1960; J. C. Stevens and S. S. Stevens, 1960).

Two points are immediately evident.   All the data in Fig. 11 approximate power functions—straight lines in log-log coordinates— and the slopes stand in the same order as the values of the exponents listed in Table 2.   Less obvious but even more interesting is the exact numerical relation between the slopes determined by matching with handgrip and those determined by matching with numbers (that is

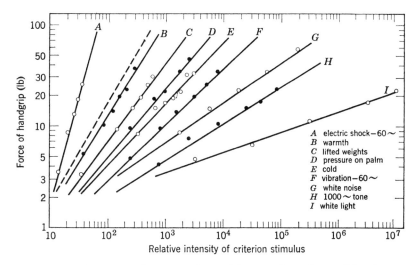

Fig. 11. Equal-sensation functions obtained by matching force of handgrip to various criterion stimuli. Each point stands for the median force exerted by 10 or more observers to match the apparent intensity of a criterion stimulus. The relative position of a function along the abscissa is arbitrary. The dashed line shows a slope of 1.0 in these coordinates.

Table 3.   The Exponents (Slopes) of Equal-Sensation Functions,
as Predicted from Ratio Scales of Subjective Magnitude,
and as Obtained by Matching with Force of Handgrip

Ratio scale		Scaling by means of handgrip		
Continuum	Exponent of power function	Stimulus range	Predicted exponent	Obtained exponent
Electric shock (60-cycle current)	3.5	0.29–0.72 milliampere	2.06	2.13
Temperature (warm)	1.6	2.0–14.5°C above neutral temperature	0.94	0.96
Heaviness of lifted weights	1.45	28–480 grams	0.85	0.79
Pressure on palm	1.1	0.5–5.0 pounds	0.65	0.67
Temperature (cold)	1.0	3.3–30.6°C below neutral temperature	0.59	0.60
60-cycle vibration	0.95	17–47 db re approximate threshold	0.56	0.56
Loudness of white noise	0.6	55–95 db re 0.0002 dyne/cm^2	0.35	0.41
Loudness of 1000-cycle tone	0.6	47–87 db re 0.0002 dyne/cm^2	0.35	0.35
Brightness of white light	0.33	56–96 db re 10^{-10} lambert	0.20	0.21

to say, magnitude estimation). Since the exponent for handgrip is approximately 1.7, we should expect that the exponent for a given continuum in Table 2 would be about 1.7 times as large as the slope of the corresponding line in Fig. 11. How nearly this expectation is fulfilled is shown by the comparisons in Table 3. Despite the variability inherent in experiments of this sort, the agreement between the obtained and predicted exponent is generally satisfactory. This agreement testifies with a certain eloquence to the basic validity of the ratio scales of sensory magnitude.

The Stimulus Scale

Except for the two continua, warm and cold, the physical stimuli of all the continua discussed above have been measured on the ordinary physical scales of amperes, grams, dynes, and so on. This practice is sufficiently accurate for most purposes, but when we look more closely we see that the general form of the power law is

$$\psi = k(\phi - \phi_0)^n$$

where ϕ_0 is a constant value corresponding to "threshold." For ranges of stimuli well above the minimum detectable level, the value of ϕ_0 is usually negligible, but it assumes larger proportions when subjective scales are extended downward toward very low values.

Temperature provides a clear and dramatic example of the importance of measuring stimuli in terms of the ratio scale of *distance* from threshold (J. C. Stevens and S. S. Stevens, 1960). The threshold for warmth when aluminum stimulators are applied to the inside of the forearm is about 305.7° above zero on the absolute scale (Kelvin). Compared to the short range of tolerable thermal stimuli, this is indeed a high threshold.

As shown in Fig. 12 (log-log plots), when apparent temperature is scaled by magnitude estimation and the results plotted against the Kelvin scale, the data fall on a curve that is sharply concave downward. When plotted in terms of degrees above the neutral or threshold value, however, the data fit a power function with an exponent of about 1.6. From these measurements it follows that the power-function formula for subjective warmth ψ_w is

$$\psi_w = k(T_K - 305.7)^{1.6}$$

where T_K is absolute temperature.

In a similar type of experiment (aluminum stimulators applied to the arm), the formula for cold ψ_c turned out to be

$$\psi_c = (304.2 - T_K)^{1.0}$$

The difference between the two values, 305.7 and 304.2, corresponding to ϕ_0 is of no particular significance. It presumably means that the observers' average skin temperature was different in the two experiments. On most other continua the value of the additive

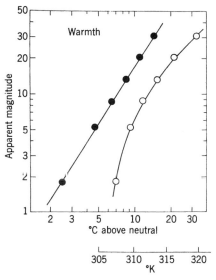

Fig. 12. Magnitude estimation of apparent warmth. Each point is the geometric mean of 36 estimates (12 observers). The upper abscissa, for the filled points, is a log scale of the difference in temperature (Celsius) between the stimulus and the "physiological zero." The lower abscissa, for the unfilled points, is a log scale of the absolute temperature (Kelvin).

constant ϕ_0 is small relative to the usable stimulus range. Nevertheless, in two instances a revision of the stimulus scale designed to take explicit account of ϕ_0 has transformed otherwise wayward data into well-behaved power functions. The scale for tactile vibration (60 cps) applied to the arm was corrected in this manner (Stevens, 1959d), and a similar treatment was applied to the loudness scale by Scharf and J. C. Stevens (in press). As Luce (1959) has pointed out, the use of an additive constant to bring the zero of the physical scale into coincidence with the zero of the psychological scale is a proper generalization of the power-function law. (Differences on a ratio scale constitute a ratio scale, as do also differences on an interval scale.)

In calling ϕ_0 the "threshold" value, we raise a problem concerning precisely what is meant by the term threshold. It is not necessarily the threshold as measured in some arbitrary manner under arbitrary conditions. Rather it should probably be thought of as the "effective" threshold that obtains at the time and under the conditions of the experiment in which the magnitude scale is determined. Needless to say, this "effective" threshold cannot be measured very precisely. Consequently, it becomes expedient to take as the value of ϕ_0 the constant value whose subtraction from the stimulus values succeeds in rectifying the log-log plot of the magnitude function. Provided the constant value so chosen is a reasonable threshold value, this procedure seems justified. At any rate, it has worked well for the four continua, vibration, loudness, warmth, and cold.

Variability

Needless to say, the responses people make to sensory intensity are variable. Although an exemplary picture of this variability is shown in Figs. 9 and 10, a further statement is in order about it. The statement will be brief, however, because the author confesses to a certain lack of enthusiasm for elaborate statistical analyses.

By and large, the interquartile range encountered when groups of observers undertake magnitude estimation on intensitive continua is of the order of 0.2 to 0.3 log unit. (It may, of course, be lower for continua that are easy to judge.) This variability contains certain obvious components, however, the most important of which appear to be the following.

1. *Variability due to the observer's modulus, that is, his conception of the "standard."* Since we are concerned only with the *form* of the magnitude scale, this source of variability is of no concern. When desired, it can be partialed out in one way or another, with a consequent reduction in the over-all variability.

This component of variability is especially evident in those experiments in which each observer is allowed to choose his own modulus (Stevens, 1956b). It also plays a prominent role in cross-modality matches (Stevens, 1959a; J. C. Stevens, Mack, and S. S. Stevens, 1960). Each observer, for example, has his own conception of what force of handgrip matches what level of loudness, but the absolute values chosen by the observer are irrelevant so far as the form of the equal-sensation function is concerned.   It is only the relative values that matter.   Concretely, our main concern is with the slopes and not the intercepts of the functions in Figs. 9 and 10.

2. *Variability due to the observer's conception of a subjective ratio.* In a method like fractionation or magnitude estimation, each person must make up his own mind about what he considers "half as bright," say, and not all observers arrive at the same conclusion.   (A plot showing an example of the variability encountered in halving and doubling may be found in Stevens, 1957a.)   Nothing much can be done about this source of variability, except perhaps to try to avoid biases and constraints in the conditions of observation (see Stevens, 1956b).

3. *Variability due to differing sense-organ operating characteristics.* This source of variability is biologically the most interesting, but it is probably of only minor magnitude in a group of "normal" observers. Nevertheless, in the hard-of-hearing ear or the night-blind eye, it may be a factor of considerable consequence.   (The state of a sense organ like the eye is also changed, of course, in the process of adaptation.) In order to evaluate this "sense-organ" factor, it may prove useful to apply a battery of cross modality matching tests   A sufficient battery of these tests should prove capable, for example, of distinguishing the person with auditory recruitment from the person whose conception of a subjective loudness ratio is merely atypical.

An interesting problem related to variability concerns the question of departures from the power-function law.   Can we expect that the power law will always hold rigorously (provided, of course, no errors arise in our measurements), or should we look for second-order deviations from it?   The data of particular experiments sometimes depart from the power law, but in most instances it is not easy to determine whether these defections are due to artificial biases of one kind or another.   Nevertheless, since the possibility of genuine departures from the power law is a problem of basic moment, an effort should be made to devise procedures of sufficient accuracy to settle the question.   The fact that the power law is closely approximated by so many data in so many different sense modalities adds interest and significance to any authentic departures from the power-function form.

The Role of Transducers

The foregoing suggestion of a method for determining the individual loudness function in a hard-of-hearing ear assumes that the nature of the sensory transducer largely determines the form of a magnitude function. An opposite assumption has often been made, however, to the effect that the magnitude function merely reflects how observers have learned in the past to associate sensory impressions with some known aspect of the physical stimulus. The "learning" explanation has recently been revived (Warren, 1958), and, under the name "physical correlate theory," it is alleged to provide a "basis for Stevens' empirical law" (Warren, Sersen, and Pores, 1958). If this theory were correct, it would presumably explain why the psychophysical law is a power function, but the evidence that learning accounts for all the exponents in Table 2 is mostly nonexistent. Familiarity with the stimulus may be a factor in people's judgments on some kinds of continua (although how one would prove it is hard to see), but many of the continua in Table 2 are quite unfamiliar to the typical observer—at least as regards measures of stimulus intensity. And especially difficult to conceive is how familiarity with the physical stimuli—even if the observers had such familiarity—could account for the results of cross-modality matching like those shown in Fig. 11.

It seems rather more probable that the exponents are what they are because of the nature of the sensory transducers. It is likely, for example, that the exponents for light and sound are smaller than 1.0 because these sensory transducers behave essentially as "compressors"—a characteristic that enables them to handle the enormous dynamic ranges of stimulation to which they are subjected. At the other extreme, in the transduction process involved with electric current applied to the fingers, there is an operation of "expansion" in the sense that the psychological magnitude grows as an accelerating function of stimulus intensity, that is, the exponent is greater than 1.0. It seems quite improbable that the form of this function was "learned" by the observer.

It is an interesting question whether electrical stimulation of nerves other than those in the fingers would also exhibit an accelerating transduction characteristic. In a study of the "electrophonic effect" (Jones, Stevens, and Lurie, 1940), patients lacking tympanic membranes were stimulated by means of an electrode placed inside the middle-ear cavity. Although some patients heard pure tones, seven of the group heard only a buzzing noise whose quality was more-or-less independent of the frequency of the stimulating current. Since it is certain that other nerves (for example, the facial and the vestibular) were occasionally stimulated in the course of these experiments, it seems safe to conclude that the auditory nerve was also sometimes directly affected by the current. Direct, unpatterned stimulation of the auditory nerve fibers would account for the patient's hearing only a noise.

Some of the patients noted a large change in loudness when only a small change was made in the stimulating current. This effect was so striking that an attempt was made to measure the loudness change by comparing it with a sound in the opposite (normal) ear. The outcome is shown in Fig. 13 (for further explanation, see Stevens, Carton, and Shickman, 1958). Apparently if the auditory end organ

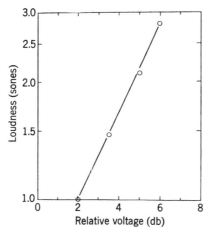

Fig. 13. Showing the steep growth of loudness with increasing electric current applied to the auditory nerve of a patient whose eardrum had been removed. The current was delivered by an electrode placed in the middle-ear cavity. The exponent is about four times as large as the exponent obtained with acoustic stimulation.

is bypassed, and if a stimulating current acts directly on the eighth nerve, a new transducer process becomes involved, and the function describing the growth of loudness acquires a radically different exponent. The implication is that the "compression" observed when the normal ear is stimulated by sound waves is a function of the sense organ, not of some higher center in the nervous system.

When it is discovered that two continua presumed to be rather similar are governed by different exponents (Table 2), one suspects that there may be basic differences in the transducer systems involved. Two pairs of such continua are especially interesting: warmth and cold, and taste and smell. Warmth and cold are interesting because the same stimulating device, applied to the same place on the arm produces two scales of sensory intensity, one for stimulus temperatures below neutral and one for temperatures above neutral (J. C. Stevens and S. S. Stevens, 1960). Not only do warmth and cold produce sensations of different quality, they also appear to do it by means of transducers with different operating characteristics. The temperature sense is also unique in that the neutral or threshold point is the bottom of one stimulus scale (warmth) and the top of another (cold). Cold increases as the stimulus value decreases.

Taste and smell are often classed together as chemical senses. The mode of action of the stimulus for smell has been such a mystery, however, that mechanisms other than the bathing of end organs by chemical solutions have been hypothesized from time to time. This state of affairs gives added significance to the obvious difference between the operating characteristics of the olfactory and gustatory systems. For the several substances thus far tested, the exponents of the power functions for olfactory intensity have run from about 0.5 to 0.6 (Jones, 1958; Reese and Stevens, 1960). Earlier experiments on taste, with the method of fractionation, gave exponents of the order of 1.0 (Beebe-Center and Waddell, 1948). The exponents for taste listed in Table 2 were obtained by Mary McLean in some exploratory experiments with the method of magnitude estimation. This work is still in progress, but there is little doubt that the exponents for taste are generally about twice as large as the exponents for smell. Does this difference in the dynamics of apparent intensity mean that two wholly different mechanisms underly the transduction processes in taste and smell?

Summary

A central issue in the struggle to transform psychology into a science has long concerned the quantification of sensory magnitudes. Fechner's creation of psychophysics was a monumental attempt to measure sensation; but the model as well as the method by which he hoped to achieve its quantification has turned out to be a vigorous stride in a wrong direction. Other models, envisaging other methods, have lately achieved success in the mapping of quantitative relations on sensory continua.

The measurement of sensation in a manner useful to science has followed upon the clarification of several sticky issues that formerly thwarted the proper investigation of quantitative relations. Progress has been made on the following points:

1. Sensation has ceased to be regarded as a private mentalistic entity comprising one-half of a metaphysical dualism, and has joined the class of other natural constructs, defined in terms of the operations used to denote and measure it. We study sensation by noting the behavior of organisms under the impact of energetic configurations (stimuli).

2. The concept of measurement has been broadened to include much more than the operation of counting, which was all there was to primitive measurement. Measurement is now conceived as the assignment of numbers to objects or events according to rule. The distinguishable rules, and the invariances consistent therewith, entail four classes of scales: *nominal, ordinal, interval,* and *ratio.* A fifth class, called *logarithmic interval,* is of some academic but little practical interest.

3. A basic distinction can be made between two kinds of sensory continua, prothetic and metathetic. The distinction rests on a set of functional criteria, but it is useful to think of the prothetic continua as those concerned with *how much* (intensity) and the metathetic continua as those concerned with *what* or *where* (quality). At the physiological level, many prothetic continua (loudness, brightness, and so forth) involve the addition of excitation to excitation, whereas some of the metathetic continua (pitch, apparent azimuth, inclination) involve the substitution of excitation for excitation.

4. Ratio scales of sensation, permitting a direct determination of the input-output characteristics of sensory systems, have been erected on prothetic continua by means of a variety of simple operations. The principal methods have involved (a) direct numerical estimations by the observer and (b) cross-modality comparisons in which the apparent strength of a sensation in one modality is made equal to a given sensation in another modality.

The outcome of these developments is a simple and attractive power law relating psychological magnitude to physical stimulus.

The formula is

$$\psi = k(\phi - \phi_0)^n$$

where ϕ_0 is the effective threshold value. On some two dozen continua the exponent n has been found to range from 0.33 for visual brightness to 3.5 for the subjective strength of electric current applied to the fingers. The generality and utility of this quantitative law may open new vistas in psychophysics and provide new insights into the roles played by the sensory transducers.

References

Beebe-Center, J. G., and D. Waddell. A general psychological scale of taste. *J. Psychol.*, 1948, **26**, 517–524.

Békésy, G. v., and W. A. Rosenblith. The mechanical properties of the ear. Chapter 27 in S. S. Stevens (Editor), *Handbook of Experimental Psychology*. New York: Wiley, 1951.

Jones, F. N. Scales of subjective intensity for odors of diverse chemical nature. *Amer. J. Psychol.*, 1958, **71**, 305–310.

Jones, R. C., S. S. Stevens, and M. H. Lurie. Three mechanisms of hearing by electrical stimulation. *J. acoust. Soc. Amer.*, 1940, **12**, 281–290.

Luce, R. D. On the possible psychophysical laws. *Psychol. Rev.*, 1959, **66**, 81–95.

Reese, T. S., and S. S. Stevens. Subjective intensity of coffee odors. *Amer. J. Psychol.*, 1960, **73**, 424–428.

Scharf, B., and J. C. Stevens. The form of the loudness function near threshold. *Proc. 3rd int. Congr. Acoust.* (in press).

Stevens, J. C., and J. D. Mack. Scales of apparent force. *J. exp. Psychol.*, 1959, **58**, 405–413.

Stevens, J. C., J. D. Mack, and S. S. Stevens. Growth of sensation on seven continua as measured by force of handgrip. *J. exp. Psychol.*, 1960, **59**, 60–67.

Stevens, J. C., and S. S. Stevens. Warmth and cold: Dynamics of sensory intensity. *J. exp. Psychol.*, 1960, **60**, 183–192.

Stevens, S. S. On the theory of scales and measurement. *Science*, 1946a, **103**, 677–680.

Stevens, S. S. The two basic mechanisms of sensory discrimination. *Fed. Proc.*, 1946b, **5**. [Abstract]

Stevens, S. S. Mathematics, measurement, and psychophysics. In S. S. Stevens (Editor), *Handbook of Experimental Psychology*. New York: Wiley, 1951.

Stevens, S. S. Decibels of light and sound. *Physics Today*, 1955a, **8**(10), 12–17.

Stevens, S. S. The measurement of loudness. *J. acoust. Soc. Amer.*, 1955b, **27**, 815–829.

Stevens, S. S. Calculation of the loudness of complex noise. *J. acoust. Soc. Amer.*, 1956a, **28**, 807–832.

Stevens, S. S. The direct estimation of sensory magnitudes—loudness. *Amer. J. Psychol.*, 1956b, **69**, 1–25.

Stevens, S. S. Concerning the form of the loudness function. *J. acoust. Soc. Amer.*, 1957a, **29**, 603–606.

Stevens, S. S. On the psychophysical law. *Psychol. Rev.*, 1957b, **64**, 153–181.

Stevens, S. S. Problems and methods of psychophysics. *Psychol. Bull.*, 1958, **54**, 177–196.

Stevens, S. S. Cross-modality validation of subjective scales for loudness, vibration, and electric shock. *J. exp. Psychol.*, 1959a, **57**, 201–209.

Stevens, S. S. Measurement, psychophysics, and utility. In C. W. Churchman and P. Ratoosh (Editors), *Measurement: Definitions and Theories*. New York: Wiley, 1959b.

Stevens, S. S. On the validity of the loudness scale. *J. acoust. Soc. Amer.*, 1959c, **31**, 995–1003.

Stevens, S. S. Tactile vibration: Dynamics of sensory intensity. *J. exp. Psychol.*, 1959*d*, **57**, 210–218.

Stevens, S. S., A. S. Carton, and G. M. Shickman. A scale of apparent intensity of electric shock. *J. exp. Psychol.*, 1958, **56**, 328–334.

Stevens, S. S., and E. H. Galanter. Ratio scales and category scales for a dozen perceptual continua. *J. exp. Psychol.*, 1957, **54**, 377–411.

Stevens, S. S., and J. Volkmann. The quantum of sensory discrimination. *Science*, 1940, **92**, 583–585.

Warren, R. M. A basis for judgments of sensory intensity. *Amer. J. Psychol.*, 1958, **71**, 675–687.

Warren, R. M., E. A. Sersen, and E. B. Pores. A basis for loudness-judgments. *Amer. J. Psychol.*, 1958, **71,**, 700–709.

Zwicker, E., G. Flottorp, and S. S. Stevens. Critical bandwidth in loudness summation. *J. acoust. Soc. Amer.*, 1957, **29**, 548–557.

2

GÖSTA EKMAN
Psychological Laboratory, University of Stockholm

Some Aspects
of Psychophysical Research

A Note on Measurement

Although the problems studied in psychophysics are rather varied, one thing is central to all the work subsumed under this name. That is the measurement of subjective variables. In the present stage of psychophysical development we are concerned in particular with quantifying human experience on a relatively simple level, for example, measuring subjective brightness or pitch or perceived skin pressure. This is not necessarily the whole thing, but it is fairly true for the present situation.

The measurement technique, which is used in most of this work, has been developed mainly by S. S. Stevens and his co-workers over a number of years. Without going into any detail, I shall briefly characterize this sort of scaling. I call it *direct* scaling of subjective variables, because the essential steps of the scaling procedure are implied in the experimental situation. You may present a tone to the subject and tell him that the loudness of this standard tone is called 100. Now, in relation to this, what is the loudness of another tone of equal pitch? You may obtain the estimate 62, and this datum is one observation of the scale value of the second tone. The two scale values are, by definition, on a ratio scale: the ratio 62/100 should, according to the instructions, be equal to the subjective ratio of the second loudness to the loudness of the standard. Of course, before the final scale can be constructed, you have to collect many observations and do some computational tricks with them, but fundamentally the situation is about as simple as outlined here.

Even if this procedure seems to be quite straightforward, one might still be in some doubt about the metric properties of the resulting scale. In an investigation carried out by Goude (1959) in this laboratory, some fundamental properties of scales obtained by direct scaling methods were studied. Results from two of Goude's experiments are shown in Fig. 1. These experiments were designed to test the *additivity* of scale values. In one part of the experiments (empty circles) the subjects were comparing single stimuli two at a time, as is commonly done in scaling. In another part of the experiments (filled circles) the subjects were working with subjective *sums* of the

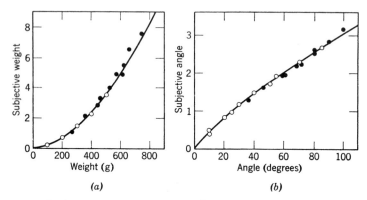

Fig. 1. Results from an experiment on weight lifting (*a*) and another experiment on subjective magnitude of angles (*b*), carried out in order to test the additivity of scales constructed by direct ratio methods (Goude, 1959). Unfilled circles represent scale values for single stimuli. Filled circles represent scale values obtained when the subjects were working with subjective *sums* of subjective magnitudes. In each experiment both sets of data exhibit the same trend that is closely approximated by the power functions represented by the curves.

subjective magnitudes under consideration. The two sets of data thus obtained in each experiment clearly exhibit the same general trend and may well be approximated by the same curves, as shown in the graphs. This is probably the first time that the fundamental metric property of additivity has been demonstrated in scales obtained by direct psychophysical methods of measurement. Results like these should increase our confidence in these methods.

Three Types of Problems

Stimulus-response relations

It was Fechner's idea to investigate the relation between a subjective intensity and the intensity of the physical stimulation evoking the subjective response. On the basis of Weber's law (which usually holds over a wide range) and the assumption that just noticeable stimulus differences are subjectively equal (which in general they are definitely not, as we now know), he derived his famous logarithmic law which has ever since continued to appear in textbooks without ever being verified.

In recent years, however, the *power function* has been found to describe the psychophysical relation very adequately for about twenty continua so far investigated. Convincing evidence has been presented, especially by Stevens (1957) and Stevens and Galanter (1957). There is hardly a really convincing exception known to this rule. We usually write the power function in the form

$$R = c(S + a)^n \qquad (1)$$

We have already seen two power functions fitted to the data in Figs. 1*a* and *b*. As a further illustration of this psychophysical law, I have chosen an experiment from our laboratory concerning the subjective brightness of monochromatic light of various wave lengths (Ekman, Eisler, and Künnapas, 1960). In order to avoid confusion, the data have been divided into two graphs of Fig. 2. The curves

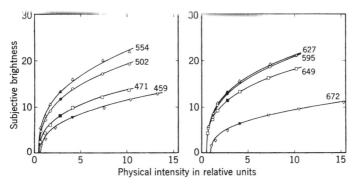

Fig. 2. Brightness as a function of stimulus intensity for eight wave lengths. The circles and other symbols represent the scale values with a common unit of measurement. In order to avoid confusion between nearly coinciding curves, the data are plotted and the curves drawn in two separate graphs. The curves represent power functions. For the various wave lengths the following functions are obtained:

$$R_{459} = 4.93(S - 0.68)^{0.381}$$
$$R_{471} = 6.20(S - 0.60)^{0.358}$$
$$R_{502} = 8.94(S - 0.55)^{0.343}$$
$$R_{554} = 9.96(S - 0.50)^{0.356}$$
$$R_{595} = 9.71(S - 0.52)^{0.343}$$
$$R_{627} = 10.02(S - 0.51)^{0.334}$$
$$R_{649} = 8.30(S - 0.54)^{0.318}$$
$$R_{672} = 4.24(S - 0.95)^{0.363}$$

represent power functions fitted to the data, and there is a rather good agreement between these theoretical curves and the experimental data, which represent brightness values in terms of a ratio scale constructed by one of the direct methods mentioned above (magnitude estimation).

The parameter descriptive of the curvature of the function is the exponent. This is about one-third for all wave lengths, which is about the same order of magnitude as the exponent for white light. The parameter c describes the relative brightness of a light of constant intensity and varying wave length. And finally, the parameter a can be considered to be related to the absolute threshold, although this relation may be complex (Ekman, 1959).

Psychophysiological relations

By far the greater part of modern psychophysical work has been devoted to the classical problem of stimulus-response relations. The main outcome has been the power law with its rather remarkable generality. This law is empirical and descriptive, and *per se* it does not tell us anything about the mechanisms responsible for the transformation it describes.

In general, these mechanisms will have to be investigated on the psychophysiological level, but very little has been done so far. It is well known that peripheral response processes often bear a logarithmic relation to stimulation. Since the final subjective outcome is a power function of stimulation, we have to expect that the central process generating the subjective response is an antilogarithmic function of the peripheral process. No such function has even been plotted, which I think should be a challenge to research workers in the neighboring fields of neurophysiology and psychophysics.

However, we have some information about differential sensitivity which will have to be incorporated in any theory of the psychophysiological transformations of a stimulus intensity into a subjective magnitude such as brightness or heaviness or pain. This information may be briefly summarized as follows.

1. Differential sensitivity, as measured on the stimulus continuum, is usually adequately described by the modified form of Weber's law

$$\Delta S = k_s S + a \qquad (2)$$

This has long been known and needs no particular illustration here.

2. Differential sensitivity, as measured on the subjective continuum,

seems to follow the same principle

$$\Delta R = k_R R + b \tag{3}$$

This is less well known and deserves a few comments. It was probably first pointed out by Harper and Stevens (1948) that the subjective counterpart of a just noticeable difference tends to increase with S. In some recent studies from our laboratory we have tried to measure the just noticeable difference in terms of R, and Fig. 3 shows the results for three sets of data. The graphs verify Eq. 3; they also suggest that the constant b may be negligible, but that question requires further investigation.

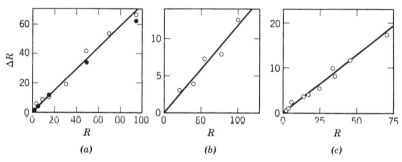

Fig. 3. Just noticeable differences in subjective units ΔR, plotted against subjective magnitude R: (a) values computed from data for lifted weights in two experiments by Oberlin (1936) transformed to a subjective scale; (b) results from an experiment on visually perceived velocity; and (c) results from an experiment on aurally perceived time. (Reproduced from Ekman, 1959.)

Let us for the moment accept both of the propositions stated above (Eqs. 2 and 3) and see what we can do with them. I think that the empirical information we now have may appear a little more intelligible if we look at it from the following points of view.

1. The *discriminative response* appears to be essentially a peripheral process. This would follow from the modification of Weber's law in Eq. 2 and the demonstration that peripheral response is a logarithmic function of stimulus intensity. When considered together, these relations mean that a just noticeable difference on the stimulus continuum has a counterpart in a constant increment on the continuum of *peripheral response*. In other words, the derivation of Fechner's law would be valid at the level of peripheral response.

2. The constant increment of peripheral response, however, corresponds to a subjective increment that is linearly related to the total

subjective magnitude. This indicates a multiplicative transformation of peripheral response into *subjective response,* and this is precisely the antilogarithmic transformation we need to come out with the power function relating subjective magnitude to stimulus intensity.

Perhaps we may tentatively conclude that *discrimination* and perception of *subjective magnitude* are psychophysiologically quite different processes, the latter being the final outcome of a step-by-step process of transformations.

Intrasubjective relations

Finally, I shall illustrate a type of problem that is more purely psychological than the two problems we have considered so far. Until now we have been investigating psychological variables as functions of physical or physiological variables. In the present context we are going to study relations *between* psychological variables.

One such problem which has been studied for some time in our laboratory is concerned with the experience of *similarity.* What function of their subjective attributes is subjective similarity between two stimuli? So far we have investigated this problem only for unidimensional continua, that is to say, we have allowed the two stimuli to vary only in one subjective dimension. I shall illustrate this work by our first experiment, in which we studied the similarity of pure tones of varying pitch and equal loudness. In one part of the experiment a pitch scale was constructed, and in the second part of the experiment a scale of subjective similarity was obtained. It was now possible to investigate similarity as a function of pitch (Eisler and Ekman, 1959). The result may be expressed by the equation

$$s_{ij} = \frac{R_i}{(R_i + R_j)/2}, \qquad (R_i \leqq R_j) \qquad (4)$$

where s_{ij} varies from 0 (complete absence of similarity) to 1 (identity). Data from this experiment on pitch are shown in Fig. 4, where the experimental measures of similarity are plotted against the theoretical values of similarity computed by Eq. 4 from the experimental measures of pitch. In subsequent experiments the same equation has been shown to describe the mechanism of similarity also in the continua of brightness (the subjective continuum was defined as darkness) and visual area (Ekman, Goude, and Waern, 1961) as well as heaviness (Eisler, 1960).

Since Eq. 4 has been verified in four quite different perceptual continua, it appears to possess a certain degree of generality. It is one of the few strictly quantitative principles or "laws" known to describe a psychological mechanism. An outstanding feature of Eq. 4 is the absence of empirical constants. One thing I want to stress here is that the simplicity of this relation could be discovered only because it was investigated exclusively *within the subjective domain*. Suppose that we had tried to investigate subjective similarity in terms of tonal frequency, other stimulus variables being equal. At best we could have arrived at a rather complex empirical equation which it would

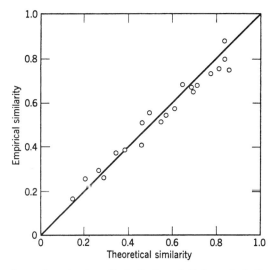

Fig. 4. Experimental estimates of similarity plotted against theoretical values computed according to Eq. 4 from a ratio scale of pitch. The data exhibit only a random fluctuation around the diagonal line representing perfect agreement. Similar results have been obtained for brightness, visual area, and heaviness. (Reproduced from Eisler and Ekman, 1959.)

not have been possible to "understand" immediately, because pitch is a nonlinear function of frequency and because loudness would have been a second subjective variable in the experiment. I think that these considerations are important, because they seem to indicate the possibility of discovering psychological mechanisms that may turn out to be rather simple when they are studied *on the proper level*.

Let me add a few concluding comments on the similarity principle. My first guess in planning these experiments was that similarity might be a simple ratio between the two scale values ($s_{ij} = R_i/R_j$, where $R_i \leq R_j$). When the data were plotted, we immediately had to reject this hypothesis, and then Dr. Eisler had a hard time fitting various functions until he found one that yielded acceptable results. It was first written as $1.98R_i/(R_i + R_j)$. We were fortunate in that 1.98 is pretty close to 2, and we could write Eq. 4 without empirical constants, letting the factor 2 represent an averaging operation.

Written in this form, Eq. 4 tells us that the subjective similarity between two percepts varying in one subjective dimension is the ratio between the lower scale value and the arithmetic mean of the two scale values. This interpretation is represented in the left-hand part of Fig. 5. It is certainly simple and, in a sense, intelligible. But perhaps we can "understand" the perceptual process even better by writing the same equation in the form

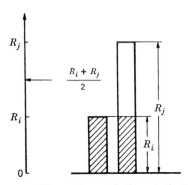

$$s_{ij} = \frac{2R_i}{R_i + R_j} \qquad (4a)$$

Fig. 5. Illustrations of the similarity principle expressed by Eq. 4. The left-hand part of the graph represents Eq. 4 and the right-hand part the alternative form, Eq. 4a. For explanations, see text.

which is represented in the right-hand part of Fig. 5. Let us see how it can be interpreted. There are two different percepts of intensities R_i and R_j, both present simultaneously as they are in the experimental situation. The similarity-generating process does two things with this material: (1) it counts off the part of R_j that corresponds to R_i (shadowed areas), and (2) it computes the *ratio between this common subjective magnitude* $2R_i$ *and the total subjective magnitude* $R_i + R_j$ present in the experimental situation.

Mathematically the two forms of Eq. 4 are identical, but personally I now prefer the latter form, since it implies a more concrete model of the operation performed in the similarity-generating process of a human nervous system. What this operation would look like from the point of view of a brain physiologist I do not know, but it would certainly be interesting to know.

Extensions of Psychophysics

All the work I have discussed in this paper has been carried out by means of those direct psychophysical methods that I mentioned in the first section. The introduction of these methods by Stevens and his co-workers initiated the rapid development of modern psychophysics.

These methods can be used only in experimentation with human subjects, since they require the subject to estimate relations between subjective magnitudes. This becomes a serious limitation as soon as we want to extend our work to include animal experimentation. There are many reasons for expecting such extensions to become important within the near future. Let me mention just two of them: (1) Some of the psychophysiological work that seems desirable, from a psychophysical point of view, will necessarily have to be done with animals. (2) Many intervening variables or hypothetical constructs (or whatever names they may go by) in learning theory are just the sort of things we are measuring every day in our human laboratories: subjective variables like pain and similarity. For well-known reasons we usually want to investigate learning as dependent on such variables on the animal level, and accordingly we would need techniques for measuring them in animals too.

For these reasons I would like to draw attention to another psychophysical tradition, initiated long ago by Thurstone (1927a, b) on the basis of test theory and developed in quite different directions. These methods may be called *indirect* methods, since they are based on a set of assumptions intervening between the experimental data and the final scale. Experimentally these methods are sometimes even better adapted to animal than to human experimentation. The kind of data for which they have been developed is in the form of right or wrong solutions, or of preference in a choice situation. This is precisely the sort of data you obtain in maze performance and other situations in which animal behavior is studied. The usefulness of these methods for measuring certain "underlying," not directly observable variables (response strength) in experiments on human learning has recently been convincingly demonstrated by Björkman (1958). The point I want to make here is that these methods might be used for measuring "subjective" variables (or whatever you may prefer to call them) in animals which cannot be asked what they experience (with human subjects it would be superstitious *not* to ask them). We shall have to make *some* assumptions, for example, that the similarity principle is valid for the rat, and then we can proceed to investigate psychological mechanisms. Let me mention here once more the possible *simplicity* of psychological relations when we investigate them on their proper (that is, psychological) level. The important thing to do in order to discover a simple relation that may exist between performance and motivation is to measure, say, amount of *thirst*—not hours of water deprivation.

One obvious step now would be to scale subjective variables, with human subjects, by the techniques of both sorts of psychophysical measurement, the direct and the indirect methods. If in a number of such situations we obtain the same results, then we should feel more confidence in using the indirect methods in those cases where only these methods *can* be used.

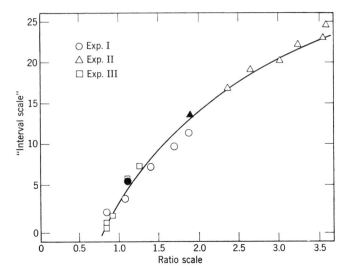

Fig. 6. A comparison between two scales obtained for the esthetic value of handwriting specimens. The "interval scale" obtained by one of the indirect methods (pair comparisons) is represented by the ordinate, and the ratio scale obtained by a direct method (ratio estimation) is represented by the abscissa. The curve represents the logarithmic relation to be expected if variability is proportional rather than constant.

Only few experiments of this sort have been undertaken. Those I know about were carried out in our laboratory (Björkman, 1959; Ekman, Künnapas, and Berglund, 1961). Some unpublished data from the latter study may serve as an illustration. Figure 6 shows results from experiments in which the esthetic value of handwriting specimens was measured. In this graph the interval scale y constructed by an indirect Thurstone method (pair comparisons) is plotted against the ratio scale x from a direct Stevens method (ratio estimation; scale values computed according to Ekman, 1958). The "interval scale" was constructed on the assumption that variability is constant (Thurstone's Case V). If, instead, it is proportional to the average scale value of the stimuli, then it may be shown that $y \approx a + b \log x$

(cf. Björkman, 1960). This function is graphed in Fig. 6, and the fit of the data shows that the same scale (apart from an arbitrary zero point) may be constructed by the indirect method (on the revised assumption of proportional variability) as well as by the direct method. Several studies are now being performed with the aim of investigating the general relations between direct and indirect scaling methods. The ultimate purpose of these studies would be to find a common rationale, so that essentially the same scale can be constructed by either type of method.

When more experiments of this sort have been carried out, we may proceed, with some confidence, to *measure* subjective pain, and thirst, and sexual drive, and brightness and—why not—esthetic value in infrahuman species. Whether, on the animal level, you prefer to call them "underlying" or "subjective," is of minor concern.

Finally it may be mentioned that even in human perception there are many psychological variables that cannot be directly observed. The variable intensity of that process that manifests itself in certain figural reversals has recently been successfully studied by Künnapas (1961) by means of an indirect method. Similar studies of the continuous, more-or-less rhythmical process which is responsible for "fluctuations of attention" are now in progress.

Summary

This paper includes an introductory discussion of the metric properties of the presumed ratio scales obtained by "direct" psychophysical scaling methods. Reference is made to some recent experiments demonstrating the additivity of such scales.

Three main types of problems, which may be investigated by psychophysical methods, are outlined:

1. Stimulus-response relations, where response is a subjective variable, constitute a classical psychophysical problem formulated long ago by Fechner. The general applicability of the power function to this kind of relation, which has been demonstrated especially by Stevens and his co-workers, is pointed out and illustrated by a new experiment concerning the brightness of monochromatic light.

2. The possibility of investigating certain psychophysiological relations is discussed. On the basis of existing information it is tentatively concluded that stimulus discrimination and, on the other hand, perception of subjective magnitude may be psychophysiologically quite

different processes, the latter being the final outcome of a step-by-step process of transformations.

3. Attention is drawn to the possibility of investigating certain intrasubjective relations, relations between subjective variables. A series of experiments concerning subjective similarity is chosen as an illustration of this approach. It was demonstrated that similarity is a simple mathematical function—without empirical constants—of the subjective properties of the percepts being compared. These results seem to indicate the possibility of discovering psychological mechanisms which may turn out to be rather simple when they are studied on the proper—that is to say, subjective—level.

Finally some possible extensions of psychophysical work are discussed with special regard to animal experimentation. Measurement of "subjective" or perhaps rather "underlying" variables in animals will necessitate the use of "indirect" psychophysical methods of the type introduced by Thurstone, and it thus becomes an interesting task to study the relation between these and the "direct" methods generally used in human experimentation. Results from a preliminary experiment of this type are reported.

References

Björkman, M. *Measurement of Learning*. Stockholm: Almqvist & Wiksell, 1958.

Björkman, M. An experimental comparison between the methods of ratio estimation and pair comparisons. *Reports, Psychol. Lab. Univ. Stockholm*, 1959, No. 71.

Björkman, M. Variability data and direct quantitative judgment for scaling subjective magnitude. *Reports, Psychol. Lab. Univ. Stockholm*, 1960, No. 78.

Eisler, H. Subjective similarity in the continuum of heaviness, with some methodological considerations. *Scand. J. Psychol.*, 1960, **1**, 69–81.

Eisler, H., and G. Ekman. A mechanism of subjective similarity. *Acta psychol.*, 1959, **16**, 1–10.

Ekman, G. Two generalized ratio scaling methods. *J. Psychol.*, 1958, **45**, 287–295.

Ekman, G. Weber's law and related functions. *J. Psychol.*, 1959, **47**, 343–352.

Ekman, G., H. Eisler, and T. Künnapas. Brightness scales for monochromatic light. *Scand. J. Psychol.*, 1960, **1**, 41–48.

Ekman, G., G. Goude, and Y. Waern. Subjective similarity in two perceptual continua. *J. exp. Psychol.*, 1961, **61**, 222–227.

Ekman, G., T. Künnapas, and A.-M. Berglund. Measurement of aesthetic value: a comparison between two psychophysical methods. *Reports, Psychol. Lab. Univ. Stockholm*, 1961.

Goude, G. On direct measurement in psychology. Thesis, University of Stockholm, 1959 [in Swedish].

Harper, R. S., and S. S. Stevens. A psychological scale of weight and a formula for its derivation. *Amer. J. Psychol.*, 1948, **61**, 343–351.

Künnapas, T. Measurement of the intensity of an underlying figural process. *Scand. J. Psychol.*, 1961, **2**, 174–184.

Oberlin, K. W. Variations in intensitive sensitivity to lifted weights. *J. exp. Psychol.*, 1936, **19**, 438–455.

Stevens, S. S. On the psychophysical law. *Psychol. Rev.*, 1957, **64**, 153–181.

Stevens, S. S., and E. H. Galanter. Ratio scales and category scales for a dozen perceptual continua. *J. exp. Psychol.*, 1957, **54**, 377–411.

Thurstone, L. L. Psychophysical analysis. *Amer. J. Psychol.*, 1927a, **38**, 368–389.

Thurstone, L. L. A law of comparative judgment. *Psychol. Rev.*, 1927b, **34**, 273–286.

3

J. C. R. LICKLIDER
Bolt Beranek and Newman Inc., Cambridge, Massachusetts

On Psychophysiological Models

Preface

This is not a written version of the talk I gave at the Symposium. In that talk, I summarized experimental data on information-processing capabilities of human beings and described a model for categorizing patterns of stimulation. The data were intended to set boundary conditions on the communicative capabilities of the over-all human organism. The model was intended to focus attention on problems that arise in learning to handle redundant messages. It was evident from the discussion that, in order to bring the data and the model into resonance with the thinking of many of the members of the group, I would have to relate both data and model more closely than I did to actual, observable physiological processes. For the material presented, however, physiological interpretations are still to be formulated.

Instead of insisting on the propaedeutic values of those data and that model, therefore, I am turning, in these written pages, to a smaller area in which it is possible to suggest fairly specific relations between parts of a model and observable neural processes. The conviction that guides my effort remains the same as it was before: a conviction that there are value and promise in psychophysiological models.

Psychophysiological models seem to me to offer the possibility of bringing together in productive interaction (1) quantitative and qualitative findings of experimental and (even) clinical psychology, (2) substantive facts of neurophysiology and neuroanatomy, (3) abstract theorems and principles of mathematics and logic, and (4) insights and formulations achieved in the course of developing the technology of information processing and communication. Perhaps because I value that possibility very highly, it seems to me that it currently commands relatively little effort—relatively little, that is, as compared with efforts in purely psychological and purely physiological research or in psychophysiological research that does not involve mathematical models or ideas from communication theory, information theory, or the young field of "artificial intelligence." Happily, however, several papers presented at the Symposium and several parts of the discussion contributed strongly toward the development of, and, I hope, an ascendency for, psychophysiological models.

Procedure and Experience

Audio analgesia is a phenomenon only recently discovered and not yet well understood (Gardner and Licklider, 1959a–c). The essential fact is that a procedure involving acoustic stimulation but no conventional anesthetic or analgesic agents has made it possible for patients who previously had required nitrous oxide or a local anesthetic to undergo dental operations without serious pain or unpleasantness.

The procedure is briefly this: The patient wears earphones and controls his own acoustic stimulation through a control box held in his lap. It has two control knobs, one for music and the other for a rushing, roaring sound derived from "white" noise. At the beginning of the session, the patient selects the music he wants to hear—a stereophonic tape recording—and adjusts it to a volume suitable for ordinary listening. When the dentist starts to work, or when his work causes any discomfort, the patient turns up the volume of the music. As soon as there is a trace or forewarning of pain, the patient turns the noise knob. It controls the level of the rushing, roaring "waterfall" sound. The over-all sound pressure of the noise may be set as high as 116 decibels above 0.0002 microbar. The intense noise drowns out or suppresses the pain in a great majority of cases.

There is no longer any question of the efficacy of the procedure in dental operations. Dr. Wallace J. Gardner, of Cambridge, has used the procedure with more than 1000 patients who previously required chemical agents. The audio analgesia was completely effective with 65 per cent, and effective enough with another 25 per cent that no other agent was required and the acoustic procedure was preferred for subsequent visits. Ten other dentists have now had fairly extensive experience with the procedure (more than 5000 patients in all), and their experience has paralleled Gardner's very closely. The operations have included cavity preparation (drilling), grinding, and extraction. Gardner has had about 200 successful extractions—including several multiple extractions—and no failures. In short, in the dental office, the acoustic procedure appears genuinely to suppress pain, to provide analgesia in the sense of standard dictionary definitions of that word.

Thus far, there is not enough experience with the procedure in nondental clinical situations to justify conclusions. The available evidence, provided by physicians who have used the same sound-producing equipment as was employed in the dental observations, is partly positive, partly negative. The procedure was effective in several minor surgical operations, but not in all. It was wholly sufficient in five cases of labor and childbirth but required supplementation in three others. It was ineffective in suppressing intractable pain associated with cancer.

In the experimental laboratory, it is difficult to make a dramatic demonstration of suppression of pain by sound, but it is easy to create a discernible effect. On the basis of preliminary experimentation, it appears that the factor of tension-relaxation must be introduced into the laboratory situation before the degree of effectiveness observed in the dental office can be achieved (Weisz, 1959). The same statement can be made, however, with reference to morphine and other pain-relieving drugs. There is little or no difference in effectiveness between morphine and placebo in a typically objective, analytical, laboratory context (Hill, et al., 1952).

Psychological Characteristics of Audio Analgesia

Experience thus far gained has revealed several characteristics of audio analgesia for which one would like to have an explanation. The characteristics are by no means precisely quantified; they are largely qualitative and clinical:

1. In the absence of acoustic stimulation and conventional analgesic or anesthetic agent, pain tends to build up regeneratively during intervals of strong nociceptive stimulation.

2. When nociceptive stimulation is withdrawn, after an interval of pain in the absence of acoustic stimulation and the conventional analgesic or anesthetic agents, the magnitude of the pain sensation decreases, but there is often a residual pain sensation, an after-pain.

3. When an intense auditory stimulus is turned on and held for an interval at constant sound-pressure level, the loudness rises rapidly to a maximum, then equilibrates or recedes to a relatively steady level ("perstimulatory fatigue").

4. Ordinarily, after the first 100 milliseconds, there is no positive after-sensation in hearing, but a depression in sensitivity can be measured for a period following moderate or intense acoustic stimulation.

5. In at least some situations, intense acoustic stimulation has a definite suppressive effect on the subjective magnitude of pain and on overt pain reactions.

6. In general, effectiveness in suppressing pain increases with sound intensity, decreases with the intensity of the noxious stimulation.

7. Relaxed patients experience more relief from pain than do tense patients. The main role of the music in the procedure is to relax the patient before the operation is begun. Except for hard-of-hearing patients and children too young to manipulate the controls effectively, the patients with whom the audio procedure has been least effective have been patients who entered the situation in extreme tension or anxiety.

8. In the experimental laboratory, in a psychophysical context, one can often detect a reduction in the pain produced by an electric shock or other noxious stimulus, but it is only a modest reduction, much less dramatic than the effects observed in clinical situations or even in laboratory situations in which the psychophysical context and set for careful observation are replaced by conditions favoring relaxation and diversion of attention.

9. The procedure is more effective if the noise is turned up before pain develops than it is if the noise is withheld until pain is clearly present.

10. Stimulation by intense noise appears to have a more-or-less persistent after-effect on pain. In some instances, after long or intense exposure, it has been possible for patients to undergo ordinarily painful operations without further presentation of noise. In other instances, patients have been able to follow the gradual reappearance of pain during the course of seconds following presentation of a burst of noise.

11. Pain may have some reciprocal effect on the loudness of noise or music, but it is not as marked as the effect of sound on pain.

12. Once a pain has been suppressed, it can sometimes be kept suppressed by a weaker sound than was at first required to overcome it.

13. Quite a few patients have fallen asleep while listening to music or noise during a dental operation. Many have fallen asleep while still listening to music or noise in the interval immediately following an operation.

14. Individual differences in use of the acoustic procedure are very great. Some patients require intense noise; others use only moderate

levels and report that pain suppression is fully effective. Some patients rely heavily on the music, turning it up to full volume instead of using noise. Others use the noise control at every suggestion of pain.

15. The several dentists who have used the audio procedure have had approximately the same percentage of success.

16. Random noise is more effective in suppressing pain than other sounds tested, with the exception of random noise plus music.

17. Although the high-frequency components of "white" noise must be attenuated if one is to produce a sound that most patients will accept and willingly adjust to a high volume level, the upper cutoff frequency of the noise must not be made too low or effectiveness in suppressing pain will be lost.

18. Some patients have reported that, while using the audio procedure, they felt something they could call pain, but that it "didn't hurt the way pain ordinarily does."

19. Against certain types of pain, such as the pain produced by an electrodepilatory needle, photic stimulation appears to be more effective than acoustic stimulation (Baruch and Fox, 1959).

20. Distraction of attention from pain to something else appears to reduce pain.

The items just listed do not provide a basis for curve fitting, for the calculation of coefficients of correlation, for any of the activities usually associated with the development and testing of mathematical models. Their qualitative nature may appear, indeed, to bar them from the realm of models. Yet it seems worth while to try to organize the picture in some other way than merely to list the 20 statements.

"Explanations" of Audio Analgesia

Almost surely, several factors operate together in producing audio analgesia. Their relative weights or importances vary from patient to patient and situation to situation. These seven seem usually to be the main ones: (1) direct suppression of pain sensation, either the primary pain or the secondary elaboration of, or reaction to, the primary pain [reference is made here to Beecher's (1957) distinction]; (2) relaxation, with consequent reduction of the elaboration reaction; (3) distraction of attention from pain, with consequent reduction of the apparent severity of the pain; (4) masking of the dentist's drill (in operations in which a drill is used), with consequent reduction in the anxiety conditioned to the drill through earlier painful experiences; (5) improved communication with the dentist (through the changing levels of music and noise, which the dentist hears through his own earphones or monitor loudspeaker), with consequent reduction in fear, anxiety, or feeling of helplessness; (6) active role for the patient (who now can control a massive part of his experience in the dental chair), with consequent reduction in fear, anxiety, or feeling of helplessness; and (7) suggestion.

The seven factors have been cited, individually and collectively, as "explanations" of audio analgesia. Clearly, they are not explanatory in the reductionist sense; they do not describe any mechanism or process. Instead, they relate characteristics of the audio procedure to generalities that cover other situations than audio analgesia, and thus perhaps lend some plausibility to what might otherwise seem wholly implausible. This kind of "explanation" seems unsatisfactory, but it is about all that can be achieved without recourse to hypotheses that involve the neurophysiological substratum of behavior.

Neurophysiological and Neuroanatomical Constraints on Models of Intersensory Interaction

Neurophysiological and neuroanatomical researches are making such rapid strides toward an understanding of interactions among sensory modalities in the reticular formation and related centers of the brain that it may soon be possible to specify precisely the loci involved in the suppression of pain by sound. For the time being, however, we may concentrate on process and neglect considerations that point to particular regions or nuclei, noting only that there is abundant opportunity, at levels from the upper boundary of the medulla to the cerebral cortex, for sound signals and pain signals to come together. What are some of the general neurophysiological and neuroanatomical considerations that may serve as guide lines or constraints in the development of a model of the interactive process?

1. The organization of the nervous system is characterized by focalized excitation or facilitation and diffuse inhibition or suppression. (The papers presented at this Symposium by Ratliff and Mountcastle provided beautiful examples of this principle of organization.) Where two or more sensory systems come into a common region but do not map themselves, point-for-point, into a common projection, therefore, we may expect the effects of one system on another to be inhibitory or suppressive rather than excitatory or facilitory.

2. It is necessary for the sake of stability that, on the average, inhibitory processes have higher gains than excitatory processes at high levels of activity, and that excitatory processes have higher gains than inhibitory processes at low levels of activity. Otherwise, the brain would either "run down" into quiescence or undergo a regenerative "chain reaction" up to saturation.

3. The temporal parameters of excitatory processes are usually shorter than those of inhibitory processes. This fact may be related to the focalization of the former and the diffuseness of the latter.

4. There is measurable activity in the neural substratum (for example, "spontaneous activity") even in the absence of contemporary stimulation and in the absence of (reliable) sensation. (The word "reliable" is introduced to avoid argument here about thresholds.) It seems reasonable to postulate an underlying pain process that has continuous existence, though at low levels, during intervals in which there is ordinarily no report of pain.

A Psychophysiological Model of the Process Underlying Audio Analgesia

The interests of simplicity and generality pull in opposite directions. Because the facts to explain are so few and so qualitative, there is justification, at best, for only a very simple model. On the other hand, we may expect—we know, in fact, that there are—interactions among all the sense departments, and it seems restrictive to limit consideration to only hearing and pain, or to only a given pattern of facilitory and inhibitory coefficients. The compromise has been to formulate a general model, one with a variable number n of channels, each channel having variable parameters, but to exercise in the trials to be described here only a few specific realizations in which n equals 2. The effect, therefore, is clearly on the side of simplicity. It is, in fact, deliberate oversimplification, to represent a neural mechanism that involves several cascaded networks, each containing thousands of nerve cells characterized in part by all-or-none response, by a mere pair of channels involving only a few continuous variables. The object, however, is not to construct a brain. It is to construct a conceptual network (1) that is easy to understand, (2) that bears a recognizable (and potentially improvable) resemblance to part of the nervous system, and (3) that will behave in ways that manifest the characteristics of audio analgesia.

The model incorporates some, but by no means all, of the ideas

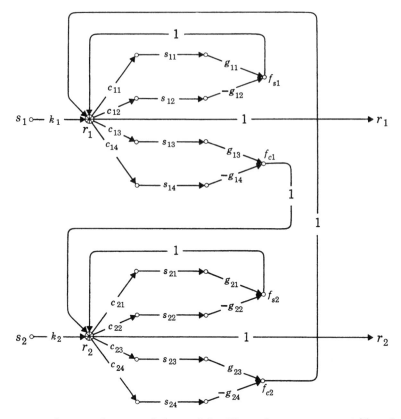

Fig. 1. Schematic diagram of the model. The nodes represent variables, the branches operations. The stimuli are s_1 (sound) and s_2 (pain). The responses are r_1 (loudness process) and r_2 (pain process). The variables f_{s1}, f_{s2}, f_{c1}, and f_{c2} are self- and cross-feedbacks derived from r_1 and r_2 through the successive operations of clipping (c_{ij}), smoothing (s_{ij}), amplification ($+g_{ij}$), and addition. The over-all effect is to produce two stimulus-response channels, each subject to both positive and negative feedback from its own response and from the response of its neighbor.

about the pain mechanism that have been described by Hardy, Wolff, and Goodell (1952), Beecher (1957), and Melzack, Stotler, and Livingston (1958).

The model is illustrated schematically in Fig. 1. The stimulus or input is s_1. The response or output is r_1. The action of the network can be described most readily if we assume, for the moment, that the node marked with superposed + and × signs is simply an adder, as all the other nodes in fact are.

The response r_1 is formed by adding together the stimulus s_1, the self-feedback signal f_{s1}, and the cross-feedback signal f_{c2}. The self-feedback signal f_{s1} is formed from r_1 by clipping (c_{11} and c_{12}), smoothing (s_{11} and s_{12}), amplifying (g_{11} and $-g_{12}$), and adding (at the node labeled f_{s1}). The characteristic of a typical clipper is

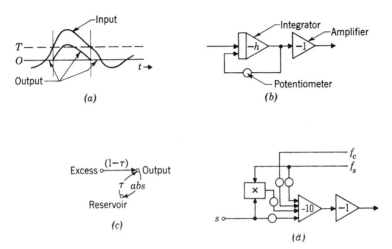

Fig. 2. Details of operations in the model. (a) Behavior of a clipper that passes only the parts of the signal that exceed a preset threshold T. (b) Conventional analog integrator with negative feedback, equivalent to a resistance-capacitance smoothing filter, with time constant determined by the setting of the potentiometer. The amplifier following the integrator makes the gain of the network positive. (c) A digital smoothing circuit. The output is the weighted sum of the input (weight $1 - \tau$) and the quantity (weight τ) accumulated in the reservoir. The quantity in the reservoir is a remnant of past values of the input. The result is closely similar to that provided by (b). (d) The network that provides a weighted sum of addition and multiplication at the nodes marked by superposed $+$ and $\times$ signs in Fig. 1. Appropriate adjustment of the potentiometers provides the response $r = a(s + f_s + f_c) + (1 - a)(s \cdot f_s)$.

shown schematically in Fig. 2a. The threshold T_{12} of c_{12} is higher than the threshold T_{11} of c_{11}. That gives the positive-feedback path the advantage at low signal levels. An analog smoother is illustrated in Fig. 2b, and a digital smoother is illustrated in Fig. 2c. In the latter, τ is a multiplicative coefficient, and $(1 - \tau)$ is its complement; *abs* is an operator that delays the signal one interval of calculation time and then replaces the signal by its absolute magnitude. Both Figs. 2b and c approximate the action of a simple resistance-capacitance smoothing filter. The time constant of s_{12} is longer than that of s_{11}. That makes the negative feedback more sluggish than the positive feedback. Finally, g_{12} is greater than g_{11}. That gives the advantage of greater gain to the negative feedback paths when the signal level is much greater than the clipping thresholds.

The signals f_{s2}, f_{c1}, and f_{c2} are derived in ways that correspond directly to the one just described. We have in sum, therefore, both positive and negative feedback from each channel to itself and from each channel to every other. The only thing that remains to be described is the node with the superimposed $+$ and $\times$ signs.

Consider two neural channels that come together and impinge upon a third. The response z of the third is a function of the levels of activity x and y of the incoming two: $z = F(x, y)$. In what is perhaps the simplest concept, F is the sum: $z = x + y$. In another interpretation, almost equally familiar, F is the product: $z = x \cdot y$. To achieve a helpful but limited degree of generality, we may combine the two notions linearly and let the equation $z = F(x, y)$ be $z = a(x + y) + (1 - a)(x \cdot y)$, with $0 \leq a \leq 1$. That is essentially the interpretation of F used in the model. In the model, however, three "neural channels," not two, impinge upon the outgoing channel.

The interpretation actually employed in the model is shown in Fig. 2d. The three incoming signals are s, f_s, and f_c. The response r of the outgoing channel is $a(s + f_s + f_c) + (1 - a)(s \cdot f_s)$. That limits the nonlinear (multiplicative) action to the self-feedback signal f_s and the input s. To a degree controlled by $(1 - a)$, the self-feedback signal turns a valve through which the input flows to yield the response. The input variable s is never negative.

Identifications between Variables of the Model and Variables of Audio Analgesia

The interpretation of the model is direct. The two channels are parts of the auditory and pain systems. The input variables s_1 and s_2 are the strengths of acoustic and nociceptive stimulation, respectively. The output variables r_1 and r_2 are the strengths of the sound and pain *processes*, respectively. The subjective magnitude of the sound (loudness in sones) is a monotonic, nondecreasing function of r_1, and the subjective magnitude of the pain (subjective painfulness in dols) is a monotonic, nondecreasing function of r_2. For present purposes, we may say that loudness L is equal to $r_1 - c_1$ when $r_1 \geq c_1$, and L is equal to 0 when $r_1 \leq c_1$, and we may say that subjective painfulness P is equal to $r_2 - c_2$ when $r_2 \geq c_2$, and P is equal to 0 when $r_2 \leq c_2$. To simplify things further, we may let $c_1 \cong 0$. For the pain channel, however, it seems better to assume that there is no ordinarily reported pain until the pain process exceeds a threshold significantly above zero.

The thing in the model that corresponds most directly to "level of tension" in a patient is the sum of the feedback signals. Note that this is not a unitary thing. Tension, as the word is used in clinical contexts, is a global term. In the model, even oversimplified as it is, there are two "tensions" in the auditory system and two more in the pain system. One member of each pair ("anxiety," perhaps) has a direct effect upon the gain of the channel; the other has only an indirect effect.

The clipping threshold T_{21} associated with c_{21} in the model bears a relation to tolerance for pain in the actual patient. If T_{21} is high, the regenerative growth of pain does not begin until the level of subjective pain is fairly high; if T_{21} is low, the regenerative build-up begins at once. The thresholds associated with the other clippers may be given corresponding interpretations.

Behavior of the Model in Response to Stimulation

The model has been realized in two forms: a digital computer program and an analog computer setup. Because the digital computer is a small, slow machine, the digital program has been run in only a few configurations, all with $a = 0$, which specifies linearly additive combination at the nodes at which stimuli and feedbacks are combined. The digital program, however, can be changed from two channels to twenty by writing "20" on a typewriter, and it thus has the advantage of growth potential. The analog computer set-up produces results much faster. It has been run with both linearly additive and partially multiplicative combinations at the crucial nodes.

In describing the simulation tests, we shall refer to the variables as acoustic and nociceptive stimuli, sound and pain responses, and so on. That seems preferable to the alternative of designating them always by letters because (1) it brings to the foreground the linkages between variables of the model and variables of the situation represented and (2) there seems to be little danger of confusion between the two realms of discourse.

Behavior of the digital simulation

The behavior of the digital program in a long run with three modes of stimulation is shown in Fig. 3. The parameters of the model for this run are given in Table 1.

For Fig. 3a, only nociceptive stimulation was employed. The painful stimulus was applied first at a weak level which we may call 1 unit, then at 2 units, then at 4 units, and finally at 8 units. The

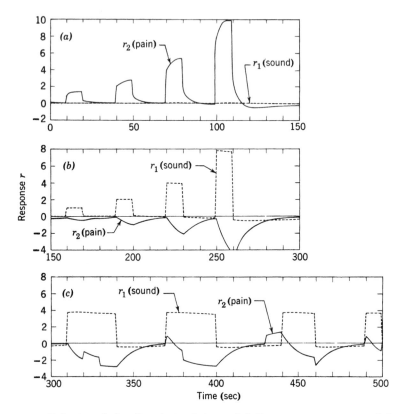

Fig. 3.   Behavior of the digital simulation. (a) Responses to a series of four noxious stimuli. The stimuli occur at the intervals 9–19, 39–49, 69–79, and 99–109 sec. Their relative strengths are 1, 2, 4, and 8. (b) Responses to a corresponding series of four acoustic stimuli. Note the depression of the already subliminal pain process. In (c) acoustic and nociceptive stimuli are presented together. The acoustic stimulus is on at strength 4 during intervals 309–339, 369–399, 439–459, and 489–499 sec. The noxious stimulus is on during intervals 319–329, 369–379, 429–459, and 489–499 sec.

responses of both the auditory and pain channels are shown. The pain response rises rapidly as soon as the nociceptive stimulus is applied, continues to build up during the course of stimulation, and drops rapidly, but not to zero level, when the stimulus is removed. The

Table 1.   Parameters of Digital Simulation for Tests Shown in Fig. 3

i	a	j	k_j	T_{ij}	τ_{ij}	g_{ij}
1	0	1	1.0	0.0	0.7	3.0
		2		4.0	0.975	5.0
		3		1.0	0.9	0.3
		4		4.0	0.925	0.5
2	0	1	1.0	0.0	0.9	1.0
		2		1.0	0.975	5.0
		3		1.0	0.9	0.5
		4		0.5	0.9	10.0

nociceptive stimlus has a small effect on the loudness process. A corresponding effect of actual pain on actual loudness might be made noticeable in a suitably designed psychoacoustic experiment. In the dental situation, however, such a small suppressive effect would not be noticed.

For Fig. 3*b*, only acoustic stimulation was employed. The applications are graded, as they were in Fig. 3*a*. While the stimulus is on, the loudness process holds almost a steady level. When the stimulus is terminated, the loudness process falls to subzero values. This is analogous to the "poststimulatory fatigue" observed in auditory tests (Caussé and Chavasse, 1947; Gardner, 1947; Lüscher and Zwislocki, 1947; de Maré, 1939).

For Fig. 3*c*, both acoustic and nociceptive stimulation were employed. We may examine the effect of the sound on the pain by comparing Fig. 3*c* with 3*a*. It is the temporal course, rather than the magnitude, of the suppressive effect that is of primary interest. [The effect can be made as marked as is desired, of course, by manipulating parameters of the model. The magnitude of the effect observed with this particular configuration of parameter values (Table 1) is characteristic of a very successful application of the audio procedure to a favorable subject.] It is noteworthy that the effectiveness of the acoustic stimulation in suppressing pain is greatest when the acoustic stimulus is applied before the painful stimulus, and that there is a small "twinge" of pain at the beginning of stimulation when the two stimuli are applied simultaneously. Corresponding effects are noted repeatedly in dental operations.

Behavior of the analog simulation

Figure 4 shows a sample of the behavior of the analog simulation. Table 2 gives the corresponding parameters. Throughout the record,

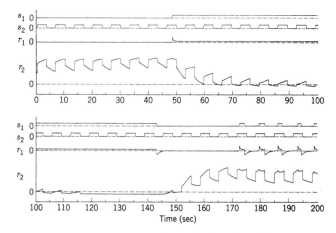

Fig. 4. Behavior of the analog simulation. The chart shows two contiguous 100-sec intervals of stimulation and response. The acoustic and nociceptive stimuli are marked s_1 and s_2, respectively, and the corresponding responses, r_1 and r_2.

Table 2. Parameters of Analog Simulation for Tests Shown in Fig. 4

i	a	j	k_i	T_{ij}	τ_{ij}	g_{ij}
1	0.1	1	1.0	0.5	0.5	1.0
		2	1.0	4.0	1.0	1.0
		3	1.0	0.5	0.5	0.5
		4	1.0	4.0	1.0	4.0
2	0.1	1	1.0	0.5	1.0	1.0
		2	1.0	4.0	10.0	1.0
		3	1.0	0.5	1.0	0.0
		4	1.0	4.0	10.0	0.5

a fairly strong nociceptive stimulus is turned on and off periodically. A relatively weak acoustic stimulus is turned on for a prolonged interval covering the middle portion of the sample and, again, for short intervals near the end of the sample. With the parameters of Table 2, the loudness process equilibrates rapidly and then holds a steady level. A corresponding temporal course has been observed for actual loudness by Hood (1955) and Egan (1955), in psychoacoustic experiments. When the stimulus is terminated, the loudness process falls to subzero values.

The suppressive effect of the acoustic stimulation appears on two quite different time scales. The gradual, progressive suppression produced by the long presentation of the acoustic stimulus is clearly evident in the middle part of the record. A more immediate, but less marked, effect can be seen at the onset of acoustic stimulation. The latter effect is clearest in the traces near the end of the record, where the sound is turned on and off during each of several bursts of pain

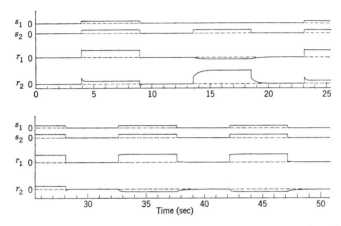

Fig. 5. Behavior of the analog simulation. The layout is similar to that of Fig. 4. Compare the second pain response, during quiet, with the first and third, during noise. The fourth and fifth pain responses are entirely suppressed.

stimulation. The great difference between the effect producible in a moment and the effect that can build up over a period of time should be borne in mind.

Figure 5 shows another sample of the behavior of the analog simulation, this time with the parameters of Table 3. In Fig. 5, the acoustic

Table 3. Parameters of Analog Simulation for Tests Shown in Fig. 5

i	a	j	k_{ij}	T_{ij}	τ_{ij}	g_{ij}
1	0.1	1	1.0	0.5	0.5	1.0
		2	1.0	4.0	1.0	1.0
		3	1.0	0.5	0.5	0.5
		4	1.0	4.0	1.0	5.0
2	0.1	1	1.0	0.5	1.0	1.0
		2	1.0	4.0	10.0	1.0
		3	1.0	0.5	1.0	0.0
		4	1.0	4.0	10.0	1.0

stimulus is not strong enough, relative to the painful stimulus, to suppress the pain process rapidly.

The painful stimulus and the acoustic stimulus are presented exactly together in every stimulus cycle but the second. In the second, the acoustic stimulus is withheld. The pain is clearly worse in the absence than in the presence of the sound. In the latter part of the record, the successive presentations of the sound gradually suppress

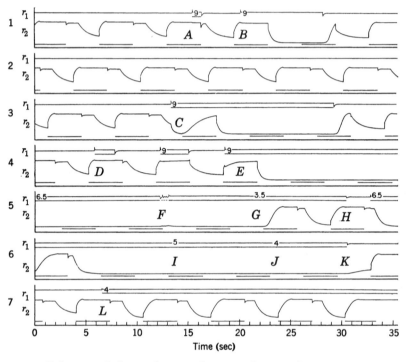

Fig. 6.   Behavior of the analog simulation. The stimulus presentations are marked, in this figure, by lines below the responses. The intensity of the painful stimulus, when on, is constant. The relative strengths of the acoustic stimuli are specified by numbers superposed upon the record. In addition to effects shown in earlier figures, this figure illustrates how the effectiveness with which sound suppresses pain depends upon the state of relaxation-tension at the time the sound is applied.

the pain process but never entirely overcome it. The pain process has a discernible effect upon the auditory process, but again it is not marked.

Essential characteristics of audio analgesia not yet discussed are exhibited in the behavior of the analog model with the parameters given in Table 4. Figure 6 shows the results of a long run with those parameters. Throughout the record, a very intense nociceptive stimu-

Table 4.   Parameters of Analog Simulation for Tests Shown in Fig. 6

i	a	j	k_{ij}	T_{ij}	τ_{ij}	g_{ij}
1	0.3	1	1.0	0.5	0.5	1.0
		2	1.0	4.0	2.0	1.0
		3	1.0	0.5	0.5	0.5
		4	1.0	4.0	2.0	4.0
2	0.3	1	1.0	0.5	2.0	1.0
		2	1.0	4.0	10.0	1.0
		3	1.0	0.5	2.0	0.0
		4	1.0	4.0	10.0	0.0

lus goes on and off periodically.   Its intensity while on is constant. An acoustic stimulus also goes on and off, but at irregular intervals. The intensity of the acoustic stimulus is varied several times during the run.

At the beginning of line 1 in Fig. 6, the acoustic stimulus is off.   The pain process is going through cyclic variations in which it builds up to quasi saturation when the noxious stimulus is on and recovers partially while the noxious stimulus is off.   The quasi saturation level is determined by the interplay of the positive and negative feedbacks. Each time the painful stimulation is removed, the pain process drops suddenly, but only for a short time.   Then for a short time it holds a high level (again, determined by the interplay of the feedback signals), and then it declines exponentially toward its quiescent operating point.

A brief burst of intense sound (intensity 9) at A in Fig. 6 has a barely discernible immediate effect on the pain process; the effect on the recovery phase is somewhat more evident.

The intense sound stimulus beginning at B has only a minor effect during the pain-stimulus interval because of the great intensity of the pain stimulus.   As soon as the pain stimulus is removed, however, the sound drives the pain process down rapidly.   Note, particularly, that, once the pain process has been suppressed, even the strong nociceptive stimulus is unable to break through the suppression provided by the sound.   Only when the sound is turned off does the pain process rise.   Then, by the second cycle, it is back to its quasi-saturation-and-partial-recovery behavior.   That behavior continues through eight pain cycles (lines 2 and 3) without further acoustic stimulation.

When the sound is turned on again at C (intensity 9), the painful stimulus is off.   The sound rapidly suppresses the pain process, but the painful stimulus comes back on before full suppression is achieved.

The burst of pain stimulation elevates the pain process but does not succeed in raising it to quasi saturation because it is working against the influence of the sound. During the second pain-stimulusfree interval, the suppressive effect of the sound fully depletes the pain process, and there is no more pain activity until the sound is turned off.

Beginning at *D*, the record illustrates effects of bursts of sound presented during the intervals of painful stimulation. As before, the effects are noticeable mainly in the recovery intervals.

Thus far, during the run, the acoustic stimulus has been either at intensity 9 or at intensity 0. At *E*, shortly after an onset of the pain stimulus, the acoustic stimulus is turned on again at intensity 9. By the second pain cycle, it has fully suppressed the pain process. Then, at the beginning of line 5, the strength of the acoustic stimulus is reduced to 6.5. Even at the reduced level, it is effective in preventing the rise of the pain process during the following cycles. During one of them, at *F*, there was difficulty with an "earphone lead," and the pain process almost was able to resume activity. The difficulty was corrected quickly, however, and the pain process subsided.

At *G*, the acoustic stimulus was again reduced in strength, to 3.5. At this new level, the suppressive effect was not great enough to hold down the pain process, which returned to its cycle of quasi saturation and partial recovery.

Next, at *H*, the acoustic stimulus was turned off, and then back on again at intensity 6.5. As before, it gradually overcame the pain and reduced the pain process to near-zero level. Again the sound intensity was reduced—to 5 at *I* and then to 4 at *J*—and again the weaker acoustic stimulation was effective in holding the pain process in abeyance through several cycles. When the weaker acoustic stimulus was then turned off at *K*, the pain process resumed its familiar course.

Note, particularly, now, at *L*, that when the weaker acoustic stimulus was turned on again, it was unable to have much effect on the pain. It could hold the pain process down when it caught the pain process in a suppressed state (*J*), but it could not overcome the pain process when the latter was fully active (*L*). The run ends with several cycles of quasi saturation and partial recovery despite the presence of the not-quite-sufficient acoustic stimulation.

Evaluation of the Model

The model is admittedly preliminary, tentative, and oversimplified. Little comes out of it that was not put into it other than the demonstration that the network actually does what intuition says it must do. Of course, the model can relate the behavior to the parameters more accurately than an intuitive grasp of the process alone can do, but that is not now an important capability because individual differences among actual patients correspond to wide ranges of parameter values. If we count all the possibly variable parameters, we find that the two-channel model with addition plus multiplication at the critical node has about 26, which is greater than the number of facts we have against which to check the behavior of the model. From one point of view, that is a disastrous situation. From another, one sees that most of the 26 parameters need never be varied and that the variations that are made are constrained by predetermined patterns, but those observations do not wholly overcome the objection that there are "too many degrees of freedom."

Is such a model, then, worth while? Does the fact that characteristics of its behavior parallel characteristics of actual audio analgesia give it any value or significance?

In approaching that question, it may be helpful to review the correspondence between model and actual characteristics. Of the twenty characteristics we listed, the behavior of the model directly displays the first twelve. Implicit in the simple, two-channel version of the model are places for three more of the characteristics (Nos. 13, 14, and 15). Increasing the number of channels would accommodate four more (Nos. 16, 17, 18, and 19). An extension would take care of the remaining characteristics (No. 20).

Some of the correspondences are obvious. Others require interpretive notes:

7. Effectiveness of suppression increases with relaxation, decreases with tension. In the model, relaxation-tension is identified with the feedback signals, which can be monitored individually. In versions of the model involving multiplication at the critical node, the gain of the pain channel depends upon the level of self-feedback. That level, in turn, depends upon the level of cross-feedback from the auditory channel.

8. The immediate suppressive effect (which corresponds to the effect that can be observed in a psychophysical context) is less marked than the effect that develops gradually.

12. When the pain process and the feedback signals are at low levels, the gain of the pain system is low, and the suppressive influence of the sound has less to overcome.

13. If we identify sleep with extremely low levels of tension

(Jacobson, 1938), and if we focus attention on the pain channel, we see that, in a few instances during the stimulation runs, the audio procedure induced sleep. That, of course, is an oversimplification, but it corresponds well with experience in the dental office, and it may go even further. Preliminary observations indicate that infants may be put to sleep by noise of the type used in the audio procedure.

14. Varying the parameters changes the amount of sound required to suppress a given amount of pain. It seems reasonable to suppose that there are large individual differences among people in respect of the neural parameters that correspond to the simulation-network parameters.

15. The dentist or experimenter enters the picture only through his manipulation of stimulus variables and of such parameters as relaxation-tension. In the actual, clinical situation, that is an over-simplification, of course. Nevertheless, the fact that approximately the same degree of success was achieved by the several dentists, all of whom were aware of the importance of relaxing their patients, is just what would be expected on the basis of the model, and not at all what would be expected if subtle suggestion, or even strong suggestion bordering on hypnosis, were essentially involved.

16–17. In a version of the model with many channels, several channels could be identified as auditory and several others as pain channels. To achieve effective suppression, with such a system, it would be necessary to stimulate many or all of the auditory channels. The advantage claimed by random noise because of its continuous coverage of the frequency scale would then be reflected by the behavior of the model. The advantage inherent in temporal continuity is, of course, reflected by the behavior of even the two-channel model.

18. In a multichannel model, the parameters would vary among the several pain subchannels. Suppression should then change the quality of experienced pain if it did not abolish the pain entirely.

19. In a multichannel model in which parameters varied among the channels, one sense modality would be likely to have an advantage against one pain subchannel, another sense modality against another pain subchannel. That would yield a parallel to the behavior observed with acoustic and photic stimulation.

20. To handle distraction of attention, it seems desirable to introduce a mechanism that will provide an analog of selective interest. It would be best to work with a multichannel model. The responses r_i would then constitute a pattern of nondegenerate dimensionality. Networks could be arranged to provide positive feedback in response to certain patterns, negative feedback in response to others. Such a

system would have a crude analog of attention, and its "attention" would be distractible. We may return, now, to the question of the value of such a model as the one we have been discussing.

The main value, it seems to me, stems from the role of the model as an organizer. The model pulls a variety of facts together into a compact diagram, and it interrelates them, one to another. The model consolidates the experience thus far obtained and makes it easy to go on to extensions and ramifications without losing one's grasp on the things one has already dealt with.

A second value that is important in the context of audio analgesia stems from the fact that, although the model displays rather subtle and complex behavior, it is mechanistic. That might not be important in a reductionist, laboratory-research context, but it is important in the clinical context. The model is useful, for example, in providing a nonmentalistic rationale for the part of the procedure intended to facilitate relaxation. The model even offers some help in clarifying the question of the role of suggestion: it is often suggested that audio analgesia is a form of hypnosis. Hypnosis and audio analgesia share the involvement of relaxation and the suppression of pain without the aid of chemical agents. However, audio analgesia involves no special relation between patient and dentist, and it appears to be effective for a larger percentage of patients. The model serves to turn the question of the relation between audio analgesia and hypnosis from semantics to experimentation. One begins to see how the model might be extended to make a place for suggestion and to yield behavior similar to some that is observed under hypnosis. One begins to ask not "Is audio analgesia hypnosis?" but "What specifiable processes underlie analgesia, and what specifiable processes underlie hypnosis?"

That takes us immediately to the third hoped-for value of the model: that it will be useful in formulating experiments and in leading to new observations. As mentioned earlier, the model suggests that we look at intersensory effects in general and try to relate to the model the many published results of that field. That will be a difficult and probably frustrating task. More specifically, the model suggests an experiment to determine the effect of pain on loudness. It makes connection with ongoing experiments on effects of sound on sleep and in inducing sleep. And, finally, the model demands quantitative data on audio analgesia. It makes it clear why experiments carried out in a traditional psychophysical context are unlikely to lead to interesting results. In doing so, it may possibly introduce new variables into psychophysics. It seems likely, for example, that the subject's state of tension-relaxation may have a significant effect upon subjective magnitudes such as loudness, though perhaps not upon sensitivity to weak signals.

There is another side of the question of value, however, and it disturbs me deeply. My primary impression of the Symposium was that many neurophysiologists are so engrossed in their substantive, purely neurophysiological work, and that it is reinforcing them so strongly through advances in understanding of the nervous system, that they do not have time for models. There is no way of escaping the cost. Even to set up and exercise such a simple model as the one described here required dozens of hours on the digital computer and several on the analog computer. It is not wholly clear to me that the time would not have been better spent with electrodes in a brain or even earphones on ears.

My ingrained predispositions toward the realm of models and toward the effort to correlate psychological and physiological data were given a sudden shock excitation by a recent letter from Vernon Mountcastle (1959), whose superb presentation at the Symposium greatly clarified my impressions of the reticular formation and the posterior group nuclei, where networks of the kind postulated here may be presumed to lie. Mountcastle has found cells, both in the posterior group nuclei and in the cerebral cortex, which respond to nociceptive stimulation and whose responses are suppressed by acoustic stimulation. His finding raises the question, whether models can lead to experiments as fast as experiments can lead to models.

Summary

"Audio analgesia" is acoustically induced suppression of pain. The phenomenon is genuine but complex and not yet well understood. Most of our knowledge about it comes from clinical (especially dental) experience.

The present aim is to clarify understanding of audio analgesia by bringing the established characteristics of the phenomenon into relation with one another, and with a few basic psychophysiological ideas, through the agency of a mathematical (or computer) model. The model is a very simple one. It involves two channels ("hearing" and "pain"), each with positive and negative feedback paths to itself and to the other channel. The dynamic behavior of the model is determined by amplification, biasing, and smoothing parameters associated with the feedback paths. With values of those parameters that are reasonable on a priori grounds, the behavior of the model reflects many of the characteristics of audio analgesia.

Acknowledgment

Dr. Wallace J. Gardner, of Cambridge, with whom I have been working these last two years, is, in my opinion, primarily responsible for advances that have been made in developing the audio procedure for dental applications and in understanding the phenomenon. Dr. Alexander Weisz has performed exploratory laboratory experiments that have clarified several issues. Dr. Ronald Melzack has contributed to the understanding of audio analgesia in relation to neurophysiological studies of pain. Dr. Martin Orne has contributed toward meaningful interpretation of the relation between audio analgesia and suggestion and hypnosis. To these men, and to the several dentists and physicians who have studied audio analgesia in clinical contexts, I express great indebtedness and great appreciation. No less great are my appreciation and indebtedness to Edward Fredkin for tutoring and guidance in digital computing and for help with both the digital and analog simulations.

Preparation of this chapter was supported by the Air Force Office of Scientific Research under Contract USAF 49(638)–355. The report number is AFOSR–TN–60–1190.

References

Baruch, J. J., and H. L. Fox. Personal communication, May, 1959.

Beecher, H. K. The measurement of pain. *Pharmacol. Rev.* 1957, **9,** 59–209.

Caussé, R., and P. Chavasse. Etudes sur la fatigue auditive. Note Technique No. 1057, Centre National d'Etudes des Télécommunications, Paris, 10 Dec. 1947, pp. 29ff.

Egan, J. P. Independence of the masking audiogram from the perstimulatory fatigue of an auditory stimulus. *J. acoust. Soc. Amer.*, 1955, **27,** 737–740.

Gardner, M. B. Short duration auditory fatigue as a method of classifying hearing impairment. *J. acoust. Soc. Amer.*, 1947, **19,** 178–190.

Gardner, W. J., and J. C. R. Licklider. Serendipitous effects of masking noise upon sensations produced by the dentist's drill. *J. acoust. Soc. Amer.*, 1959a, **31,** 117 [Abstract].

Gardner, W. J., and J. C. R. Licklider. Follow-up report on audio analgesia. *J. acoust. Soc. Amer.*, 1959b, **31,** 850 [Abstract].

Gardner, W. J., and J. C. R. Licklider. Audio analgesia. *J. Amer. dental Soc.*, 1959c, **59,** 1144–1149.

Hardy, J. D., H. G. Wolff, and H. Goodell. *Pain Sensations and Reactions.* Baltimore: Williams and Wilkins, 1952.

Hill, H., C. Kornetsky, H. Flanary, and A. Wikler. Effects of anxiety and morphine on discrimination of intensities of painful stimuli. *J. clin. Invest.,* 1952, **31,** 473–480.

Hood, J. D. Auditory fatigue and adaptation in the differential diagnosis of end-organ disease. *Ann. Otol. Rhinol. Laryngol.,* 1955, **64,** 507–518.

Jacobson, E. (Editor), *Progressive Relaxation.* (2nd Ed.)   Chicago: University of Chicago Press, 1938.

Lüscher, E., and J. Zwislocki. The decay of sensation and the remainder of adaptation after short pure-tone impulses on the ear. *Acta oto-laryng., Stockh.,* 1947, **35,** 428–445.

de Maré, G. Audiometrische Untersuchungen. *Acta oto-laryng., Stockh.,* 1939, Suppl. **31.**

Melzack, R., W. A. Stotler, and W. K. Livingston. Effects of discrete brain stem lesions in cats on perception of noxious stimulation. *J. Neurophysiol.,* 1958, **21,** 353–367.

Mountcastle, V. Personal communication, November, 1959.

Weisz, A. Z. Personal communication, November, 1959.

4

FRANK A. GELDARD
University of Virginia

Cutaneous Channels of Communication

It is conventional, when life scientists come together to focus on communication processes, to mark off three domains, the relations among which define their chief problems. Briefly specified, they are: S, the world of stimuli; ϕ, the physiological organism, especially its neural mechanisms; and ψ, the response system. There is much talk of "black boxes," some feeling safer when they get inside them whatever the complexity encountered, others concerned with more remote terms, either as a matter of preoccupation with broad psychophysical relations or in the pretense that the box is empty. To a person interested in the communication process, broadly conceived, all sets of relations—those between S and ϕ, ϕ and ψ, and the larger ones joining S and ψ—are insistent ones. In addition, some new emphases become important, for the conceptions of "stimulus" and "response" most useful for dealing with physiological changes within the organism prove to be too limited in scope and implications to meet the demands of communication problems.

At the stimulus end of the chain, it is necessary to think not only of energetics—luminous flux or acoustic power—but also in terms of "things," objects in the world, manufactured articles. It is with the aid of these that we communicate. It is also most frequently *about* these that we communicate.

At the response end of the sequence there is likewise an expansion possible, indeed inevitable, for we have not come far as yet in the systematic description of responses. More often than not, when we attempt to specify a response, we do so in terms of its significance, utilitarian value, or purpose. We have never evolved a really first-rate set of descriptive categories that will permit us to keep close to the physiological actions involved in behavior. Where we should perhaps be talking in terms of "envelopes of contraction and relaxation," or some such, we find ourselves saying, "The animal obtained food," or "The subject made a judgment." We tend to classify responses in relation to social significance. Even the statement, "The rat pressed the bar," implies a little teleology.

The broadest relations available for study and analysis, then, are those between "things" in the world and the complex behavioral sequences they ultimately evoke—what perhaps had best be subsumed under the term, "conduct." This set of relations between objects and conduct defines the domain of behavior theory. It also sets the stage for the first approach to any new communication problem, for when, now, we ask about the potentialities of a little-used sense channel for mediating messages, the broad relation between things and conduct has to be faced.

Two radically opposed approaches may be made to the assessment of cutaneous communication possibilities. If a person is interested in the "things" that go into conventional communication systems and in trying to utilize them in novel ways, he may ask, for example, what the skin's capacity is for feeling the disturbances present in a telephone receiver or other electromechanical transducer, and interpreting them. This, in fact, is what was asked years ago by those who got interested in "hearing through the skin" programs. Their approach was reasonable, given as a starting point the considerations that the skin is quite capable of apprehending mechanical vibrations and that the world contains vast quantities of hardware capable of transducing speech and music into mechanical vibrations of a power level that can be felt by the skin. The argument was an evolutionary one—the eardrum, which does so well at picking up and transmitting the fine dance of the air molecules present in speech, is a descendant of a cruder but not dissimilar tissue also capable of behaving with some efficiency in acoustic systems. Why not simply train the skin to do substantially what the tympanic membrane does? There were many patient experiments, and enough came out of them to encourage persistent effort. Human speech, suitably amplified, was applied to the fingers. After a sequence of 28 half-hour training periods, in one experiment, the feeler could learn to judge, with about 75-per-cent accuracy, which one of 10 short sentences had been delivered to the skin. After some 30 hours of practice, single words, presented without context, could be recognized about half the time. However, even these poor levels of performance deteriorated if the talker changed his rate of presentation or if another person sat in for him. The efforts at getting the skin to encompass a language not natural to it eventually petered out; the approach by way of adapting existing hardware had, for all practical purposes, failed.

If one attack is to force a receptor system to perform in an unnatural way by trying to adjust it to the world's hardware, the other involves asking how the world and the things in it might be modified to get the most out of the senses. What discriminations are possible for the skin? What is the stuff out of which a cutaneous language has to be built if it is to be optimally utilizable by the skin? Having ascertained what sensory dimensions there are, one is in a position to catalogue the possible discriminations, relative and absolute, that might be made along such continua. This should yield building blocks, collocations of stimulus properties, for coding. One would, indeed, be in possession of the stuff out of which all possible cutaneous communication systems could be constructed, provided only that the dimensional analysis were to be exhaustive and systematic. Then it would depend mainly on what had to be communicated—whether the "message" was a warning, directional or rate information, a more sophisticated and elaborate formal "language," or whatnot—what stimulus arrays were selected out for coding and use.

There are but a few stimulus dimensions of the first order available. They are easily listed: locus, intensity, duration, and frequency, if one begins with repetitive mechanical impacts, that is, vibration. This is not the only possibility—the skin responds to thermal, chemical, and electrical stimuli as well—but mechanical vibration is the most promising one if continuous signaling is contemplated. There are also some derived dimensions, and we shall come to them later. What of the first-order ones? This is not the place to review all the psychophysical data. They are the joint products of a succession of workers in the Virginia laboratory, and most of them have been published (Geldard, 1957). But we must see where we stand.

First-Order Dimensions

Locus has never been systematically investigated, though a good deal is known about it through the studies of Spector and Howell (cf. Geldard, 1957). The question of how many places on the skin can be utilized is not settled by extrapolation of the results of two-point esthesiometer measurements. Unlike static pressure, mechanical vibration applied to the skin does not stay "within bounds" unless special steps are taken to prevent its spread (Békésy, 1959). This means that the two-point limen for vibration is many times greater than the static one for a given region. Howell found that seven vibrators could be spaced on the ventral rib cage with 100-per-cent identifiability of locus, under his conditions. This is probably about the limit for a practical cutaneous communication system. The chest accommodates five conveniently, and it is tempting—until one finds that it will not work—to consider the combinations of vibrators (31 signals for 5 contactors) that might be coded. The difficulty is that two or more simultaneously acting vibrators feel no different from one, once the static pressure of each has adapted out, and provided the vibratory pattern is set up in all of them with the same onset. A split-second temporal differential is all that is needed to restore two local impressions, but this leads to the further complication that such a manipulation also provides the essential conditions for synthetic movement, tactual "phi." This, in turn, could be coded, of course, and we shall encounter it again later as belonging in the family of "derived" phenomena.

The failure to explore the entire body for locus as a codable cue is not the result of neglect, nor yet of lack of interest in the outcome. In this electronic age so many things have to await technological advance. We simply have not had a transducer with the right properties to make the experiment feasible. When Dr. Howell had an array of eight vibrators on his subject's chest, he was literally trapped in a forest of concrete-rooted supports, flexible goosenecks, and long, flat springs that were necessary, with the vibrators then in use, to ensure independence of vibratory generation and to preserve a uniform static background pressure. It is, therefore, a considerable satisfaction to report at this time a technological "breakthrough" of recent weeks whereby my colleague, Dr. R. C. Bice, has succeeded in modifying radically a transducer of the hearing-aid variety to provide strong, low-frequency vibrations which are not readily damped by the skin. He has also found an ingenious way of utilizing fabric fasteners to couple the vibrator firmly to any chosen site of stimulation. The instruments are sufficiently small and light and their electrical properties are such that they yield high powers without undue heating. For the first time the systematic exploration of the dimension of locus seems to be in sight.

What of *intensity?* There have to be some limits. Each part of the integument has its own absolute threshold, and each part has its susceptibility to both discomfort and damage by the trip-hammer action of powerful vibrators. In the region of the chest, a useful range of stimulus amplitudes is bracketed by the values, 50 to 400 microns, the former because it is safely above the 100-per-cent absolute threshold, the latter because it falls well below the threshold of discomfort. Between these limits the average observer can, under laboratory conditions and with the use of a careful psychophysical procedure, detect about 15 intensive steps. On an absolute recognition basis, unless one were to select subjects or train them, it would be unsafe to include more than three steps, widely spaced over this range. The intensive dimension is, in fact, the least exploitable of all the first-order dimensions. An analysis of errors in a communication system that codes locus, intensity, and duration shows nearly all mistakes to be made along the dimension of intensity. There appear to be two chief reasons for this: (1) a fixed amplitude applied to different loci varies considerably in its "feel," owing perhaps to accidents of local innervation, perhaps to reinforcing or damping variations in underlying tissue; (2) it is not easy, in the face of breathing motions and those coming from circulatory events, to maintain a strictly invariant relation between a mechanical vibrator and the skin surface on which it rests.

Duration of a continuous vibratory "package" is judged with some precision over the entire useful range. The range must, of course, be selected on extraneous considerations. In our experiments we have chosen not to deal with any durations under 0.1 second, this on the ground that a "buzz" much shorter than this is likely to be mistaken for a "nudge" or a "poke." At the other end of the scale, we have set 2.0 sec as a limit beyond which we are unlikely to wish to code signals; a communication system employing units lasting more than 2 sec is certainly a ponderous one. Between 0.1 sec and 2.0 sec, then, there is a durational continuum within which the average observer can make about 25 distinctions, the steps being of the order of 0.05 sec at the low end and 0.15 sec at the high end of the range. This is again the relatively precise Δt of the psychophysical experiment. Absolute identifications with 100-per-cent accuracy yield four or five considerably more widely dispersed levels, and, if neither selection of subjects nor training of them is intended, it is safer to use only three.

It is clear from the foregoing that three sets of building blocks of cutaneous communication systems are in our possession—a limited number of absolutely discriminable steps of locus, intensity, and duration. There is, of course, a fourth primary dimension. This is *frequency*. The story of frequency discrimination in the vibratory realm is not a simple one. There is a history, and also some recent experiments have had interesting outcomes. The history we may summarize by saying that failure to control for differences in subjective intensity, when frequencies were being compared, and for contaminating transients at the onset and offset of the stimulus envelope has invalidated all measures prior to those recently obtained by Genevieve Goff in our laboratory (Goff, 1959). She overcame these defects by first assembling a band of equal-loudness stimuli, differing

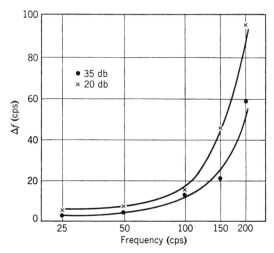

Fig. 1. Differential frequency discrimination of mechanical vibration applied to the finger tip (Goff, 1959). The curves are for a single subject.

in frequency, and then measuring Δf systematically throughout the obtainable frequency range. Her basic results are shown in Fig. 1.

It is clear that, at very low frequencies, judgments of vibratory "rate" are quite good, but it is equally clear that discriminability fades rapidly as the frequency scale is ascended. In the region best for speech sounds the skin does very badly indeed, and this finally explains why the "hearing through the skin" programs, alluded to earlier, yielded such disappointing results. A direct comparison between the skin and the ear is made possible in Fig. 2; the auditory data are the classical curves obtained in the Bell Telephone Laboratory (Shower and Biddulph, 1931) at a roughly comparable intensity level.

There is one additional difficulty where vibratory frequency is concerned; were it not for this, it might conceivably be possible to transpose audible frequencies downward into the tactile range. The difficulty is this—the correspondence between vibratory frequency and perceived "pitch" is a tenuous and uncertain one. Vibratory pitch proves to be a joint function of both frequency and amplitude.

To be sure, this is also the case, for much of the audible range at least, in hearing as well, but intensity is only a very minor determinant of auditory pitch; making pure tones louder or softer can move pitch about only a little. Frequency is very nearly in absolute control. In the cutaneous sphere things are different. Increase the amplitude of a moderately loud 40-cycle sinusoidal vibration applied to the

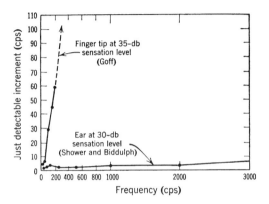

Fig. 2. Discriminability of vibratory frequency (Δf) compared with the auditory pitch discrimination data of Shower and Biddulph, 1931 (after Goff, 1959).

finger tip, and it undergoes a marked downward shift in pitch. Decrease the amplitude, and its rate goes up perceptibly. Békésy has recorded shifts of the order of three octaves (Békésy, 1959). It is obvious that frequency would have to be handled gingerly in a communication system, especially if intensity were simultaneously manipulated as a variable.

A Workable Vibratory System

In our initial efforts to form a system of vibratory signals capable of being coded, only the three obviously useful dimensions of locus, intensity, and duration were employed. The code shown in Fig. 3 was devised and applied successfully by Howell, so successfully that a subject who had invested a total of 30 hours in learning the alphabet of the "vibratese" language could, after a further training period of only 35 hours, receive sentences with 90-per-cent accuracy when transmitted at the rate of 38 five-letter words per minute (Howell, 1956).

This performance in no wise represents the optimal attainable.

Using the same rules as those of international Morse code with respect to interword and intraword spacing (0.1 sec and 0.05 sec,
respectively), the system requires only 0.79 sec to transmit the
"average" five-letter English word.   This means that the ceiling
transmission rate is 67 words per minute, a speed well over three

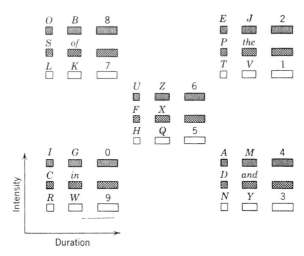

Fig. 3.   Code of the "vibratese" language (Howell, 1956).

times that of proficient Morse.   To approach this rate, however, there
would be required an extensive and somewhat elaborate engineering
job to devise an automatic coder—perhaps of the tape variety—that
would initiate signals faster than our present homemade machine, a
manually operated typewriter that triggers "flip-flop" circuits (for
time) and a bank of potentiometers (for intensity).   The closing of
the gap between 38 and 67 words per minute we leave to those interested in establishing world's records.

Future Potentialities

Meanwhile what other building blocks are available?   We have
considered only the first-order dimensions of the vibratory stimulus.
There are some derived dimensions, a few of which have already
received some attention.   It would be helpful to have additional
codable cues, if only to be in a position to add redundant elements.
Much has been written about purifying and simplifying languages by
reduction of redundancy.   Where intelligibility is less than optimal

there is much to be said for making the language more, rather than less, redundant (Miller, 1951, p. 103), and the vibratese language would doubtless benefit from this kind of doctoring.

What are the candidates here? Intensity variations as a function of time present one set of possibilities. A signal may be imposed on the skin quite abruptly or more gradually, just as in music one may have variations in "attack," "hitting" a tone or "sliding into" it. Sys-

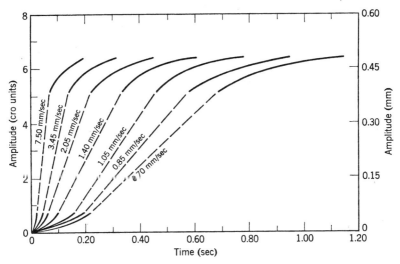

Fig. 4. Variable rates of rise from zero to a vibratory amplitude of 480 μ. The seven curves have been drawn to mark off six successive jnd steps (Howell, 1958).

tematic manipulation of this variable has been carried out by Howell. Figure 4 serves to describe both how the envelope shapes were made to vary and how discriminability of rate of onset changed with decreasing rate of attack. For a growth of amplitude from zero to 480 microns—the largest that it was practicable to work with—and for rise times that ranged from the shortest that the system could produce without transients to the longest that might serve for signaling purposes, there prove to be six discriminable steps. As with all other similar functions, these come from the conventions of psychophysics. Absolute identifications are, of course, not spaced this closely. Indeed, if 100-per-cent recognizability of "attack rate" is demanded, over this range of onset slopes there are but two: the "magical number" proves to be 7 minus 5! There are presumably another two steps associated with offsets.

A second possibility, in the realm of derived dimensions, is wave form. This has not been investigated yet, though it is clear that wave-form variations should be discriminable if the basic frequency is low enough. Still other moment-to-moment variations should be detectable. Perhaps it would be possible to introduce variations within signal envelopes other than those of onset and offset, a gradually increasing or decreasing frequency, say. This is an unexplored field.

Another patterning of stimuli involves space as well as time. As in vision, the successive stimulation of two separated receptive areas, provided the temporal relations are right, leads to perceived movement. Unlike vision, the critical time relation for tactual "phi" is the absolute interval between the beginnings of the two successive exposures, not the duration of the silent interval between them. Indeed, there need be no silent interval at all. The exposures may overlap and still yield good movement if the onsets are properly spaced. Cutaneous phi has been surveyed, not so much with a view to coding it and incorporating it into a communication system, as to studying its essential conditions as a phenomenon, but clearly it offers some possibilities.

It is doubtful whether vibratory phi should be coded into any language of the type described above. Indeed, in any rapid succession of signals applied to spatially discrete loci, the crucial conditions for synthetic movement are already given. Movement has to be overlooked, not apprehended in this situation, of course. On the other hand, the phenomenon of cutaneous movement is an extraordinarily powerful one; it really commands attention. Place a ring of vibrators around the body, for example, three across the front and three across the back of the thorax, energize them successively with a 0.1-sec temporal separation, and the observer feels a vivid "swirling" motion, entirely novel in his experience, because he seems to be at the center of it! This effect is a completely prepossessing one and, accordingly, it is ideal for coding in a different fashion, say, as a warning signal to be infrequently used but capable of highly significant coding. It would do valiant service attached to a "panic button."

This raises the important question of the ultimate place of the cutaneous channels in the total communication picture. Coding to letters and numerals is really a quite pedestrian way of getting meanings into tactile patterns. There are, to be sure, obvious ways of making such a system "fly" at a faster rate. One way would be to code the vibratory signals to phonemes. We have not attempted it because of the prodigious investment entailed in learning the phonemes themselves, but it ought to be tried. It is also possible that there may come out of hiding an entirely novel cutaneous shorthand, one capitalizing on distinctively tactile properties. Serious study of basic cutaneous perceptual phenomena, an area dignified by the devotion of not more than a dozen first-rate minds in the whole of recorded history, might turn up such a linguistic development.

Vibratory Tracking

The possibilities of cutaneous communication are by no means confined to conventional language, of course. Other kinds of information may be imparted tactilely. Rates, amounts, directions—anything falling on unidimensional or bidimensional continua could presumably be represented to the skin by means of suitably patterned mechanical impacts or sequences of them. One of these possibilities has already been exploited in our experiments. Vibratory tracking by compensatory pursuit has been carried out by lining up three vibrators across the chest and letting them be successively energized to give the impression of continuous movement in one direction or the other (through utilization of "phi"). The "arrowhead" was always pointed" toward the target, and the vibratory sequences were temporally spaced to indicate degrees of urgency in getting back "on-target." The subjects manipulated a steering wheel and attempted to eliminate all cutaneous signals by promptly neutralizing all off-target indications.

Ten subjects performed in response to these tactile signals. Ten others were presented the visual analogue, with lights substituted directly for the vibrators. Both groups learned rapidly, and the vibratory performance was in no wise inferior to the visual. Comparative learning curves are shown in Fig. 5. Although the visual conditions are not optimal for this sense—the target was "traveling" at the rate of only 3.5 degrees per second, and the eye, of course, can handle speeds many times as great—the tracking task imposed on the subjects was one that would keep all but the speediest vehicles comfortably on course, and the skin was handling the assignment fully as well as the eye.

Subsequently, this experiment underwent simplification. Synthetic movement was taken out of the display, the three vibrators were reduced to two, and even the "urgency" feature was eliminated. Now there occurred only a simple "nudge," a brief burst of 60-cycle vibration to indicate direction off the target to the right or left. Performance showed no significant degeneration with this removal of redundancy in the signal. Very little is needed to give directional information that is adequate to fairly complex performance.

Currently there is being tested a bidimensional vibratory display designed to give both "right-left" and "up-down" deviations from the target (glide path? trajectory?), and there is every indication that the system will do what is expected of it. The value for situations in which vision and hearing are pre-empted, for one reason or another, is obvious.

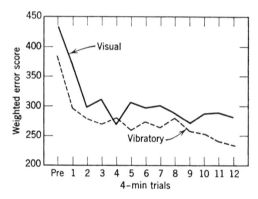

Fig. 5. Comparison of visual and vibratory tracking performances. The error scores were weighted for distance off target (Bice, 1953).

Another whole domain of possibilities opens up when we consider that only one form of energy, the mechanical, has thus far entered our calculations. Not much is to be hoped for from the chemical and thermal forms of stimulation. They are both too ponderous in their operation to be of much use in communication; at best they could provide only the analogues of smoke signals. But there is the whole important realm of electrical stimulation. The skin responds with lively patterns to both direct and alternating current. Indeed the heart of the problem is that the patterns are, in general, somewhat too lively for comfort. We have devoted several years of intensive effort to finding the conditions of electrical stimulation that will yield codable vibratory patterns, bereft of pain, and that can be reproduced on demand. The problem has turned out to be a slippery one. Electricity is the great "nonadequate" stimulus; it triggers everything, as physiologists well know. However, the important stimulus parameters are few in number, and my colleague, Dr. John F. Hahn, has been able to isolate the really significant one in skin stimulation (Hahn, 1958). As Fig. 6 taken from his work shows, where square waves are employed and are systematically varied in frequency and duration, thus obviating any influence of the change in rate of current increase such as occurs with alternating current of variable frequency, it turns out that absolute threshold is related to duration only.   This

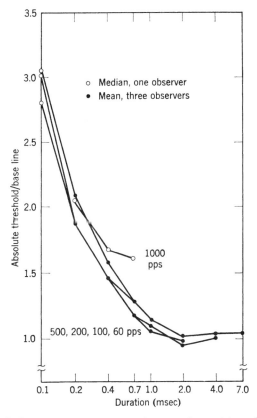

Fig. 6. Strength-duration curves at several rates of repetition of direct square-wave stimulation of the skin. Absolute threshold/base line is the ratio of absolute threshold to the mean of thresholds with pulse durations of 2.0 msec or longer; pps is pulses per second (Hahn, 1958).

fundamental discovery permits us to narrow our search for the basic medium of reception in electrical sensitivity of the skin. The great stumbling block, thus far, is the omnipresent pain. Although under some reproducible conditions pain tends to adapt out in continuous signaling, leaving behind a not too unpleasant tingle, it is doubtless asking too much, in practical communication situations, to expect that even transient discomfort from a transducer will be tolerated.

Summary

A really comprehensive approach to communication problems entails not only the domains implied by the terms, "stimulus," "physiological changes," and "response," and the three sets of relationships into which they enter, but the much more remotely related terms, "physical objects" at the beginning of the series, and "conduct" at the end of it. The position one takes on these relations has much to do with how questions concerning communication are to be asked.

There are reviewed the currently available facts concerning cutaneous sensibilities, insofar as they have bearing on the communication problem. The possible elements of a cutaneous language are identified as those collocations of stimulus properties that are recognizable in the absolute. The first-order dimensions of mechanical vibration (locus, intensity, duration, frequency) provide the major possibilities, though interactions between frequency and intensity are such as to render one or the other of these difficult of utilization. A few steps of locus, intensity, and duration provide enough codable signals to make rapid and accurate communication possible.

In addition to the primary dimensions, there are some secondary or derived ones. Intensity variations as a function of time ("attack") constitute one such possibility. A second dimension is the derivative of frequency, wave form. A third involves both spatial and temporal variations, synthetic cutaneous movement. The first and third have already received some attention; the second has not been much investigated as yet. Movement (cutaneous "phi") is especially useful in unusual warning situations because of its vividness. It has also served as an intensifier of signals in tracking situations.

A second contender for prominence in future cutaneous communication systems is electrical stimulation. The possible stimulus parameters are few in number, and recent work has shown pulse duration to be the important one in sensitivity measurements. Direct electrical stimulation of the skin minimizes the transducer problem and is ultimately to be preferred over mechanical systems if the ubiquitous problem of pain can be circumvented.

References

Békésy, G. v. Synchronism of neural discharges and their demultiplication in pitch perception on the skin and in hearing. *J. acoust. Soc. Amer.*, 1959, **31**, 338–349.

Bice, R. C.   Vibratory tracking.   In Progress Report No. 21, ONR Project NR140–598, Aug. 29, 1953.

Geldard, F. A.   Adventures in tactile literacy.   *Amer. Psychologist,* 1957, **12,** 115–124.

Goff, Genevieve D.   Differential discrimination of frequency of cutaneous mechanical vibration.   Doctoral dissertation, University of Virginia, 1959.

Hahn, J. F.   Cutaneous vibratory thresholds for square-wave electrical pulses. *Science,* 1958, **127,** 879–880.

Howell, W. C.   Training on a vibratory communication system.   Master's thesis, University of Virginia, 1956.

Howell, W. C.   Discrimination of rate of amplitude change in cutaneous vibration. Doctoral dissertation, University of Virginia, 1958.

Miller, G. A.   *Language and Communication.*   New York: McGraw-Hill, 1951.

Shower, E. G., and R. Biddulph.   Differential pitch sensitivity of the ear.   *J. acoust. Soc. Amer.,* 1931, **3,** 275–287.

5

IRWIN POLLACK

Operational Applications Laboratory, Air Force Cambridge Research Center
Bedford, Massachusetts

Selected Developments in Psychophysics, with Implications for Sensory Organization

The purpose of this chapter is to outline some selected recent developments in psychophysics that may be relevant to problems of sensory organization. In this discussion it is assumed that psychophysics is concerned primarily with "terminal" activities, in that specified sensory inputs are related to specified behavioral outputs, without direct observation of the dazzling complexity of intermediate neural events. Psychophysical data are thus necessary, but not sufficient, for a complete model of sensory organization. From this point of view, psychophysics will be considered to suggest crude alternative models, or broad strategies of action, for approaches to the sensory organization of the human observer, rather than crucial details. All references to the nervous system in the chapter refer to a conceptual nervous system that serves between the terminal activities of stimulus and response.

Psychophysics and Information Measurement

The outstanding impact of information theory—more accurately, of information ·measurement—upon psychophysics has been the specification of the stimulus (Quastler, 1956). Instead of the specification of a particular energy configuration, information measurement has called for the specification of the entire *ensemble* of possible events to which the organism might be exposed in a particular situation. Although minor contextual effects have long been known to psychophysics, the concept of the ensemble of possible stimuli far transcends second-order effects. In the light of information measurement, the entire ensemble of events becomes, in effect, a superordinate stimulus.

With the same impetus, the psychophysical behavior of interest becomes the stimulus-response *confusion matrix* which relates each of the possible responses to each of the possible stimuli (Miller and Nicely, 1955). The stimulus-response confusion matrix in turn is crucial in testing alternative models of sensory organization. These models consider the representational mapping of stimulus objects on the relevant portion of the conceptual nervous system. This relevant portion will be termed the *discrimination space*. In particular, two extreme alternative models will be further considered in detail.

The fixed-map model

The fixed-map model of sensory organization implies a complete lack of interaction among the alternative stimulus objects of the ensemble (Luce, 1959). The model implies that the mapping of any specific stimulus object onto the conceptual discrimination space is invariant with the other members of the stimulus ensemble. That is to say, corresponding to each stimulus object there is a location in the discrimination space, and the location associated with a particular stimulus object does not depend upon the other members of the ensemble.

In terms of psychophysical results, the model suggests that any stimulus of the ensemble is effective only insofar as the stimulus contributes responses to the total pool of responses. Removal of the stimulus and redistribution of its corresponding responses within the remaining stimulus-response confusion matrix leave the ratio among specific stimulus-response confusions unchanged. Hence, the predictive instrument for the fixed-map model has been termed the constant-ratio rule (Clarke, 1957). It should be noted that the model makes no assumption of the distribution, or of the scalar properties, of the objects of the stimulus ensemble.

Rubber-band model

An alternative model of sensory organization assumes a high degree of interaction among the members of the stimulus ensemble. It assumes a variable mapping of the stimulus ensemble onto the appropriate discrimination space, depending upon the available range of stimulus objects in the ensemble. The discriminative *density*, in turn, is inversely proportional to the range. The change in density with range may be illustrated by marking off equal distances on a rubber band, and then counting the number of marks as the band is stretched and contracted, with reference to a fixed length. From this illustration, the model gets its name: the rubber-band model of sensory organization (McGill, 1958). In terms of the model, any specific pair of items may be highly discriminable if the entire range under presentation is narrow, but poorly discriminated if the entire range is wide.

The rubber-band model, however, does not yield exact quantitative prediction unless we assume that the discrimination space is always stretched to accommodate the entire range of the stimulus ensemble. This assumption is unreasonable, for it would result in the prediction that the discriminability between any two stimuli, of an ensemble of only two stimuli, would be independent of the differences between the stimuli. This is certainly not the case. Further, whereas the rubber-band model may reasonably be applied to stimulus ensembles that may be arranged along a single physical continuum, it is not evident how the rubber-band model would accommodate stimulus ensembles that vary along several physical continua. These points suggest that predictions based on the rubber-band model might be successful only for stimulus ensembles that vary with respect to a single physical variable, or a small number of variables.

Confusion-matrix results

Several empirical studies have now shown that, for stimulus ensembles of a high order of physical complexity (for example, words presented in noise, letters presented in a brief flash), the constant-ratio rule is an excellent predictive instrument, even for relatively small confusion matrices (Clarke, 1957; Clarke and Anderson, 1957). Said otherwise, with these stimulus ensembles, the fixed-map model appears to lead to precise prediction of psychophysical data.

Stimulus ensembles that vary with respect to only one variable do not yield clear evidence. Preliminary data (Clarke, 1959) suggest that large deviations from the constant-ratio rule (in the direction predicted by the rubber-band model) are observed for stimulus ensembles in which only a single variable is manipulated. However, these deviations are observed only with small confusion matrices and only when the stimulus objects of the ensemble are highly discriminable. For larger confusion matrices, or for small confusion matrices in which the stimulus objects of the ensemble are highly confusable, the constant-ratio rule is an excellent predictor of the psychophysical findings, even for stimulus ensembles that vary with respect to a single variable (Hodge and Pollack, in preparation).

Thus, the concept of the stimulus ensemble, derived from information measurement, with the associated stimulus-response confusion matrix, permits examination of alternative models of sensory organization. The weight of evidence favors a model that assumes that stimulus objects are represented in a discrimination space by mapping that is relatively invariant with other objects of the ensemble. This conceptual picture may require modification, particularly for ensembles in which a small number of stimulus objects vary with respect to a single physical variable. For the latter case, it may be necessary to consider conditions under which transformation or remapping may take place.

Other implications for sensory organization

The concept of the discrimination space upon which stimulus objects are mapped—or even the concept of the stimulus ensemble—is, of course, not original with information measurement as an approach to psychophysics. However, information measurement provided the reorientation of thinking necessary to focus attention on a class of problems previously neglected (Frick, 1953).

The mapping of the ensemble onto a fixed representational discrimination space directly leads to a familiar result. Consider a two-dimensional discrimination space in which the area around each point is the indeterminacy of that point. As we pack more and more points into the fixed discrimination space, the greater is the overlap among the points. The representational picture leads to the following experimental prediction: the larger the stimulus ensemble, the greater will be the confusion among the items of the ensemble. And, indeed, many experiments bear out this prediction (for example, Miller, Heise, and Lichten, 1950).

Examination of a large number of ensembles has turned up a result that may have interesting implication for sensory organization. When the members of the ensemble differ with respect to only a single physical variable, the amount of information transmitted is approximately constant, irrespective of the variable chosen. In order to transmit more information, it is necessary that the members of the ensemble differ with respect to more than one variable. The larger the number of variables that require specification, the higher will be the information transmitted (Miller, 1956). If we insist that stimulus ensembles are mapped with respect to their corresponding physical variables in discrimination space, then we must be prepared to accept the concept of a higher-order representational space in the conceptual nervous system.

The relative constancy of information transmission suggests a functional psychophysical capacity. But, in order to match effectively a functional capacity with an unstructured environment in which events occur at irregular intervals, a buffer storage is most desirable. Several lines of evidence point to functional temporary buffer storage for the human operator (Broadbent, 1958). If we assume an effective buffer storage between the environment and subsequent information processing, an interesting question arises. Are selective operations imposed upon sensory information before, or after, buffer storage? That is to say, is there filtering of information at the receptor level prior to later information processing? A crude type of selective filtering is feasible by simply orienting the receptors, as in opening and closing the eyes, but can we ascribe a functional location for additional selective operations? Experimental tests on the reception of messages in noise suggest strongly that the primary locus of selective filtering is after, rather than before, buffer storage. This evidence is based on the finding that restriction of the stimulus ensemble after presentation of a message is often as effective as restriction of the ensemble before presentation (Pollack, 1959).

One class of experiments deserves brief mention in that they show that the size of the ensemble is not critical for performance. In short-term memory, the number of items recalled appears to be independent of the size of the ensemble: our memory capacity seems to be limited more by the number of items retained than by the information per item. In George Miller's descriptive terms, memory is limited by the number of "chunks" rather than the "bits per chunk" (Miller, 1956). In order to store more information, it is necessary to repackage more information per "chunk." This repackaging, or recoding, has been demonstrated experimentally. A comprehensive treatment of sensory organization should include a treatment of recoding.

In partial summary, the stimulus-response confusion matrix, stemming from information measurement, has focused interest on the stimulus ensemble and the mapping operations between objects of the ensemble and their associated loci in the (conceptual) nervous system.

Statistical Detection Theory

We turn now to a second development in psychophysics which appears to have important implications for sensory organization: the application of statistical decision theory to psychophysics (Tanner and Swets, 1954; Licklider, 1959). The particular form of the application has been termed statistical detection theory.

Consider, for the moment, the behavior of an electronic detection system in the context of a psychophysical experiment. On 50 per cent of the trials, a controlled brief flash is presented above a controlled background. On the other 50 per cent, there is no additional flash. The detection system consists of a photocell sensor and a response-threshold detector, such that if the output of the photocell sensor reaches a critical response level, the system responds "flash"; and, if the sensor fails to reach the critical response level, the system responds "no flash." Let us assume that, on successive occasions, the response-threshold detector may be set to different critical response levels.

If the critical response level of the threshold detector is set extremely low, the photocell system "sees" nearly all the flashes, that is, it registers the presence of the flash upon nearly all the trials in which the flash has been presented. In the language of signal detection, a high "hit" probability has been achieved. The price that must be paid for setting the critical response level low is that the photocell system may register often the presence of a flash on trials in which the flash was not presented. In the language of signal detection, a high "false-alarm" probability would be obtained. Thus, when the critical response level is low, a high hit probability is associated with a high false-alarm rate.

Similarly, if the critical response level of the threshold detector is set high, the system may fail to report many of the flashes on trials in which the flash was, indeed, presented. And, in the same manner, the system would rarely register the presence of a flash on trials in which the flash was not presented. That is, when the critical response level is high, a low hit probability is associated with a low false-alarm rate.

Note that, with the same photocell sensor, a wide range of hit probabilities, with an associated wide range of false-alarm probabilities, may be obtained simply by varying the critical response level of the threshold-response detector operating upon the information furnished by a sensor of the same photocell system. Should we now conclude that under these conditions the photocell sensor is changing in its sensitivity or discrimination of the flash above the background illumination? No, it is not reasonable to infer changes in the behavior of the photocell sensor, for its characteristics are determined by its own construction relative to the sensing environment under examination. It is rather more reasonable to assume that the discriminability of the photocell sensor is constant, and that the operating level of the decision part of the system is changing. This distinction between the *discriminability* of the sensory part of the system and the *operating level* of the decision components of the system, acting upon the sensory information, is crucial.

What has been said for the electronic photocell system can also be said for the human discriminator. Under a given set of viewing conditions of observation, a wide range of operating levels may be obtained by varying the instructions to the observer or by varying the relative values and costs of missed detections and false alarms. In fact, it is possible to show that, under a specific signal and noise condition, the discriminability of the human observer is invariant over a wide range of operating levels (Egan, Schulman, and Greenberg, 1959; Pollack and Decker, 1958; Tanner and Swets, 1954).

What are the implications of the signal-detection approach for sensory organization? The primary implication is that, whereas a *response* threshold detector is reasonable, a *sensory* threshold detector is not tenable. The experimental evidence indicates strongly that the sensory information about the environment is preserved in a form such that a continuous range of information is available for subsequent decision making.

The negation of the sensory threshold is, of course, a radical notion— so radical as to invite immediate suspicion. In the comment, "Is There a Quantal Threshold?" (in this volume), Stevens takes an opposite position. He points out that, in every experiment carried out within the framework of signal detection, noise has been introduced with the signals to the observers. He notes that the introduction of noise is alone sufficient to mask the "sensory neural quantum" (Miller, 1947). Experiments with pure tones (Stevens, Morgan, and Volkmann, 1941) may, or may not, yield the sensory quantum. Further, it is argued that the forced-choice procedure of the signal-detection approach places an unnecessary burden on the observer which tends to mask the sensory quantum. Much heat can be generated by arguing these points. At present, however, it seems more profitable to determine whether we can set up an experimental test between the statistical-detection and the sensory-quantum approaches. The following set of experiments might provide such a test if a number of assumptions can be met.

First we recreate the conditions of the initial Stevens, Morgan, and Volkmann experiment, in which the "neural quantum" was observed. We assume that we are successful in obtaining "a fortunate selection of observers capable of maintaining an unwavering attention over extended periods of time" so that quantal functions can be obtained. We also assume that we can get the listener to make a confidence rating of each of his responses without disturbing the quantal function. (For a discussion of the use of confidence ratings, see Egan, Schulman, and Greenberg, 1959; Pollack and Decker, 1958.) And, finally, we assume that we can interpose a small number of "catch trials," in which the signal is not presented, without disturbing the quantal functions of the observer.

If the neural model is represented by a discriminative continuum, the listener's confidence ratings should reflect his accuracy of judgment on the critical trials. On the other hand, the sensory-quantum approach predicts that the listener's confidence ratings will not reflect any difference, owing to the quantal nature of the changes. In this regard, it may be noted, the original article reported that large intensity increments often "sounded louder" than small ones (Stevens, Morgan, and Volkmann, 1941, p. 334).

Of course, the experiment cited above may not be feasible. As Stevens and others (Miller and Garner, 1944) have noted, any disturbance of the observer often results in the failure to obtain the neural quantum. Adding the confidence response may vitiate the entire experiment. Still the approach may warrant a serious attempt.

The theory of signal detection has another—and potentially more important—implication for sensory organization. The theory defines an *ideal observer*. The ideal observer is a performance model that is able to extract all available information from the signal and noise environments (Tanner and Birdsall, 1958). The ideal observer sets an upper limit on the discriminability of the observer under examination—whether human or electronic.

Experiments demonstrate that the human observer performs poorly relative to the ideal observer in the detection of weak signals in noise. Here the ideal observer can use phase information and other classes of information that are not employed effectively by the human observer. But the human observer's performance approaches that of the ideal observer for the recognition of large signals, that is to say, at favorable signal-to-noise ratios (Tanner, 1957). It may be noted that the memory requirements imposed upon the observer for strong signals are substantially less than for weak signals. The full implications of these findings for sensory organization are not yet clear. There is an undercurrent of excitement that suggests that powerful implications will follow clarification.

And, finally, the author would be remiss if he did not attempt to capture the flavor of the signal-detection approach to present-day psychophysics. In this framework, the trained human observer becomes an active goal-seeking discriminator with the ability (within limits) to vary characteristics of his sensory equipment and of his decision and operating levels. Though this picture is too molar to permit resolution of details of sensory organization, it may be profitable to make room for such flexibility in considering the molar properties of the sensory organization of the human observer.

Summary

Two recent contributions to psychophysics—information theory and signal-detection theory—have been briefly considered insofar as they may point toward molar properties of sensory organization of the active human observer. Psychophysical data, which reflect only stimulus-response correlations of the molar observer, by their very nature cannot be expected to reveal the details of the actual neural substratum of sensory organization.

References

Broadbent, D. E. *Perception and Communication.* New York: Pergamon Press, 1958.

Clarke, F. R. Constant-ratio rule for confusion matrices in speech communication. *J. acoust. Soc. Amer.,* 1957, **20,** 715–720

Clarke, F. R. Proportion of correct responses as a function of the number of stimulus-response alternatives. *J. acoust. Soc. Amer.,* 1959, **31,** 835 [Abstract].

Clarke, F. R., and C. D. Anderson. Further test of the constant-ratio rule in speech communication. *J. acoust. Soc. Amer.,* 1957, **29,** 1318–1320.

Egan, J. P., A. I. Schulman, and G. Z. Greenberg. Operating characteristics determined by binary decisions and by ratings. *J. acoust. Soc. Amer.,* 1959, **31,** 768–773.

Frick, F. C. Some perceptual problems from the point of view of information theory. In *Current Trends in Information Theory.* Pittsburgh: University of Pittsburgh Press, 1953. Pp. 76–91.

Hodge, M., and I. Pollack. Confusion matrix analyses and the constant-ratio rule (in preparation).

Licklider, J. C. R. Three auditory theories. In S. Koch (Editor), *Psychology: A Study of a Science. Vol. I–I. Sensory, Perceptual and Physiological Foundations.* New York: McGraw-Hill, 1959. Pp. 41–144.

Luce, R. D. *Individual Choice Behavior: A Theoretical Analysis.* New York: Wiley, 1959.

McGill, W. J. Modern psychophysics. Paper presented at 1958 meeting of American Psychological Association.

Miller, G. A. Sensitivity to changes in the intensity of white noise and its relation to masking and loudness. *J. acoust. Soc. Amer.,* 1947, **19,** 609–619.

Miller, G. A. The magical number seven, plus or minus two: Some limits on our capacity for processing information. *Psychol. Rev.,* 1956, **63,** 81–97.

Miller, G. A., and W. R. Garner. Effect of random presentation on the psychometric function: Implications for a quantal theory of discrimination. *Amer. J. Psychol.,* 1944, **57,** 451–467.

Miller, G. A., G. A. Heise, and W. Lichten. The intelligibility of speech as a function of the context of the test materials. *J. exp. Psychol.,* 1950, **41,** 329–335.

Miller, G. A., and P. E. Nicely. An analysis of perceptual confusions among some English consonants. *J. acoust. Soc. Amer.,* 1955, **27,** 338–352.

Pollack, I. Message uncertainty and message reception. *J. acoust. Soc. Amer.,* 1959, **31,** 1500–1508.

Pollack, I., and L. R. Decker. Confidence ratings, message reception, and the receiver operator characteristic. *J. acoust. Soc. Amer.,* 1958, **30,** 286–292.

Quastler, H. (Editor). *Information Theory in Psychology: Problems and Methods.* Glencoe, Ill.: Free Press, 1956.

Stevens, S. S., C. T. Morgan, and J. Volkmann. Theory of the neural quantum in the discrimination of loudness and pitch. *Amer. J. Psychol.,* 1941, **54,** 315–335.

Tanner, W. P., Jr. Large signal methods for the study of psychoacoustics. *J. acoust. Soc. Amer.,* 1957, **29,** 766 [Abstract].

Tanner, W. P., Jr., and T. G. Birdsall. Definitions of d' and η as psychophysical measures. *J. acoust. Soc. Amer.,* 1958, **30,** 922–928.

Tanner, W. P., Jr., and J. A. Swets. A decision-making theory of visual detection. *Psychol. Rev.,* 1954, **61,** 401–409.

6

COLIN CHERRY

Imperial College, University of London

Two Ears — but One World

The Formation of Binaural *Gestalten*

I believe it was Epictetus the Stoic who was reported as saying, "God gave man two ears, but only one mouth, that he might hear twice as much as he speaks."

This wishful thought may be empty, so far as human wisdom is concerned, but it is curiously near truth on the plane of psychophysics. For the possession of two ears gives us greatly enhanced powers of aural discrimination: we can the better separate a single voice in a buzz of conversation and attend to it; or we can single out a voice from the sounds of traffic, or of the wind, and all the myriad of disturbing noises. The brain takes maximum advantage of the slight differences between the signals that reach the two ears—differences in timing, in intensity, and in microstructure—and by processes of inductive inference breaks down the complex of sounds into separate coherent images, *Gestalten*, which becomes projected to form the subjective "spatial world" of sound.

The basic fact about any one of these images, say that of a voice, is that it is single; with two ears we hear only one world. By what logical processes does the brain examine and analyze the sense data reaching the two ears, so as to achieve this fusion? Understanding of the fusion process seems to be fundamental to an understanding of directional hearing. Understanding of the logic of the processes whereby voices in a crowd can be separated is a parallel problem; for we *can* separate voices, and other *Gestalten,* with one ear alone but with reduced effectiveness.

In this chapter we shall be concerned only with this basic phenomenon of fusion, discussing it as a psychological problem and its description as a logical (mathematical) problem. Physiology will not be mentioned, for we shall not look inside the head at all. On the contrary, we shall try to set up a model to describe what "must" be there.   Before you physiologists charge me with arrogant presumption in saying this, I should ask you to remember two things.   First, we are not saying what physiological *mechanism* must be there, but only by what logical principles it can be described as carrying out its functions (for example, whether it adds or multiplies or integrates); second, if you disagree with any aspects of our mathematical model, it is up to you to show that the logical operations are *physiologically* impossible.   If you succeed, we on our side will retract and re-examine our logic (or, more likely, our premises).

The model of binaural fusion, to which I refer here, has been described elsewhere in complete detail (Sayers and Cherry, 1957).   It is largely the result of the extensive experimental and computational work of Sayers while he was a student in my group at Imperial College, London.

To Fuse, or Not to Fuse?

In order to study fusion, as opposed to directional hearing (Cherry and Sayers, 1956), we eliminate the effects of head turning by fitting the subject with headphones.   If the two earphones are driven with identical signals of any kind, the subject hears a single fused image located in the center of his head.   If the interaural time difference T_e, or the relative intensity A_L/A_R, of the left- and right-hand signals is varied, the image appears to move across the head laterally in a line between the two ears.   The image does not appear to pass outside the head and has no angular direction, as in real life; it has only lateral position, left to right.

We control the lateralization by varying the interaural time interval T_e only, driving each earphone from a separate reproducing head on a special magnetic-tape recorder, running at high speed.   The interaural time interval T_e can be set to any value to an accuracy of less than 20 microseconds.

The interval T_e is set to a succession of random values (using a table of random numbers), and the subject is required to guess whether the fused image appears to lie to the right or left—a dichotomous judgment, R or L.   The forms of such lateralization "judgment curves" fall broadly into two classes: (1) If the source is predominantly random, or quasi-random, the form is like that shown in Fig. 1a.   When T_e is large (say 1 to 5 milliseconds), the subject answers 100 per cent correctly; when small, he makes errors; when zero, he is (by chance) correct about 50 per cent of the time.   (2) If the source is

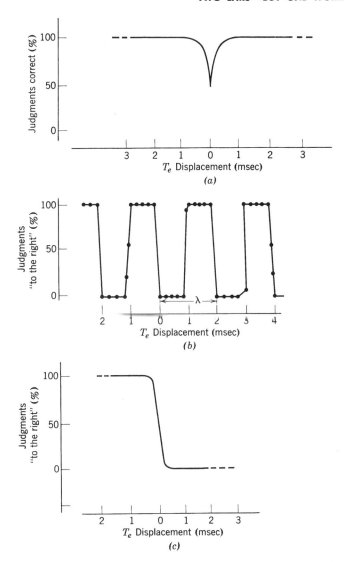

Fig. 1. Fusion judgment curves for random and periodic sources: (*a*) lateraliza-tion for a white-noise source, speech source, or a similar quasi-random source; (*b*) lateralization for a periodic 800-cps source; (*c*) replotting of curve (*a*) in terms of percentage of judgments to the right.

predominantly periodic, as with pure tones or multiple tones, the curve itself is periodic, as shown in Fig. 1*b* for a pure 800-cycle-per-second sine wave. These two forms of curve represent extreme tendencies;

real-life sources of sound contain both (quasi-) periodic and stochastic components.

We shall concentrate first on sources of type 1—for example, speech or noise.

This downward dip in Fig. 1a represents the subject's uncertainty, which has at least two origins: (1) since the fused image has a finite subjective size, its central point is uncertain, and (2) the subject is uncertain of his own midplane (intracranial) about which he is making acoustic judgments.

The arbitrariness between zero values of T_e on the magnetic tape and the central (intracranial) plane is removed by replotting Fig. 1a in terms of percentage of right-hand judgments, as in Fig. 1c, thereby skewing the curve. This method is adopted hereafter for all forms of "judgment curve." The fact that $T_e = 0$ now is not to be taken as "an applied central image." Note that Fig. 1b is plotted this way also.

So much, briefly, for the case of identical aural stimuli. If, however, the left and right signals, $S_L(t)$ and $S_R(t)$, are utterly different (statistically independent), say, two different voices, or different noises, then no fusion takes place (Cherry, 1953). What we are here regarding as a "fusion mechanism" is not triggered. This mechanism is then operated, or not operated, as determined by the relation between $S_L(t)$ and $S_R(t)$.

It has been our purpose to examine experimentally the nature of this critical relation and to find a model for the analytical processes whereby the brain determines it, as a control for the "fusion mechanism."

Obviously it is far more interesting to see what happens when $S_L(t)$ and $S_R(t)$ are neither identical nor totally unlike, but only *partially* alike. In what sense is "alike" to be defined? What does the brain look for in its running analysis?

If we were to confine ourselves to the use of simple determinate signals, like sine waves, clicks, and so forth, we might oversimplify the problem. For in real life the aural stimuli, such as human speech, street noises, and so on, are stochastic. "Stochastic" means "probabilistic"; the brain cannot know in advance exactly what is coming to the ears next, in microscopic detail, instant by instant. Further than this, the two signals $S_L(t)$ and $S_R(t)$ are never *exactly* alike, in detail, in real life. They differ not only in timing T_e and average intensity A_L/A_R, owing to the spatial directions of sound sources, but in other ways, owing to head sound-shadows, reflections and reverberations, and random sound contributions from the wind on our faces.

Again, in real life, an aural image stays fused and does not hop about as the two aural signals fluctuate. It must be average, statistical, invariant properties that control this "fusion mechanism," and several different storages may be used, for different averaging times.

Mathematically speaking, the measures that assess the average (statistical) degree of dependence, or independence, of stochastic sources are the *correlation functions*. There are different forms of such measures, but certain other real-life conditions, under which the brain must operate, narrow down the possibilities. Such functions have found a considerable place in models of other aural phenomena, notably in the work of Licklider (1951, 1956), especially in connection with pitch perception and spatial separation, but not, so far as we can find, in models of binaural fusion prior to our 1956 paper (Cherry and Sayers, 1956). We shall be returning to such theoretical aspects in a later section.

Experimental Control of the Statistical Independence of the Two-Ear Signals

Before theorizing further, let us refer to some experiments. Figure 2 shows the same binaural stimulus arrangement as resulted in the curves of Fig. 1, except that now band-limited white noise is added to the signal at one ear only. The magnitude of this noise $N(t)$ in relation to that of $S(t)$ controls the statistical independence of the two-ear stimuli to any degree we wish.

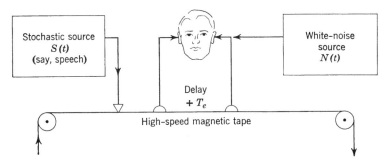

Fig. 2. Control of the statistical dependence between the two-ear stimuli.

Figure 3*a* shows a typical judgment curve (percentage of left judgments) as T_e is set to successive random values when $S(t)$ is male speech. This curve differs from that of Fig. 1 in two ways: (1) Dips appear, showing less certainty of left lateralization of the image at certain values of time T_e. Fusion is now only *partial* over various regions of interaural delay T_e. (2) A total left-right dissymmetry shows *total fusion* when the pure speech signal leads in time but *partial*

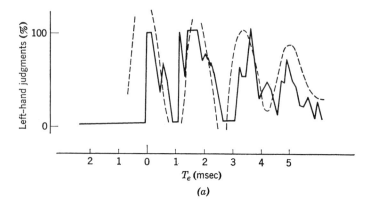

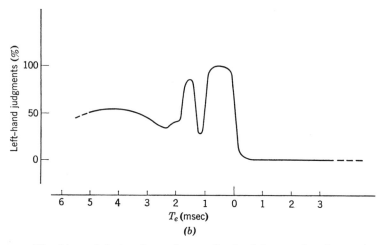

Fig. 3. The binaural fusion "precedence effect": (*a*) typical judgment curve using male speech masked by noise in left ear only [$S_R(t)$ = pure speech; $S_L(t)$ = noisy speech]; (*b*) typical judgment curve, using noise for left ear and noisy noise for right ear (theoretical curve dashed).

fusion when the noisy speech leads. This we have referred to as a type of *precedence effect* (Cherry and Sayers, 1956).

This new "precedence effect" seems to us to have important implications, and it will be referred to again in the next section.

My colleague Dr. Sayers has made extensive measurements of such binaural "judgment curves," each the result of one to two thousand separate right-left judgments with randomly set time delays. Partly from such curves the nature of the analysis performed by the brain (in the "fusion mechanism") has been exposed. This analysis is required, in part, to determine the degree of statistical independence of the signals that arrive at each ear in real life, since this assessed degree of independence is a measure of the likelihood that these two signals come from a single external source of sound. If this likelihood is higher than some threshold (not necessarily constant), fusion is enforced; if lower, separate images are set up at each side of the head.

If the logic of such argument is accepted, we are led to the conclusion that the "fusion mechanism" can be described as carrying out some type of correlation analysis, though there can be many forms of such measures. To determine the precise form used by the "fusion mechanism," further experiments were performed using not only stochastic signals like speech and noise but determinate ones such as sinusoids. We shall return to these experiments later, in the section on Binaural Fusion of Speech. But first let us return for a moment to the curious "precedence effect," already noted.

The "Precedence" Assumed by a Pure Sound Source at One Ear over a Noisy or Distorted Source at the Other Ear

There are other "precedence effects" in hearing, but the one referred to earlier and illustrated in Fig. 3 has not been reported, so far as we are aware, prior to our 1956 and, in more detail, our 1957 papers (Cherry and Sayers, 1956; Sayers and Cherry, 1957).

The effect was mentioned only in passing in these papers, and the writer would like now to take the opportunity of calling the attention of psychologists and neurophysiologists to it, since there may be important implications.

With reference to Fig. 3 the effect is this: if the "pure" signal $S_R(t)$ (say, speech) is set earlier in time than the noisy signal $S_L(t) + N(t)$, then total fusion takes place, with zero uncertainty concerning its right-left localization, over a wide range of signal-to-noise ratios, provided the delay T_e is greater than some residual amount (about 100 microseconds in practice); but if the noisy signal is earlier in time, fusion is quite varied and uncertain. This total dissymmetry is illustrated by the judgment curve of Fig. 3a.

In our earlier publications (1956, 1957) no explanation of this effect was offered. It was examined over a wide range of conditions, however, with not only white-noise masking in one ear but also various distortions (for example, severe amplitude clipping).

What distinguishes a "pure" signal from a noisy or distorted one, thereby giving precedence to the former? Our first reaction was to assume that, in the case of speech, the listener's great experience of speech gives him prior information (and thus a set toward "pure" speech). But experiments soon showed the effect to exist, equally strongly, with sources other than speech (such as combinations of sine waves of different frequency, and many others), of which we humans have far less prior knowledge. Finally, we can destroy all effects of prior knowledge by using, as the binaural source $S(t)$, another wide-band white-noise source, independent of the added noise source $N(t)$.*

Figure 3b shows a typical curve, which is quite asymmetrical about $T_e = 0$. Notice, in this case, that the "noisy noise" $S(t) + N(t)$ has taken precedence over the "pure noise" $S(t)$.

Clearly, then, the inference process by which the brain (fusion mechanism) discriminates between the two signals does not necessarily require prior probabilities (though with speech or other common signals such prior knowledge may conceivably enhance the effect). Recently H. B. Voelcker (of our laboratory) has given a complete theoretical explanation of the effect; unfortunately, at the time of writing, this has not appeared in print, so no reference can be given. Without anticipating its appearance, we might merely point out that the signal $S_R(t)$ is "contained within" the noisy or distorted signal $S_L(t) + N(t)$, but that the converse is not true. Voelcker has examined the nature of the relation "contained within" mathematically, concluding that this precedence effect requires only a simple correlation-analysis mechanism, which can be of the type that we have described as the basis of binaural fusion (Sayers and Cherry, 1957), and which is outlined in the section on The Model of the Binaural Fusion Process in the present paper. In other words he observes that the processes adopted in our model are all that are strictly needed to explain the effect.

Binaural Fusion of Speech

A great deal of auditory analysis is carried out with simple determinate signals, like clicks and pure sine waves. Speech, which is the

---

* Perhaps we could give a psychological definition to white noise, as being "that signal of which the subject can have no prior microscopic knowledge (but only statistical knowledge), which prevents him from making predictions for a time ahead greater than the Nyquist Sampling Time."

class of signal of greatest importance in real life, has a very complex structure, but this does not preclude its use for such experiments as we are considering here.   On the contrary, we regard as essential the use of speech and other stochastic signals, when attempting to build models of aural mechanisms—which, after all, have evolved around such natural sources.

The usual and most effective way of dealing with such complex "stochastic" sources as speech is to use statistical (average) methods and measures.   The method we have adopted, for studying binaural fusion, is of the type illustrated by Figs. 1 and 3; our original paper (Sayers and Cherry, 1957) contains many more.

It is from the parameters of such statistical data that the parameters of the binaural-fusion mechanism may be inferred.   But such parameters must not be expected necessarily to remain invariant as the type of signal is changed (or the noise of the environment or other conditions).   This variation we have found to be the case, in certain ways. But it is not the fusion *mechanism* that seems to change; rather it is extremely flexible in the way it handles limitless variety of binaural signals—speech, with its transients, its quasi-random breath noises, its quasi-periodic formants as well as all the myriad of other sound sources that surround us in daily life.   Rather it is the *data* that the fusion mechanism recognizes as being "in common" between the two ears that are found to vary according to circumstances; if one class of data is not available in the stimuli, the mechanism finds another.

This kind of flexibility is, of course, indicative also of correlation processes being used by the brain.   Before examining for the exact processes, let us illustrate this flexibility in the case of speech.

The data in speech controlling the binaural fusion process

In this series of experiments, intoned vowels were used (to remove transients).   They were recorded on tape, which was subsequently closed into a loop to provide a continuous source.

When such an intoned vowel is presented binaurally to a subject, the resulting "judgment curve" is exactly the same as that of Fig. 1a for a random or quasi-random source (or replotted so as to eliminate arbitrariness of zero delay T_e as in Fig. 1c).   This is indicative of the fact that the random breath sounds, closely similar in both ears, control the fusion process; the formant periodicities are not apparent here.

But we can easily destroy the similarity of the breath sounds arriving at each ear by adding to one side a white-noise generator, as in Fig. 2.   What does the mechanism fuse on now?   Figure 4a shows one example, with an average signal-to-noise ratio of −9 decibels;

the difference now is that dips have appeared at values of T_e that indicate that fusion is taking place on the periodic components (formants) of the intoned vowel source (which have known frequencies). Thus the fusion mechanism is operated by common breath sounds, under quiet conditions, but partly by the common formants when breath tones are rendered useless.

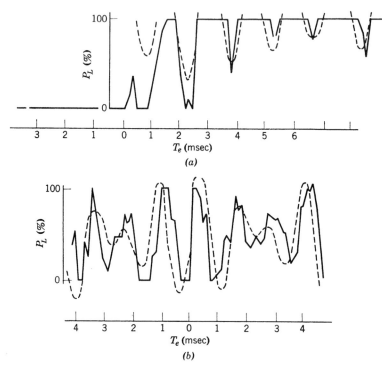

Fig. 4. Judgment curves for (a) the intoned vowel a, noise-masked in the left ear only, with signal-to-noise ratio -9 db (theoretical curve dashed); (b) two sinusoids (600 and 800 cps, equal level).

We shall not repeat details of the calculations here but merely state that the correspondence between the dips in the curve of Fig. 4a and the formant frequencies of the intoned vowel has been examined by Sayers (Sayers and Cherry, 1957); his calculated "judgment curve" is shown dashed on Fig. 4a. Note the coincidence of the peaks. Such a calculation, based on the mathematical model to be described in the next section, is possible with some accuracy because the formant frequencies and decrements can be measured accurately in this case of a steadily intoned vowel.

The asymmetry of the curve of Fig. 4*a* representing our "precedence effect" should also be noted.

The Model of the Binaural Fusion Process

The few curves we have illustrated here (Figs. 1, 3, 4) are typical of a very large number resulting from measurement with many different types of sound source and different listening subjects (Sayers and Cherry, 1957).

Let us look now at the steps involved in building up a theoretical model, by which such fusion "judgment curves" may be calculated for all sorts of sound sources—pure tones, chords, intoned vowels, noise, running speech.

The first "simple model" of binaural fusion

The simplest interaural correlation process is illustrated by Fig. 5. Briefly, we argue that the L and R signals, $S_L(t)$ and $S_R(t)$, are cross-correlated and that the correlation function $R_{12}(t, \tau)$ is represented upon a "conceptual surface," as shown by the dashed curve. Then the "judgment mechanism" decides whether the function lies, on an average, more to the left or more to the right of the mid- ("intracranial") line. The measure used does not appear to be critical, since only a dichotomous judgment is needed; for our calculations, we have used the normalized areas lying, left and right, under the curve $R_{12}(t, \tau)$. One additional operation is necessary at this stage; before cross-correlation, the signals $S_L(t)$ and $S_R(t)$ need to be combined with their own mean values A_L and A_R (the method of combination used tentatively at this stage is simple linear addition). The reason for this is that these mean intensities can themselves also control the R-L judgment and so must appear upon the "conceptual surface." This oversimplified operation is clarified in our final model, as described in a later section.

Conventional cross-correlation is represented by the function

$$\Phi_{12}(\tau) = \lim_{T \to \infty} \frac{1}{2T} \int_{-T}^{+T} f(t) \cdot g(t + \tau) \, dt \tag{1}$$

but clearly this is not usable, since the aural processes are proceeding in real time, and integration over the future is unrealistic. Again, the brain cannot integrate indefinitely over the past, because judgments are made moment by moment, as sources of sound move about. Clearly, the process can only be one of short-term *running correlation*,

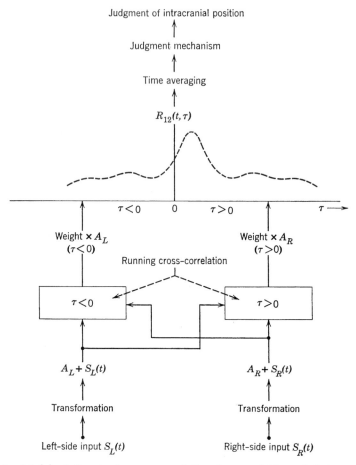

Fig. 5. Model of the simple cross-correlation theory of binaural fusion. Note that the running cross-correlation unit has been divided in two parts only for clarity of representation.

represented most generally by

$$R_{12}(t, \tau) = \int_{-\infty}^{+\infty} f(t - T)g(t - T - \tau)W(T) \, dT \qquad (2)$$

where $W(T)$ is a weighting function, corresponding to a short-term memory. The exact form of this function is not critical, we have found, but it is convenient to assume that it is exponential:

$$W(T) = \epsilon^{-KT}\Big|_{T>0} \qquad (3)$$

Licklider (1951) used a similar running correlation function in his model of pitch perception. We have estimated the time constant K here as approximately 6 msec. This running cross-correlation is illustrated by Fig. 6.

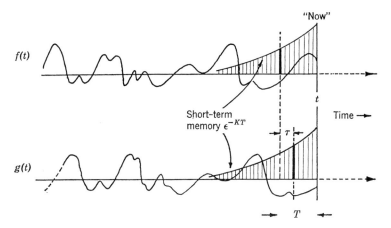

Fig. 6. Short-term running correlation, as from Eq. 8.

Theoretical results, based upon the "simple model"

The key to the exact form of the correlation process was first obtained from fusion measurements based upon single and multiple pure tones (see, for example, Figs. 1b and 4b). Such curves appeared to be closely similar to the correlation functions of the signals themselves (for such steady-state signals no short-term memory ϵ^{-KT} was involved).

Notice one important thing about these curves, typical of all our measurements; they are "sawn-off" sharply at the 100-per-cent and 0-per-cent levels, corresponding to the listener's absolute certainty of L or R. We shall cite only one example here: Fig. 4b shows (dashed) the calculated curve based upon the "simple model" when 600 and 800 cps are applied together binaurally. This does not, of course, recognize the existence of these 100-per-cent and 0-per-cent limits, since the model contains no "thresholds of absolute confidence."

The complete model of binaural fusion

The "simple model" of Fig. 5 was found to be inadequate for several reasons. Predominant is the fact that the L-R lateral movement of a fused sound image is controlled also by the relative *intensities* of the two ear signals A_L and A_R, as well as by their timing in relation to one another. We have made many measurements of fusion "judg-ment curves," using, in particular, multiple sinusoids and intoned vowels, and have found the following a most surprising result.

Briefly, if the *form* of the signals $S_L(t)$ and $S_R(t)$ is held constant, but their relative mean intensities A_L/A_R are varied, the *form* of the

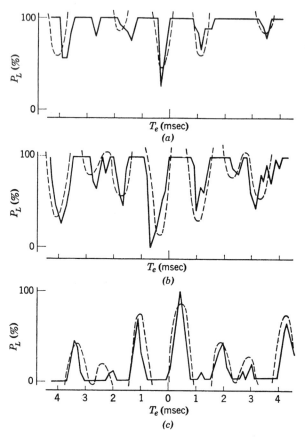

Fig. 7. Judgment curves for two-sinusoid binaural signal, showing influence of interaural amplitude difference (600-cps component, 2 db with respect to 800-cps component in each channel). (*a*) $S_L(t)$ above $S_R(t)$ by 10 db; (*b*) by 6 db; and (*c*) by −6 db. Theoretical curves dashed.

judgment curve itself also remains invariant; however, this judgment curve is moved bodily, up and down, between the "sawn-off" 100-per-cent and 0-per-cent limits. Figure 7 shows one typical result, using two sinusoids, 600 and 800 cps, as in Fig. 4*b* also. It is as though the 100- and 0-per-cent limits form a mask, behind which the judgment curve slides up and down, as controlled by the ratio A_L/A_R.

Such a result is very simply accounted for if the relative signal amplitudes A_L and A_R at the left and right ears are assessed, by some
averaging process, and used to weight the L and R sides of the correlation function $R_{12}(t, \tau)$ (see Fig. 5) accordingly, before the final
judgment process decides whether this function lies predominantly
L or R of the mid- (intracranial) line.

Such an assessment of A_L and A_R might well result from an additional *autocorrelation* process, operating on the signals at each ear
separately. Again, if this is short term, as for the cross-correlation
$R_{12}(t, \tau)$ in Eq. 2, running average values are assessed, $A_L(t)$ and
$A_R(t)$, so that these are themselves functions of time. Such a method
of assessment provides, inherently, a theoretical basis for other binaural
phenomena, to be referred to later.

The complete model upon which calculations have been based is
now shown in Fig. 8 and is described in detail in our 1957 paper.
Briefly, the signals at each ear $S_L(t)$ and $S_R(t)$ are first subjected to
separate running autocorrelation, and the two resulting functions are
then cross-correlated.

This cross-correlation function $R_{12}(t, \tau)$ is subsequently represented
on a "conceptual surface" but, before the "judgment mechanism" decides whether this function lies predominantly L or R of the midline,
the two parts of the function (L and R) are weighted by a long-term
time average of the two autocorrelation functions, $R_{11_L}(t, \tau)$ and
$R_{11_R}(t, \tau)$. There are, therefore, three "conceptual surfaces" involved,
one at each ear and one in the center.

Such a model is, of course, not neurophysiological, but *functional;*
it has been used for calculating L-R lateral "judgment curves," with
many forms of binaural signal, either identical at each ear or noise-
masked at one ear. We cannot cite many here (other than Figs. 3a,
4b, and 7) but would refer the interested reader to the original (Sayers
and Cherry, 1957).

Some Further Aural Effects Predicted by the Model

1. When the left or right ear only, but not both, is stimulated, no
centrally fused subjective image is formed. In the model, only the
L or R "conceptual surface," and not the central one, is energized.
Similarly, when the L and R signals are from statistically independent
sources the same result obtains (Cherry, 1953).

2. When a complex source of sound applied binaurally has no frequency components below approximately 1200 to 1500 cps, it is an
experimental fact that only the *envelopes* of the L and R signals fuse

into a binaural image; their microstructures do not operate the fusion mechanism (Leakey, Sayers, and Cherry, 1958). But when the signals

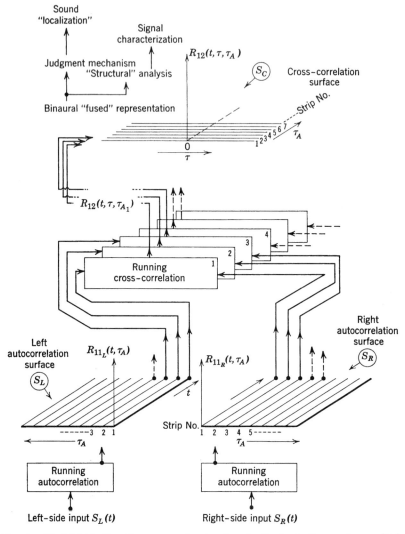

Fig. 8. Model including the proposed extensions to the basic theory, from which calculations have been made.

are identical (or different) *pure* tones above about 1500 cps, with *no* varying envelopes, no fused image forms. This phenomenon assists our directional hearing of signals, all of whose components of wave length are shorter than twice the distance between the ears. Our fusion model does the same, since, under these conditions, the running autocorrelation processes at each ear automatically pass on only the envelopes (running averages) to the central cross-correlation process.

3. When listening binaurally to multiple sinusoids, the subject can, at will, listen to the whole fused chord, or to any one, or any group, of fused tones as distinct *Gestalten*. Again, by independent time-delay controls, each of these different fused images can, subjectively, have independent lateral movement. Our model also enables us to calculate the "judgment curves" of lateral image positions for each of the images separately (Sayers and Cherry, 1957), since the preliminary autocorrelation processes perform the necessary periodicity discriminations.

4. If, with binaural stimulation, one ear is, in addition, masked by noise, or if other perturbation is introduced, a kind of "precedence effect" arises which we noted in an earlier section. Voelcker has shown that this can be explained only in terms of the short-term auto- and cross-correlation processes used in the model, but his work is not yet published.

The relations between our model and the proposals made by Licklider (1956) concerning pitch perception might be noted here. We have adopted the same running-correlation function as Licklider and, like him, we require both auto- and cross-correlation (although the order is reversed in our model). But our premises and experimental data are utterly different in kind. Licklider is concerned with neurophysiological facts; we are not. Again, it is on pitch perception that he concentrates, rather than upon the various fusion phenomena. We have been concerned solely with building a functional model of fusion processes, that is, with a set of mathematical operations whereby we can *calculate numerically* the probability of binaural fusion— nothing else.

It cannot be overstressed that models such as ours are to be regarded as "one of a class of models"; it is always possible to take the mathematical processes cited and recast them, or transform them, into alternative processes. Thus there is no *need* to restrict our processes to correlations; correlations can well be interpreted by filtering processes instead. And, once again, it is not neurophysiology that concerns us here, but representations of behavior—entirely an "outside view" of the human black box.

A great many perception phenomena concern extraction of *statistical invariants* from among the raw sense data; for *Gestalten* are recognized under many forms of transformation and perturbation—both determinate and statistical, for example, noise. Again, the perceptions operate in real time, moment by moment.

Short-term correlation thus seems to be a logical choice for .the extraction of such invariants. We end then with a somewhat vague suggestion that perhaps the various hierarchical processes of perception may be described by hierarchical successions of such correlations. The question is whether short-term correlations may not be the bricks of which part of the house of perception is built.

Summary

In this paper a mathematical model is developed which is descriptive of the binaural fusion process. Fusion of all types of signal can be handled: sinusoids, noise sounds, intoned vowels, running speech, and so forth.

It is entirely stimulus-response *behavior* that is described by the model, and no reference whatever is made to the physiology or anatomy of hearing. With the model calculations can be made, in complete detail, of the probability with which a human listener will hear sounds as lying to the *right* or *left* side of his intracranial plane, when binaural stimuli are presented with some mutual time delay. The two signals (left-ear, right-ear) need not be identical; fusion still occurs when they are different, as in fact they always are in real life. Again, the mathematical model will assess the probability of fusion. Only if the two signals are totally unlike (statistically independent) are we assured of total failure of fusion.

The details of the model have already been published elsewhere. In the present paper attention is paid particularly to the arguments underlying the development of the model. Its relation to models of other hearing processes is clarified.

Finally, some typical examples are cited, comparing the experimental and the calculated results, and some general implications of the model for binaural perception are drawn.

References

Cherry, E. C. Some experiments upon the recognition of speech, with one and with two ears. *J. acoust. Soc. Amer.*, 1953, **25,** 975–979.

Cherry, E. C., and B. McA. Sayers. Human "cross-correlator"–A technique for measuring certain parameters of speech perception. *J. acoust. Soc. Amer.*, 1956, **28,** 889–895.

Leakey, D. M., B. McA. Sayers, and E. C. Cherry. Binaural fusion of low- and high-frequency sounds. *J. acoust. Soc. Amer.,* 1958, **30,** 222–223.

Licklider, J. C. R. A duplex theory of pitch perception. *Experientia,* 1951, **7,** 128–134.

Licklider, J. C. R. Auditory frequency analysis. In C. Cherry (Editor), *Proceedings of the Third London Symposium on Information Theory.* London: Butterworth, 1956.

Sayers, B. McA., and E. C. Cherry. Mechanism of binaural fusion in the hearing of speech. *J. acoust. Soc. Amer.,* 1957, **29,** 973–987.

7

HALLOWELL DAVIS

Central Institute for the Deaf, St. Louis, Missouri

Peripheral Coding
of Auditory Information

Neurophysiological Specifications for the Form of the Auditory Code

The most strategic spot at which to examine and assess the information that enters the central nervous system through its sensory input systems is at the level of the primary afferent neuron. In some systems, such as the cutaneous and the muscular, the afferent neurons are grouped into many separate afferent nerve bundles, but in others, notably the visual, olfactory, vestibular, and auditory systems, a single pair of cranial nerves carries the entire sensory input of that particular sense modality. In the auditory nerve it is possible to record the action potentials of peripheral neurons that have not yet made synaptic connections. It is particularly appropriate to use the auditory system as an example of neurological coding because information theory, which is obviously the background for our thinking about this problem, has developed primarily around auditory communication. Perhaps the auditory system can serve as a model in this regard for some of the other sensory systems.

The auditory nerve in man, and in mammals generally, is composed of myelinated nerve fibers of moderate and quite uniform size (3 to 5 microns in diameter). The cell bodies lie peripherally in the spiral ganglion inside the cochlea (Figs. 1, 2). The cells are bipolar cells, and no synaptic connections in this ganglion are recognized. (There is also a relatively small efferent bundle, the olivocochlear bundle or tracts of Rasmussen, with cell bodies in both the ipsilateral and the contralateral olivary complex. We really do not know its function, and I shall disregard it deliberately.) In man the total number of afferent fibers, based on ganglion cell counts, is between 25,000 and 30,000 in each ear. The number is about the same as the number of sensory cells (hair cells), and the distribution of both ganglion cells and sensory cells along the length of the basilar membrane of the cochlea is approximately, although not absolutely, uniform (Davis, 1957, 1959; Stevens and Davis, 1938).

The relation of one set of sensory cells, the internal hair cells, to its nerve supply is fairly simple. As a first approximation, we may think

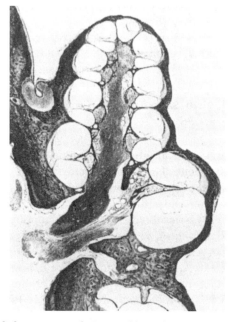

Fig. 1. Mid-modiolar section of the cochlea of a guinea pig. The cochlear canal is cut across nine times. The auditory nerve passes down the center of the modiolus and through the internal auditory meatus to the brain. The cell bodies lie in the spiral ganglion between the nerve proper and the organ of Corti on the cochlear partition. In this particular animal the nerve cells and the organ of Corti of the first turn had degenerated, following exposure to a high-intensity tone of high frequency.

of two or three internal hair cells as connecting to each fiber, and of each hair cell as receiving innervation from two or three nerve fibers (Fig. 3). A similar relation apparently holds for one class of fibers, the radial fibers, to the external hair cells, but each of the more numerous external spiral fibers innervates many cells, and each cell receives innervation from a number of fibers (Fernández, 1951; Stevens and Davis, 1938).

In the overlapping innervation of hair cells, we find what may be analogous to the peripheral nerve net of the cutaneous system and to the extrafoveal regions of the retina, where a considerable amount of overlap of innervation seems to be the rule.   The overlap may be simply part of nature's margin of safety or redundancy in the structure of the nervous system.   (I tell my students, "No single neuron,

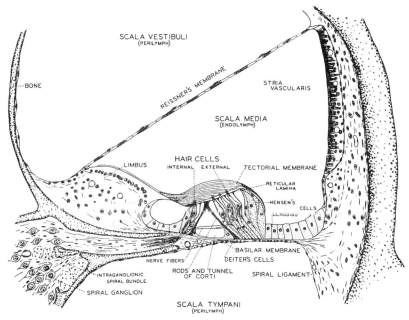

Fig. 2. Camera lucida drawing of a cross section of the cochlear partition in the second turn of the cochlea of a guinea pig.   The bipolar nerve cells of the spiral ganglion appear at the lower left.   The nerve fibers have no myelin sheath in the organ of Corti.   The sensory cells are the internal and external hair cells.

unless it be an anterior horn cell, is ever completely indispensable.") The overlap of innervation also provides at least the anatomical possibility of some interaction among sensory cells prior to the setting up of the familiar all-or-none impulses of medullated nerve fibers.   The peripheral endings of the sensory nerve fibers are nonmedullated throughout the organ of Corti.   They are completely naked, without even a Schwann cell sheath.   They should be regarded as a dendritic system.   Conduction by electrotonus only or by decremental conduction is common in dendritic systems.   We have no right to assume that all-or-none impulses followed by refractory periods arise until the auditory fibers acquire their myelin sheaths in the habenula perforata.

In the auditory nerve where it passes through the internal auditory meatus, however, we find all-or-none impulses, separated in a given fiber by time intervals of at least 1 millisecond and usually somewhat longer (Tasaki, 1954). This form of conduction, typical of peripheral nerves and of white matter in the central nervous system, constitutes a very severe limitation for the coding of incoming sensory information.

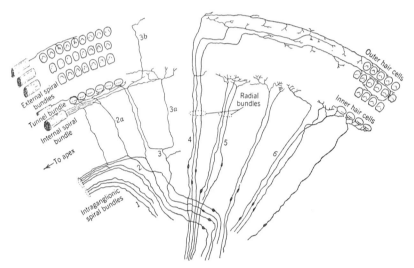

Fig. 3. Plan of innervation of the hair cells, according to Lorente de Nó (in Wever, 1949). The intraganglionic spiral bundle is composed of efferent fibers. It arises in the medulla and does not have its cell bodies in the spiral ganglion.

The dominant problem of auditory theory for many years has been to determine how the physical dimensions of the monaural acoustic stimulus, namely, frequency and intensity, are coded in the form of all-or-none impulses, separated by silent intervals and traveling in a large number of parallel, insulated channels. A second problem, which has attracted more and more attention in recent years, has been the coding of the information that forms the basis of sensing the direction from which sound arrives. Binaural differences in time become very important in this problem.

The Classical Rules and Possible Mechanisms for Coding Auditory Information

From the work of Békésy and many others, we know that the cochlea acts as a mechanical acoustical analyzer, and that the position of maximum mechanical activity along the basilar membrane changes as a function of frequency (Békésy and Rosenblith, 1951; Davis, 1958a). It seems inescapable that an important part of the information concerning the frequency of the incoming sound waves, particularly those of high frequency, is carried by the built-in assignment of certain channels (nerve fibers) to certain parts of the basilar membrane. This arrangement takes advantage of the differential response to frequency of different sections of the cochlea. This is the *place principle* of coding of frequency (pitch) information.

It is also well established that for lower frequencies (below 2000 cycles per second or thereabouts) nerve impulses in the auditory nerve tend to group into volleys because each sound wave acts as a separate stimulus. Volleys of nerve impulses following one another at the frequency of the original sound waves have been observed in the pattern of nerve impulses as they pass up the auditory nerve (Békésy, 1959b; Békésy and Rosenblith, 1951). It has been abundantly proved in psychophysical experiments that information concerning the frequency of a low tone (its pitch) can be transmitted by nerve fibers other than those that are excited by a pure tone of that frequency. A high-frequency noise that is periodically interrupted causes us to hear a low-frequency pitch—its "periodicity pitch"— in addition to its high-frequency spectral pitch. This is the so-called *frequency or volley principle* of auditory coding (Davis, Silverman, and McAuliffe, 1951; Licklider, 1951; Miller and Taylor, 1948).

In some sensory systems the *frequency* of discharge of impulses in each afferent fiber is an important method of coding information concerning the *intensity* of the stimulus. This form of coding seems to be of little significance in the auditory system. Experimentally it appears that the dynamic range over which the frequency of discharge in a given fiber is also a function of intensity is very limited: not more than 20 or 25 decibels (Galambos and Davis, 1943; Katsuki, et al., 1958). Very likely the information concerning the intensity of an acoustic stimulus is more directly represented by the total number of fibers that are active than by the total number of nerve impulses in a given length of time.

A third method of conveying (coding) intensity information is by a systematically graded set of thresholds among the receptor units and a correspondingly graded contribution per fiber to the ultimate sense of loudness. In the ear there are two dimensions of systematic grading, both of them probably significant. First, the different rows of sensory cells probably have different thresholds and perhaps different importance centrally. Second, for a sound of a given frequency there is a point of maximum sensitivity somewhere along the basilar membrane. Receptors that are located between this point of maximum sensitivity and the basal end of the organ of Corti can also be stimulated by this same tone, but its intensity must be increased. Thus there is, for every audible frequency, a systematic grading of thresholds lengthwise along the organ of Corti.

For the lateralization of the source of a sound, the two parameters of binaural difference in intensity and binaural difference in time of arrival of corresponding sound waves are both known to be significant. The importance of earlier arrival (the precedence effect), even by a small fraction of a millisecond, is well recognized. The fixed anatomical separation of the two ears with the intervening acoustic baffle of the head gives a physical base on which nature has built into the nervous system a particular code, based on differences in time and differences in intensity appropriate to low and to high frequencies, respectively, to carry information relative to the direction of arrival of incoming sounds. Considerable attention has been given to the "trading relation" between intensity and time. There is a peripheral trading relation between intensity and time, because neural latency depends not only on the physical time of arrival of a sound but also on its intensity.

It is not so well recognized that there must inevitably be certain small-scale time differences (differences in latency) among the responses of the nerve fibers of an individual ear. The traveling waves on the cochlear partition impose certain necessary differences in the time of arrival of the crests of the waves at different points along the cochlea (Fig. 4). This is true for acoustic transients as well as for sustained tones. These time differences are a function of the frequency of the sound waves in question (Tasaki, Davis, and Legouix, 1952). If it is true (and I believe that it is a good general rule) that the central nervous system never fails to make some use of every bit of information that is presented to it, the consequent systematic differences in time of arrival of impulses in different nerve fibers are presumably utilized—although for what purpose we do not know. Some other observations, however, allow us to speculate concerning possible roles for these systematically delayed impulses and also for the synchronized volleys of impulses described earlier.

We do know, as a matter of experimental observation, that individual elements in the midbrain and cortex are more "sharply tuned" than are the neurons of the auditory nerve (Katsuki, et al., 1958). It seems clear that inhibitory processes are at work and bring about the

selective tuning by suppressing the excitatory effects of a given frequency of acoustic sound wave except in a particular channel or set
of channels (Galambos, 1944). It has been generally assumed that
the suppression is determined in some way by differences in intensity
of stimulation at different points along the cochlear partition. It is
unlikely, however, that the differences in intensity are great enough
to account for the sharpness and precision of frequency discrimination. My suggestion is that systematic differences in time of arrival,
due to the delay in the traveling wave, may also enter into this par-

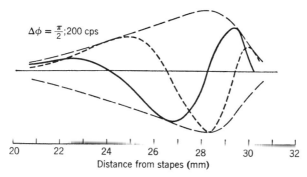

$$\Delta\phi = \frac{\pi}{2}; 200 \text{ cps}$$

Distance from stapes (mm)

Fig. 4. Diagram of the traveling-wave pattern of movement of the cochlea partition near the apex in a human ear. Note that the successive crests of the
short-dashed curve are separated by about 5 mm. The time of travel over this
distance for a wave of this frequency corresponds to 1 cycle. Its duration is
1/200 sec, or 5 msec. The maximum amplitude of the traveling waves is
reached between 28 and 29 mm from the stapes. (From Békésy and Rosenblith, 1951.)

ticular code for sharpening frequency discrimination. Such a code,
however, would have the curious feature that impulses arriving later
from more apical regions would produce excitatory effects in the
"tuned" units at higher levels, whereas the earlier impulses from the
more basal region traversed by the traveling wave would not excite
these particular units but would somehow be suppressed.

The pitch discrimination demonstrated by psychophysical methods
could easily require a code of this complexity. There is also considerable evidence to suggest that the earlier synchronized impulses in
each volley from the more basal regions are not merely suppressed
but carry information about the time of arrival of the individual low-
frequency sound wave or burst of high-frequency sound. This contributes to lateralization if nothing else. Actually this may be the
principal function of all well-synchronized neural volleys.

The Place of the Auditory System in Phylogenetic Development

Some insight into the coding and the nature of organization of each sensory system may be obtained from the generalizations formulated by G. H. Bishop (1959), elaborating the concepts of Judson Herrick, in relation to fiber size and to phylogenetic development. Bishop points out that the central nervous system seems to have evolved by a process of successive accretion. In the simplest, earliest vertebrates the primitive system was composed of relatively small nerve fibers that conduct rather slowly. From the sensory point of view the system was sensitive, and from the organizational point of view it was well integrated. Responses tended to involve large parts of the body in rather complicated patterns of movement for orientation, escape, food taking, and so forth. The reticular formation of the central nervous system represents part of this primitive organization and so do some touch and pain fibers. The cutaneous nerve net is probably a survival of this primitive form of organization.

To the original primitive nervous system, which nevertheless still persists in the higher and more complicated forms, has been added a second and in some cases a third system. These new systems supplement but do not displace the old. The newer systems are characterized by larger, faster fibers and by discrete innervation. The latter feature makes possible finer discrimination, better localization, and "mensuration." The absolute thresholds of the newer systems may not be so low, but the differential thresholds are smaller.

The existence of two sets of receptors, fibers, and even centers for a given modality is in itself a form of coding, but we should not try to interpret this situation without looking at the organism as a whole. I suspect that the key to this duality may be an ability on the part of the organism to accept or reject the information from one or the other of these systems as a whole, for example, by what we call "paying attention," or perhaps to use one set of information for one purpose and another set for another purpose or set of purposes.

In the special case of the auditory system, the concept of old versus new does not turn out to be immediately or obviously helpful. As nearly as we can tell, the cochlear system is a relatively late development in the evolutionary sequence. It is certainly quite primitive in birds and reptiles and is fully developed only in the mammals. The cochlear fibers are considerably smaller than the large fibers of some parts of the (muscular) proprioceptive system. The labyrinthine system seems to have a component belonging to this large fiber system, namely, the type I hair cells in the nonauditory labyrinth (Wersäll, 1956). These cells are nearly surrounded by a nerve ending in the form of a cup or chalice that is connected to a single fiber about 9 microns in diameter.

In the ear the external hair cells and the internal hair cells look somewhat different, but the fibers are all of the same diameter (Wersäll, 1959). The external hair cells seem to be innervated by a nerve net, whereas the inner hair cells have a more discrete, but not quite one-to-one, innervation and a higher threshold of stimulation. [This statement about threshold is supported by the association of the summating potential, with its higher threshold, with inner hair cells, and also by the higher threshold for action potentials in ears whose external hair cells had been eliminated by streptomycin while the internal hair cells remained intact. These ears had much elevated thresholds for all electric responses, but they responded fairly well to the higher levels of stimulation (Davis, Deatherage, et al., 1958b).] It is tempting to compare the external hair cells to the rods and the inner hair cells to the cones of the retina (Davis, 1958b), but the differences do not seem to be so great among the auditory hair cells.

Bishop's reasoning suggests another line of thought. The cochlea represents a relatively recent addition to the more primitive nonauditory labyrinth. The saccule (and utricle) are certainly sensitive to strong alternating linear accelerations. The question is not whether the saccule is or is not sensitive to "vibration," but rather what its threshold may be for various frequencies of vibration. The contribution of the saccule to hearing in man and animals has been the subject of controversy for decades. Perhaps the saccule-mediated and even the cutaneously mediated sense of vibration bears the same relation to hearing that Head's protopathic sense of touch bears to epicritic touch. This comparison may have some validity in spite of the greatly extended frequency range of hearing and its lower rather than higher absolute threshold of intensity.

Within the cochlea we must recognize the important differences between the possibilities for low-frequency detection and coding and the possibilities for high-frequency detection and coding. The two overlap in a large middle-frequency range from 500 cps (or lower) to perhaps 2000 or 3000 cps. One important difference is that the frequency principle can and presumably does operate in the low-frequency range. This part of the auditory mechanism is the direct extension of the sense of vibration. All that has been added is a more sensitive detector and some more nerve channels, with the result that intensity discrimination is not sacrificed for an extended dynamic range. Stimulation is still substantially wave by wave, and the refractory period still imposes an upper limit of the range of frequencies that can be transmitted directly.

The mechanical frequency analyzer of the cochlea was a biological breakthrough. It has certainly made possible the detection and the discrimination of frequencies above 2000 cps, which is above the effective range of the turtle and some birds such as the pigeon (Wever and Vernon, 1956). Only the place principle operates above 4000 cps (and perhaps above 2000 cps).

In the high-frequency range, successive sound waves apparently can summate their excitatory effects (Davis, Fernández, and McAuliffe, 1950). The envelope of the stimulating pattern becomes of direct importance, and a second mechanical detector system, the d-c summating-potential mechanism of the inner hair cells, detects the envelope directly before neurological coding has taken place (Davis, Deatherage, et al., 1958a). The summating potential is relatively small in the apical turns, but it is very prominent in the basal turn. It seems particularly important for the detection of high-frequency signals.

We can also speculate that the mechanical frequency analyzer of the cochlea has considerably improved the ability of the organism to resolve low-frequency tones. Superimposed on volleys which can be "counted" are neural impulses that are systematically dispersed in space and time as a consequence of the traveling-wave pattern of motion in the basilar membrane. When Békésy imposed a similar pattern of motion on the skin, he found that the principal sensation is localized at a place where adjacent areas show large phase differences. Sensation is minimal or absent in the areas that are stimulated in phase. Also the difference limen for the vibratory sensation is smaller when place and "pitch" are observed together than when the "pitch" alone of a vibrating rod is observed (Békésy, 1959a, b).

To sum up this line of thinking, we say that hearing resembles other senses in showing a duality of mechanisms but the duality is not clearly related to phylogeny according to the Bishop fiber-size rule. Hearing is unorthodox, in that a primitive coding system—that of the sense of vibration—remains as part of the auditory mechanism for middle and low frequencies, but, in addition, two entirely new mechanical analyzing devices have developed in the cochlea. One of these is the frequency analyzer, the other is the envelope detector. These peripheral analyzers allow a much greater variety and latitude in peripheral neural coding. The most complex bio-acoustic development is probably the echo-location system of bats and porpoises, which involves very good discrimination of frequency, of envelope, and of time differences of high-frequency signals. The new channels and codes of the auditory system do not replace but are added to more primitive systems. One consequence of this is that we may expect the psychoacoustics of low-frequency sounds to be different from and probably more complicated than the psychoacoustics of high-frequency sounds.

Analysis of the Action Potentials of the Auditory Nerve of the Guinea Pig

For several years in our laboratory we (D. H. Eldredge, B. H. Deatherage, D. C. Teas, S. P. Diamond, G. Pestalozza, and H. Davis) have measured the action potentials of the auditory nerve of the guinea pig as a function of the intensity, the frequency, and certain other characteristics of the stimulus. The analysis of some of these data throws some additional light on the coding of certain aspects of auditory information and raises some other interesting questions.

As we have pointed out, the auditory nerve is composed of a large number of parallel fibers of uniform diameter. These fibers pass through an opening, the internal auditory meatus, in a relatively non-conducting bony partition. An electrode in the cochlea and another on the neck (effectively an intracranial electrode) give a situation closely resembling the classical fluid electrodes of Keith Lucas. In this situation it is reasonable to assume that every fiber contributes approximately equally to the potential difference measured by the electrodes. If so, when a group of fibers discharges synchronously, the peak voltage of the composite action potential is proportional to the number of active fibers.

In order to measure the action potential satisfactorily, we must eliminate the unwanted cochlear microphonic. This can be accomplished fairly effectively by using, as one electrode for the action-potential input circuit, the mid-point on a Wagner bridge between two intracochlear electrodes opposite one another in the scala tympani and the scala vestibuli. We are not, however, able to achieve reliably rejection ratios of better than 40 db. Some of our measurements at high input intensities and low frequency may therefore be contaminated with a small but significant amount of cochlear-microphonic artifact. Other limitations on our ability to estimate the number of active fibers are (1) at low intensities the physiological noise from action potentials of the neck muscles, respiratory muscles, and so forth, of the animal, and (2) asynchrony of discharge of the volley of impulses. The asynchrony is due partly to the Békésy traveling wave (Davis, 1958a) when the frequency is low, and partly to systematic variations in latency with the intensity of stimulation. This is significant because some fibers in each volley are always near their threshold of response. There are also spontaneous fluctuations in latency, and in the guinea pig, at least, there is systematically a greater length of fiber from the apical than from the basal region.

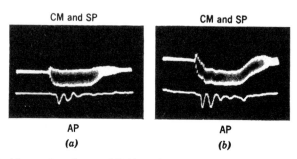

Fig. 5. Cochlear microphonic (CM) and summating potential (SP) recorded by intracochlear electrodes from the basal turn of a guinea pig. The stimulus was a 7000-cycle tone burst with a 1-msec rise time. The stimulus for the oscillogram in (a) was just maximal for CM; for the oscillogram in (b) it was 10 db stronger. Note that the CM is smaller in (b), whereas the SP (the d-c displacement of the upper trace) and the action potentials (downward "spikes" in the lower trace) are larger. (From Davis, Deatherage, et al., 1958a.)

We have concentrated on two forms of stimulus that evoke well-synchronized volleys of action potentials from known regions of the cochlea (Fig. 5). The first is the onset of a tone burst at 7000 cps with a rise time of 1 millisecond. (We use 1-msec rise time in order to minimize the scattering of acoustic energy to other frequencies. The frequency, 7000 cps, is selected as being high enough to give "summated" stimulation, without the volley effect, that is determined by the shape and amplitude of the envelope and quite independent of the individual sound waves; at the same time 7000 cps is low enough to allow us a wide dynamic range within the limits of our transducer; and finally it stimulates at a point on the cochlea that is anatomically accessible for the placement of electrodes.) The pattern of action-potential response to such a stimulus is the familiar spike (N_1) followed by successively smaller spikes (N_2 and N_3) at intervals of about 1 msec. This interval represents the refractory period of the auditory nerve fibers. Incidentally, we should think of 7000 cps to the guinea pig as the equivalent of 3000 or 4000 cps to the human ear. Man's ear is tuned at least an octave lower than the guinea pig's. The guinea pig's locus for 7000 cps is about one-third the distance along the basilar membrane from the oval window, and it is reasonable to assume that just about one-third of the cochlea is stimulated by a strong tone burst at this frequency.

Our second acoustic stimulus is the onset of a tone pip of about 500 cps with a rise time of three wave lengths. Each of these three waves acts as a separate stimulus, each about 3 db stronger than its predecessor. The whole basal turn of the cochlear partition moves substantially in unison in response to this stimulus and delivers a well-synchronized volley of impulses (Deatherage, Eldredge, and Davis, 1959). Of course, the more apical regions are also stimulated and even more effectively, but the action potentials are spread out in time because of the delay of the traveling wave. This "tail" of delayed impulses adds an uncertain amount to the synchronized basal-turn response to the second wave, and the "tail" from the second wave adds with the volley produced by the third wave. As a first approximation, however, we shall disregard these overlaps and also the possible contamination from "uncanceled" cochlear microphonic, mentioned above.

The frequency of 500 cps has its maximum of activity in the third and apical turns, but it is also a very effective stimulus for the basal turn (Deatherage, Eldredge, and Davis, 1959). It stimulates the entire length of the organ of Corti with a gradient of effectiveness that increases from the basal end through the third turn and is still quite effective near the apex. The synchronized spike (the N_1 for each wave) is contributed by the basal turn and perhaps part of the second turn. Beyond the middle of second turn, the impulses are too long delayed. (The actual delay can be measured directly by comparing the cochlear-microphonic response from a second pair of electrodes in the second or third turn with the response from the basal turn.)

We estimate, as an order of magnitude, that there are about 15,000 fibers in the auditory nerve of a guinea pig. The figure of about 25,000 (not over 30,000) is fairly well established for man. The sensory cells and pattern of innervation seem to be much the same for the two species, but the length of the partition is shorter in the guinea pig than in man. Thus the basal half of the guinea-pig cochlea, which can be excited synchronously, contains approximately 7500 fibers, and the basal third contains about 5000.

Input-output relations of N_1 in response to 7000-cycle tone bursts

Viewed on the oscilloscope, the threshold of action potential is quite sharp. As the stimulus increases in strength, the latency of response

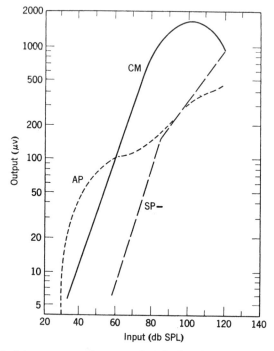

Fig. 6. Typical input-output functions for the basal turn of guinea pig stimulated by 7000-cycle tone bursts with 1-msec rise time. Recording by intracochlear electrodes (Tasaki, Davis, and Legouix, 1952). The cochlear microphonic (CM) is a linear function of sound pressure level up to nearly 80 db SPL but goes through a maximum between 100 and 105 db SPL, though without peak clipping. The summating potential (SP−) is also a linear function up to about 85 db SPL but then becomes a power function with an exponent of about 0.5. It never shows a maximum like that of CM but increases up to injurious sound pressure levels. The action potential (AP) output, measured from base line to the peak of N, is described in the text. The curves for CM and SP− represent the typical form and median values of about thirty experiments. The AP curve represents one particular experiment selected as typical. Both SP − and AP are more variable from experiment to experiment than is CM.

shortens appreciably over a range of 10 to 20 db. The amplitude grows rapidly at first and then, in terms of microvolts of response per decibel of increase, more slowly, and then faster and faster over a very wide dynamic range (Fig. 6). At a sound pressure level of

120 db, N_1 reaches an amplitude of nearly 1 millivolt. At this level we can assume that nearly all the fibers of the basal third of the cochlea are being stimulated. We have established, therefore, an approximate equivalence of about five or six (not more than ten) nerve fibers per microvolt of peak voltage. This relation implies in turn that the usual threshold of visible response on our oscilloscope is of the order of five to eight fibers. In fact, at threshold in favorable preparations with low physiological noise, we believe that we have occasionally seen abrupt jumps in both latency and amplitude, as though we were just detecting the responses of individual fibers.

If the intensity of the 7000-cycle tone burst is increased above 120 db, the action potential continues to grow, but we have reason to believe that the more apical regions of the cochlea and perhaps even the nonauditory portions of the labyrinth now contribute to the total discharge. We shall not consider further this population of nerve fibers with very high threshold. The dynamic range, before the new components confuse the measurements, is 90 to 100 db. The input-output curves show, on a double logarithmic plot, a sharp initial rise and then a very roughly linear trend corresponding to a power function with an exponent of 0.25 to 0.20. The slope is rather strongly determined by the size of a population of fibers with low threshold that centers at about 40 db sound pressure level. These low-threshold responses are easily reduced or eliminated, either by surgical clumsiness, a bout of anoxia, or a blast of high-intensity noise.

The experiment illustrated in Fig. 6 was chosen as representative of our more sensitive (and therefore presumably more normal) preparations. In Fig. 7 the data have been replotted on linear-versus-logarithmic coordinates and the cumulative response curve of action potentials has been broken down to increments in voltage (and therefore numbers of fibers) per 5-db increase in the strength of stimulation. It seems quite clear that in this animal we encountered one small, sensitive population that determined the threshold of response and a larger population with increasingly higher thresholds that generated the large voltages seen at high intensities of stimulation.

We thought at one time that the population with low threshold was the group of fibers that innervates the external hair cells and that the high-intensity group was the fibers from the internal hair cells (Davis, Deatherage, et al., 1958b). The relative sizes of the two populations are, however, such as to make this interpretation incomplete. Furthermore, the internal hair cells should begin to respond when the summating potential reaches a few microvolts, which is only some 10 or 15 db above the corresponding level for the cochlear microphonic. On the other hand, a prominent effect in ears partly dam-

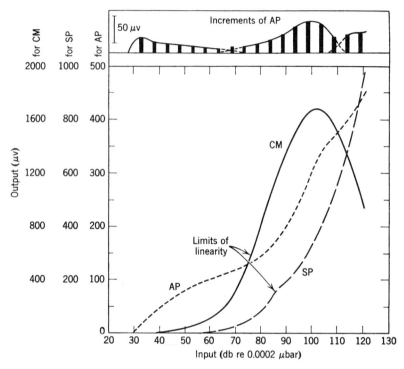

Fig. 7. The data of Fig. 6 are here replotted in semilogarithmic ⁻coordinates. Above are shown the increments of AP voltage for successive 5-db increases in sound pressure level. Three populations of neurons seem to be present. The population with very high thresholds (above 110 db SPL) undoubtedly contains many neurons that do not arise in the basal turn. The calibration bar for voltage (50 μv) is estimated to correspond to about 500 nerve fibers.

aged by streptomycin, and in which the damage seemed chiefly confined to the external hair cells, was the disappearance of just this low-threshold population and with it much of the cochlear-microphonic response. We look favorably on the suggestion of Engström (1960) that the inner row of external hair cells, which seems to be clearly differentiated morphologically from the two more peripherally located rows, may be physiologically specialized for low threshold.

Input-output relations of N_1 for 500-cycle tone pips

The dynamic range of simple response to a tone pip at 500 cps is narrower than for a high-frequency tone burst. Above 110 db the cochlear microphonic shows serious distortions of wave form, and the action-potential "spikes" become confused and reduced in height.

At 110 db the voltage of the largest N_1 is about 1 mv or perhaps a little more (Fig. 8). We estimate, on the basis of information given above, that at 110 db a 500-cycle pip is evoking synchronized action potentials from approximately half the cochlea or about 8000 fibers in all. As for the high-frequency burst, the voltage is about 1 mv, so we find again that 1 μv corresponds to less than ten fibers.

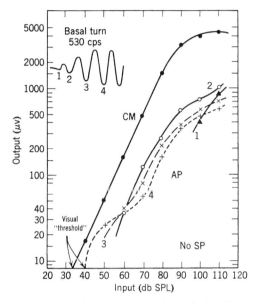

Fig. 8. Input-output functions for CM and AP from the basal turn for a 530-cycle tone pip of approximately the form shown in the sketch. There is no measurable SP in the basal turn at this frequency. The CM wave form shows peak clipping above 110 db. The crossing of the output curves for the successive waves is explained in the text and in Fig. 9.

The threshold of action potentials at 500 cps is likely to be between 40 and 45 db SPL, so that the dynamic range from threshold to saturation may be only 60 or 70 db. The rate of increase, if we confine our attention to the height of the largest N_1 spike, whether it is evoked by the first, second, third, or fourth wave, is also roughly a power function, but with an exponent between 0.5 and 0.4.

For what it may be worth, the psychophysical rule (Stevens, 1955) that the loudness of a tone is doubled for each increase of 10 db in the intensity of the stimulus corresponds to a power function with an exponent of 0.3. This is intermediate between our high-frequency trend and our low-frequency trend. The more rapid increase that we have found at 500 cps than at 7000 cps is similar to the low-frequency "recruitment" in human hearing.

The largest N_1 spike is not always elicited by the same wave of the tone pip. At threshold, in the experiment shown, it was elicited by

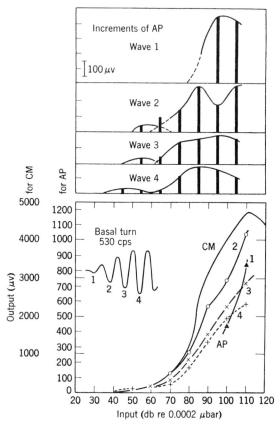

Fig. 9. The data of Fig. 8 are here replotted in semilogarithmic coordinates, and the increments in voltage of the AP spike per 10-db step in SPL are shown for each wave. A small low-threshold and a large high-threshold population of neurons can be identified. When there is a large increment in voltage for one wave, the increase in the voltage of the following wave tends to be smaller than expected. The vertical calibration bar corresponds to approximately 1000 nerve fibers.

the largest wave in the group, namely, the fourth. When the stimulus was made a few decibels stronger, the previous wave (No. 3 in Fig. 9) reached threshold strength. The axons that responded to this stimulus, however, were not all re-excited by the fourth wave, to which they had been previously responding, and consequently the later wave grew less than it would otherwise have. With still further increase in the strength of the stimulus, the smaller earlier waves successively reached threshold and evoked earlier responses, always to some extent at the expense of the size of the following waves. At 110 db the relatively very small first wave evoked a nearly maximal action-potential spike. When the stimuli were well above threshold, each of the sound waves evoked not only a major spike (N_1) but also a smaller N_2 at the usual 1-msec interval.

The meaning of "latency" for a low-frequency tone pip is obviously equivocal. In some contexts the latency of the very first small response may be decisive, in others the latency of the largest spike. If the tone pip is made to start more abruptly (with the first wave the largest, as in the "low-frequency clicks" used in some psychoacoustic experiments), its frequency spectrum is significantly broadened. Here we encounter the indeterminacy that is inescapable when we try to specify time and frequency simultaneously. Gabor (1947) pointed out this dilemma very clearly. No cleverness in peripheral coding can circumvent this basic reciprocal relation inherent in the *logon.*

Some Final Speculations

When, in response to a 500-cycle pip or tone, the auditory nerve is saturated or nearly saturated with activity from extensive excitation of the basal as well as the apical turn, the question may well be asked how the listener can still identify the pitch of the tone. I repeat that possibly, under these circumstances, the temporal sequence of discharge due to the traveling-wave pattern becomes significant in helping to determine somehow the inhibitory effects that sharpen the tuning of the individual elements higher in the nervous system. (We need not assume that pitch and loudness are discriminated at the same level in the auditory system, nor certainly by the same mechanism.) Before speculating further, however, it might be well to do some appropriate psychophysical experiments and determine to just what extent the recognition of pitch actually does deteriorate at these high intensity levels.

Incidentally, if the effects of saturation can be demonstrated in psychophysical experiments, they may help us decide how much of the adaptation of the sense of loudness (perstimulatory fatigue) is peripheral and how much of it takes place centrally. Adaptation of the frequency of impulses in individual units in the cochlear nucleus and in the auditory nerve has been known experimentally in animals for a long time (Galambos and Davis, 1943). It is not so clear, however, that the marginal fringe of fibers, which are stimulated only enough to discharge at a low average rate, undergo adaptation sufficiently to drop out of activity. It is probable that during adaptation to an intense tone the total number of active fibers may not diminish very much, although the total number of impulses that they deliver per second may fall off considerably. It is still uncertain whether the sense of loudness corresponds more nearly to the total number of peripheral neurons that are active or to the rate of arrival of impulses arriving at some center, regardless of the particular fibers over which they came, that is, the "mass" of neural activity.

All in all, we see that the peripheral mechanism for coding auditory information is highly complex. The same nerve impulses serve in more than one way—according to the channels in which they travel, according to their number, and according to the time relations among them. The volley principle contributes to the sense of pitch, but the time differences between volleys from the two ears contribute to localization of the source, and perhaps time differences between different fibers in the same ear assist in the operation of the place principle for pitch. Dr. Eldredge suggests that an appropriate comparison to a dual electronic system would be to color television. Here the color information has been inserted in the frequencies between the picture frequencies. The code is so complex that the signals, as seen in the transmission channel, appear chaotic, but the information is successfully decoded and the system works. Our broadest generalization concerning the peripheral code for auditory information is that it is not a single but an elaborate, multiple set of codes. There is not only one but several acoustic transducer mechanisms, each with its own code, and in the central nervous system there must be at least as many distinct neural receiving mechanisms, and also a mechanism or mechanisms for resynthesis.

Summary

The classical theories of the coding of auditory information ascribe pitch partly to the activation of particular nerve fibers (place principle) and partly to the repetition rate of volleys of nerve impulses (frequency principle). We suggest that systematic time differences between impulses in basal and apical fibers due to the delay of the traveling waves on the cochlear partition may also assist in pitch discrimination of low frequencies.

Interaural time differences between corresponding impulses in corresponding fibers in the basal turn are probably the chief basis of lateralization. Intensity discrimination is probably based on some combination of the total number per second of nerve impulses arriving at the nerve centers and the total number of nerve fibers activated, plus perhaps a weighting of the central effect of certain fibers as a function of the thresholds of their receptor cells.

The cochlea is a phylogenetically new organ. The fibers of its nerve are all of similar diameter, and we do not find a new, large-fiber, mensurating system (Bishop, 1959) opposed to an old, small-fiber, integrating system. The otolithic organs and semicircular canals are the old system. The former are sensitive to low-frequency vibration as well as to steady linear accelerations, and the high-frequency detector and analyzer systems of the cochlea are the new developments. One aspect of the cochlear detector system is that successive high-frequency sound waves can summate their excitatory effects.

Input-output functions for synchronized initial volleys of action potentials elicited by high-frequency (7000 cps) and by low-frequency (530 cps) acoustic signals have been established for guinea pig. The over-all trends are very approximately power functions with an exponent of 0.25 to 0.20 for 7000 cps and between 0.5 and 0.4 for 530 cps. The peak voltage of these initial synchronized volleys is presumably proportional to the number of fibers activated. Analysis of the growth of the input-output functions in these terms shows two fairly distinct but overlapping populations of nerve fibers, a small one with low thresholds and a larger one with thresholds broadly distributed over the higher range of intensities.

In general, the auditory system depends on an elaborate multiple set of codes for transmitting the information concerning frequency, intensity, and time differences that is contained in the incoming acoustic signals. The multiplicity of codes implies an equal number of central receiving or decoding mechanisms to make central integration possible.

References

Békésy, G. v. Similarities between hearing and skin sensations. *Psychol. Rev.*, 1959a, **66**, 1–22.

Békésy, G. v. Synchronism of neural discharges and their demultiplication in pitch perception on the skin and in hearing. *J. acoust. Soc. Amer.*, 1959b, **31**, 338–349.

Békésy, G. v., and W. A. Rosenblith. The mechanical properties of the ear. Chapter 27 in S. S. Stevens (Editor), *Handbook of Experimental Psychology*. New York: Wiley, 1951.

Bishop, G. H. The relation between nerve fiber size and sensory modality: Phylogenetic implications of the afferent innervation of cortex. *J. nerv. ment. Dis.*, 1959, **128**, 89–114.

Davis, H. Biophysics and physiology of the inner ear. *Physiol. Rev.*, 1957, **37**, 1–49.

Davis, H. Transmission and transduction in the cochlea. *Laryngoscope*, 1958a, **68**, 359–382.

Davis, H. A mechano-electrical theory of cochlear action. *Ann. Otol. Rhinol. Laryngol.*, 1958b, **67**, 789–801.

Davis, H. Excitation of auditory receptors. In J. Fields (Editor), *Handbook of Physiology. Neurophysiology* I. Washington: American Physiological Society, 1959.

Davis, H., B. H. Deatherage, D. H. Eldredge, and C. A. Smith. Summating potentials of the cochlea. *Amer. J. Physiol.*, 1958a, **195**, 251–261.

Davis, H., B. H. Deatherage, B. Rosenblut, C. Fernández, R. Kimura, and C. A. Smith. Modification of cochlear potentials produced by streptomycin poisoning and by extensive venous obstruction. *Laryngoscope*, 1958b, **68**, 596–627.

Davis, H., C. Fernández, and D. R. McAuliffe. The excitatory process in the cochlea. *Proc. Nat. Acad. Sci.*, 1950, **36**, 580–587.

Davis, H., S. R. Silverman, and D. R. McAuliffe. Some observations on pitch and frequency. *J. acoust. Soc. Amer.*, 1951, **23**, 40–42.

Deatherage, B. H., D. H. Eldredge, and H. Davis. Latency of action potentials in the cochlea of the guinea pig. *J. acoust. Soc. Amer.*, 1959, **31**, 479–486.

Engström, H. Electron micrographic studies of the receptor cells of the organ of Corti. In G. L. Rasmussen and W. F. Windle (Editors), *Neural Mechanisms of the Auditory and Vestibular Systems*. Springfield, Ill.: C. C Thomas, 1960.

Fernández, C. The innervation of the cochlea (guinea pig). *Laryngoscope*, 1951, **61**, 1152–1172.

Gabor, D. Acoustical quanta and the theory of hearing. *Nature*, 1947, **159**, 591–594.

Galambos, R. Inhibition of activity in single auditory nerve fibers by acoustic stimulation. *J. Neurophysiol.*, 1944, **7**, 287–303.

Galambos, R., and H. Davis. The response of single auditory-nerve fibers to acoustic stimulation. *J. Neurophysiol.*, 1943, **6**, 39–57.

Katuski, Y., T. Sumi, H. Uchiyama, and T. Watanabe. Electric responses of auditory neurons in cat to sound stimulation. *J. Neurophysiol.*, 1958, **21**, 569–588.

Licklider, J. C. R. A duplex theory of pitch perception. *Experientia*, 1951, **7**, 128–134.

Miller, G. A., and W. G. Taylor. The perception of repeated bursts of noise. *J. acoust. Soc. Amer.*, 1948, **20**, 171–182.

Stevens, S. S. The measurement of loudness. *J. acoust. Soc. Amer.*, 1955, **27**, 815–829.

Stevens, S. S., and H. Davis. *Hearing: Its Psychology and Physiology*. New York: Wiley, 1938.

Tasaki, I. Nerve impulses in individual auditory nerve fibers of guinea pig. *J. Neurophysiol.*, 1954, **17**, 97–122.

Tasaki, I., H. Davis, and J.-P. Legouix. The space-time pattern of the cochlear microphonics (guinea pig), as recorded by differential electrodes. *J. acoust. Soc. Amer.*, 1952, **24,** 502–519.

Wersäll, J. Studies on the structure and innervation of the sensory epithelium of the cristae ampullares in the guinea pig. *Acta oto-laryngol.*, 1956, Suppl. 126.

Wersäll, J. Personal communication, 1959.

Wever, E. G. *Theory of Hearing.* New York: Wiley, 1949.

Wever, E. G., and J. A. Vernon. Auditory responses in the common box turtle. *Proc. Nat. Acad. Sci.*, 1956, **42,** 962–965.

8

LLOYD M. BEIDLER
Division of Physiology, Florida State University

Mechanisms of Gustatory and Olfactory Receptor Stimulation

All information concerning the external world passes through our sense organs. Since the neural output of these organs is the input to the rest of the nervous system, it is important to consider the characteristics of the receptors and their relationship with the external environment. It has been shown in other sensory areas, particularly in vision, that many of the simple properties of the sensory receptor can account for some of the psychophysical data. Thus, a better understanding of the receptor mechanism will enable us better to understand the behavioral response to external stimuli.

The gustatory and olfactory receptors usually act as monitors during the intake of food into the oral cavity and air into the nasal cavity. There are many other chemoreceptors, internal to the organism, that will not be considered in this brief presentation. Chemoreceptors are even more important to lower animals, where their activity is intimately related to reproduction, nutrition, food finding, and so forth.

Food taken into the oral cavity of man may stimulate both the gustatory and olfactory receptors. Natural foods are usually cellular in organization and, therefore, very complex. The chemical components of flavor may not be uniformly distributed throughout the food, and in some cases flavor may be released only upon chewing. The flavor may normally be present in the cell and then released during the grinding process of mastication, or a precursor of the flavor may exist, which is acted upon by an enzyme, a "flavorase," released during mastication. This latter concept has led to the development of dramatic effects in the food industry.

In order to understand the mechanism of taste and smell, it is necessary to simplify the stimulus by using pure chemical compounds of known concentration. The remainder of this paper will be based upon observations using such simple stimuli. It should be remembered, however, that touch, temperature, pain, and other sensory qualities, in addition to the flavorous components, are important in the appreciation of food.

Taste

The taste bud, the organ of taste, was first discovered in fish, where taste buds are distributed over the surface of the body, as well as in the mouth. In mammals, most of the taste buds are located on the surface of the tongue and are associated with the circumvallate, the

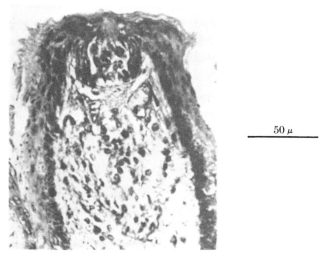

50 μ

Fig. 1. A single taste bud is located near the surface of each fungiform papilla of the rat. Calibration line, 50 μ.

foliate, and the fungiform papillae. The chorda tympani nerves innervate all the fungiform papillae, whereas the glossopharyngeal nerves innervate the taste buds on the circumvallate and foliate papillae at the posterior of the tongue. The taste receptor characteristics described in this paper refer to those taste buds found on top of the fungiform papilla. The rat has only one taste bud per fungiform papilla, although most other mammals have more than one. Each taste bud consists of about 10 to 15 single taste cells arrayed in a budlike organ much like the segments of an orange (see Fig. 1). These taste cells have small villi-like structures called taste hairs which project through the taste pore into the aqueous medium covering the surface of the tongue. These villi are about 4 microns in length and about 0.2 micron in diameter. It is commonly thought that the chemical stimuli of taste interact with the cell at the site of the villi. Branches of a single nerve fiber may innervate several cells or a single cell several times.

Histologists have observed mitotic division in the taste bud which led many to believe that the taste cells continually multiply and degenerate much like most other epithelial cells lining the oral surface. The presences of phagocytes lent credence to this idea. Recent work with colchicine, which blocks mitotic division during metaphase, confirms the concept of a continual replacement of taste cells within the bud. One may also assume then that the fine endings of the taste fibers also move about as the cells they innervate disappear and reappear. Thus, we have a dynamic picture of the anatomical structures concerned with the taste process. The functional importance of a dynamic structural state in the physiology of taste is not at present known. The taste buds on the anterior part of the tongue are near the surface and are continually insulted by various strong agents in the food, whereas those at the rear of the tongue are situated in deep grooves and are, therefore, better protected and less often stimulated. Perhaps the average age of taste cells at the anterior of the tongue is quite different from that of those at the back. This difference, if it exists, may be responsible for functional differences between the two parts of the tongue. It is well known that in man the taste buds at the tip of the tongue are more sensitive to sweet substances than are those at the back of the tongue, and those at the back of the tongue are in turn more sensitive to bitter substances. Only future experiments will reveal the functional importance of variations in rate of cell division among the cells of the taste bud.

The response of a single taste cell to taste substances may be recorded by inserting a very fine microelectrode into the interior of the cell. A change in potential, usually a depolarization, may then be measured as various taste stimuli are placed on the surface of the tongue. From such studies one finds that the individual taste cells of a rat may respond to one or more of the four basic taste stimuli. That is, some cells may respond to sugars, salts, and acids, whereas others may respond only to salts and acids. Furthermore, the sensitivity of each single taste cell to a given stimulus differs; some cells are very sensitive, others respond only sluggishly. Thus, the population of receptors concerned with taste consists of cells with varying sensitivities. However, the information is transmitted to the central nervous system along taste nerve fibers, each of which innervates more than one taste cell. Each fiber may therefore respond to one or more stimuli representing different taste qualities, as was shown many years ago by Pfaffmann (1941). How the central nervous system decodes these mixed-up messages concerning the taste quality is not known (Pfaffmann, 1959).

The single taste fibers are about 4 μ in diameter and have small action potentials that are generated at an irregular frequency. A measure of the response of the population of taste receptors to chemical stimulation may be obtained by recording the total electrical

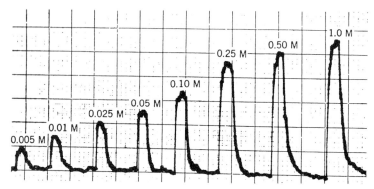

Fig. 2. Summated electrical activity of chorda tympani nerve in response to NaCl solutions, flowed over rat's tongue, of concentrations: 0.005, 0.01, 0.025, 0.05, 0.10, 0.25, 0.50, and 1.0 M. Test solutions of 7 to 10 sec duration of application are interspersed with water rinses. Time scale: 1 large division = 20 sec.

activity from the chorda tympani taste nerves that innervate the front two-thirds of the tongue. The electrical discharge is amplified, electronically summated, and then displayed on a Sanborn recorder (see Fig. 2). The response of the taste cells declines in magnitude during the first 2 seconds of stimulation and then maintains a steady state of electrical activity that is proportional to the concentration and the type of chemical stimulus. This steady-state response is maintained for at least 10 minutes during continuous application of NaCl to the rat's tongue. In man, the salty sensation in response to a continuous application of NaCl on the surface of the tongue decreases in intensity until the saltiness is completely absent within 2 minutes. It is very probable that this adaptation is due to central neural processes and not peripheral. The response of the rat chorda tympani nerve declines steadily during application of CaCl₂ to the tongue, and may attain complete adaptation within several minutes.

It is interesting to note that a particular single fiber may show different rates of adaptation when different chemical stimuli are applied to the tongue. In the rat, the response to low concentrations of potassium benzoate declines during application of the stimulus, and a positive response is not attained until the potassium benzoate is washed from the tongue with water. The large positive response during the rinse rapidly declines to the initial steady level. Thus one notes that the temporal sequence of response depends upon the particular chemical stimulus. The latency of response, the rate of rise of response, the magnitude of the response, the rate of adaptation, and the rate of decline of the response after the stimulus is removed all are peculiar to the particular stimulus chosen.

The cation plays a predominant role in stimulation by salts, and the anion has a slight inhibitory effect. Compared to KCl, NaCl provokes a strong response in the rodent chorda tympani. The reverse is true in carnivores. The guinea pig and hamster respond well to sucrose, whereas the cat responds very poorly. Thus, species differences exist even with most simple kinds of taste stimuli.

The taste receptors are stimulated within 50 msec after the chemical stimulus flows over the surface of the tongue. The wide variety of chemicals and the large range of concentration that can stimulate the taste cells suggest that the stimulus does not need to enter the cell in order to stimulate. Furthermore, the response declines rapidly after the tongue is rinsed with water. A theory of the mechanism of taste stimulation has been proposed (Beidler, 1954). If a monomolecular reaction between the taste stimulus and some part of the cell is assumed, the reaction may be described as

$$\begin{array}{ccc} \mathbf{A} + & \mathbf{B} & \rightleftharpoons \mathbf{AB} \\ C & N - Z & Z \end{array}$$

where **A** is the stimulus, **B** are unfilled receptor sites, **AB** are filled receptor sites, C is the concentration of stimulus, N is the total number of sites available, and Z the number of sites filled at concentration of stimulus C.

The equilibrium constant K may be written as

$$K = \frac{(Z)}{(C)(N - Z)} \tag{1}$$

If it is assumed that the magnitude of response is proportional to the number of sites filled, and that the maximum response occurs at a high concentration of stimulus when all the sites are filled, then

$$R = \alpha Z \quad \text{and} \quad R_s = \alpha N$$

where R is the magnitude of response and R_s the magnitude of maximum response. Inserting this information into Eq. 1, one arrives at the fundamental taste equation:

$$\frac{C}{R} = \frac{C}{R_s} + \frac{1}{KR_s} \qquad (2)$$

Notice that all the parameters can be directly measured except the equilibrium constant K. If the theory is correct, then the plot of the experimental data in the form of C/R versus C should reveal a straight line. Figure 3 shows agreement of the data with the theory, and

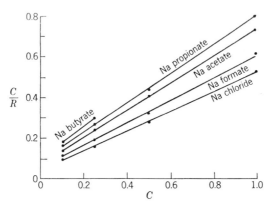

Fig. 3. The ratio of the molar concentration of the stimulus and the magnitude of the integrated response of the taste receptors is plotted against the molar concentration of the stimulus.

one may conclude that the mathematical expression in Eq. 2 is sufficient to describe the quantitative experimental data. This does not necessarily mean that the assumed model is a correct one, however. In order to obtain more information concerning the interaction of the stimulus with the receptor, the equilibrium constant K can be found from the graph in Fig. 3. This will provide a measure of the binding force of the stimulus with the receptor. The small equilibrium constants found (see Table 1) are in agreement with the concept that

Table 1. Calculated Equilibrium Constants

Sodium salts	Equilibrium constant K
Sodium chloride	9.80
Sodium formate	9.00
Sodium acetate	8.55
Sodium propionate	7.58
Sodium butyrate	7.72

the stimulus is adsorbed to the cell surface and that the reaction is not an enzymatic one. One finds similar magnitudes of equilibrium constants in the adsorption of salts to proteins.

The magnitude of response of the taste receptors to salts does not change when the temperature changes from 20 to 30°C. Thus the reaction is independent of temperature in this range and agrees with the notion that the process is a physical one and not an enzymatic reaction. Also, ΔH is equal to 0 in the expression

$$\Delta F = \Delta H - T \Delta S$$

Since ΔF can be calculated from the equilibrium constant by the equation

$$\Delta F = -RT \ln K$$

one notes that the ΔF, or change in free energy of the reaction, is negative, and thus, the change in entropy, ΔS, is positive. This means either that the water of hydration of the ion decreases when the ion is adsorbed to the receptor surface, or else that the receptor molecule changes shape slightly when the ions are adsorbed to it. One also finds small positive changes in entropy when ions are bound to proteins.

Properties of the binding may also be studied by changing the pH of the solution. The magnitude of response to 0.1 molar NaCl is not changed appreciably when the pH of the solution is varied from 3 to 11. This implies that most carboxyl units of proteins are not involved in stimulation, but other groups, such as phosphate groups, may be involved. In the present theory of taste, one may think of the receptor molecule as a polyelectrolyte containing a large number of charged side chains that make up the sites.

Although this theory was developed for the interaction of salts with a receptor, it has also been found useful to describe stimulation by means of acids, sugars, and bitter substances. Certainly the type of binding to sugars is different from that to salts, but the over-all monomolecular reaction of adsorption is still maintained. The acids stimulate by means of their hydrogen ions, although the adsorption to the surface is also dependent upon the amount of un-ionized acid in the solution and the ionic strength. Thus, a weak acid does not stimulate as well as an equimolar concentration of a strong acid. On the other hand, the weak acid stimulates more effectively than a strong acid of equal pH (see Table 2). Furthermore, if the ionic strength is changed by the addition of salt common to the weak acid, the subjective sourness does not decrease to the extent that would be expected from the change in the pH. This can be accounted for by the increase in ionic strength, which tends to increase the amount of acid adsorbed at a given pH.

Table 2.   Molar Concentration of Total Acid and Hydrogen Ions Necessary to Elicit a Response Equal in Magnitude to that to 0.0050 M HCl

Acid	Molarity (mM)	mM H^+
Sulfuric	2.2	4.2
Hydrochloric	5.0	5.0
Dichloroacetic	9.0	7.7
Formic	11.6	1.43
Lactic	15.6	1.47
Acetic	64.0	1.06
Propionic	130.0	1.33
Butyric	150.0	1.51

Small changes in spatial arrangement of the molecules making up the surface of the receptor may result in large changes in response to a given stimulus, since the binding forces involved are very small. This means that differences in sensitivity as well as in response to the four taste qualities may differ from one taste cell to the next in a given population of taste receptors.   Furthermore, differences in response to stimuli are to be expected from one species to another. Since the response to a stimulus of large molecular size may be dependent upon the molecular shape, taste specificity is not uncommon. The taste of such molecules as phenylthiourea and other related molecules is genetically determined.

The formulated taste equation can be applied to psychophysical data if one uses a unit of jnd (just noticeable difference) as a measure of response.   If one assumes that a jnd unit represents an equal increment of neural activity in response to any given taste stimulation, then the plot of the accumulative number of jnds versus concentration can be described by the taste equation.   Furthermore, if the concentration divided by the cumulative number of jnds is plotted against the concentration, then a straight line should be obtained according to Eq. 2.   Figure 4 shows how well the theoretical curve conforms with the experimental points.   The value of the equilibrium constant of the taste reaction can be obtained from psychophysical measurements by measuring the slope and intercept of the plot in Fig. 4. Once this is obtained, many other predictions concerning the relationship of stimulus to taste sensation can be found.

Weber's ratio $\Delta C/C$ can be determined by calculating from Eq. 2

the change in concentration necessary to produce a given change (assigned as unity) in neural response. The theoretical prediction of a $\Delta C/C$ versus C plot agrees well with the experimental data.

One may conclude, therefore, that taste intensity discrimination is related in a simple way to the magnitude of neural output from the receptors. This does not necessarily mean, however, that all psychophysical measurements of taste sensation can be so simply described.

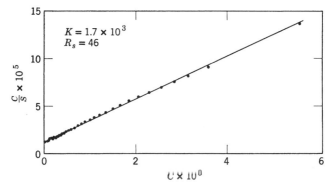

Fig. 4. The ratio of the molar concentration of the stimulus and the number of consecutive taste jnds (ΔS) for man is plotted against the molar concentration of the stimulus (data from Lemberger. 1908).

Olfaction

Anatomy

The olfactory receptors are limited to the posterior region of the septum and turbinates of the nose. In the rabbit there are about 100 million olfactory receptors in the small olfactory area. Nine to ten receptors are arranged in a symmetrical pattern around the periphery of each columnar supporting cell with a density of 127,000 receptors per square millimeter. The receptor is a bipolar neuron with a cell body 5 to $10\,\mu$ in diameter. Immediately below the nucleus, the cell forms a centrally directed process of about $0.2\,\mu$ in diameter, the olfactory nerve fiber. There is one nerve fiber from each receptor cell connecting the receptor directly to the olfactory bulb, or a total of about 100 million olfactory nerve axons in the rabbit. The peripheral portion of the cell forms a rodlike structure, about $1\,\mu$ in width and about 20 to $90\,\mu$ in length, that projects to the surface of the olfactory epithelia. The rod terminates in a swelling about $2\,\mu$ in diameter, which contains a marginal ring to which are attached, in radial symmetry, about 9 to 16 cilia (see Clark, 1956). These cilia are about 0.1 to 0.2 μ in diameter, and electron microscopy reveals the usual fibrillar structure common to all cilia. Whether these cilia are motile is not known. The rapid response of the receptors to an odor of minute concentration passing over the aqueous medium lining the olfactory epithelium is not easily understood if diffusion alone operates to carry the odor molecules to the receptors. Bloom (1954) observed in the toad olfactory cilia 50 μ in length. The ends of these cilia were broken, suggesting that the length may approach 100 to 200 μ, as reported by earlier workers using the light microscope. Gasser (1956) indicates that the cilia extend 1 to 2 μ and then form a fine process that may extend to 100 μ in length. These long cilia are believed to turn at right angles and to lie tangential to the aqueous surface covering the olfactory epithelium. It is thought that the cilia play a large role in the olfactory process. If one assumes a diameter of 0.15 μ, a length of 100 μ and 13 cilia per receptor, then these cilia present a total surface area for adsorption of odor of about 600 cm^2.

There appears to be little definite proof either for or against the concept that single receptor cells have great selectivity. A small number of different but specific olfactory receptors could account for the discrimination of a large number of different odors. On the other hand, it is possible that each odor has its own spatial pattern of adsorption on the olfactory area, and thus the quality of odors could be differentiated by a distinctive spatial and temporal adsorption pattern over a large population of receptors having uniform properties.

Many olfactory receptors in rabbit are contained in a thin sheet of epithelium lining the septum. Parallel to the septum is another thin sheet of epithelium on the turbinate, where receptor cells are also located. The distance between these two parallel sheets of epithelia is usually 1 to 2 mm. Since the tissue is heavily vascularized, the separation between the two receptor areas can easily be diminished upon vasodilation. This in effect tends to decrease the amount of odor that can pass between the sheets of epithelium and thus stimulate the olfactory region. Vasodilation and constriction are controlled by the autonomic nervous system. The anatomical structure of the nose is very important because, as will be shown later, the olfactory response is dependent upon flow rate and thus upon sympathetic neural activity reaching the olfactory region.

Experimental procedures in study of olfaction

A good proportion of the animals kept under normal laboratory conditions have various forms of rhinitis, and such animals are unsuitable for experiments in olfaction. In a suitable animal, strands of olfactory nerve can be dissected free and the neural activity recorded electrophysiologically. Unfortunately, such small strands contain a large number of small olfactory fibers, and single-fiber analysis in the periphery has not yet been attained. Some useful information has been obtained from the fiber bundles, however.

The concentration and flow rate of applied odor is maintained by the use of an olfactometer. After purification by means of charcoal and silica gel, room air is odorized by passing purified air through liquid odorous material until the stream is saturated with the odorous vapor. This odor stream is then mixed with a purified air stream in the proper proportion to attain any given concentration of odor. The flow rate of the mixed stream is calibrated and controlled by a flow meter. The odor can be passed through a plastic breathing chamber at a flow rate higher than the highest instantaneous rate that can be measured flowing through the nares during inspiration. This assures that the animal always breathes pure odor when the stimulus is presented. During some experiments, the trachea of the animal can be cannulated and a syringe used on the trachea to pull the air through the nares at a predetermined rate. In other experiments, the nares can be cannulated by pushing a very small polyethylene tubing through the nares to the olfactory area. The air stream may then be withdrawn by means of a syringe attached to a hole drilled through the maxillary sinus. The trigeminal nerves, which innervate the same general area of tissue, also respond to odors, and the neural activity can be recorded simultaneously with the olfactory activity. The olfactory response in the rabbit declines steadily and is completely eliminated about 10 minutes after the blood flow is stopped. The trigeminal response, on the other hand, is maintained for about 1 hour after stoppage of the blood flow.

Characteristics of olfactory receptor responses

The olfactory response is dependent upon the type of chemical stimulus applied to the olfactory area. For example, amyl acetate produces a large olfactory response, whereas phenylethyl alcohol usually produces a small response. The olfactory receptors respond to amyl acetate at lower concentration than do the trigeminal fibers. However, the trigeminal nerves often respond to the phenylethyl alcohol at a concentration lower than that necessary to initiate an olfactory response. At high concentrations of chloroform odor a large olfactory response is produced during the first inspiration, but thereafter the response declines rapidly until all olfactory activity disappears.

The concentration of stimulus is very important. Both the olfactory and trigeminal responses increase with concentration. The magnitude of olfactory response is limited by adaptation at high concentrations; that is, successive inspirations of odor produce a decrease in response until a steady level is reached (see Fig. 5). At high concentrations of odor the steady level of the olfactory response may be zero. When the trachea is cannulated and the odor is pulled continuously through the nares by means of an attached syringe, the process of adaptation occurs at a much lower concentration of odor. There

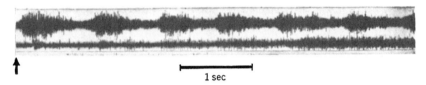

↑ |————————|
1 sec

Fig. 5. The olfactory nerve activity (first trace) in response to an odor of medium concentration decreases with each successive inspiration to a steady level. The trigeminal response (second trace) slowly increases with time. Time trace = 1 sec.

is initially a large response that rapidly declines to a steady level. Adaptation is seen at a much lower concentration when the odor is continually applied than when it is rhythmically breathed during inspirations and washed during expirations. The rate of adaptation depends upon the manner in which the stimulus is applied to the olfactory area. The olfactory receptors respond well during each inspiration, and the response declines rapidly during each expiration. The trigeminal response, on the other hand, usually commences later (and at higher concentrations) than the olfactory, and it builds up to a steady level that shows little decrease during expiration. Thus, the trigeminal responses are more sluggish than the olfactory.

The flow rate of the odor over the olfactory area is also very important. When air is pulled through the nares by means of a syringe attached to the trachea, one observes an increase in magnitude of neural response to a given concentration of odor when the flow rate is increased. No additional increase in response is noticed after the flow rate attains a value of one-half liter per minute through the nares of a rabbit. The maximum instantaneous flow rate through the nares during a normal inspiration is several times greater than that necessary for a maximum olfactory response. If a plastic cannula is inserted through the nares to the olfactory area and a hole drilled through the side of the nose through the maxillary sinus to the olfactory area, then the odor can be pulled through the cannula, over the olfactory area, and out the exit hole.   In this case, the magnitude of response is maximum at a flow rate of about 1 cc per second.

In a cannulated preparation, the nose of the rabbit can be completely filled with a given concentration of odor.   In this instance the olfactory response is maximal during the first fraction of a second when the stimulus is applied, and then adaptation decreases the response to a very low level.   If the flow rate is stopped, the olfactory response disappears.   A maximal response is again observed as soon as the flow rate is continued, even though no air has entered the nares during this time.

Many authors use the criterion of selective adaptation, that is, adaptation to one odor but not to another, as a proof of receptor specificity.   However, if the olfactory receptors have characteristics of adaptation similar to those observed with taste receptors, then it is possible for a single receptor to have different rates of adaptation to different odors.   The properties of adaptation of single olfactory receptors to different odors is unknown because the activity from a single olfactory receptor has never been recorded.   It was stated above that during normal breathing the olfactory response is maximum during the first inspiration at medium to high concentrations of odor.   If the concentration of odor stimulus is increased from one level to another, a transient overshoot in response above the new steady response level is again observed.   However, if the concentration of odor is decreased to a lower level, then a transient undershoot in response occurs followed by an increase in olfactory activity until the new steady level of response to the lower concentration of odor is attained.   These overshoots and undershoots during a change in concentration of odor would tend to enhance the ability for the discrimination of changes in intensity of stimulus.

The olfactory response is also dependent upon the amount of activity reaching the nasal area by way of the sympathetic fibers in the ethmoidal nerve.   The magnitude of sympathetic activity to the olfactory area can be recorded from the ethmoidal nerve of the narcotized rabbit.   Loud noises, flashes of light, or pinching the toe of the animal will all increase the amount of sympathetic activity.   As this activity is increased, the magnitude of olfactory response to a given odor is also increased.   This can easily be demonstrated by observing the increased magnitude of olfactory response recorded from the olfactory nerve bundle during electric stimulation (once every 2 seconds) of the cervical sympathetic nerve in the neck of the animal (see Fig. 6). It is not known what the mechanism is for the sympathetic control of the olfactory output. It may merely be a control of nasal constriction which determines the flow rate of odor reaching the olfactory area. Such a mechanism would be similar to that of the pupil in the visual system. However, it should be noted

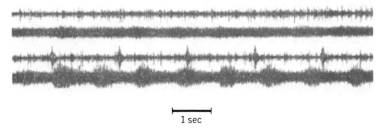

1 sec

Fig. 6. The olfactory response to a given odor may be greatly enhanced by sympathetic stimulation. First and third traces show sympathetic activity of ethmoidal nerve before and after electrical stimulation at ½ sec. Second and fourth traces indicate olfactory activity before and after sympathetic stimulation. Time trace = 1 sec.

that researchers on other receptors have noticed an influence of the sympathetic system upon the receptor itself. For example, Loewenstein (1956) reported an increase in response of the frog mechanoreceptors innervated by the dorsal cutaneous nerve when the sympathetic nerve was stimulated.

The direct recording of neural activity from the primary olfactory nerve is difficult. Some of the parameters important in olfaction have been illustrated, and others may remain unknown. Unless all parameters are considered and controlled, a quantitative study of olfaction has little meaning (Beidler, 1960).

Summary

The first event in the mechanism of taste stimulation can be considered to be a weak binding of the ion or molecule of the taste substance to the molecular structure of the taste cell surface. The response of the taste receptors can then be described by applying the mass-action law to the adsorption of the chemical stimulus to the receptor surface. A simple but fundamental equation is obtained which describes the magnitude of taste response in terms of the concentration of the stimulus, the equilibrium constant of the reaction, and the maximum response obtainable at high stimulus concentrations.

This equation adequately describes the taste data obtained electrophysiologically from animal experimentation. By assuming that an equal increment of neural activity is associated with each just noticeable difference, one may use the taste equation (Eq. 2) to describe adequately human psychophysical data such as absolute and differential thresholds. Considerable specificity is observed in the binding of many sweet and bitter substances.

The magnitude of olfactory receptor response in mammals depends upon the odor concentration, the rate of flow of odor across the olfactory epithelium, the amount of sympathetic neural activity coming to the nasal region, and the length of time the stimulus is applied. Adaptation of the neural activity to olfactory stimulation occurs at high but not at low concentrations of odor. Other functional characteristics of the olfactory receptors may be studied by recording the neural activity associated with the primary olfactory receptor in response to different odors at various concentrations. Information concerning the relation of response to concentration is at present inadequate for the formation of a theory of olfactory receptor stimulation.

Acknowledgments

The author would like to acknowledge the assistance of Mr. Don Tucker in the study of olfactory processes.

The research from the author's laboratory was supported bv USPH Grant B-1083, NSF Grant G-3865, and Contract Nonr 589(00) between Florida State University and the Office of Naval Research.

References

Beidler, L. M. A theory of taste stimulation. *J. gen. Physiol.*, 1954, **38,** 133–139.
Beidler, L. M. Physiology of olfaction and gustation. *Ann. Otol., Rhinol., Laryngol.*, 1960, **69,** 398–409.
Bloom, G. Studies on the olfactory epithelium of the frog and the toad with the aid of light and electron microscopy. *Z. Zellforsch.*, 1954, **41,** 89–100.
Clark, W. E. LeG. Observations on the structure and organization of olfactory receptors in the rabbit. *Yale J. Biol. Med.*, 1956, **29,** 83–95.
Gasser, H. S. Olfactory nerve fibers. *J. gen. Physiol.*, 1956, **39,** 473–496.
Lemberger, F. Psychophysische Untersuchungen über den Geschmack von Zucker und Saccarin. *Arch. ges. Physiol.*, 1908, **123,** 293–311.
Loewenstein, W. R. Modulation of cutaneous mechanoreceptors by sympathetic stimulation. *J. Physiol.*, 1956, **132,** 40–60.
Pfaffmann, C. Gustatory afferent impulses. *J. cell. comp. Physiol.*, 1941, **17,** 243–258.
Pfaffmann, C. The sense of taste. In *Handbook of Physiology. Neurophysiology* I. Washington, D. C.: American Physiological Society, 1959. Pp. 507–533.

9

HESSEL DE VRIES and **MINZE STUIVER***
Natuurkundig Laboratorium, Groningen, The Netherlands

The Absolute Sensitivity of the Human Sense of Smell

A perpetually interesting aspect of the sense organs is their high sensitivity. The sense cells of the eye respond to one quantum of light with a macroscopic reaction. The ear and the other organs of the labyrinth have a sensitivity that comes close to Brownian motion. Until now, olfaction has been neglected; it will be shown below that the olfactory cells respond to one single odorous molecule, or at least to a very small number. Taste and the sense of touch are less sensitive; however, the calculation of the absolute sensitivity of the tactile receptors is fairly complicated. It is difficult to estimate what part of the energy applied actually comes to the receptor.

The high sensitivity of the sense organs offers special problems. At the present time we can only say that a "trigger reaction" occurs, in which the stimulus triggers metabolic energy. Hardly anything is known with certainty about these trigger reactions, however.

Finally, the high sensitivity of the sense organs not only is interesting in itself but may also be a guide to an understanding of the mechanism involved. Confining ourselves to olfaction, we find that, if a sense cell reacts to macroscopic amounts of the odorous compound, we have to look for "normal" macroscopic chemical reactions. Actually, the cell reacts to one or a few molecules, which means that it is just a triggering of metabolic energy. Needless to say, this brings us straight to an unsolved problem, but a problem of general importance.

Up until a few years ago, quantitative studies in olfaction were rare and no complete set of data was available with which we could calculate the sensitivity of single olfactory receptors. Recently Neuhaus (1955) proved that the olfactory cells of a dog could react to single molecules of fatty acids. The data and calculations that follow have been taken from the work of Stuiver (1958).

* Now at Geochronometric Laboratory, Yale University.

159

The following six steps are involved in the calculation of the sensitivity of individual olfactory cells.

1. The absolute threshold of the organ as a whole, that is, the minimum perceptible number of molecules N_0 entering the nose under optimal conditions.

2. The fraction f_1 of the inhaled air that passes through the olfactory slit.

3. The fraction f_2 of odorous molecules left in the air when it reaches the olfactory slit. This factor f_2 is smaller than unity, because part of the molecules will be adsorbed on the mucous membrane that covers the nasal cavity.

4. The remaining molecules that pass through the olfactory slit will be effective only insofar as they hit the epithelium (fraction f_3). The number N of odorous molecules reaching the epithelium is then given by

$$N = N_0 f_1 f_2 f_3 \qquad (1)$$

5. From N and the number of sense cells in the epithelium it is possible to calculate the average number of molecules per sense cell. A simple statistical analysis gives the minimum number n of molecules to which a cell must respond. If one were to assume a high value of n, stimulation at the actually observed threshold would "never" occur.

6. Finally, the steepness of the frequency-of-smelling curve gives information about the minimum number of molecules involved in the process (see below).

A similar line of approach was followed in the analysis of visual threshold data. It has been mentioned already, however, that none of the factors in Eq. 1 was known for olfaction.

1. The absolute threshold for human smell

The threshold, expressed in number of molecules entering the nose, depends on time of presentation of the stimulus, concentration of the odorous substance, and rate of flow of the carrier gas (generally air). The effect of the rate of flow on the threshold could be correlated with its influence on the adsorption of the odorous substance on the mucous membranes (see below). In most of the experiments the rate of flow for normal breathing has been used (250 cubic centimeters per second per nostril). The effect of the time of presentation is illustrated by Figs. 1 and 2. Figure 1 gives the threshold expressed in number of molecules. For short stimuli it is nearly constant; for longer stimuli it becomes proportional to the time of presentation. This means that

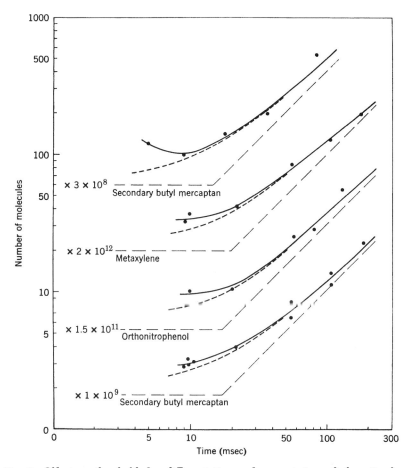

Fig. 1. Olfactory threshold for different times of presentation of the stimulus. Logarithmic scales. Absolute values indicated for each odor. Dotted curve: see text. (From Stuiver, 1958.)

the *concentration* becomes the controlling factor (see also Fig. 2). For very short pulses the threshold tends to go up (Fig. 1). This is because the volume of air involved becomes very small; consequently the volume of the nasal cavity that has to be filled before the odorous substance reaches the sensory epithelium becomes relatively more important. Making a correction for this loss, one obtains the dotted line in Fig. 1.

The curves of Fig. 1 were obtained by the blast-injection technique. Normal breathing is more pleasant, however, and for this reason the greater part of the threshold measurements on different substances and observers were made with normal breathing. This means a presentation time of 1 second. Thus Fig. 1 can be used to convert the results for normal breathing to the results that would have been obtained if short stimuli had been used.

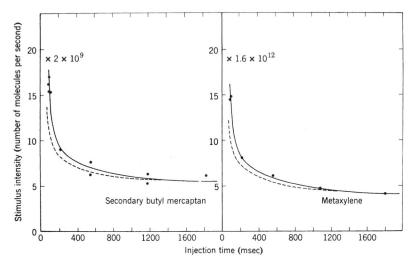

Fig. 2. Absolute threshold as concentration ("intensity") versus time of presentation. The same data as Fig. 1, but converted into concentration (number of molecules per cubic centimeter). Dotted curves: see text. (From Stuiver, 1958.)

2. The fraction of the inhaled air that passes through the olfactory slit

Though the geometry of the nasal cavity looks fairly complicated, it can be represented by a simple model. The two nasal cavities are separated by a wall, the *septum*. In each half there are three folds, the conchae. The final geometry is such, however, that the "cavity" is more like a slit whose width varies from 1 to 2 millimeters. Unfolding the conchae gives the model shown in Fig. 3. This model is, of course, only an approximation. Individual differences occur, as well as appreciable variations in the lumen of the slit, due to variations in the blood content of the conchae or excretion of mucus.

The sensory epithelium lies in the upper part of the nasal cavities (see Fig. 3) on both sides of the lumen. This part is sometimes called the olfactory slit. The total area (on each side) is about 2.5 square centimeters.

The flow pattern in the model of Fig. 3 was studied by observing aluminum particles suspended in water flowing through it; the rate of flow of the water has to be 1/15 times the rate of flow of air in order to give the same pattern.   For normal breathing between 5 and 10 per cent of the total flow was found to pass through the olfactory slit. For higher flow rates this fraction increases up to at most 20 per cent. In the lower part of the nasal cavity no flow of water was observed. Needless to say, individual deviation from these figures may occur because of differences in the geometry of the nose.

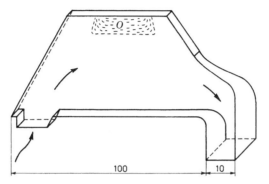

Fig. 3.   Model of the human nasal cavity. O = olfactory epithelium.   Area O about 2.5 cm^2.   (From Stuiver, 1958.)

3. Loss of odorous substance by adsorption to the wall before the olfactory slit is reached

It is well known that odorous substances tend to adhere to surfaces. This is one of the reasons that it is not possible to dilute an odorous mixture in a well-defined ratio by adding a small amount of the mixture to clean air in a larger vessel.   As far as we can see there is no special reason why odorous substances alone should have this strong adsorption.   The reason it shows up in this case is probably because the concentrations used are so low that the amounts of odorous compound are not even large enough to leave a monomolecular layer on the surface of the container.   So a relatively large fraction will be adsorbed, partly on "active" spots on the surface, partly just by a random distribution between gas phase and the adsorbed phase.

For reasons to be given below, it will be assumed that each molecule colliding with the mucous membrane or the olfactory epithelium will not return in the air.   It may come off much later, or it may be absorbed by the cells and transported by the blood.   If we assume complete loss after collision with the wall, the amount a left after a time t is easily calculated as:

$$a = a_0 e^{-D\pi^2 t/d^2} \qquad (2)$$

In this formula d stands for the width of the slit, and D for the coefficient of diffusion of the odorous substance. Inserting for d 1 mm and for D the value 0.08 gives

$$a = a_0 e^{-80t} \qquad (2a)$$

The rate of decrease is the same, whether the gas is at rest or is flowing laminarly.

For a rate of breathing of V cm³ per second the time of transport from the entrance of the nose to the epithelium is about $2/V$ sec. The average width d is 1 mm. Inserting this in Eq. 2 gives

$$a/a_0 = f_2 = e^{-(80 \times 2)/V} \qquad (3)$$

For normal breathing the rate of flow V (per nostril) is 250 cm³ per second, giving

$$f_2 = e^{-0.64} \approx 0.5 \qquad (4)$$

4. Molecules escaping from adsorption in the olfactory slit

Equation 2 also gives the fraction of molecules escaping from adsorption in the olfactory slit. Inserting for the width d of the slit 1 mm, the fraction adsorbed is

$$f_3 = 1 - e^{-80t} \qquad (5)$$

For normal breathing the result is

$$f_3 = 0.5 \qquad (6)$$

Inserting the values obtained for f_1, f_2, and f_3 in Eq. 1 gives, for the fraction of inhaled odorous material adsorbed in the olfactory slit, a value of 2 per cent for normal breathing. The other 98 per cent may be lost by adsorption to the mucous membranes, by passing through another part of the nose, or by passing the olfactory slit without hitting the epithelium.

The values of f_2 and f_3 have been calculated from the assumption that each odorous molecule hitting the mucous membrane adheres there for a fairly long time (of the order of a second or more). Unfortunately the experimental evidence supporting this assumption is somewhat indirect. The first indication is the fact that the olfactory sensation disappears "nearly immediately" after the stream of air stops. It is difficult to evaluate this quantitatively. A more quantitative experiment, carried out by Stuiver (1958) was not conclusive either.

Further evidence for immediate adsorption is obtained from the measurements of the effect of the rate of flow on the absolute threshold. According to Eqs. 4 and 6, f_2 and f_3 depend on the rate of flow V. The variation of the absolute threshold with V was in agreement with the calculated variation. (For further details see Stuiver, 1958.) However, it is not possible to conclude unambiguously from these experiments that adsorption is complete, and further experiments are planned. For the present discussion it is quite unimportant. The effect of adsorption on the final amount reaching the epithelium affects f_2 and f_3 in opposite ways, and the product is nearly independent of the adsorption for the rates of flow normally used. Moreover, the values f_1, f_2, and f_3 do not affect appreciably the conclusions to be drawn in the following sections.

5. The threshold of one sense cell

According to section 1, the threshold for mercaptans is of the order of 10^9 molecules entering the nose. Since about 2 per cent reach the epithelium, 2×10^7 molecules are effectively adsorbed. They are distributed over all sense cells, that is, about 4×10^7 cells; consequently the average number per cell is $\frac{1}{2}$. If one assumes that n or more molecules are necessary to excite a cell, the probability is easily calculated that at least one cell will be excited. Obviously at least one cell *has* to be excited in order to produce a sensation. If n is set too high, no sensation will occur. A simple statistical analysis shows that n is at most 8 (Stuiver, 1958). This low value suggests that it is probably 1.

A few additional remarks have to be made. The first concerns the assumption that the odorous molecules are evenly distributed over the sense cells. It can be shown (see Stuiver, 1958) that diffusion (through the mucus) is too slow to concentrate the molecules on special receptors which might be especially sensitive to the odor used. So the number of molecules is certainly not larger than assumed. It may be smaller, since molecules may be adsorbed on ineffective sites. Secondly, not all cells may be effective for the odor; this too will lead to a lower value of n. Since the chance of excitation varies strongly with n, the value derived under modified assumptions will not be much below 8. Finally, it was tacitly assumed that the cells and the nerve connections respond in an all-or-none way. The discussion would become more complicated if we accepted a graded response.

6. Number of molecules involved in producing a sensation

On the analogy of results well known from the analysis of visual data, it is of interest to study the steepness of the frequency-of-smell-

ing curve, that is, the curve relating the frequency of a sensation with
the intensity of the stimulus.   A few curves are shown in Fig. 4.   The
steepness gives information about the number of molecules involved
in producing a sensation.   If one calls this number m and assumes
that no other factors are involved, the frequency-of-response curve

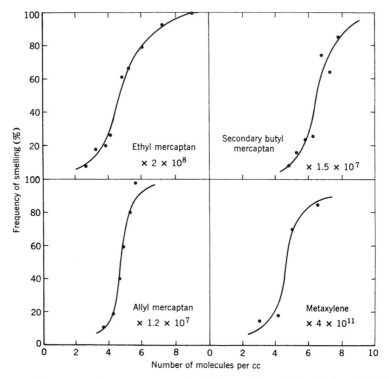

Fig. 4.   Frequency-of-response curves for different odors.   (From Stuiver, 1958.)

can be calculated: it becomes steeper with increasing m.   Other factors
—physiological and psychological variation—tend to make the curve
less steep, so the value of m derived is the minimum value.   The
curves obtained (Fig. 4) indicate m to be 40.   This may mean that
40 cells have to cooperate, in order to produce an olfactory sensation,
each responding to one molecule.   This figure can be used in section 5; it means that a smaller value of n must be chosen to give 40
excitations instead of only one.   The new value of n is 6.   A further
reduction of n is possible if it is assumed that the 40 receptors have
to be close together.

Summary

From an analysis of olfactory threshold data it is derived that the threshold of one human olfactory cell is at most 8 molecules for appropriate odorous substances.   Analysis of frequency-of-response curves shows that at least 40 molecules are necessary to produce a sensation.

References

Neuhaus, W.   Die Unterscheidung von Duftquantitäten bei Mensch und Hund nach Versuchen mit Buttersäure.   Z. vergl. Physiol., 1955, **37,** 234.
Stuiver, M.   Biophysics of the sense of smell.   Thesis, Groningen, 1958.

10 W. A. H. RUSHTON
Trinity College, Cambridge

Peripheral Coding in the Nervous System

The Nature of the Nerve Message

The verbal content of a telephone message may be transmitted exactly by telegraphing it in Morse code, but the conversational overtones, the excitement, or the indignation will be lost, and we cannot help feeling that communication with our fellows would indeed be drab if it had to be restricted to a series of clicks. Yet a series of "clicks" is in fact all that our brains ever receive—at any rate from nerves. What necessity underlies this impoverishment in the potentialities of communication?

A telephone wire can transmit a message because conduction along the wire is good and in other directions is bad, so nearly all the electricity flows down the wire and may thus be detected at a distance. But nerves are "wires" made not of metal but of dilute salt solution, with a resistance of some 25 megohms per millimeter. An ordinary telephone wire of this resistance would stretch across a continent, and after traveling so far the signal would need "boosting." Nerve signals therefore need boosting every 1 mm, and in fact the anatomical continuity of the nerve fiber is interrupted every 1 mm by a *node of Ranvier*, which electrophysiology has shown to be a boosting station. So along the stretch of nerve that runs between the fingers and the spinal cord in man there are some 800 nodes or boosting stations. Now if the telephone message is to preserve its quality, each node should restore to the signal exactly what was lost in traveling along the cable from the last node. Suppose that restoration was not perfect but only 99-per-cent complete at each node. Then after 800 nodes the signal would be reduced to

$$(0.99)^{800} = 1/3000$$

of its original size.   If on the other hand restoration was overdone by 1 per cent, the signal would reach 3000 times its initial size—or in practice the system would grossly overload and hence become saturated.   So the minutest fluctuation from a perfect restoration will result in a signal that is either *all or none*.   Amplitude modulation of signal is impossible: the only reliable code is by unit (saturated) change.

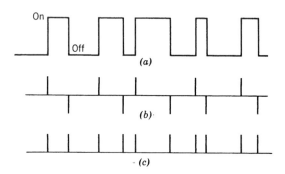

Fig. 1.   Possible types of all-or-none signal.

The most general form of this signal is thus given by Fig. 1*a*, where the *on* and the *off* of the unit response occurs in any desired time sequence.   The information content of this signal in the long run is the same as that of its differential (Fig. 1*b*), and since negative and positive pulses alternate strictly, the change of sign carries no information.   So the series of positive pulses in Fig 1*c* carries all the information of the most general form (Fig. 1*a*) but expends far less physical energy.   This is the signal of choice, and it is also the one actually found in peripheral nerve.

The exact form of the brief pulse is irrelevant since it is not the form but the presence or absence of pulse at any instant that defines the message.   But it is important that two successive pulses should not gradually fuse into one as they approach each other, for the distinction between one and two is crucial to the code.   Overtaking and fusing is prevented in nerve by the existence of the *refractory period*—a short time following each pulse during which the nerve is unresponsive and cannot transmit a second pulse.   It is seen then that the telegraphic *all-or-none* pulse code follows from the poor conductivity of a cable made of salt solution and the relatively great distances over which an impulse has to be conducted.   This argument loses its force if the conduction distance is only a fraction of a millimeter, and experiment suggests that over such short distances all-or-none conduction does not in fact obtain.

It is not easy to define the maximum amount of information that a single nerve fiber can carry using this pulse system, nor is it important to do so, for an estimate much lower than the maximum is still enormously greater than that which the central analyzer can handle. A human nerve can easily transmit impulses at 300 per second, so if a single second of time is subdivided into 300 equal intervals there is the possibility that each interval could either contain or not contain at least one impulse. In 0.1 sec this involves 30 *bits* of information, that is, the capacity to distinguish one out of a thousand million situations. Our experience does not at all support the view that a single nerve fiber could in 0.1 sec carry so much discrimination. The whole of the fovea centralis of the human retina contains some 30,000 nerve fibers and would be capable of transmitting in 0.1 sec a million bits. But all we can handle centrally is about 4 bits. Clearly the flow of information is limited not by the transmission lines but by the coding or decoding of the messages transmitted.

The Coding of Impulses

The pattern of nerve pulses may be thought of as arising in the sense organs, sometimes spontaneously, but mainly as a result of external stimuli. Sense organs are in fact energy filters through which special kinds of environmental change are allowed to pass and generate trains of impulses in particular nerves, which may result in specific and appropriate action. Light is a form of energy that can be precisely controlled, so a good place to look for the way in which a stimulus is encoded as a pulse pattern might be in the optic nerve of some animal in which a single nerve fiber makes contact with the photoreceptor itself, without the complexity of nerve junctions being interposed on the way. This requirement, unfortunately, excludes all vertebrate eyes (where both retinal structure and impulse patterns are complex), but it is satisfied by the lateral eye of the arthropod *Limulus*.

Figure 2 shows some results from experiments by Fuortes (1959). The optic nerve ganglion (eccentric cell) was impaled by a microelectrode arranged in a bridge circuit so as both to measure the membrane potential and to pass current which would polarize the cell membrane. When the current had the value indicated on the horizontal axis (Fig. 2), the membrane potential (ordinates) was as

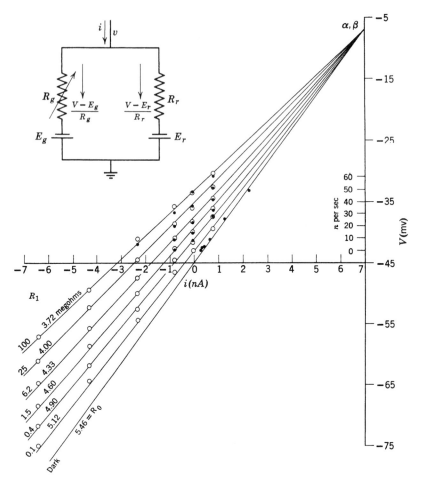

Fig. 2. Intracellular records from an optic nerve cell of *Limulus* (after Fuortes, 1959). Abscissa plots depolarizing currents; ordinates plot membrane potential in millivolts (circles) or frequency of impulses per second (dots). For a fixed light intensity I (numbers at left of curves), the circles and dots lie upon the same fixed straight line, whose slope represents the membrane resistance. The whole structure of concurrent lines follows from the electrical representation of the membrane shown inset, where R_g is dependent upon light, but R_r is not.

shown by the open circles. It is seen that, for any fixed light intensity I, the circles lie upon a straight line whose gradient is obviously equal to the resistance of the cell membrane. Thus in the dark or in any fixed light the membrane resistance is constant and unaffected by the passage of current in or out. An increase in the light, however, decreases the slope of the line, and thus is associated with a lower membrane resistance. This structure of concurrent lines is precisely what would be expected if the cell membrane had the electrical constitution represented by the conventional diagram inset in Fig. 2, where R_g is dependent upon light and R_r is not.

Now the eccentric cell whose properties are studied in Fig. 2 is not itself the photoreceptor, and it contains no visual pigment. Light, of course, excites the photoreceptors, and it might have been thought that these generated an electric current which in turn stimulated the eccentric cell. But this cannot be the case. For we have seen that current excites without any change of membrane resistance, whereas light produces a resistance drop. So the photoreceptors must act by producing some agent that lowers the resistance of the cell membrane. This concept is familiar in studies of synaptic transmission, and the action of acetylcholine upon the myoneural junction has been particularly well investigated.

Now a steady depolarization of the eccentric cell is accompanied by a steady rhythmic discharge of impulses in the optic nerve whose frequency is shown by the black dots in Fig. 2. The fact that the dots fit the same pattern of concurrent lines as do the membrane potentials (circles) means that a given membrane potential generates impulses at a fixed frequency. The fact that the frequency scale (n per second) and the membrane potential (millivolts) are both uniform scales means that they are linearly related to each other and hence to the change in membrane resistance ($R_0 - R_I$). The effect of steady light intensity upon impulse frequency is thus given by Fig. 3, which plots ($R_0 - R_I$) against log I. The experimental points lie very close to the curve which is the mathematical function

$$\log_{10}(1 + I)$$

suitably displaced along the horizontal axis. If we assume with Barlow (1956) that there is "noise" in the receptor even in the dark and that this has the same effect as light of intensity I_D, then a logarithmic transform of the total light ("dark" plus bright) will be

$$\log(I_D + I) = \log[1 + (I/I_D)] + \log I_D$$

Thus the good fit in Fig. 3 between points and curve can be described as follows. The resistance is always proportional to the logarithm of $(I + I_D)$, the total "light"; hence the change in resistance from the dark value is $\log[1 + (I/I_D)]$ where I_D is about -1.4 log units.

Now the change in membrane resistance, as we have seen, is produced by the liberation of some transmitter chemical, so the fact that the resistance is always proportional to log $(I_D + I)$ must be linked to the rate of production of this chemical. The simplest suggestion is that production is proportional to log $(I_D + I)$ and decay is exponential, so that the concentration is always closely proportional

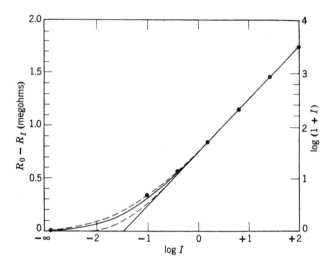

Fig. 3. $R_0 - R_I$, the drop in membrane resistance due to light I, is plotted against log I (dots). The smooth curve represents the mathematical function log $(1 + I)$ slid horizontally to fit the points.

to log $(I_D + I)$. Then if the drop in membrane resistance is proportional to the concentration of the transmitter, the observed relations will obtain. All this is a bit too facile, but it leads to some interesting and very simple conclusions.

The Transmission of the Code

As we have seen, it is plausible to suppose that the transmitter chemical C_1 (Fig. 4) has a concentration proportional to log $(I_D + I)$ and that it causes a proportional drop both of membrane resistance and membrane potential, and a proportional rise in the frequency n_1 of impulses generated in the nerve.

Now at the other end of the nerve a similar (or perhaps a somewhat different) chemical C_2 (Fig. 4) will be liberated at a rate proportional to n_1, since each impulse releases one small "packet" of transmitter substance. If C_2 is continuously removed according to monomolecular kinetics, its mean level will be proportional to n_1 and hence will be linear with C_1. It may therefore be suggested that the main function of nerve transmission is to set up rapidly at a distant place a concentration C_2 of chemical which is linearly related to C_1 in quantity, though possibly quite different from it in chemical constitution.

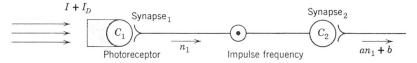

Fig. 4. Diagram of single unbranched transmission line.

But C_2 is also a chemical transmitter and may be expected similarly to generate in the efferent nerve a train of impulses whose frequency is a linear function of C_2. So a chain of m one-to-one nerve junctions as in Fig. 4 should set up a final concentration C_m which is a linear replica of the initial concentration C_1.

However, one-to-one synaptic connections are unusual (and as a transmission line are inferior to a single long nerve fiber). The familiar synaptic condition is as in Fig. 5, where several afferent

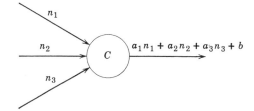

Fig. 5. Diagram of nerves converging upon a synapse.

fibers converge upon one outgoing nerve. If then each incoming fiber secreted the same kind of transmitter substance, the frequency of outgoing impulses would be linearly related to the weighted mean of the incoming frequencies, as indicated in Fig. 5.

If some of the afferent nerve fibers were inhibitory, they might be regarded as secreting a chemical that neutralized some of the exciting transmitter substance—a situation which is formally represented by giving a negative sign to the appropriate coefficient a_1 (Fig. 5). So the effect of an impulse train n_1 in an inhibitory nerve should be to change the frequency in the efferent nerve by an amount linear with n_1. The elegant experiments of Hartline and Ratliff (1957; see also

Ratliff in this volume) show exactly this linear relation between the frequencies of excitation and inhibition in the optic nerve of *Limulus*.

Adaptation

We have been considering only steady states in nerve fibers that respond with steady rhythms to a steady excitation. But in fact most nerves are not like this; they adapt, that is to say, their discharge frequency declines rapidly or slowly when a constant stimulus is applied and maintained. The value of this property appears to lie in the fact that the linear relation between transmitter chemical and impulse frequency has only a limited range. It is probable that if all but one of the excitatory nerves converging upon a synapse were maximally active, the synapse would saturate and could not signal changes in the remaining nerve. But if the strong stimulus were maintained and the nerves adapted, then the level of transmitter at the synapse would soon fall to a point upon the linear range where the contribution of the remaining nerve would be perfectly transmitted. Adaptation is therefore rather like placing a condenser in series with the input of an amplifier so that it is not pushed outside its working range. Sensory systems are in fact for the most part a-c detectors, and this is effective since it is the *change* in the environment that is biologically important.

Transmitter decay

It has been assumed that the transmitter chemical is liberated in a fixed "packet" by each impulse and decays exponentially at a rate slow compared with the interval between impulses; consequently, a steady pulse frequency will build up a steady concentration proportional to that frequency. However, it certainly is not always true that the transmitter decays as slowly as this. At the (vertebrate skeletal) myoneural junction, for instance, acetylcholine has mainly disappeared during the refractory period, so that no chemical build-up occurs during a burst of nerve impulses. This allows any nerve rhythm to be transmitted to the muscle practically unchanged and with no loss of information except at high frequencies. If conservation of information were the main consideration, *rapid* removal of the transmitter would be the optimum condition. But a synapse that did no more than this were better replaced by the through conduction of a (branched) axon.

If many converging afferent neurons are to interact in a way that does not depend dramatically upon the exact instant at which impulses arrive, it will be necessary that the decay constant of the transmitter chemical should be far longer than the refractory period, though this will limit severely the rate at which it is possible to transmit information in practice. It seems plausible to suggest that at the periphery the decay constant may be rather fast, and in the brain centers rather slow.

The Logarithmic Transformation of Stimuli

As appears from Fig. 3 the frequency of impulses in the optic nerve of *Limulus* is a linear function of log $[1 + (I/I_D)]$. It is unlikely that this logarithmic transformation is a peculiarity of the mechanism by which light energy is turned into nerve impulses, for the same transformation holds with mechanical receptors (for example, stretch and pressure) and seems to be a property of sense organs generally. It rather looks as though very different modalities of sensation have evolved this code for intensity because of its special advantages. These have been analyzed with respect to the vertebrate eye by D. M. MacKay in a private communication from which I shall largely draw in what follows.

All highly evolved sense organs are required at the lower end of their intensity range to detect signals so weak as to approach the prevailing noise level. With strong signals, on the other hand, the organ's task is not merely to detect; it must make fine discrimination with regard to time course, spatial location, and quality (for example, color, pitch, sliminess) and do so over a range of 50 to 100 decibels.

Now the minimum noise (power) of a sense organ will depend directly upon the number of degrees of freedom of output, and this corresponds more or less to the resolving power. So, as the output intensity is reduced the signal-to-noise power ratio cannot be maintained unless the resolving power is also reduced in the same proportion. If this is achieved, however, there need be no loss of information, for as the resolving power is diminished the integrating power may be increased in inverse proportion. Thus at any intensity level I, the resolving power will vary directly as I, and by integration the "gain" of the system will vary inversely as I, so that $\Delta I/I$ will have the same output Δn at all intensity levels. This is the well-known Weber's law which holds approximately for the increment thresholds of so many modalities of sensation. It is here seen to be the relation expected of a sense organ that wasted no information at any level of intensity, and maintained a steady signal-to-noise discrimination.

The long-distance receptors (eyes, ears, and nose) are receptive over an enormous span of intensities, but in all this range what the animal needs to detect is not so much the absolute intensity as the *pattern* in which intensity is distributed over the appropriate continuum (space, pitch, or the olfactory gamut). When light, etc., becomes fainter, all parts of the pattern become fainter in the same proportion; hence the invariant feature by which the pattern may be recognized is the *ratio* of the intensities of each part in comparison with the rest. So if I is the mean intensity and $(I + \Delta I_p)$ the actual intensity at any point p in the continuum, the pattern of $\Delta I_p/I$ will be the same at all levels of I.

But this invariance of $\Delta I/I$ with intensity level is precisely the relation derived above for maintaining a constant signal-to-noise ratio without loss of information. So we see a double significance in the logarithmic transform from intensity of stimulus I to frequency of impulses n. For if

$$a(n - n_0) = \log I$$
$$a \, \Delta n = \Delta I/I$$

Now $a \, \Delta n$ is that aspect of the signal which, as we saw earlier, tends to be transmitted without change through the neuron chain. So the invariant in transmission is that feature of the response pattern that has a constant signal-to-noise ratio and is unchanged by great alterations in the mean level of the stimulus. When the daylight fades, the general response pattern therefore remains unchanged; only the resolution of detail declines.

There is one further feature of the logarithmic transform that may be important if the central nervous system performs sensory discrimination by obtaining the cross-correlation function between neighboring sensory inputs (see Cherry in this volume). An important step in this process lies in the multiplication of the corresponding input energy functions with various time delays.

Now the fine structure of the cerebral cortex seems well suited to give delays of various magnitudes, and the pooling of transmitted chemicals from the various paths will result in an output giving the sum of the (delayed) input frequencies. Since these are logarithmic transforms of the stimulus intensities, those intensities will be delayed and multiplied in the output signal. However, in order to complete the cross-correlation before integration can be effective, we need an exponential transform of the frequency pattern, and this has yet to be discovered in the central nervous system.

The Power Law of Sensation (Stevens)

It is possible to compare directly the magnitudes of sensations of different kinds (see Stevens in this volume). For instance, two lights may be adjusted in brightness so that their ratio appears to have the same value as the tension ratio of two squeezes. It is not clear what is the physiological process upon which this comparison is based, but it must be one that is the same for both modalities, and one thing that *is* the same is a train of impulses.

According to the logarithmic transform, the difference in impulse frequency with two lights of ratio I/I_0 is given by

$$n - n_0 = a \log (I/I_0)$$

Similarly with two squeeze tensions of ratio T/T_0

$$n - n_0 = b \log (T/T_0)$$

If the sensation ratios are judged the same in magnitude when $(n - n_0)$ is the same for both, there will be a linear relation between $\log I$ and $\log T$, as Stevens has found experimentally. This, of course, is identical with a power relation between the corresponding intensities, the exponent a/b being the linear slope of the logarithmic plot. But not all Stevens' measurements were made like this. Instead of the magnitudes of one kind of sensation being compared with those of another, many subjects were asked to compare their sensations with what they understood as a uniform scale of sensations. It is not easy to see a priori what can be the basis of such an understanding, for a uniform increase of sensation certainly does not correspond to a uniform increase either in the strength of the stimulus or in the number of its just detectable increments (jnd). But whatever the nature of this subjective magnitude S, it possesses the property that its logarithm is linear with the logarithm of the external stimulus I, whose resulting sensation is judged.

The objective basis of S may thus be some special kind of stimulus, such as "repetition rate," which is found to have an exponent of 1.0 for all modes of sensation (light, sound, touch, or shocks). According to this idea the quantity "rate" would be encoded in an impulse train in some center by means of the logarithmic transform, and the coefficient of this transform would give the scale of sensation.

Nerves, Chemicals, and Behavior

The simplest animals have no nerves; yet they react with purpose, seeking and avoiding. Life for them presumably lies in their sensitivity to the chemical environment and the flow of protoplasm or thrash of flagella by which they can move in it, in the change of permeability with the mixing of cell ingredients, and in the secretion and removal of the hormones by which activity is controlled.

At the dawn of life, urges and efforts must have been chemical— have they ever been otherwise?

Nerves do not replace chemicals: they secrete them—instantly, exactly, and at a distant location. Nerves are the biological response to the needs of an animal who thinks with its hormones but has grown so large that diffusion can no longer distribute them fast enough nor precisely enough. We do not need to wait for adrenalin to be delivered even by an efficient blood flow, for we may secrete it immediately and intimately from a thousand nerve endings, and we possess other nerves that are faster and discharge their hormones more particularly.

In the long and faltering journey of evolution one great stride permitted animals to increase in size and specialization without loss of a unified chemical control—the appearance of a system of nerves whose essential function was to respond to a hormone here and at once to secrete a linear replica of it exactly there. Indeed the new transmission line was so much more capable of handling information than the chemical system it served that, as we have seen, the nerve message is largely redundant and nerves remain virtually idle. A species that could process nerve information without referring back at every stage to the wasteful atavism of chemical secretion would win the world. Would it also lose its soul? What is the material basis of feeling and purpose?

Impenetrably close are interwoven the cause-effect relations between the chemicals and nerves. But two chinks appear in that screen. The first lies in the long continuity of the urges of life—the will to live and to reproduce; those antedated nerves. The second is the dramatic and often highly integrated changes of personality that result from altering the chemical environment of the brain by giving or withholding hormones or drugs.

Slight as are these hints they lend a faint color to the view that the material correlate of cold thought or hot passion may be the play of chemicals in a chemical playground.

Summary

1. Nerves are conductors of such high resistance that they require hundreds of relays. This makes them unsuited to transmit amplitude-modulated signals. The most efficient signal is a train of brief pulses, separated from each other by a refractory phase. The information content of this channel could be vastly greater than is actually used. It is limited by the coding-decoding system.

2. In the encoding of sensory messages there is a more or less linear relation between the impulse frequency n and the logarithm of the stimulus intensity I. Thus Δn is proportional to $\Delta I/I$, the Weber fraction. This means that the *contrast* in a sensory ensemble is preserved by a fixed pattern in Δn independent of the absolute intensity level.

3. Nerve trains are generated by transmitter chemicals and in turn secrete transmitter chemicals, and there is a linear relation between the concentrations of the two chemicals involved. Converging nerves upon a synapse may be thought of as adding (or with inhibition subtracting) their chemicals. Most nerves adapt, so the effects of steady stimuli are not maintained.

4. It is suggested that the sole function of nerve is to secrete rapidly at a distant spot a hormone whose concentration is a linear replica of the mean input concentration. Upon this view we think and feel with our hormones. The suggestion is almost as groundless and implausible as that we do it by trains of nerve impulses.

References

Barlow, H. B. Retinal noise and absolute threshold. *J. opt. Soc. Amer.*, 1956, **46**, 634–639.

Fuortes, M. G. F. Initiation of impulses in the visual cells of *Limulus*. *J. Physiol.*, 1959, **148**, 14–28.

Hartline, H. K., and F. Ratliff. Inhibitory interaction of receptor units in the eye of *Limulus*. *J. gen. Physiol.*, 1957, **40**, 357–376.

11

FLOYD RATLIFF
The Rockefeller Institute

Inhibitory Interaction and the Detection and Enhancement of Contours

The interplay of excitatory and inhibitory influences over interconnections within the retina yields patterns of optic-nerve activity that are more than direct copies of the pattern of external stimulation. Certain significant information is selected from the immense detail in the temporal and spatial pattern of illumination on the receptor mosaic, enhanced at the expense of less significant information, and only then transmitted to the central nervous system.

Among the most significant features of a pattern of illumination are the loci of transitions from one intensity to another and from one color to another. Indeed, if only these contours are represented—as in a line drawing or cartoon—much of the significant information is retained. This paper describes an integrative neural mechanism which plays a role in the detection and enhancement of such contours. A quantitative experimental analysis of the purely inhibitory interaction among retinal elements in the lateral eye of *Limulus* is reviewed in detail, and the physiological significance of inhibitory mechanisms is discussed briefly.

Inhibitory Interaction in the Eye of *Limulus*

Fundamental properties of excitation and inhibition

The lateral eye of *Limulus* is a compound eye containing approximately 1000 ommatidia (Fig. 1a). Nerve fibers arise from the ommatidia in small bundles and come together to form the optic nerve. A plexus of nerve fibers interconnects these bundles immediately behind the layer of ommatidia (Fig. 1b).

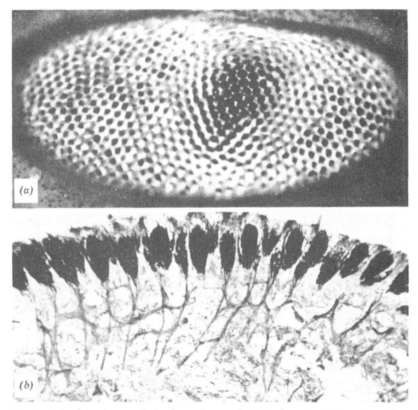

Fig. 1. The lateral eye of the horseshoe crab, *Limulus*. (*a*) Corneal surface. In a medium-sized adult each eye forms an ellipsoidal bulge on the side of the carapace, about 12 mm long by 6 mm wide. Each ommatidium has an optical aperture about 0.1 mm in diameter; the facets are spaced approximately 0.3 mm apart, center to center, on the surface of the eye. The optical axes of the ommatidia diverge, so that the visual fields of all those in one eye cover, together, approximately a hemisphere. The optical axes of the dark circular facets near the center of the eye are oriented in the direction of the camera. (*b*) Photomicrograph of a section of the compound eye of *Limulus* taken perpendicular to the plane of (*a*) at a slightly higher magnification. Samuel's silver stain. The cornea has been removed. The heavily pigmented sensory parts of the ommatidia are at the top of the figure. The silver-stained nerve fibers originating in the retinular cells and the eccentric cell of each ommatidium emerge as a bundle and join with similar bundles from other ommatidia to form the optic nerve. Small lateral branches of the nerve fibers form the network, or plexus, of interconnections immediately below the receptors. (Figure prepared by W. H. Miller.)

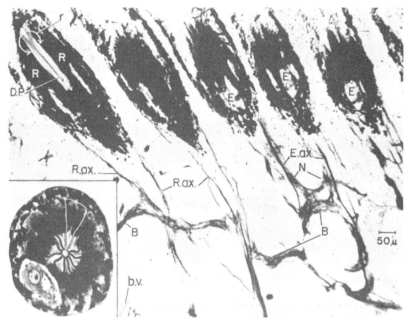

Fig. 2. Photomicrograph of a longitudinal section through several ommatidia. Samuel's silver stain. In one ommatidium (upper left) the dendritic distal process (D.P.) of an eccentric cell is seen extending up the axial canal of the rhabdom (r). This canal is formed at the junction of the medial portions of the retinular cells (R). The cell body of the eccentric cell of this ommatidium is not visible, but the eccentric cells (E) of other ommatidia to the right may be seen. The thicker, more densely stained fibers arising from the ommatidia are eccentric cell axons (E.ax.). The thinner, less dense fibers are retinular cell axons (R.ax.). Both types of axons give off small branches which form the bundles (B) of lateral interconnections that make up the plexus. These fine branches converge in regions of neuropile (N) in close proximity to eccentric cell axons (E.ax.). A portion of a blood vessel (b.v.) is shown at the bottom of the section.

The insert is a cross section of one ommatidium at the level of the eccentric cell body. Fixed in OsO_4 and stained by Mallory's aniline-blue method. One of the eleven retinular cells is outlined in white ink. Its axial border rests against the distal process of the eccentric cell which, at another level in the ommatidium, is continuous with the eccentrically placed cell body shown in this section. The spokelike structures constitute the rhabdom, probably the site of the photosensitive pigment. Electron micrographs (Miller, 1958) show that the rhabdomeres are formed by microvilli projecting from the boundaries of the radially arranged retinular cells. (Figure prepared by W. H. Miller.)

Each ommatidium (Fig. 2) contains approximately a dozen cells: a cluster of wedge-shaped retinular cells and one bipolar neuron, the eccentric cell (Miller, 1957). Both the eccentric cell and the retinular cells have axons that together make up the small bundle arising from the ommatidium; both types of axons branch profusely, and these branches constitute the plexus of interconnecting fibers (Miller, 1958; and Ratliff, Miller, and Hartline, 1958).

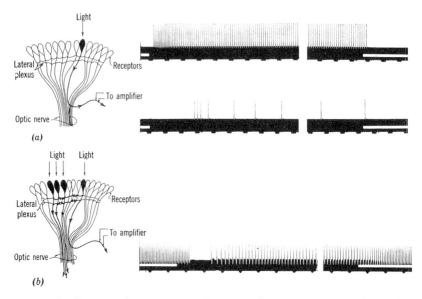

Fig. 3. Oscillograms of action potentials in a single optic nerve fiber of *Limulus*. The experimental arrangements are indicated in the diagrams at the left. (*a*) Response to prolonged steady illumination. For the upper record the intensity of the stimulating light was 10,000 times that used for the lower record. Signal of exposure to light blackens out the white line above the ⅕-sec time marks. Each record interrupted for approximately 7 sec. (Records from Hartline, Wagner, and MacNichol, 1952.) (*b*) Inhibition of the activity of a steadily illuminated ommatidium in the eye of *Limulus*, produced by illumination of a number of other ommatidia near it. The oscillographic record is of the discharge of impulses in the optic nerve fiber arising from one steadily illuminated ommatidium. The blackening of the white line above the ⅕-sec time marks signals the illumination of the neighboring ommatidia. (Record from Hartline, Wagner, and Ratliff, 1956.)

A small bundle containing a single active nerve fiber may be dissected from the optic nerve and placed on electrodes to record the action potential spikes (Hartline and Graham, 1932). Hartline, Wagner, and MacNichol (1952) have shown that the impulses recorded in this preparation originate in the eccentric cell, which seems to be a neuron rather than a true receptor (see also Waterman and Wiersma, 1954). This cell (Fig. 2) sends a dendritic distal process into the center of the rhabdom, which is made up of a dozen or so retinular cells and is in close juxtaposition with the specialized portions of the retinular cells (rhabdomeres) which are believed to be the photoreceptors (Miller, 1957). This whole assembly of cells appears to function as a "receptor unit."

The activity of one of these fibers in response to stimulation of the ommatidium from which it arises (Fig. 3a), is typical of the responses of many sensory nerves: there is a sizable latent period after the stimulus comes on before the first impulse is discharged; the frequency of discharge is relatively high at first; subsequently the frequency settles down to a lower steady level which may be maintained for long periods of time; and the frequency of discharge, particularly in this steady state, depends primarily on the intensity of stimulation. Responses to an abrupt increase or decrease in a steady level of stimulation are somewhat more complex, as will be shown later.

Illumination of other ommatidia near the ommatidium whose activity is being recorded produces no discharge in the eccentric cell axon of this test ommatidium: activity in any one optic nerve fiber can be elicited by illumination of only the one specific receptor unit from which that fiber arises. Nevertheless, the sensory elements in this eye do exert an important influence on one another by way of the plexus of lateral interconnections. This interaction is purely inhibitory. The frequency of discharge of impulses in an optic nerve fiber from a particular ommatidium is decreased (Fig. 3b), and may even be stopped altogether, by illuminating neighboring areas of the eye (Hartline, 1949).

This inhibitory effect may be summarized as follows. The ability of an ommatidium to discharge impulses in the axon of its eccentric cell is reduced by illuminating other ommatidia in its neighborhood: the threshold to light is raised, the number of impulses discharged in response to a suprathreshold flash of light is diminished, and the frequency with which impulses are discharged during steady illumination is decreased. The magnitude of the inhibition, measured in terms of decrease in frequency, has been shown to depend upon the intensity, area, and configuration of the pattern of illumination on the retina: (1) the greater the intensity on neighboring receptors, the greater the inhibition they exert on the test receptor; (2) the greater the number of neighboring receptors illuminated, that is to say, the larger the area of illumination, the greater the inhibition exerted on the test receptor; (3) illumination of neighboring receptors near the test receptor results in greater inhibition than does illumination of more distant receptors (Hartline, Wagner, and Ratliff, 1956).

Mutual inhibition

These inhibitory influences are exerted mutually among the receptors in the eye of *Limulus:* the activity of each ommatidium influences, and is influenced by, the activity of its neighbors. If activity is recorded from two optic nerve fibers coming from two ommatidia not too widely separated in the eye, the frequency of their maintained dis-

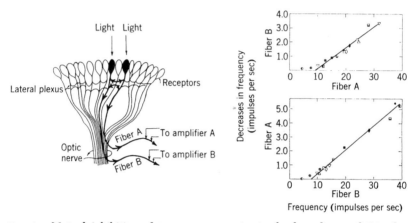

Fig. 4. Mutual inhibition of two receptor units in the lateral eye of *Limulus.* Action potentials were recorded from two optic nerve fibers simultaneously, as indicated in the diagram of the experimental arrangement. In each graph the magnitude of the inhibitory action (decrease in frequency of impulse discharge) of one of the ommatidia is plotted on the ordinate as a function of the degree of concurrent activity (frequency) of the other on the abscissa. The different points were obtained by using various intensities of illumination on ommatidia A and B, in various combinations. The data for points designated by the same symbol were obtained simultaneously. In the upper graph the slope of the line gives the value of the inhibitory coefficient of the action of receptor A on receptor B, $K_{B,A} = 0.15$; the intercept of the line with the axis of abscissas gives the value of the threshold $r^0{}_{B,A} = 9.3$ impulses per second. From the lower graph, $K_{A,B} = 0.17$; $r^0{}_{A,B} = 7.8$ impulses per second. (Reproduced from Hartline and Ratliff, 1957.)

charges of impulses—in response to steady illumination—is lower when both ommatidia are illuminated together than when each is illuminated by itself. The magnitude of the inhibition of each one has been shown to depend only on the degree of the activity of the other; thus the activity of each is the resultant of the excitation from its respective light stimulus and the inhibition exerted on it by the other. Furthermore, it has been shown that, once a threshold has been reached, the inhibition exerted on each is a linear function of the degree of activity of the other (Fig. 4).

The responses to steady illumination of two receptor units (ommatidia) that inhibit each other mutually may thus be described quantitatively by two simultaneous linear equations that express concisely all the features of the interaction (Hartline and Ratliff, 1957):

$$r_1 = e_1 - K_{1,2}(r_2 - r^0{}_{1,2})$$

$$r_2 = e_2 - K_{2,1}(r_1 - r^0{}_{2,1})$$

The activity of the receptor unit—its response r—is to be measured by the frequency of discharge of impulses in its axon. This response is determined by the excitation e supplied by the external stimulus to the receptor, diminished by whatever inhibitory influences may be acting upon the receptor as a result of the activity of neighboring receptors. (It should be noted that the excitation of a given receptor is to be measured by its response when it is illuminated by itself, thus lumping together the physical parameters of the stimulus and the characteristics of the photoexcitatory mechanism of the receptor.) The subscripts are used to label the individual receptor units. In each of these equations the magnitude of the inhibitory influence is given by the last term, which is written in accordance with the experimental findings as a simple linear expression. The "threshold" frequency that must be exceeded before a receptor can exert any inhibition is represented by r^0. It and the "inhibitory coefficient" K are labeled in each equation to identify the direction of the action: $r^0{}_{1,2}$ is the frequency of receptor 2 at which it begins to inhibit receptor 1; $r^0{}_{2,1}$ is the reverse. In the same way, $K_{1,2}$ is the coefficient of the inhibitory action of receptor 2 on receptor 1; $K_{2,1}$, the reverse.

Spatial summation of inhibitory influences

The quantitative description given thus far is concerned only with the interaction of two elements. To extend the description to more than two elements, it is necessary to know how the inhibitory influences from different elements combine with one another. The spatial summation of inhibitory influences was analyzed by measuring the inhibition exerted on a test receptor separately by each of two small groups of ommatidia near it, and then by these two groups together (Hartline and Ratliff, 1958). It may be anticipated that in general the results of such an experiment will depend on the amount of inhibitory interaction between the two groups. A special case in which there is little or no interaction could easily be achieved experimentally since the interaction between ommatidia is less, the greater their separation. Consequently it was possible to choose two regions of the eye, on either side of the test receptor, that were too far apart to affect each other appreciably but were still close enough to the test receptor to inhibit it significantly. Under these conditions the inhibitory effects exerted by these two widely separated regions on the test receptor were undistorted by their own mutual inhibition, and the experimental results were quite simple. The inhibitory effects, when measured in terms of the decrease in frequency of the test receptor, combine in a simple additive manner: the arithmetical sum of the

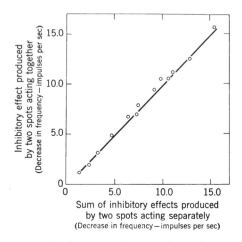

Fig. 5. The summation of inhibitory effects produced by two widely separated groups of receptors. The sum of the inhibitory effects on a test receptor produced by each group acting separately is plotted as abscissa; the effect produced by the two groups of receptors acting simultaneously is plotted as ordinate. (Reproduced from Hartline and Ratliff, 1958.)

inhibitory effects that each group produces separately equals the physiological sum obtained by illuminating the two groups together (Fig. 5).

These results permit the extension of the quantitative description to include any number of interacting elements by expressing the total inhibition exerted on any one receptor as the arithmetical sum of the individual inhibitory contributions from all the others (Hartline and Ratliff, 1958). Consequently, the activity of n interacting receptors may be described by a set of simultaneous linear equations, each with $n - 1$ inhibitory terms combined by simple addition:

$$r_p = e_p - \sum_{j=1}^{n} K_{p,j}(r_j - r^0_{p,j})$$

where $p = 1, 2, \cdots, n$; $j \neq p$; and $r_j \nless r^0_{p,j}$. In each such equation the magnitude of the inhibitory influence is given by the summated terms, written in accordance with the experimental findings as a simple linear expression.

These equations have been applied to experimental results obtained by illumination of three receptors whose optic-nerve responses could be recorded simultaneously. In these experiments (unpublished), the six thresholds and six coefficients of inhibitory action were first determined by illuminating the receptors by pairs, as in the experiment illustrated in Fig. 4. On the basis of these experimentally determined constants, the responses expected from each member of the group of three, when illuminated together, were calculated. The observed responses of the group of three receptors illuminated simultaneously agreed satisfactorily with that predicted from the interaction observed when the receptors were illuminated in pairs.

Diminution of inhibition with distance

It will be noted that the equations given above lack an explicit expression for the effects of distance. Fortunately for the simplicity of the quantitative treatment, no such expression is required. The effects of distance are already implicit in the equations: it has been found (Ratliff and Hartline, 1959) that the threshold of inhibitory action increases with increasing distance between the units involved, and that the coefficient of inhibitory action decreases with increasing distance (Fig. 6). Thus, in the quantitative formulation, the diminution

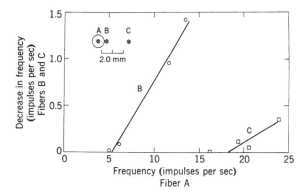

Fig. 6. The dependence of the magnitude of inhibition on distance. The inhibition (measured in terms of decrease in frequency) exerted by a small group of receptors (A) on two other receptors (B and C) is plotted as ordinate. As abscissa is plotted the concurrent frequency of the discharge of impulses of one of the receptors in group A. The geometrical configuration of the pattern of illumination on the eye is shown in the insert. The locations of the facets of the receptors whose discharges were recorded are indicated by the symbol ⊗. (Reproduced from Ratliff and Hartline, 1959.)

of the inhibitory effect with increasing distance may be ascribed exactly to the combined effects of increasing thresholds ($r^0{}_{p,j}$) and decreasing inhibitory coefficients ($K_{p,j}$) that accompany increasing separation of the interacting elements p and j.

The dependence of the mutual inhibition among receptors on their separation greatly affects the quantitative outcome of experiments with various configurations of interacting elements. Indeed, in many cases the law of spatial summation would seem to be called into question: often the inhibitory effect produced by the combined influence of several groups of receptors is *less* than the sum of the inhibitory effects produced by each such group illuminated alone (Hartline and Ratliff, 1958). But this occurs simply because the inhibitory influence

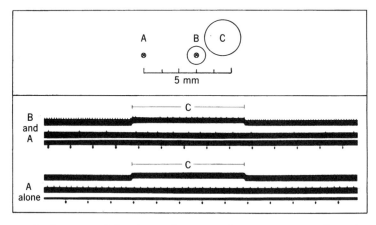

Fig. 7. Oscillograms of the electrical activity of two optic nerve fibers showing disinhibition. In each record the lower oscillographic trace records the discharge of impulses from ommatidium A, stimulated by a spot of light confined to its facet. The upper trace records the activity of ommatidium B, stimulated by a spot of light centered on its facet, but which also illuminated approximately 8 or 10 ommatidia in addition to B. A third spot of light C was directed onto a region of the eye more distant from A than from B. The geometrical configuration of the pattern of illumination is sketched above. Exposure of C was signaled by the upward offset of the upper trace. Lower record: activity of ommatidium A in the absence of illumination on B, showing that illumination of C had no perceptible effect under this condition. Upper record: activity of ommatidia A and B together, showing (1) lower frequency of discharge of A (as compared with lower record) resulting from activity of B, and (2) effect of illumination of C, causing a drop in frequency of discharge of B and concomitantly an increase in the frequency of discharge of A, as A was partially released from the inhibition exerted by B. Time marked in ⅕ sec. The black band above the time marks is the signal of the illumination of A and B, thin when A was shining alone, thick when A and B were shining together. (Records from Hartline and Ratliff, 1957.)

exerted by a receptor depends on its own *activity*. And since the amount of receptor activity in each of several groups close to each other is less when they are illuminated together than when they are illuminated separately (because of the mutual inhibition), the inhibitory effect produced by the combined influence of these several groups acting together should be less than the sum of the inhibitory effects produced by each group illuminated alone.

This interpretation, as well as the quantitative description, is given even stronger support by experiments in which a third spot of light is used to provide additional inhibitory influences that can be controlled independently of the two interacting receptor units whose activity is being measured. When additional receptors are illuminated in the vicinity of an interacting pair, too far from one ommatidium to affect it directly but near enough to the second to inhibit it, the frequency of discharge of the first increases as it is partially released from the inhibition exerted on it by the second (Fig. 7). Such "disinhibition" simulates facilitation: illumination of a distant region of the eye results in an increase of the activity of the test receptor. This observed result is a direct consequence of the principle of interaction that was established above: the inhibitory influences exerted by a receptor depend on its own activity, which is the resultant of the excitatory stimulus to it and whatever inhibitory influences may, in turn, be exerted upon it.

Responses to simple spatial patterns of illumination

Although the activity of a system of interacting elements can conveniently be described without making explicit reference to their relative locations in the receptor mosaic and to the spatial pattern of illumination (since the dependence of the inhibitory influences on distance is implicit in the values of the thresholds and inhibitory coefficients), it is nevertheless clear that the strong dependence of the inhibitory thresholds and coefficients on the separation of the elements introduces a topographic factor that must be of considerable significance in retinal function. Any complete description of the spatial characteristics of the inhibitory interaction must, therefore, provide an explicit statement of the relations between these inhibitory parameters and corresponding distances on the receptor mosaic. At the present time, however, a sufficient number of measurements has not been made, covering the wide variety of locations, directions, and distances, to formulate exactly such a law. Nevertheless, on the basis of a quantitative analysis of the inhibitory interaction, one can predict the general form of the patterns of response that will be elicited from the

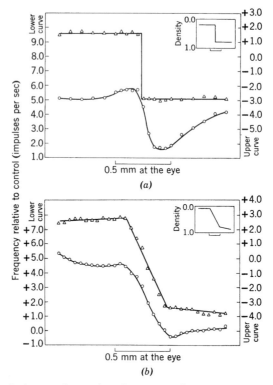

Fig. 8. The discharge of impulses from a single receptor unit in response to simple patterns of illumination in various positions on the retinal mosaic.

(*a*) "Step" pattern of illumination. The demagnified image of a photographic plate was projected on the surface of the eye. The insert shows the relative density of the plate along its length as measured, prior to the experiment, by means of a photomultiplier tube in the image plane where the eye was to be placed. The density of the plate was uniform across its entire width at every point. The upper (rectilinear) graph shows the frequency of discharge of the test receptor, when the illumination was occluded from the rest of the eye by a mask with a small aperture, minus the frequency of discharge elicited by a small control spot of light of constant intensity also confined to the facet of the test receptor. Scale of ordinate on the right. The lower (curvilinear) graph is the frequency of discharge from the same test receptor when the mask was removed and the entire pattern of illumination was projected on the eye in various positions, minus the frequency of discharge elicited by a small control spot of constant intensity confined to the facet of the receptor. Scale of ordinate on the left.

(*b*) A simple gradient of intensity (the so-called Mach pattern). Same procedure as in (*a*). (Reproduced from Ratliff and Hartline, 1959.)

elements of the receptor mosaic by various simple spatial patterns of illumination.

Contrast effects, for example, may be expected to be greatest at or near the boundary between a dimly illuminated region and a brightly illuminated region of the retina. A unit within the dimly illuminated region, but near this boundary, will be inhibited not only by dimly illuminated neighbors but also by brightly illuminated ones. The total inhibition exerted on such a unit will be greater, therefore, than that exerted on other dimly illuminated elements that are farther from the boundary; consequently its frequency of response will be less than theirs. Similarly a unit within but near the boundary of the brightly illuminated field will have a higher frequency of discharge than other equally illuminated units that are located well within the bright field but are subject to stronger inhibition since all their immediate neighbors are also brightly illuminated. Thus the differences in activity of elements on either side of the boundary will be exaggerated, and the discontinuity in this pattern of illumination will be accentuated in the pattern of neural response.

The ideal experimental test of these qualitative predictions would be to record simultaneously the discharge of impulses from a great number of receptor units in many different positions with respect to a fixed pattern of illumination on the receptor mosaic. Since such a procedure is impractical, the discharge of impulses from only one receptor unit near the center of the eye was measured, and the pattern of illumination shifted between measurements, so that this one receptor unit assumed successively a number of different positions with respect to the pattern (Ratliff and Hartline, 1959). Two simple patterns of illumination were used: an abrupt step in intensity, and a simple gradient between two levels of intensity (the so-called Mach pattern). In each case (Fig. 8), transitions in the pattern of illumination are accentuated in the corresponding pattern of neural response: maxima and minima appear in the frequency of receptor discharge at the sides of the transitions.

Temporal aspects of inhibitory interaction

It is a fundamental characteristic of most receptors that not only do they signal the information about steady-state stimulus conditions, but they also respond vigorously to temporal changes in stimulus intensity. Both the excitatory and the inhibitory components of activity in the lateral eye of *Limulus* are marked by large transient responses to stimulus changes. The inhibitory transients may best be understood if the excitatory transients are examined first. MacNichol and

Hartline (1948) found that a steady discharge to constant illumination by a single receptor unit in the eye of *Limulus* is modulated in the following manner by changes in the level of illumination: in response to a small step increase in the intensity of illumination, a large transient increase in frequency was produced—the frequency eventually subsiding to a steady level slightly greater than that preceding the change in illumination; a similar small decrease in intensity like-

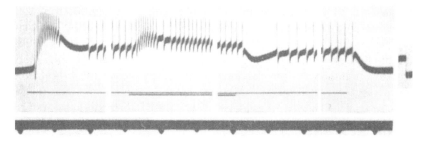

Fig. 9. Slow "generator" potentials and propagated action potentials arising within an ommatidium in response to transients in illumination. Recorded between a micropipette inserted into the ommatidium and an indifferent electrode in the solution covering the eye. Upward deflection of the trace indicates increasing positivity of pipette. Calibration signal, 10 mv; ⅕-sec time marks. Approximately 1 sec cut from each section of the record. Illumination is indicated by the dark line above the time marks. The shorter dark line near the center of the record indicates a step increment in the intensity of illumination. Under steady illumination the frequency of the spikes recorded by means of the microelectrode, as well as their concomitant propagated impulses, depends linearly upon the level of the slow potential (cf. MacNichol, 1956). During transients the frequency is less simply related to the slow potential changes; both the absolute level and the rate and direction of change of the slow potential determine the momentary rate of response. Any abrupt increase or decrease in the slow potential produced by a change in illumination is accompanied by a marked increase or decrease in the frequency of response.

wise produced a large decrease in frequency, followed by a gradual recovery to a level slightly below that preceding the change in illumination. Thus any change in the level of illumination is marked by a large transient change in frequency that is much greater than the comparable change from one steady-state frequency to the other. Transients such as these (Fig. 9) appear to depend upon both the rate of change and the absolute level of the slow "generator" potentials within the ommatidium (unpublished experiments).

These marked excitatory transients produce a similar but opposite effect in the frequency of response of neighboring elements on which the excitation has not been changed. Consequently, if the frequency of response of a particular element is compared with that of its neighbors, the excitatory transients will seem even larger, relatively speaking, since they produce this opposite effect on the neighboring elements. A typical example of such excitatory and inhibitory transients,

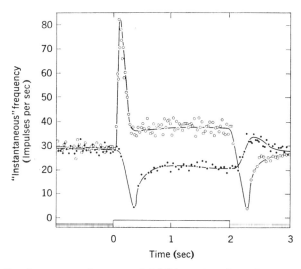

Fig. 10. Simultaneous excitatory and inhibitory transients in two adjacent receptor units in the lateral eye of *Limulus*. One receptor unit, black dots, was illuminated steadily throughout the period shown in the graph. The other unit, open circles, was illuminated steadily until time 0, when the illumination on it was increased abruptly to a new steady level where it remained for 2 sec and then was decreased abruptly to the original level. The added illumination produced a large transient increase in frequency of the second receptor, which subsided quickly to a steady rate of responding; the subsequent decrease in illumination to the original level produced a large transient decrement in the frequency of response, after which the frequency returned to approximately the level it had prior to these changes. Accompanying these marked excitatory transients are large transient inhibitory effects in the adjacent, steadily illuminated receptor unit. A large decrease in frequency is produced by the inhibitory effect resulting from the large excitatory transient; during the steady illumination the inhibitory effect is still present but less marked; and finally, accompanying the decrement in the frequency of response of the element on which the level of excitation was decreased, there is a marked release from inhibition.

recorded simultaneously from two optic nerve fibers, is shown in Fig. 10. Corresponding to the excitatory transient in one element, but slightly later in time, there is a marked inhibitory transient in the other element. Corresponding to the decrement in the frequency of response of the element on which the level of excitation was decreased, there is a release from inhibition evident in the response of the other element, again slightly later in time than the excitatory transient. Such an increase in frequency due to this release from inhibition cannot exceed appreciably the level of response that this element would have shown if there had been no inhibition at all. That is to say, the uninhibited response of this element is the maximum response, since there are no excitatory interactions in this eye. Occa-

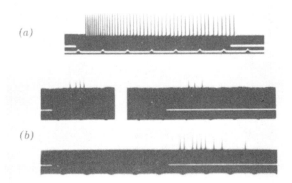

Fig. 11. Oscillograms of diverse types of impulse discharge patterns in single fibers of *Limulus* optic nerve. (*a*) Typical sustained discharge in response to steady illumination. (*b*) Upper record: a synthetic "on-off" response (approximately 1 sec was cut from the middle of this record). Lower record: a synthetic "off" response. Time is marked in ⅕ sec. Signal of exposure to light blackens the white line above the time marker. Fibers whose activity is shown in the two records of (*b*) gave a sustained discharge like that shown in (*a*) when the ommatidia from which they arose were illuminated alone. (Reproduced from Ratliff and Mueller, 1957.)

sionally, however, there has appeared a very slight suggestion of an overshoot above the uninhibited level of response following inhibition, but such overshoots are always of extremely small magnitude. Although the inhibitory transients parallel the excitatory transients, they are probably not entirely dependent upon them. Tomita (1958) has shown that a large transient inhibitory effect can be produced by antidromic impulses of constant frequency. Apparently the initial phases of the inhibitory effects are inherently somewhat greater than the subsequent effects.

It is evident that the relatively simple discharge pattern typically observed in fibers of the optic nerve of *Limulus* (Fig. 3a) may be greatly modified by the combined action of the excitatory and inhibitory transients to produce a relatively complex response (Fig. 10). If this response is further modified by other means—such as exposure time, intensity of stimulus, and state of adaptation—it is possible to generate "on-off" and "off" responses in the individual fibers of the *Limulus* optic nerve (Ratliff and Mueller, 1957). These "synthesized" transient responses (Fig. 11) have the properties of the analogous responses in the vertebrate eye (Hartline, 1938): the "on-off" responses are characterized by a burst of activity when the light is turned on, no further activity as the light stays on, and a final burst of activity after the light is turned off; the "off" responses do not appear until after the light goes off, and they may be inhibited by reillumination. These experiments lend considerable support to the view (Granit, 1933; Hartline, 1938) that "on-off" and "off" responses may be the result of the complex interplay of excitatory and inhibitory influences by showing that the experimental manipulation of these influences can, indeed, yield such transient responses.

The role of inhibitory interaction in generating specialized neural responses to *temporal* changes in stimulation may well be of greater physiological significance than the better known part it plays in the enhancement of contrast under steady-state conditions.

Discussion

Inhibitory interaction among neural elements is a fundamental neural mechanism common to many sense modalities, integrative levels, and species (cf. the brief review by Brooks, 1959); and its principal functional properties appear to be much the same wherever it is found. The eye of *Limulus* provides an especially favorable preparation for the quantitative analysis of these functional properties: the interaction is purely inhibitory; the population of interacting elements is relatively small; and the pattern of stimulation on the receptor mosaic may be controlled with considerable precision. Experiments with this eye furnish a model of a relatively simple integrative process that may be useful in understanding the more complex integrative processes in other parts of the nervous system and in other species.

The influence of inhibitory interaction among neural elements on their patterns of activity is clearly illustrated in the eye of *Limulus*. The discharge of impulses in any one optic nerve fiber depends not only upon the stimulus to the specific receptor unit from which that fiber arises but also upon the spatial and temporal distribution of the stimulation of the entire population of interacting elements. These interactions accentuate contrast at sharp spatial and temporal gradients and discontinuities in the retinal image: borders and contours become "crisp" in their neural representation. Thus, the pattern of optic-nerve activity that results is by no means a direct copy of the pattern of stimulation on the receptor mosaic; certain information of special significance to the organism is accentuated at the expense of less significant information.

Interaction in the vertebrate retina is more complex; it comprises both inhibitory and excitatory influences, which result in great diversity, and often lability, of the patterns of optic-nerve activity. The one influence often obscures the contribution of the other. Nevertheless, it has been possible to parcel out some of the separate contributions of excitatory and inhibitory influences to the patterns of optic-nerve activity in the vertebrate visual system. For example, Hartline (1939) found that, in the eye of the frog, an "off" response elicited in a single fiber by illuminating one group of receptors may be inhibited by illuminating another group of receptors in the same receptive field. It has since been shown (Barlow, 1953) that light falling entirely outside but close to the receptive field of a particular fiber also has an inhibitory effect on that fiber's response. Kuffler (1953) has shown that in the eye of the cat certain areas within a single receptive field make either a predominant "on" or "off" contribution to the discharge pattern; and when two opposed areas within the receptive field interact, the responses of both through their common neuron become modified; the interaction is mutual. The possible role of inhibition in the modification of these responses in the retina is discussed in some detail by Barlow, FitzHugh, and Kuffler (1957). Further modifications of the pattern of nerve activity take place at higher integrative levels in the visual pathway; for example, Baumgartner and Hakas (1959) have observed inhibitory interaction among neural units in the visual cortex of the cat.

Inhibitory interaction is undoubtedly the basis of a number of well-known visual phenomena such as brightness contrast and color contrast. Indeed, early psychophysical studies of these phenomena presaged the discovery of inhibitory interaction by electrophysiological methods. For example, Mach's quantitative formulation of the interdependence of retinal areas in the human eye (1866), based entirely on psychophysical observations, contains most of the important features of inhibitory interaction subsequently revealed by electrophysiological studies on lower animals nearly a century later! The visual significance of the contrast effects, too, has long been known. For hundreds of years artists have utilized these effects to brighten or subdue colors, or to alter their apparent hue, and—especially—to emphasize lines and contours. Indeed, it seems almost instinctive for the artist to accentuate the contours of an object he is representing; and he does this—as the eye does—at the expense of accuracy of representation of less significant features.

Summary

This paper reviews a quantitative analysis of the inhibitory interaction among receptor units (ommatidia) in the lateral eye of *Limulus*.

Activity in any one optic nerve fiber of the *Limulus* eye, isolated by dissection, can be elicited only by illumination of the receptor unit from which that fiber arises (there is no excitatory convergence in this eye). But the discharge of nerve impulses by any given receptor unit, nevertheless, is influenced by illuminating other receptor units in its neighborhood. This influence is purely inhibitory.

These inhibitory influences are exerted mutually among the interacting receptors; the activity of each is the resultant of the light stimulus to it and the inhibition exerted on it by the others. Under steady illumination the inhibition exerted by one receptor unit on a second receptor unit in its neighborhood is a linear function of the frequency of discharge of the first receptor unit. When several receptor units act on a given unit in their vicinity, the total inhibition they exert is determined quantitatively by the inhibitory influences of each, combined by simple addition. As a consequence, the responses of n receptors interacting with one another may be described by a set of n simultaneous equations, linear in the frequencies of the interacting units.

The inhibitory interaction between two receptors in the eye of *Limulus* is stronger, the closer they are to one another in the receptor mosaic. This factor has an important effect on the patterns of optic-nerve activity elicited by various spatial configurations of light on the receptor mosaic: contrast is accentuated in the vicinity of steep gradients in the retinal image. These effects resemble closely the analogous effects in human vision (Mach bands and border contrast).

The transient, dynamic phases of the inhibitory interaction accentuate the optic-nerve response to temporal changes in illumination. Indeed, by suitably pitting the excitatory influences of light stimulation of a receptor unit against inhibitory influences from some of its neighbors, transient discharges of impulses can be produced that resemble very closely the "on-off" and "off" bursts of impulses so characteristic of various optic nerve fibers of the vertebrate retina.

These experiments on the eye of *Limulus* furnish a model illustrating one mechanism that may contribute to spatial and temporal contrast effects in more highly organized visual systems, and that may be useful in understanding complex integrative processes in other parts of the nervous system.

Acknowledgment

The quantitative analysis of inhibitory interaction reviewed here was supported by a research grant (B864) from the National Institute of Neurological Diseases and Blindness, Public Health Service, and by Contract Nonr1442(00) with the Office of Naval Research.

References

Barlow, H. B.   Summation and inhibition in the frog's retina.   *J. Physiol.*, 1953, **119**, 69–88.

Barlow, H. B., R. FitzHugh, and S. W. Kuffler.   Change of organization in the receptive fields of the cat's retina during dark adaptation.   *J. Physiol.*, 1957, **137**, 338–354.

Baumgartner, G., and P. Hakas.   Reaktionen einzelner Opticusneurone und corticaler Nervenzellen der Katze im Hell-Dunkel-Grenzfeld (Simultankontrast).   *Pflügers Arch. ges. Physiol.*, 1959, **270**, 29.

Brooks, V. B.   Contrast and stability in the nervous system.   *Trans. N. Y. Acad. Sci.*, 1959, **21**, 387–394.

Granit, R.   The components of the retinal action potential and their relation to the discharge in the optic nerve.   *J. Physiol.*, 1933, **77**, 207–240.

Hartline, H. K.   The response of single optic nerve fibers of the vertebrate eye to illumination of the retina.   *Amer. J. Physiol.*, 1938, **121**, 400–415.

Hartline, H. K.   Excitation and inhibition of the "off" response in vertebrate optic nerve fibers.   *Amer. J. Physiol.*, 1939, **126**, 527.

Hartline, H. K.   Inhibition of activity of visual receptors by illuminating nearby retinal elements in the *Limulus* eye.   *Fed. Proc.*, 1949, **8**, 69.

Hartline, H. K., and C. H. Graham.   Nerve impulses from single receptors in the eye.   *J. cell. comp. Physiol.*, 1932, **1**, 277–295.

Hartline, H. K., and F. Ratliff.   Inhibitory interaction of receptor units in the eye of *Limulus*.   *J. gen. Physiol.*, 1957, **40**, 357–376.

Hartline, H. K., and F. Ratliff.   Spatial summation of inhibitory influences in the eye of *Limulus*, and the mutual interaction of receptor units.   *J. gen. Physiol.*, 1958, **41**, 1049–1066.

Hartline, H. K., H. G. Wagner, and E. F. MacNichol, Jr.   The peripheral origin of nervous activity in the visual system.   *Cold Spring Harbor Sympos. Quant. Biol.*, 1952, **17**, 125–141.

Hartline, H. K., H. G. Wagner, and F. Ratliff.   Inhibition in the eye of *Limulus*.   *J. gen. Physiol.*, 1956, **39**, 651–673.

Kuffler, S. W. Discharge patterns and functional organization of mammalian retina. *J. Neurophysiol.*, 1953, **16**, 37–68.

Mach, E. Ueber den physiologischen Effect räumlich vertheilter Lichtreize, II. *Sitzber. Akad. Wiss. Wien.* (Math.-nat. Kl.), Abt. 2, 1866, **54**, 131–144.

MacNichol, E. F., Jr. Visual receptors as biological transducers. In *Molecular Structure and Functional Activity of Nerve Cells.* Amer. Inst. Biol. Sci. Publ. No. 1, 1956, 34–53.

MacNichol, E. F., and H. K. Hartline. Responses to small changes of light intensity by the light-adapted photoreceptor. *Fed. Proc.*, 1948, **7**, 76.

Miller, W. H. Morphology of the ommatidia of the compound eye of *Limulus*. *J. biophysic. biochem. Cytol.*, 1957, **3**, 421–428.

Miller, W. H. Fine structure of some invertebrate photoreceptors. *Ann. N. Y. Acad. Sci.*, 1958, **74**, 204–209.

Ratliff, F., and H. K. Hartline. The responses of *Limulus* optic nerve fibers to patterns of illumination on the receptor mosaic. *J. gen. Physiol.*, 1959, **42**, 1241–1255.

Ratliff, F., W. H. Miller, and H. K. Hartline. Neural interaction in the eye and the integration of receptor activity. *Ann. N. Y. Acad. Sci.*, 1958, **74**, 210–222.

Ratliff, F., and C. G. Mueller. Synthesis of "on-off" and "off" responses in a visual-neural system. *Science*, 1957, **126**, 840–841.

Tomita, T. Mechanism of lateral inhibition in the eye of *Limulus*. *J. Neurophysiol.*, 1958, **21**, 419–429.

Waterman, T. H., and C. A. G. Wiersma. The functional relation between retinal cells and optic nerve in *Limulus*. *J. exp. Zool.*, 1954, **126**, 59–86.

12

YNGVE ZOTTERMAN

Department of Physiology, Veterinärhögskolan, Stockholm

Studies in the Neural Mechanism of Taste

We all learned at school that our gustatory sensations could be analyzed introspectively into four basic tastes, sweet, acid, bitter, and salt. In 1891 Öhrwall at Uppsala demonstrated that some taste buds reacted only to salt and acid but not to sweet and bitter, and vice versa. Since Blix had a decade earlier discovered the cold and warm spots in the skin, the theory developed that every sensory quality was subserved by specific receptors which discharged into specific afferent nerve fibers, and that these specific fibers sent signals to specific cells in the cortex of the brain. Recent electrophysiological studies of the impulse traffic in the taste nerves have in general confirmed this conception. The response of single taste fibers in amphibians, as well as in mammals, has revealed that the taste fibers seem to a great extent to be specific, but also that this specificity is not always quite strict. Thus Pfaffmann (1941) found that in cats strong acid solutions stimulated not only specific "acid fibers" but also "salt fibers." The salty taste would thus be discriminated from acid taste only by the absence of impulses in the specific taste fibers.

Further it was found that amphibians, such as the frog and toad, possess taste fibers that respond to the application of pure water to the tongue (Zotterman, 1949, 1950), and this has been amply confirmed by Japanese workers (Koketsu, 1951; Kusano and Sato, 1957). Originally I thought that these fibers that responded specifically to water (Andersson and Zotterman, 1950) served a particular purpose in the regulation of the water intake in those animals that live mainly in fresh water. This finding, however, raised the old question, whether mammals including man are equipped with specific taste organs for water.

In 1954 Liljestrand, confirming earlier investigations, found in experiments on himself that the threshold value for NaCl solutions lay between 0.009 and 0.002 molar (M), that is, about 0.05 to 0.01 per cent NaCl (Liljestrand and Zotterman, 1954). This was just between the concentrations where I had found that the specific water taste fibers began to respond. Liljestrand also found that Stockholm tap water (with a dry residue of 0.04 per cent) could be distinguished from distilled water (dry residue of 0.0004 per cent) with a certainty of 100 per cent. This also held after the tap water had been boiled for 5 minutes and then cooled. In addition, von Skramlik (1926) had reported that, to men, salt solutions below 0.03 M taste sweet and that saltiness appears only with concentrations above 0.03 M. For that reason Liljestrand and I (1954) decided to study the effect of water upon the taste fibers of the chorda tympani in some mammals.

Water Fibers in Mammals

By picking up the action potentials from the entire chorda tympani nerve in the cat, dog, and pig, we were able to demonstrate that these animals responded positively to the application of water to the tongue. By splitting up the nerve into fine strands, we were able to find preparations that responded to the application of water to the tongue but not to 0.5 M NaCl solution, and also strands that responded to 0.5 M NaCl but not to water (Liljestrand and Zotterman, 1954). In a further investigation on single taste fibers from the chorda tympani of the cat, Cohen, Hagiwara, and I (1955) found that the activity of the water fibers was depressed by various inorganic salt solutions with a concentration above 0.03 M. Some of the water fibers—but not all—were found to be stimulated by quinine as well as by acids of pH below 2.5. In addition to this, fibers were found that responded only to quinine but not to water or salt, and only to a very small degree to acids. In confirmation of Pfaffmann's findings (1941), many salt fibers in the cat responded to acids as well. No fibers in the cat were found to respond specifically to sucrose or other substances that taste sweet to men. Sucrose dissolved in water did not inhibit the response of the water fibers.

As will be seen in Table 1, there are only two test solutions which in the cat stimulate only one kind of taste fiber, that is, water and NaCl. It is interesting to note that NaCl is the only salt that elicits a pure salty sensation in man. That the cat does not possess any fibers that respond to sweet-tasting substances like sucrose (Zotterman, 1935) was confirmed by Pfaffmann (1941). Recent investigations have also revealed that the calf and the lamb either lack sweet fibers

Table 1.   Response of Various Types of Fibers to Four Solutions

| | Type of Response | | | | |
Stimulus	Water fiber	Salt fiber	Acid fiber	Quinine fiber	Sensation evoked
H_2O (salt <0.03 M)	+	0	0	0	water
NaCl >0.05 M	0	+	0	0	salt
HCl (pH 2.5)	+	+	+	0	sour
Quinine	+	0	0	+	bitter

in their chorda tympani or have very few such fibers.   Sweet-tasting solutions elicit positive responses in the rat, the dog, and the pig, but a very sweet-tasting saccharine solution, 0.02 M, fails to elicit any response in these animals.

In contrast to the mammals previously tested, the rat did not give any positive response to water.   The application of water to the rat's tongue caused only an immediate decrease in the spontaneous activity (which could be abolished completely), but the activity returned slowly within 10 seconds of the cessation of the water flow.   Immediately after a water rinse, a solution even as weak as 0.003 M NaCl caused an obvious and persisting response in the rat.   The high level of spontaneous activity in the rat's chorda tympani is most likely an expression of the high sensitivity of its taste receptors to NaCl.   In other mammals such as the cat, the dog, and the pig, the salt receptors seem to adapt quite completely and quickly to a 0.01 M NaCl solution.   After the mouth is rinsed with water, however, there occurs in the cat a transient phasic response of salt fibers to a solution as weak as 0.002 M NaCl.

The very high sensitivity of the rat to weak NaCl solutions found in these experiments (Zotterman, 1956) accords with Pfaffmann's experiments on rats and can in itself explain the high discriminatory ability of these animals.   There is as yet no information about salt discrimination in the cat, dog, or pig.   If these animals depended heavily on their salt fibers, their discrimination of salty solutions would be rather poor.   But these species possess water fibers that may play a part in discrimination between water and weak salt solutions.

Taste Fibers in Birds

Electrophysiological investigations of taste in birds have been made recently by R. Kitchell, L. Ström, and the writer (1959) and include so far only chickens and pigeons. In the chicken, taste fibers that run to both the tongue and the pharynx were found in the glossopharyngeal nerve. Responses were observed following the application of distilled water, salt, glycerine, ethylene glycol, quinine, and acetic acid solutions to the tongue. Sucrose and saccharine solutions did not produce any positive response in the chicken.

In the pigeon also, there was a strong response to distilled water, as well as to salt, glycerine, ethylene glycol, and acetic acid, but no response to sucrose. In about half the number of pigeons investigated, however, we noticed a positive response to saccharine. No response was observed to quinine. Both birds have rather numerous specific cold fibers ending in their tongue, but no receptors were found that responded to warming the tongue until after the tongue temperature had been raised to above 45°C. This indicated the absence of specific warm receptors and suggested that the activity observed was due to stimulation of nerve endings subserving pain.

Our findings are in good general agreement with the studies of the behavior of the bird in relation to a particular substance, as will be seen from Table 2. In two instances, however, there appear to be

Table 2. Comparison of Electrical Responses from Taste Fibers (TR) and Behavioral Responses (BR) in Birds

	Chicken		Pigeon	
Solution	TR	BR[*]	TR	BR
NaCl	+	R[1,2]	+	R[1]
Sucrose	—	P[1,2,3]	—	A[1]
Glycerine	+	R[1]	+	R[1]
Ethylene glycol	+	O	+	O
Saccharine	—	R[1,2,3]	+	R[1]
Quinine	+	R[1,2]	—	A[1]
Acetic acid	+	R[1]	+	R[1]

[*] Behavioral responses: R, reject; A, do not discriminate; P, prefer; O, not determined—according to [1]Engelmann (1934); [2]Kare, Black, and Allison (1957); [3]Jacobs and Scott (1957).

differences in the results. Three studies report that chickens seem to prefer sucrose solutions to water (Engelmann, 1934; Kare, Black, and Allison, 1957; Jacobs and Scott, 1957), whereas no response to sucrose was observed in our studies. The degree of preference observed in the behavioral studies was small in all instances. Jacobs and Scott suggested that the preference could be due to a difference in viscosity rather than to a taste response. The results of behavioral studies show that chickens prefer water to saccharine, whereas in our experiments no taste response to saccharine in Ringer's solution was observed. Saccharine in distilled water produced a response indistinguishable from the response to distilled water alone. Our observations do not suggest the possible sources of the afferent inflow that enables chickens to discriminate against saccharine, but we can state that it probably does not originate in any taste receptors in the tongue.

It appears very odd that in our experiments 50 per cent of the pigeons responded positively to saccharine although they all lacked fibers that responded to sucrose and quinine. In behavioral tests the pigeons reject saccharine, but they do not discriminate between quinine solutions and pure water. Thus it is not in jest that I venture to suggest that saccharine may taste salty to these birds.

Taste Fibers in the Monkey

In a recent series of experiments on rhesus monkeys (Gordon, et al., 1959), quite strong responses were recorded from the chorda tympani when water was applied on the tongue, as will be seen in Fig. 1. After the tongue has been rinsed with Ringer's solution, the response to Ringer's is very small, but after a rinse with water there is a transient response (Fig. 1). It will be seen from Fig. 1 that the integrated response to 0.5 NaCl solution is smaller than the response recorded after the application of an equal amount of distilled water. The relation between the responses to water and salt varies with the individual. This relation does not necessarily depend upon the relative number of specific fibers in action but may depend on the size of the fibers, since the recorded spike height varies with the diameter of the fiber. Consequently, a quantitative comparison of the integrated responses is not a safe way to ascertain the relative numbers of the various specific fibers unless the size of the fibers is known. Figure 2 shows some records from a fine strand of the chorda. This preparation contained one fiber that gave fairly large spikes when NaCl was applied to the monkey's tongue. The spontaneous activity of the salt fiber giving large spikes was suppressed by the application of distilled water. When the tongue was irrigated with sucrose, saccharine, glycerine, or quinine solutions, a massive response of smaller spikes appeared.

In Fig. 3 records are presented from another preparation of the same chorda tympani that responded vigorously to salt but not to sugar or quinine. The threshold of the receptor of this fiber lay some-where between 0.02 and 0.01 M NaCl. This particular salt fiber also responded, but to a slight degree, to 0.2 M acetic acid. In more recent experiments, however, we have found several single salt fibers that did not respond at all to an acid stimulus.

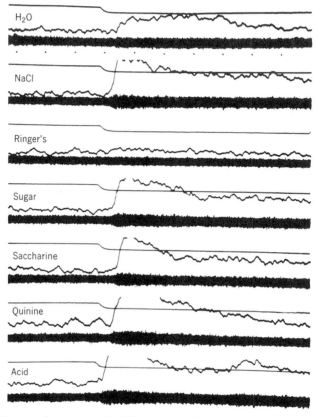

Fig. 1. Integrated responses (middle tracings) from the intact chorda tympani of rhesus to the application of various sapid solutions on the tongue. Upper tracings signal the moment of application; lower tracings, the direct response. Time: 1 per second.

The chorda tympani of the monkey apparently contains a large number of fibers that respond specifically to sweet-tasting substances such as sucrose, glycerol, and ethylene glycol, and even to saccharine. So far, the monkey is the only animal I have found having specific fibers that respond to sucrose and that also respond positively to saccharine. In dogs and pigs, animals that both possess specific sweet fibers—saccharine does not produce a positive effect on the sweet receptors. In the dog, strong concentrations of saccharine seemed to stimulate the bitter fibers (Andersson, Landgren, et al., 1950). Some preparations with a small number of fibers were ob-

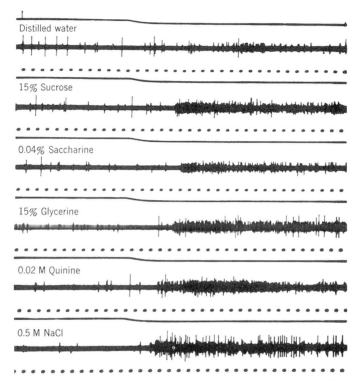

Fig. 2. Records from a strand of the chorda tympani of rhesus responding to the application upon the tongue of various substances dissolved in water. Note the large spike response to NaCl. Time: 10 per second.

tained from the monkey's chorda tympani in which the sweet fibers did not respond to any other sapid solutions except those that taste sweet to man. An example of this is seen in Fig. 4, in which the records were obtained from the same nerve strand as Fig. 2 after successive subdivisions. By thus reducing the number of fibers, we succeeded in getting rid of those fibers that had in the original preparation responded to salt, quinine, and water (see Fig. 2).

The response to acetic acid obviously derived from a specific acid fiber (see Fig. 4). This fiber produced much bigger spikes than the sweet fibers in this nerve preparation and did not respond to any other sapid stimulus. Thus we can conclude that the rhesus monkey has specific fibers for salt, sweet, bitter, and acid solutions, as well as for water.

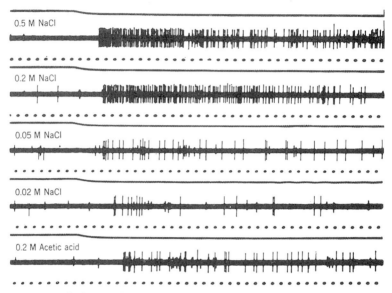

Fig. 3. Records from a fine strand of the chorda tympani of rhesus. Note the large spike response to NaCl. Threshold about 0.01 M NaCl. Time: 10 per second.

Experiments on Human Subjects

A few years ago Dr. C. Åhlander and I made our first attempts to place leads on the chorda tympani of man during operation on the middle ear, in the Ear, Nose and Throat Department of Södersjukhuset (Southern Hospital, Stockholm). Out of ten trials in which the surgeon applied the electrodes to the exposed chorda tympani in the cavum tympani, we obtained very weak signals in only two cases in response to cold and gustatory stimulation of the tongue. The responses were only just audible in the loudspeaker, however, and could not be recorded.

In recent attempts made in August and September 1958 in the Ear Clinic of Karolinska Sjukhuset, Stockholm, in collaboration with Dr. H. Diamant, we were able to record the response from the chorda tympani during operations performed in order to mobilize the stapes in the middle ear (Diamant and Zotterman, 1959). In three out of

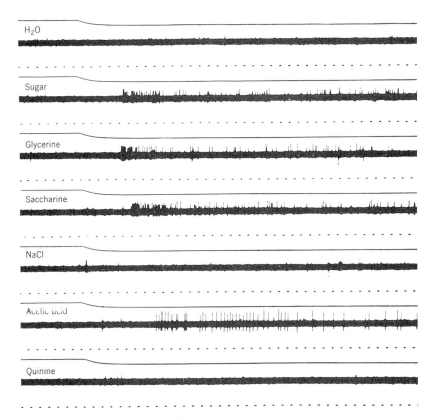

Fig. 4. Action potentials from the chorda tympani of rhesus showing response to different solutions applied to the tongue. Upper trace signals opening of stopcock. Time: 10 per second.

eight cases tried so far, we have been able to record the integrated electrical response of the nerve when the tongue was stimulated by touch and solutions of various flavors. As will be seen in Fig. 5, there was a good response to 0.5 M NaCl solutions and also to 15 per cent sucrose, 0.04 per cent saccharine, 0.02 M quinine sulphate, and 0.2 M acetic acid. The application of water to the tongue, however, was followed by a reduction in the spontaneous activity in the nerve in exactly the same fashion as we had previously found in the rat, which does not possess any taste fibers that respond positively to water. Thus man seems to lack a specific water taste, in contrast to the cat, the dog, the pig, and a mammal as high on the evolutionary scale as the rhesus monkey. The evidence so far obtained is that three persons definitely lack water taste. Future experiments will tell us whether this holds for us all or whether, in addition to great species differences, there also exist individual differences in taste sensation

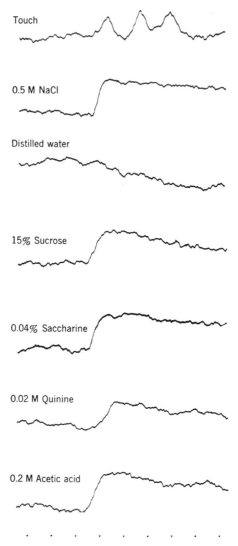

Fig. 5. Integrated responses of the whole chorda tympani nerve of man to various solutions flowed over the tongue. Time: 1 per second.

for water. Introspectively, I rather doubt whether water elicits any positive taste sensation. The action of water on our gustatory receptors, and particularly on the salt fibers, must be of a negative nature, in that water abolishes or decreases the resting activity of our taste fibers. A high sensitivity of the salt fibers, which can be contrasted with the negative effect of water, would thus be the instrument for our discrimination between dilute salt solutions and pure water. It seems to us rather odd that we should lack specific water fibers, whereas our next relatives the monkeys have such fibers. This fact does not make it easier to understand what physiological purpose the water taste fibers may serve.

Summary

The electrical response to the application of various sapid solutions on the tongue has been studied in amphibians, birds, and mammals, including the rhesus monkey as well as man.

Positive response to the application of water was obtained from frogs, chicken and pigeons, cats, dogs, pigs, and rhesus monkeys—but not from rats and humans. Great species differences were also noticed in the response to different sapid stimuli.

A study of the response of individual taste fibers in the rhesus monkey revealed that each fiber has a specific pattern of sensitivity to the various sapid substances. Certain gustatory fibers responded very specifically to one class of substances only, for example, to salt, acid, or quinine. Fibers responding to sucrose almost always responded to saccharine as well, but the "sweet" fibers of the dog's chorda tympani did not respond to saccharine.

References

Andersson, B., and Y. Zotterman. The water taste in the frog. *Acta physiol. scand.*, 1950, **20**, 95–100.

Andersson, B., S. Landgren, L. Olsson, and Y. Zotterman. The sweet taste fibres of the dog. *Acta physiol. scand.*, 1950, **21**, 105–119.

Blix, M. Experimentella bidrag till lösning av frågan om hudnervernas specifika energi. *Upsala läkaref. Förh.*, 1882, **18**, 87–102.

Cohen, M. J., S. Hagiwara, and Y. Zotterman. The response spectrum of taste fibres in the cat: A single fibre analysis. *Acta physiol. scand.*, 1955, **33**, 316–332.

Diamant, H., and Y. Zotterman. Has water a specific taste? *Nature, Lond.*, 1959, **183**, 191–192.

Engelmann, C. Versuche über den Geschmack von Taube, Ente und Huhn. *Z. vergl. Physiol.*, 1934, **20**, 626–645.

Gordon, G., R. Kitchell, L. Ström, and Y. Zotterman. The response pattern of taste fibres in the chorda tympani of the monkey. *Acta physiol. scand.*, 1959, **46**, 119–132.

Jacobs, H. L., and M. L. Scott. Factors mediating food and liquid intake in chickens. I. Studies on the preference for sucrose or saccharine solutions. *Poultry Sci.*, 1957, **36**, 8–15.

Kare, M. R., R. Black, and E. G. Allison. The sense of taste in the fowl. *Poultry Sci.*, 1957, **36**, 129–138.

Kitchell, R. L., L. Ström, and Y. Zotterman. Electrophysiological studies of thermal and taste reception in chickens and pigeons. *Acta physiol. scand.*, 1959, **46**, 133–151.

Koketsu, K. Impulses from receptors in the tongue of a frog. *Kyushu Mem. med. Sci.*, 1951, **2**, 53–61.

Kusano, K., and M. Sato. Properties of fungiform papillae in frog's tongue. *Jap. J. Physiol.*, 1957, **7**, 324–338.

Liljestrand, G., and Y. Zotterman. The water taste in mammals. *Acta physiol. scand.*, 1954, **32**, 291–303.

Öhrwall, H. Untersuchungen über den Geschmacksinn. *Skand. Arch. Physiol.*, 1891, **2**, 1–69.

Pfaffmann, C. Gustatory afferent impulses. *J. cell. comp. Physiol.*, 1941, **17**, 243–258.

Skramlik, E. v. *Handbuch der niederen Sinne.* Bd. 1. Leipzig: Thieme, 1926.

Zotterman, Y. Action potentials in the glossopharyngeal nerve and in the chorda tympani. *Skand. Arch. Physiol.*, 1935, **72**, 73–77.

Zotterman, Y. The response of the frog's taste fibres to the application of pure water. *Acta physiol. scand.*, 1949, **18**, 181–189.

Zotterman, Y. The water taste in the frog. *Experientia*, 1950, **6** (2), 57–58.

Zotterman, Y. Species differences in the water taste. *Acta physiol. scand.*, 1956, **37**, 60–70.

13

H. B. BARLOW

Physiological Laboratory, Cambridge University

Possible Principles Underlying the Transformations of Sensory Messages

A wing would be a most mystifying structure if one did not know that birds flew. One might observe that it could be extended a considerable distance, that it had a smooth covering of feathers with conspicuous markings, that it was operated by powerful muscles, and that strength and lightness were prominent features of its construction. These are important facts, but by themselves they do not tell us that birds fly. Yet without knowing this, and without understanding something of the principles of flight, a more detailed examination of the wing itself would probably be unrewarding. I think that we may be at an analogous point in our understanding of the sensory side of the central nervous system. We have got our first batch of facts from the anatomical, neurophysiological, and psychophysical study of sensation and perception, and now we need ideas about what operations are performed by the various structures we have examined. For the bird's wing we can say that it accelerates downwards the air flowing past it and so derives an upward force which supports the weight of the bird; what would be a similar summary of the most important operation performed at a sensory relay?

It seems to me vitally important to have in mind possible answers to this question when investigating these structures, for if one does not one will get lost in a mass of irrelevant detail and fail to make the crucial observations. In this paper I shall discuss three hypotheses according to which the answers would be as follows:

1. Sensory relays are for detecting, in the incoming messages, certain "passwords" that have a particular key significance for the animal.
2. They are filters, or recoding centers, whose "pass characteristics"

can be controlled in accordance with the requirements of other parts of the nervous system.

3. They recode sensory messages, extracting signals of high relative entropy from the highly redundant sensory input.

These hypotheses are presented in order of increasing sophistication, and in the following pages most space is given to the last one, for the simple reason that it requires more thought—and has certainly consumed more of mine recently. I have omitted the idea that sensory relays are mere accidents of embryological or evolutionary development whose sole function is to pass on information without transforming it significantly, but this uninteresting possibility should probably be borne in mind, especially when considering the earlier relays. I am using the term "sensory relays" rather loosely, and I intend it to include synapses at the highest levels.

As with the bird's wing, the summaries are in physical rather than biological language, but before discussing them in greater detail two explanations and an apology are needed. First, it is unlikely that sensory relays perform just one operation of such outstanding importance that one can say it is *the* most important function, regarding all others as subsidiary in the same way that one can legitimately regard signaling, or sheltering young, as subsidiary functions of a bird's wing. Hence the present hypotheses are not mutually exclusive, nor do they exclude other theories about the important operations of sensory relays. Second, these are really orientating ideas, not detailed hypotheses about mechanism of action. The appropriate test for them is whether they help to make sense of the facts already known about "sensory integration," and whether the further investigations they prompt one to make are fruitful. Correct or not, I feel sure that ideas of this sort are needed. A bird's ability to fly is certainly an important fact, but it might easily be missed by someone concentrating his attention too narrowly on the anatomy and physiology of wings.

The apology is for the absence of a discussion of the experimental evidence bearing on these ideas. Nevertheless they do come from puzzling over experimental facts, not from abstract speculation. The "password" idea came from the realization that a frog's retina had an organization that made it quite unsuitable for the kind of task we use our own eyes for. The recoding idea came from recognizing that the retinal organization (in the cat in this case) was not only rather complicated but could also vary with the state of adaptation of the eye. It seemed to me that one could only hope to understand the complex, variable transformations the retina was imposing on the sensory messages if one knew what they were directed toward, or what part they played in the whole animal. It may be too ambitious to try to answer this question, but at least I want to make it clear that I do not regard these ideas as moulds into which all experimental facts must be forced. They are just attempts to make some sense out of what would otherwise be a muddle.

Password Hypothesis

In studying sensory physiology many of us start with the idea that what we discover will be simply related to the subjective sensations of which we are aware by introspection. This is of course naive: the primary effect of the sensory messages an animal receives is not to enrich its subjective experience of the world but to modify its behavior in such a way that it and its species have a greater chance of survival. Accordingly it would be one step less naive to expect that, when sensory messages are transformed at sensory relays, they are being organized in accordance with the responses that the initiating stimuli would have produced in the normal animal. The subjective sensations they would produce in ourselves may or may not be relevant. Cutaneous stimuli that elicit flexion and withdrawal in the spinal cat are probably roughly congruent to those we call "painful," and those that elicit a scratch reflex may be analogous to those we call "tickling"; thus having these categories in mind is as helpful as thinking of the responses themselves. But we have no subjective category that adequately describes the class of stimuli that elicits the snapping response in frogs, though this is obviously an important category to have in mind when investigating the frog's visual system.

These preliminary remarks should have indicated what is meant by the "password" hypothesis. Specific classes of stimuli act as "releasers" and evoke specific responses; these classes of stimuli are thought of as "passwords" which have to be distinguished from all other stimuli, and it is suggested that their detection may be the important function of sensory relays. Looking at the case of flexion withdrawal, one sees that here the discrimination is mainly achieved, not at a sensory relay, but by having a class of sensory fibers that respond to potentially harmful stimuli. One knows little about the sensory discriminatory mechanism for the scratch reflex, except that it lies in the spinal cord. Probably no-one has recorded from a cell that performs the operation of distinguishing scratchworthy from unscratchworthy cutaneous stimuli, but it is worth asking whether, if one were picked up, its function would be spotted during an ordinary physiological investigation. There is an objection that could be raised here. It might be held that the decision whether to scratch or not cannot be taken without considering the state and requirements of the rest of the animal. In this case you could not expect to find such a discriminating unit at a low level in the nervous system, but only at a level where all necessary information has been brought together. This applies to a unit that decides whether to scratch or not, but it does not apply to a unit that does the preliminary sorting of cutaneous stimuli into a class that should be scratched, and a class that should not. It is units doing this preliminary classification that one is led to expect if one bears in mind the responses ordinarily elicited by the stimuli employed.

Take the visual system of the frog as a specific example. The range of visual responses is rather limited. A small moving object elicits a sequence of reactions consisting of alerting, turning toward the object, hopping toward it if necessary, and finally hopping and snapping at it. Frogs also follow, with eye, head, and body movements, a moving object in the visual field, but a large moving object, especially if it is in the upper part of the visual field, may provoke an escape reaction in which the frog dives under a stone or into the deepest part of the pond. Yerkes (1903) was unable to get any evidence that frogs used vision to locate themselves in their habitat, nor did he find evidence of form discrimination or learned visual reactions. To some extent the neurophysiology fits in with this. The fact that it is predominantly change of retinal illumination which elicits discharges is obviously related to the fact that it is movement which is most effective in eliciting behavioral responses. One may be able to go further and identify the "on-off" units as the detectors of snapworthy objects (Barlow, 1953), for their properties are such that they respond vigorously to the type of stimulus that is particularly effective in eliciting the hunting sequence. Lettvin, et al. (1959) have recorded responses from the frog's optic tectum that seem to fit in with the behavioral requirements in a most striking manner, and it seems possible that the neurophysiology of the frog's hunting and feeding habits will become comprehensible in some detail.

If there is a moral to be drawn from the password hypothesis, it is as follows. We know that specific stimuli elicit specific responses, and it is reasonable to look out for the physiological mechanisms responsible for the preliminary classification of "releasers," even at early stages in the sensory pathways. To do this one needs some knowledge of the behavioral results of the stimuli one employs—and one must use stimuli that have specific behavioral results.

Controlled Pass-Characteristic Hypothesis

The idea that the incoming flow of sensory impulses is regulated or controlled at sensory relays is fashionable and has been experimentally fruitful, but the existence of such control raises many further points of interest. For instance, it is obvious that the control may be much more specific than is implied by the analogy of a volume or gain control. Sensitivity to one type of stimulus might be increased while another is decreased, or, combining this with the previous hypothesis, the whole characteristic of the relay might be changed, so that, in effect, the "password" is altered.

Another point is that it is not always obvious or easy to assess the significance of even a simple form of control, particularly if one fails to take into account more than the sensory pathway itself. To illustrate this, let us consider an example in the periphery. The γ efferents control the range of muscle length over which the discharge of the spindles shows finest gradation in accordance with changes in that length; since they appear to act as a zero adjustment, it was natural to think that the function of this control was to adjust the muscle spindles so that they could continue to give finely graded discharges at whatever length the muscle happened to be. The incompleteness of this picture of their function emerges when one takes into account the fact that afferent impulses from muscle spindles evoke a reflex discharge down the α efferents, causing powerful contraction in the muscle fibers lying in parallel with the spindles. Clearly activation of the γ efferents will bring about a reflex shortening, in the manner described by Eldred, Granit, and Merton (1953). In comparison with contractions produced by direct α-efferent excitation, the amount of shortening occurring in such servo-assisted contractions will be relatively independent of changes in the externally applied load and will be affected only slightly by moderate losses of muscle power resulting from fatigue. The task of controlling movement is thereby greatly simplified, and in understanding this we have gained considerable insight into the way the nervous system manages its affairs. If one is to gain comparable insight into the significance of controlled transmission at sensory relays, one must look beyond the effect of the control upon the afferent impulses themselves and consider what part these impulses play in the behavior of the intact animal.

Another point comes from theorists considering how to make a machine capable of learning to recognize complex patterns. In two schemes that have been offered (Lee, 1959; Selfridge, 1959), feedback is required from higher centers to points early in the pathway of incoming information. The basic idea is to have elements early in the pathway that can change their transmission characteristics. They change in this way only when the feedback from above signals lack of success (for example, that the recognition problem has not been solved), but when success is signaled, the transmission characteristics being used are held unchanged.

Desirable or effective transmission characteristics thus survive as a result of a selective process rather analogous to natural selection acting on genes and causing evolutionary adaptation of species to their environment. It seems just possible that control fibers entering sensory relays might be exerting such a selective action, and because of its interesting implications this possibility might be worth exploring. The semipermanent change of "set" of the relays which this idea suggests needs to be looked for by experimental techniques rather different from those used to investigate continuous, moment-to-moment control of the type usually considered.

Redundancy-Reducing Hypothesis

The first hypothesis postulated preset mechanisms for detecting and passing on restricted classes of key signals, rejecting messages that did not fit into these classes. One can liken this to permanent editorial policy: for instance, one periodical only publishes information about sporting events and personalities, another rejects everything except original scientific papers. The second hypothesis suggested that the acceptance or rejection of messages might be controlled from elsewhere, either to make a temporary change in the type or amount of information passing, or to make a more permanent adjustment to the accept-reject criteria of the sensory relays. Pursuing the editorial analogy, one can liken the temporary control to rejection on the grounds of lack of space or to suit an editorial whim, the more permanent control to the long-lasting influence an editor can exert on the preliminary selection of news by his reporters. Now it is clear that there is one important editorial criterion for acceptance or rejection that is not included in either of these broad categories. Is this *news?* Has it been said before, or has it been said elsewhere? If so, it is redundant and can be rejected.

The idea that sensory relays try to ensure that what they pass on really is news is close to the basic one behind the third hypothesis. But one is liable to several misinterpretations if one thinks of the hypothesis solely in terms of the analogy. For this reason I have used the language of information theory to state the hypothesis, together with certain simplifying assumptions. Then I have given a brief reminder of the meaning of terms such as redundancy and information, and following this an account of the sort of recoding the hypothesis leads one to expect, and the sort of predictions it leads one to make. I think the statement of the assumptions and hypothesis are precise and accurate, but they demand an accurate understanding of the meaning in information theory of the terms used; the brief reminder given here may not be sufficient to prevent misconceptions suggested by the editing analogy or by phrases like "stripping the sensory messages of their redundancy," and the only way to avoid these is to read an authoritative exposition of information theory (for example, Shannon and Weaver, 1949; Woodward, 1953).

The idea that reduction of redundancy is an important operation in the handling of sensory information is not a new one. Attneave (1954) argued that it made sense of many of the psychological facts of perception, and the points of view set out by MacKay (1956) and Craik (1943) are certainly closely related. I have written about it elsewhere (1959, in press) from a physiological point of view, and much further back in time one finds the idea, applied to much higher mental processes, clearly expressed in the writings of Ernst Mach (1886) and Karl Pearson (1892): their argument was that concepts, hypotheses, and laws of nature serve the purpose of bringing order and simplicity to our complex sensory experiences in order to achieve "economy of thought"; this seems to be the same idea as recoding to reduce the redundancy of our internal representation of the outer world.

Simplifying assumptions

1. For present purposes sensory pathways can be treated as noiseless systems using discrete signals.

2. The discrete signals are single impulses in particular nerve fibers in particular time intervals. For any one fiber and time interval, an impulse is either present or absent, so the code is binary.

3. The constraints on the capacity of a nerve pathway are the number of fibers F, the number of discrete time intervals per second R, and the average number of impulses per second per fiber I. The average number of impulses per fiber is assumed to be a variable constraint.

These are quite specific assumptions which cannot be justified by experimental results, nor are they really essential parts of the redundancy-reducing hypothesis. This would gain in generality if one dispensed with them entirely, but it is very difficult to discuss coding in the nervous system without making some assumptions about what variables in a nerve message are used to convey information, and it seemed wise to make those assumptions both explicit and as simple as possible. The assumptions adopted here are certainly oversimple in some respects. For instance, the first one side-steps the question of intrinsic neural noise, such as might be caused by random perturbations in transit time of impulses or by chance interruptions in synaptic transmission. This is not because I want to deny the importance or existence of these effects, but because the present hypothesis has something interesting to say about how the nervous system handles certain extrinsic properties of nerve messages—properties that are inherent in the physical stimuli impinging on the sense organs themselves. From this point of view, intrinsic noise, which is added to the messages at or after the sense organs, is a complicating factor that might obscure the issue, and so it seems best to neglect it at this stage.

Another point on which the assumptions might be criticized is that they fail to state some additional restraints that one feels pretty sure nerve fibers and synapses work under. For instance, FitzHugh (1957) has produced evidence that it is not the presence or absence of a single impulse in a particular short time interval that matters in a nerve message, but the aggregate number of impulses in a longer time interval. This additional restraint greatly decreases the capacity of a nerve fiber; consequently, if it holds in the higher parts of the nervous system, as well as in the simpler situation investigated by FitzHugh, my assumptions allow too much information to be passed down a nerve fiber. This has been done deliberately, because the safe course here is to assume that the nervous system is efficient. If it is clearly demonstrated that the nervous system is inefficient in some particular well-defined way, this can quite easily be incorporated into the hypothesis and its implications correspondingly modified, whereas our whole frame of thought might be undermined if it turned out that the nervous system was more efficient than we had supposed.

In fact the assumptions are simple; they suggest what we should look out for if the nervous system is smarter than we are inclined to think; and they define a communication system that will be helpful in discussing the hypothesis; but physiologically they are certainly oversimple and unproved, and they may be quite wide of the mark.

Hypothesis

The hypothesis is that sensory relays recode sensory messages so that their redundancy is reduced but comparatively little information is lost. To clarify this, what is meant by "information," "redundancy," "message," and so on, must first be explained.

A "message" is a set of "signals"; for example, it might be the particular pattern of impulses that arrives along a set of 10 fibers during an interval of 1/10 seconds. These signals are carried into the relay along a set of fibers that constitute the "input" channel, and they generate impulses in other neurons that are the "output" signals in the output channel. If one writes down all the different input messages that occur, and for each input the output message that results, this will constitute the "code" relating input to output.

"Information" is a quantitative attribute of a message if the prior probability of receiving it is known. This usually means that it belongs to an ensemble or population of mutually exclusive and statistically independent messages whose frequency distribution is known. If P_m is the probability of the message m in such an ensemble, then the information attributed to m is $H_m = -\log P_m$; the average information of all messages is

$$H_{\mathrm{av}} = -\sum_m P_m \log P_m$$

summed for all members of the ensemble. The rate of flow of information is H_{av}/T, where T is the average duration of messages from the ensemble, weighted for frequency of occurrence, that is,

$$T = \sum_m P_m T_m$$

The "capacity" C of a channel is equal to the greatest rate of flow of information that can be passed down it. This is calculated from its physical limitations and the constraints on the way it is used. For instance, with the constraints assumed under Simplifying Assumptions 3 above, the capacity of a nerve pathway is

$$C = -FR\left[\frac{I}{R}\log\frac{I}{R} + \left(1 - \frac{I}{R}\right)\log\left(1 - \frac{I}{R}\right)\right]$$

If messages of average information content H and duration T are passing down a channel of capacity C, the "relative entropy" of the messages is the ratio of rate of flow of information to capacity H/CT. The "redundancy" is 1 minus this ratio $[1 - (H/CT)]$; it can be thought of as the fraction of the channel capacity that is not occupied by the message it is being used to transmit.

Returning to the code that relates input to output, if there is a one-to-one relation, it is clear that there will be no loss of information, because the probability of each output is the same as that of its corresponding input. At first it might be suspected that the redundancy would also be unchanged, but this is not so; redundancy depends upon the capacity of the channel as well as the amount of information it is passing; so if the output has a lower capacity, and carries the same information, the redundancy must be less. The task achieved by a redundancy-reducing code is, in fact, to pack the messages more neatly, so that they can be passed down a smaller channel, with less unused space.

Now according to the assumptions, the only restraint on the output that can be varied is the average frequency of impulses. The capacity C is maximum for $I = R/2$; I is normally below this, so that if the sensory relay decreases the redundancy, it must do so by decreasing still further the average frequency of impulses being used to convey the input messages. We may suppose that the relay has a range of possible codes relating input to output: the hypothesis says that, for a given class of input message, it will choose the code that requires the smallest average expenditure of impulses in the output. Or putting it briefly, it economizes impulses; but it is important to realize that it can only do this on the average; the commonly occurring inputs are allotted outputs with few impulses, but there may be infrequent inputs that require more impulses in the output than in the input.

There is an important difference between this and the two previous hypotheses. They considered possible principles for selecting some sensory messages while rejecting others; and it was taken for granted that those selected would be rather infrequent, whereas those rejected would, in effect, all be classified alike. Consequently, both these hypotheses involve discarding a large fraction of the incoming information. In contrast, the emphasis in the present hypothesis is on the preservation of information: it is the redundancy that is discarded, and although an incidental loss of information may result this is not an essential feature of a redundancy-reducing code. If the morning paper fails to state that the sun set last night, one does not conclude that it did not happen, because one knows that this is the kind of event, which, though important, is omitted. In the same way, impulses can be economized without misleading the more central parts of the nervous system.

Recoding two binary inputs

The effect of coding to reduce redundancy is not just the elimination of wasteful neural activity. It constitutes a way of organizing the sensory information so that, on the one hand, an internal model of the environment causing the past sensory inputs is built up, while on the other hand the current sensory situation is represented in a concise way which simplifies the task of the parts of the nervous system responsible for learning and conditioning. In order to illustrate this, consider the very simple optimal code finder which has been described by the author and Mr. P. E. K. Donaldson (Barlow, 1959). Suppose there are two fibers A and B entering a relay, and two fibers X and Y leaving it. Take a time interval such that only one impulse can arrive in a fiber; now the possible input messages are impulse in A alone (Ab), impulse in B alone (aB), impulses in both (AB), or impulses in neither (ab). (Capitals symbolize impulses present, small letters impulses absent.) There are also four possible output messages, and these can be related one-to-one to the inputs in factorial $4 = 24$ ways, each of which is a reversible code in which no information is lost. To choose the appropriate code according to the hypothesis, it is necessary to measure the relative frequencies of the input messages. As soon as the commonest has been determined, it should be allotted to the output xy, since this is cheapest in terms of impulses. As soon as the rarest is known, this should be allotted to XY, the most expensive output. This narrows down the choice of code to 2 out of the 24. A choice between these two could be made if impulses in X and Y were valued differently; otherwise either would do.

The first claim for the code was in relation to the kind of neural model of the environment which Craik (1943) talked about. If one tries to think what is meant by such a neural model, it is clear that this must consist of a store of the frequencies of occurrence of a myriad of combinations and sequences of sensory stimuli. Excluding genetic factors, there is nothing else causally connected with the environment from which our internal representation of it could be constructed. It is the relative frequencies of the input states that are required for choosing the code, and so, conversely, the code chosen reflects these relative frequencies. Thus it acts as a store or model, but it is, of course, defective to the extent that it has only the rank order of input events, not their actual frequencies.

The fact that a redundancy-reducing code orders the input messages in accordance with their frequency of occurrences is also the basis of the second claim, that such coding is a useful preliminary to the learning and conditioning tasks performed by the nervous system. There are, of course, some operations that it will not assist but will hinder.   Take, for instance, the control of pupil diameter; this seems to require information about the average amount of light entering the pupil, and it is precisely such average properties that the code tends to subtract from the messages.   On the other hand, in learning and conditioning, the animal does not act upon a predetermined feature of the sensory input but has to find sensory correlates of the rewards, punishments, and unconditional stimuli it receives before it can act on them.   This is no simple task, for with input fibers numbered in millions the number of possible states of the input is more than just astronomical, it is meaninglessly large.   Yet it would seem to be necessary to separate and inspect individually a rather large fraction of these possible states in order to have a reasonable chance of finding the required sensory correlate.   This would be a formidable task.

After they have been coded, the messages are arranged according to their prior probabilities.   Those containing a small number of impulses are the commonly occurring ones and lie at one extreme. Those containing many impulses occur infrequently and lie at the other extreme.   In deciding which of the possible states one should inspect one would be greatly helped by this arrangement, for one could avoid allotting neural machinery to the task of discriminating between the vast numbers of possible states that contain many impulses and, therefore, occur infrequently or not at all.   One could start the search with the possible states that contain few impulses and therefore include the states that occur most often, and by this means a vast curtailment of search effort would seem to be possible.

One can go a stage further along these lines; the requirement is to find which of the incoming messages are correlated with a particular event, such as a punishment, a reward, or the receipt of an unconditional stimulus.   Now the frequency of this event can itself be determined, and then one could avoid wasteful searching among the possible messages that occur too frequently as well as too rarely; one could confine the search to those possible states that contain the appropriate number of impulses.   If a rat runs a maze once a day, it should search for the key to the correct turning among patterns of sensory stimulation that also occur, very roughly, once a day.   Obviously the frequency matching must not be too accurate, or the rat will be foiled by any nonregularity in the environment or in the experimental design, but a degree of frequency matching does seem to make the correlate-finding problem much more feasible.

Returning now to the two-binary-input recoder, we can illustrate the following property of redundancy-reducing codes: an elementary signal in the output may correspond to an unexpected and less simple feature of the input. Suppose that the inputs A and B are both fairly infrequent, and that very nearly all instances of B are accompanied by A. This means that the input state aB is the rarest, and it must be allotted the output XY: that is, on the rare occasions it does occur, it causes both outputs to fire. Now since both A and B are fairly infrequent, the commonest input state is ab, which is accordingly allotted to xy. It then turns out that, if the code is reversible, one of the output fibers must correspond to the situation in which A and B are different from each other, so that it is active when A fires without B, or B without A. In other words, this output fiber signals when A and B hold this relation to each other, and it could not be understood or described adequately in terms of the responses to either input alone. This example, incidentally, illustrates the way in which the code is typically incomplete as a model of the environment, for it tells us that the input "B without A" has occurred less often than any other input, but it does not tell us whether it has never occurred, or occurred only very rarely. This does not mean that the code is unlike the nervous system's model, for that too is incomplete, and this type of omission might perhaps be characteristic here also.

Recoding more complex inputs

When one tries to consider a recoder for a more complicated input, an interesting situation arises. There are $(2^2)! = 24$ possible codes for two binary inputs; for n inputs, each with m discriminable levels of activity, there are $(m^n)!$ possible codes, a number that obviously gets impossibly large when m and n increase. It would be unreasonable to assume that the nervous system was able to choose any one of this number, so the range is presumably restricted by genetic and purely chance factors and possibly also by the "engineering difficulties" of arranging certain codes. The magnitude of $(m^n)!$ emphasizes that there is plenty of scope for such factors in limiting the choice of code; all the present hypothesis requires is that there should be a considerable range left to be selected from on the basis of frequencies of past sensory messages.

The way in which a code can act as a model of the environment can be brought out more directly in a more complicated example. The principle of recoding is to find what messages are expected on the basis of past experience and then to allot outputs with few impulses to these expected inputs, reserving the outputs with many impulses for the unusual or unexpected inputs. Imagine that the incoming pattern of impulses forms a sensory "image" which can be likened to the pattern of light intensity in an optical image. To determine the expectations, take a time exposure, and develop the negative; the blackening then indicates the expected intensity in each part of the image. Now look at the sensory image at present being received through the negative. Any regions in the image that have not changed since the time exposure was started are reduced to a uniform gray, but regions that have changed stand out by being lighter or darker than their background. The procedure thus emphasizes the unusual at the expense of the usual.* In the same way, a redundancy-reducing code in the nervous system cuts down the impulse traffic from expected messages, whereas any sequence or combination of inputs that is unexpected on the basis of previous experience requires more impulses, and so stands out from the background. The code must be superior to the photographic negative in taking account of ordered sequences (that is, movement), but like the negative it is a representation of the environment—the fact that it is a negative model is not important.

This picture of the operation of a redundancy-reducing code also brings out its close relation to the "matching response" described by MacKay (1956). His conception was that a nervous center produces an outgoing signal that is an attempt to match the incoming signal. The "error" between incoming signal and matching response indicates how successful the attempt has been, and a second-stage matching response could be made to this error signal, and so on. Since the matching response must correspond to a redundant feature of the original signal, the effect of the operation is to recode the signal without this redundant element.

If one thinks of reducing redundancy as "economy of impulses" and "emphasizing the unusual," it will be seen that such recoding tends to impart a dual character to a nerve impulse. On the one hand, it signals the occurrence of a specific, but not necessarily simple, feature of the input. On the other hand, it also contributes to the sum total of impulses required to convey the information and thus helps to indicate how improbable the current sensory input is. Impulses are reserved for the unusual so that they carry more information indi-

---

* See illustration of a similar redundancy-reducing photographic process in Comment on Lateral Inhibition, p. 783 in this volume.

vidually, and the total number required in the output increases with the improbability of the input.

Predictions and speculations

Three predictions following from the hypothesis are, first, that impulse frequencies in response to "usual" stimuli should be decreased; second, that the type of transformations should change according to the probabilities of the stimuli being passed through the sensory relay; and, third, that the outputs may correspond to rather complex features of the inputs, not to properties that are simple in physical or anatomical terms. I think it is probably worth looking for evidence on these points with present techniques, though it should be realized that, even if the principle is correct, there are a great many more specific details to be given before one really has a working hypothesis on organization of the sensory input. For instance, one does not know how rapidly to expect the changes in code to follow a change in input. Reducing redundancy defines a strategy, but the tactics by which this objective is attacked are all-important, and nothing has been said about this. Now the object of this paper was to set out possible strategies for sensory integration, so I am going to confine my speculation about tactics of redundancy reduction to two points.

The first is that the coding out of redundancy is an operation that lends itself to subdivision. One can suppose that small parts of the input, selected by spatial and temporal contiguity, are dealt with in isolation before being brought together for coding out the more complex forms of redundancy. If the coding was successful in the first stage, the capacity of the channel acting as input to the next stage would be reduced. This serial reduction seems to be advantageous, because the difficulty in selecting a code is related to the number of possible codes, which in turn is dependent upon channel capacity, not the amount of information the channel is carrying. Of course, the recognition of a complex pattern, to which a complex recoding operation is comparable in difficulty, can also be conceived as a subdivided series of acts, but in this case it is not at all clear how the earlier acts take one toward the accomplishment of the complete recognition. In contrast, coding out redundancy must almost inevitably lead to the situation in which a single impulse in the output corresponds to a complex feature of the input, and this complex feature will be one that enables a concise, nonredundant description of the sensory situation to be given. One thus sees a way of breaking down pattern recognition into a succession of autonomous stages that do not need controlling from above.

The second point about tactics concerns the form of the output, which has not been specified either by the assumptions or by the hypothesis itself. These would allow the number of fibers conveying the information either to increase or to decrease, or to stay the same, but on this point anatomical clues are available. The situation is that the number of fibers may increase as one goes toward the cortex (see, for example, Galambos, 1954, for auditory pathway), or it may decrease (see Walls, 1953, for visual pathway). But whatever may happen in the subcortical relays, in the sensory areas of the cortex itself, there is a vastly greater number of cells than in the incoming fibers. Now it follows from the formula for the capacity of a nerve channel (given on p. 223 according to the admittedly oversimplified assumptions used here), that the aggregate of impulses required to carry the same information at the same redundancy is lower in a large channel than in a channel with fewer fibers; not only is I, the mean impulse frequency per fiber, less, but also IF, the total number of impulses in the whole pathway. Combined with reduction of redundancy, an enormous decrease in the number of impulses required seems to be possible without the loss of any information. One consequence is that impulses in some units may occur so rarely that it is possible to conceive that a response, such as salivation or the raising of a forepaw, would become linked in direct fashion to the occurrence of an impulse in that unit alone, rather than to impulses occurring in a particular combination of units. Expansion of channels works in the same direction as redundancy reduction; they both increase the informational value of single impulses at the higher levels.

It is amusing to speculate on the possibility that the whole of the complex sensory input we experience is represented, at the highest level, by activity in a very few, and perhaps only a single, neural unit at any one instant. At first this seems a monstrous suggestion, but consider how complex a sensory situation a skilled writer can evoke with a very small number of words. These words are taken at, say, 4 per second, and are chosen from a vocabulary of the order of 10^4; with an impulse chosen from a neural vocabulary of 10^6 cells, and occurring at an average rate of, say, 1 per $1/10$ second (an average of about 1 per day per fiber) a representation of the current sensory situation should be possible which would be as complete as what we actually experience.

Now the process of reducing the redundancy must be stopped at some point, and instead the nervous system must disseminate its representation of the sensory input to all parts of the nervous system that require it: having edited its newspaper, it must print it and distribute it, and this is of course a redundancy-increasing process.   The present speculation is that the sensory image that is thus disseminated consists of very few impulses and perhaps only a solitary one, in a very large array of nerve fibers.   But whether this particular suggestion is right or not, it offends one's intuition, and one's experience of the efficiency and economy of naturally evolved mechanisms, to suppose that sensory messages are widely disseminated through the nervous system before they have been organized in a fairly non-redundant form.

Summary

This paper is not a discussion of the physiological mechanisms of sensory pathways, but an attempt to formulate ideas about the operations these mechanisms perform.   "What are sensory relays for?" is the question posed, and three hypotheses are put forward as answers.

The first—the "password" hypothesis—really says that, since animals respond specifically to specific stimuli, their sensory pathways must possess mechanisms for detecting such stimuli and discriminating between them: one might therefore look for such mechanisms in neurophysiological preparations.

The second hypothesis is the fashionable one that relays act as control points at which the flow of information is modulated according to the requirements of other parts of the nervous system.   It is pointed out that such control might have more interesting consequences than are suggested by the analogy of a simple gain or sensitivity control.

Most space is given to discussion of the third hypothesis, that reduction of redundancy is an important principle guiding the organization of sensory messages and is carried out at relays in the sensory pathways.   Some simplifying assumptions about the information-carrying variables of nerve messages are made, followed by a statement of the hypothesis and an explanation of the terms used.   Examples of recoding are described to illustrate its consequences, and predictions (which might be experimentally testable) and speculations (for entertainment only) are made.

To strip the redundancy from the preceding pages, what I have said is this: it is foolish to investigate sensory mechanisms blindly—one must also look at the ways in which animals make use of their senses.   It would be surprising if the use to which they are put was not reflected in the design of the sense organs and their nervous pathways—

as surprising as it would be for a bird's wing to be like a horse's hoof.

References

Attneave, F. Informational aspects of visual perception. *Psychol. Rev.*, 1954, **61,** 183–193.

Barlow, H. B. Summation and inhibition in the frog's retina. *J. Physiol.*, 1953, **119,** 69–88.

Barlow, H. B. Sensory mechanisms, the reduction of redundancy, and intelligence. In National Physical Laboratory Symposium, *Mechanisation of Thought Processes*. London: H.M. Stationery Office, 1959.

Barlow, H. B. The coding of sensory messages. In W. H. Thorpe and O. L. Zangwill (Editors), *Current Problems in Animal Behaviour*. Cambridge: University Press (in press).

Craik, K. J. W. *The Nature of Explanation.* Cambridge: University Press, 1943.

Eldred, E., R. Granit, and P. A. Merton. Supraspinal control of the muscle spindles and its significance. *J. Physiol.*, 1953, **122,** 498–523.

FitzHugh, R. The statistical detection of threshold signals in the retina. *J. gen. Physiol.*, 1957, **40,** 925–948.

Galambos, R. Neural mechanisms of audition. *Physiol. Rev.*, 1954, **34,** 497–528.

Lee, R. J. *Self-Programming Information and Control Equipment* (SPICE). Falls Church, Virginia: Melpar Inc., 1959.

Lettvin, J. Y., H. R. Maturana, W. S. McCulloch, and W. H. Pitts. What the frog's eye tells the frog's brain. *Proc. Inst. rad. Engrs.*, 1959, **47,** 1940–1951.

Mach, E. *The Analysis of Sensations.* Trans. by C. M. Williams from 1st German ed., 1886. Chicago and London: Open Court, 1914.

MacKay, D. M. Towards an information-flow model of human behaviour. *Brit. J. Psychol.*, 1956, **47,** 30–43.

Pearson, K. *The Grammar of Science.* London: Walter Scott, 1892.

Selfridge, O. G. Pandemonium: a paradigm for learning. In National Physical Laboratory Symposium, *Mechanisation of Thought Processes*. London: H.M. Stationery Office, 1959.

Shannon, C. E., and W. Weaver. *The Mathematical Theory of Communication.* Urbana: Univ. of Illinois Press, 1949.

Walls, G. L. The lateral geniculate nucleus and visual histophysiology. *Univ. Calif. Publ. Physiol.*, 1953, **9,** 1–100.

Woodward, P. M. *Probability and Information Theory with Applications to Radar.* London: Pergamon Press, 1953.

Yerkes, R. M. The instincts, habits, and reactions of the frog. I. The associative processes of the green frog. *Psychol. Monogr.*, 1903, **4,** 579–597.

14

CLINTON N. WOOLSEY

Laboratory of Neurophysiology, Department of Physiology
Medical School, University of Wisconsin

Organization of Cortical Auditory System

In an earlier symposium (Woolsey, 1961), we reviewed the present state of knowledge concerning organization of the cerebral cortical auditory system. As a result of that review and analysis we drew certain deductions and proposed a new synthesis. Since then some of these deductions have been checked by new experiments, and we are now able to offer substantiating evidence for them. The present account expands the argument offered in that symposium by including these new findings.

Topical Projection of Cochlear Nerve to A I and A II

In 1942 Woolsey and Walzl presented results, obtained by recording electrical potential changes evoked locally in the cerebral cortex, on stimulating electrically small bundles of cochlear nerve fibers in the spiral osseous lamina. Two auditory areas were thus defined below the suprasylvian sulcus on the lateral aspect of each hemisphere. In the more dorsal one, occupying middle ectosylvian cortex, the cochlear nerve was so projected that responses to stimulation of fibers in the basal turn appeared in the rostral part, whereas those produced by stimulation of apical fibers occurred in the caudal part of the area. The focal zone of this response field extended from the upper end of the anterior ectosylvian sulcus to a position just behind the upper end of the posterior ectosylvian sulcus. In the second area, lying immediately ventral to the first, and extending across the anterior ecto-sylvian, the pseudosylvian, and the posterior ectosylvian gyri, the order of cochlear representation was reversed. Figure 1 shows these relations as illustrated in a diagram drawn especially for Davis' chapter in Stevens' *Handbook of Experimental Psychology* (1951).

The figure shows the areas of response to stimulation of the base and apex of the cochlea crossing each other between the two ectosylvian sulci, to produce the reverse order of representation in the lower as compared with the upper part of the cortical auditory region. The terminology, auditory areas I and II (AI and AII), was suggested in a paper by Woolsey and Fairman (1946). A point of importance for

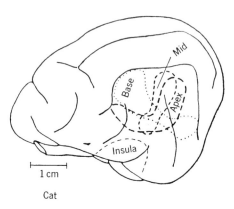

Cat

Fig. 1. Areas of response to focal electrical stimulation of cochlear nerve of cat, in basal, middle, and apical turns of cochlea, according to results of Woolsey and Walzl (1942). Prepared by C. N. W. for Davis (1951). AI above; AII below. Note reversal of cochlear representation.

what follows is the relatively great difference in the actual lengths of AI and AII, as indicated in this figure. The total length of AI from center of basal to center of apical focus is between 10 and 11 millimeters. In AII the distance is much greater, since to the longer surface measurement must be added twice the depths of the anterior and posterior ectosylvian sulci, giving a total distance of at least 33 to 35 millimeters.

Ades' Secondary Auditory (Association) Area

While the paper of Woolsey and Walzl was in press, Ades (1943) submitted for publication a study in which he described a "secondary" acoustic ("association") area in the posterior ectosylvian gyrus of the cat, defined by recording activity induced in this gyrus by click stimulation when strychnine was applied to the "primary" auditory area. Ades' "secondary" area thus included that portion of the second auditory area of Woolsey and Walzl that responded to electrical stimulation of the basal end of the cochlea, but it did not include the rest of AII.

Tunturi: Confirmation and Doubt; A III

In 1944 Tunturi confirmed for the dog the results obtained by cochlear-nerve stimulation in the cat and for the first time succeeded in clearly demonstrating localization to tonal stimuli, by using tone bursts with gradually rising amplitude at onset, thus avoiding click artifacts. Tunturi had difficulty in following the frequency-localization pattern from the anterior ectosylvian to the posterior ectosylvian gyrus, perhaps, as he then thought, because this portion of AII is rolled into the continuous ectosylvian sulcus of the dog. Later, as the result of strychnine studies, Tunturi (1950) came to have doubts about the reality of this central sector of AII, because evoked potentials there were incapable of tripping strychnine spikes. He therefore questioned whether the whole second area of Woolsey and Walzl could be considered a single system. In addition, Tunturi (1945) described a third auditory response area in the dog, well separated from AI and AII, beneath the anterior end of the suprasylvian gyrus, within or overlapping the face subdivision of the second somatic sensory area (Pinto Hamuy, Bromiley, and Woolsey, 1956).

Order of Frequency Representation in Anterior Ectosylvian Gyrus

Another feature of Tunturi's (1952) results, which may be pointed out at this time, is illustrated in Fig. 2a. This concerns the order of frequency representation in the anterior ectosylvian gyrus. The figure shows that frequency *increases* as one moves *forward*, rather than the reverse, as would be the case if Woolsey and Walzl's view of the organization of the second auditory area were correct. Similar, and perhaps more striking, because the frequency range is greater, are the observations of Hind (1953), illustrated in Fig. 2b. On the basis of his findings, Hind drew attention to the existence of discontinuities in the frequency gradient in the "second" auditory area of the cat and stated: "These discrepancies suggest that additional data (especially latency measurements) are desirable to support the concept of functional unity implied in the use of a single term to designate this relatively broad expanse of cortex."

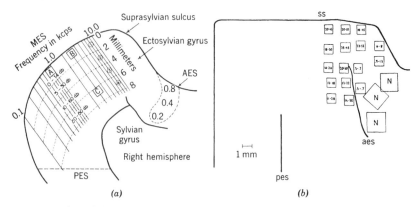

Fig. 2. Order of frequency representation in anterior ectosylvian gyrus. (*a*) *Dog*: 0.2, 0.4, 0.8 kcps (from Tunturi, 1952); (*b*) *Cat*: 0.3 to 0.7; 0.5 to 1.3; 6 to 8 kcps (from Hind, 1953).

Evidence for Dependence of A II-Ep upon A I for Activation

In 1953 Bremer, concurring with Ades' concept that the posterior ectosylvian gyrus is a secondary area in the sense of "association cortex," presented evidence that in his opinion not only substantiated Ades' view but also allowed him to conclude that the area "extended ahead of the posterior ectosylvian gyrus to form a narrow strip on the ventral border of the primary area, thus coinciding with the 'second auditory field' described by Woolsey and Walzl." Bremer offered as support for this conclusion the general characteristics of responses in his secondary area: "Thus by comparison with the response in the primary area, the higher threshold, the longer and more variable latency—indicating a more complicated synaptic transmission—the slower onset and still slower decay of the surface-positive wave, the smaller voltage, the usual monophasicity—that is, the absence of a surface-negative wave following the positive one—and finally the different sensitivity to local strychninisation are all suggestive of a relayed response. That the sensory responses of the posterior ecto-sylvian gyrus are actually mediated by the auditory area is made likely by the following facts (Ades, 1943, 1949; Arteta, Bonnet, and Bremer, 1950): the response is increased after a local strychninisation of the projection area; the spontaneous strychnine spikes of the primary area are followed, on the secondary one, by potentials of a latency and shape similar to those recorded in response to a click; such repercussions are absent in the non-strychninised part of the primary area and in the upper part of the posterior ectosylvian gyrus;

the response of the secondary area disappears when the primary area is destroyed or made momentarily unresponsive by the local application of a depressing agent; a superficial incision of the brain—presumably interrupting short association fibers by which the two areas are interconnected—has the same effect."

Mickle and Ades (1953), studying the spread of evoked cortical activity before and after ablations in area I and incisions between AI and AII, also supported the view that the cortex of the pseudosylvian and posterior ectosylvian gyri is dependent upon corticocortical activation from AI.

Lilly and Cherry (1954), however, interpreted their findings on the movements of "response figures" over the auditory cortex as reflecting the magnitudes and the timing of parts of a preformed afferent figure (moving up to the cortex from below) exciting cortex at different places at different times.

Cytoarchitectural Parcellation of Auditory Region

Some of these preceding studies were influenced in part by another study pertinent to our discussion, that of Rose (1949) on the cytoarchitectural characteristics of the auditory region of the cat (Fig. 3).

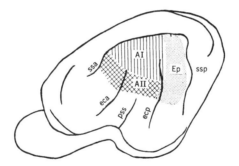

Fig. 3. Cytoarchitectonic fields of auditory region according to Rose (1949). Suprasylvian fringe sector lies buried in suprasylvian sulcus along dorsal border of AI; ssa, anterior suprasylvian; ssp, posterior suprasylvian; eca, anterior ectosylvian; pss, pseudosylvian; ecp, posterior ectosylvian.

According to this study, "the auditory region consists of a central sector and a peripheral belt divisible in turn into three sectors. The dorsally situated sector of the belt lies hidden in the depth of the suprasylvian sulcus and adjoins the central sector throughout its extent.

The central sector, roughly triangular in shape, will be referred to as the first auditory area (AI). The oroventral sector of the peripheral belt will be designated as the second auditory area (AII), while the posterior portion of the belt will be called the posterior ectosylvian area (Ep). The dorsal fringe sector will be referred to as the suprasylvian fringe sector (SF). Its functional relation to the auditory area of the cat is not clear."

Relation of Architectural Areas to Evoked Activity

Rose and Woolsey (1949) undertook to relate these cytoarchitectural differentiations to evoked electrical activity and to thalamic connections. The chief difficulty encountered concerned the relation of the basal end of the "second" auditory response field (AII) to the posterior ectosylvian area. They wrote: "There is a discrepancy between the finding of Woolsey and Walzl that the second auditory area extends into the posterior ectosylvian gyrus and the architectonic definition which suggests that it ends in front of the posterior ectosylvian sulcus." This discrepancy between histological and electrical data suggested the need to reinvestigate the relations of the posterior ectosylvian area to the rest of AII and its relation, via corticocortical connections, with AI.

Re-Examination of Cochlear-Nerve Projection to A I, A II-Ep

A re-examination was undertaken in 1953 by Downman and Woolsey (see Downman, Woolsey, and Lende, 1960); they repeated the experiments on cochlear-nerve stimulation, studied the effect upon responses in AII-Ep of removing AI, and examined the problem of corticocortical relations using the techniques of electrical stimulation and recording.

In the nerve-stimulation experiments they were able to repeat the original observations of Woolsey and Walzl concerning localization in AII, but highly significant additional information was secured. Stimulation of the basal end of the cochlea (Figs. 4a–c) gave results in good agreement as to the locus of the response field in AI, but in the second area the response field was sometimes ahead of the posterior ectosylvian sulcus (Fig. 4a), sometimes astride it (Fig. 4b), and sometimes two definite response areas were seen (Fig. 4c), one ahead of and one behind the sulcus. Responses in the posterior ectosylvian gyrus could further be characterized by a longer latency. This was also pointed out by Perl and Casby (1954). These results suggested the possibility of differentiating one high-frequency focus in middle AII and another in Ep, thus bringing the physiological data into better

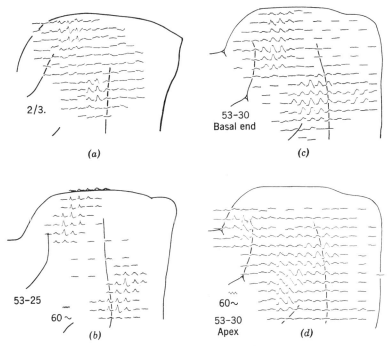

Fig. 4. Areas of response to focal stimulation of cochlear nerve fibers: (a–c) 2 mm from basal end of cochlea in three different cats, showing similarity of locus in AI, but variation of locus in AII-Ep; (d) in apex of cochlea; same animal as in (c). (From Downman, Woolsey, and Lende, 1960.)

relation with Rose's anatomy of these parts of the original second auditory area. The depth of the posterior ectosylvian sulcus is such (6 to 7 mm) that one or the other of these foci might escape notice in any particular experiment, depending on the cochlear site stimulated and on the variability of sulci with respect to the total cortical auditory region.

Restriction of A II to Mid-Portion of Original A II

We have already called attention to doubts concerning the validity of including in a single AII system the anterior ectosylvian low-frequency area and have emphasized particularly evidence that the frequency sequence appears to be running in the wrong direction (Tunturi, 1952; Hind, 1953). If the anterior ectosylvian and the posterior ectosylvian portions of AII must be removed from the original second auditory system, there is left to it only the central portion, which lies between the anterior and the posterior ectosylvian sulci. That this still contains a complete representation for the cochlea is indicated by Fig. 4d, which gives the results of stimulating the apex of the cochlea in the same animal for which responses to stimulation of the basal end are given in Fig. 4c. Figure 4d shows three foci of maximal response, one astride the upper posterior ectosylvian gyrus— a little less prominent than usual—a second, well isolated in the anterior ectosylvian gyrus, and a third just above the pseudosylvian sulcus. The latter overlaps to some extent the adjacent basal response area but is clearly centered more rostrad. Existence of a nonresponsive zone between the anterior ectosylvian and the pseudosylvian response foci appears significant. A less striking example of this may be seen in Fig. 7b of Woolsey and Walzl (1942). In that case the two foci were thought to be one, with response amplitude attenuated by fluid along the anterior ectosylvian sulcus. These focalized responses to stimulation of the apical and basal coils in middle AII also, I think, dispose of the doubt raised by Tunturi (1950) concerning the validity of auditory responses in this part of the original second auditory area. Response patterns such as those shown in Figs. 4c and d cannot be accounted for on the basis of volume-conductor artifact. This conclusion is also supported by the experiments of Kiang (1955), who confirmed Tunturi's finding that responses in middle AII are incapable of tripping strychnine spikes but established that small isolated islands of cortex in middle AII still yield evoked responses which are then abolished by undercutting the island.

Activation of A II-Ep Independently of A I

A second contribution of the study with Downman was the demonstration that the middle part of AII and the posterior ectosylvian area can be activated by cochlear-nerve stimulation and by clicks after acute or chronic bilateral removal of AI, thus establishing the existence of afferent connections to these parts of the auditory region which are independent of corticocortical connections from AI. Kiang (1955), on isolating small islands of cortex by removing surrounding cortex, independently established the same conclusion somewhat more elegantly.

Corticocortical Interconnections between A I and A II-Ep

The demonstration of pathways to AII-Ep, which are independent of relays from AI, does not deny the existence of corticocortical activation of AII-Ep nor does it invalidate the evidence for such activation presented by Bremer (1953), Ades (1943, 1949), or Mickle and Ades (1953). It indicates that two ways of activating AII-Ep exist.

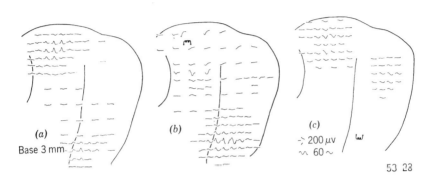

Fig. 5. (*a*) Surface-positive responses in AI and Ep to electrical stimulation of cochlear nerve, 3 mm from basal end of cochlea. (*b*) Surface-positive responses elicited in Ep by electrical stimulation at center of response area in AI. (*c*) Surface-negative responses in AI and upper Ep to electrical stimulation of response focus in lower Ep. (From Downman, Woolsey, and Lende, 1960.)

A third aspect of the study by Downman and Woolsey was the demonstration, by electrical stimulation and recording, of the existence of specific interconnections between regions of cortex focally activated by cochlear-nerve stimulation. The three potential plots of Figs. 5a–c give the results of one such experiment. Figure 5a shows two areas responsive to stimulation of cochlear nerve fibers in the bony spiral, 3 mm from the basal end, one in AI and a second in Ep. In Fig. 5b the AI focus has been stimulated with single electric shocks and the activity induced thereby has been mapped. The main effect is a family of well-developed, initially surface-positive responses, occurring in that part of the posterior ectosylvian gyrus that had been activated by cochlear-nerve stimulation. Figure 5c shows recordings made in AI and in upper Ep when the lower posterior ectosylvian focus was stimulated with the same strength of stimulus used in AI. Responses occur in the AI focus and less prominently in upper Ep. The responses, however, are now surface-negative. The significance of the negativity (whether dromic or antidromic) is not clear, but it probably means activity in elements nearer the surface than those giving rise to surface-positive potentials. In any case, specific connections from AI to Ep may be inferred from these results.

Attempts were also made to demonstrate similar relations between the apical focus in AI and the apical focus of AII in the anterior ectosylvian gyrus. No such relationships could be shown. This can be taken to mean that the relations between AI and anterior ectosylvian AII are different from those between AI and Ep and again suggests that anterior ectosylvian AII and Ep do not belong to a single functional subdivision of the auditory region. Kiang (1955) came to a similar conclusion on the basis of other properties of the low-tone anterior ectosylvian area, which led him to classify low frequency AII with AI.

Collapse of the Concept That A II Is a Single Functional Area

We are thus brought to the point where the second auditory area of Woolsey and Walzl, in the sense in which it was originally conceived, has been destroyed by the force of additional data. We are left with (1) a central remnant (middle AII) in which frequency representation is complete and the order is still the reverse of that in AI, (2) an orphaned low-frequency anterior ectosylvian area, and (3) a similarly dissociated high-frequency Ep area, which has specific corticocortical connections from high-frequency AI (Downman and Woolsey) but which, according to strychnine effects, can be activated from all parts of AI (Ades, Bremer, Kiang), and from the anterior ectosylvian low-frequency area as well (Kiang). We are thus led to consider whether information concerning the posterior ectosylvian and the anterior ectosylvian areas is complete, since we are faced with what appears to be a special status for high frequencies in the former and for low frequencies in the latter.

Complete Representation of Cochlea in Ep

Recently in our laboratory Sindberg and Thompson (1959) have carried out a series of experiments, under chloralose anesthesia, involving the posterior ectosylvian gyrus. They have shown under these conditions that Ep can be activated by click as far ventrad as the lower end of the posterior ectosylvian sulcus (Fig. 6) and that responses in the ventral region are similar in characteristics to those in more dorsal Ep. The more ventral extension was still responsive to click after acute removal of AI and AII (as originally defined). In some experiments electrical stimulation of the cochlear nerve indicated that ventral Ep was predominantly apical in representation, whereas basal representation was more dorsal, as is already known. Thus it appears that the posterior ectosylvian gyrus may contain another complete representation of the cochlea.

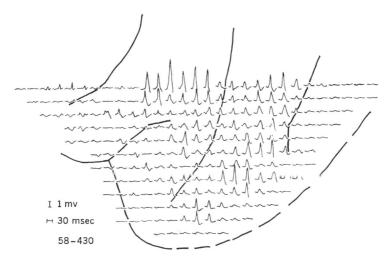

I 1 mv

⊢ 30 msec

58–430

Fig. 6. Potential plot of click responses in posterior ectosylvian gyrus (Ep) and in middle AII (brain not explored above area mapped). Note extension of Ep response field toward temporal tip. (From Sindberg and Thompson, *J. Neurophysiol.*, 1962.)

It should be noted that the ventral part of the posterior ectosylvian gyrus extends considerably below the limit of the posterior ectosylvian field of Rose. However, Rose remarked that changes in characteristics of Ep "take place so gradually that sharp separation of the posterior ectosylvian field from what is, apparently, the rest of the temporal cortex seems artificial." Failure to report evoked responses in this region previously may be attributed only in part to anesthesia, although it is well known (Woolsey and Walzl, 1942; Kiang, 1955) that Ep is activated only with difficulty by click stimuli under pentobarbital anesthesia. Electrical stimulation of basal cochlear nerve fibers is quite effective, but the ventral part of the region has never been examined in any previous experiment during stimulation of the apical coil.

Frequency Representation in the Suprasylvian Fringe Sector

We may now return to the anterior ectosylvian low-frequency area. We have already called attention to the reversed frequency sequence exhibited in this area in experiments of Tunturi (1952) and of Hind (1953); see also Katsuki, Watanabe, and Maruyama (1959) and Hind, et al. (1960). The possible significance of this struck us only recently when we attempted to account for another finding that was not in keeping with the dual-projection schema of Woolsey and Walzl. This finding was the occurrence of responses evoked by high-frequency tone pulses in an area just above the low-frequency focus of AI, close to the suprasylvian sulcus. This was originally noted by Hind (1953) in strychnine experiments but has recently been confirmed by him and his co-workers (unpublished) in experiments mapping the fields of response evoked by tone pulses under pentobarbital or chloralose anesthesia. Hind, et al. (1959, 1960) also found single units in this locus which responded optimally to high-frequency tone pulses. Downman, Woolsey, and Lende (1960) in one experiment elicited well-localized surface-negative potentials at this site (separate from negative potentials in AI) when the basal focus in Ep was stimulated electrically.

If we now consider the low-frequency anterior ectosylvian area, in which the frequency represented increases as the suprasylvian sulcus is approached, together with the high-frequency locus near the suprasylvian sulcus above the low-frequency focus of AI, we may bring them into one family by supposing that they may belong to Rose's suprasylvian fringe sector. Thus we may deduce that another complete frequency representation probably exists along the suprasylvian border of AI, in which the order of frequency representation is the reverse of that in AI.

Experimental test of this deduction by Woolsey, Hind, Benjamin, Welker, Sindberg, and Ladpli yielded results shown in Fig. 7 and summarized and interpreted in Fig. 8. In these experiments the lateral bank of the anterior and middle portions of the suprasylvian sulcus were exposed by removal of the suprasylvian and lateral gyri. Responses were then evoked by tone pulses of various frequencies. The cortex of the inferior bank of the suprasylvian sulcus and of the auditory region on the exposed lateral surface of the hemisphere were mapped in millimeter steps (Fig. 7). As can be seen in Fig. 8, the frequency sequence along the suprasylvian sulcus increases from 0.7 kcps at the anterior to 32 kcps at the caudal end, and the response

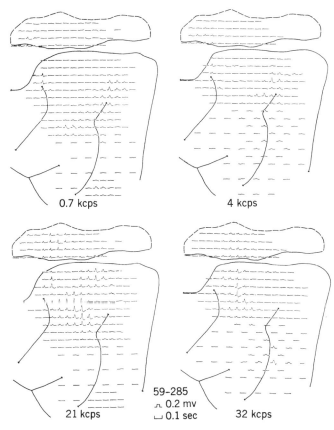

Fig. 7. Distribution of responses recorded from lateral bank of suprasylvian sulcus and lateral surface of hemisphere to tone pulses of 0.7, 4, 21, and 32 kcps. The responses in the suprasylvian sulcus are plotted as seen in a plane different from that of the lateral surface.

areas show the usual shifting overlap. Response areas in AI are shown for these same frequencies, and in AII and Ep for 0.7 kcps and 32 kcps.

Three additional sets of facts concerning sound-induced responses in the cerebral cortex must still be mentioned.

Tunturi's Third Auditory Area (A III) in Cat

We have already drawn attention to the fact that in the dog Tunturi's (1945) third auditory area is situated within or overlaps the anterolateral margin of the second somatic sensory face area (Pinto Hamuy, Bromiley, and Woolsey, 1956) at some distance from AI and AII. Bremer (1953) has illustrated this area in the cat as located near the anterior suprasylvian sulcus, adjacent to apical AII,

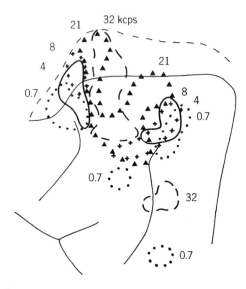

Fig. 8. Areas of response to tone pulses of 0.7, 4, 8, 21, and 32 kcps. At upper left is shown overlapping distribution of response areas on lateral bank of suprasylvian sulcus and their extensions onto the free lateral aspect of the cortex. The frequency increases from low to high in a rostrocaudal sequence. In the upper central part of the figure, the sequence in AI is the reverse of that in the suprasylvian fringe sector. In the posterior ectosylvian area, foci for 32 and 0.7 kcps are shown. Extension of the 32-kcps focus into AII may indicate a high-frequency focus of AII on the anterior bank of the posterior ectosylvian sulcus. A focus for 0.7 kcps in AII is shown.

in Mickle and Ades' (1952) polysensory area. This would bring it in relation to the leg and tail representations of SII. Actually the face subdivision of SII in the cat is inaccessible unless the orbital surface of the brain is exposed. When this is done, evidence for AIII in this locus is forthcoming (see Fig. 13 for locus).

An Insular Auditory Area

In 1953 Carl Pfaffmann and I were attempting to do some experiments on cortical taste centers in the cat and came by accident on an area in insular cortex which was responsive to auditory stimulation. This is illustrated in Fig. 9, in a map prepared by R. A. Lende in 1954. Recently, in our laboratory, Loeffler (1958) added information on this area in experiments employing cochlear-nerve stimulation.

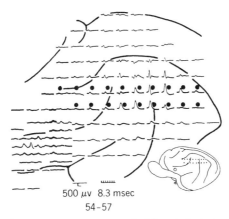

500 μv 8.3 msec
54-57

Fig. 9. Insular auditory response area, marked by circle in inset diagram. (Unpublished map prepared by R. A. Lende, 1954.)

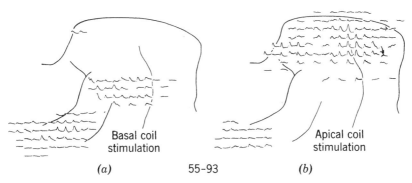

Basal coil
stimulation

Apical coil
stimulation

(a)   55-93   (b)

Fig. 10. Responses in insular cortex to electrical stimulation of cochlear nerve in basal (a) and apical (b) coils. Partial mapping of rest of auditory area. (From Loeffler, 1958.)

Figure 10a shows responses elicited in this area by electrical stimulation of basal cochlear nerve fibers. Some evidence was obtained that responses to apical stimulation are focalized more ventrally (Fig. 10b), indicating some degree of spatial differentiation in projections to this area. Desmedt and Mechelse (1959a; see also Desmedt, 1961) have independently reported the occurrence of auditory responses in this area and have described an efferent influence originating in this part of the cortex and acting to control ascending auditory volleys (1959*b*). They also found that the insular area could be activated by photic stimuli and that interaction occurred between visual and auditory responses. Desmedt and Mechelse refer to this as a fourth auditory area (AIV); we shall call it the insular area. Goldberg, Diamond, and Neff (1957) have presented behavioral evidence of the importance of insular and temporal cortex for auditory function, and Rose and Woolsey (1958) and Diamond, Chow, and Neff (1958) have described sustaining projections from auditory thalamus to these cortical areas.

Auditory Responses in Suprasylvian and Anterior Lateral "Association" Areas, in Sensorimotor Cortex, and in Second Visual Area

Finally, I should like to mention responses to auditory stimulation which occur not only outside all the cortical areas so far discussed but also in the complete absence of all these areas.

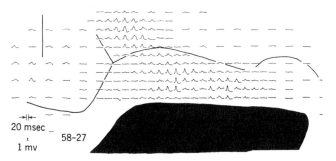

Fig. 11. Responses to click stimulation (latency, 15 msec) in suprasylvian and anterior lateral gyri and in sensorimotor cortex. All cortex between suprasylvian and rhinal sulci removed. (From Thompson and Sindberg, 1960.)

Some years ago Tunturi (1950) reported that certain potentials recordable from the suprasylvian gyrus of the dog were presumably volume-conductor artifacts, because they were incapable of tripping strychnine spikes. In recent years, however, Buser and co-workers (1959) in Paris have conclusively established the validity of auditory responses in this area (see their 1959 papers and their lists of references). Figure 11 shows some of the results obtained in our laboratory by Thompson and Sindberg (1960) with chloralose anesthesia after removal of all cortex between the suprasylvian and the rhinal sulci. There are two peaks of response to click stimulation in the suprasylvian gyrus, one anteriorly in the lateral gyrus and still another in the sensorimotor region, with maximal amplitude in the "precentral" motor area. At all these sites the response characteristics are very similar: a surface-positive wave with latency of approximately 15 milli-

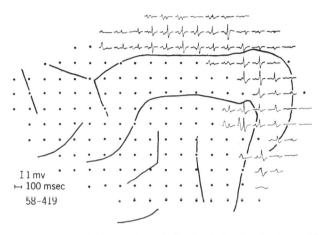

I 1 mv
⊢ 100 msec
58-419

Fig. 12. Late responses (100 msec) to click stimulation in visual area II. (From Thompson and Sindberg, 1960.)

seconds, up to 3 millivolts in amplitude, followed by a longer, more variable negativity. Electrical stimulation of apical and basal portions of the cochlear nerve, while yielding the same pattern of localization in AI and AII (original meaning) as under pentobarbital, gave no evidence of topical projection to the suprasylvian, anterior lateral, and precentral areas. Variation of stimulus repetition rate demonstrated that the response in these areas is all or none rather than graded, as in AI. The responses at all four foci were highly correlated with one another in occurrence and amplitude, but they were not at all correlated in amplitude with response amplitude in AI. It was concluded that all four foci receive input from some common subcortical system not involving the medial geniculate body. According to Buser, Borenstein, and Bruner (1959), the input to the suprasylvian areas is through the pulvinar.

In addition to the four response foci with 15-msec latency just discussed, a late response to auditory stimulation, with a latency of about 100 msec, was found to occur in an area apparently coinciding quite well with the extent of Talbot's visual area II (Fig. 12; see Woolsey and Fairman, 1946). Aside from latency, this area also exhibited characteristics similar to those described for the 15-msec "association" responses. Auditory responses in this locus were previously noted by Bremer (1953).

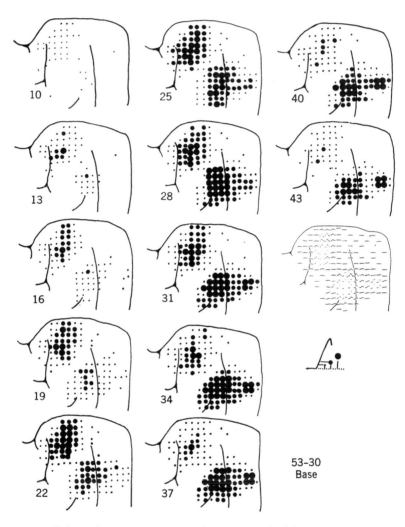

Fig. 13. Voltage-time sequence map of activity evoked by contralateral basal-turn stimulation. Actual responses shown at right and in Fig. 4c. Response size, at intervals of 3 msec, has been expressed as falling within three ranges. These have been charted as spots of three arbitrary sizes. Time of each map, taking stimulus artifact as zero, is shown in milliseconds at lower left of each map. (From Downman, Woolsey, and Lende, 1960.)

Temporal Sequence of Cortical Activation by Auditory Stimulation

Bremer (1953), Mickle and Ades (1953), and Lilly and Cherry (1954) have given particular attention to the temporal sequence in which evoked responses appear in various parts of the cortical auditory region. We have pointed out above some of the time differences observed in activity evoked outside the main auditory region. Figure 13, taken from Downman, Woolsey, and Lende (1960), shows a series of time-voltage maps illustrating the manner in which responses (lower right) to electrical stimulation of cochlear nerve fibers in the basal end of the cochlear spiral activate the cortex in place and time. The voltage of each response was measured at intervals after stimulation noted at the lower left of each map. These values were then represented by dots of three different arbitrary sizes to show the state of the "response figure" throughout auditory areas AI, AII, and Ep for each time interval selected. Three distinct centers of activity in these three subdivisions are seen. Activity in AII began later than in AI and continued longer; activity in Ep began even later and lasted longer still.

Summary

Figure 14 summarizes the story that has been set forth above. It shows that cortical auditory response mechanisms are organized in a much more complex way than was encompassed by the concept of a dual cortical auditory area. Within the central auditory region there appear to be four complete representations for the cochlea, which correlate in some degree with the four principal cytoarchitectural divisions of Rose.

1. The *suprasylvian fringe area* (SF) begins on the anterior ectosylvian gyrus, then disappears in the depth of the suprasylvian sulcus, and emerges again above the posterior ectosylvian sulcus. In this area low frequencies (A) are represented anteriorly, high frequencies (B) posteriorly. Nothing is known specifically about the connections of this area with the thalamus, but it is possible that they come from *pars principalis*, thus helping to explain one difficulty found by Rose and Woolsey (1958) in relating frequency localization in the medial

geniculate body (experiments of Woolsey, Rose, Lende, and Ullrich) to retrograde degenerations produced by lesions in AI.

2. *Auditory area I* (AI) coincides with Rose's area AI and includes the focal band of the first auditory area of Woolsey and Walzl, in which high frequencies (B) are represented at the rostral end, low frequencies (A) at the caudal end. It is the only cortical auditory response area known to receive essential projections from the thalamus (Rose and Woolsey, 1949, 1958).

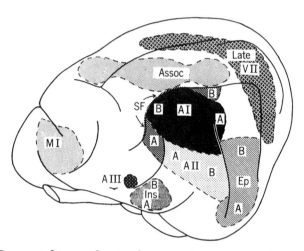

Fig. 14.   Summary diagram showing four central areas with cochlea represented anteroposteriorly from apex (A) to base (B) in the suprasylvian fringe sector (SF), from base to apex in AI; from apex to base in AII. In Ep representation is base above, apex below. In insula (Ins) evidence suggests base represented above, apex below. AIII is Tunturi's third auditory area. "Association" cortex (Assoc) and precentral motor cortex (MI) gave responses to click with 15-msec latencies under chloralose. Visual area II (Late) gave responses with 100-msec latency, also under chloralose. (From Woolsey, 1961.)

3. Ventral to AI is *auditory area II* (AII) lying between the ectosylvian sulci. In this remnant of the original AII of Woolsey and Walzl, frequency representation is less sharply differentiated than in AI, but the order of representation is again reversed with low frequencies (A) forward, high frequencies (B) back. The thalamic connections to this area have been described as sustaining (Rose and Woolsey, 1958).

4. *Area Ep* occupies the lower two-thirds of the posterior ectosylvian gyrus, with higher frequencies (B) represented above and lower frequencies (A) below. Small lesions in this area do not produce any marked changes in the thalamus, but larger ones undoubtedly

produce degeneration in the caudal part of *pars principalis* (Rose and Woolsey, 1958). According to Diamond, Chow, and Neff (1958), the projection is a sustaining one.

The upper part of the posterior ectosylvian gyrus has not been included as a part of the auditory region in this summary, since the responses recorded here by Woolsey and Walzl (1942) may have been the result of electrotonic spread. However, Lilly and Cherry (1954) described activity in this area, which traveled across the cortex at a much slower rate than activity in AI. On the basis of Rose's anatomical study, this should be a part of Ep. Further study of it is needed.

Outside the central region there is the *insular area* (Ins) with some degree of frequency localization suggested (A, B). The thalamic connections for this area are not known. Above it, rostrally, lies the *third auditory area* (AIII) of Tunturi in the head subdivision of somatic sensory area II. This area apparently receives connections from the posterior nuclear group (Rose and Woolsey, 1958). The insular and AIII areas both yield short latency responses (8 to 10 msec).

In the *suprasylvian and anterior lateral gyri* (Assoc) and in the *precentral motor area* (MI) are the response areas with 15 msec latency. These areas can also be activated by visual and somatic stimulation (Buser and Borenstein, 1959; Buser, Borenstein, and Bruner, 1959; Thompson and Sindberg, 1960). The thalamic connections, in part at least, are from the pulvinar (Buser, Borenstein, and Bruner). The area of late, 100-msec responses (Late) occupies the *second visual area* (VII). Its activation pathway is not known.

The functional contributions of these many connections to the cortex from the auditory receptors have, for the most part, yet to be determined (see Neff in this volume). It seems clear from what has been said about them that the temporal sequence of their activations and the different paths by which they are activated, in addition to their cortical locations, serve to characterize them as anatomically and functionally differentiated parts of a complex mechanism, to which a great deal of experimental study must be devoted before their contributions to the physiology of the organism can be understood.

Acknowledgment

The research described in this chapter was supported by grants from the NINDB (B-896), the Alfred Laukhuff Trust, and the Research Committee of the University of Wisconsin out of funds provided by the Wisconsin Alumni Research Foundation.

References

Ades, H. W.   A secondary acoustic area in the cerebral cortex of the cat.   *J. Neurophysiol.*, 1943, **6,** 59–63.

Ades, H. W.   Functional relationships between the middle and posterior ectosylvian areas in the cat.   *Amer. J. Physiol.*, 1949, **159,** 561 [Abstract].

Arteta, J. L., V. Bonnet, and F. Bremer.   Répercussions corticales de la réponse de l'aire acoustique primaire: l'aire acoustique secondaire.   *Arch. int. Physiol.*, 1950, **57,** 425–428.

Bremer, F.   *Some Problems in Neurophysiology.*   London: Athlone Press, 1953.

Buser, P., and P. Borenstein.   Réponses somesthésiques, visuelles et auditives, recueillies au niveau du cortex "associatif" suprasylvien chez le chat curarisé non anesthésié.   *EEG clin. Neurophysiol.*, 1959, **11,** 285–304.

Buser, P., P. Borenstein, and J. Bruner.   Etude des systèmes "associatifs" visuels et auditifs chez le chat anesthésié au chloralose.   *EEG clin. Neurophysiol.*, 1959, **11,** 305–324.

Davis, H.   Psychophysiology of hearing and deafness.   Chapter 28 in S. S. Stevens (Editor), *Handbook of Experimental Psychology.*   New York: Wiley, 1951.

Desmedt, J. E.   In G. L. Rasmussen and W. F. Windle (Editors), *Neural Mechanisms of the Auditory and Vestibular Systems.*   Springfield, Ill.: C. C Thomas, 1961.

Desmedt, J. E., and K. Mechelse.   Mise en évidence d'une quatrième aire de projection acoustique dans l'écorce cérébrale du chat.   *J.· Physiol.*, 1959a, **51,** 448–449 [Abstract].

Desmedt, J. E., and K. Mechelse.   Corticofugal projections from temporal lobe in cat and their possible role in acoustic discrimination.   *J. Physiol.*, 1959b, **147,** 17–18P [Abstract].

Diamond, I. T., K. L. Chow, and W. D. Neff.   Degeneration of caudal medial geniculate body following cortical lesion ventral to auditory area II in cat.   *J. comp. Neurol.*, 1958, **109,** 349–362.

Downman, C. B. B., and C. N. Woolsey.   Inter-relations within the auditory cortex.   *J. Physiol.*, 1954, **123,** 43–44P [Abstract].

Downman, C. B. B., C. N. Woolsey, and R. A. Lende.   Auditory areas I, II and Ep: Cochlear representation, afferent paths and interconnections.   *Bull. Johns Hopkins Hosp.*, 1960, **106,** 127–142.

Goldberg, J. M., I. T. Diamond, and W. D. Neff.   Auditory discrimination after ablation of temporal and insular cortex in cat.   *Fed. Proc.*, 1957, **47,** 204 [Abstract].

Hind, J. E.   An electrophysiological determination of tonotopic organization in auditory cortex of cat.   *J. Neurophysiol.*, 1953, **16,** 473–489.

Hind, J. E., R. M. Benjamin, P. W. Davies, J. E. Rose, R. F. Thompson, W. I. Welker, and C. N. Woolsey.   Unit responses in auditory cortex.   *Fed. Proc.*, 1959, **18,** 68 [Abstract].

Hind, J. E., J. E. Rose, P. W. Davies, C. N. Woolsey, R. M. Benjamin, W. I. Welker, and R. F. Thompson.   Unit activity in the auditory cortex.   In G. L. Rasmussen and W. F. Windle (Editors), *Neural Mechanisms of the Auditory and Vestibular Systems.*   Springfield, Ill.: C. C Thomas, 1961.

Katsuki, Y., T. Watanabe, and N. Maruyama. Activity of auditory neurons in upper levels of brain of cat. *J. Neurophysiol.*, 1959, **22**, 343–359.

Kiang, N. Y-S. An electrophysiological study of cat auditory cortex. Doctoral dissertation, University of Chicago, 1955.

Lilly, J. C., and R. B. Cherry. Surface movements of click responses from acoustic cerebral cortex of cat: Leading and trailing edges of a response figure. *J. Neurophysiol.*, 1954, **17**, 521–532.

Loeffler, J. D. An investigation of auditory responses in insular cortex of cat and dog. M.D. Thesis, University of Wisconsin, 1958.

Mickle, W. A., and H. W. Ades. A composite sensory projection area in the cerebral cortex of the cat. *Amer. J. Physiol.*, 1952, **170**, 682–689.

Mickle, W. A., and H. W. Ades. Spread of evoked cortical potentials. *J. Neurophysiol.*, 1953, **16**, 608–633.

Perl, E. R., and J. U. Casby. Localization of cerebral electrical activity: The acoustic cortex of cat. *J. Neurophysiol.*, 1954, **17**, 429–442.

Pinto Hamuy, T., R. B. Bromiley, and C. N. Woolsey. Somatic afferent areas I and II of dog's cerebral cortex. *J. Neurophysiol.*, 1956, **19**, 485–499.

Rose, J. E. The cellular structure of the auditory area of the cat. *J. comp. Neurol.*, 1949, **91**, 409–439.

Rose, J. E., and C. N. Woolsey. The relations of thalamic connections, cellular structure and evocable electrical activity in the auditory region of the cat. *J. comp. Neurol.*, 1949, **91**, 441–466.

Rose, J. E., and C. N. Woolsey. Cortical connections and functional organization of the thalamic auditory system of the cat. In H. F. Harlow and C. N. Woolsey (Editors), *Biological and Biochemical Bases of Behavior*. Madison: Univ. of Wisconsin Press, 1958. Pp. 127–150.

Sindberg, R. M., and R. F. Thompson. Auditory responses in the ventral temporal cortex of cat. *Physiologist*, 1959, **2**, 108–109 [Abstract].

Sindberg, R. M., and R. F. Thompson. Auditory response fields in ventral temporal and insular cortex of cat. *J. Neurophysiol.*, 1962, **25**, 21–28.

Thompson, R. F., and R. M. Sindberg. Auditory response fields in association and motor cortex of cat. *J. Neurophysiol.*, 1960, **23**, 87–105.

Tunturi, A. R. Audio frequency localization in the acoustic cortex of the dog. *Amer. J. Physiol.*, 1944, **141**, 397–403.

Tunturi, A. R. Further afferent connections of the acoustic cortex of the dog. *Amer. J. Physiol.*, 1945, **144**, 389–394.

Tunturi, A. R. Physiological determination of the boundary of the acoustic area in the cerebral cortex of the dog. *Amer. J. Physiol.*, 1950, **160**, 395–401.

Tunturi, A. R. A difference in the representation of auditory signals for the left and right ears in the iso-frequency contours of right middle ectosylvian auditory cortex of the dog. *Amer. J. Physiol.*, 1952, **168**, 712–727.

Woolsey, C. N. Organization of cortical auditory system: A review and a synthesis. In G. L. Rasmussen and W. F. Windle (Editors), *Neural Mechanisms of the Auditory and Vestibular Systems*. Springfield, Ill.: C. C Thomas, 1960.

Woolsey, C. N., and D. Fairman. Contralateral, ipsilateral and bilateral representation of cutaneous receptors in somatic areas I and II of the cerebral cortex of pig, sheep, and other mammals. *Surgery*, 1946, **19**, 684–702.

Woolsey, C. N., and E. M. Walzl. Topical projection of nerve fibers from local regions of the cochlea to the cerebral cortex of the cat. *Bull. Johns Hopkins Hosp.*, 1942, **71**, 315–344.

15

WILLIAM D. NEFF

Laboratory of Physiological Psychology, University of Chicago

Neural Mechanisms of Auditory Discrimination

In an earlier paper, a rational procedure for a program of research on the neural mechanisms of auditory discrimination was described and the results of a series of initial experiments were summarized (Neff and Diamond, 1958). More detailed accounts of some of the experiments have been given in other published reports (Neff, et al., 1956; Butler, Diamond, and Neff, 1957; Diamond and Neff, 1957; Diamond, Chow, and Neff, 1958).

The present paper brings up to date the summary of results obtained in studies of localization of sound in space and of discriminations of changes in frequency, pattern, and duration of tones. A comparison is presented of the findings for the cat and the monkey, and related clinical studies on man are noted. Finally, an attempt is made to construct a simple, neural model that will explain the differences between discriminations that can and cannot be made after ablation of auditory areas of the cerebral cortex.

Auditory Areas of Cerebral Cortex of the Cat

Evidence from anatomical and electrophysiological investigations has indicated that the areas shown in Fig. 1 may have auditory function. According to a number of different experimenters, electrophysiological changes may be elicited from areas AI, AII, the ventral part of Ep, SII, anterior SS, and parts of I-T.

Studies of retrograde degeneration in the thalamus after cortical ablation have shown that AI, AII, Ep, and I-T receive direct projection from the thalamic center of the primary auditory afferent system, the medial geniculate body (GM). The projection from GM to the cortex is not the same for all the subareas. As indicated in Fig. 2, AI receives dense projection from the anterior part of GM; ablation of AI results in severe degeneration in GM (Diamond and Neff, 1957;

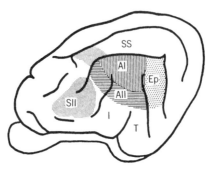

Fig. 1. Auditory areas of the cerebral cortex of the cat. AI, auditory area I; AII, auditory area II; Ep, posterior ectosylvian; I, insular; T, temporal; SII, somatic area II; SS, suprasylvian.

Rose and Woolsey, 1958). The insular-temporal region receives a much less dense projection from the posterior part of GM; ablations in I-T produce definite but more diffuse degeneration of cells in GM

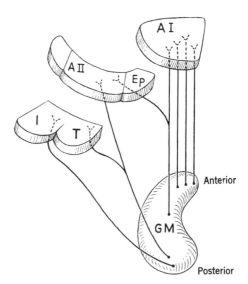

Fig. 2. Diagram illustrating projection of medial geniculate body (GM) upon different areas of auditory cortex.

(Diamond, Chow, and Neff, 1958). Auditory area II and the ventral part of Ep apparently receive only collateral projection from GM;

ablation of these areas does not result in noticeable degeneration in GM, but ablation of AII and Ep, in combination with AI, results in more widespread degeneration of GM than does ablation of AI alone; the region of degeneration is extended in a caudal direction (Diamond and Neff, 1957; Rose and Woolsey, 1958). Anatomical evidence also suggests that greater degeneration of the posterior part of GM is produced when ablation of I-T is extended to include parts of AII and Ep (Diamond, Chow, and Neff, 1958).

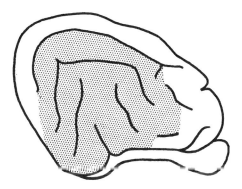

Fig. 3. Largest region of cortex ablated in animals tested for auditory discriminations.

At least two definitions of auditory cortex may be given: (1) all cortical areas from which electrical responses may be evoked by acoustic stimuli or by direct electrical stimulation of auditory nerve endings; or (2) those cortical areas, ablation of which causes total degeneration of the medial geniculate body. In most of our experiments, we have made ablations according to both definitions. The largest ablations we have made are shown by the stippled region in Fig. 3. All cortex defined by either (1) or (2) above is included in such an ablation.* In some cases, we have also limited our ablations to as little as one subarea of auditory cortex.

Behavioral Studies

Localization of sound in space

Cortical ablations and subcortical transection of main afferent pathways. A severe deficiency in ability to localize sound in space is

---

* Results reported at this symposium by Woolsey indicate that under certain conditions, for example, when chloralose is used, responses may be elicited in some other areas such as the lateral gyrus and sensorimotor areas.

shown by the cat after bilateral ablations of auditory areas AI, AII, and Ep (Neff, et al., 1956; Neff and Diamond, 1958). Unilateral ablation of these areas has little or no effect on localizing behavior. Experiments are now being done in our laboratory by N. Strominger in an effort to answer these further questions:

1. Can a deficit in localizing ability be produced by a large unilateral cortical ablation or by unilateral transection of the main afferent pathways at a subcortical level?

2. Is there a subarea of cortex or a total region that is particularly critical for localization?

The results at present available indicate that a small loss in ability to localize may occur after a large unilateral ablation that includes areas AI, AII, Ep, SII, I-T, and SS, as shown in Fig. 1. The capacity to localize is retained, but there may be an increase in the size of the smallest angle that can be discriminated. For example, one animal with a preoperative threshold of 3 degrees had a postoperative threshold of about 11 degrees. In other cases, little or no change has been noted. The differences between animals that show a deficit and those that do not may become clear when the brains are examined post mortem and the cortical damage is assessed.

An increase in the threshold angle that can be discriminated has also been found after unilateral transection of the brachium of the inferior colliculus (BIC). From tests made on three animals, it appears that a more severe deficit may occur after complete unilateral BIC transection than after the largest unilateral cortical ablations.

Almost complete loss of localizing ability is found after bilateral cortical ablations that include at least the whole of AI, AII, and Ep. Tests have been made on animals with: (1) bilateral ablation of AI, AII, Ep, SII, I-T, and SS; (2) bilateral ablation of AI, AII, Ep, and I-T, and (3) bilateral ablation of AI, AII, and Ep. In all cases, ability to make discriminations at an angle of 90 degrees was little better than chance.

Small increases in threshold angle discriminated have been found in animals with the following bilateral ablations: (1) AI; (2) AII, Ep, I-T, and SII; (3) AII and Ep; and (4) I-T. No loss was seen after bilateral ablation of anterior and middle suprasylvian gyrus. We were interested in this suprasylvian region because the experimental results of Lambroso and Merlis (1957) suggest that it might have to do with some kind of binaural interaction.

Transection of commissural pathways. G. C. Nauman (1958) has completed experiments in which the main commissural pathways (the corpus callosum, the commissure of the inferior colliculus, and the trapezoid body) were transected, singly or in combination. Transection of the corpus callosum, of the commissure of the colliculus, or of both has little or no effect on the ability of the cat to localize sound in space.

Complete transection of the trapezoid body is difficult to accomplish. In one cat, a complete or nearly complete transection was made; a threshold shift from 4 degrees preoperatively to 9 degrees occurred. Results were similar for a second animal whose trapezoid body was partially sectioned in a second operation after the corpus callosum had been transected: threshold before any operation, 4.0 degrees; after transection of corpus callosum, 4.5 degrees; after transection of corpus callosum and trapezoid body, 8.8 degrees.

From the results of Nauman's study, it may be concluded that for accurate localization of sound in space it is essential that the nerve impulses from the two ears interact at some center in the lower brain stem. When results of anatomical and electrophysiological studies are considered (Stotler, 1953; Galambos, Schwartzkopff, and Reipert, 1959), the medial superior olivary nucleus appears to be the critical center.

Frequency discrimination

Butler, Diamond, and Neff (1957; see also Neff and Diamond, 1958) reported that cats were able to make fine discriminations of frequency after complete bilateral ablations of cortical areas AI, AII, and Ep and in some cases after ablations that included not only these areas but somatic area SII as well. On the other hand, Meyer and Woolsey (1952; see also Rose and Woolsey, 1958) reported loss of ability to discriminate frequency changes in cats in which somatic area SII, in addition to AI, AII, and Ep, had been ablated. Later experiments by Goldberg, Diamond, and Neff (1958) and by Thompson (1958) have shown that, with the training and testing procedures described by Butler, Diamond, and Neff, cats can learn to discriminate small changes in frequency after bilateral ablation of AI, AII, Ep, SII, and cortex in the insular-temporal region, ventral to SII and Ep. Moreover, Goldberg, Diamond, and Neff (1958) have also found that frequency discrimination can be relearned after bilateral ablation of areas AI, AII, Ep, SII, I-T, and SS (see Fig. 4). Goldberg has also found that frequency discriminations can be relearned by animals with not only these large cortical ablations but ablation of the inferior colliculi as well.

Goldberg has also tested frequency discrimination in a series of nine animals in which the brachium of the inferior colliculus was sectioned bilaterally. The extent of the lesion was varied intentionally. The amount of postoperative impairment of the learned habit was clearly related to the completeness of transection of auditory pathways. Two animals were unable to perform above a chance level.

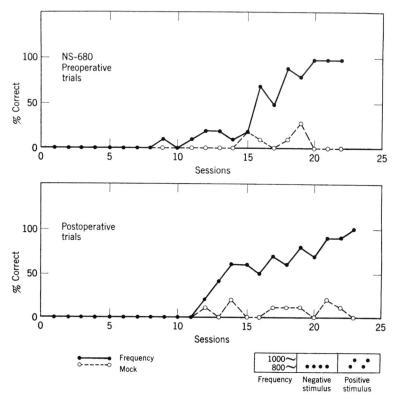

Fig. 4. Pre- and postoperative learning curves for a frequency discrimination. The region of cortex shown in Fig. 3 was ablated bilaterally in cat NS-680. Mock trials indicate level of performance resulting from spontaneous responses. They are trials during which no change is made in the stimulus; a correct response is scored if the animal responds during a period of time equal to that allowed for the positive stimulus in the frequency-discrimination test.

When evoked responses to click stimulation were recorded from the cortex at the time the animals were sacrificed, some responses were obtained in the seven animals that either relearned or performed better than chance postoperatively; none were found in the two animals that failed to relearn.

Intensity discrimination

Raab and Ades (1946) and Rosenzweig (1946) have reported that cats, with bilateral ablation of temporal cortex that includes auditory areas AI, AII, and Ep, can learn to discriminate changes in intensity of pure tones and that the difference limens are not changed. Raab and Ades further found that discriminations of intensity differences could be made after bilateral ablations of the inferior colliculi as well as AI, AII, and Ep; in these animals, the difference limens were increased (for example, approximately 10 decibels at 1000 cycles per second).

Oesterreich and Neff have repeated and extended the observations of Raab and Ades. We have shown that the cat can discriminate changes in intensity and that difference limens are little affected by bilateral ablation of a cortical region that includes AI, AII, Ep, SII, I-T, and anterior and middle SS (see Fig. 3). When this cortical region and the inferior colliculi are both ablated, intensity discriminations can still be made, but the difference threshold may be increased by 5 to 7 db. After bilateral section of the brachium of the inferior colliculus, intensity discriminations can also be made. The greatest threshold change that has been observed after brachium section was 10 db. No evoked responses could be recorded from the auditory cortex of this animal when it was tested at time of sacrifice.

Pattern discrimination

Cats with bilateral ablation of cortical areas that includes at least AI, AII, and Ep are unable to discriminate changes in temporal patterns of tones—for example, a change from low, high, low to high, low, high (Diamond and Neff, 1957). As might be expected, capacity to make pattern discriminations is also absent after bilateral section of the brachium of the inferior colliculus.

Goldberg, Diamond, and Neff (1957) have found that a severe deficit in discrimination of temporal patterns occurs after bilateral ablation of the insular-temporal region in the cat. The results for one animal are shown in Fig. 5. Cat G-536 lost the learned pattern discrimination habit after removal of insular-temporal cortex and was unable to relearn it to preoperative criterion in 40 sessions (10 trials per session). Its performance at times was above the chance level, but there was no consistent improvement. Other animals had even poorer postoperative records, performing at chance level; they were, however, able to learn a frequency discrimination.

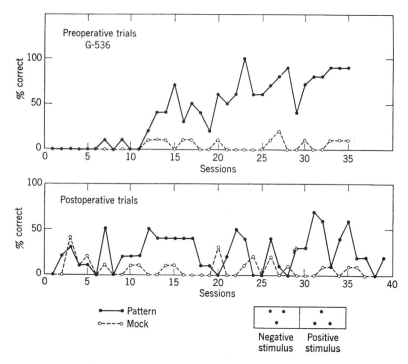

Fig. 5. Pre- and postoperative curves for a pattern discrimination. Insular-temporal cortex was ablated bilaterally in cat G-536.

Duration discrimination

For the reasons discussed in the section on A Proposed Neural Model, Scharlock and Neff tested the discrimination of changes in duration of a pure tone in the cat after bilateral ablations of auditory cortex. A severe deficit was found after removal of areas AI, AII, Ep, SII, and I-T. Removal of AI, AII, and Ep or removal of I-T produced loss of the learned discrimination; postoperative relearning required nearly as long as initial learning before the operation.

Control ablations

In a number of experiments it has been shown that impairment of ability to discriminate such auditory attributes as pattern, duration, and localization of sound in space is the result of damage to auditory areas of cortex, and that no deficit occurs after ablations of other areas of the cortex such as somatic areas SI and II, suprasylvian gyrus (Wegener, 1954), or suprasylvian gyrus plus temporal area ventral to Ep and entorrhinal cortex (Deatherage and Neff).

Comparative Studies

Anatomical studies

It was noted in the section on Auditory Areas of the Cerebral Cortex of the Cat that the region called insular-temporal cortex (Figs. 1 and and 2) receives direct projection from the posterior portion of the medial geniculate body.   In the monkey, the thalamocortical pro-

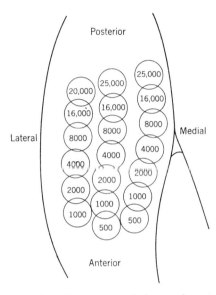

Fig. 6.   Tonotopic map of auditory cortex of the monkey.   This is a view from above of the superior surface of the superior temporal convolution of the right cerebral hemisphere.   The circles indicate regions of maximal sensitivity for the frequencies noted, the low frequencies in the anterior part of the area, the higher frequencies in more posterior parts.   (From Kennedy, 1955.)

jection of the auditory system appears to be similar to that for the cat.   The projection area as usually described lies in the posterior half of the superior surface of the superior temporal convolution (Poliak, 1932; Walker, 1938; Clark, 1936).   From this region large evoked potentials of short latency may be recorded when the ear is stimulated by clicks or by tonal pulses.   The tonotopic organization of this area has been described by Licklider and Kryter (1942), by Walzl (1947), and, in greater detail, by Kennedy (1955) (see Fig. 6).   Ablation of auditory cortex, so defined, results in retrograde degeneration of the anterior two-thirds or more of the medial geniculate body.   As in the cat, the posterior part of GM usually remains intact after ablation of the cortex which has commonly been called auditory projection area.

Bucy and Klüver (1955), in their report on the anatomy of the brain of one monkey from the series in which they observed behavioral changes after temporal lobectomy, noted that degeneration occurred in the posterior tip of GM, although the temporal-lobe ablation had extended very little, if at all, into the auditory projection area. They suggested that the ablation may have invaded projection fibers to the auditory area. Recently, Akert, et al. (1959) have looked for an area of cortex in the monkey which might, in terms of its anatomical connections with GM, be considered homologous to the I-T region in the cat. Such a region has been found. It lies on the superior temporal surface rostral to the classical auditory area. It is anterior to the ventral tip of the central sulcus. Ablation of this region produces severe degeneration in the posterior pole of GM.

Behavioral studies

In a preliminary study, Oder (1959) has tested monkeys on discriminations of auditory and visual patterns before and after ablation of (1) parietal-temporal-preoccipital cortex and (2) anterior temporal cortex. In some cases the anterior temporal ablations included part, at least, of the region described by Akert, et al., as receiving projection from the posterior pole of GM. Results for two animals of the group examined by Oder are shown in Figs. 7 and 8. It may be seen that monkey HO-9 with a rather extensive ablation of parietal-temporal-preoccipital areas retains the ability to make an auditory pattern discrimination after both unilateral and bilateral removal of cortex (Fig. 7, upper curves); it is unable to make a visual pattern discrimination after the second operation and fails to relearn in twice the number of trials required for initial learning. Since visual tests were made in a two-choice situation, chance level of response is 50 per cent (Fig. 7, lower curves). Monkey HO-11 with ablation of a small area of cortex on the superior surface of the superior temporal convolution, unlike monkey HO-9, fails to retain the auditory habit and relearns at about the initial rate; it retains the ability to make visual discriminations (Fig. 8).

Similar results have been found by other investigators. Konorski reports a deficit in auditory discrimination in the dog after insular-temporal ablations and Stepién, Cordeau, and Rasmussen have found, in the monkey, a loss in discrimination of auditory patterns with no change for visual patterns after removal of the anterior portion of the

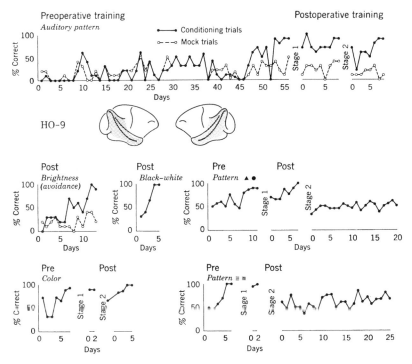

Fig. 7. Pre- and postoperative learning curves for monkey HO-9. The bilateral ablation of cortex, as shown in the diagram, included parietal, preoccipital, and temporal areas of the cortex. Surgery was done in two stages, cortex of one side being removed in each operation. The curves at the top of the figure show performance on an auditory discrimination; those at the bottom, on a series of visual problems.

superior temporal convolution, and deficit in visual discrimination with no change in auditory after ablations of infero-temporal cortex.

Clinical studies of man

The results of clinical studies of patients with temporal-lobe damage or removal cannot, in most cases, be compared readily with the results of animal studies, because auditory tests used in the clinic and in the animal laboratory are usually quite different. Tests of speech perception are often used in the clinic, and if an impairment in perception is found it cannot be easily ascertained whether it is due to inability to recognize the complex patterns of sound or inability to associate meanings with the sounds.

In a number of clinical investigations, ability to localize sound in

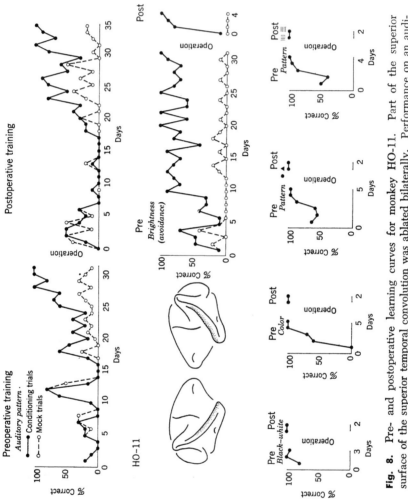

Fig. 8. Pre- and postoperative learning curves for monkey HO-11. Part of the superior surface of the superior temporal convolution was ablated bilaterally. Performance on an auditory discrimination is shown in the curves at the top of the figure. Performance on a series of visual problems is shown in the two lower sets of curves.

space or to perceive binaural time differences has been measured. The results reported are not in agreement. Two recent studies will serve as examples. Sanchez-Longo and Forster (1958) report a loss in localizing ability in patients with unilateral temporal-lobe lesions; they found a deficit in the accuracy of localizing sounds in the auditory field contralateral to the side of brain damage. Walsh (1957), on the other hand, reports that localization in the horizontal plane, at least for discriminations based on binaural time differences, is unaffected by unilateral temporal-lobe damage. Localization in the vertical plane may be disturbed. In view of the clear demonstration of loss in the localizing ability of animals with temporal-lobe damage, further studies on man are in order.

Other recent studies (for example, those by Bocca, 1958; Bocca, Calearo, and Cassinari, 1957; and de Sa, 1958) have reported impairment in the understanding of speech after temporal-lobe damage. Bocca used tests of filtered, interrupted, and accelerated speech and found that intelligibility scores were lower for the ear contralateral to the side of brain damage than for the ipsilateral ear. And de Sa reported that, compared to patients with cochlear damage, patients with damage to the central nervous system more often made mistakes in perceiving nasal sounds such as *an, en, in, on,* and *un.* In order that a direct comparison may be made with the results of animal studies, it is desirable that tests of temporal patterns of tones or other acoustic signals, excluding speech, be carried out with human patients. It seems likely that impairment in perception of complex sequences of nonspeech sounds may be seen.

A Proposed Neural Model

At the time when our experiments had shown that the cat suffers a complete loss of capacity to discriminate temporal patterns of tones after ablations of auditory cortex, but that it can relearn responses to onset of tones, change in intensity, and change in frequency, a first attempt was made to describe, in terms of events in the auditory afferent pathways and centers, the differences between those discriminations that can be made after cortical ablations and those that cannot.

In a discrimination that requires response to the onset of a tone, we can assume that there is no activity in the neural units of the afferent auditory system during the intervals between presentations of the positive stimulus.* In Fig. 9*a*, units *A* to *F* are at rest. When the positive stimulus (onset of a tone) is presented, impulses are aroused in neural units *A* and *B*.

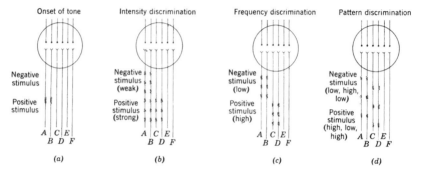

Fig. 9. Diagram of response to positive and negative stimuli.

In an intensity discrimination, based on the procedure of Raab and Ades (1946), Rosenzweig (1946), and Oesterreich and Neff, namely, a pulsing tone of medium intensity as neutral or negative stimulus and an increase in intensity of the tone as positive stimulus

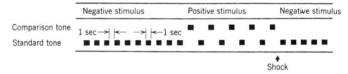

Fig. 10. Diagram to show sequence of tonal stimuli used in frequency, intensity, and duration discriminations. In frequency discrimination, the negative stimulus is a pulsing tone of a given frequency, say, 800 cps, and the positive stimulus is a pulsing tone of a different frequency, say, 1000 cps. The positive stimulus may consist of the new frequency alternated with the old, as shown in the diagram, or of the new frequency presented alone. For intensity discrimination, the comparison tone differs in intensity from the standard tone. For duration discrimination, the comparison tone is longer or shorter than the standard tone.

(see Figs. 10 and 11 for explanation of stimulus sequence), the neural events may be described by Fig. 9*b*. During the negative stimulus, neural units *A* and *B* are excited. When the stimulus intensity is increased, not only *A* and *B* are excited but also units *C* and *D*. From numerous studies of cochlear action and of electrically recorded events

---

* For the purposes of the present argument, "spontaneous" firing of impulses in some neural units will be disregarded.

in neurons of the auditory system, we know that when the intensity of an acoustic stimulus is increased there is more widespread activity in the cochlea and new neural units are caused to fire.

In a frequency discrimination, based on the test procedure of Butler, Diamond, and Neff (1957), the neural events may be pictured by Fig. 9c. The negative stimulus, a low tone, elicits nerve impulses in neural units A and B. The positive stimulus, a high tone, excites units C and D. Again we know from many experiments that a shift

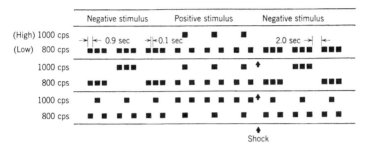

Fig. 11. Diagram showing sequences of tonal stimuli used in pattern discriminations. Pulsing tones are presented in groups of three. The sequence shown in the top line requires only a discrimination of frequency. The sequences of the two lower lines require discriminations of temporal patterns.

in the place of maximal response in the cochlea and a shift in the place of excitation in neural units of the auditory system occur when the frequency of an acoustic signal is changed.

In a pattern discrimination such as that used in the experiments of Diamond and Neff (1957), the sequence of neural events may be represented by diagram d of Fig. 9. The negative stimulus (low, high, low) gives rise to bursts of nerve impulses, first in units A and B, then in C and D, and again in A and B. The positive stimulus excites the same units but in the sequence C D, A B, C D.

From this analysis of events in a simplified model of the auditory afferent system, a difference between those discriminations (onset, intensity, and frequency) that can be made after ablation of auditory cortex and one (pattern) that cannot becomes apparent immediately. For each of the discriminations that can be made, the positive stimulus produces excitation in new neural units, that is to say, units that were not excited by the negative stimulus. In the pattern discrimination, on the other hand, the positive stimulus and the negative stimulus excite the same neural units, and the difference is in the order or sequence of excitation or, in the case illustrated in Fig. 9d, in the total amount of excitation produced in certain units; during a given time interval the positive stimulus (high, low, high) produces more excitation in units C and D than does the negative stimulus.

With this picture of neural events in mind, we next looked for another auditory discrimination in which, like pattern, the negative and positive stimuli would excite the same neural units, but in which the difference would be one simply of amount of stimulation. The obvious choice is discrimination of duration. When the stimulus is a pulsing tone, a learned response can be established in the experimental animal to a change in duration of the tone, with the frequency and intensity held constant. Figure 12 shows the sequence of neural events. It was predicted that discrimination of change in duration would, like discrimination of change in pattern, be affected by ablation of auditory cortex. As noted earlier in the section on Duration Discrimination, this prediction was upheld; a severe deficit was found after large bilateral ablations of auditory cortex.

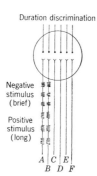

Duration discrimination

Negative stimulus (brief)

Positive stimulus (long)

$A \quad C \quad E$
$\quad B \quad D \quad F$

Fig. 12. Diagram of response to positive and negative stimuli: duration.

The hypothesis has been made and supported by experimental evidence that the critical difference between those auditory discriminations that can be made after ablation of auditory cortex and those that cannot is that, in the former, new neural units in the afferent pathways are excited by the change from negative to positive stimulus. A question that might be asked next is: what implications does this explanation have for the special functional features of the auditory areas of the cortex? The auditory cortex (for the present, the critical areas appear to be AI, AII, Ep, and I-T; ablation of these areas produces complete retrograde degeneration of the medial geniculate body) or, perhaps better, the thalamocortical section of the primary auditory system appears to be of particular importance in differentiating between two sets of neural events that differ only in their temporal patterning. In order to make such a differentiation, a short-term memory (between a fraction of a second and a few seconds) would seem to be essential.

A number of other questions may be asked about the neural model that has been presented and its implications. A few of them, which are most directly related to the experimental results summarized, will be considered briefly.

Are frequency, intensity, and other similar discriminations, which can be learned after destruction of the primary thalamocortical portion of the auditory system, handled by subcortical centers or by other cortical centers? Evidence upon which to base an answer is still far from adequate. In our studies on frequency and intensity discriminations after bilateral section of the brachium of the inferior colliculus, there is the suggestion that cortical responses may be evoked by acoustic signals after complete transection of BIC, and that in such cases frequency and intensity discriminations can still be learned. When the section of BIC is extended so as to eliminate all evoked cortical responses, there may be a complete loss of capacity to learn frequency discriminations but not intensity discriminations. These results, which must be considered as tentative, suggest that a thalamocortical circuit other than the primary afferent one (medial geniculate–auditory projection areas) is important for frequency discrimination, but that subcortical centers may subserve intensity discrimination. If additional evidence supports these findings, then our neural model can be extended to account for the difference between frequency and intensity discriminations. An obvious difference is that, with change in the intensity of tonal stimuli, there is a change in the total flow of nerve impulses. When intensity is increased, total flow is increased; when intensity is decreased, total flow is decreased. With change in the frequency of tonal stimuli, no such simple relation holds between the direction of change and the total flow of nerve impulses. It might be expected, therefore, that neural centers capable of serving for intensity discrimination may not be capable of serving for frequency discrimination, or that centers able to subserve frequency discrimination may not be able to subserve intensity discrimination.

Like pattern discrimination, the capacity to localize sound in space is lost or at least severely affected by ablation of the cortical projection areas of GM. Can this deficit in auditory function be accounted for by the neural model constructed to account for discriminations of intensity, frequency, duration, and pattern? There is certainly evidence that temporal events in the auditory nervous system are of primary importance in the localization of sound in space. It might be expected, therefore, that, as a center equipped to deal with temporal events, the auditory cortex would be essential for accurate discriminations involving sound localization. The orders of time involved in the pattern and duration discriminations described above, and in localization of sound in space, are quite different; for the former, seconds or relatively large fractions of a second, for the latter, microseconds. It may be an oversimplification, therefore, to postulate the same neural mechanisms for both.

A number of predictions suggested by the neural model described above have been mentioned elsewhere (Neff, in press). These would include the suggestion that the cortical projection areas of other sensory systems, for example, the visual and somesthetic, may, for discriminations in their particular modalities, play a role similar to the one the auditory cortex plays for hearing.

Summary

After bilateral ablation of those areas of the cerebral cortex that receive projection from the medial geniculate body (AI, AII, Ep, and I-T), the cat is able to learn to respond to the onset of tones, changes in intensity, and changes in frequency. It suffers severe impairment, or complete loss, of ability to make discriminations involving localization of sound in space, change in temporal patterns of tones, and change in duration of tones.

When somatic area SII and the suprasylvian gyrus as well as AI, AII, Ep, and I-T are ablated, responses to onset of tone, change in intensity, and change in frequency can still be learned.

After bilateral section of the auditory pathways at the level of the brachium of the inferior colliculus, so complete as to eliminate all evoked cortical responses to auditory signals in the anesthetized animal, there is some evidence suggesting that ability to learn frequency discriminations may be completely lost but that ability to learn intensity discriminations remains. Both frequency and intensity discriminations can be made after less extensive BIC transections, although these lesions appear, on post-mortem examination of stained sections, to destroy all the primary auditory fibers ascending to the medial geniculate bodies.

Insular-temporal cortex in the cat has been shown to receive projection from the posterior part of the medial geniculate body. Bilateral ablation of the insular-temporal region produces a severe deficit in capacity to discriminate changes in patterns of tones.

In the monkey, bilateral ablation of a region of cortex anatomically homologous to the insular-temporal region in the cat, in that it receives projection from the posterior end of the medial geniculate body, produces a behavioral deficit similar to that found for the cat, namely, impairment of pattern discrimination.

Auditory discriminations that can be made after bilateral ablations of the medial geniculate projection areas of the cortex differ from those that cannot be made, in that for the former the positive stimulus causes new neural units in the auditory afferent system to be excited. In discriminations that cannot be made after the cortical ablations, the same neural units are excited by both negative and positive stimuli; the difference between the two, in terms of neural activity, is in the sequence of excitation or in total amount of excitation in some units during a given time interval. The geniculocortical part of the auditory system appears to provide a mechanism for short term "memory" or storage.

Acknowledgments

The Laboratory of Physiological Psychology is supported by the Office of Naval Research, the Air Force Office of Scientific Research, and Air Force Cambridge Research Center, as well as by grants from the Sonics Research Foundation and the Wallace C. and Clara A. Abbott Memorial Fund of the University of Chicago.

References

Akert, K., C. N. Woolsey, I. T. Diamond, and W. D. Neff. The cortical projection area of the posterior pole of the medial geniculate body in Macaca mulatta. *Anat. Record*, 1959, **134**, 242.

Bocca, E. Clinical aspects of cortical deafness. *Laryngoscope*, 1958, **68**, 301–309.

Bocca, E., C. Calearo, and V. Cassinari. La surdité corticale. *Rev. Laryngol. Otol. Rhinol.*, 1957, **78**, 777–856.

Bucy, P. C., and H. Klüver. An anatomical investigation of the temporal lobe in the monkey (*Macaca mulatta*). *J. comp. Neurol.*, 1955, **103**, 151–252.

Butler, R. A., I. T. Diamond, and W. D. Neff. Role of auditory cortex in discrimination of changes in frequency. *J. Neurophysiol.*, 1957, **20**, 108–120.

Clark, W. E. Le Gros. The thalamic connections of the temporal lobe of the brain in the monkey. *J. Anat. Lond.*, 1936, **70**, 447–464.

Deatherage, B. H., and W. D. Neff. Unpublished experiments.

Diamond, I. T., K. L. Chow, and W. D. Neff. Degeneration of caudal medial geniculate body following cortical lesion ventral to auditory area II in cat. *J. comp. Neurol.*, 1958, **109**, 349–362.

Diamond, I. T., and W. D. Neff. Ablation of temporal cortex and discrimination of auditory patterns. *J. Neurophysiol.*, 1957, **20**, 300–315.

Galambos, R., J. Schwartzkopff, and A. Reipert. Microelectrode study of superior olivary nuclei. *Amer. J. Physiol.*, 1959, **197**, 527–536.

Goldberg, J. M. Unpublished experiments.

Goldberg, J. M., I. T. Diamond, and W. D. Neff. Auditory discrimination after ablation of temporal and insular cortex in cat. *Fed. Proc.*, 1957, **16**, 47.

Goldberg, J. M., I. T. Diamond, and W. D. Neff. Frequency discrimination after ablation of cortical projection areas of the auditory system. *Fed. Proc.*, 1958, **17**, 55.

Kennedy, T. K. An electrophysiological study of the auditory projection areas of the cortex in monkey (*Macaca mulatta*). Doctoral dissertation, University of Chicago, 1955.

Konorski, J. Personal communication.

Lambroso, C. T., and J. K. Merlis. Suprasylvian auditory responses in the cat. *EEG clin. Neurophysiol.*, 1957, **9**, 301–308.

Licklider, J. C. R., and K. D. Kryter. Frequency localization in the auditory cortex of the monkey. *Fed. Proc.*, 1942, **1**, 51.

Meyer, D. R., and C. N. Woolsey. Effects of localized cortical destruction on auditory discriminative conditioning in cat. *J. Neurophysiol.*, 1952, **15**, 149–162.

Nauman, G. C. Sound localization: the role of the commissural pathways of the auditory system of the cat. Doctoral dissertation, University of Chicago, 1958.

Neff, W. D. Role of the auditory cortex in sound discrimination. In G. L. Rasmussen and W. F. Windle (Editors), *Neural Mechanisms of the Auditory and Vestibular Systems.* Springfield, Ill.: C. C Thomas (in press).

Neff, W. D., and I. T. Diamond. The neural basis of auditory discrimination. In H. F. Harlow and C. N. Woolsey (Editors), *Biological and Biochemical Bases of Behavior.* Madison: Univ. of Wisconsin Press, 1958. Pp. 101–126.

Neff, W. D., J. F. Fisher, I. T. Diamond, and M. Yela. Role of auditory cortex in discrimination requiring localization of sound in space. *J. Neurophysiol.*, 1956, **19**, 500–512.

Oder, H. E. Functions of the temporal lobe in monkey (*Macaca mulatta*). Master's thesis, University of Chicago, 1959.

Oesterreich, R. E. Unpublished experiments.

Oesterreich, R. E., and W. D. Neff. Unpublished experiments.

Poliak, S. *The Main Afferent Fiber Systems of the Cerebral Cortex in Primates.* Berkeley: Univ. of California Press, 1932.

Raab, D. H., and H. W. Ades. Cortical and midbrain mediation of a conditioned discrimination of acoustic intensities. *Amer. J. Psychol.*, 1946, **59**, 59–63.

Rose, J. E., and C. N. Woolsey. Cortical connections and functional organization of the thalamic auditory system of the cat. In H. F. Harlow and C. N. Woolsey (Editors), *Biological and Biochemical Bases of Behavior.* Madison: Univ. of Wisconsin Press, 1958. Pp. 127–150.

Rosenzweig, M. R. Discrimination of auditory intensities in the cat. *Amer. J. Psychol.*, 1946, **59**, 127–136.

de Sa, G. Audiologic findings in central nerve deafness. *Laryngoscope*, 1958, **68**, 309–317

Sanchez-Longo, L. P., and F. M. Forster. Clinical significance of impairment of sound localization. *Neurology*, 1958, **8**, 119–125.

Scharlock, D. P., and W. D. Neff. Unpublished experiments.

Stepién, L. H., P. Cordeau, and T. Rasmussen. Personal communication.

Stotler, W. A. An experimental study of the cells and connections of the superior olivary complex of the cat. *J. comp. Neurol.*, 1953, **98**, 401–432.

Thompson, R. F. Function of auditory cortex of cat in frequency discrimination. *Fed. Proc.*, 1958, **17**, 163.

Walker, A. E. *The Primate Thalamus.* Chicago: Univ. of Chicago Press, 1938.

Walsh, E. G. An investigation of sound localization in patients with neurological abnormalities. *Brain*, 1957, **80**, 222–250.

Walzl, E. M. Representation of the cochlea in the cerebral cortex. *Laryngoscope*, 1947, **57**, 778–787.

Wegener, J. G. The role of special cerebral and cerebellar systems in sensory discrimination. Doctoral dissertation, University of Chicago, 1954.

16

JERZY E. ROSE, LEONARD I. MALIS, and CHARLES P. BAKER
Departments of Physiology and Psychiatry, The Johns Hopkins
University School of Medicine;
Mount Sinai Hospital, New York City; and
Departments of Medicine and Physics, Brookhaven National Laboratory

Neural Growth in the Cerebral Cortex after Lesions Produced by Monoenergetic Deuterons

I wish to present some of our results pertaining to laminar lesions caused by monoenergetic, heavy particles in the cerebral cortex of the rabbit. It is readily possible to produce, by these means, sharply delimited zones of destruction completely devoid of nerve cells, which display only moderate numbers of glia cells after long survival periods. A full account of this work will be given elsewhere (Malis, et al.; Rose, et al.). Here I shall present only some of these data which show that, if a laminar lesion is produced, there soon appear nerve fibers within the zone that is completely deprived of nerve cells. Most if not all of these fibers must be interpreted as newly grown. With time the fibrilloarchitectonic pattern of the cortex displaying a laminar lesion may resemble rather closely a normal picture. I shall discuss the implications of these findings, for a possibility exists that the growth observed under experimental conditions may actually be but an expression of a normal continuous growth of neurons in the central nervous system.

The radiation beam used in all the experiments was the beam produced by the 60-inch cyclotron in the Brookhaven National Laboratory. This beam consists of reasonably monoenergetic 20-mev deuterons, or 10-mev protons, or 40-mev alpha particles. Since 20-mev deuterons penetrate about 2.5 millimeters deep into the brain and since the ranges of the available protons and alpha particles are very nearly half the mentioned value, deuterons were usually employed, and all the lesions to be shown were produced by deuterons.

I shall not discuss here any details of the irradiation technique. However, the way heavy particles ionize may be mentioned briefly, for it is the ionizations that are responsible for the destruction of the tissue, and it is the differential in the number of ions produced by the beam along the different sectors of its path that makes a laminar lesion possible.

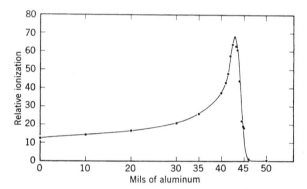

Fig. 1.   Relative ionizations produced by the beam of deuterons in a thin ionization chamber after passage through aluminum foil of varying thickness. The energy of the deuterons in the beam is close to 20 mev. Ordinate: amount of ionization current in arbitrary units per constant number of incident deuterons. Abscissa: thickness of aluminum foil in mils.

Figure 1 illustrates a typical Bragg ionization curve when 20-mev deuterons, as produced by the cyclotron used, penetrate aluminum foil. In the actual experimental set-up the beam passed through a testing ion chamber, then through an aluminum foil of stated thickness, and finally through a recording ion chamber. For each point on the curve the quantity of the ion-chamber current in the testing chamber (and hence the number of deuterons per square centimeter) is the same. The corresponding quantity of the ion-chamber current in the thin recording chamber (and thus the number of ions produced) varies as shown when aluminum foil of a stated thickness is placed in the beam path between the two ion chambers.

It is apparent that the number of ions produced per unit length of path rises initially quite slowly and then increases abruptly over a very short distance by a factor of about 5 near the end of the path of the particles. If then the radiation dose is so chosen that only ionization levels near the end of the path are destructive to the tissues, it should be possible to produce narrow lesions in depth without substantial damage to the cortical layers above the lesion. It may be added that 1-mil (25.4-micron) aluminum foil is equivalent in its stopping power to somewhat more than 50 microns of the cortical

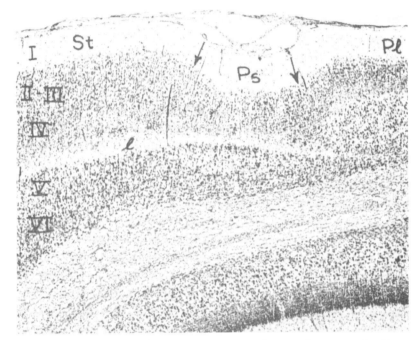

Fig. 2. Laminar lesion (l) produced in the zone of maximal ionization by 20-mev deuterons. The zone of destruction lies mainly in the fifth cortical layer of the striate and peristriate field. Rabbit 78 RP, 65 days after irradiation. Note that there are no nerve cells in the zone of maximal ionization and that there is a slight cell loss just above this zone. Layers I to III appear substantially intact. Notice further that the glial content of the laminar lesion is about the same as that of the first cortical layer. The laminar lesion is narrower in the peristriate than in the striate field, owing to scatter of the beam near the edge of the irradiated region. Peak dose, 33,000 rads. Pl, posterior limbic region; Ps, peristriate area; St, striate area. I, first cortical layer or zonal lamina; II to VI, cellular cortical layers. Thionin stain, 30×.

tissue. Thus it is possible to shorten the range of particles in the tissue and elevate the zone of maximal ionization by introducing into the beam path absorbers of suitable thickness.

Figure 2 illustrates a typical laminar lesion in the cerebral cortex of a rabbit. The peak dose, that is, the dose delivered at the height of ionization, was 33,000 rads. It may be mentioned that the dose delivered at the surface of the cortex (and which remained biologically ineffective) was here 7000 rads; the dose averaged over the entire depth of tissue that was penetrated by the beam was 12,000 rads. These doses resulted from a bombardment by 8.8×10^9 deuterons/cm^2. For all other illustrations only the peak dose in rads will be given, since it is this dose that is obviously most pertinent for the evaluation of the laminar destructions.

It will be noted that the laminar lesion is sharply delimited from its surroundings. No nerve cells at all survived within the zone of maximal ionization, and the number of glia cells present is but moderate. It is a striking fact that the glia cell content of the older laminar lesions is essentially the same as that of the first cortical layer, which is normally almost devoid of nerve cells.

Laminar lesions such as those shown in Figure 2 result from peak doses of about 15,000 to 35,000 rads. The peak dose may be increased to about 45,000 rads, and the outcome may still be only a heavy laminar destruction. In general, the higher the dose, the broader is the laminar lesion, at least up to a certain point. Peak doses higher than about 45,000 rads tend to produce necrotic foci, and if the dose is increased still further, a typical radiation necrosis of the entire irradiated region results (Malis, et al.).

A laminar lesion presents a quiescent histological picture if it is examined some months after irradiation. In the early stages, however, there is a marked vascular reaction even with mild radiation doses, and this reaction is violent if the dose is high. Moreover, a very marked early glia reaction is present for some weeks after exposure. These facts must be kept in mind if one examines laminar lesions at different times after irradiation.

If one irradiates with peak doses that are nearly maximal for the production of heavy laminar lesions and studies the irradiated brains at various time intervals, remarkable histological findings present themselves.

Figure 3 shows a cell preparation of a very heavy laminar lesion 14 days after irradiation by a peak dose of 40,000 rads. The reaction of the tissue is violent. Most of the nerve cells have already disappeared in the zone of maximal ionization (l). There is a heavy vascular reaction, and dense nests of glial cells are apparent. The limits of the laminar lesion do not stand out as sharply from the surroundings as they do in older lesions. Since the radiation dose is high, there are marked signs of damage immediately above the zone of maximal ionization. However, most nerve cells above the lesion are preserved.

Fig. 3. Very heavy laminar lesion 14 days after irradiation with a peak dose of 40,000 rads. Normal cortical sector (to the right of arrow) stands out sharply from the irradiated region to the left. Note the heavily infiltrated zone (*l*) of total nerve-cell destruction, the violent vascular reaction in the irradiated field, and the nests of glia cells in the zone of maximal ionization. Observe, however, that most of the nerve cells above the zone of destruction are preserved. Rabbit 241 L, striate cortex. Thionin preparation, 50 ×.

The adjacent section stained for fibers (Figs. 4 and 5) reveals that fibers were damaged still more severely. The site of the laminar lesion (*l*) is clearly devoid of any normal fibers. In fact, the fibrillar pattern in the entire irradiated field above the zone of maximal ionization appears virtually eliminated as well.

Figure 6 shows a very heavy laminar lesion 49 days after irradiation. The animal was irradiated on the same day and in the same experimental run as was the rabbit whose lesion is illustrated by Figs. 3 to 5. The peak dose here was 39,000 rads. The cell picture is consistent with the presumption that radiation doses for lesions shown in Figs. 3 and 6 were about the same, since the differences in histological appearance are certainly due to the different survival times. Most of the acute reactions seen in Fig. 3 subsided. The laminar lesion (*l*)

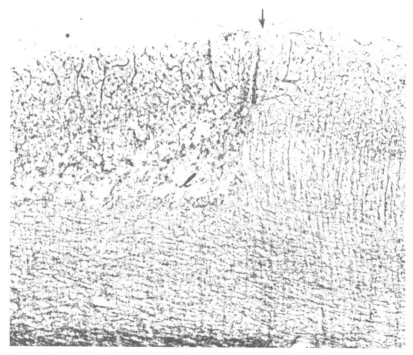

Fig. 4.   Section adjacent to that shown in Fig. 3.   Schultze's silver stain for fibers.   Normal cortical sector is to the right of arrow.   Note that there are no normal fibers stained in the zone of maximal ionization (l) and that the fibrillo-architectonic pattern over the entire irradiated field is not recognizable.   Note the vascular reaction in the irradiated field, $60 \times$.   (Since the sections treated with silver stains shrink much more than do thionin preparations, they are shown under 20 per cent higher linear magnification.)

stands out as a broad band.   The width of the lesion, the marked destructions above the zone of maximal ionization, and the accumulation of glia cells at the edge of the laminar lesion all testify to the severity of the dose.

The adjacent fiber preparation (Figs. 7 and 8) is of great interest. It is immediately apparent even under low magnification that the zone of maximal ionization (l) is filled with a network of nerve fibers that run predominantly in the horizontal direction and follow the undulations of the laminar lesion itself, forming an abnormal, rather dense striation.   Even though most of these fibers are oriented along the long axis of the lesion, some run more-or-less perpendicular to the lesion and ramify within it.   Many fibers clearly enter the lesion from below.

Some conclusions can be drawn immediately. First of all, it appears that the fibers seen in the zone of maximal ionization cannot be simply fibers that were normally present in this zone and somehow escaped destruction. This cannot be so because these fibers run in an abnormal direction and form an abnormal striation not present in the normal

Fig. 5. Edge of the radiation lesion shown in Fig. 4 under magnification of $150 \times$.

cortical sectors. They appear, moreover, in the zone that received the highest radiation dose of all the zones of the irradiated cortical field, but which is topographically closest to the normal cortical sectors. The obvious implication is that these fibers must be, at least predominantly, *newly grown fibers* which grow into the zone of maximal ionization from the adjoining normal sectors. This implication is harmonious with the finding that no fibers at all were seen in the zone of maximal ionization two weeks after irradiation with an almost identical dose and also with the fact that the fibrillar pattern above the laminar lesion still appears grossly abnormal.

Figure 9 shows a cell preparation in rabbit 242 RA, 78 days after irradiation with a peak dose of 37,000 rads. The animal belongs to

Fig. 6. Very heavy laminar lesion 49 days after irradiation with a peak dose of 39,000 rads. The laminar lesion (l) stands out as a broad band. Observe the marked cellular changes for some distance above the zone of maximal ionization. Normal cortical sector to the right of arrow. Note especially the irregular edge of the laminar lesion which sometimes occurs as a consequence, of shielding. Rabbit 239 L, striate cortex, thionin preparation, 50×.

the same series as rabbits 239 and 241. Despite a slightly smaller peak dose, the laminar lesion is very heavy and is in all essentials similar to that shown in Fig. 6. Again the laminar lesion is broad, there is considerable destruction for some distance above the zone of maximal ionization, and nests of glial cells are apparent in the laminar lesion.

Figures 10 and 11 show the adjacent fiber preparation. The laminar lesion (l) is not strikingly apparent at the first glance. A closer inspection reveals that the zone devoid of nerve cells is everywhere filled with nerve fibers, and that the fibrillar pattern of the entire irradiated region has been restored to a considerable extent.

The lesions shown are all lesions that lead only to a laminar destruction, however heavy such a laminar destruction may have been. It should be emphasized that if a radiation dose is so intense that a

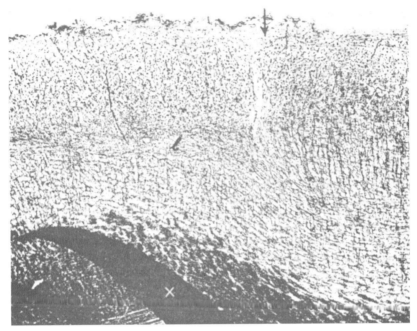

Fig. 7. Section adjacent to that shown in Fig. 6. Schultze's silver stain for fibers. Normal cortical sector to the right of the arrow. Note the dense network of fibers which form an abnormal striation in the zone of maximal ionization (l). Observe that the abnormal striation is coincident with the limits of this zone. X, technical artifact. 60 ×.

necrotic focus results, which leads to a cavitation or a glial scar, no such growth is observed. On the other hand, if the laminar lesion is light or moderate, abundant numbers of fibers are seen in the laminar lesion very much earlier than is the case when the radiation dose is larger. Once fibers appear in the laminar lesion they persist there, as far as we know, indefinitely; they can certainly persist up to about eighteen months after irradiation, which is thus far the longest time any of the animals examined have survived.

Figure 12 shows a laminar lesion in rabbit 239 R, 49 days after irradiation with a peak dose of 28,000 rads. The lesion was placed in the sixth cortical layer of the postcentral region in the right hemisphere. The rabbit is the same animal whose lesion on the left side in the striate cortex is shown in Figs. 6 to 8. The laminar lesion is, of course, markedly milder here than the one in Fig. 6. The adjacent fiber section (Figs. 13 and 14) indicates that the laminar lesion is traversed by fibers, and that the fibrillar pattern of the cortex is actu-

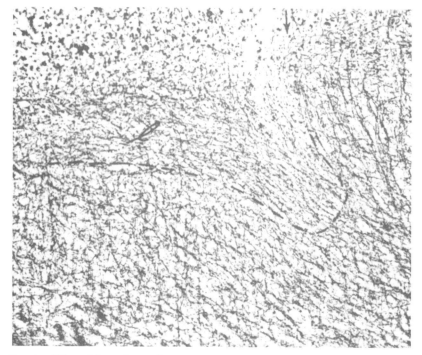

Fig. 8.   Edge of the radiation lesion in Fig. 7 shown under magnification of 150 ×. The approximate ventral limits of the zone of maximal ionization that corresponds to the laminar lesion in Fig. 6 is indicated by bars.   Normal cortical sector to right of arrow.

ally surprisingly close to the norm.   A horizontal striation and a denser fiber network are, however, apparent in the laminar lesion and the orderly radial arrangement of the fiber bundles is clearly disturbed by the lesion.   Nevertheless, there is no doubt that fibers in abundant numbers cross the zone of nerve-cell destruction, and that a remarkable restitution toward normal organization has indeed taken place.

Figure 15 shows a moderate laminar lesion 204 days after irradiation. The peak dose here was 27,000 rads.   Figures 16 and 17 show the adjoining section stained for fibers.   The fiber content of the laminar lesion and the general fibrillar appearance of the cortical field are substantially the same as those illustrated in Figs. 13 and 14.   It would appear then that, with a moderate dose of radiation, at least a temporary morphological end point has already been achieved at the end of seven weeks, and in this case the result was maintained without noticeable changes for seven months.

For all the fiber sections shown, Schultze's silver method was em-

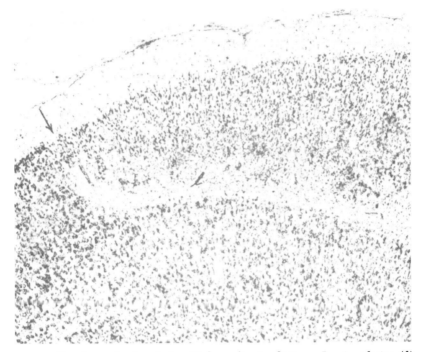

Fig. 9. Very heavy laminar lesion 78 days after irradiation. Laminar lesion (*l*) is broad. Marked destruction above the lesion is evident. Normal cortical sector is to the left of the arrow. The animal was irradiated in the same experimental runs as were rabbits 239 and 241. Rabbit 242 RA, postcentral region. Peak dose, 37,000 rads. Thionin stain, 50 ×.

ployed. This technique was chosen because it permitted us to stain dependably the cortical network of fibers in thick (50 to 60 microns) frozen sections. The defect of the technique is that the silver precipitate is rather crude, and for that reason the sections are not well suited for studies under high resolution. However, studies under higher power with other silver techniques yield results identical to those shown, as we have illustrated elsewhere (Rose, et al.). Here I shall mention briefly only two other observations of interest before I consider the implications of these findings.

If one concedes that nerve fibers in large numbers actually grow from the normal sectors into the zone of maximal ionization, some rather spectacular findings are easily interpreted.

Figure 18 illustrates one such observation. The animal was irradiated with a dose chosen to destroy the superficial cortical layers. The radiation dose was high, and a necrotic focus is present over some distance in the irradiated region. Over a considerable portion of the irradiated field, however, the cortex consists essentially of only the sixth and part of the fifth cortical layers. The remarkable fact is that adjoining the fifth layer dorsally there is a viable strip of tissue, com-

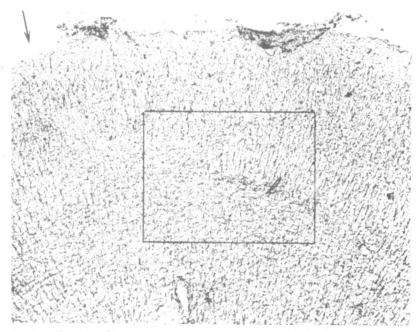

Fig. 10. Section adjacent to that shown in Fig. 9. Schultze's silver stain for fibers. The laminar lesion (*l*) is everywhere filled with nerve fibers. The fiber network is in places heavier than normal. Note that the fibrillar pattern above the lesion approaches the normal pattern in the right half of the field where the laminar lesion lies deep within the cortex and the number of surviving nerve cells above the lesion is large. 60 ×.

pletely devoid of nerve cells, which is quite obviously the region in which maximal ionization took place, and which with milder radiation doses is the site of the laminar lesion. In other words, the site that received a very much higher dose (by a factor of 2 or more) than did the cortical layers above it displays a viable tissue, whereas the layers that received a much smaller dose underwent a complete necrosis. If fibers grow into the zone of maximal ionization, such a finding is, of course, only to be expected. It would be quite puzzling otherwise. It can be shown that nerve fibers are actually present in the cellfree strip (Rose, et al.) and thus take part in creation of an "artificial zonal lamina."

The correspondence of an artificial zonal lamina or of any laminar lesion in general to the first cortical layer does not end with the recognition that both contain nerve fibers. For older laminar lesions regularly contain also numerous dendrites emitted by cells lying just below

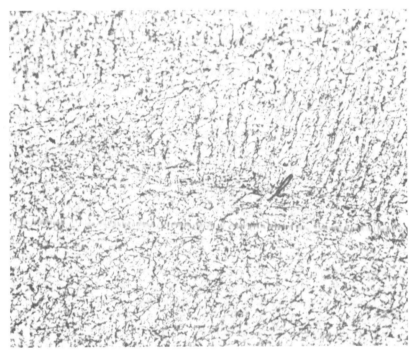

Fig. 11. The sector enclosed by rectangle in Fig. 10 shown under magnification of 150×.

the lesion itself. The presence of dendrites in a laminar lesion can be easily demonstrated in silver preparations, but they are seen equally clearly in suitable thionin sections.

Figure 19 shows one such lesion. The laminar lesion here is located in the fifth cortical layer, and apical dendrites of the large cells are stained for a considerable distance. A closer scrutiny of the laminar lesion even under low magnification discloses the presence of a large number of apical dendrites which arise from the cells just below the lesion. Figure 20 illustrates these facts under higher magnification. The apical dendrites in the lesion are oriented in the usual radial direction. In contrast to the evidence in regard to the nerve fibers, there is no compelling reason to assume that the dendritic processes seen in the laminar lesion must be new sprouts, although it is likely that such an interpretation is correct.

Let me summarize our results to date. Our data, based on the study of nearly 400 radiation lesions, imply that when a laminar lesion is produced nerve fibers in large numbers are present after some time in the acellular zone. These fibers appear sooner if the radiation dose

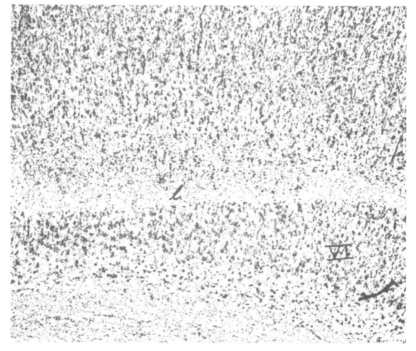

Fig. 12. Laminar lesion (l) in the sixth cortical layer of the postcentral field 49 days after irradiation with a peak dose of 28,000 rads. Rabbit 239 R, thionin stain, 50×.

was relatively mild, later if it was high. The findings apply to every appropriately stained section in every lesion of the material, provided the radiation dose led only to laminar destruction.

It seems obligatory to conclude that most (or all) of these fibers must be new sprouts. This must be concluded because the fibers in the laminar lesion run, initially at least, predominantly in an abnormal direction following the long axis of the zone of maximal ionization, thus forming a conspicuous abnormal striation. Furthermore, they form, for many months at least, dense and irregular networks in the zone of maximal ionization that are foreign to the normal pattern. In addition, if the radiation dose is relatively high, the first normally staining fibers to appear in the irradiated region are those in the zone of maximal ionization; if an artificial zonal lamina is formed, the *only* fibers to appear in the irradiated field are those in this zone, which is the zone of maximal radiation damage. This must be so if the fibers

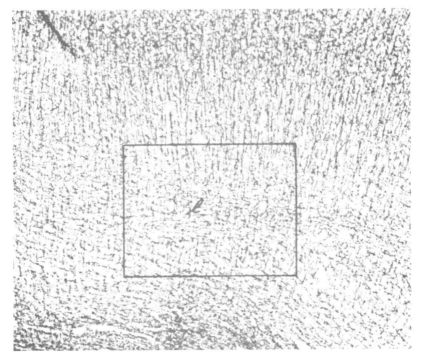

Fig. 13. Section adjacent to the one shown in Fig. 12 in Schultze's silver stain for fibers. Laminar lesion (*l*) stands out rather indistinctly from the surroundings. Heavier fiber network and horizontal striation in the lesion are apparent, however. Note that the laminar lesion interrupts the orderly arrangement of radial fibers in the field. The general fibrillar appearance of the cortical field is remarkably close to the norm. 60 ×.

grow, since it is the zone of maximal ionization that always adjoins the nonirradiated cortical sectors.

It is apparent from these remarks that the fibers would have to grow into the laminar lesion, even if one could assume that the original, normal fiber content there remained undisturbed by radiation. There is, however, no reason at all to make such an assumption, for with radiation doses that lead to a heavy laminar lesion the original fiber pattern can be shown to undergo complete destruction in the

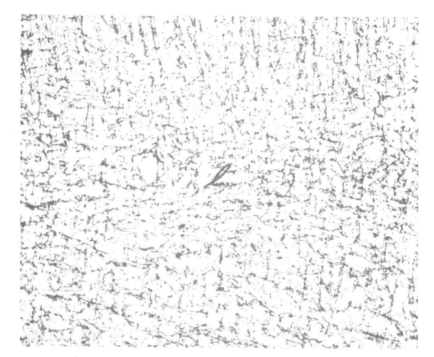

Fig. 14. The sector enclosed by a rectangle in Fig. 13 shown under magnification of 150×.

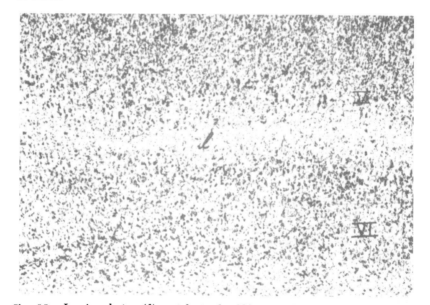

Fig. 15. Laminar lesion (l) mainly in the fifth cortical layer of the postcentral region 204 days after irradiation with a peak dose of 27,000 rads. Rabbit 206 L; thionin stain, 50×.

zone of maximal ionization. The available evidence is also conclusive that, even with a mild radiation dose, most if not all of the normal fibers in this zone are destroyed as well (Rose, et al.).

If then one concludes that all or most of the fibers seen in the laminar lesion must of necessity be new sprouts, one can consider the question, how to interpret these findings. One interpretation could,

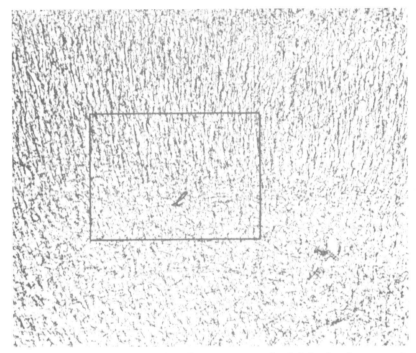

Fig. 16. Section adjacent to that shown in Fig. 15. Schultze's silver stain for fibers. Note that the laminar lesion (l) is almost identical in appearance to the lesion in Fig. 13. Note the elegant fibrilloarchitectonic pattern of the field despite the lesion. 60 ×.

of course, be that under the conditions of our experiments a massive regeneration of central neurons is actually a demonstrable phenomenon. If this were so, the regeneration observed would differ in two important respects from the usual findings after surgical lesions. First, the growth is truly of massive character, and second, it may lead to a reconstruction of the fiber pattern in the cortical field which can be considered a reasonable facsimile of the original picture. I shall not dwell here on the problem of regeneration since we consider it elsewhere (Rose, et al.), and since I wish to discuss here an alternative interpretation which, though it does not preclude the regeneration of central neurons, makes such an assumption superfluous. The interpretation is that the growth observed is not due to regeneration (if by regeneration is meant a growth that occurs only after injury has been inflicted) but is actually the result of normal, continuous growth

Fig. 17. The sector enclosed by a rectangle in Fig. 16 shown under magnification of 150×.

of the fibers that is still possible after an appropriate radiation dose and can be observed, since all the fibers in the zone of maximal ionization have been clipped, as it were, synchronously over a very short distance.

In order to make such an interpretation acceptable, it is mandatory to assume that the growth of the central neurons takes place only at the terminal or paraterminal region of the axon, since only such an assumption is concordant with the massive growth present with laminar lesions and the scarcity or absence of growth with lesions produced by necrotizing radiation doses or by surgical means.

One could assume then that all central neurons normally have a capacity for continuous growth in the paraterminal region of every axonic collateral. It would also seem reasonable to believe that the activity of the neuron itself is a factor in modulating the rate and extent of the growth. It follows from these assumptions that all synaptic endings have to be metabolized, and, therefore, that any given ending

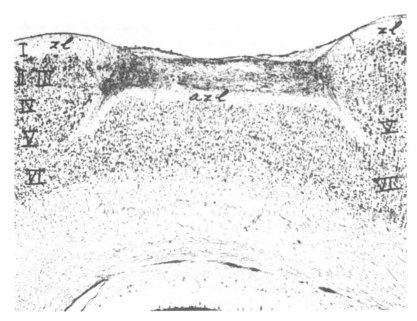

Fig. 18. Destruction of upper layers in the striate field of rabbit 275 L, 122 days after irradiation. Only the sixth and part of the fifth layers are preserved. A viable zone free of nerve cells forms the free surface. Creation of an "artificial zonal lamina" (*azl*). Peak dose not certain, probably above 50,000 rads. *zl*, first cortical layer or zonal lamina. II to VI, cellular cortical layers.

could be expected—in contrast to the classical views—to be not a permanent fixture given for life but only an evanescent morphological structure.

As you all know, the concept of continuous growth of nerve fibers has recently been formulated by Weiss and Hiscoe (1948) on the basis of their experiments on the *peripheral* nerves. It certainly seems possible to interpret some older experimental work and some histochemical observations, of which but a few may be cited here (Cook and Gerard, 1931; Hydén, 1943, 1950; Parker and Paine, 1934), as supporting or compatible with this view, and Gerard (1950), Sanders (1948), and Young (1945) all endorse this formulation. Clearly, if

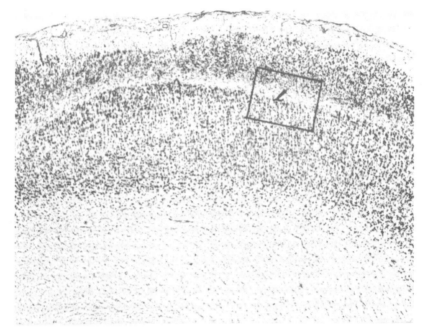

Fig. 19. Laminar lesion (l) in the fifth layer of the postcentral region in rabbit 67 LA, 132 days after irradiation with a peak dose of 48,000 rads. Closer inspection reveals that numerous dendrites enter the laminar lesion. Thionin stain, 30×.

it is true that continuous growth occurs in the peripheral fibers, it would be only reasonable to expect that the same holds true for the central axons. As far as the central neurons are concerned, there seems to be no anatomical objection to the acceptance of the proposition of continuous growth, once it is conceded that central neurons can be shown to grow at all in large numbers. I shall not argue here the merit of the concept but may mention that the evidence concerning the sprouting of the intact dorsal root axons (Liu and Chambers, 1958; McCouch, et al., 1958) and the retrograde and transneuronal degeneration would fit such a concept quite well. It appears, moreover, that the often contradictory and puzzling observations concerning the behavior of nerve fibers after surgical transections in the central nervous system could be harmoniously interpreted if a continuous growth of central neurons in their paraterminal region were indeed a reality (Rose, et al.).

It is not my purpose here to suggest that a continuous growth of central axons necessarily occurs under physiological conditions. The

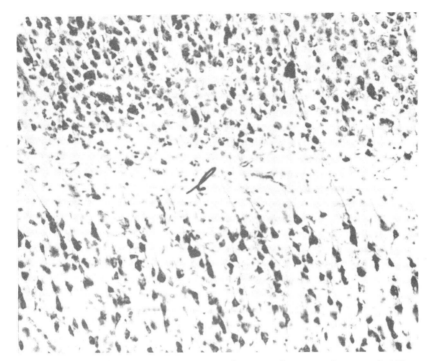

Fig. 20. The enclosed area in Fig. 19 shown under magnification of 200×. Notice the apical dendrites entering the laminar lesion.

point to be made is merely that, under the conditions of our experiments, a massive growth of nerve fibers can actually be shown to take place in the cerebral cortex and that the available evidence appears *compatible* with the presumption that this growth may in fact be due to a physiological perpetual growth of central neurons.

Summary

It is possible to produce, by means of monoenergetic deuterons, narrow lesions within the cerebral cortex in which all nerve cells are destroyed. Such lesions, which are termed laminar lesions, can be restricted to a single cortical layer and can be placed at any desired depth within the cortex.

If a laminar lesion is produced, the zone of total destruction of nerve cells sooner or later contains nerve fibers in great numbers. The available evidence appears conclusive that many or all nerve fibers in the laminar lesion *must be new sprouts* which invade the zone of maximal radiation damage from the adjacent normal sectors. The growth is abundant and the end result is a remarkable reconstitution of the fibrillar pattern in the affected cortical field. Apical dendrites of cells that lie just below the zone of maximal radiation damage are also regularly found in older laminar lesions. It appears probable that a nerve cell may regrow or reconstitute its apical dendrite that has been destroyed or damaged by radiation.

The implications of the findings are discussed. It seems possible that massive growth of nerve fibers into and across the laminar lesion may represent a *regeneration phenomenon*. It seems more likely, however, that the growth is actually due to a *normal, continuous growth of central neurons*.

Acknowledgment

This research was supported by Contract No. AT(30-1)-2033 with the United States Atomic Energy Commission.

References

Cook, D. D., and R. W. Gerard. The effect of stimulation on the degeneration of severed peripheral nerve. *Amer. J. Physiol.*, 1931, **97,** 412–425.

Gerard, R. W. Some aspects of neural growth, regeneration, and function. In P. Weiss (Editor), *Genetic Neurology.* Chicago: University of Chicago Press, 1950. Pp. 199–207.

Hydén, H. Protein metabolism in the nerve cell during growth and function. *Acta physiol. scand.*, 1943, **6,** Suppl. 17, 5–136.

Hydén, H. Spectroscopic studies on nerve cells in development, growth and function. In P. Weiss (Editor), *Genetic Neurology.* Chicago: University of Chicago Press, 1950. Pp. 177–193.

Liu, C. N., and W. W. Chambers. Intraspinal sprouting of dorsal root axons. *Arch. Neurol. Psychiat.*, 1958, **79,** 46–61.

Malis, L. I., C. P. Baker, L. Kruger, and J. E. Rose. Effects of heavy, ionizing, monoenergetic particles on the cerebral cortex. I. Production of laminar lesions and dosimetric considerations. *J. comp. Neurol.* (in press).

McCouch, G. P., G. M. Austin, C. N. Liu, and C. Y. Liu. Sprouting as a cause of spasticity. *J. Neurophysiol.*, 1958, **21,** 205–216.

Parker, G. H., and V. L. Paine. Progressive nerve degeneration and its rate in the lateral-line nerve of the catfish. *Amer. J. Anat.*, 1934, **54,** 1–25.

Rose, J. E., L. I. Malis, L. Kruger, and C. P. Baker. Effects of heavy, ionizing, monoenergetic particles on the cerebral cortex. II. Histological appearance of laminar lesions and growth of nerve fibers after laminar destructions. *J. comp. Neurol.* (in press).

Sanders, F. K.   The thickness of the myelin sheath of normal and regenerating peripheral fibers. *Proc. roy. Soc.*, 1948, **B135**, 323–357.

Weiss, P., and H. B. Hiscoe.   Experiments on the mechanism of nerve growth. *J. exp. Zool.*, 1948, **107**, 315–396.

Young, J. Z.   Structure, degeneration and repair of nerve fibers. *Nature, Lond.*, 1945, **156**, 132–136.

17

WERNER REICHARDT

Max-Planck-Institut für Biologie, Tübingen

Autocorrelation, a Principle for the Evaluation of Sensory Information by the Central Nervous System

Many animals react to optical stimulation by moving the eyes, the head, or even the whole body. These reactions are called "optomotor responses." They are continuously graded functions of optical stimulation.

Figure 1 illustrates how optomotor reactions can be elicited (Hassenstein and Reichardt, 1956a, b, 1959). The animal (Chlorophanus) sits inside a hollow cylinder which is composed of perpendicular black and white stripes. When the cylinder rotates, the animal tries to follow the movement. For the insect, this response to the movement observed reduces the relative speed of the surroundings to a residual speed—the slip. Insect and surroundings together form a feedback loop.

The direction and intensity of the optomotor responses have been used as an indicator of the perception processes involved in the central nervous system (CNS) of the experimental animal. In the experiments the feedback loop has been cut off by fixing the animal in the center of the cylinder in such a way that its optomotor reactions could be observed by the experimenter without influencing the position of the animal itself in relation to its optical environment. The experimental procedure is shown in Fig. 2 (Hassenstein, 1951). The beetle's back is glued to a piece of cardboard which is held by a clip, and the clip is fixed to a stand. The beetle is thus freely suspended in air. Then the beetle is given the Y-maze globe which it carries of its own free will. The Y-maze globe consists of six pieces of curved straw that join at four points to form Y-like junctions. When the beetle starts to walk it remains fixed, but the Y-maze globe performs the negative of the movements the beetle would perform if it were walking freely. After a few steps the beetle reaches a Y junction, or rather a Y junction reaches the beetle, and it has to choose right or left. After passing the junction, the animal is in the same

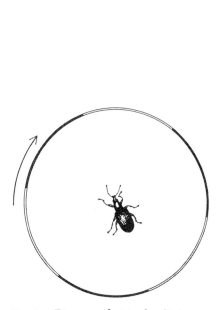

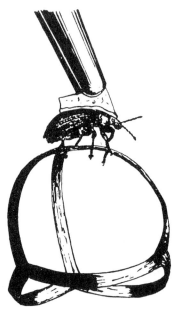

Fig. 1. Diagram of striped cylinder experiment in which optomotor responses can be elicited.

Fig. 2. The animal (Chlorophanus) during the experiments.

situation as before. After the next few steps it has to choose again, and so on. For the beetle the Y-maze globe is an infinite Galton probability apparatus. In a given situation of optical stimulation, the ratio of choices has been proved to be a sensitive quantitative measure of optically induced (optomotor) turning tendencies.

The intensity of the reaction has been measured by the ratio:

$$R = \frac{W - A}{W + A} \tag{1}$$

where W indicates the number of choices "with" and A the number of choices "against" the direction of cylinder motion during the experiments. Statistical considerations have shown that R is practically a linear function of the turning tendency as long as R is smaller than 0.7.

The number of choices that has to be taken into account depends on the intensity of reaction. This has been explained in full detail elsewhere (Hassenstein, 1951, 1958a).

In order to check the relations between stimuli and reactions, successions of narrow linelike light stimuli were delivered to the eyes of the experimental animal. Figure 3 presents one of the experimental arrangements used in the various tests (Hassenstein, 1958b). The set-up consists of three concentric cylinders. During the experiments the beetle with the Y-maze globe is suspended in the center of the cylinders. The inner cylinder is fixed and contains perpendicular slits. The outer cylinder is made up of black and white stripes so arranged that—looking from the beetle's position—the background of each slit consists of either a black or a white field. A "rotating cylinder" of separated gray screens is located between the inner and outer cylinders. If the gray screens are rotated, their leading and

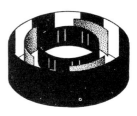

Fig. 3. Experimental arrangement for the production of sequences of light change.

trailing edges generate an alternating sequence of running light changes in the slits of the inner cylinder. If light changes from dark to lighter are designated by a plus sign (+) and those from lighter to darker by a minus sign (−), and the slits are labeled $X, Y, Z, \cdots$, the arrangement of Fig. 3 produces sequences + in X, − in Y, + in Z, or − in X, + in Y, − in Z, and so on. With similar arrangements, programs involving nearly any light change can be produced and used for different stimulations of single ommatidia in the beetle's eyes.

If we use $A, B, C, D, E, \cdots$ to designate adjacent ommatidia in a horizontal row of the facet eye, and if again the plus sign represents an illumination change from darker to lighter, we can describe with the formula $S_{AB}^{++} (t_1, t_2)$ a succession of two stimuli in adjacent ommatidia. The first stimulus is received by ommatidium A at time $t = t_1$, the second stimulus by ommatidium B at $t = t_2$. The reaction of the animal to this stimulation S_{AB}^{++} is represented by R_{AB}^{++}.

The results of stimulation to the ommatidia are the following:

1. The most elementary succession of light changes that is able to release an optomotor response consists of two stimuli in adjacent ommatidia.

2. In generating optomotor responses, the stimulus received by an ommatidium can interact only with the stimulus received by the immediately adjacent ommatidia or by those once removed. No physio-

logical interaction exists between ommatidia separated by more than one unstimulated ommatidium.

3. The maximum reaction is elicited with a time interval $\Delta t = t_2 - t_1$ of about $\frac{1}{4}$ sec between the two stimuli. The strength of reaction decreases with both greater and smaller time intervals. The maximum time interval between stimuli producing small reactions was found to measure slightly more than 10 seconds. The physiological interaction takes place between the after-effect of one stimulus and the effect of a following one.

4. The combined stimulus S_{AB}^{++} leads to the reaction $+R_{AB}^{++}$. This means that the animal follows the direction of stimulus successions.

5. Stimulus S_{AB}^{--} produces R_{AB}^{--}. We have found that $+R_{AB}^{--} = +R_{AB}^{++}$.

6. Successions of stimuli in adjacent ommatidia produce the reaction $R_{ABCD}^{++++} \vdots \vdots$. This reaction turns out to be the sum of the partial reactions R_{AB}^{++}, R_{BC}^{++}, R_{CD}^{++}, $\cdots$ and R_{AC}^{++}, R_{BD}^{++}, R_{CE}^{++}, $\cdots$. More precisely, $R_{ABCDE}^{+++++} \vdots \vdots = R_{AB}^{++} + R_{BC}^{++} + R_{CD}^{++} \cdots + R_{AC}^{++} + R_{BD}^{++} + R_{CE}^{++} \cdots$.

7. Stimulating the facet eye with alternating light sequences as S_{AB}^{+-} or S_{AB}^{-+} leads to reactions *opposite* to the direction of stimulus successions. In other words, R_{AB}^{+-} is equal to R_{AB}^{-+} is equal to $-R_{AB}^{++}$ is equal to $-R_{AB}^{--}$: R_{AB}^{+-} and R_{AB}^{-+} result in "negative" optomotor responses.

The experimental results reported in 5 and 7 show clearly that the relation between stimulus input and reaction output is in accordance with an algebraic sign multiplication.

Table 1. Algebraic Multiplication

	S_A^+	S_A^-
S_B^+	$+R$	$-R$
S_B^-	$-R$	$+R$

8. A cylinder of gray stripes on a white background releases weaker optomotor responses than a cylinder of black stripes on a white background that rotates at the same angular speed. The intensity of optomotor reactions depends not only on the speed of moving patterns but also on the absolute amount of the individual light changes of which the stimulus situation consists. In one of the experiments, the time intervals of stimulus successions were constant, and the stimulus intensities were varied by using screens of different shades of gray (Fig. 4). This was done in such a manner that the sum of stimulus intensities was also kept constant. When ommatidium A received the stimulus amount x, B received the amount $-(1-x)$.

The result of the experiment, as plotted in Fig. 4, showed that the reaction is a quadratic function of stimulus intensities, namely, R is proportional to $-x(1-x)$.

The experimental results of 5, 7, and 8 have the following consequence: physiological mechanisms must exist in the beetle's CNS which cause the sensory inputs and the motor output to be linked together by a process working in accordance with the mathematical operation of multiplication.

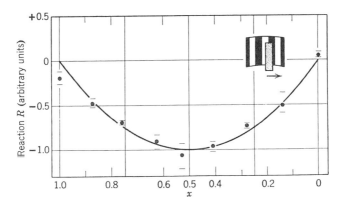

Fig. 4. The intensity of optomotor reactions to movements of different gray screens in front of black and white stripes. On the ordinate, the maximum intensity is equal to -1. On the abscissa, x represents differences in reflected light intensities between moving screens and black cylinder fields; the black-white interval equals 1.

From the findings reported under 1 through 8, we have designed a minimum mathematical model that describes the relations between stimulus inputs and reaction output (Reichardt, 1957; Reichardt and Varjú, 1959). The model, which is presented in Fig. 5, enables us to understand the corresponding evaluation principle in the beetle's CNS and in addition makes it possible to predict reactions to known stimulations. The model contains only two light-sensitive receptors, A and B, representing two adjacent visual elements of the facet eyes (ommatidia). This takes into account that the optomotor response is the sum of the partial reactions to stimulations of adjacent ommatidia and those once removed. The receptors A and B transform the space- and time-dependent processes of the optical surroundings into the time functions L_A, L_B. When we shift the light pattern in front of the receptors from left to right (see Fig. 5), the light-intensity values of the pattern are received first by receptor A and, after a time interval Δt, by receptor B. If Δs designates the angular distance between A and B, the time interval is connected with the velocity of the pattern by the relation $\Delta t = \Delta s/w$. The time functions L_A and L_B feed information channels which are linked together in the multi-

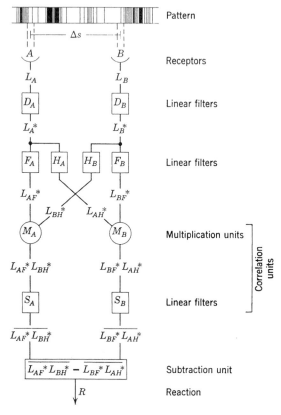

Fig. 5. Mathematical model describing stimulus-reaction relations of optomotor responses in Chlorophanus.

plier units M_A, M_B. The time functions L_A, L_B are linearly transformed by the units D, F, and H. We call these units linear filters in accordance with expressions used in control engineering (see, for instance, Laning and Battin, 1956).

The multiplier and low-pass filter units M_A, S_A and M_B, S_B process the transformed time functions $L_{AF}^* L_{BH}^*$ to the time average $\overline{L_{AF}^* L_{BH}^*}$ and $L_{BF}^* L_{AH}^*$ to the average $\overline{L_{BF}^* L_{AH}^*}$. Since multiplication and time averaging of two time functions is called first-order

correlation (see, for instance, Wiener, 1949), we speak of the correlator units. The outputs of the correlator units are subtracted from
each other in the subtraction unit. The output of the last-mentioned
unit controls the motor output of the animal (see Fig. 5).

The symmetry of optomotor reactions to pattern movement from
left to right, and vice versa, requires symmetry of the model, which
means that

$$D_A = D_B = D; \qquad F_A = F_B = F; \qquad \text{and} \qquad H_A = H_B = H \qquad (2)$$

We now consider, as an example, the movement of a one-dimensional light pattern in Fig. 5 from left to right. This pattern is thought
to be built up of perpendicular stripes with shades whose reflection
values are measured in the center of the cylinder and are called L.
The value L covers a range of $0 \leq L \leq L_{max}$. The distribution of
L values can be of very different types. In the irregular case, for
instance, L is a stochastic variable. The time functions L_A, L_B are
considered to be the sum of the time average C and the fluctuating
light component $G(t)$:

$$L_A = C + G(t) \qquad L_B = C + G(t - \Delta t) \qquad (3)$$

In other words, G is defined in such a way that its average value
is zero. The time functions L_A and L_B are transformed by the filters
D, F, and H. Since the experimental results have shown that these
filters are linear ones, the transformed time functions L^* can be
written as convolution integrals of L. We consider the input functions of the filters to be built up of narrow pulses. If a unit pulse
(δ-function) with the properties

$$\delta(t - t_0) = \begin{cases} \infty & t = t_0 \\ 0 & t \neq t_0 \end{cases} \qquad \int_{-\infty}^{+\infty} \delta(t - t_0) \, dt = 1 \qquad (4)$$

stimulates a filter input, the output responds with the time function
$W(t)$. (W is the weighting function of the filter.) For an arbitrary
input function $F(t)$, the output of the filter can be written as a convolution integral

$$F^*(t) = \int_0^{+\infty} W(\xi) F(t - \xi) \, d\xi \qquad (5)$$

If we call W_{DF} the weighting function of the vertical and W_{DH} the
weighting function of the cross channels in Fig. 5, then we obtain at
the correlator inputs the time functions

$$L_{AF}^* = \int_0^{+\infty} W_{DF}(\eta) L_A(t - \eta) \, d\eta \quad \text{and similarly} \quad L_{BH}^*, L_{BF}^*, L_{AH}^* \qquad (6)$$

After straightforward calculations the output R was found to be

$$R = \int_{\eta=0}^{+\infty} W_{DF}(\eta)\, d\eta \int_{\xi=0}^{+\infty} W_{DH}(\xi)\Phi_{GG}(\eta - \xi - \Delta t)\, d\xi$$

$$- \int_{\eta=0}^{+\infty} W_{DH}(\eta)\, d\eta \int_{\xi=0}^{+\infty} W_{DF}(\xi)\Phi_{GG}(\eta - \xi - \Delta t)\, d\xi \qquad (7a)$$

In Eq. 7a Φ_{GG} designates the autocorrelation function of $G(t)$. This function is defined as follows:

$$\Phi_{GG}(\xi) = \lim_{T \to \infty} \frac{1}{2T} \int_{-T}^{+T} G(t)\, G(t - \xi)\, dt \qquad (8)$$

The relation in the frequency domain equivalent to Eq. 7a was determined as

$$R = \tfrac{1}{2} \int_{-\infty}^{+\infty} Y_D Y_D{}^*[Y_F Y_H{}^* - Y_F{}^* Y_H] S(\omega)\, e^{+i\omega \Delta t}\, d\omega \qquad (7b)$$

In the equation Y_D, Y_F, and Y_H are the transfer functions of filters D, F, and H (they are related to the weighting functions by Fourier transforms); Y^* designates the conjugated complex of Y; $S(\omega)$ is the spectral density of $G(t)$; $S(\omega)$ and $\Phi_{GG}(\xi)$ are Fourier pairs of each other; ω is connected with the wave length λ of a sinusoidal light pattern and its speed w by the relation $\omega = (2\pi/\lambda)w$; i designates the imaginary unit $\sqrt{-1}$.

Equations 7a and 7b describe the relation between stimuli and reactions as far as the model has been determined by the experiments reported here. But there remain still to be investigated quantitatively the linear transformations of filters D, F, and H.

In order to analyze these transformations, we have studied experimentally the reaction R as a function of cylinder speed w to a sinusoidal light pattern with wave length $\lambda = 4.7\Delta s$ (Hassenstein, 1959). The results of this experiment are shown by the points in Fig. 6. With rising speed the intensity of optomotor reactions increases, reaches a maximum, and finally falls again to zero. In the semilogarithmic plot, the reaction curve is symmetrical with respect to its maximum. From this reaction curve we have determined the filter transformations. This analysis (Reichardt and Varjú, 1959) can be presented here in rough outline only.

The spectral density of the sinusoidal component $G(t)$ was determined as

$$S(\omega) = \sigma^2 \left[\delta\left(\omega - \frac{2\pi}{\lambda} w\right) + \delta\left(\omega + \frac{2\pi}{\lambda} w\right) \right] \qquad (9)$$

In this equation σ^2 equals $A^2/2$; A designates the amplitude of the sinusoidal component. Taking into account Eq. 9, we obtain as the result of integration in Eq. 7b

$$R = i|Y_D|^2(Y_FY_H{}^* - Y_F{}^*Y_H)\sigma^2 \tag{10}$$

where

$$\omega = \frac{2\pi}{\lambda}\,w = \frac{2\pi}{4\Delta s}\,w$$

Since the transformations in the filters D, F, and H have been proved to be linear and constant in time, the input-output relations of the filters can be described only by linear differential equations with constant coefficients.

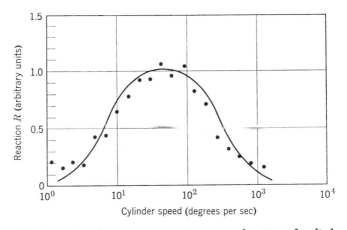

Fig. 6. The intensity of optomotor reactions as a function of cylinder speed w. In this experiment a sinusoidal pattern with wave length $\lambda = 4.7\Delta s$ was used. On the ordinate, the maximum intensity is equal to $+1$.

In principle, these equations can be of the ordinary or partial type. The results of the analysis—on the basis of the optomotor responses indicated in Fig. 6—have shown that the transforming properties of the F and H filters can be described by first-order linear differential equations, whereas the D filters respond in accordance with a partial differential equation of the one-dimensional diffusion type. From these findings we derived the filter transfer functions of the model, which are in the Laplace domain.

$$Y_F = \frac{\pm b_{oF}}{a_{oF} + p} \tag{11a}$$

$$Y_H = \frac{\pm b_{oH}}{a_{oH} + p} \tag{11b}$$

$$Y_D = \pm b_{oD}\sqrt{p}\,e^{-\sqrt{b_{1D}p}} \tag{11c}$$

In Eq. 11, p designates the variable of the Laplace domain. Finally, the time constants of the filter transformations have been determined from the measurements plotted in Fig. 6. They are

$$\tau_F = \frac{1}{a_{oF}} = 1.6 \text{ sec}, \quad \tau_H = \frac{1}{a_{oH}} = 0.03 \text{ sec}, \quad \tau_D = b_{1D} \leq 10^{-4} \text{ sec} \quad (12)$$

The solid curve in Fig. 6 was calculated by Reichardt and Varjú (1959) from Eqs. 10 and 11 and the time constants τ_F, τ_H, and τ_D.

Up to this point the results of experiments reported here have been used to design a mathematical model that describes in quantitative terms the relations between stimuli and observed optomotor responses. It remains as a challenge to prove that this model (Fig. 5) enables us to predict optomotor responses of Chlorophanus to movements of any patterns mathematically analyzable.

We can proceed here in the following way. A one-dimensional cylinder pattern is selected and mathematically described; more precisely, the $G(t)$ function is determined. Introducing $G(t)$ into Eq. 7a or 7b and considering Eq. 11, we can predict the intensity of optomotor responses to different speeds of the selected pattern. The results are compared with experiments carried out with the same pattern.

This procedure can be combined with an even more rigorous test of the model. We have already shown that the optomotor responses depend on the autocorrelation or spectral density of $G(t)$ and not on $G(t)$ itself, as is expressed in Eqs. 7a, b which describe the relations between stimuli and responses in the time and frequency domain. The one-dimensional light patterns and, in addition, the time functions $G(t)$ can be decomposed into components of Fourier series. Each sinusoidal component is determined by its frequency, amplitude, and phase shift. Since the transformations by the filters D, F, and H are linear, the Fourier components interact with each other only in the correlator units of the mathematical model. It is well known that the output of the correlator unit is not influenced by phase shifts between sinusoids at its inputs. This property of a first-order correlation process has an important consequence (Reichardt and Varjú, 1959) for optomotor responses of Chlorophanus: the class of all patterns that differ from each other only in the different phase relations of their Fourier components should produce the same optomotor responses for any pattern velocities.

In order to test this predicted property of the beetle's CNS, we have selected the two periodic cylinder patterns of Figs. 7a and b. The patterns consist of equally spaced black and white stripes with wave lengths $\lambda = 90$ degrees and $\lambda = 22\frac{1}{2}$ degrees. We superimposed (Varjú, 1959; Hassenstein, 1959) the two patterns in two ways. (1) Each contour edge of the 90-degree pattern was placed in the middle of two contour edges of the $22\frac{1}{2}$-degree pattern (shown in Fig. 7c). (2) The contour edges of the two patterns are superimposed without any shift (Fig. 7d). The patterns of Figs. 7c and 7d differ, therefore, by only a 90-degree phase shift of the short wave-length pattern.

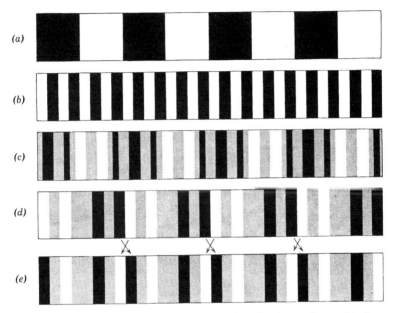

Fig. 7. Periodic cylinder patterns: (a) wave length of periodicity, 90 degrees; (b) wave length of periodicity, $22\frac{1}{2}$ degrees; (c and d) constructed by superposition of patterns (a) and (b), in accordance with the mixing rules for pigment colors (black + black = black, white + white = white, black + white = gray); pattern (e) was derived from pattern (d) by the reversal of one pair of black and white stripes in each period.

Figure 8 shows optomotor reactions to movements of Figs. 7c and 7d (Hassenstein, 1959). The open circles designate measurements taken with pattern 7c, the black circles, measurements taken with pattern 7d. The reactions show clearly the predicted effect: phase shifts between periodic components of the pattern do not change the optomotor responses. In other words, the beetle is phase-blind with respect to the optomotor reactions. The solid response curve in Fig. 8 was calculated from Eqs. 7b and 11, and the time constants of Eq. 12. The calculations were carried through with consideration of the three lowest Fourier components that are contained in the patterns of Figs. 7c and 7d. In order to demonstrate that only the phase shift is responsible for the generation of the same response curves to patterns 7c and 7d, we have reversed a pair of black and white stripes in each period of Fig. 7d. The new pattern is presented in Fig. 7e. The measured (Hassenstein, 1959) optomotor responses to movements of pattern 7e are contained in Fig. 8 and are designated by crosses. The corresponding broken-line reaction curve

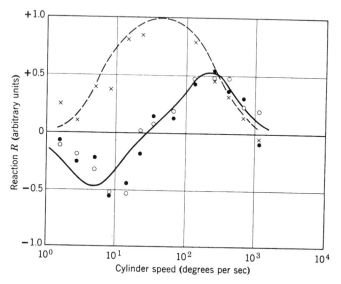

Fig. 8. The intensity of optomotor responses as a function of cylinder speed w. The reactions were measured with patterns in Figs. 7c, d, and e. The maximum intensity of the dashed curve equals $+1$.

was calculated (Varjú, 1959) from Eqs. 7b, 11, and 12, and the lowest five Fourier components of pattern 7e. It is obvious from the reaction curve for pattern 7e, that the slight alteration in pattern 7d resulted in a drastic change in the reaction curve. This is because the reversal of one black and white stripe per period changes not only the phase relations in pattern 7d but also the amplitude distribution of the Fourier components. Since the predicted reaction curves to movements of patterns 7c, d, and e are in accordance with the measurements, we conclude that the mathematical model of Fig. 5 describes the perception processes in the beetle's CNS with respect to optomotor reactions.

Finally, we shall consider another property of the optomotor response predictable from the model in Fig. 5. Up to this point we have studied only responses to moving periodic patterns. These patterns are rich in redundancy, since one period of light distribution determines the whole pattern. But the natural environment of the beetle is not made up of pure periodicities; it consists of a distribution of "spots" with different degrees of reflection. These distributions demand a statistical description (Reichardt, 1957). The limiting case of such a surrounding is given if the degree of reflection of each spot

(a)

(b)

Fig. 9. Cylinder patterns consisting of random light sequences. (*a*) Width of stripes, 7 degrees; (*b*) width of stripes, 1 degree.

is statistically independent of the degrees of reflection of all other spots. A one-dimensional pattern of this type can be built up by purely random sequences of small stripes with different degrees of reflection. Figure 9 contains samples of random patterns. The angular width of the equally spaced stripes in Fig. 9*a* was selected as 7 degrees, whereas in the pattern of Fig. 9*b* the width of the stripes amounts to 1 degree. Both are random sequences, but pattern 9*b* is a much better approximation to the case that involves "white noise," since the smaller the stripes, the closer we approach a constant spectral density.

Figure 10 contains optomotor responses measured (Hassenstein, 1959) with the patterns of Fig. 9. The black points were obtained from measurements using pattern 9*a* and the crosses from pattern 9*b*. The solid curve was calculated from Eqs. 7*b*, 11, and 12 under the assumption that the randomly constructed pattern is "white noise." This reaction curve reveals a curious property: at very low pattern velocities the optomotor response turns out to be opposite to the direction of pattern movement. This predicted effect (Reichardt and Varjú, 1959) is found experimentally with pattern 9*b* which approximates the white-noise limiting case much better than does pattern 9*a*, since the angular width of stripes is only 1 degree. With increasing velocities the experiments done with pattern 9*b* (crosses) follow the calculated reaction curve until the optomotor response reaches a maximum. For higher velocities the responses decrease more rapidly than the predicted values. This is because the random pattern was glued on a cylinder. Therefore the white-noise assumption is violated

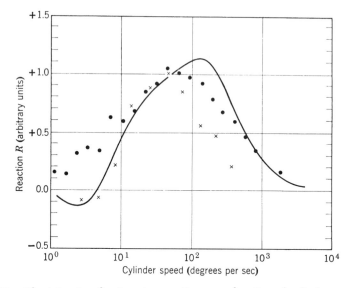

Fig. 10. The intensity of optomotor reactions as a function of cylinder speed w. Points, reactions to pattern in Fig. 9a; crosses, reactions to pattern in Fig. 9b. Maximum intensity of measured values equals $+1$.

at high velocities, since the longest wave length in the pattern is 360 degrees and not infinity. We have shown that the negative part of the reaction curve at low velocities is due to the kinetics of the D filters (Reichardt and Varjú, 1959).

Summary

Like many other animals, the beetle Chlorophanus responds with optokinetic reactions to relative movement of its optical surroundings. These reactions are elicited not only from the movement of figures distinguishable against their backgrounds but also from the movement of randomly constructed patterns of shades from white through black. The most elementary succession of light changes that is able to release an optomotor response consists of two stimuli in adjacent ommatidia of the facet eye. A stimulus received by one ommatidium can interact only with a stimulus received by the adjacent ommatidia or by those once removed. The stimuli received by the ommatidia are linearly transformed, and their interaction in the central nervous system is in accordance with the principle of first-order correlation. As a consequence of this evaluation principle in Chlorophanus, the class of all optical surroundings that differ from each other by different phase relations of their Fourier components produce the same optomotor responses for any pattern velocities.

References

Hassenstein, B.   Ommatidienraster und afferente Bewegungsintegration.   Z. vergleich. Physiol., 1951, **33**, 301–326.

Hassenstein, B.   Die Stärke von optokinetischen Reaktionen auf verschiedene Mustergeschwindigkeiten. Z. Naturforsch., 1958a, **13b**, 1–6.

Hassenstein, B.   Über die Wahrnehmung der Bewegung von Figuren und unregelmässigen Helligkeitsmustern. Z. vergleich. Physiol., 1958b, **40**, 556–592.

Hassenstein, B.   Optokinetische Wirksamkeit bewegter periodischer Muster. Z. Naturforsch., 1959, **14b**, 659–674.

Hassenstein, B., and W. Reichardt.   Functional structure of a mechanism of perception of optical movement. Proc. I int. Congr. Cybernet., Namur, 1956a, 797–801.

Hassenstein, B., and W. Reichardt.   Systemtheoretische Analyse der Zeit-, Reihenfolgen- und Vorzeichenauswertung bei der Bewegungsperzeption des Rüsselkäfers Chlorophanus. Z. Naturforsch., 1956b, **11b**, 513–524.

Hassenstein, B., and W. Reichardt.   Wie sehen Insekten Bewegungen? Die Umschau, 1959, **10**, 302–305.

Laning, I. H., and R. H. Battin. Random Processes in Automatic Control. New York: McGraw-Hill, 1956.

Reichardt, W.   Autokorrelationsauswertung als Funktionsprinzip des Zentralnervensystems. Z. Naturforsch., 1957, **12b**, 447–457.

Reichardt, W., and D. Varjú.   Übertragungseigenschaften im Auswertesystem für das Bewegungssehen. Z. Naturforsch., 1959, **14b**, 674–689.

Varjú, D.   Optomotorische Reaktionen auf die Bewegung periodischer Helligkeitsmuster. Z. Naturforsch., 1959, **14b**, 724–735.

Wiener, N.   Extrapolation, Interpolation, and Smoothing of Stationary Time Series. New York: Technology Press and Wiley, 1949.

18

WOLF D. KEIDEL, URSULA O. KEIDEL, and MALTE E. WIGAND

Physiologisches Institut, Universität Erlangen

Adaptation: Loss or Gain of Sensory Information?

Adaptation was known first as a threshold shift of the visual system, after exposure to sudden changes in illumination level, to a new steady state. For a long time it seemed to be not much different from fatigue, a fading out of excitability of the sensory organs, a state in which the imperfection of the constituents of the living tissue was revealed in quantitative research, inconvenient for the observer, but inevitable. In time it was found to be a feature of almost all sensory systems, perhaps with the exception of a few, such as the nociceptive system.

The time course of any adapting part, within the total chain of information transfer in response to a step rise in stimulus intensity, has generally been found to be a "proportional-differential" block, such as is used in those technical servomechanisms that need high speed for operation. This means, physiologically, that the firing rate of single units is smaller in the steady state of adaptation than during the transient period (the "overshoot"). The same is true for the size of the compound action potential recorded from a population of cells or fibers. Thus, as a consequence of adaptation, the number of units used in the coded information is smaller than the number used at the first step rise in the stimulus. Therefore, in the classical sense, adaptation means a *loss* of sensory information. In spite of this, however, my question still remains: is this true in all cases where sensory adaptation occurs? More specifically, I shall discuss three sets of experimental data in which the classical definition of adaptation as a loss of sensory information does not hold.

Visual System: Simultaneous Difference Limen

The first set of experiments concerns the eye, and the research to be reported was carried out in our laboratory by Dr. Ranke (1952) in cooperation with Dr. Kern (1952). Dr. Ranke based his concept on some measurements on the ear by Dr. von Békésy (1929). Békésy had presented a tone of high intensity and long duration to a subject's ear, and immediately after the end of that intense stimulation he measured the so-called curves of equal loudness shown in Fig. 1. The result—an increase in the number of just noticeable differences (jnd) for intensity in the steady state of adaptation—was clearly

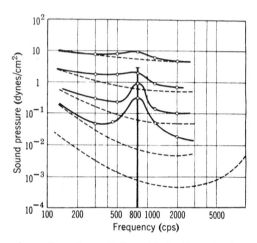

Fig. 1. Curves of equal loudness before (dashed) and after (solid) the ear was adapted to a tone of 800 cps. Steps of equal loudness correspond to smaller steps of sound pressure in the adapted than in the unadapted ear. Thus the number of jnd for a given upward step in stimulus intensity is increased when the steady state of adaptation is reached. (From Ranke, 1952, after Békésy, 1929.)

mentioned by Békésy as early as 1929. Ranke's question was whether this result was unique for the ear, or whether there was imbedded a general principle common to all sensory systems. In order to investigate this question Ranke used a large white screen that illuminated the entire visual field of a subject and thus determined the adaptation state of the subject's eye. The state of adaptation depended, therefore, only upon the level of illumination of the screen ("total illumination level"). The subject's task was to decide, at each illumination level, whether the two halves of a small circular test area (34 minutes, 0.5 second) had the same or a different brightness. The difference between the two stimuli was measured and then the ratio of that difference (minimum separabile) to the control stimulus intensity was calculated. This ratio $\Delta I/I$ was then inverted to $I/\Delta I$, which is a measure of the "excitability."

Ranke now plotted the ratio, so defined, as a measure of the actual sensitivity of the eye, against both the total illumination level and the ratio of the illumination level of test area to the total illumination level. This is shown in Fig. 2. Since the total illumination level is the adapting light intensity, this diagram means that the number of jnd, as a correlate of the available information, reaches a maximum only when the intensity of the test stimulus is equal to the adapting intensity. Or, in other words, since the number of jnd is the integral of the

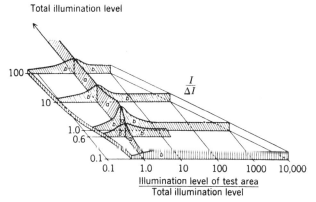

Fig. 2. The vertical dimension represents $I/\Delta I$ (the ratio of control intensity of test area to that difference in intensity that is needed in order for simultaneous stimulation to increase the brightness by 1 jnd). The abscissa is the ratio of illumination level of test area to the total illumination level. The third dimension (depth) is the adapting (total) illumination level. It is clear that, for the adapted steady state, the ratio plotted vertically reaches a maximum. The unit of illumination level is the apostilb. (After Ranke, 1952, and Kern, 1952.)

ratio $I/\Delta I$, the steepness of the number of jnd plotted against the test stimulus intensity is also greatest for the steady state of adaptation, much like the results reported by Békésy on the ear. This is illustrated in the two graphs of Fig. 3. The curves in the lower figure are the five shaded functions of Fig. 2, and the upper graph shows the integrals of those five functions.

Although a particular photographic emulsion usually reveals the same steepness of gradation along the total range of its sensitivity to light, these experiments make it evident that this is not the case for the visual system. Here the number of jnd for each single information line (a receptor-neuron unit) is quite restricted; according to Ranke the total number of jnd is at most 35. This means that, from a critical flicker frequency around 20 cycles per second up to a maxi-

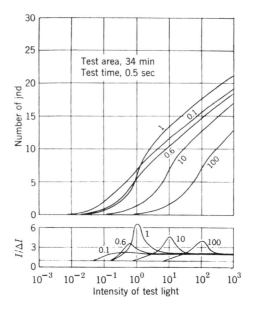

Fig. 3. The five shaded functions of Fig. 2 (lower curves) are integrated and represent the number of jnd as a function of the test light intensity (upper curves). The parameter is the adapting illumination level of the screen. It is clear that the steepest parts of the integral curves are around those values of test light intensities where they are equal to the adapting illumination level. (After Ranke, 1952.)

mal firing rate of a single fiber of about 800 pulses per second within the optic nerve, the difference limen (DL) would be about 10 per cent, which is in good agreement with other data. Since the total range of intensity discrimination of the visual system is about 10^5, there are two ways of conveying that information, one of which uses a monotonic function, as in photographic emulsion. If the conveying system were of this kind, the size of each DL would be three times too high, that is to say, about 3 db (100 db divided by 35 steps). The alternative is to cut out of the total range of intensity a small strip with enlarged sensitivity. This means that an increased number of jnd would lie within this small strip. Now if this strip should migrate with the continuously changing state of adaptation level, the result would be precisely the kind of thing that happens in the visual system. This is what Ranke called the *Bereichseinstellung der Sinnesorgane*, and in this restricted sense this is, of course, a real gain in information through the adaptation of the eye. This is my first example. Now let us return to the auditory system.

Auditory System: Directional Hearing

In accordance with the Hornbostel-Wertheimer theory of directional hearing, experiments were carried out with 23 cats to determine how the information for directional hearing is stored at the cortical level during recording of the evoked slow potentials that result from the presentation of two clicks to the two ears.

Our main interest here was first to find whether the size of the cortical potentials is varied by the two parameters for directional hearing—the attenuation of the ear that is turned away from the sound source, and the time delay. If the size of these potentials is correlated with the direction of the sound source, a second problem is to determine whether such a change in size itself depends on the state of adaptation, much like what has just been described for the DL for intensity in the visual system.

All cats were anesthetized with Dial and kept at a given level of anesthesia. The left auditory area was then exposed and, by means of platinum wire gross electrodes, the slow evoked potentials were recorded (from area AI and in a few cases from AII). Clicks at durations of 100 to 200 microseconds and at repetition rates of 1 to 10 per second were delivered to both ears by means of two sound pressure chambers, using plastic tubes to the meatus of the two ears, so that the cross-attenuation was of the order of 60 decibels. The clicks could be presented simultaneously. Their intensity ratio could be adjusted by means of a decade attenuator. In addition, the second click could be delayed. For this purpose the built-in delay line of a Tektronix oscilloscope (type 535) was used to trigger a second pulse generator whose output pulse after amplification was matched in shape, duration, and intensity to the first one. It differed only in time. The records usually consisted of ten superimposed traces. In some cases an electronic averaging machine (Keidel, 1959) was used in order to measure the total size of the potentials with an accuracy of 1 per cent. For routine processing of the data the average size of ten traces was measured in microvolts. The animals were kept in a sound-proofed room under constant climatic conditions (temperature held constant at 20° C and relative humidity at 60 per cent by means of an air conditioner). After the dura mater was cut, the cortex was covered with mineral oil, and saline solution was given subcutaneously at about 2-hour intervals.

In the first series of experiments the clicks were delivered simultaneously. Only the ratio of intensities between the left and right ears was varied, over a range corresponding to the ratio that occurs under physiological conditions for directional hearing in the cat. The attenuation of the clicks by the cat's head, measured by means of a sound pressure meter, was found to be of the order of 10 to 15 db, which corresponds to a shift of the sound source from 90 degrees to the midline of the head. In the first set of experiments the sound received by the ipsilateral ear (with respect to the cortex with recording electrode) was attenuated in steps up to 14 db, while the contralateral ear received up to a 4-db increase in sound pressure. The procedure was then reversed and carried out on opposite ears. The total schedule of measurements was repeated for different absolute intensities up to 80 db above threshold. These experiments yielded the result that, over the range 20 to 50 db above threshold, the ipsilateral ear usually contributes less to the compound slow evoked potential of one side of the cortex than does the contralateral ear. This finding holds for a small range of absolute intensities between 20 and 50 db above threshold; for both higher and lower intensities, the curves are flat. This means then that both ears contribute about equally to the compound cortical potential. In addition, for both ears the curves of equal sound energy and different ratio of intensities between the left and right ears change their steepness from zero through a maximum value and back to zero again as the intensity is increased. This is shown in Fig. 4.

If one considers that both cortices (in corresponding cell groups of the auditory area) differ only in that this steepness is positive for the one and negative for the other, then it is highly probable that this difference in steepness between the left and right cortical auditory areas is correlated with the information about directional hearing at the cortical level. This assumes that the threshold for directional hearing should be lowest where the steepness is greatest, that is, for intensities around 40 db above threshold. On the other hand, this threshold should rise as it approaches the absolute threshold and the very high sound intensities. We carried out some psychophysical experiments using the same stimulus equipment, the results of which seem to lead in this direction. For all kinds of difference limens this is well-known behavior. In the psychophysical experiments a given delay was compensated by an increase in intensity ratio to the opposite ear.

The next step was to observe the influence of adaptation on the steepness of our curves. On the basis of earlier experiments (Keidel, et al., 1957, 1958, 1960), we used different repetition rates in order to be able to compare the transient and the steady state of the adapta-

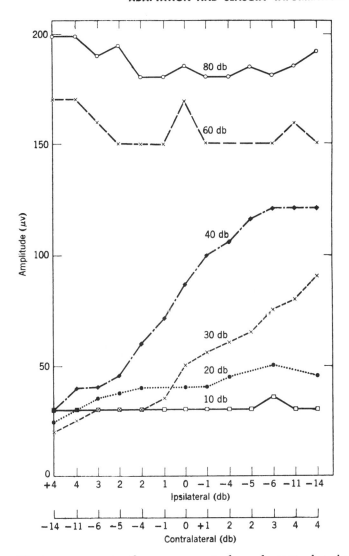

Fig. 4. The two ears receive the same amount of sound energy, but the ratio of intensities between the left and right ears is varied by attenuating the intensity to the one while an increment is fed to the other ear. The parameter is the absolute intensity. The steepness of these curves runs through a maximum as the absolute intensity increases. Recording from left auditory area AI.

tion time course. Since a first response cannot know what repetition rate (rr) follows it, the responses to very slow repetition rates, around 1 per second and below, are representative of the nonadapted state.

As the repetition rate rises to 5 to 10 per second, the transient state follows until finally, after a few clicks, the steady state of adaptation is achieved. All the following responses to clicks are representative of the adapted state at the given repetition rate. The results of these experiments—with repetition rate as parameter—are shown in Fig. 5.

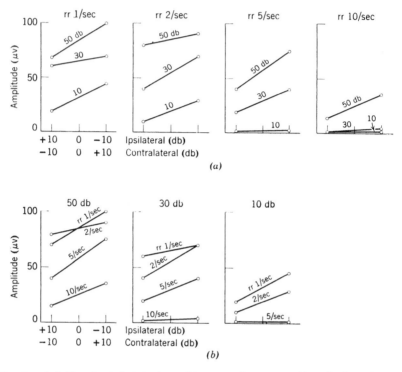

Fig. 5. (*a*) Nonadapted state (repetition rate, 1 per second) and adapted state (steady-state responses, rr 10 per second) are compared for different intensities, and (*b*) the influence of repetition rate on the size of the cortical responses is plotted for three given intensities. The curves in (*b*) demonstrate that increasing adaptation not only diminishes the absolute size of the potentials but also affects the steepness of the curves. Recording from left auditory area AI.

The results leave no doubt that adaptation alters the size of the cortical responses in a clear-cut manner. Here again, the steepness of the curves is altered by the process of on-going adaptation, and the effect of adaptation is not limited to a mere decrease in size, as seems to happen in the usual curves in which size of potential is plotted against repetition rate.

There is a further complication when time delay of the second click is introduced in addition to attenuation. All these data are compared in Fig. 6. The curves in (a) show the effect of attenuation alone, in (b) that of time delay alone, and in (c) a comparison of the effects of the two factors on the size of evoked slow cortical potentials.

In the physiological range of the two parameters (τ and i), the difference in effectiveness of the two ears when interacting to one cortex (or, alternatively, the difference in size between potentials from corresponding areas on the two sides of the auditory cortex) increases as the intensity ratio increases and as the delay increases. In both cases, this means that the difference between the ears grows with increasing distance of the sound source from the midline. Thus the steepness of these curves seems to be related to the number of units for a given information content about the locus of the sound source. So it is quite understandable that both parameters, τ and i, yield curves that are symmetrical around the y axis (up to 150 μsec delay and 20 db attenuation). Since the number of units for a given sound direction increases with the transition from the transient to the steady state of adaptation (or as the repetition rate increases from 1 to 10 per second), again adaptation means an actual gain in sensory information.

There is included, however, one assumption for which we have no experimental evidence. That is the assumption that the difference in excitation state of the two auditory cortices is evaluated by means of another part of the cortex, for example, an association area or something of that sort, and moreover that this is done in the form of a difference in percentage. There is also the possibility that such "processing areas" are only small parts within the total auditory cortex. We have found points within auditory area AI as well as AII just a few millimeters apart with quite different features: some cell groups behaved like those described above; others revealed a dependence upon intensity ratio such as is shown in Fig. 7a, or upon time delay, as in Fig. 7b. Since Fig. 7 contains averaged data of more than 100 potentials, there can be no doubt that there exist fields within the auditory cortex that are independent of the binaural delay. We feel that these observations actually do not contradict each other if one assumes that the delay is evaluated far below the cortical level, say, by the interaction of the trapezoid bodies. Then the information about the processed input information, not the delay itself, would be led to the cortical level for storage, and at this place the information, so coded, would be evaluated a second time. This dual handling of information would also explain loci with and without dependence on delay between the two stimuli. In any case, the first processing of τ

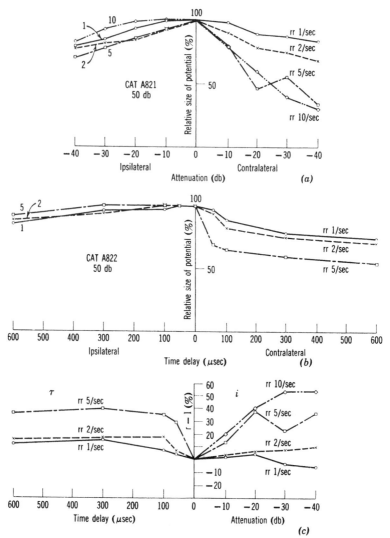

Fig. 6. (*a*) Effect of attenuation *i* of the ipsilateral and contralateral ears on potentials recorded from the auditory cortex; obviously the contralateral ear contributes more to the compound slow evoked potential. (*b*) Effect of time delay *τ* on the size of potential compared for delay of the ipsilateral and the contralateral ears; here too the contralateral ear contributes more. (*c*) Differences between the right and left sides of diagrams (*a*) and (*b*) are compared (right minus left). Up to about 150 μsec delay and 20 db attenuation, the curves are grouped symmetrically around the *y* axis. This makes it probable that the information for directional hearing is stored in this way at the cortical level. Greater steepness then means more units for a given *τ* or *i*. Greater steepness is recorded at higher repetition rates. Therefore transition to the steady state of adaptation, from a repetition rate of 1 to 10 per second, means a clear gain in information.

has to be done at as low a level as possible, first, because the accuracy must be of the order of 30 μsec and, second, because the loss of time before the first motor reaction must be kept to a minimum (as when an animal in an emergency must rotate his head toward a sound

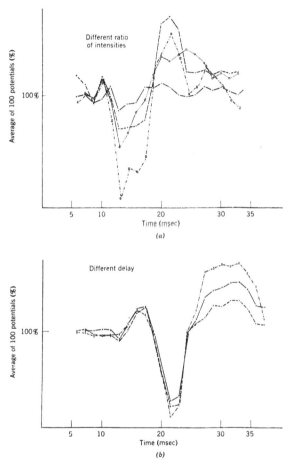

Fig. 7. Averaged data of 100 evoked slow potentials showing that there are cell groups within auditory areas AI and AII whose potentials (*a*) vary with the intensity ratio of left to right ear, or (*b*) are independent of the binaural delay. The 100% level on the ordinate indicates the response to a single click.

source). Later on, there is time enough to evaluate the two kinds of information, intensity ratio and delay, probably in a processed and precoded manner, and then to decide, at the cortical level, the next motor step. We are now occupied by experiments in directional feeling, and we hope in this way to come to a clearer knowledge of the underlying neural circuits by comparison.

Peripheral and Cortical Levels: Time Course of Adaptation

In our third set of experiments we started with the question, how is the total amount of information about just one parameter of a multi-dimensional sensory input, namely, intensity, changed along its path through the chain of synapses up to the cortex? Two levels were recorded, (1) at the first neuron and (2) at the projection area of the cortex. The whole problem was restricted to a not-too-complicated kind of stimulus which nevertheless made it possible to record evoked potentials, that is, trains of clicks whose parameters are intensity and repetition rate. Duration of pulses was kept constant.

Twenty-seven cats were prepared the same way as described above for recording the evoked slow potentials. In a few animals, in addition, the bulla was opened and the compound action potential of the eighth nerve was recorded near the round window, according to the technique described by Rosenblith and Rosenzweig (1951, 1953). The experiments described here were carried out in the Communications Biophysics Laboratory at the Massachusetts Institute of Technology (Keidel, et al., 1957, 1958, 1960). In addition, some data were obtained from microelectrode studies on single fibers of the maxillary nerve of guinea pigs when their whiskers were stimulated. These experiments were carried out in 1958 by Keidel, Ichioka, and Trincker at our Institute in Erlangen.

Within the auditory system the drop in size of the compound action potential to a lower steady-state value, as a consequence of the time course of adaptation, starts at the first neuron at repetition rates somewhere around 100 per second (Keidel, et al., 1958). At the cortical level, however, this drop starts at much lower repetition rates, somewhere around 2 per second. This effect is shown in Fig. 8.

Single-fiber responses are pretty well synchronized with the pulses in the trains and adapt only by a decrease in the rate of responses per second. This takes the form of increased staggering, since these responses obey the all-or-none law. We obtained data for this kind of response only in the somesthetic system, by means of pulls on the whiskers, recording from single A and C fibers of the maxillary nerve. However, according to experiments of Tasaki and Davis (1955), single fibers of the auditory nerve should respond similarly when stimulated by trains of clicks. Our results are shown in Fig. 9 for typical single

A and C fibers. At the cortical level, however, a special time course of the envelope of the responses to the clicks in the trains was observed: at medium and high intensities, not at very low ones, a periodicity of about 3 to 5 sec appears. This effect depends upon intensity as well as repetition rate. It is shown in Fig. 10. This periodicity con-

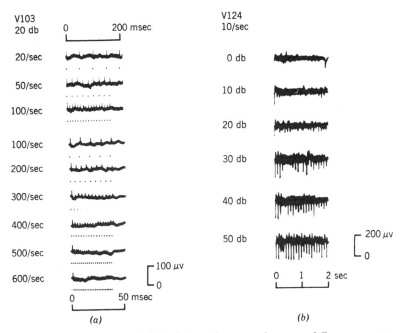

Fig. 8. (*a*) With trains of clicks delivered to a cat's ear at different repetition rates, the compound action potential of the auditory nerve is recorded. The typical time course of adaptation (drop of envelope to a lower steady-state value) starts somewhere around a repetition rate of 100 per second. (*b*) A comparison of intensities for a given repetition rate of 10 per second shows that the typical time course of adaptation at the cortex starts at much lower repetition rates (somewhere around 2 per second).

sists of two phenomena: (1) a periodic increase in the size of potentials, and (2) an appearance of repetitive responses with the same period. The two phenomena may occur simultaneously or alternately. We looked for a possible quantitative relation between intensity and period and for this purpose compared measurements of this period for more than 1000 single potentials with the so-called intensity curves of the same potentials. As a result we found that there is a clear reciprocal relation between the period on the one hand and the two intensity functions on the other. This is shown

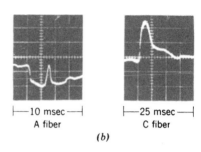

|—10 msec—| |—25 msec—|
A fiber   C fiber

(b)

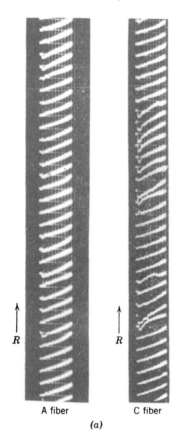

Fig. 9. (a) Staggering of single A and C fibers as adaptation proceeds; trains of pulls on guinea pigs' whiskers (repetition rate, 10 per second; pulse duration, 600 μsec; intensity, 35 db above threshold); R, onset of train. (b) Single responses; here upward deflection means negative potential.

R

R

A fiber   C fiber

(a)

in Fig. 11 (mean value of the first three potentials labeled "dynamic"; mean value of the following 100 potentials labeled "steady state").

The origin of this periodicity is not clear. Obviously it is related to the two-click experiments of Rosenzweig and Rosenblith (1953), although in the two-click experiments the period of changing excitability is much shorter, of the order of 100 to 200 msec (see Goldstein, Kiang, and Peake, in press). This difference in length of period from that of our experiments may be reduced to the fact that a train contains many, and not just two, clicks in a row. It is tempting, of course, to look not only for local reasons, but also for centrifugal feedback loops or something of that sort. This is especially tempting, since in our experiments this periodicity fades out more and more with increasing depth of anesthesia. In addition, there is a systematic difference between the auditory system and the somesthetic system in which the second, third, and so forth, maxima, if observable, never exceed the size of the first few potentials—which they may well do in the auditory system if it is stimulated with high enough intensities.

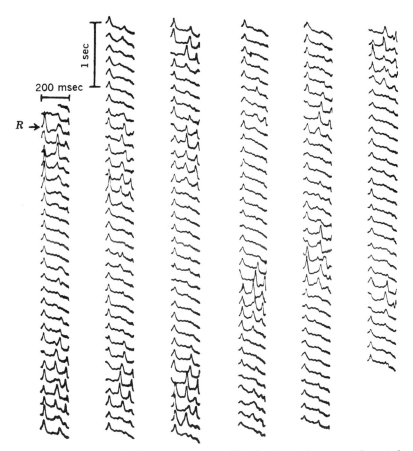

Fig. 10. Time course of adaptation at cortical level. Original traces with periodic enhancement and repetitive responses (auditory cortex area AI; trains of clicks; repetition rate, 5 per second; duration, 200 μsec; R, onset of train; intensity, 80 db above threshold).

Finally, all the data collected for the third problem of this paper are put together in Fig. 12, which gives an idea of the variation of information about a single parameter of a multidimensional sensory input, namely, intensity. Let us try to point out the similarities and differences for the peripheral and cortical levels. (This does not, of course, mean that these differences have their origin in the cortex; on the con-

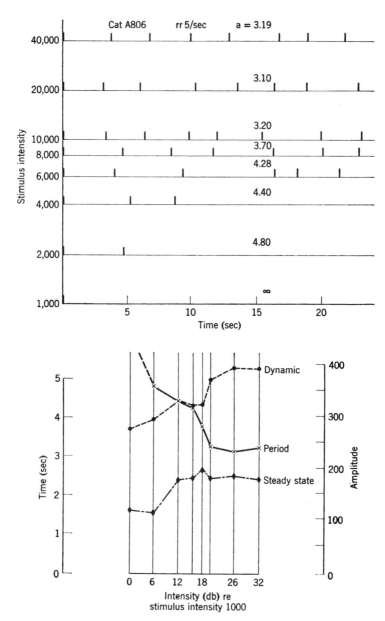

Fig. 11. Upper: Occurrence of potentials at various levels of stimulus intensity (arbitrary units). The periodicity of repetitive responses (a) is indicated for each intensity. Lower: Periodicity of repetitive responses (left-hand scale) as a function of stimulus intensity. This curve has a falling characteristic; the "dynamic" and "steady-state" intensity curves are included for comparison (right-hand scale). Stimuli, trains of clicks; repetition rate, 5 per second; recorded at auditory area I.

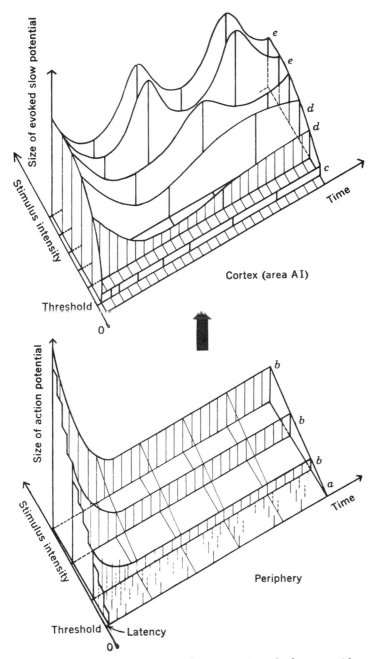

Fig. 12. Similarities and differences in the time course of adaptation (during a train of clicks) at the peripheral and cortical levels. The curves labeled (*a*) to (*e*) refer to experimental data from the following figures or references: (*a*) Fig. 9*b*; (*b*) Fig. 8*a*; (*c*) Mountcastle (1957); (*d*) Fig. 8*b*; and (*e*) Fig. 10.

trary, the cortex is only the locus for recording.   Feedback loops may play a role over wide parts of the brain.)

First of all, it is quite clear that in the periphery, as well as in the cortex, trains of clicks show about the same time course, with only slightly different time constants, as do sinusoidal stimuli delivered to any sensory system.   In a word, in the periphery the transition function of the adaptation shows the features of a proportional-differential block, which includes an initial overshoot and a final steady state. The synchronized, but staggered, responses to single clicks within the train with the highest firing rate at the beginning obviously add up to an initial overshoot in size and a full line in periodicity by combining several fibers.   This principle is shown in the lower drawing of Fig. 12, in which three single-fiber responses are distinguished by solid, broken, and dotted lines.   For each fiber the all-or-none law is valid.   This behavior is in accordance with the volley theory.   The same principle holds for recording from the maxillary nerve in the somesthetic system.   By this means the drop in size of the compound action potential by adaptation can be explained.   As a consequence, the dynamic and the steady-state intensity functions are the upper limits of the left and right cross section of the three-dimensional graph. It is obvious that the dynamic curve is steeper than the steady-state curve: this is nothing but the above-mentioned loss of information (measured as number of units for a given state) by the adaptation. In principle the same is true for the corresponding records at the cortical level: the dynamic and steady-state curves show a roughly similar difference in steepness (but with a different shape).   Only the envelopes of the evoked slow potentials are clearly different in shape from those of the periphery, since they have not a constant but a fluctuating "steady state," or in other words a cyclical component, dependent for each period on intensity and repetition rate.

According to the definition given, this again means a gain in information through adaptation for those periods in which maxima of the envelopes are reached (and therefore the steepness is considerably greater than at lower intensities).   For then too, the number of units for a given intensity step is greater.   This holds even more when the number is compared with the corresponding number at the peripheral level.   However, until we know more about the underlying mechanisms, principles, and circuits, we cannot determine whether this is a true gain in information (perhaps via other systems) or just an amplification of potential size by local processes, which may be caused by a local shift in the d-c potential.   However, the difference in this periodicity that can be observed quantitatively, at least for two different sensory systems (auditory and somesthetic), promises that it may be more than a merely local phenomenon.

Summary

Three sets of experiments have been discussed, relating to (1) the simultaneous difference limen of the visual system, (2) directional hearing in the auditory system, and (3) the time course of adaptation at the peripheral and cortical levels of the auditory system, with a few excursions into the somesthetic system. All three sets of experimental data lead to the conclusion that adaptation does not mean a mere loss of sensory information, as is usually thought, but in quite a few situations in sensory systems adaptation may result in an actual gain in sensory information. This is true at least for discrimination measured in DL. Consequently adaptation can be considered a very basic principle, like facilitation, potentiation, sensitization, and inhibition within the total system of sensory communications.

Acknowledgments

It is a pleasure to thank the Deutsche Forschungsgemeinschaft for the grants that made available nearly all the equipment used for this study.

References

Békésy, G. v. Zur Theorie des Hörens. Über die eben merkbare Amplituden- und Frequenzänderung eines Tones. Die Theorie der Schwebungen. *Physik. Z.*, 1929, **30**, 721–745.

Goldstein, M. H., Jr., N. Y-S. Kiang, and W. T. Peake. Interpretation of responses to repetitive acoustic stimuli. *Proc. III. int. Congr. Acoust., Stuttgart* (in press).

Keidel, W. D. Elektronisches Rechenwerk zur Mittelwertsbildung statistisch streuender periodischer bioelektrischer Potentiale. *Z. Biol.*, 1959, **111**, 54–66.

Keidel, W. D., U. O. Keidel, and N. Y-S. Kiang. Cortical and peripheral responses to vibratory stimulation of the cat's whiskers. *Quart. Prog. Rep., Res. Lab. Electronics, MIT*, July 1957, pp. 135–139.

Keidel, W. D., U. O. Keidel, and N. Y-S. Kiang. Cortical and peripheral responses to vibratory stimulation of the cat's vibrissae. *Arch. int. Physiol. Biochim.*, 1960, **68**, 241–262.

Keidel, W. D., U. O. Keidel, N. Y-S. Kiang, and L. S. Frishkopf. Time course of adaptation of evoked responses from the cat's somesthetic and auditory systems. Quart. Prog. Rep., Res. Lab. Electronics, MIT, January 1958, pp. 121–124.

Kern, E. Der Bereich der Unterschiedsempfindlichkeit des Auges bei festgehaltenem Adaptationszustand. Z. Biol., 1952, **105,** 237–245.

Mountcastle, V. B. Modality and topographic properties of single neurons of cat's somatic sensory cortex. J. Neurophysiol., 1957, **20,** 408–434.

Ranke, O. F. Die optische Simultanschwelle als Gegenbeweis gegen das Fechnersche Gesetz. Z. Biol., 1952, **105,** 224–231.

Rosenblith, W. A., and M. R. Rosenzweig. Electrical responses to acoustic clicks: influence of electrode location in cats. J. acoust. Soc. Amer., 1951, **23,** 583–588.

Rosenzweig, M. R., and W. A. Rosenblith. Responses to auditory stimuli at the cochlea and at the auditory cortex. Psychol. Monogr., 1953, **67,** No. 363.

Tasaki, I., and H. Davis. Electric responses of individual nerve elements in cochlear nucleus to sound stimulation (guinea pig). J. Neurophysiol., 1955, **18,** 151–158.

19

D. M. MacKAY

King's College, University of London*

Interactive Processes
in Visual Perception

Several writers in this symposium have discussed physiological evidence of interaction between neighboring areas of the visual field. This paper is an interim report of some related psychological observations that fall into two classes, the perception of pattern and the perception of motion.

Our group's interest in interactive processes began with a study of the integration of directional information in the setting of vernier lines.† It is well known that the accuracy of alignment of two straight lines can considerably exceed the resolution of the retinal mosaic, indicating some integration of evidence either from several receptors or from the same receptor in several positions, or both. Since, with a nearly stationary retinal image, vernier acuity is not initially impaired (Ratliff, 1952), the first process seems to be the more important.

Briefly, what we found was that, when R. A. Fisher's (1935) measure of information content was applied, the amount of information in the alignment increased linearly with the length of the vernier line over a certain range (of the order of 50 cone diameters), and more slowly thereafter.

One inference was that directional information was abstracted by the visual system in such a form that samples from separate areas could be combined. As a result, we became interested in the response of the visual system to stimuli that were highly enriched in directional information. These turned out to evoke an unexpectedly striking class of effects, the first example of which was encountered accidentally in a BBC studio lined with slotted wallboard. This can be seen in Fig. 1 as a curious rapid "trickling" of wedge-shaped shadows up and

---

* Now at Department of Communication, University College, Keele, Staffs., England.

† To be published elsewhere in collaboration with Miss S. J. Todd.

down the blank strips between columns of parallel lines. Investigation of this phenomenon, which was well known to broadcasting personnel as a nuisance, led to the discovery—in many cases, alas, the

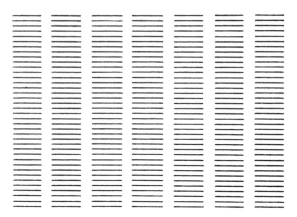

Fig. 1. BBC wallboard pattern.

rediscovery—of a wide range of effects associated with the perception of regular, spatially repetitive patterns. It is these patterns that will occupy us in the first half of this paper.

The Perception of Pattern

In terms of selective information content, one of the simplest stimuli—much simpler than a single spot of light—is an infinitely extended field of regularly spaced parallel lines, since it is invariant under all translations in one direction, and under all translations differing by a multiple of the pattern spacing in the other direction. It is probably not surprising, therefore, that such stimuli evoke a whole family of striking effects—some relevant, others irrelevant to our topic, and many familiar in the older literature.

Helmholtz (as usual) was one of the first in the field (1856) with a report of the wavy appearance of the lines of a grating near the limit of acuity. Though he attributed this to the granularity of the retinal mosaic, we shall see that it may be connected with other interesting phenomena of a similar kind, but requiring more sophisticated explanation. These may be grouped into four classes:

1. "Moiré" effects due to the superposition of successive images as the eyes move. These appear as flickering shadows which can be

reproduced by superimposing a slightly displaced duplicate pattern on the original (see, for example, Figs. 2a and 3). They disappear if the retinal image is stabilized (MacKay, 1958a).

2. Faint, superimposed, colored fringes, and patterns, extensively studied by Erb and Dallenbach (1939) and others, who believed them

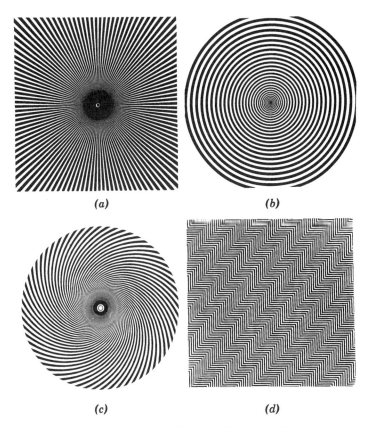

Fig. 2.   Patterns that induce complementary images.

to result (as in Benham's top) from on-off modulation of light falling on receptors near black and white boundaries—here as a result of eye movement. It is far from certain that all the effects they studied could be explained in this way, but they will not be discussed here.

3. A "streaming phenomenon," first noticed over 50 years ago (Pierce, 1901), as if "fine dust" were drifting over the pattern at right angles to the lines. We have found this to be especially noticeable at lower levels of illumination.

4. A curious shimmering after-effect, seen for a few seconds on a blank background as a scurry of wavy lines or shadows moving roughly at right angles to the direction of the stimulus lines. When I happened on this effect a few years ago (MacKay, 1957*a*), I suggested that it be called the *complementary after-image,* because of its geometrical relation to the stimulus. Though I later discovered that

Fig. 3. "Moiré" effect of superimposing two patterns of type described.

some aspects of this phenomenon had been reported by W. S. Hunter (1915) over 40 years earlier, I think the name (abbreviated to CAI) may be worth retaining to distinguish it from the ordinary negative after-image. Like the latter it is limited to the area of the field stimulated.

To a physicist the appearance of waves inevitably suggests the possibility of standing-wave phenomena, and Fig. 2*a* was constructed to see what happened when the wave pattern was closed. As expected, the CAI comprised concentric systems of wavy lines, sometimes contrarotating, but often organized to form a single rotating pattern with the general appearance of Fig. 4. At first it was hoped that the direction of rotation might be related to the handedness of the subject, but a sample of over 100 observers failed to show any uniform correlation (MacKay, 1957a).

Experiments with other patterns, of which Figs. 2b to d are typical, gave similar results.

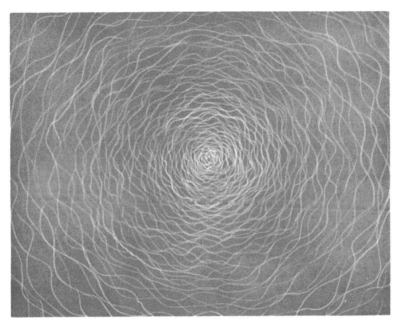

Fig. 4. An impression of the complementary after-image (CAI) evoked by Fig. 2a.

Effects of eye movements

Although the geometrical *form* of the CAI poses a separate problem from its *motion*, one of our first questions concerned the possible role of eye movements. Two types of test convinced us that eye movements were irrelevant to the streaming phenomenon and the CAI. First, the stimulus was stabilized using a simple adaptation of the method devised by Ditchburn (Ditchburn and Ginsborg, 1952), and by Riggs, et al. (1953). A contact lens of 60-diopter power was used, so that the usual mirror and optical system could be dispensed with and the stimulus figure viewed directly, supported by a light dural stalk, 1.5 centimeters long, cemented to the lens (MacKay, 1957b). Homatropine was used to immobilize the ciliary muscle; a 3-millimeter artificial pupil, formed during the molding of the lens, eliminated trouble from the resulting pupillary dilation.

Both the streaming and the CAI are unaffected by this stabilization. The stabilized image fades irregularly and spasmodically, and on each fade-out it is replaced by a moving complementary image (CI).

As a second test, an electronic photoflash lamp (rated at 120 joules) was used to impress a stationary image on the retina. After an initial fraction of a second the CAI appears, in rapid motion, and persists for 1 or 2 seconds before the appearance of the normal, but positive, after-image* which follows powerful flash stimulation (MacKay, 1958a). It may safely be concluded that motion of the retinal image is not necessary for either the formation or the motion of the CAI. It should be said, however, that the CAI can be somewhat enhanced in intensity by sweeping the contours of the stimulus to and fro on the retina, for example, by slightly rocking the ray figure (Fig. 2a) about its center. This, plus the fact that the CAI is observed no matter what part of the figure is fixated, eliminates any idea that the effect might be due to some periodic structure in the retinal mosaic happening to coincide with the periodicity in the figure.

Evocation by stroboscopic illumination

In view of the oscillatory "shimmering" of the CAI, it was thought worthwhile to use stroboscopic illumination and look for possible resonance effects. What was discovered, however, was something quite different. At rates of the order of 5 flashes per second, and energies of the order of 1 joule per flash, *stationary* fragments of the CI can be seen distributed over the figure, usually as wispy blue lines, roughly (but never exactly) normal to the stimulus contours, and interspersed with broader wisps of orange or pinkish brown. No sharply critical flash rate has been observed, and the blue lines disappear gradually as the rate is increased to 15 or 20 per second, without any unusual effects in the magical region of 10 cycles per second. (There are, of course, traces of the "honeycomb" and other evoked images familiar to clinicians, but these are present even with unpatterned stroboscopic stimulation.)

Evocation by visual "noise"

The second and perhaps most interesting discovery was made through a fortunate combination of circumstances. In order to provide structurally neutral moving stimuli in research on the after-image of seen movement, we use patterns of randomly distributed dots

* First observed to my knowledge by R. L. Gregory of Cambridge, England.

(actually photographs of sandpaper). Viewing one of these patterns through a transparency of the ray figure (Fig. 2a), I saw traces of the complementary image in the "random" background. Moving the background did not displace this CI but rather enhanced it.

This suggested that a background random in both space and time might evoke a complementary image continuously. A television set with a high-gain amplifier was used as a visual noise source, and it was indeed found that when any of the regular stimulus figures were superimposed on the "noise" (whether as a transparency, by projection on the screen, or by superposition of images in a half-silvered mirror) the "Brownian movement" which is normally seen in a noise source was replaced by a highly organized and persistent pattern having the form and directions of motion of the complementary image.

Since the characteristics of noise on a television screen are not widely controllable (in respect of spot size, for example), J. P. Wilson of our laboratory has more recently been using a closed loop of motion picture film constructed by repeated photography of confetti scattered over a black background. This gives qualitatively similar results but makes possible a much more systematic study, which is now in progress. To mention just one result, it appears that the optimal diameter of a dot is comparable (as might perhaps be expected) with the spacing of the stimulus lines.

Dependence on parameters of stimulation

At the outset it seemed possible that these phenomena might be related to the lateral inhibitory and other effects in the retina described by Kuffler (1953). (The suggestion has recently been made also by W. K. Taylor, 1958.) This possibility led us to investigate the effects of altering such parameters as line spacing, intensity, and the like. Although the study (chiefly by J. P. Wilson) is still in progress,* a few preliminary results may be of interest.

1. Contrary to Taylor's findings (1958), the CAI has been observed with *line spacings* varying over a range of several hundred to one, the lower limit being near that for grating acuity. At large spacings the CI has the appearance of wisps of smoke rather than waves, but in general character it resembles a magnified version of the CI seen at normal spacings.

2. The critical parameter is not the spacing but the *number* of parallel lines, a minimum of between four and ten being required, irrespective of spacing, if the CAI is to be seen.

* Now published as: The response of the visual system to spatially repetitive stimuli. Doctoral dissertation, University of London, 1960.

3. The optimal *ratio of black to white spaces* turns out to be one to one, though the curve has a broad maximum—estimated in terms of the duration of the CAI.

4. With many normal subjects, considerable *defocusing* of the stimulus has relatively little effect on the duration of the CAI. Occasionally, however, observers suffering from eye defects such as astigmatism have reported difficulty in seeing the CAI. This curious paradox needs further investigation.

5. The CAI can be seen over a wide range of *intensity,* and at low levels (especially near the threshold of intensity for evocation of CAI) the CI is visible continuously, drifting over the stimulus figure. (Is this related to the fact that here the signal-to-noise ratio is poorer, producing the same effect as when noise is added to a figure of normal brightness?) The threshold intensity for the production of the CAI depends on the spacing of the stimulus lines, and the curve of threshold versus spacing has the same form as the grating acuity curve, at a level about 100 times higher in intensity.

6. The CAI is rapidly reduced in intensity and duration if the stimulus lines are randomly deflected from parallelism; but randomness of *spacing* between lines that are perfectly parallel is found to be equally destructive of it. Systematic changes in direction or spacing (as in Figs. 2a to c) seem, however, to be tolerated.

Laws of form of the complementary image

As has been mentioned, the fine structure of the CI usually has a wavelike appearance. Over the whole range of line spacings, the average "wave length" (which is never too well defined) seems to be proportional to spacing, being approximately three or four times the interval from line to line.

As to its over-all form, the CI appears to try to differ as much as possible in direction from the original stimulus. Horizontal lines give a roughly vertical CI, radial give roughly circular, circular roughly radial, and so forth. The term "roughly" is meant to allow for the wavy character referred to, which may result—for example, with radial or circular stimuli—in the formation of petaloid patterns of overlapping left- and right-handed spirals. When two stimuli are combined—for example, with horizontal and vertical lines overlapping to form a grid—the CAI has its main lines at 45 degrees to the stimulus lines, as if they were the resultant of vector summation of the two component CAIs. Similarly, if parallel lines are viewed for a few seconds after radial lines, the CAI is elliptical—again as if the individual circular and linear CAIs have combined like vector fields.

Physiological location

A few clues toward the physiological location of these oscillatory after-effects have been gathered.

In the first place, although they occur with either monocular or binocular viewing, the binocular CAI is twice as lasting. This would hardly be expected if the oscillatory after-effect were in the retina, unless we postulate complex mutual reinforcement via centrifugal fibers. It looks rather as if the incoming signals evoke the oscillatory responses after binocular fusion has occurred.

Further evidence to the same effect comes from the fact that the CAI transfers to some extent from one eye to the other, in that, while viewing a regular pattern with one eye, one can see traces of its CI in a random-noise source viewed in the other eye. In prolonged monocular viewing (unstabilized) with one eye covered, the pattern tends to fade periodically—presumably owing to binocular rivalry— and then is *replaced* by the CAI on a dark background (MacKay, 1957a). Moreover, as has been mentioned, we have found no sign of a critical spacing of the stimulus lines on the retina, as might have been expected if the effect depended on the range of lateral interaction at that level. Again, since blurring of the edges of large-scale patterns may be tolerated over an area which would be covered by 100 lines of the smallest patterns, it seems clear that retinal mechanisms of contour detection or enhancement can hardly be essential to its occurrence.

On the other hand, there is a variety of evidence that the effect is not too far central. The CAI is confined, for example, to the area of the visual field stimulated. Moreover, it may sometimes be detected when the subject does not consciously resolve the details of the stimulus pattern—for example, with some moving stimuli, or after prolonged fixation of a pattern near the limit of grating acuity.

One further result tends in the same direction (MacKay, 1957a). If the subject, after watching a rotating neutral pattern (see section on Evocation by visual "noise") until he has developed a strong and lasting after-impression of rotation, observes the "target" figure (Fig. 2b) for a few seconds, the radial CAI that he sees has the appearance of rotating. This suggests that, whatever may be the transformation responsible for the after-impression of rotation, it occurs *after* the stage at which the CAI is generated, since it operates on the CAI and could have no effect on the circularly symmetrical stimulus.

At the National Hospital, Queen's Square, I have had an opportunity to study the effects of certain brain injuries on the CAI. The work is difficult for obvious reasons; but one subject with a homonymous quadrantal defect due to a vascular lesion showed an interesting and suggestive response (MacKay, 1957a). With the "ray" figure (Fig. 2a) he reported only a semicircular CAI occupying the unimpaired half-field. With the "target," however (Fig. 2b), he could see the radial CAI over all but the damaged quadrant, which was visible as a "nick" in the petaloid figure. This seems to fit with an interpretation of the process as a type of standing-wave phenomenon in the neural network, requiring an undamaged path in the direction of wave propagation.

After saturation with the rotating neutral pattern, this subject reported that he could see the CAI "rotating past the nick," which remained stationary. It is regrettable that the exact locus and extent of this lesion (presumed occipital) is unknown, as, fortunately for the subject, he required no surgical treatment. More evidence of this sort may go far to locate the oscillatory activity that generates the CAI.

Inferences

At this stage any inferences must be tentative, but it seems likely that we are here dealing with a neural network sufficiently richly interconnected to be virtually a continuous medium. The fact that two parallel lines added to a pattern of, say, four may make all the difference between seeing and not seeing a CAI (which covers the whole extent of the pattern) indicates a powerful type of long-range interaction; this moreover must take place in a system that can ignore changes of scale of several hundred to one. The fact that the CI can be "shock-excited" continuously by visual noise or stroboscopic lighting seems to indicate that a "state of strain" is present in the network concerned during all the time that a regular figure is viewed, being revealed by the noise in much the same way that magnetic lines of force are revealed by iron filings.

A further possible inference would be that the *direction of a contour* is signaled in a particularly simple code, such that the presence of many contours with the same direction results in what may be called "directional satiation" (MacKay, 1957a), with a resulting hypersensitivity to contours in the complementary direction. That this is not the whole story is shown by the dependence of the effect also on regularity of line spacing. It is our feeling that, if the type of code that shows such dependencies could be discovered, we should be a long way toward understanding the mechanism of pattern perception.

The Perception of Motion

The waterfall illusion

Evidence of neural interaction in the perception of motion is familiar in the ancient "waterfall illusion," in which, after observing a moving surface for some time, one sees whatever one looks at as "moving" in the opposite direction—without, of course, changing its position. Any idea of explaining this by nystagmic eye movements or the like is ruled out by the fact, familiar to those who look directly up or down the track from a moving train, that radial motion is equally effective in evoking such after-impressions; and two simultaneous contrarotating stimuli, for example, produce two contrarotating after-images.

The inference seems reasonable that the visual system at some stage incorporates detectors of motion as such, which become adapted, just as our sensory systems for discrimination of brightness and color do, and for some time afterwards adopt a "zero level" that is negative.

An illusion of instability in stroboscopic illumination

A remarkably powerful illusion, which has a bearing on this type of interactive process, was discovered accidentally while stroboscopic lighting was being used in a room containing self-luminous objects (MacKay, 1958*b*). When the head was turned suddenly, objects such as the glowing cathodes of valves were distinctly seen to leave their glass envelopes for a moment and then return. In a more systematic experiment it was then found that, if a card carrying a grid of reference lines and a steadily lighted point or cross at its center was lit stroboscopically at rates lower than some 15 flashes per second, irregular movement of the card resulted in a dramatic displacement of the central light within its coordinate framework, or even out of the plane if motion was perpendicular to the latter.

By gentle pressure on the corner of the eyelid, sufficient to rotate the eyeball slightly to and fro, it proved easy to see that the sensitivity to displacement of the retinal image under these conditions is much smaller for the flash-lit background than for the steady light source. By increasing the flash rate above about 15 per second, one can see the sensitivity to background motion increase to "catch up" with the sensitivity to motion of the steady source.

At these higher frequencies the illusion of displacement within the framework also disappears, the transition (around 15 cps) being a relatively sudden one of a "*Gestalt*-forming" sort, whereby one ceases to perceive the light and its background as two independently mobile entities.

It is known that a reduction in the intensity of retinal stimulation causes an increase in the latency of response (witness the familiar Pulfrich effect). Some of the effects just described can in fact be counterfeited by using a sufficiently faint *steady* source to illuminate the background: the brightly lighted central cross then appears to move appreciably ahead of its framework when the retinal image is displaced.* In order to verify that this effect could be discounted as an explanation of the stroboscopic phenomenon, we have used a wide range of intensities both of the stroboscopic lighting (0.3 to 3 joules per flash) and of the steady source illuminating the central cross. It turns out that the dissociation of movement persists even when the average (stroboscopic) background intensity is considerably greater than that of the cross, whereas in steady light of the same subjective intensity as the stroboscopic source (and indeed at much lower levels) no trace of dissociation can be seen. This seems to confirm that the explanation lies in the discontinuity of the background illumination, rather than its intensity. Its color has also been found to make no qualitative difference to the effect.

Effects of eye-hand coordination

A striking indication of one factor operative in the illusion can be had by allowing the subject to move the coordinate framework himself. At once the apparent dissociation of the cross from the background is reduced. Although the amount remaining depends largely on the regularity of the movement and on its frequency, reductions of 50 to 100 per cent are normal.

If, on the other hand, the subject displaces the retinal image by rocking a viewing mirror instead of moving the framework, little if any reduction is observed in the degree of dissociation, although the subject knows quite well what movement he will produce. Yet again, if the experimenter moves the framework, but the subject is allowed to hold or even touch it as it moves, almost as large a reduction is observed as if the subject were responsible for the movement.

The inference would seem to be that eye movements, when coordinated by positional feedback from hand movements, reduce the displacement of the retinal image and hence reduce the disparity between the perceived motions of cross and background. A further clue pointing in the same direction may be mentioned briefly. It comes from

---

* Dr. Janders of the Max-Planck Institute, Seewiesen, has drawn my attention to a similar effect visible with a bright star near the horizon at twilight.

a preliminary test* of a subject with a lesion of the dorsal column which impaired the sense of position of her hands and wrists. The subject reported no change in the degree of dissociation of cross and background on touching, holding, or moving the card. Further work on subjects with impaired sense of position is required before firm conclusions can be drawn, and this will be reported elsewhere. The evidence fits in, however, with the other data just mentioned and suggests that the illusion arises from a relative deficit of retinal information on the retinal velocity of the flash-lit background. Where the retinal velocity of the image can be reduced as a result of proprioceptive cues, the illusion is reduced in strength. Where the movement of the stimulus is sufficiently regular (for example, where it is viewed by reflection in a *mechanically* rocked mirror which gives it a circular motion), the illusion is also much reduced, since the amount of information (in the technical sense) that is required to enable the eyes to follow such circular motion is so small that proprioceptive cues are unnecessary. As soon as the eyes are prevented from following, by fixation on a stationary spot, the illusion reappears.

The foregoing evidence would seem to rule out the otherwise tempting explanation that the dissociation is due solely to the time lag between successive samples of the flash-lit background, which thus appears to lag behind the cross. This explanation would not alone account for the differences between cases of regular and irregular movement, nor for the effects of holding, touching, or controlling the moving framework.

If, however, these differences are put down to the effects of following eye movements, the evidence all indicates that dissociation occurs only when the *retinal image* is displaced. May we not then conclude that the time lag between successive samples of *retinal* image-position causes the effect? If the image is stationary on the retina, successive samples are identical, and no illusion arises. If the image moves over the retina, the self-luminous cross is always *ahead* of the last sample of the background, by an average of half the distance between successive samples.

This would certainly seem to be a large part of the story. There is, however, a further complication, in the evidence already mentioned that small displacements of the retinal image, easily detectable in steady light, may go undetected in stroboscopic light. To most observers, moreover, the total amplitude of movement of the flash-lit framework looks considerably smaller than that of the cross, irrespective of the timing of the light impulses in relation to the motion

* Arranged by kindness of Sir Russell Brain.

(that is, in a way that cannot be completely explained by accidental "missing" of the extreme positions by the stroboscope). On both counts, then, we seem to have to reckon with a genuine diminution of sensitivity to movement seen intermittently—presumably for lack of "velocity-detecting" neural interactions normally set up by the displacement of continuous stimuli. When the interval between samples is less than 1/15 sec, the information held in transient storage from one sample is apparently still sufficient, when the next sample arrives, to enable velocity detection to occur.

A parallel to this may be seen in cutaneous sensitivity to change in the position of intermittent and continuous stimuli. Some tentative experiments, suggested by the foregoing results, have shown that movement of a point pressed continuously against the skin can be detected, and its direction sensed, more readily than the same movement of the same object tapping intermittently on the skin. These experiments will be reported elsewhere, but they supply an analogy here that seems suggestive. We might epitomize the difference between the retinal images of the cross and the flash-lit background by saying that, whereas the first "strokes" the retina, the second "taps" its way over it. Our inference from the stroboscopic illusion is that the visual mechanism has developed an enhanced sensitivity to "stroking" by the retinal image, as distinct from a mere displacement, in virtue of interactive processes whose nature remains to be elucidated.

The "rubber-sheet" illusion

A final sample of evidence on interactive processes in the perception of velocity takes us back to Figs. 2a and b and their like. If Fig. 2a, for example, is viewed with one eye, and the eyeball is gently rotated to and fro by finger pressure, a curious non-Euclidean distortion is observed (MacKay, 1958b). The rays pointing in the directions in which the image is displaced seem to move apart and together fanwise, as if on a rubber sheet that is being locally distended and contracted. (The apparent contraction occurs on the side *toward* which the center is moving.)

If, however, a line (such as a thread) is placed across the pattern on one side of the center, perpendicular to the direction of motion, the "rubber" distortion in its neighborhood is eliminated, although on the opposite side it continues unaltered.

Similar effects are found with all patterns of this sort. Their interest lies in the evidence they supply on the order of priorities in perceptual organization. One might have supposed, on grounds of "parsimony," that the *form* of the perceived figure should be preserved invariant under involuntary as much as under voluntary movements of its retinal image. In fact, however, the parsimony is exercised at a lower level. Each local area of the perceptual field appears to have metrical autonomy, constrained only by the topological necessities of linking up with its neighbors. Each area responds to a change by making the minimal alteration justified by the evidence—irrespective of the *conceptual* improbability of the perceptual result.

Thus, since only the velocity component normal to a contour generates information, the rays in the direction of motion are seen to rotate only toward or away from one another. An object (such as a thread perpendicular to the rays) that supplies evidence of the complementary velocity component enables its neighborhood to be perceived in motion as a whole, undistorted.

This effect, like the others accompanying forced rotation of the eyeball, can readily be shown to have nothing to do with eyeball distortion, though subjectively this might be the most tempting hypothesis to fit the impression one gains. Rocking a viewing mirror, or even rapidly moving the stimulus pattern itself, can evoke the same impression. They are in fact what one would expect if, as I have argued elsewhere (MacKay, 1959), the perceptual mechanism functions on the principle of Fisher's null hypothesis—taking *stability* as the norm, demanding adequate evidence before making any change in the world-as-perceived, and in every case making the minimal change (however surprising) that will match such new evidence as may arise.

Conclusion

It will be clear that this survey has been more in the nature of a progress report than a finished article. What it indicates is not a set of firm conclusions so much as a number of clues to the kind of theoretical thinking and model-making that would seem to be appropriate, at least in relation to visual perception.

Predominant has been the emphasis on *lateral interaction*, both short-range and long-range, apparently present at all levels from the retina inwards. This interaction is evidently so strong that oscillatory states can be set up by suitably periodic spatial stimuli and can persist for several seconds after their removal. The characteristic geometrical patterns in which these oscillatory states "crystallize" present a problem as yet unsolved. But these patterns suggest a neural model whose interactive processes make it specifically sensitive to *direction of contour*, so that "saturation" with contours in one direction induces a hypersensitivity to contours in the complementary direction. Some evidence mentioned makes it unlikely that this is a purely retinal process.

The new stroboscopic illusions, coupled with more familiar evidence, indicate a further role of neural interaction, not only in the abstraction of velocity as a basic variable, but also in the computation of position as the integral of velocity. This latter principle seems to apply on a local basis, so that (as with the rubber-sheet illusion) local changes of perceived position may occur in response to suitably misleading evidence regardless of the metrical distortions that result. A general inference that I have discussed elsewhere (MacKay, 1960) is that our model-making at this level may find it useful to borrow (or develop) analogies with continuous structures (such as elastic membranes or solids) as an adjunct to our thinking in terms of discrete neural networks.

Summary

Interaction between widely separated areas of the visual field appears to be essential to many perceptual processes. Some fresh evidence of its nature and extent is provided by recent investigation of anomalies in the perception of pattern and of motion.

1. In vernier alignment, information from a considerable length of the line contributes to the accuracy achieved, the (metrical) information content of the alignment increasing linearly with length over a restricted range.

2. With spatially periodic stimuli, a dynamic form of interaction sets in, manifesting itself in a vigorously moving, transient "complementary after-image," suggestive of standing-wave patterns of activity in the stimulated area. Shock excitation by superimposed random visual noise, or by brief flashes of illumination, shows that even during inspection these spatially periodic stimuli set up a state of strain in the visual system, the main strain lines (rendered visible by the shock excitation) running roughly at right angles to the stimulus lines. A variety of evidence (including some preliminary clinical data) suggests that this phenomenon is not entirely (if at all) retinal in origin. Stabilization of the retinal image and the use of flash illumination have also shown it to be quite distinct from effects due to eye movement.

3. A striking indication of the role of interactive processes in the perception of motion is provided by the relative stability of the visual

world in stroboscopic lighting. It is found that, with a field containing both self-luminous and stroboscopically lit objects, the acuity for displacement of the retinal image is much lower for the latter. With sufficiently irregular displacement of the whole field, moreover, the self-luminous objects may appear to move quite independently of their flash-lit background. Information supplied by allowing normal subjects to touch or move the stimulus area for themselves reduces the apparent dissociation of the steadily lit and flash-lit objects, whereas one subject, who lacked position sense in wrist and arm, reported no change in similar circumstances. At flash rates above some 15 cps, the dissociation disappears. The inference is that an interactive mechanism, similar to that indicated by the familiar "waterfall illusion," normally may enable the *velocity* of the image over the retina to make a contribution essential to the veridical perception of displacement.

References

Ditchburn, R. W., and B. L. Ginsborg. Vision with a stabilized retinal image. *Nature, Lond.*, 1952, **170**, 36.

Erb, M. B., and K. M. Dallenbach. Subjective colors from line patterns. *Amer. J. Psychol.*, 1939, **52**, 227–241.

Fisher, R. A. *The Design of Experiments.* London: Oliver and Boyd, 1935.

Helmholtz, H. von. *Physiologische Optik*, Vol. II. 1856.

Hunter, W. S. Retinal factors in visual after-movement. *Psychol. Rev.*, 1915, **22**, 479–489.

Kuffler, S. W. Discharge patterns and functional organization of mammalian retina. *J. Neurophysiol.*, 1953, **16**, 37–68.

MacKay, D. M. Moving visual images produced by regular stationary patterns. *Nature, Lond.*, 1957a, **180**, 849–850.

MacKay, D. M. Some further visual phenomena associated with regular patterned stimulation. *Nature, Lond.*, 1957b, **180**, 1145–1146.

MacKay, D. M. Moving visual images produced by regular stationary patterns, II. *Nature, Lond.*, 1958a, **181**, 362–363.

MacKay, D. M. Perceptual stability of a stroboscopically-lit visual field. *Nature, Lond.*, 1958b, **181**, 507–508.

MacKay, D. M. The stabilisation of perception during voluntary activity. *Proc. 15th int. Congr. Psychol., Brussels*, 1959, 284–285.

MacKay, D. M. Modelling of large-scale nervous activity. In Soc. Exp. Biol. *Symposium on Models and Analogues in Biology.* Cambridge: University Press, 1960. Pp. 192–198.

Pierce, A. W. *Studies in Space Perception.* 1901.

Ratliff, F. The role of physiological nystagmus in monocular acuity. *J. exp. Psychol.*, 1952, **43**, 163–172.

Riggs, L., F. Ratliff, J. C. Cornsweet, and T. N. Cornsweet. The disappearance of steadily fixated visual test objects. *J. opt. Soc. Amer.*, 1953, **43**, 495–501.

Taylor, W. K. Visual organization. *Nature, Lond.*, 1958, **182**, 29–31.

20

TIMOTHY H. GOLDSMITH
The Biological Laboratories, Harvard University

The Physiological Basis of Wave-Length Discrimination in the Eye of the Honeybee

The world of the honeybee is not the world that we know. First and foremost, this is because of the profound differences in the sensory apparatus of bees and men. The wave lengths to which bees respond, the optics of the eye, and the neural organization of the visual system all are different. Yet at the level of primary receptive processes the bee and man have much in common.

In the pages that follow we shall consider the color vision of bees and the methods that have been employed in its study. The first section of this paper is an examination of the ability of bees to discriminate different wave lengths, as revealed by behavioral experiments. The following sections treat what is known of the underlying physiological and biochemical mechanisms.

The "Psychophysics" of the Color Vision of Honeybees

I should like to begin this essay by considering some of the work of Daumer (1956). Daumer was not the first to demonstrate color vision in the honeybee; that was accomplished by von Frisch, about forty-five years ago; however, Daumer has drawn on the extensive experience of others in training honeybees, and he has provided some remarkable data on the ability of bees to distinguish colors.

Daumer's apparatus consists of a round feeding platform on which are four quartz feeding dishes. Any or all of the dishes can be illuminated from below by monochromatic lights or mixtures of monochromatic lights, selected with interference filters from the spectrum of a xenon arc. Bees, individually marked for identification, are trained to fly to the apparatus and collect sugar syrup at a particular color. The training color is then put in competition with other lights, and as a trained bee returns to the feeding platform its choice of feeding dish is noted. Daumer has shown that the lights are selected on the basis of wave length, rather than either position or intensity.

The relative stimulative efficiency of different wave lengths

In one set of experiments Daumer trained bees to particular wave lengths. He then reduced the intensities by steps until the bees went with equal frequency to unilluminated feeding dishes. The reciprocal of the energy just necessary to attract 60 per cent of the bees to the training light was defined as the stimulative efficiency. The relative stimulative efficiency of several wave lengths is given in Table 1.

Table 1.   The Relative Stimulative Efficiency of Several Wave Lengths in Attracting Trained Bees to an Illuminated Feeding Dish

(Data of Daumer, 1956.)

Wave length ($m\mu$)	Reciprocal of the relative energy required, at threshold, to attract bees to the training color
360	5.6
440	1.5
490	0.5
530	1.0
588	0.8
616	0.3

Table 1 shows that the maximum stimulative efficiency lies in the near ultraviolet, with a second small peak in the green. A similar result was obtained by Bertholf (1931a, b) in studying the positive phototaxis of the bee.

The ability of bees to distinguish different regions of the spectrum as different colors

In many experiments bees were trained by Daumer to one wave length; their choices of feeding dish were then tabulated when two or three other wave lengths were presented simultaneously with the training light. Figure 1 summarizes the results of several of these experiments.

Each curve in Fig. 1 is drawn through three or four points, corresponding to the wave lengths present at the feeding platform during the test period. The number over each curve is the wave length, in millimicrons, to which the bees had previously been trained; each curve is based on an average of 170 observations of individual bees. In (*a*), (*b*), and (*c*) the relative energies of the test and training lights were adjusted for equal stimulative efficiency (Table 1); in

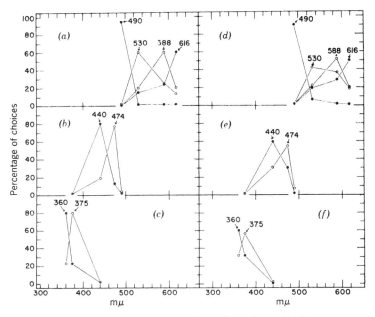

Fig. 1. The wave-length discrimination of the honeybee. Each curve represents the distribution of choices—single visits of individual bees—in percentages among three or four illuminated feeding dishes, following training to the wave length indicated over the curve. In (*a*), (*b*), and (*c*) the energies of test and training lights were adjusted for "equal stimulative effect" (see Table 1); in (*d*), (*e*), and (*f*) the intensities of the test lights remained the same, but the training light was reduced by a factor of 4. (After Daumer, 1956.)

(*d*), (*e*), and (*f*) the energies of the *training* lights were reduced by a factor of 4. The similarities between (*a*) and (*d*), (*b*) and (*e*), and (*c*) and (*f*) indicate that the bees did not distinguish the different feeding dishes on the basis of brightness.

Figures 1*a* and *d* show that bees distinguish blue-green (490 mμ) very sharply from the broad band of wave lengths encompassing green, yellow, and red. There is, however, a significant ability of bees to distinguish colors within the broad spectral region, 500 to 650 mμ. At shorter wave lengths, (*b*) and (*e*) show that bees distinguish blue (474 mμ) and violet (440 mμ) lights from each other fairly well but do not confuse either of these two wave lengths with the neighboring spectral regions, blue-green (490 mμ) and near ultraviolet (375 mμ). And similarly, (c) and (f) indicate that, within the ultraviolet, bees differentiate 360 mμ and 375 mμ from each other, but they distinguish both ultraviolet wave lengths from blue-violet (440 mμ) with even greater reliability.

Such experiments as are shown in Fig. 1 outline the system of color vision of the honeybee. Throughout the spectrum visible to the bee— from the near ultraviolet to the red—bees can be trained to alight on feeding dishes illuminated by any wave length. The ease with which two wave lengths are differentiated, however, depends not only on how close together they lie in the spectrum but also on their absolute positions in the spectrum. The spectrum can be divided into four bands between which bees exhibit a minimum of confusion—between which color differentiation is most highly developed: near ultraviolet (300 to 400 mμ), blue (400 to 480 mμ), blue-green (480 to 500 mμ), and green-yellow-red (500 to ca. 650 mμ).

"Bee-purple"

When wave lengths from the ends of the visible spectrum are combined, the resulting light appears a different color from any spectral region. For man, combinations of red and blue give rise to purples; similarly for bees, mixtures of yellow and near ultraviolet produce "bee-purples." Table 2 shows how sharply one bee-purple—consisting of 10 per cent 360 mμ and 90 per cent 588 mμ—is differentiated from all regions of the spectrum.

Table 2.   Distribution of Visits of Bees to Daumer's Illuminated Feeding Dishes Following Training to a Light Consisting of 10 Per Cent Ultraviolet and 90 Per Cent Yellow

Test and training lights were of equal energies; thus the 360-mμ test light should have had a greater "stimulative effect" than the training light (cf. Table 1). Note, however, the bees were not confused; more than 90 per cent visited the training light.

Wave lengths mμ		Percentage of visits
Training light ("bee-purple")	$\begin{cases} 10\% \ 360 \\ 90\% \ 588 \end{cases}$	91.5
Test lights	360	2.5
	474	<1
	490	<1
	588	4.5

Daumer has shown that mixtures of from 2 to 50 per cent ultra-violet (360 mμ) with yellow (588 mμ) produce bee-purples. Two of these bee-purples are differentiated from each other with better than 90-per-cent accuracy, three more with better than 70-per-cent accuracy.

"Bee-white" and complementary colors

It is common experience for us that a continuum of wave lengths from 400 to 750 mμ elicits a sensation associated with white light. It is equally well known that the same sensation can be evoked by combining just a few widely separated wave lengths; when only two wave lengths suffice, the two colors are said to be complementary. An analogous phenomenon is observed with honeybees. A continuum of wave lengths from the red or orange through the near ultraviolet produces for bees a new color that is not confused with any spectral region, and which it is appropriate to call "bee-white." The experiments of Daumer, which are summarized in Fig. 2, show the kind of data on which this conclusion is based.

In each of the experiments of Fig. 2 bees had a choice between a single test light and a light to which they had previously been trained. The proportion of bees going to each light is shown on the ordinate as a function of the spectral composition of the test light. The energies of the test and training lights were equal in the experiments of Fig. 2.

In (a) the training light was bee-white (the full spectrum of a xenon arc); the test light was obtained by adding known proportions of spectral 360 mμ to bee-white from which the ultraviolet had been filtered. When about 15 per cent of the energy of the test light was at 360 mμ, the bees selected the test and training dishes with equal frequency. Bee-white must, therefore, include some ultraviolet; however, the continuum of near-ultraviolet wave lengths in the xenon arc can be replaced in a "color match" by an appropriate amount of monochromatic 360 mμ.

The same experiment is repeated in (b), except that the broad band of wave lengths longer than 400 mμ, which was present in the test light of (a), has been replaced with monochromatic 490 mμ. Thus 490 mμ and 360 mμ are complementary; a mixture of about 15 per cent 360 mμ and 85 per cent 490 mμ is confused by bees with bee-white. Furthermore, these experiments imply that light that is white for us but contains no ultraviolet has something of the character of spectral 490 mμ for bees (cf. Hertz, 1939).

In (c) the bees were trained to white from which the ultraviolet had been removed; the test light was 440 mμ plus 588 mμ, mixed in various proportions. The bees confused the test and training lights when the test light contained 35 per cent blue and 65 per cent yellow.

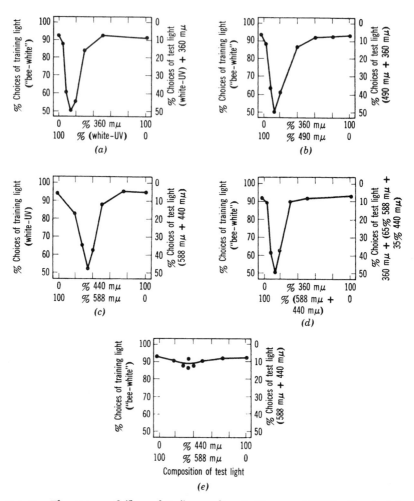

Fig. 2. The nature of "bee-white," according to Daumer (1956). (See text for explanation.)

Experiment (c) suggests that a mixture of 35 per cent 440 mμ and 65 per cent 588 mμ can be substituted for 490 mμ. Experiment (d) is a repeat of the second experiment (b), with the 490 mμ in the test light replaced by this mixture of blue and yellow. The resulting curve is identical to those of (a) and (b).

Finally, (c) confirms that no mixture of blue and yellow is equivalent to bee-white. Bee-white must include some ultraviolet.

The relationships in Fig. 2 can be summarized by the following shorthand notations, which should be read as ordinary colorimetric equations. From (d):

$$0.55 \ (588 \ m\mu) + 0.30 \ (440 \ m\mu) + 0.15 \ (360 \ m\mu)$$
$$= \text{bee-white} \quad (1)$$

From (b):

$$0.85 \ (490 \ m\mu) + 0.15 \ (360 \ m\mu) = \text{bee-white} \quad (2)$$

And from (a) through (d), considered together:

$$0.65 \ (588 \ m\mu) + 0.35 \ (440 \ m\mu)$$
$$= (490 \ m\mu) = (\text{bee-white} - \text{UV}) \quad (3)$$

The wave lengths 360 $m\mu$ and 490 $m\mu$ are complementary. Figure 2d shows, however, that the complementary to 440 $m\mu$ is a bee-purple. The composition of this bee-purple can be calculated from Eq. 1—79 per cent 588 $m\mu$, 21 per cent 360 $m\mu$.[*] The question remains, what is complementary to 588 $m\mu$? We might guess from Eq. 1 that some wave length lying between 360 $m\mu$ and 440 $m\mu$ will serve. Daumer has not put this question to direct experimental test. He has, however, shown that a mixture of 440 $m\mu$ and 360 $m\mu$ is equivalent to the one intermediate wave length available with his equipment:

$$0.80 \ (360 \ m\mu) + 0.20 \ (440 \ m\mu) = (375 \ m\mu) \quad (4)$$

Mixtures of ultraviolet and yellow are not equivalent to any intermediate wave length; the existence of such bee-purples shows clearly that two variables do not suffice to describe the color-vision system of the bee as it is revealed in these "color-matching" experiments. The simplest interpretation of these behavioral experiments—the interpretation offered by Daumer—is that the bee has three receptor mechanisms excited maximally by different regions of the spectrum. The data of Figs. 1 and 2 suggest that the three receptor systems are centered in the near ultraviolet (300 to 400 $m\mu$), blue-violet (400 to 480 $m\mu$), and green-yellow-red (500 to $ca.$ 650 $m\mu$).

The narrow band of wave lengths 480 to 500 $m\mu$ produces an excitation that is as sharply defined as the effects of the three broader spectral regions. Because of the equivalence of stimuli expressed in

[*] $\dfrac{0.55}{0.55 + 0.15} = 0.79; \quad \dfrac{0.15}{0.55 + 0.15} = 0.21$

Eq. 3, however, it is likely that the behavioral responses associated with blue-green light occur when stimulation is confined (or virtually so) to the blue-violet and green-yellow-red receptors. This interpretation explains why 490 mμ is complementary to 360 mμ (Eq. 2) and makes it unnecessary to postulate a fourth, blue-green receptor. This argument suggests an analogy between the place of blue-green in the color-vision system of the honeybee and yellow in the color vision of man. In each system, this is the spectral region in which wave-length discrimination is best developed, and these are the wave lengths that are the least saturated. Daumer has shown that, after training to either 490 mμ or 360 mμ, in order for a bee-white test light to assume the character of the training color and attract the bees, about ten times as much blue-green as ultraviolet must be added to the bee-white.

Concerning the Physiological Substratum

The ability of an animal to distinguish different colors has its correlates in the physiological properties of the animal's photoreceptors. As we have just seen, the behavioral responses of trained honeybees imply the presence of receptor mechanisms responding to ultraviolet, blue, and green-yellow-red. It is a matter of some interest, therefore, if other evidence indicates the presence of receptors with peaks of sensitivity in these regions of the spectrum.

Some anatomical considerations

The honeybee possesses two kinds of eye. Two large compound eyes cover the lateral surfaces of the head; three small, simple eyes—ocelli—form the corners of a triangle on the dorsum of the head.

The compound eyes consist of several thousand ommatidia. Each ommatidium is an elongate, cylindrical structure, optically isolated from adjacent ommatidia by a sleeve of dense pigment, and capped on its distal end by a transparent corneal facet (Fig. 3). Within each ommatidium are eight long retinular cells; each retinular cell has a differentiated medial border, the rhabdomere. Collectively the rhabdomeres form the rhabdom, a long, slender body which lies on the axis of the ommatidium.

The rhabdoms of species representing five orders of insect have been examined with the electron microscope (Odonata, Goldsmith and Philpott, 1957; Orthoptera, Fernández-Morán, 1958; Lepidoptera, Fernández-Morán, 1958; Diptera, Goldsmith and Philpott, 1957;

Wolken, Capenos, and Turano, 1957; Danneel and Zeutzschel, 1957; Fernández-Morán, 1958; Yasuzumi and Deguchi, 1958; and Hymenoptera, Goldsmith, unpublished), and all, like the rhabdoms of *Limulus* (Miller, 1957), possess a similarly oriented fine structure of microtubules. This fine structure suggests an analogy in function between rhabdoms and the outer limbs of vertebrate rods and cones and is a principal reason why the rhabdom is believed to be the site of the primary process of photoreception.

Each retinular cell of the bee is a primary sensory neuron, sending a fiber to the central nervous system. The fibers pass through the basement membrane of the ommatidial layer, and many, if not all, synapse immediately in the *lamina ganglionaris* of the optic lobe of the brain (Kenyon, 1897; Phillips, 1905).

It is generally believed that single ommatidia respond as units. This certainly appears to be true in *Limulus*, where one cell, the eccentric cell, occupies a unique position with respect to the rhabdom and is the only cell from which it is possible to record the discharge of propagated spikes (MacNichol, 1958); however, it has yet to be proved that the ommatidia of insects respond as units. Although the visual acuity of the honeybee correlates with the interommatidial angle (Hecht and Wolf, 1928–29), the possibility remains that, for other visual functions, retinular cells behave as independent units. This possibility perhaps is more likely in Diptera, where the rhabomeres of each ommatidium are not fused into a single structure.

The ocelli are inconspicuous by comparison with the compound eyes, and their function in the life of the bee is not known. Most of what can be said about them is negative; they do not, for example, resolve

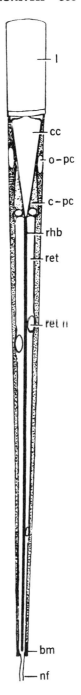

Fig. 3. Diagram of a single ommatidium from the compound eye of a worker honeybee (*Apis mellifera*), after Phillips (1905): 1, corneal lens; cc, crystalline cone (composed of 4 cells); o-pc, outer pigment cell (12 cells); c-pc, corneal pigment cell (2 cells); rhb, rhabdom; ret, retinular cell (8 or 9 cells); ret n, retinular cell nucleus; bm, basement membrane; nf, nerve fibers (one from each retinular cell). Each ommatidium thus consists of 26 or 27 cells.

images and, by themselves, have never been shown to mediate any behavioral response. As we shall see, however, they possess two different types of photoreceptor responding in different regions of the spectrum, and so their role in the visual process remains a perplexing problem.

Electrophysiological observations of spectral sensitivity

1. *The compound eye of the worker bee.* When light strikes the compound eye of a worker bee, an electrode in the superficial layers of the retinulae records a monophasic, negative retinal action potential (Fig. 4). This response is complex in origin. Its main com-

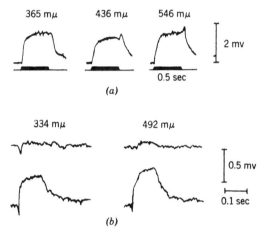

Fig. 4. Electroretinograms from the dark-adapted (*a*) compound eye and (*b*) ocellus of the worker bee. A direct-coupled amplifier was used to obtain both sets of records; an upward deflection signifies negativity of the electrode in the illuminated eye. (*a*) Retinal action potentials from the compound eye in response to 0.5-sec test flashes. With such long test flashes it is possible to see that the off effect increases in prominence at long wave lengths. Lower trace indicates light stimulus. (*b*) Threshold and suprathreshold responses of an ocellus, showing the positive on effect which is present at threshold only in the near ultraviolet. In both the compound eyes and ocelli the differences between retinal action potentials evoked by long and short wave lengths are enhanced by light adaptation.

ponent is a sustained negativity that persists as long as the stimulus and that is believed to arise from the retinular cells. Other transient components—on and off effects—may be present; these are thought to originate in the optic ganglion. The shape of the response can vary somewhat with the wave length of the stimulus and the state of light adaptation of the eye (Goldsmith, 1959–60).

The magnitude of the retinal action potential is a useful measure of excitation. Measurements of spectral sensitivity have been made on the compound eye by determining the energy (quantized) required to elicit retinal action potentials of equal magnitude in response to stimuli of different wave lengths. The test stimuli lasted 0.1 sec; with

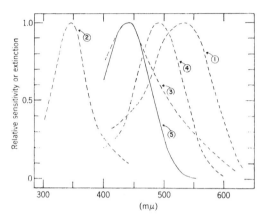

Fig. 5. Spectral-sensitivity functions from the compound eyes and ocelli of honeybees (broken curves) and the absorption spectrum of a photosensitive pigment extracted from the heads of bees (solid curve). The spectral-sensitivity functions are based on reciprocals of number of quanta required to elicit constant-size retinal action potentials in response to 0.1-sec test flashes. (1) The spectral sensitivity of the dark-adapted compound eye of the worker in the visible region of the spectrum. This curve appears to describe a green-sensitive receptor system. (2) The spectral sensitivity of the ultraviolet receptor system of the compound eye of the worker, as revealed during adaptation to a bright yellow-green light. Within the uncertainties of the available measurements, the spectral sensitivities of the ultraviolet receptor systems of the ocellus and the compound eye of the drone are the same. (3) Dark-adapted compound eyes of drone bees in which the sensitivity in the visible region of the spectrum appears to be determined principally by a blue-sensitive receptor system. (4) The blue-green receptor of the ocellus of the worker. (5) Absorption spectrum of the photosensitive retinene-protein pigment extracted from the heads of bees.

this duration it was often possible to obtain approximate matches of the wave forms of responses of the dark-adapted eye at different wave lengths.

The spectral-sensitivity function of the dark-adapted compound eye of the worker honeybee has a broad peak centered at about 535 mμ (Fig. 5, curve 1). Only a low shoulder or, occasionally, a second minor maximum appears in the near ultraviolet, in marked contrast to the "relative stimulative efficiency" of different wave lengths in evoking behavioral responses (cf. Table 1). Often it is assumed that the capacities of the sense organs of "lower" animals are directly measurable in the animals' behavior. The relatively great effectiveness of ultraviolet wave lengths in stimulating phototaxis appears, however, to involve selection of information in the central nervous system of the bee (Goldsmith, 1959–60).

The green receptor system is somewhat sensitive in the near ultraviolet; however, it is possible to reveal the presence of a separate ultraviolet receptor system in the compound eye of the worker honeybee by adapting the eye to green, yellow, or red light. These long wave lengths depress the sensitivity of mechanisms functioning maximally at long wave lengths but spare the ultraviolet receptor. The spectral sensitivity of the eye, measured in the presence of a bright yellow-green adapting light, is also shown in Fig. 5 (curve 2). (The data for curves 1 and 2 can be found in Goldsmith, 1959–60.) The peak of sensitivity appears to be at 345 mμ and is probably correct to 5 or 10 mμ. This curve does not extend to longer wave lengths, because in such a light-adapted eye the retinal action potentials differ appreciably in wave form at the two ends of the spectrum; only through the violet and near ultraviolet was it possible to match responses at different wave lengths. However, sensitivity of the ultraviolet receptor system decreases in the visible (Goldsmith, 1959–60).

Because the retinal action potential arises from different sites, measurements of its height reflect the activities of both retinular cells and associated neurons. It is convenient, therefore, to speak of these spectral-sensitivity curves as the spectral sensitivities of receptor systems, with the realization that these may not describe just the properties of primary receptor cells. The anatomical basis of the receptor systems includes both retinular cells and neurons in the optic ganglion but is otherwise unknown; the receptor systems are at present really defined by the spectral-sensitivity functions. Actually, to identify the experimentally measured spectral-sensitivity functions with the properties of discrete receptor systems further assumes that the receptor systems contribute independently to the retinal action potential. There is some preliminary evidence for this: the spectral sensitivity of the green receptor system is not much altered in different states of light adaptation (Goldsmith, 1959–60).

2. *The compound eye of the drone.* The compound eye of the drone bee is about three times larger than that of the worker. Like the worker's, it contains receptor systems for green and ultraviolet. It is possible, however, to observe the ultraviolet receptor system without selectively adapting the eye to long wave lengths; an ultraviolet maximum is present in the spectral-sensitivity curve of the

dark-adapted eye. In different preparations this peak varies in height, relative to the peak in the visible part of the spectrum. There is another striking difference between the two castes; in the drone the peak of sensitivity in the visible is not constant in position or shape in different preparations. This appears to be because at least three receptor systems contribute prominently to the retinal action potential of the drone, the two described for the worker and a third in the blue-violet. In different preparations, the responses of blue-violet and green receptor systems are combined in varying proportions in the retinal action potential, probably depending on the distribution of receptors in the immediate vicinity of the tip of the electrode. When the mechanisms sensitive to long wave lengths appear to contribute relatively little to the retinal action potential, maximum sensitivity of the eye is at about 440 mμ (Fig. 5, curve 3). This is probably the position of maximum sensitivity of a blue-violet receptor system. The retinal action potential of the compound eye of the drone appears to be more complex than that of the worker, and the drone has not yet been studied by selective adaptation.

0. *Tho ooolli of workers.* The ocelli of worker bees also respond to light stimulation with a retinal action potential that is primarily negative (Fig. 4). Like the retinal action potential of the compound eye, the wave form depends on the wave length of stimulation. In the visible, the threshold response is entirely negative; in the near ultraviolet the threshold retinal action potential begins with a rapid positive component.

It is possible to construct spectral-sensitivity curves by matching either the positive or negative waves of the retinal action potential. In either case the spectral-sensitivity function has two maxima, at 490 mμ and at about 340 mμ. The relative heights of the two peaks depend on which component of the retinal action potential the spectral-sensitivity curve is based on (Goldsmith and Ruck, 1957–58).

These experiments demonstrate the existence of two receptor mechanisms in the ocellus. The ultraviolet receptor very likely has the same spectral sensitivity as the ultraviolet receptor system of the compound eyes; however, the 490-mμ receptor is known only in the ocellus. The sensitivity of the 490-mμ receptor is shown in Fig. 5 (curve 4).

It was mentioned above that the ocelli, by themselves, do not mediate any behavioral responses. Furthermore, the relatively great "stimulative effect" of ultraviolet (Table 1) does not require excitation of the ultraviolet receptor of the ocellus (Goldsmith, 1959–60). It is probably also true that the wave-length discrimination demonstrated in the behavioral experiments of Daumer does not involve the 490-mμ receptor of the ocellus; however, this has not been tested by experiment.

The Photochemical Basis of Vision in the Bee

The electrophysiological measurements of the spectral sensitivities of compound eyes and ocelli provide evidence for photoreceptor mechanisms functioning in the spectral regions postulated by Daumer. It is tempting to assume that the broken curves of Fig. 5 (curve 3 excepted) reflect the properties of single receptor systems, perhaps with spectral sensitivities not much different from the primary receptors. This view could be strengthened considerably if it were possible to show that the broken curves of Fig. 5 correspond to absorption spectra of visual pigments extracted from the honeybee. Unfortunately, only one such comparison is possible at this time.

The visual pigments of all animals that have been examined are formed of protein (opsin) conjugated to retinene (Wald, 1958; Hubbard and St. George, 1957–58; Wald and Hubbard, 1957; Brown and Brown, 1958). Retinene, the aldehyde of vitamin A, is an unsaturated molecule of 20 carbon atoms, and it can exist in several geometrical isomers. Only one of these isomers, neo-b (11-*cis*), is found in combination with opsins to form visual pigments (Hubbard and Wald, 1952–53; Hubbard and Kropf, 1958). Neo-b retinene has an absorption maximum in the near ultraviolet at about 376 to 379 mμ; however, visual pigments (or chromopsins) absorb maximally anywhere from the blue (see, for example, Denton and Warren, 1956) through the yellow-green (Wald, Brown, and Smith, 1954–55), depending on the opsin.*

The heads of honeybees contain retinene bound to protein. This retinene is probably restricted to the compound eyes and ocelli, for no retinene is present in the thoraces and abdomens. When the heads of dark-adapted honeybees are ground in neutral buffer, most of the retinene-protein complex comes into solution. After partial purifica-

* This account is somewhat simplified. Actually there are two retinenes— retinene$_1$ and retinene$_2$, the latter with an extra double bond—and two vitamins A. When combined with a given opsin, the retinene$_2$ pigment has absorption maximum 20 mμ or more to longer wave lengths than the corresponding retinene$_1$ complex. To date only retinene$_1$ chromopsins have been found in invertebrates, and the present discussion is concerned with retinene$_1$ pigments only. However, Wald, Brown, and Smith (1953) have described a red-sensitive retinene$_2$ cone pigment.

tion, this buffer extract can be shown to contain a photosensitive pigment with peak of absorption at about 440 mμ. On exposure to light the pigment bleaches, liberating retinene (Goldsmith, 1958).

The absorption spectrum of this photosensitive retinene-protein complex is shown by the solid curve (5) in Fig. 5. This curve was obtained from a difference spectrum made in the presence of potassium borohydride. The retinene was therefore reduced to vitamin A, with peak of absorption at 328 mμ but negligible absorption at wave lengths longer than 400 mμ. The unbroken curve of Fig. 5 therefore corresponds closely to the absorption spectrum of the unbleached pigment.

There is approximate correspondence between the absorption spectrum of this blue-sensitive pigment and the spectral sensitivity maximum in the blue reported for the compound eye of the drone bee. The spectral sensitivity function of the blue-sensitive mechanism shown in Fig. 5 (curve 3) is almost surely contaminated with contributions from the green receptor system; however, the coincidence of the maxima suggests that the 440-mμ photopigment is the visual pigment of the blue-violet receptor system.

There is as yet no electrophysiological evidence for the existence of a blue receptor system in the compound eye of the worker bee. The behavioral experiments of Daumer and the occurrence of a blue-sensitive pigment in the heads of worker bees both suggest, however, that such a blue-violet receptor is present.

The spectral sensitivities of the green and ultraviolet receptor systems correspond fairly well to the maxima of the curves of relative stimulative efficiency obtained in behavioral experiments (Table 1; Bertholf, 1931a, b; see also Goldsmith, 1959–60). Although to date only one photosensitive pigment has been identified in the extracts of bees, there is reason to hope that further work will reveal others. After the partial purification of the buffer extracts of dark-adapted bees' heads, only about half the total retinene originally present can be accounted for by the 440-mμ pigment.

The occurrence of a receptor with maximum sensitivity in the near ultraviolet poses new problems. Free neo-b retinene absorbs maximally at about 376 to 379 mμ, and the combination of retinene and opsin is customarily associated with a shift in absorption maximum to longer wave lengths. The ultraviolet receptor mechanisms of the bee, however, are maximally sensitive at *shorter* wave lengths, 340 to 345 mμ. Nevertheless, the apparently universal utilization of retinene as the chromophore of other animal visual pigments leads one to suspect that the bee is playing a variation on a familiar theme.

As Professor G. Wald has pointed out to the author, any process that removes the terminal double bond from the conjugated system of the retinene molecule will result in a shift in spectrum to shorter wave lengths. (The reduction of retinene to the alcohol, vitamin A, is an example; the shift in this case is about 50 mμ.) Wald and Brown (1951–52, and in a personal communication) have shown that the aldehyde group of retinene combines spontaneously in solution with the sulfur-containing amino acids cysteine and homocysteine to form complexes with absorption maxima between 335 and 340 mμ. Such reactions show that the variety of possible combinations of retinene with protein is, in principle, sufficiently great to construct a visual pigment for the ultraviolet receptors of the bee.

Summary

The spectrum visible to the honeybee extends from about 300 mμ in the near ultraviolet to at least 650 mμ in the red; beyond these limits sensitivity steadily decreases. Bees can be trained to associate

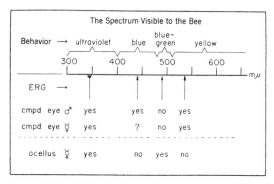

Fig. 6. Above the scale of wave lengths: the spectral regions which, in behavioral experiments, appear to be relatively distinct colors for bees (see von Frisch, 1914; Kühn, 1927; Daumer, 1956). Below: the positions of maxima of spectral sensitivity that are revealed by analysis of electroretinograms (vertical arrows), and the occurrence of these spectral-sensitivity functions in several photoreceptors of the honeybee.

food with a great number of wave lengths; however, Fig. 6 shows diagrammatically the several spectral regions that bees can be trained to distinguish with greatest reliability. The approximate extents of these spectral bands are as follows: near ultraviolet, 300 to 400 mμ; blue-violet, 400 to 470 mμ; blue-green, 470 to 510 mμ; and green-yellow-red, 510 to 650 mμ.

An analysis of complementary colors and of "bee-purples" leads to the conclusion that the visual system of the bee is built upon receptors maximally sensitive in the near ultraviolet, blue, and green (or yellow) regions of the spectrum. Electrophysiological measurements of spectral sensitivity, based on the energies necessary to evoke retinal action potentials of equal size, support this view. The positions of the spectral-sensitivity maxima determined by this electroretinographic method are shown in Fig. 6 by the vertical arrows.

An ultraviolet receptor system is present in the three photoreceptors of the bee that have been studied, the compound eyes of workers and drones, and the ocelli of workers. Maximum sensitivity appears to be at about 340 or 345 mμ.

A blue-violet receptor system contributes to the retinal action potential of the drone; there is at this time no conclusive electrophysiological evidence for its presence in the compound eye of the worker. It is likely based on the blue-sensitive retinene-protein pigment with λ_{max} at 440 mμ that has been extracted from the heads of workers and mixtures of workers and drones. It is suggested that the blue-violet receptor system may also occur in the compound eye of the worker.

A receptor maximally sensitive in the blue-green at about 490 mμ is present in the ocelli of worker bees. However, it is thought that the bees' ability to distinguish sharply between blue-green and all other wave lengths is mediated by the receptors of the compound eyes. The function of ocelli is still unknown.

A green receptor system with a broad maximum of sensitivity centered at about 535 mμ occurs in the compound eyes of workers; a receptor system with similar, if not identical, spectral sensitivity is also present in the compound eyes of drones.

Acknowledgment

The experiments of the author were supported in part by a grant B-2077 from the Institute of Neurological Diseases and Blindness, U. S. Public Health Service. The author is a member of the Society of Fellows of Harvard University.

References

Bertholf, L. M. Reactions of the honeybee to light. I. The extent of the spectrum for the honeybee and distribution of its stimulative efficiency. *J. agric. Res.*, 1931a, **42**, 379–419.

Bertholf, L. M.   The distribution of stimulative efficiency in the ultra-violet spectrum for the honeybee.   *J. agric. Res.*, 1931*b*, **43**, 703–713.

Brown, P. K., and P. S. Brown.   Visual pigments of the octopus and cuttlefish. *Nature, Lond.*, 1958, **182**, 1288–1290.

Danneel, R., and B. Zeutzschel.   Über den Feinbau der Retinula bei *Drosophila melanogaster*.   *Z. Naturforsch.*, 1957, **12b**, 580–583.

Daumer, K.   Reizmetrische Untersuchung des Farbensehens der Bienen.   *Z. vergleich. Physiol.*, 1956, **38**, 413–478.

Denton, E. J., and F. J. Warren.   Visual pigments of deep sea fish.   *Nature, Lond.*, 1956, **178**, 1059.

Fernández-Morán, H.   Fine structure of the light receptors in the compound eyes of insects.   *Exp. Cell Res.*, 1958, Suppl. **5**, 586–644.

von Frisch, K.   Der Farbensinn und Formensinn der Biene.   *Zool. Jahrb., Abt. Allg. Zool. Physiol. Tiere*, 1914, **35**, 1–188.

Goldsmith, T. H.   The visual system of the honeybee.   *Proc. Nat. Acad. Sci.*, 1958, **44**, 123–126.

Goldsmith, T. H.   The nature of the retinal action potential, and the spectral sensitivities of ultraviolet and green receptor systems in the compound eye of the worker honeybee.   *J. gen. Physiol.*, 1959–60, **43**, 775–799.

Goldsmith, T. H., and D. E. Philpott.   The microstructure of the compound eyes of insects.   *J. biophysic. biochem. Cytol.*, 1957, **3**, 429–440.

Goldsmith, T. H., and P. R. Ruck.   The spectral sensitivities of the dorsal ocelli of cockroaches and honeybees—an electrophysiological study.   *J. gen. Physiol.*, 1957–58, **41**, 1171–1185.

Hecht, S., and E. Wolf.   The visual acuity of the honey bee.   *J. gen. Physiol.*, 1928–29, **12**, 727–760.

Hertz, M.   New experiments on colour vision in bees.   *J. exp. Biol.*, 1939, **16**, 1–8.

Hubbard, R., and A. Kropf.   The action of light on rhodopsin.   *Proc. Nat. Acad. Sci.*, 1958, **44**, 130–139.

Hubbard, R., and R. C. C. St. George.   The rhodopsin system of the squid. *J. gen. Physiol.*, 1957–58, **41**, 501–528.

Hubbard, R., and G. Wald.   Cis-trans isomers of vitamin A and retinene in the rhodopsin system.   *J. gen. Physiol.*, 1952–53, **36**, 269–315.

Kenyon, F. C.   The optic lobes of the bee's brain in the light of recent neurological methods.   *Amer. Naturalist*, 1897, **31**, 369–376.

Kühn, A.   Über den Farbensinn der Bienen.   *Z. vergleich. Physiol.*, 1927, **5**, 762–800.

MacNichol, E. F.   Subthreshold excitatory processes in the eye of *Limulus*. *Exp. Cell Res.*, 1958, Suppl. **5**, 411–425.

Miller, W. H.   Morphology of the ommatidia of the compound eye of *Limulus*. *J. biophysic. biochem. Cytol.*, 1957, **3**, 421–428.

Phillips, E. F.   Structure and development of the compound eye of the honey bee. *Proc. Acad. Nat. Sci. Phila.*, 1905, **57**, 123–157.

Wald, G.   Retinal chemistry and the physiology of vision.   In *National Physical Laboratory* Sympos. No. 8.   *Visual Problems of Colour*.   London: H. M. Stationery Office, 1958.

Wald, G., and P. K. Brown.   The role of sulfhydryl groups in the bleaching and synthesis of rhodopsin.   *J. gen. Physiol.*, 1951–52, **35**, 797–821.

Wald, G., P. K. Brown, and P. H. Smith.   Cyanopsin, a new pigment of cone vision.   *Science*, 1953, **118**, 505–508.

Wald, G., P. K. Brown, and P. H. Smith. Iodopsin. *J. gen. Physiol.*, 1954–55, **38,** 623–681.

Wald, G., and R. Hubbard. Visual pigment of a decapod crustacean: the lobster. *Nature, Lond.*, 1957, **180,** 278–280.

Wolken, J. J., J. Capenos, and A. Turano. Photoreceptor structures. III. *Drosophila melanogaster. J. biophysic. biochem. Cytol.*, 1957, **3,** 441–448.

Yasuzumi, G., and N. Deguchi. Submicroscopic structure of the compound eye as revealed by electron microscopy. *J. Ultrastr. Res.*, 1958, **1,** 259–270.

21

M. A. BOUMAN

Institute for Perception RVO-TNO, Soesterberg, The Netherlands

History and Present Status of Quantum Theory in Vision

Soon after the final discovery of the corpuscular nature of light by Planck (1901), the minimum perceptible amount of light for human vision was compared with the energy of the quantum of light.

At about the same time as the discovery of the quantum, fairly accurate measurements were made by different investigators of this minimum perceptible. It was already known that the smallest values are found for bluish-green light in the dark-adapted peripheral retina. Under these conditions, 1 to 6×10^{-10} erg suffices for perception (Langley, 1889; Grijns and Noyons, 1905). More recent measurements with modern technical equipment have given similar results.

Around 1916 the first remarks about the number of quanta involved in the retinal acts at the threshold for perception appeared in the literature. One of the very first is due to Lorentz (Zwaardemaker, 1920). He called the attention of Zwaardemaker to the fact that a very restricted number of quanta—25 to 150—have to enter the eye in order for it to see. Zwaardemaker concluded that because of this small number the sensitivity of the eye is almost as great as is physically realizable. Accounting for the losses due to absorption, reflection, and scatter in the eye media, Zwaardemaker postulated that the requirements for transport of energy in the visual cycle are determined by the nervous system. Probably being aware of Sherrington's results about transmittance of action potentials in the central nervous system, he suggested speculatively in 1918 that the active energy needed in order to reach the threshold is fulfilled for the nervous system when two quanta are absorbed. More than 25 years later, van der Velden (1944, 1946) gave a more rigid base to this hypothesis with the aid of ingenious considerations of a theoretical-statistical nature and with some related experiments.

His work was preceded by that of other authors who have built on the idea that the flux of quanta at threshold level is so low that statistical fluctuations in this flux become apparent in perception (Ferree, 1913). Barnes and Czerny (1932) and Brumberg and Vavilov (1933) especially have pointed out that the scintillations of point sources seen near the threshold level in the dark-adapted eye are not attributable exclusively to variations of a physiological or psychological nature. Their studies were only of qualitative importance, partly by intention and partly because of the application of a questionable method of analysis to their experimental statistical results.

The Frequency-of-Seeing Curve

The statistical fluctuations in the number of quanta are described by the Poisson distribution. The actual number in a particular sample has the probability W of being k

$$W(k) = e^{-\bar{n}} \cdot \frac{\bar{n}^k}{k!} \tag{1}$$

in which $\bar{n}$ is the average number. In psychophysical experiments, only $\bar{n}$ can be determined in its absolute value with thermopile or photoelectric cell. The amount of light per unit of visual angle and per second incident in the eye is determined in this way. About half this light is lost on its way from cornea to retina (Ludvigh and McCarthy, 1938). A good deal of the retinal illumination perhaps does not contribute to the perception. Indeed when the total amount of photosensitive pigment that can be extracted from retinas is homogeneously distributed over an area of the same size, no more than 20 per cent of the light is absorbed (König, 1894; Hecht, 1944). The actual distribution of the pigment over the retina is not known, but 20-per-cent absorption of the incident light might be the maximum.

With the aid of these estimates, together with Lorentz's and Zwaardemaker's considerations, it was possible to conclude that the number of quanta that the photopigment would have to absorb in order for perception to take place is at least 1 and not more than 15. Indeed Hecht (1942) working along these lines came to the same conclusion, so far as the upper limit is concerned. He seemed not to be interested in the lower limit. However, the data he used also point to 1 as the possible lower limit.

These low numbers apply when a small area of the retina is exposed for a short period, say, about one-tenth of a second, and when the eye is dark adapted, and bluish-green quanta are used. Since, in such an area, several hundred receptors are available, the chance that more than one of the quanta are absorbed in the selfsame receptor is sufficiently low to warrant the conclusion that for activation of an individual receptor one absorbed quantum is sufficient. Early in World War II, Hecht (1942), van der Velden (1944), and de Vries (1943) all came independently to this conclusion. The work just referred to involves only rod vision, and I will leave the question of the cones open until a later section.

The same authors studied the problem, how many of these individually activated rods are required for perception? Hecht and van der Velden, in particular, have contributed much to the solution of this problem. Hecht's work in this respect was limited to a method that van der Velden also used, along with other methods with greater analytic power.

In short the idea is as follows. If k absorptions are required—or k activated receptors—the chance that, in an actual flash with an average number of $\bar{n}$ quanta, k successful absorptions will occur is

$$W(k) = 1 - e^{-f\bar{n}} \sum_{0}^{k-1} \frac{f\bar{n}^{\nu}}{\nu!} \qquad (2)$$

In this equation, f is the fraction of the incident light that on the average is not lost between the outside surface of the cornea and active absorption in the receptor.

If the chance for perception of such flashes should be only a physical phenomenon like the occurrence of spikes in nerve fibers (Hartline, 1959; Mueller, 1954), Eq. 2 would represent the dependence of the chance for yes responses on the flash intensity. The shape of the experimental frequency-of-seeing curve in that case can be compared with the curves represented in Fig. 1. The value of f is not relevant for this comparison. A secondary hypothesis in this procedure is the constancy of k. If k is unstable, the procedure of comparison just described does not give rigid information on the lowest possible k value.

The results obtained with this method by various authors are far from equal. Frequency-of-seeing curves were found with k values between 2 and 8. Evidently perception even under these very sophisticated conditions is far from a physical act.

Various authors have produced evidence that the frequency-of-seeing curve is highly dependent on the conditioning and motivation of the test subject. After all, the method using the frequency-of-seeing curve as the only source of information for determination of k has not produced spectacular results. The region of possible k values has narrowed from between 1 and 15 to between 2 and 8.   Recently Pinegin (1958) came again to higher values up to 12.

Anyhow 2 to 12 is a small number.   In the light of the results from the frequency-of-seeing curve, it is worth while to consider whether thermal noise in the photopigment of the receptor does not already produce a permanent supraliminal stimulus situation.   If so, the soundness of this theory should become highly suspect.

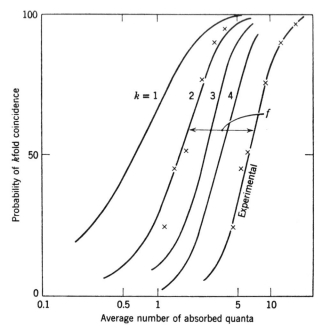

Fig. 1.   The probability $W(k)$ that in a flash with average quantum number $\bar{n}$ at least $k = 1,2,3,4$ quanta occur.   The x's around the experimental curve indicate frequency of seeing as a function of flash intensity for small short test stimuli; experimental curve represents parallel shift of $k = 2$ curve (van der Velden, 1944, 1946).

In 1948 de Vries estimated the number of rhodopsin molecules that decompose per second in the human eye by thermal movements.   He calculated 10^{-9} to 10^{-6} decomposition per rod per second.   This means at least several hundred per second in the whole retina.

If the eye behaved exactly like an ideal physical instrument, a signal containing approximately the square root of these few hundred decompositions could be detected.   This means about 50 quanta on the cornea. Actually this is the number which under suitable conditions can result in a perception. These conditions include the necessity to focus the energy in a very restricted area and to apply it in a short time. This area is about half a degree in diameter, and the time is about one-tenth of a second. Only 10^{-3} thermal spontaneous decomposition then coincides with the flash. The retinal threshold, 2 to 12 quanta, exceeds enormously the number of spontaneous decompositions per summation area and per summation time. It seems as if the demand of spatial and time coincidence of some small number of quanta by the nervous structures eliminates the thermal noise. A real profit in terms of signal-to-noise ratio!

In order to explain the behavior of visual functions, several authors have suggested the introduction of noise, a type of operational noise. They point to the fact that the nerve elements in the visual cycle from receptor to the visual cortex also have spontaneous activity (Barlow, 1956). The *"Eigengrau,"* "retinal light," "dark light"—all names for the same phenomenon—is supposed, at least by some authors, to be a symptom of this spontaneous activity. The *Eigengrau* is the subjective appearance in the visual field under complete absence of light. Barlow especially has made use of an operational noise corresponding to a retinal "dark illumination" expressed in luminous units (candles per square meter). The real existence of such a noise is evident when some mechanism can be traced back which produces effects exactly equal to these as a result of possible actual adequate activity induced by an outside stimulus. The thermal noise in the receptors is the only possibility known directly that can satisfy these conditions. Earlier it was demonstrated that this noise was probably far too small to serve this purpose.

Barlow suggested that the disagreement between Hecht's and van der Velden's results could be explained by different combinations of reliability and noise effects. Studies by Wertheimer (1953), Verplanck (1952), Howarth and Bulmer (1956), Miller and Garner (1944), and Verplanck, Cotton, and Collier (1953) demonstrated the nonrandom behavior of yes and no responses of test subjects. This is strong evidence against the possibility of finding a general frequency distribution in a combination of noise and reliability. The difference between Hecht's and van der Velden's work might be due to subtle differences in psychological by-effects.

Finally it may be said that, in the explanation and in the experimental results referring to the analysis of the frequency-of-seeing curve, great controversies still go on.

Piper's and Piéron's Laws

When the dark-adapted peripheral retina is studied, the threshold energy is found to increase with exposure time and stimulus area so long as they exceed 0.1 sec and about 10 min of arc. These facts made van der Velden think about the information they might contain for

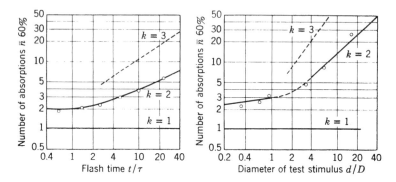

Fig. 2. Curves representing average quantum number required in a flash for a 60-per-cent chance of at least one single quantum absorption $k = 1$, at least one twofold coincidence $k = 2$, at least one threefold coincidence $k = 3$: t, flash time; d, diameter of a circular test stimulus. Open circles are 60-per-cent thresholds for green light 7 degrees nasal from the fovea; dark-adapted eye (Bouman, 1950a).

a further determination of k. He realized that this could mean that more than one activated receptor is needed for perception. Possibly these activations have to occur sufficiently close in time and space. This would mean that k successful quantum absorptions would have to coincide, within a small area, with diameter D and in a length of time τ. He calculated the possibility that, at least once, coincidence conditions are satisfied for various sizes of exposed area and for various durations of exposure. In Fig. 2 the results are represented in terms of number of absorptions required for a 60-per-cent chance that at least one twofold, threefold, etc., coincidence is reached. The actual value of f is again not relevant.

The mathematical calculations are rather difficult for inexperienced mathematicians. It is also not easy to feel the results intuitively. The problem is to calculate the chance for a twofold, threefold, etc., coincidence when the average number over the whole area and exposure time is given. Empirically, one can convince oneself that the results presented are correct by throwing a collection of small pieces of paper onto a checkerboard a great number of times, and looking for the chance that two, three, etc., pieces will fall in the same box.

The next step in van der Velden's work was to compare the curves in Fig. 2 with the dependence of threshold upon time and area. In the figure it is clear that a twofold coincidence represents the actual behavior of the threshold. When t is large and d is large, the slope of the curve is very sensitive to changes in k. With this explanation van der Velden presented a new basis for Piper's law, according to which threshold energy is proportional to the diameter of circular targets, and also for Piéron's law, which asserts that threshold energy is proportional to the square root of exposure duration. In this way both these laws found their bases in one single elementary threshold condition.

After all, the widely accepted Piper's law is an important argument in van der Velden's two-quantum mechanism. Some evidence supporting this theory can also be found from electrophysiological experiments on simple reflex arcs. No synaptical connection in these arcs can be passed by a single spike; at least a twofold coincidence is required (Sherrington, 1906). Histological studies have made it clear that signals of the individual receptors meet each other in nerve elements, in which summation of subliminal effects can occur.

This two-quantum hypothesis has been criticized widely (Pirenne and Denton, 1951, 1952). First, the discrepancy with the slope of the frequency-of-seeing curve, which in the view of some authors points to k values larger than 2, has already introduced elaborate discussion into the literature. On the one hand, there are the pure mechanistic-statistical people who tried to force experimental differences (thought to be due to effects of training, reliability, and noise) into some one frequency distribution. On the other hand, psychologists have still not succeeded in making the results of the different investigators, in themselves good and reproducible, comparable to each other.

Important for the acceptance of the hypothesis are the conditions underlying its application: in the exposed area the retina is homogeneous with respect to (1) receptor density times sensitivity of the individual receptors, (2) distance and time over which subliminal effects can cooperate, and (3) frequency distribution of the required number k. When these conditions are satisfied, Piper's law means that two quantum absorptions represent the lowest possible number that can lead to perception.

There are experimental situations in which the conditions just described are sufficiently closely realized to make by-effects negligible. Moreover, the hypothesis was checked for other than circular objects.

The way in which a two-quantum threshold behaves was calculated for line-shaped targets and moving point sources. From these calculations the dependence of threshold on length of a line, velocity of movement, brightness, and exposure time was predicted for a twofold coincidence of space and of time. An overwhelming number of actual threshold measurements all agreed with this prediction (Bouman, 1953, 1955; van den Brink, 1957; van den Brink and Bouman, 1957; Bouman and van den Brink, 1953).

Baumgardt (1953) claimed that all experimental results could also be described by the supposition that not one twofold coincidence but more—each one producing an elementary effect in the retina—is required. In a subsequent paper Bouman (1955) pointed out that to fulfill this supposition would require an unlimited possibility of summation in time and space between these elementary effects. Baumgardt afterwards placed this possibility high in the visual cycle in the brain. Explicitly said, subliminal effects, no matter where and essentially no matter when (within perhaps a few seconds over which Piéron's law is valid) initiated in the receptors, would have the possibility of cooperating to fulfill threshold conditions. There have been efforts to find such mutual sensitization between subliminal illuminations located far apart. When simple absolute thresholds are considered, and more complicated visual functions like perception of shape are not hidden in the experiments, the results aré all negative. The experimental set-up which Baumgardt used to corroborate his idea involved an identification aspect in the test object, and because of that it is, in our opinion, not conclusive. Just as Pirenne (1943) demonstrated the mutual independence of the two eyes for the absolute threshold, so did van den Brink show with clear-cut experiments the absence of summation between small short flashes, separated by more than 20 min of arc and by more than 0.15 sec in the same retina.

Summation effects that cover the whole retina are found for the pupillary reflex. So there is in the nerve system of the visual cycle a possibility of integration of widely spaced local effects. However, studies of the pupil have shown that there is summation for pupil contraction only when the visual threshold conditions of the more differentiated system of perception are satisfied (Schweitzer and Bouman, 1956; Schweitzer, 1956).

Intensity Discrimination

Independently of each other, de Vries (1943) and Rose (1948) developed arguments of a quantum-statistical nature for the behavior of contrast thresholds. They wondered why the increment ΔB on a background B at threshold level is, in a considerable region of values, proportional to $B^{1/2}$. They presented the hypothesis that ΔB is just beyond threshold when the statistical fluctuations in quantum flux B are exceeded by ΔB. These fluctuations are proportional to $B^{1/2}$. The underlying idea, at least in de Vries' work, was that the elementary events in the visual system are connected in a one-to-one relation with

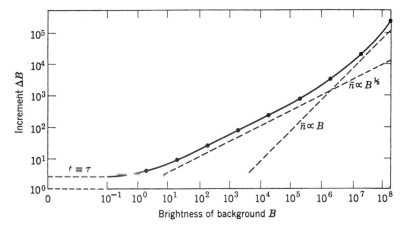

Fig. 3. Increment threshold ΔB as a function of background brightness B for small short flashes 7 degrees nasal from the fovea for green light (van den Brink, 1959; van den Brink and Bouman, 1959).

the active quantum absorptions. Deviations from this law are due either to photochemical and nervous by-effects or to spike saturation of the nerve fibers when the retina is very strongly illuminated. In the latter situation the number of elementary events is no longer proportional to the incident number of quanta (van der Velden, 1949).

The simplest situation is when the signal is small and short enough to cover only one single retinal recipient unit and only one summation time. Under this condition, deviations from the proportionality, ΔB to $B^{1/2}$, over about 6 to 7 log units are very small (van den Brink, 1957; van den Brink and Bouman, 1957; see Fig. 3). It seems that when more recipient units for a duration several times longer than the simple summation time are activated by the stimulus, the agreement with de Vries' and Rose's law begins to fail. This cannot simply be taken as an argument against this law, as various authors have suggested (for instance, Aguilar and Stiles, 1954; Bouman, 1954). Neither de Vries nor Rose was very explicit on whether the signal-to-noise ratio in one single channel through which the information from the whole retina flows is decisive for the threshold. Anyhow this idea can be easily abstracted from their papers. Probably there are, however, a great number of relatively independent units O, each of which takes samples over periods lasting T, and compares these samples with some actual value of the background illumination, or with some average value of retinal activity due to general background illumination. The size of these channels in the retina, as well as the sample period T, is only slightly dependent on the state of adaptation. The size is also different for different locations on the retina, and different for rods and cones.

It is generally accepted that the area over which Ricco's law, $IO = $ constant, holds represents also the size of the recipient units. Correspondingly, the summation time in such a unit is given by the region of t values in which Bunsen-Roscoe's law, $It =$ constant, is valid.

If the existence of independent sampling in retinal areas of relatively small size is recognized, de Vries' and Rose's law for large and long-lasting stimuli will have to be extended. Under these conditions the contrast threshold ΔB will be reached when, owing to the test flash, in at least one OT sample the fluctuations of the background per area O per time T are exceeded. If a short and small test flash must contain at least k quantum absorptions, the threshold energy ΔE is proportional to $(o, t)^{(k-1)/k}$ in which o and t are area and exposure time of the test stimulus and both are large.

The mathematical manipulations that lead to this formula are again rather complex. If a nonmathematician wants to check them, he can again throw pieces of paper onto a checkerboard and in this way study the probability that three, four, and so forth, will fall onto one box. As was mentioned earlier, if B is equal to zero, then k equals 2. In this way it is suggested that de Vries' and Rose's law be modified by letting k increase proportionally to $B^{1/2}$ instead of ΔB. When o and t are both smaller than O and T, $\Delta E \propto B^{1/2}$ as $\Delta E \propto k$. Although the formula $\Delta E \propto (o, t)^{(k-1)/k}$ may not describe the dependence of ΔE on o and t exactly, it shows that when $k \propto B^{1/2}$, ΔB increases faster (Bouman, 1953). Only with small area and exposure time will ΔB be proportional to the quantum demand of the individual retinal channels. For small and short flashes, changes due to adaptation in the span of summation in the retina and over time can hardly influence the ΔB vs. B function. Because of the small deviation from $\Delta B \propto B^{1/2}$ for small and short flashes, it can be concluded that changes in concentration of the active photopigment in this region of B occur almost not at all. From the small deviations mentioned above we have earlier calculated, for a few wave lengths of the light, what this photochemical component does (Bouman, 1950b, 1952a, b). By this component is meant the decrease in the active fraction of the incident light due to decreased quantity of photopigment in the receptor when the eye becomes light adapted. From these calculations a remarkably small change resulted. An important piece of evidence in support of de Vries' and Rose's theory is that Rushton and Cohen (1954) also concluded from direct measurements of remission from the retina that the photochemical components contribute almost nothing to adaptation.

In the literature, de Vries' and Rose's hypothesis is also put forward as an explanation of Piper's law, by suggesting that spontaneous effects in the visual cycle represent a statistical noise that impedes the signal. The relevant part of the noise should be proportional to the area of the test stimulus $a \cdot o$. The signal in this idea is just visible when it is $(ao)^{1/2}$ as Piper's law predicts. Piéron's law is also thought to be explained along this line. The threshold energy should be proportional to $t^{1/2}$ since the relevant noise is proportional to exposure time t. Because of the noise, the absolute threshold should actually be a contrast threshold to the noise, corresponding to the "dark illumination" of the retina. As has been mentioned, the spontaneous thermal decompositions of photopigment are not frequent enough to yield the required number of statistical events. In these considerations, it is thought that the threshold is determined by the signal-to-noise ratio in a single channel which collects only the results of the area covered by the test stimulus. It is hard to imagine how this channel may know that the unexposed areas of the retina are unexposed, so that the noise of these areas is not weighted in the decision about the threshold in the exposed area.

Finally, a "dark illumination" dependent on color from stimulus and background, as well as from location on the retina, is used by Barlow (1958) to explain deviations from $\Delta B \propto B^{1/2}$. He suggested $\Delta B \propto (B + B_0)^{1/2}$, in which B_0 represents the "dark light" under the actual stimulus conditions. The essential idea that the internal noise of the visual system can be projected mathematically in the outside world as dark illumination is again present in this hypothesis. It is essential also that one channel be present in which decisions on perception are made. Also we call attention to the fact that the noise projected to the outside has a color similar to the background and a geometrical extension similar to that of the test stimulus. It is evident that the basis of all these speculations is the same: there is a threshold mechanism in which signal-to-noise ratios determine perception.

Most of the threshold studies on contrast are made in the periphery. In the fovea only a very few results are available (ten Doesschate, 1942). The behavior of absolute thresholds in the fovea was also found to follow Piper's and Piéron's laws; thus the two-quantum hypothesis describes the threshold mechanism here (Bouman, 1950a). Foveal measurements on ΔB vs. B with small o and t values, which are most suitable for checking de Vries' and Rose's law, are not available. Probably a theory for contrast thresholds and absolute thresholds in the periphery would also be applicable to the fovea. An interesting complication here is that a pure contrast threshold, in terms of intensity discrimination, is impossible in the fovea. The appearance of the color changes with brightness (Hurvich and Jameson, 1951). Whether de Vries' and Rose's law has a more general meaning, so that it can be applied to combined brightness and color contrast or to pure color contrast is a wide open question.

Finally, we make a general remark about contrast thresholds and quantum statistics. The sensitivity of a physical instrument is after all determined by its intrinsic noise factors. In a photocell, for instance, these factors are the noise in the incident light, in the photoelectric flux, and in the anode current. According to elementary fluctuation theory, the noise is the largest in the component in which the number of elementary events is the smallest. The sensitivity of the system is for the greater part determined and limited by the fluctuations in this component. If the final practical accuracy that can be obtained is Δe, and the value according to physical arguments like those mentioned above is ΔE, then $\Delta E/\Delta e$ is the efficiency factor of the system. Under conditions in which de Vries' and Rose's law is proved to be valid, the efficiency factor can be determined. If, for instance, a thousand quanta is the average input per second, about 25 quanta per second are absorbed in the retina in order to contribute to perception. The fluctuation in this average value of 25 is $25^{1/2} = 5$. The relative noise is $1/5$, and by this fraction the maximal obtainable sensitivity is determined when, in subsequent phases of the visual process, influences of other nature do not decrease the number of events further.

Various authors have estimated how closely the eye reaches the physical limit. Efficiency factors between 2 and 8 are given (Rose, 1948; Bouman, 1952a, b; Jones, 1953). The value of these factors depends on the properties of the nerve system and perhaps on psychological effects. It is likely that, in photochemical reaction in the receptor by each absorbed quantum, a rhodopsin molecule is broken down so that it delivers a fraction that activates the nerve fiber connected to the receptor (Dartnall, 1957).

Adaptation

The dark-adaptation curve, representing threshold measurement as a function of time after a preceding exposure to quite high brightness, demonstrates a larger total sensitivity variation for large test objects than for small ones (Hecht, Haig, and Wald, 1935). Various authors have described and studied this phenomenon (see Fig. 4). Most of them suggest that during dark adaptation the summation area, or the

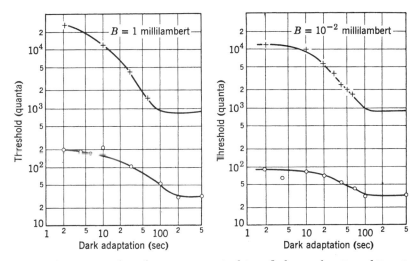

Fig. 4. Average number of quanta $\bar{n}$ required in a flash as a function of time in the dark after a preceding illumination for 10 min with $B = 1$ millilambert and $B = 10^{-2}$ millilambert. All with green light. Test stimuli: crosses, 330 min of arc; open circles, 11 min of arc (Bouman and ten Doesschate, 1953).

size of the recipient unit, increases. By this action the threshold for large targets should indeed improve more than for small ones. When the adaptation amplitude of a large object is, for instance, 400 times that of a small object, the change in size of the recipient unit must be 20. Direct estimation of the possibility for summation has been made by van den Brink and Bouman (1953, 1954; Bouman and van den Brink, 1952; van den Brink, 1957). They found that the actual change does not exceed a factor between 2 and 4 and is not large enough by far to explain the difference found between adaptation amplitudes.

However, a very simple explanation can be found in the point of view of quantum theory. When with light adaptation the number of quantum absorptions required per retinal unit per summation time of the system increases from 2 to k (and for the moment no regard is paid to possible changes in concentration of photopigment), the threshold for small short flashes is $k/2$ times as large as in a dark-adapted state. For large circular targets with diameter d, the threshold energy is approximately $fk \cdot (d/D)^{2(k-1)/k}$. In the previous section the more general formula $(ot)^{(k-1)/k}$ was given, which can also be written: threshold energy $\Delta E = fk(d^2/D^2 \cdot t)^{(k-1)/k}$. D is the diameter of the summation area which is still thought to be independent of state of adaptation. When, for instance, d/D is equal to 100—that is, when D is about 12 min of arc, d is 20 degrees of arc, and k increases from 2 to 4—the threshold for this object goes up by a factor of 10 (for a small target by only a factor of 2). When k goes from 2 to 10, these factors are 200 and 5.

In the case of an additional photochemical component, the thresholds for all target sizes rise accordingly. This component is also, according to Rushton's findings, relatively small, perhaps between 10 and 100. If we choose for it a value of 30, the factors in the example just mentioned become 6000 and 150. The diameter D of the recipient unit increases about three times, according to van den Brink and Bouman. This increase is active only for large targets and makes the two factors 42,000 and 150! Actual thresholds for a 20-degree target and a point source are indeed subject to about these factors during dark adaptation. The enormous dynamics in adaptation is in this way described by a few simple small effects, and the difference in dark-adaptation behavior for small and large targets is explained by a quantum-theoretical argument. The number of quanta needed per summation area, in order to give rise to perception, increases from 2 to higher values in the light-adapted state. At least one of the summation areas with diameter D covered by the test stimulus must have absorbed k quanta in order for perception to occur.

Unfortunately the mathematical treatment by which the formulas were found is so difficult as to make it almost impossible to evaluate their correctness intuitively. Wald concluded from our treatment and from our experiments (Bouman and ten Doesschate, 1953) that the quantum demand in the separate rods and cones increases with light adaptation. For that reason we here lay stress upon the fact that the suggested quantum-theoretical aspect is strictly a neural component in adaptation.

The required number of quanta k per summation area in the peripheral retina is still small compared to the number of receptors in it. It is our opinion that a single receptor can always be activated by absorption of one quantum. Only the number of those quantum absorptions that have to occur in one or another recipient unit of the exposed area, in order to give rise to a visible effect, increases during light adaptation from 2 to higher values.

The reader will be aware of the similarity in the description of contrast thresholds against a background B and those for the absolute threshold in the nondark-adapted state. This resemblance can, in our opinion, be easily accepted. Change in adaptation condition in our scheme is a matter of change in the organizational pattern of the nerve system of the retina. Through this pattern, action potentials can find their way and spikes can meet each other at the places of rendezvous—the synapses—in order to continue nerve action beyond these barriers. A light-adapted state in this picture means increased k, decreased summation area and summation time, and a little photochemistry. It takes at the most about a half hour in the dark to reset the organizational pattern from a light-adapted state to a dark-adapted state.

The background B in measurements of contrast threshold will bring the pattern of possible required and permitted nerve actions out of the pattern of the dark-adapted state. It means increased k, decreased D and T and, of course, decreased concentration of photopigment. When, in a particular pattern corresponding to some B value for the contrast threshold, a special number of k quantum absorptions is required for short and small flashes, there will be some phase in the dark-adaptation process with the same k value. This might make one inclined to describe this state of adaptation as a dark illumination of B. It must be recognized, however, that the patterns just mentioned for both conditions might still be different with respect to summation spans over space and time.

An interesting point is how the setting of the organization pattern occurs after a change in background brightness. There must exist in the retina a continuous effort to reset the pattern with the lowest possible k and the largest D and T values. The spontaneous activity in the visual system, which is always available, is itself thought to be the agent for it. When B is not equal to 0, we think that some balancing action in the nervous system, between this spontaneous nerve activity and the statistical fluctuation in B itself, sets the k value in direct relation to $B^{1/2}$.

In the formula that Barlow used to describe the ΔB against B function, $\Delta B \propto (B + B_0)^{1/2}$, B_0 in itself does not depend on B. This means that ΔB has to follow this formula momentarily. When B is suddenly made 0, ΔB must immediately come down to $B_0{}^{1/2}$, the minimal possible value. No dark adaptation lasting for half an hour can be deduced from this idea of intrinsic noise connected with the dark light B_0. We think that a stimulus from the outside world gives rise to a unique pattern of nerve activity which can by no means be generated by the visual system itself spontaneously. This implies that the noise in a physical stimulus and its corresponding physiological effects in the eye are of quite a different nature from the intrinsic noise of the visual system itself. It does not mean that visual illusions cannot have their basis in nervous activity in the retina. When they have their basis in the retina, they never occur isolated from the stimulus by which they are initiated, and hence they are due to noise or some other nerve activity produced by the real stimulus.

Finally, we want to make two remarks. First, in the section on contrast thresholds we suggested that contrast threshold is reached, when in at least one OT sample, by chance, a number of quanta from the test stimulus is absorbed that should have made a short and small flash visible. It can be asked whether each OT sample is compared with immediately preceding, succeeding, or neighboring samples. We left these possibilities open, but since the study of adaptation we are inclined to suggest that the parts of the retina exposed to the background as a whole adapt to the fluctuations in this background flux by the balancing action between intrinsic noise and noise from the outside world.

The second remark is that, in the suggestions made by various authors and mentioned in the previous section, the signal-to-noise ratio in one channel is decisive for perception. This channel covers the whole area of the test stimulus. We have asked already how this channel knows about the unexposed and exposed areas at the threshold. We think that the eye is always ready to resolve a complex picture according to details comparable in size with the summation areas. If the eye correlates the information present in each of these recipient units in order to decide between perception and no perception, then more than just a yes or no response is involved.

An argument against the position we have taken in these various crucial points is the evident existence of false responses. We cannot place these in our theory easily. In the absence of any better idea, we suggest that these false responses are due to some irregularities of a more general nature and not strictly to the processes in the visual pathways.

Experiments in Color Identification

There are only a very few experiments in the literature suitable for a check on quantum-theory views on color vision. In the central fovea we found that the absolute threshold obeys Ricco's, Bunsen-Roscoe's, Piper's, and Piéron's laws, much like the threshold for the rod system. This means that, if we accept van der Velden's hypothesis, for all wave lengths and hence also for all cone systems through which color vision is working, a twofold coincidence is sufficient for perception. The time in which this coincidence must occur is equal to the time for rod vision. The recipient unit, however, is much smaller. It is not certain whether this unit is the receptor itself. Ricco's region is only a few minutes of arc.

A direct approach to summation between point sources by van den Brink and Bouman proved that by-effects, such as aberrations in the focusing system of the eye and diffraction, are a real handicap to the solution of this problem. From experiments in which the energy of the test stimulus contains two wave lengths, it turned out that subliminal effects initiated by any wave length could cooperate with such effects from any other wave length. Because of the great variety of combinations of wave lengths used, and because of the differences among the fundamental response curves of the cone systems, it was concluded that each individual receptor can be activated by one single quantum, and that two of these receptors, no matter whether of the same kind or not, result in a perception.

These considerations made us anxious to know what color sensations correspond to this small number of activated receptors. Indeed when, for instance, monochromatic stimuli are used of a wave length for which two of the cone systems are equally sensitive, it is to be expected that three types of twofold coincidences will occur almost equally often: the two activated cones both belong to one or the other system, or each of them to one of the two systems. If all this should be true, it would be expected that, for stimuli near threshold, different discrete color sensations would occur with repetitions of the stimulus.

To check this we carried out color-identification experiments in the fovea with monochromatic stimuli of different sizes. The idea was that (1) around the threshold a very restricted number of quanta is active for perception; (2) instability in retinal stimulation at repetitions of the flashes occurs, owing to statistical fluctuations in number and distribution of these quanta over the different receptors; (3)

instability in retinal stimulation and in perception are possibly correlated with each other.

We tried to discover whether a special quality in the perception—for instance, reddish—is connected with a special number and with a special distribution of quantum absorptions over the retinal cones.

If success should be achieved, the following alternatives would be available: (a) a special quality would result from activation of a special cone system by two or more quanta; (b) a special quality would result from activation of a special cone system with special numbers—each amounting to at least two quanta; (c) a special quality would result from combined activation of special systems, each amounting to at least one quantum; (d) a special quality would result from combined activation of special systems with special numbers, each amounting to at least one quantum. Different combinations of these numbers would result in different qualities.

Another point in the philosophy behind these considerations is that the enumerable collection of possible patterns of retinal stimulation corresponds to an enumerable collection of possible patterns of nerve activity, and this corresponds further to an enumerable collection of color sensations which can be perceived as mutually different.

From this work, whose results are given partly (Bouman and Walraven, 1957) in Fig. 5, it was clearly shown that conditions for a color perception are not fulfilled when perception alone occurs. - In red and green there is an achromatic zone. This means that, if for perception alone a simple twofold coincidence of quantum absorption suffices, three or more absorptions are needed for a color sensation to occur; that is, a twofold coincidence gives sometimes a colored, sometimes an uncolored, sensation.

The experimental results for large targets in Fig. 6 demonstrate that with green the achromatic zone increases with the size of the test stimulus; with red it remains constant. From this it can be concluded that the green cone system gives an uncolored perception at two absorptions and a colored one at three or more. For red there seemed to be two systems. One gives an uncolored perception, the other a red one, both with a twofold coincidence. An alternative possibility is a system that sometimes gives an achromatic signal and sometimes a chromatic. The conclusions from these subtle experiments are rather provocative and far-reaching. As with rod vision, the slope of the frequency-of-seeing curve, analyzed in terms of Poisson, does not agree with the absolute threshold or color threshold.

In the section on the frequency-of-seeing curve, this discrepancy is discussed in detail. In these color-identification experiments it is striking that for green light the curve becomes shallower for large test stimuli. It is possible, of course, that a change in secondary effects occurs. Our experiments would become more conclusive if,

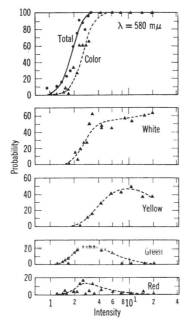

Fig. 5. Frequency-of-seeing curves for total and for color thresholds (top), and for perception of white, yellow, green, and red. All for small short foveal flashes (Bouman and Walraven, 1957).

for instance, for moving point sources also, similar results were found in terms of twofold and threefold coincidences required for special qualities.

Visual Acuity

In the foregoing sections, absolute and contrast thresholds were discussed with relation to quantum theory. Visual acuity also is related by some authors to this theory. When recognition of a simple form is involved, visual acuity refers mostly to the perception of one or a few discrete critical details, as in the Landolt ring or the Snellen optotype. When the detail is small relative to the recipient unit of the retina, perception of this element is an act independent of neighboring

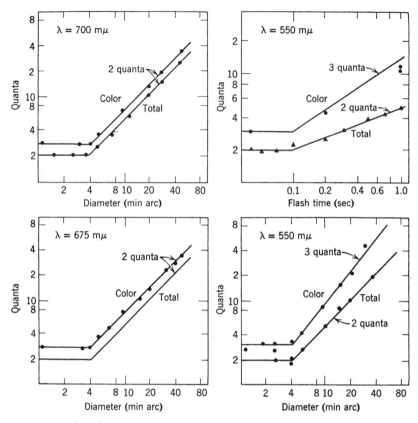

Fig. 6. Total and color thresholds in the fovea as a function of flash diameter and flash time (Bouman and Walraven, 1957).

elements of the figure. In these cases the visual acuity as a function of brightness B can be deduced from the behavior of the contrast threshold for targets of comparable sizes. By simple means it can be found that for the lowest B values the visual acuity is proportional to B (Bouman, 1950a; Bouman and van der Velden, 1947, 1948).

Recently van der Gon (1959) described visual-acuity measurements in the fovea. For higher brightness values his results agreed with calculations he made on the chance that the critical detail at the threshold contains a significantly smaller or larger number of quanta than neighboring elements of the test figure. For lower brightness values the results deviated from this concept. In order to explain this, he introduced an internal noise in the retina as Barlow did. But van der Gon did not realize that his measurements in this region of low brightness values approached nicely the asymptote predicted by two-quantum theory.

For the recognition of more complex figures the simple quantum-statistical approach has to fail. Psychological influences, such as the *Gestalt* of the target, knowledge of available choices in the collection of test figures used, and experience become the more important aspects, as they are in everyday vision.

In summary, it can be said that quantum-statistical knowledge of the nature of light, together with simple properties of nervous and photochemical mechanisms in the visual system, has led to a more-or-less rounded picture of the behavior of thresholds as a function of brightness, size, exposure time, and velocity for a wide variety of conditions, including aspects of adaptation and color identification.

A real handicap to a correct evaluation of this theory is the fact that the underlying mathematics can hardly be felt intuitively. On the contrary, almost none of the physiologists interested in these problems have succeeded in freeing themselves from the intuitive conclusion that the threshold brightness for a large target can only decrease further for a still larger one if summation can take place over the whole stimulated area. Even Baumgardt (1957), a pioneer in the quantum-statistical approach to visual functions, did not succeed completely in this. The simple two-quantum hypothesis, corroborated by Piper's law, requires only a summation over discrete retinal distances— of about 2 to 10 min of arc—in order to explain that the threshold brightness for a 20-degree target is half as great as for a 10-degree area!

The increase of k above 2 explains the much larger variation in threshold in contrast situations and in adapting processes for large targets as compared to small ones. This is also why, for small and short flashes only, ΔB is proportional to $B^{1/2}$.

Summary

A review is presented of the development of quantum-statistical theories for visual functions such as absolute thresholds for steady and moving targets, contrast thresholds, visual acuity, adaptation, and color identification. A few essential points in the discussion of these theories follow.

1. For each state of adaptation the nerve system in the retina sets an organizational pattern through which action potentials can find

their way to the higher centers. For each actual pattern, threshold perception of a stimulus occurs when k quanta are absorbed by chance in any summation area of size O covered by the test stimulus. These k quanta must be absorbed in this area within the summation time T. The crucial parameters—the value k, the size O, and the time T— change with the state of adaptation. In the dark-adapted nonilluminated eye, k is equal to 2. For short and small test stimuli the threshold energy is fk, in which f is the fraction of the incident light that is actively absorbed in the receptors. For long flashes and large targets, threshold energy is proportional to $f(ot)^{(k-1)/k}$, in which o is area, and t is duration of the test stimulus. From this formula, it is clear that an increase in k affects threshold energy for large and/or long flashes more than for small short ones. For contrast thresholds this formula has not been adequately checked (Bouman, 1953).

2. There is a continuous force available in the spontaneous intrinsic activity of the nerve system that tries to make the pattern such that k is as small as possible, O and T as large as possible. The noise in the physical stimulus from the outside world competes with this force, resulting after some time in a balance between the two, giving a pattern in which the number k is directly related to the average of the fluctuations in the O, T samples of the physical stimulus. When the latter has the brightness B, the number k becomes proportional to $B^{1/2}$. After strong illumination it takes about a half hour to bring the organizational pattern back to that of the dark-adapted state, and k drops from perhaps between 10 and 20 to 2, this being the lowest possible value of k.

3. Paragraphs 1 and 2 make the point that contrast threshold ΔB is proportional to $B^{1/2}$ only when o and t from the test stimulus are both small. Deviations from this proportionality—and indeed Piper's and Piéron's laws too—are explained by some authors by introducing some intrinsic noise into the visual system such that absolute thresholds are actually contrast thresholds against some "dark illumination" B_0. The consequences of these assumptions are discussed.

Instability phenomena in color identification and dependence of visual acuity on brightness are discussed with reference to quantum-statistical approaches.

References

Aguilar, M., and W. S. Stiles. Saturation of the rod mechanism of the retina at high levels of stimulation. *Optica Acta*, 1954, 1, 59.

Barlow, H. B. Retinal noise and absolute threshold. *J. opt. Soc. Amer.*, 1956, **46**, 634–639.

Barlow, H. B. Intrinsic noise of cones. In National Physical Laboratory Sympos. No. 8. *Visual Problems of Colour.* London: H.M. Stationery Office, 1958, p. 615.

Barnes, R. B., and M. Czerny. Lässt sich ein Schroteffekt der Photonen mit dem Auge beobachten? *Z. Physik*, 1932, **79**, 436.

Baumgardt, E. Seuils visuels et quanta lumineux. *Ann. psychol.*, 1953, **53**, 431.

Baumgardt, E. Théories quantiques de la vision. *Proc. 2nd int. Congr. Photobiol., Turin*, 1957, p. 195.

Bouman, M. A. Quanta explanation of vision. *Doc. Ophthal.*, 1950a, **4**, 23–115.

Bouman, M. A. Peripheral contrast threshold of the human eye. *J. opt. Soc. Amer.*, 1950b, **40**, 825–832.

Bouman, M. A. Peripheral contrast threshold for various and different wave lengths for adapting field and test stimulus. *J. opt. Soc. Amer.*, 1952a, **42**, 820–831.

Bouman, M. A. Mechanisms in peripheral dark adaptation. *J. opt. Soc. Amer.*, 1952b, **42**, 941–950.

Bouman, M. A. Visual thresholds for line-shaped targets. *J. opt. Soc. Amer.*, 1953, **43**, 209–211.

Bouman, M. A. Comment on Aguilar and Stiles' discussion of their increment threshold measurements. *Optica Acta*, 1954, **3**, 155.

Bouman, M. A. The absolute threshold conditions for visual perception. *J. opt. Soc. Amer.*, 1955, **45**, 36–43.

Bouman, M. A., and G. van den Brink. On the integrate capacity in time and space of the human peripheral retina. *J. opt. Soc. Amer.*, 1952, **42**, 617–620.

Bouman, M. A., and G. van den Brink. Absolute thresholds for moving point sources. *J. opt. Soc. Amer.*, 1953, **43**, 895–898.

Bouman, M. A., and J. ten Doesschate. Nervous and photochemical components in visual adaptation. *Ophthalmologica*, 1953, **126**, 222–230.

Bouman, M. A., and H. A. van der Velden. The two-quantum explanation of the dependence of the threshold values and visual acuity on the visual angle and the time of observation. *J. opt. Soc. Amer.*, 1947, **37**, 908–909.

Bouman, M. A., and H. A. van der Velden. The two quanta hypothesis as a general explanation for the behavior of threshold values and visual acuity for the several receptors of the human eye. *J. opt. Soc. Amer.*, 1948, **38**, 570–581.

Bouman, M. A., and P. L. Walraven. Some color naming experiments for red and green monochromatic lights. *J. opt. Soc. Amer.*, 1957, **47**, 834–839.

van den Brink, G. Retinal summation and the visibility of moving objects. Thesis, Utrecht, 1957; Report 1957–9 from Institute for Perception, Soesterberg, The Netherlands.

van den Brink, G., and M. A. Bouman. Variation of integrate capacity in time and space: an adaptational phenomenon. *J. opt. Soc. Amer.*, 1953, **43**, 814. [Abstract]

van den Brink, G., and M. A. Bouman. Variation of integrative actions in the retinal system: an adaptational phenomenon. *J. opt. Soc. Amer.*, 1954, **44**, 616–620.

van den Brink, G., and M. A. Bouman. Visual contrast thresholds for moving point sources. *J. opt. Soc. Amer.*, 1957, **47**, 612–618.

Brumberg, E., and S. Vavilov. Visuelle Messungen der statistischen Photonen-schwankungen. *Bull. Acad. Sci. URSS*, 1933, p. 919.

Dartnall, H. J. A. *The Visual Pigments*. London: Methuen, 1957.

ten Doesschate, J. Physiologisch-optische beschouwingen betreffende de visuele beoordeling van röntgenphotogrammen. Thesis, Utrecht, 1942.

Ferree, C. E. The fluctuations of luminal visual stimuli of point area. *Amer. J. Psychol.*, 1913, **24**, 378.

van der Gon, J. J. Denier. Gezichtsscherpte—een fysisch-fysiologische studie. Thesis, Amsterdam, 1959.

Grijns, G., and A. K. Noyons. Über die absolute Empfindlichkeit des Auges für Licht. *Arch. Anat. Physiol., Physiol. Abt.*, 1905, 25–52.

Hartline, H. K. Sight quanta and the excitation of single receptors in the eye of Limulus. *Proc. 2nd int. Congr. Photobiol., Turin*, 1959, p. 193.

Hecht, S. The quantum relations of vision. *J. opt. Soc. Amer.*, 1942, **32**, 42–49.

Hecht, S. Energy and vision. *Amer. Scientist*, 1944, **32**, 159.

Hecht, S., C. Haig, and G. Wald. The dark adaptation of retinal fields of different size and location. *J. gen. Physiol.*, 1935, **19**, 231.

Howarth, I., and M. G. Bulmer. Non-random sequences in visual threshold experiments. *J. exp. Psychol.*, 1956, **8**, 163.

Hurvich, L. M., and D. Jameson. A psychophysical study of white. *J. opt. Soc. Amer.*, 1951, **41**, 521.

Jones, R. Clark. Detectivity of the human eye. *J. opt. Soc. Amer.*, 1953, **43**, 814. [Abstract]

König, A. Über den menschlichen Sehpurpur und seine Bedeutung für das Sehen. *Sitzber. Akad. Wiss. Berlin*, 1894, 577–598.

Langley, S. P. Energy and vision. *Phil. Mag.*, 1889, **27**, Ser. 5, 1.

Ludvigh, E., and E. F. McCarthy. Absorption of visible light by the refractive media of the human eye. *Arch. Ophthal.*, 1938, **20**, 37.

Miller, G. A., and W. R. Garner. Effect of random presentation on the psychometric function. *J. Psychol.*, 1944, **57**, 451.

Mueller, C. G. A quantitative theory of visual excitation for the single photo-receptor. *Proc. Nat. Acad. Sci.*, 1954, **40**, 853.

Pinegin, N. I. Independence of wavelength of the threshold number of quanta for peripheral rod and foveal cone vision. In National Physical Laboratory Sympos. No. 8. *Visual Problems of Colour*. London: H.M. Stationery Office, 1958, p. 725.

Pirenne, M. H. Binocular and uniocular threshold of vision. *Nature, Lond.*, 1943, **152**, 698.

Pirenne, M. H., and E. J. Denton. Quanta and visual thresholds. *J. opt. Soc. Amer.*, 1951, **41**, 426.

Pirenne, M. H., and H. J. Denton. Accuracy and sensitivity of the retina. *Nature, Lond.*, 1952, **170**, 1039.

Planck, M. Über das Gesetz der Energieverteilung im Normalspectrum. *Ann. Physik*, 1901, **4**, 553.

Rose, A. The sensitivity performance of the human eye on an absolute scale. *J. opt. Soc. Amer.*, 1948, **38**, 196.

Rushton, W. A. H., and R. D. Cohen. Visual purple level and the course of dark adaptation. *Nature, Lond.*, 1954, **173**, 301.

Schweitzer, N. M. J. Threshold measurements on the light reflex of the pupil in the dark adapted eye. *Doc. Ophthal.*, 1956, **10**, 1–78.

Schweitzer, N. M. J., and M. A. Bouman.   Threshold measurements on the light reflex of the pupil.   *Ophthalmologica*, 1956, **132**, 286.

Sherrington, C. S.   *Integrative Action of the Nervous System.*   New Haven, Conn.: Yale University Press, 1906.

van der Velden, H. A.   Over het aantal lichtquanta, dat nodig is voor een lichtprikkel bij het menselijk oog.   *Physica*, 1944, **11**, 179.

van der Velden, H. A.   The number of quanta necessary for the perception of light of the human eye.   *Ophthalmologica*, 1946, **111**, 321.

van der Velden, H. A.   Quanteuse verschijnselen bij het zien.   *Ned. Tijdschr. Natuurk.*, 1949, **15**, 146.

Verplanck, W. S.   Nonindependence of successive responses in measurements of the visual threshold.   *J. exp. Psychol.*, 1952, **44**, 273–282.

Verplanck, W. S., J. W. Cotton, and G. H. Collier.   Previous training as a determinant of response dependency at the thresholds.   *J. exp. Psychol.*, 1953, **46**, 10–14.

de Vries, Hl.   The quantum character of light and its bearing upon threshold of vision, the differential sensitivity and visual acuity of the eye.   *Physica*, 1943, **10**, 553.

de Vries, Hl.   Die Reizschwelle der Sinnesorgane als physikalisches Problem.   *Experientia*, 1948, **4**, 205.

Wertheimer, M.   An investigation of the randomness of threshold measurements.   *J. exp. Psychol.*, 1953, **45**, 294–303.

Zwaardemaker, H.   *Leerboek der Physiologie.*   Haarlem: Bohn, 1920.

22

VERNON B. MOUNTCASTLE

Department of Physiology
The Johns Hopkins University School of Medicine

Some Functional Properties
of the Somatic Afferent System

In this presentation I should like to pursue two general themes. The first is that it is necessary to look to something more than anatomical connections between a given sensory sheet and the neural centers upon which it projects to understand the very highly developed capacities for discrimination which mammalian afferent systems possess. I propose to describe some of the dynamic attributes of the somatic afferent system which contribute, I believe, to its discriminative capacity, with particular emphasis on the role of afferent inhibition in its function. In doing so I shall draw upon data from studies of what I shall term the *lemniscal component* of the somatic system. It comprises first-order afferents from the periphery which project via the dorsal columns upon relay neurons of the dorsal column nuclei, and thence upon cells of the ventrobasal nuclear complex of the thalamus, and from there upon the cells of the postcentral gyrus of the cortex. I shall, however, use the term lemniscal to refer more generally to a certain set of physiological properties, as well as to a rather strictly defined anatomical entity.

This is, however, but a part of the neural substratum of general somatic sensibility. At the level of dorsal root entry there is a massive offshoot, impinging upon polysynaptic mechanisms of the dorsal horns at the segmental level, and serving reflex functions of many sorts, which yields another great ascending pathway, the anterolateral columns. My second purpose is to describe the functional properties of some of the neural elements of this system, observed at thalamic and cortical levels. Once again I shall refer to this system more in terms of a set of properties than as an anatomical entity, and shall use the terms posterior group or spinothalamic rather loosely in referring to it. These systems, divergent at the level of the spinal cord, very likely converge again upon the cerebral cortex, at least to a certain extent. How they must work synergistically in depicting sensory events is an interesting but still highly speculative matter to which I shall devote some discussion.

Place and Method of Investigation

The several series of observations concerning the lemniscal system which I wish to describe were made at its thalamic and cortical levels, in both the cat and the monkey. Figure 1 shows the cytoarchitectural subdivisions of the postcentral gyrus of the monkey, as determined in

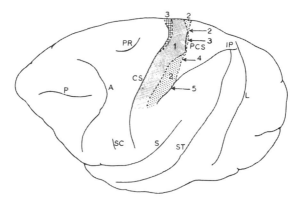

Fig. 1. Reconstruction of the three cytoarchitectural areas of the somatic sensory cortex of the macaque monkey. Numbers without arrows indicate these areas. The junctions between them are considered to be regions of sharpest gradient in cellular changes, rather than sharp lines. Abbreviations as follows: A, arcuate sulcus; CS, central sulcus; IP, intraparietal sulcus; L, lunate sulcus; P, principal sulcus; PCS, postcentral sulcus; PR, superior precentral sulcus; S, sylvian fissure; SC, anterior subcentral sulcus; ST, superior temporal sulcus. (From Powell and Mountcastle, 1959a.)

a recent study by Dr. T. P. S. Powell, of Oxford University (Powell and Mountcastle, 1959a). He has, in large measure, confirmed the much earlier observations of Brodmann (1905) that this region is divisible into three on the basis of its cellular arrangement, and we have recent evidence that these divisions are meaningful in physiological terms (Powell and Mountcastle, 1959b). It is the postcentral gyrus that receives the cortical projection of cells of the ventrobasal complex of the thalamus. We use the term ventrobasal (Rose and Mountcastle, 1952) to refer to that region of the thalamus of distinctive architecture which receives the terminals of the fibers of the medial lemniscus, and which degenerates following removal of the postcentral gyrus. It is included within but is not exactly coextensive with the regions termed n. ventralis posterolateralis and posterome-dialis by other authors (for example, Walker, 1938).

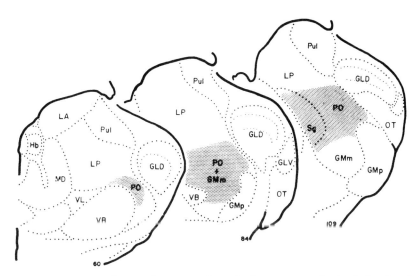

Fig. 2. Drawings of coronal sections through the posterior third of the dorsal thalamus of the cat. The sections were some 600 microns apart in the orocaudal dimension. The shaded region indicates the position of the posterior group of nuclei of the thalamus in relation to the ventrobasal complex. Abbreviations: GLD, dorsal nucleus of the lateral geniculate body; GLV, ventral nucleus of the lateral geniculate body; GMm, magnocellular division of the medial geniculate body; GMp, principal division of the medial geniculate body; Hb, habenular complex; LA, lateral anterior nucleus; LP, lateral posterior nucleus; MD, medio-dorsal nucleus; OT, optic tract; PO, posterior group of nuclei of the thalamus; Pul, pulvinar; Sg, suprageniculate nucleus; VB, ventrobasal nuclear complex of the thalamus; VL, ventrolateral nucleus.

The spinothalamic component of the anterolateral system sends only a sparse projection to the ventrobasal complex in the cat. It increases greatly in primates. It is now clear from the work of Mehler, Feferman, and Nauta (in press) that the spinothalamic tract termi-nates, at least in part, within the region of the magnocellular portion of the medial geniculate body. This nucleus is part of a larger region which has recently been defined by Rose and Woolsey (1958) as the posterior nuclear group of the dorsal thalamus. It lies posterior and dorsal to the ventrobasal complex, as can be seen from the drawings of Figs. 2 and 15, and encompasses a region of varied cytoarchitecture. This group includes parts of the posterior, lateral posterior, and supra-geniculate nuclei, as well as the magnocellular portion of the medial geniculate, referred to above. We have studied this region in order to assay the functional properties of the spinothalamic system. Some further discussion of the connections of this region will be given below.

In our investigations we have used the method of single-unit analysis of Adrian. The method is well known, and reference is made to several earlier publications in which our particular techniques have been described in some detail (Mountcastle, Davies, and Berman, 1957; Powell and Mountcastle, 1959b). With this method, it is pos-sible to observe the electrical signs of the impulse discharges of a single cortical or thalamic neuron, to the exclusion of others. One can then study the relation of this discharge to the place, intensity, frequency, and qualitative nature of the physiological stimuli that bring the cell to action. By studying *seriatim* a sufficient number of cells of a given population, under conditions as standard as possible, one can reconstruct the behavior of the population under that par-ticular set of conditions. Several examples of the primary data ob-tained are given in Figs. 5, 8, 9, and 14, to which reference will later be made.

The Lemniscal System

It is now somewhat unpopular to speak of the "successive relays" of a sensory system. In recent years we have come to realize that nuclei such as the ventrobasal complex or the geniculate bodies of the thalamus may be capable of a wider range of activity than simply relaying to higher levels the neural activity that reaches them. Some evidence does indeed suggest that other neural systems may condition this relay, and that some degree of neural integration occurs in sensory systems, at subcortical levels (for review, see Livingston, 1959). In-tuitively this is an attractive idea, tenuous as the evidence for it may be at present, for a simple relay in a system is not parsimonious of neurons. This idea should not be exaggerated, however, for studies of the lemniscal system that have been carried out so far emphasize its great capacity to reproduce at the cortical-input stage a rather faithful replica of its peripheral neural input. Such a tenacity of nuclear transfer, without marked signs of integrative action, is quite different from the properties of the spinothalamic system, which will be described later. In the lemniscal system, the functional properties of the various cells are quite similar, whether they lie in the dorsal-column nuclei, in the thalamus, or at the first stage of cortical activation. For this reason I shall consider the properties of the system as a whole, noting such differences as we have observed at successive levels.

Patterns of representation in the lemniscal system

Thanks in large measure to the continuing and persistent investigations of Dr. Woolsey and his colleagues, it is now well known that the contralateral body is represented in the postcentral region of the cortex in great detail, in a pattern worked out by Woolsey for an extensive series of mammals (Woolsey, 1952, 1958). I should like to reiterate what these workers originally emphasized, that detailed representation maps of the sort resulting from evoked-potential studies do not imply a point-to-point projection of the periphery upon the cortical surface (Woolsey, Marshall, and Bard, 1942). Perusal of the figurine drawings published by these investigators makes it clear that a given peripheral spot may, when stimulated, activate maximally a given cortical spot, but that such a stimulus activates also a rather large area of surrounding cortex, along a gradient of decreasing intensity. Reciprocally, there is a peripheral spot that activates maximally a given cortical spot, but surrounding areas of skin, and frequently quite large ones, also activate that cortical point, but at decreasing intensities as the peripheral stimulus is moved away from the center toward the edges of the peripheral receptive field.

The pattern of representation of the body form in the lemniscal transfer nucleus of the thalamus has also been determined in some detail (Mountcastle and Henneman, 1949, 1952; Rose and Mountcastle, 1952). In comparing these maps with those of the cortical pattern determined by Woolsey, Marshall, and Bard (1942) we must allow for the fact that the thalamic pattern exists in three dimensions, and that this is projected upon the two dimensions of the cerebral cortex. With this in mind, I think it is safe to say that there is no major difference between the two. There has been a somewhat greater disruption of middorsal continuity at the cortical than at the thalamic level, owing to the great expanse of tissue devoted to the apices of the limbs, and the necessity to accommodate this in two dimensions. This is not a marked difference, and in general the cortical is a rather straightforward replication of the thalamic pattern.

Data from such mapping experiments at the level of the dorsal columns and the dorsal-column nuclei are not available for the monkey. They are for the cat, however, and comparisons of patterns at successive levels of the system in that animal indicate what is a rather simple reduplication of the input pattern, a pattern that is itself a distortion of the actual body form. There appears to be an increasing degree of the divergence and convergence, to which I have referred above, as one ascends the system. But this is only an impression, and it seemed to us important to investigate the matter further by a study of single neurons of the system.

Topographic attributes of lemniscal neurons

Each postcentral cell that can be activated by stimulation of the skin is responsive to stimulation within a certain limited area of skin. Stimulation elsewhere does not excite that neuron. The sizes and positions of the receptive fields for the neurons encountered in each of several microelectrode penetrations of the postcentral gyrus of the monkey are shown in Fig. 3. The fields for all cells of a given penetration are more-or-less superimposed, and this is of some importance to the functional organization of the cortex—organization in a series of vertically oriented columns of cells (Powell and Mountcastle, 1959b). Here, however, I wish to use these observations to emphasize the point that a single spot in the periphery can activate many, in fact a multitude of, cortical cells. There is a divergence of the system from skin to cortex. The mean values for the sizes of these fields, for the monkey, range from about 3 square centimeters on the hand to some 25 square centimeters at the shoulder, and they are even larger on the trunk, abdomen, and back. This indicates that the denser the innervation of a peripheral part, the smaller the cutaneous fields that play upon the cortical cells related to it.

It would be of interest to compare in size the fields for cortical cells, which are at least fourth-order neurons in this polysynaptic system, with those innervated by the first-order sensory fibers of the dorsal roots. Data on the latter are not available for the monkey. Dr. Puletti, working in Dr. Woolsey's laboratory, has made these measurements for the cat (Puletti, 1959). When they are compared with data obtained in an earlier study of the cells of the sensory cortex of cat (Mountcastle, 1957), it turns out that the receptive fields of the cortical cells are from 15 to 100 times as large as those of the first-order elements. The exact ratio depends upon the part of the body considered and is smallest at the tip of the limb. Dr. Puletti's technique of dissecting dorsal root fibers may have selected a population of the larger cutaneous afferents. However, Dr. Iggo at Edinburgh has recently shown that even the tactile afferents of C-fiber size innervate small and spotlike receptive fields (Iggo, 1959).

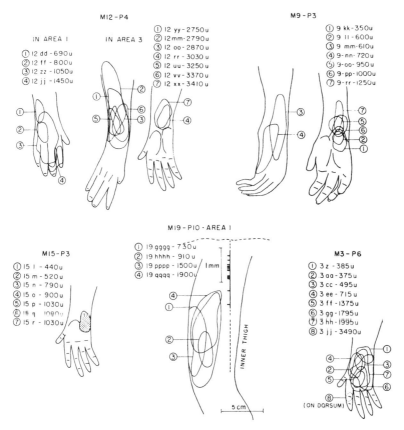

Fig. 3. Peripheral receptive fields of neurons of the cutaneous subclass, encountered in five microelectrode penetrations of the postcentral gyrus of the macaque monkey. There is a close superimposition of the fields, which was nearly complete for the seven fields isolated and studied in M15-P3. The centimeter scale refers to drawings of parts of the body; the millimeter scale to depth below the cortex for the reconstructed electrode track shown (M19-P10). (From Powell and Mountcastle, 1959*b*.)

Obviously many dorsal root fibers project via the polysynaptic chain of the system upon a single cortical cell. That is to say, not only does the system diverge from periphery to cortex, but a reciprocal convergence occurs over the same system. Thus the single-unit studies have confirmed *in extenso* the earlier observations of evoked potentials made with the gross electrodes. Now you may legitimately ask, where does this divergence-convergence occur? Our studies of receptive-field sizes of neurons in dorsal column nuclei, thalamus, and cortex indicate that it occurs to a gradually increasing degree from one "relay" to another up the system.   This is one example, then, of nonlinearity of action across relay nuclei.   The extent of this loss of a specific point-to-point relation is the basis of my earlier statement that the discriminatory capacity of the system cannot be explained by any theory of insulated connections.

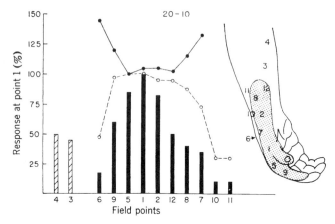

Fig. 4.   Gradations of the intensity of projection of the peripheral receptive field in the skin upon a cortical neuron of the postcentral gyrus of the macaque monkey.   The bar graph indicates the mean value of the number of impulses evoked per electrical stimulus to the skin at each point marked oñ the drawing, plotted as percentage of the mean response of the point 1 population.   Stimuli delivered at points 3 and 4, which are outside the field determined by physiological stimulation, were thought to activate nerve fibers innervating the field. Dashed lines plot the probability that a response will occur; solid lines, the mean latency of the population of responses, expressed as percentages of these values for the point 1 population. (From Mountcastle and Powell, 1959*b*.)

Gradient of projection of a peripheral receptive field.   It now becomes of some importance to examine the more dynamic aspects of the system, if we wish to understand the mechanism it employs in making the very fine spatial and temporal discriminations of which it is capable.   Before turning to observations of that sort, I would like to add one further fact concerning the static properties of the system: all the points in the peripheral receptive field of a thalamic or cortical neuron do not possess the same potency for excitation of the cell.   Examples of this gradient of projection are given for a cortical cell in Fig. 4.   It is possible, by making the reciprocal interpretation, to say that although a light touch activates a very large number of cells distributed over perhaps several millimeters of thalamus or cortex, the degree of excitation among them is not uniform.   Cells in the center of this discharge zone discharge the longest trains of impulses with the most certain probabilities at the shortest latencies. As we move from the center of the discharge zone toward its edge, cells are encountered that discharge fewer impulses, less certainly, at longer latencies, until the cells of the subliminal fringe are reached.

I shall return again to this somewhat imaginary reconstruction of the lemniscal events evoked by peripheral stimuli. First, however, I wish to discuss the modal properties of lemniscal neurons, and to describe some experiments aimed at a measure of the capacity of the system to transmit neural activity in the dimension of time. I should add that we have not seen the receptive field of a lemniscal neuron change its position, change significantly its size or shape, or become discontinuous.

Modal properties of lemniscal neurons

The weight of evidence favors the view that the several submodalities of somatic sensibility which are discriminative in nature are mediated via the lemniscal system. Yet it is not at all evident from introspection that such sensory experiences as light touch and the awareness of the position of a limb are linked together in any intimate way. During the last ten years there has been some difference of opinion over whether the first-order afferent fibers of the system are mode-specific or not. Dr. Rose and I have reviewed this problem in another place (Rose and Mountcastle, 1959), and I shall not develop it in detail here. So far as the available evidence is telling, and it is very strong indeed, one must conclude that the endings of the myelinated afferents show a differential sensitivity to one or another form of peripheral stimulation. That is, these afferents are mode-specific, even though in some areas of the body their endings may appear histologically similar! Whether smaller afferents are universally sensitive to various stimuli or whether they too are specific is still somewhat uncertain. However, Dr. Iggo at Edinburgh (Iggo, 1959) has recently produced very good evidence that at least some of the C fibers are indeed highly stimulus-specific.

The evidence concerning mode-specificity, though slowly emerging from a cloud at the level of first-order neurons, is established with some certainty for the cells of the central stations of the lemniscal system. Neurons in the dorsal column nuclei, the thalamic ventrobasal complex, and the overwhelming majority of those observed in the postcentral gyrus respond to a particular form or locus of stimulation, and not to others. They may be divided into two groups: those activated by stimulation of the skin, and those driven by stimulation of nerve endings or sensory organs located in the deep tissues. Many of the second group are selectively activated by gentle joint rotation and depict, by their activity patterns, the positions and movements of the limbs, as indicated in Fig. 5: they deal with kinesthesis (Mountcastle and Powell, 1959a).

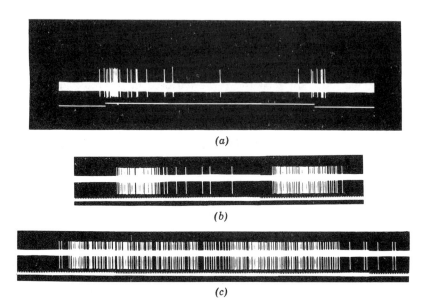

(a)

(b)

(c)

Fig. 5. (a) Impulse discharges of a neuron of the postcentral gyrus of the macaque monkey, activated only during the transient phases of movement to and from full flexion of the contralateral elbow. Onset and cessation of movement indicated by the signal line. Movement lasted 3.3 sec, which gives time. (b) and (c) Impulse discharges of another postcentral neuron, evoked by brisk lateral deviation of the contralateral toe at the metatarsophalangeal joint. Responses in (b) evoked by two brisk movements; those in (c) indicate the sustained discharge that accompanied sustained joint displacement; time, 40 per second. (From Mountcastle and Powell, 1959a.)

At the level of the cerebral cortex these mode-specific neurons are arranged in columns of cells related to one or the other submodality; this supports the view that the initial activation of the cortex by a sensory stimulus is organized in terms of a vertical module of cells, extending across the lower five layers of cells. In the monkey there is a differential distribution of these mode-specific modules in the anteroposterior direction across the postcentral gyrus, so that, although area 3 is devoted mainly to the cutaneous sense, area 2 is concerned mainly with deep sensibility, and area 1 is a gradient between them. All are organized in a single topographic pattern (Powell and Mountcastle, 1959*b*).

Mode specificity is an equally certain property of cells of the ventro-basal complex of the thalamus and the dorsal column nuclei. We do not yet know whether there is a differential distribution of cells of the submodal types within these subcortical nuclei.

Mode specificity is, at least in my experience, the *sine qua non* of the lemniscal system. It is regularly and unvaryingly observed, regardless of the experimental conditions or the level of anesthesia, and it is observed also when no anesthetic agent is used (see page 415).

Response patterns of lemniscal neurons

Relation of response to the intensity of the stimulus. I should like now to present data on the relation of the responses of lemniscal neurons to the parameters of the stimuli that excite them. The first observation is that the probability that a response will occur and its latency and degree are sensitively related to the strength of the cutaneous stimulus. With increasing stimulus intensity, a gradual increase in the response in the directions indicated takes place. Such at least is the common course of events when either mechanical or electrical stimuli are employed on the skin. Rarely, however, increasingly strong stimuli may either produce no change in the response or even cause a decrease in the number of impulses making up the evoked discharge. This latter we attribute to a recruitment of afferent inhibition by the stronger stimuli, a subject to which I shall return a bit later.

The usual response of a lemniscal neuron at the thalamic or cortical level is a brief repetitive train, even when the input in first-order afferent fibers is made as synchronous as possible by electrical stimulation of the skin. Several observers have shown that this repetition begins at first relay, in the dorsal column nuclei, and that this occurs in both anesthetized and unanesthetized animals, and over a considerable range of body temperature (Amassian and DeVito, 1957; Berman, 1957). It is important to add that repetition occurs also when intra-axonal recording in first-order fibers of the dorsal columns gives evidence that each fiber delivers but one impulse for each stimulus (Mountcastle and Poggio, 1958). The increase in discharge produced by increasing intensity of stimulation must result from a recruitment of additional first-order fibers by the stimulus. Why such a repetition should occur at all is not clear. It seems to be a general property of sensory systems, for it has been observed in all sensory systems that have been examined, at all levels. It is very likely due to some intrinsic but as yet unknown property of the synaptic relays of those systems, which seem to possess functional properties quite different from those that monosynaptically link certain dorsal root afferents to motoneurons. Here all observers agree that a single volley that occupies even a maximal number of such fibers evokes but a single impulse in each large motoneuron brought into action by the volley.

Data such as these, and those presented earlier, indicate that the latency and degree of the response of a neuron of the lemniscal system are functions of the position of the stimulus within the peripheral receptive field, of the stimulus intensity, and, as will be shown presently, of the immediate past history of the cell as well. So far as I know, there is not the slightest evidence to support the idea that space, that is, the position of the stimulus, is coded in any reliable way in terms of the temporal characteristics of the response of central neurons, though this idea is frequently put forward in discussions of the neural mechanisms of discrimination. Which cells respond is determined by the place stimulated. The temporal properties of the response, and its degree, appear to be sensitively determined by the stimulus parameters of intensity and frequency.

The capacity of the lemniscal system to transmit repetitive activity. Although analysis of the lemniscal activity evoked by single brief stimuli has proved of some value, that form of activity is in a sense unique to the experimental situation. Under more nearly normal circumstances the system is bombarded by trains of impulses in first-order afferents, trains whose frequencies reflect the steady and changing states of excitation of peripheral receptors. It is of some importance then to discover whether the system can follow with some fidelity this input or, if direct following of elements in the chain leading to the cerebral cortex is not possible, whether the input activity is coded in some reliable way by transformation in regions of nuclear relay.

One measure of this capacity in the temporal dimension is the speed with which the system recovers excitability after passage of a single volley. Marshall and his colleagues found this to be the property of the system most severely affected by barbiturate anesthesia (Marshall, 1941; Marshall, Woolsey, and Bard, 1941). An example of the prolonged recovery for a thalamic cell of the system in a deeply anesthetized cat is given in Fig. 6. Marshall and his co-workers found that, as the anesthetic state lightens, there is a remarkable shortening of these unresponsive times. By using implanted electrodes resting on the cortical surface, they were able to show that in the waking state no unresponsive time exists at all—apart from the refractoriness of peripheral nerve fiber. They found, in fact, that for the first 20 to

30 msec after the response to an initial stimulus there is facilitation of the response to a second one.

These observations we have now been able to confirm in studies of single neurons. The curves of Fig. 7 show the absence of unresponsiveness and the period of early facilitation for a postcentral neuron of a lightly anesthetized monkey (Mountcastle and Powell, 1959*b*).

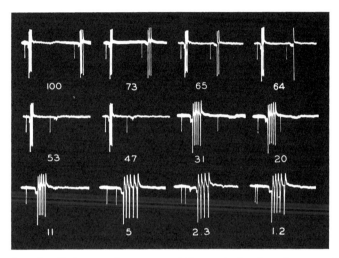

Fig. 6. Impulse discharges of a neuron of the ventrobasal nuclear complex of a cat under deep pentobarbital anesthesia. The thalamic cell was activated by light mechanical stimulation of the skin of the contralateral forepaw. Here it is driven by two electrical stimuli to the skin, delivered at a series of intervals indicated in milliseconds below each record. The record chosen for illustration at each interval of this recovery-cycle study represents the mean, in terms of the number of impulses evoked by the second stimulus, of the population of responses recorded for that interval. Relative unresponsiveness is apparent at 73 msec and is complete at 53 msec. There is no period of temporal facilitation.

More recently, we have observed this same phenomenon in the unanesthetized monkey. Since I am very much concerned about the use of unanesthetized animals in neurophysiological research, I should like to describe the methods my colleague Dr. G. F. Poggio and I have used in this series of experiments.

In a preliminary aseptic procedure the animals (monkeys and cats) have been subjected to a bilateral retrogasserian section of the trigeminal nerves, via the posterior-fossa approach. At the same time, the ascending branches of the cervical plexi are sectioned low in the neck. This results in a total denervation of the tissues of the head, except for the external auditory canal. The terminal acute experiment is carried out ten to twelve days later. Whatever operative procedures are necessary in preparation for recording can then be confined to the insensitive regions of the head. In recording from deep structures, a Horsley-Clarke instrument is used which has been modified so that the pressure points are placed on the base of the zygoma, rather than

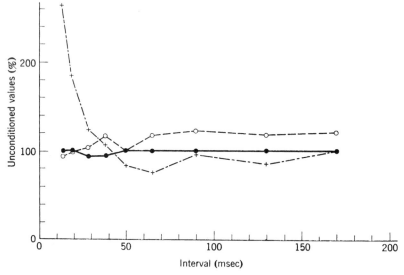

Fig. 7. Excitability of a neuron of the postcentral gyrus of the macaque monkey, studied in the same manner as in Fig. 6, but here in an animal under very light thiopental anesthesia. Plus signs and dash-dot line plot the mean number of impulses per response; open circles and dashed line, the mean latencies of the responses to a second stimulus delivered at a series of intervals after the first, as percentage of the unconditioned response; solid line, the probability that any response will occur at all. There is no sign of unresponsiveness of the system, at least beyond the very short intervals at which the two responses fuse, and there is an early period of temporal facilitation. (From Mountcastle and Powell, 1959*b*.)

in the ear. The animals are held motionless, when necessary, with Flaxedil, and are artificially respired.

In these preparations we have found that neurons at thalamic and cortical levels of the lemniscal system show no unresponsive periods, and for many cells a period of early facilitation can be demonstrated.

A second test of the capacity of the lemniscal system in the temporal dimension is to determine how faithfully cortical neurons follow repetitive peripheral stimuli. At very light anesthetic levels, cortical neurons follow cutaneous stimuli beat for beat, up to about 75 to 100 per

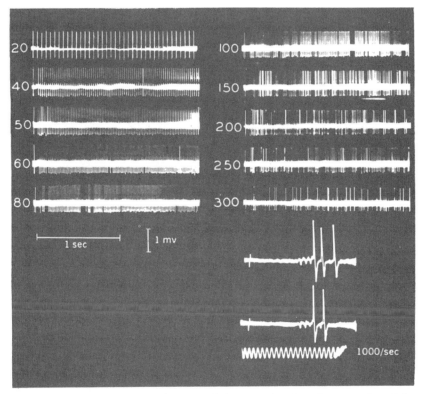

Fig. 8. Responses of a single neuron of the ventrobasal nuclear complex of the thalamus of a cat, under deep pentobarbital anesthesia, to electrical stimulation of the skin of the contralateral foreleg. The stimuli were delivered at different frequencies per second, which are indicated at the left of each record. There is equilibration of the response at higher frequencies of stimulation. Each record is for 2 sec, which gives time line. Inset records: typical responses to single stimuli. (From Rose and Mountcastle, 1959.)

second. At higher rates of stimulation the response equilibrates, that is, it comes to an over-all average response rate. An example is given in Fig. 8. During the period of equilibration the responses occur in a definite relation to certain stimuli of the train—so far as can be determined at such rapid rates of stimulation—and not to others. Although adequate stimuli and inadequate ones are randomly intermingled, the evoked discharges are not at all randomly placed.

This equilibratory response is seen regularly, however, only when the stimulus is placed near the center of the peripheral receptive field. When it is placed near the field's edge, results such as those shown in Fig. 9 are obtained. The neuron follows the stimuli up to a relatively low frequency, but when trains are delivered at higher frequencies the cell responds to the first or first few stimuli only, and thereafter it discharges only rarely if at all. Renkin, in our laboratory, has found this to be the common behavior for a large number of cortical

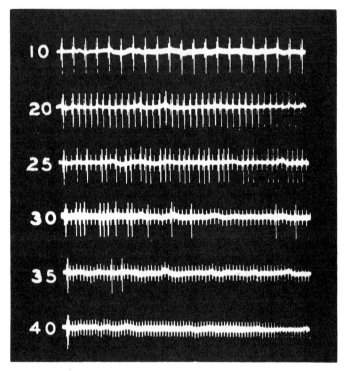

Fig. 9. Responses of a single neuron of the ventrobasal nuclear complex of the thalamus of a cat, under deep pentobarbital anesthesia, to electrical stimulation of the skin near the first digit of the contralateral forepaw, delivered at different frequencies. At frequencies of 40 per second or higher, the neuron responded to the first stimulus of a train and failed to respond thereafter. This cut-off characteristic is commonly seen when the stimuli are delivered near the edge of the peripheral receptive field. Small deflections are stimulus artifacts. (From Rose and Mountcastle, 1959.)

cells, that is, that the frequency following depends upon stimulus position (Renkin, 1959). We interpret this to mean that the synaptic connections linking the edge of its receptive field to a cortical cell are too tenuous to support activity at a high rate, but we have no explanation for the failure of transmission that ensues when rapid trains of stimuli are delivered.

This phenomenon seems to be of some importance for the following reason. If we make the reciprocal interpretation, we can postulate that slowly repeated stimuli create a discharge zone of cells in the cerebral cortex, and each cell brought above threshold follows the stimulus frequency, beat for beat. As the frequency of input increases—as the intensity of stimulation increases—the cells of the center of the discharge zone continue to follow faithfully, while those at its edge cease to respond. This mechanism tends to limit, or even to reduce, the zone of active cells as the stimulus grows. It serves to enhance the capacity of the system for spatial discrimination.

Adaptive patterns. The cells of the lemniscal system which are specific for a given mode of stimulation may nevertheless differ from one another in the temporal course of their responses to steady stimuli. For example, of the postcentral neurons activated by joint rotation, 19 per cent discharge only rapid onset transients when the exciting joint rotation is begun and adapt quickly to extinction (Fig. 5). The remaining 81 per cent behave in quite a different fashion. These neurons discharge very rapid onset transients with movement and adapt rapidly to a somewhat lower rate of discharge, which is steadily maintained so long as the joint displacement is continued (Fig. 5). The level of maintained discharge is determined by the angle of joint displacement, but we do not yet know the quantitative relation between angle of displacement and the frequency of discharge in the steady state (see Mountcastle and Powell, 1959a). Quickly and slowly adapting patterns are seen for neurons that seem to be intermingled within the discharge zone of cells set up by rotation of a given joint. Thus the active population has the capacity to signal both the occurrence and the degree of the transient movement, and the steady position of the limbs as well. How these neural patterns are used at higher levels, that is to say, how their content of information is extracted, is unknown.

Afferent inhibition

I should like now to consider a little further the capacity of this system to discriminate in the dimension of space. Behind much of what I have said is the assumption that cortical mechanisms important for localizing a point stimulated in the periphery must begin with the creation of a zone of activity in a restricted region of the postcentral gyrus. We have no inkling yet of how this is further elaborated in this and perhaps other cortical regions to result finally in a meaningful percept. If this hypothesis is true, however, the perception of two stimulated points as two and not one depends in the first instance upon the provocation by the two stimuli of two local zones of activity. These zones may be very close together and will certainly overlap, but they must not be completely superimposed, and there must remain a region of less intense neural activity between them. More complicated neural patterns must accompany peripheral stimuli with

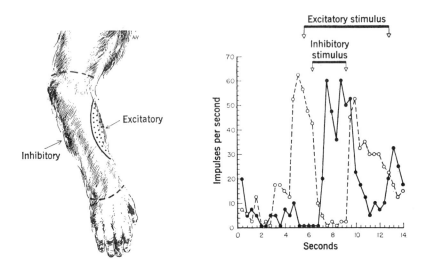

Fig. 10. Afferent inhibition in the somatic system. A neuron of the postcentral gyrus of the macaque monkey was driven from the receptive field of the skin of the contralateral preaxial forearm, as shown in the drawing, and its discharge was inhibited by light mechanical stimulation within a much larger surrounding area, the inhibitory receptive field. This stimulation excited a second neuron whose discharges were also observed in the record, and the second neuron was inhibited by stimuli within the excitatory receptive field of the first. The reciprocal behavior of the two cells is indicated by the graph of impulse frequency versus time, the first cell by the dashed line, and the second by the solid line. (From Mountcastle and Powell, 1959b.)

more complicated contours. One need only recall the phenomenal discriminations in space and in time which humans can make when reading Braille to be sure that something more than simple connections is required for mechanism.

We have observed one property of the lemniscal system which provides, I believe, a very powerful part of that something more, and that is afferent inhibition.

The phenomenon is illustrated in Fig. 10. The discharges of this postcentral cell were evoked by stimulation of the skin of the preaxial forearm. Stimulation anywhere within the large surrounding area of skin produced a cessation of the spontaneous or driven activity of the cell. The graphs of Fig. 10 show also the changes in impulse discharge provoked by these two stimuli in a closely adjacent cortical cell. The two cells behave in a reciprocal fashion, and such a reciprocity has been commonly observed in the cerebral cortex.

We have defined true afferent inhibition as that case in which a stimulus delivered to the periphery will inhibit a cortical cell but will not, at any intensity or in any temporal pattern, excite that same cell. Such a stimulus will, however, produce an evoked slow wave at the recording site. Whether this response is related to the local inhibitory process, or whether it is recorded electrotonically from an adjacent region of cortex whose cells are excited by this stimulus, is uncertain. We have observed that the inhibitory stimulus must lead the excitatory by 2 or 3 msec in order for complete inhibition to take place, and this suggests that the inhibition is accomplished by some collateral process within the cortex.

The topographical relation of excitatory and inhibitory receptive fields has always been determined with physiological stimuli. On the extremities the optimal site for inhibition is sometimes axially opposed across the limb from that for excitation. Most commonly, the pattern is that of an excitatory center and an inhibitory surround, though occasionally the inhibitory field is located eccentrically to the excitatory.

This pattern of central excitation and surround inhibition of the receptive fields appears to be the basic one for many afferent systems. It is true for the retinal ganglion cells (Kuffler, 1953), and a similar topographical arrangement has been described by Hartline and his colleagues for the lateral eye of *Limulus* (Hartline and Ratliff, 1957, 1958; Hartline, Wagner, and Ratliff, 1956; Ratliff and Hartline, 1959). Such a relation has recently been observed by Hubel and Wiesel for single neurons of the striate cortex of the cat (1959). A picture comparable to surround inhibition, in terms of the topography of the peripheral sensory sheet, is well known in the auditory system, where inhibition of central neurons by tones just off the best frequency for their excitation has been observed at each level of the system.

Afferent inhibition possesses several of the properties of afferent excitation: (1) temporal summation of the inhibitory effects produced by repetitive stimuli to a single site; (2) spatial and temporal summation of the inhibition caused by stimuli delivered to separate sites in the field; (3) adaptation to steady inhibitory stimuli—for some neurons. On the other hand, inhibition differs from excitation in its susceptibility to the depressing effects of anesthetic agents. Inhibition is never seen under deep anesthesia, appears with increasing frequency as the anesthetic state lightens, and thus is likely to be an attribute of a large proportion of the postcentral neurons related to the skin.

One of the most consistent observations we have made is that, whenever a neuron is inhibited by stimulation of the skin, that inhibitory stimulus is in fact excitatory for cortical cells that are closely adjacent. Adjustment of the position of the electrode tip then makes it possible to observe simultaneously the impulse discharges of two units that display reciprocal behavior. An example of this "afferent reciprocal innervation" is given in Fig. 10.

The conclusion I draw from these observations is that the group of cortical cells excited by stimulation at a single peripheral site is surrounded by a ring of cells inhibited by this same stimulus. The fact that no first-order afferents are concerned with inhibition alone suggests that some collateral mechanism may be involved. Whether this is true is uncertain, but the functional meaning of the surround inhibition is, I think, quite clear. It is to limit in a spatial sense the zone of cells thrown into action by a peripheral stimulus.

Intermediate summary: Properties of the lemniscal system

A transient light touch to the skin sets up a zone of active cells in the cerebral cortex. The cells of this zone do not all respond in an identical manner. Those at or near the center of the group of cells discharge short trains of impulses, and the cells of the center respond with the greatest security and at the shortest latencies. Cells just off the center discharge fewer impulses, at longer latencies, and with less certain probabilities. The response parameters continue to change gradually from cell to cell between the center and the edge of the discharge zone; those at its edge respond insecurely, whereas those just outside receive only subthreshold synaptic excitations. This description follows from the reciprocal interpretation of the studies of the gradient of projection of a cutaneous peripheral receptive field upon the related cortical cell.

This pattern of response is exquisitely dependent upon the strength of the peripheral stimulus, that is, upon the numbers of first-order afferents activated, and the durations and frequencies of the trains of impulses set up in them. Which cells of the cortex are activated is determined by the place in the periphery stimulated. The intensity of the stimulus is indicated by the spatial and temporal patterns of activity of those cells.

Such a description seems reasonable for single brief stimuli. But the physiological stimuli of ordinary life are scarcely ever so brief, for they involve stimulus patterns of duration and complexity. Thus the capacity of the system to transmit messages composed of a variety of temporal patterns could be a limiting factor in its discriminatory function. It is important to say, therefore, that our studies of the recovery of excitability under conditions of the lightest anesthesia show that no true unresponsive period exists. In fact, following a single activation, and of course after recovery from nerve-fiber refractoriness, there is a very powerful temporal facilitation. And when studies are made of the responses of cortical cells to repetitive trains of stimuli, an equilibrating response is found at a very high level—a level of the same order as, or higher than, the usual input frequencies from cutaneous receptors.

The remarkable precision of this system in the temporal dimension makes it likely, it seems to me, that the limiting factor in the transmission of information in the time dimension is the peripheral receptor itself. The frequency pass-band, as it were, of the central neural relays seems to be wider than that of the relevant receptors. Little input escapes, for even the most intense or rapidly shifting stimulus pattern that the receptors can follow is relayed centrally with considerable fidelity. At the other end of the scale, this great security of transmission has another implication: the true threshold for perception, in the alert and attending individual, is probably the threshold of the relevant receptors. In the limit, it is likely that a single impulse in a single peripheral nerve fiber will provoke a discharge of cortical cells. It is not necessary to push the analogy so far, however, for it is difficult to think of any stimulus in normal life—for any afferent system—whose effects could be confined to a single afferent nerve fiber.

All the discussion presented above pertains to the lemniscal component of the somatic afferent system, as observed at its region of thalamic transfer, and at its cortical terminus in the postcentral gyrus. It is a system of great synaptic security, poised for action in rapid cadence, of wide frequency-carrying capacity and open amplification. It is a system of precise anatomical connections, although over and above its anatomical specificity—even here there is considerable divergence—there are several physiological attributes which tune it for discriminatory functions as precise as any known in nature. These are temporal facilitation, fringe failure, and afferent surround inhibition, and doubtless there are others yet to be discovered. Recent findings of both anatomists and physiologists indicate that the system is governed by descending systems that modulate relay, which thus brings this system into that general framework of servo control so common in the nervous system.

The Spinothalamic System

I should like now to consider the spinothalamic system, the second component of the somatic afferent system. Perhaps anterolateral system is a better term, for it impinges upon many regions of the brainstem and mesencephalon, as well as on the dorsal thalamus. The system takes origin from cells of the dorsal horn of the spinal gray, cells that are activated either directly or after interneuronal relay by first-order afferents of the dorsal roots.

Some of the functional properties of dorsal-horn cells have recently been studied by several investigators (Kolmodin, 1957; Hunt and Kuno, 1959). Hunt and Kuno found these cells to serve rather large receptive fields, which may include more than one limb. Many can be activated from either side of the body. Although the majority are sensitive to mechanical stimulation of peripheral tissues, many discharge only when the stimulus is definitely injurious in nature. These properties are quite different from those of the lemniscal system at a comparable second-order level, in the dorsal column nuclei.

In my own studies of this system I first attempted to define the thalamic terminus of the spinothalamic projection, using ipsilateral input as the means of identifying transmission in this system, for the medial lemniscus is well known to be contralateral in its projection upon the thalamus. Responses to ipsilateral stimuli did occur in the thalamus; they were clustered in a region medial and mediodorsal to the medial geniculate body, posterior to the ventrobasal complex. This region, continuous above and behind with the pulvinar, has been designated by Rose and Woolsey (1958) as the posterior group of nuclei; they showed that it projects upon the cortex rather widely, in a sustaining fashion, and upon a region that includes the second somatic area. Dr. G. F. Poggio and I have recently undertaken a study of this region with the method of single-unit analysis (Poggio and Mountcastle, 1960), and I should like now to discuss those results and to compare the functional properties of the cells of this region with those of the lemniscal relay nucleus which lies more to the anterior.

Properties of the cells of the posterior thalamic group— the spinothalamic relay nucleus

The neurons of the posterior group differ in many of their functional properties from those of the lemniscal system, which I have described above. In the first place, their receptive fields may be very

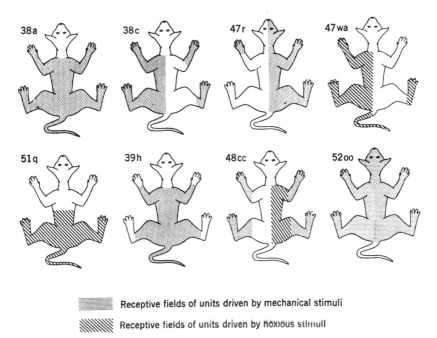

Receptive fields of units driven by mechanical stimuli

Receptive fields of units driven by noxious stimuli

Fig. 11. A representative sample of the peripheral receptive fields of neurons of the posterior nuclear group of the thalamus of the cat.   Five of these cells (38a, 38c, 47r, 39h and 52oo) were activated by light mechanical stimuli.   Two were responsive only to noxious stimuli (47wa, 51q), and one (48cc) was responsive to either form of stimulation, but the receptive fields for the two forms were not identical.   The ipsilateral side of the body, relative to the recording electrode, is the right-hand side of each figurine drawing.   (From Poggio and Mountcastle, 1960.)

large, and they may cover parts of both sides of the body and, in the limit, the entire body surface.   The fields of eight cells of this region are shown in Fig. 11, which serves also to introduce two other properties.   First, some of these cells may be activated by stimuli that are clearly destructive of tissue.   Two of the cells whose receptive fields are shown in Fig. 11 were activated only by such stimuli.   Second, the cells of the posterior group are not mode-specific.   For example, cell 48cc of Fig. 11 was activated by mechanical stimulation of any one of the four paws.   In addition, this cell responded to noxious stimulation over a much wider receptive field.   Thus it appears that there is a very wide convergence upon these thalamic cells, in terms of both topography and modality.   These attributes contrast markedly with the very high degree of place and mode specificity of the lemniscal system.

This contrast in the topographic properties of neurons of the thalamic relay nuclei for the two systems can be seen by comparing the results obtained in microelectrode penetrations of the two regions. The anatomical reconstruction of the course of a typical penetration

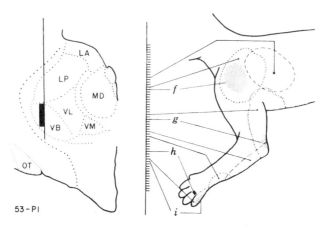

Fig. 12. Data obtained in a microelectrode penetration which passed through the ventrobasal nuclear complex of the thalamus of a cat. The reconstruction of the penetration is presented at the left; in the region marked by a thick line neurons were observed that were activated by somatic sensory stimuli delivered to the contralateral side of the body. In the drawing to the right the expanded scale (50-micron intervals) represents the responsive region of the ventrobasal complex. The areas indicated by lines and shading on the figurine drawing of the cat indicate the positions and extents of the peripheral receptive fields of the neurons observed; at positions *f, g,* and *i* neurons were isolated which were activated by light mechanical stimuli delivered to the skin. At position *h* a neuron was isolated which was activated by gentle rotation of the metacarpophalangeal joint of the third toe of the contralateral forepaw. The other fields represented are those of the several neurons active together at the levels indicated. The entire sequence of receptive fields is typical of the rather precise topographical pattern of representation of the contralateral body surface in the ventrobasal nuclear complex. Abbreviations: LA, lateral anterior nucleus; LP, lateral posterior nucleus; MD, mediodorsal nucleus; VB, ventrobasal nuclear complex; VL, ventrolateral nucleus; VM, ventromedial nucleus; OT, optic tract. (From Poggio and Mountcastle, 1960.)

through the lemniscal relay region is shown in Fig. 12. In its course active and drivable cells were isolated and studied along the segment indicated. The receptive fields of some of these neurons are shown on the drawing of the cat's foreleg. They were restricted in size and, as the electrode descended the fields for the units encountered, were successively more distally placed. This conforms to the pattern of representation in this region determined with gross-electrode methods

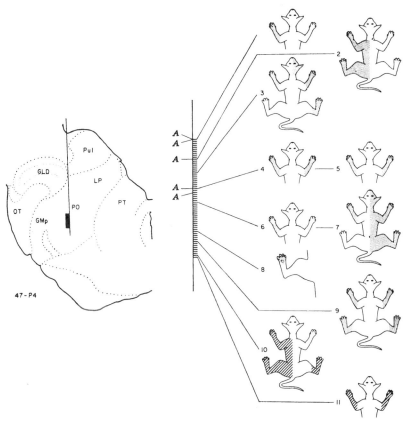

Fig. 13.   Data obtained in a microelectrode penetration that passed through the posterior nuclear group of the thalamus of a cat.   The penetration is reconstructed in the drawing at the left.   In the region marked by the heavy line, active and drivable neurons were observed that displayed the "posterior group properties" described in the text.   The responsive region is indicated by the expanded scale (20-micron intervals) shown at the right.   As indicate levels at which neurons were isolated that were activated only by sound.   Figurine drawings 1 through 9 show the peripheral receptive fields of cells that were activated by light mechanical stimuli delivered to the body surface; the cells whose receptive fields are shown by drawings 10 and 11 could be activated only by stimuli that were destructive of tissue.   It is important to note that, in contrast to the data shown in Fig. 12, the sequence of receptive fields here composes no topographical pattern.   In many other penetrations of the posterior group the auditory, nociceptive, and mechanoreceptive cells, as well as those responsive to all three forms of stimulation, were indiscriminately mixed in the dorsoventral direction.   (From Poggio and Mountcastle, 1960.)

(Mountcastle and Henneman, 1949).   On the other hand, the penetration illustrated in Fig. 13 passed through the posterior nuclear group. In its passage nine neurons were isolated and studied; their receptive fields are shown to the right. They indicate another and in my experience a unique property of an afferent system: there is no regular pattern of representation of the body form in this region. In addition, other cells were observed in this region that were sensitive to sound, and we have observed some cells that were sensitive to both tactile and auditory stimuli.

Responses to noxious stimuli

We have regularly observed, at the very lightest levels of anesthesia, that many cells of the posterior group are activated by noxious stimuli. In fact, 60 per cent were sensitive to only such destructive stimuli, whereas others were sensitive to this form of stimulation in only part of their receptive field but responded to light mechanical stimulation delivered to other parts of the field. Figure 14 shows representative

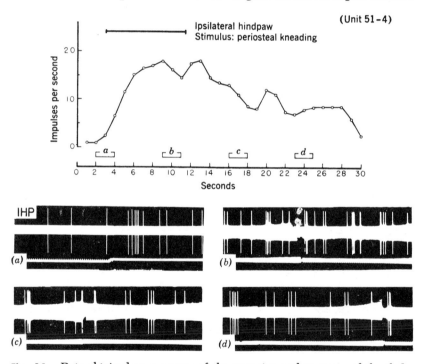

Fig. 14. Data obtained on a neuron of the posterior nuclear group of the thalamus of a cat, a cell which was activated only by noxious stimulation of the deep tissues of any part of the body except the face. The graph indicates the slow recruitment of activity during periosteal kneading and the prolonged afterdischarge that followed it. Records (a), (b), (c), and (d) are samples taken at the times indicated on the graph above. Note the slow recruitment of activity and the prolonged discharge after cessation of stimulation. (From Poggio and Mountcastle, 1960.)

records below and the graph of impulse frequency versus time above for a posterior group cell driven by destructive stimulation. The graph shows the typical pattern of a slow recruitment of activity and a prolonged after-discharge. This pattern is very different from the rapid onset transients and dead-beat endings of the discharges of cells of the lemniscal system.

Our observations upon this group of cells of the posterior group have led to the conclusion that this region is likely to play an essential role in the neural mechanisms for the perception of pain. I am well aware of the difficulty one faces in terming painful any stimulus delivered to an experimental animal, for no introspective report is available. Nevertheless, when a stimulus by its nature destroys tissue, provokes a defense or escape reaction accompanied by signs of the appropriate emotional change in the normal waking animal, and provokes painful sensations when applied to man, it seems reasonable to regard the stimulus as painful to both animal and man. To label such stimuli descriptively and not call them painful seems to me a semantic triviality. There is certainly no a priori reason to believe that in evolution the perception of pain appears as a wholly new sensory phenomenon in man. Many recent observations indicate that very widespread areas of the brainstem and diencephalon may be activated by painful stimuli. The suggestion that, of these regions, the posterior thalamic group is more likely to be concerned with the perception of pain rests upon the fact that it is the only one known to project in any direct way upon the cerebral cortex.

Inhibition

For cells of the posterior group the topography of the excitatory and inhibitory receptive fields is very complex. Cells activated from one side of the body may be inhibited from the other side, and the fields may be homologously distributed. Other cells, activated from more restricted fields, may be inhibited from almost all the rest of the body. These patterns of inhibition are quite similar to those of the interneurons of the dorsal horn studied by Hunt and Kuno (1959), and they are very different indeed from those of the lemniscal system, described above. The functional meaning of these very complex relations is obscure.

Anatomical correlations

Figure 15 shows reconstructions of nine electrode tracks that passed through the region of the posterior nuclear group. It is clear from these drawings that there is a rather precise correlation between the functional properties of a thalamic cell—lemniscal or spinothalamic as

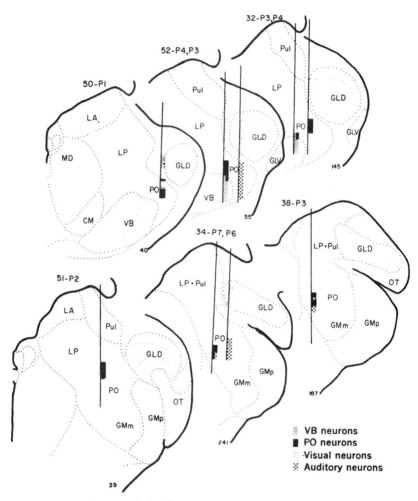

Fig. 15. A correlation of the functional properties of thalamic cells with their position in the thalamus. The drawings above represent the reconstructions of nine microelectrode penetrations of the cat thalamus. The properties of the cells observed in each penetration are indicated by the symbols. Note especially penetrations 52-P4 and 32-P3, in which the change of properties from those termed "posterior group" to those termed "lemniscal" (see text) correlates with the passage of the electrode tips from posterior group above to the ventro-basal complex below. Abbreviations as in Fig. 2. (From Poggio and Mount-castle, 1960.)

I have defined them—and the position of the cell in the thalamus. This is made very clear by the results obtained in two of the recon-structions illustrated, 52-P4 and 32-P3. In each case the properties of the cells observed changed suddenly in the course of the penetra-

tion, and the reconstructions show that this region of change correlates very well with movement of the electrode from PO above to VB
below. There is some intermingling of cells of the two types at the
cytoarchitectural junction. The symbols indicate also that many
neurons of PO, particularly in its caudal portion, were observed which
responded to sound, and it is here that we have observed interaction
of sound and somatic stimuli upon the same single cells.

Intermediate summary: Properties of the spinothalamic system

The neurons of the posterior thalamic group are related to very
large peripheral receptive fields, which may cover the entire body
surface. No pattern of representation of the body form exists in this
region. Posterior group cells may be driven by mechanical stimuli,
or by noxious stimuli, or by both, and many of these cells are affected
also by auditory stimuli—in either an excitatory or an inhibitory way.
These cells are specific to neither mode nor place. Their properties
are different in almost every respect from those of cells of the lemniscal
portions of the somatic afferent system.

It is this posterior region that Whitlock and Perl (1959) activated
by nerve volleys, after transection of the dorsal columns. It is, I
believe, this same region that Knighton (1950) stimulated to produce
local responses in the second somatic area of the cerebral cortex. It
is indeed this same region that Rose and Woolsey (1958) found to
degenerate when a removal of the second somatic area is added to
that of the adjacent auditory areas of the cortex. One is tempted by
this evidence to accept the hypothesis that the lemniscal and spinothalamic systems are quite separate and distinct, projecting upon separate thalamic relay nuclei, which in turn are related to different areas
of cortex. It is well known, however, that the spinothalamic system
has a component that projects upon the lemniscal relay nucleus, as
well as upon the posterior group (Mehler, Feferman, and Nauta, in
press), and that this component, rudimentary in the cat, grows
disproportionately at an accelerating rate in primates. And other
experimental results already available render the idea untenable. In
presenting them I should like first to consider what is known of the
second somatic area of the cortex.

The Second Somatic Area

This region of the cortex was described independently by Adrian
(1941) and Woolsey (1943). Woolsey has since shown that it exists
in each of a large series of mammals. It is located in the anterior

ectosylvian gyrus of the cat, immediately below the postcentral homologue.

There is evidence that this region possesses functional properties both spinothalamic and lemniscal in nature. Dr. Alvin L. Berman, working in our laboratory, has confirmed Woolsey's earlier observation that the region possesses as detailed and precise a pattern of representation as does the postcentral homologue. There is one difference, that the projection into the second somatic area is from both sides of the body, in two coextensive and completely overlapping patterns. Berman (1957) has shown that the auditory responses that can be recorded in this region (Tunturi, 1945; Bremer, 1952) interact with those evoked by somatic stimulation. His results show that in the second somatic area the leading somatic stimulus prevents the response to a succeeding auditory one, and that the reverse holds for the adjacent auditory cortex. Double interaction occurs at the region of transition. The precise pattern of representation described above is a lemniscal property. The somatic-auditory interaction is reminiscent of the similar interaction we have observed for cells of the posterior nuclear complex of the thalamus.

Recently, Carreras and Levitt (1959) have carried out a study of the second somatic area in the cat using the method of single-unit analysis. They found that the columnar pattern of organization described for the first somatic area holds also for the second. They also observed many cells that were specific to mode and place. However, Carreras and Andersson have continued this study and have observed some cells of the region, with very wide receptive fields, that could be activated only by noxious stimulation. Thus it is possible to say, on several counts, that the second somatic area possesses the functional properties of both the lemniscal and the spinothalamic components of the somatic afferent system.

In spite of all the data I have presented, it is not possible to assert that the postcentral region possesses exclusively lemniscal properties, though it certainly does to an overwhelming extent. In the postcentral gyrus of monkey Dr. Powell and I found 12 cells of a total population of nearly 1400 studied that had unusual properties (Mountcastle and Powell, 1959b). These cells had extremely large receptive fields, they could be activated by ipsilateral stimuli, and they could then be inhibited from homologously located stimuli to the contralateral side. Some could be driven only by stimuli clearly injurious to the skin and when so driven could be inhibited by light mechanical stimuli delivered to the same peripheral locus. Each cell, of course, did not display all these properties. The observations made upon this small group of cells are, I think, reliable. If so, they carry an important implication. That is, that the postcentral area expresses a range of functional properties comprising those of both the lemniscal and the spinothalamic systems.

Summary

It must be apparent by now that it is not yet possible to present an orderly and well-integrated synthesis concerning the neural correlates of somatic sensibility. Certain generalizations begin to stand out, however. The first is that the dorsal column–medial lemniscal system is one designed to serve the discriminative forms of somatic sensibility, and certain of its properties endow it with an exquisite and precise capacity to present to higher levels of the brain neural transforms concerning the position, form or contour, and change with time of the peripheral stimulus. This fits with clinical observations upon human patients with lesions of this system, and with its avalanching phylogenetic development.

The spinothalamic system, on the other hand, seems to be concerned with much more general aspects of sensation and to transmit information concerning the qualitative nature of peripheral events, rather than place, pattern, or temporal cadence. The very widespread projection of some components of the ascending systems of the anterolateral columns of the spinal cord upon the reticular formations of medulla and midbrain, and upon the intralaminar nuclei of the thalamus, predicts its prepotent role in arousal and in what might loosely be called the vegetative functions. That we have observed cells of this system in the posterior thalamic group to be activated by noxious stimuli is, I think, clear. This observation does not mean that pain projects only upon this region of the thalamus. What it may imply, with some reason, is that this is the pathway to the conscious perception of pain, for the region projects upon the cerebral cortex. Other more medial regions, similarly activated, are likely to be concerned with reflex responses, the alteration of other motor phenomena, and the evocation of that emotional overtone that is the common attribute of the painful experience in man.

At the cortical level it is not yet possible to ascribe a very reasonable meaning to the duality of projection, to somatic areas I and II, and the meaning of the duality—indeed of the multiplicity—of representation in all the afferent systems remains a puzzling and important problem. The evidence indicates that somatic areas I and II are not exclusively representative of the lemniscal and the spinothalamic components of the somatic afferent system, respectively. They are so, however, to a considerable extent, a fact confirmed in part by behavioral studies, for it is only when removals of the postcentral homologue are made that animals show deficits in the discriminative sphere.

Finally, there is one other impression I have from our work on the somatic sensory areas of the cortex, using the method of single-unit analysis. That is, that the first response of cortical cells is a fairly simple re-representation of the temporal and spatial patterns of activity set up in first-order afferent fibers by the peripheral stimulus. It is true that there are some signs of integrative action—for example, the role of afferent inhibition in limiting discharge zones—but in general the patterns of activity we have observed are far too simple, at least to my way of thinking, to be the neural essence *per se* of sensory perception. It is as if we observed an intermediate level, in the sense of Hughlings Jackson. Yet we know that a monkey using only the postcentral gyrus among parietal areas is capable of very high-order sensory discriminations and integrations. This suggests an important next avenue of study: to observe and define the transformations of activity patterns that occur sequentially in time in the cerebral cortex—successive patterns of activity occupying both the original cells activated and perhaps other populations so far not brought under study. We may then approach some of the problems of first importance, and particularly among them that one so fascinating to all those interested in the function of the brain: how does the brain—or how do we—extract or derive universal generalizations from sensory stimuli?

References

Adrian, E. D. Afferent discharges to the cerebral cortex. *J. Physiol.*, 1941, **100,** 159–191.

Amassian, V. E., and J. L. DeVito. La transmission dans le noyau de Burdach (nucleus cuneatus). *Colloq. int., Centre nat. Rech. sci.,* 1957, No. 67, 353–393.

Berman, A. L. Auditory and somatic interaction in the anterior ectosylvian gyrus of the cerebral cortex of the cat. Doctoral dissertation, The Johns Hopkins University, 1957.

Bremer, F. Analyse oscillographique des résponses sensorielles des écorces cérébrale et cérébelleuse. *Rev. Neurol.,* 1952, **87,** 65–92.

Brodmann, K. Beiträge zur histologischen Lokalisation des Groshirnrinde. 3. Die Rindenfelder der niederen Affen. *J. Psychol. Neurol.*, 1905, **4**, 177–226.

Carreras, M., and S. A. A. Andersson. Unpublished observations, Laboratory of Physiology, The Johns Hopkins University, 1960.

Carreras, M., and M. Levitt. Microelectrode analysis of the second somatosensory area in the cat. *Fed. Proc.*, 1959, **18**, 24.

Hartline, H. K., and F. Ratliff. Inhibitory interaction of receptor units in the eye of *Limulus*. *J. gen. Physiol.*, 1957, **40**, 357–376.

Hartline, H. K., and F. Ratliff. Spatial summation of inhibitory influences in the eye of *Limulus*, and the mutual interaction of receptor units. *J. gen. Physiol.*, 1958, **41**, 1049–1066.

Hartline, H. K., H. G. Wagner, and F. Ratliff. Inhibition in the eye of *Limulus*. *J. gen. Physiol.*, 1956, **39**, 651–673.

Hubel, D. H., and T. N. Wiesel. Receptor fields of single neurones in the cat's striate cortex. *J. Physiol.*, 1959, **148**, 574.

Hunt, C. C., and M. Kuno. Background discharge and evoked responses of spinal interneurones. *J. Physiol.*, 1959, **147**, 364–384.

Iggo, A. A single unit analysis of cutaneous receptors with C afferent fibers. In G. E. W. Wolstenholme and M. O'Connor (Editors), *Pain and Itch*. Ciba Study No. 1. London: Churchill, 1959. Pp. 41–56.

Knighton, R. S. Thalamic relay nucleus for the second somatic receiving area of the cerebral cortex of the cat. *J. comp. Neurol.*, 1950, **92**, 183–191.

Kolmodin, G. M. Integrative processes in single spinal interneurones with proprioceptive connections. *Acta physiol. scand.*, 1957, **40**, Suppl. 139. Pp. 89.

Kuffler, S. W. Discharge patterns and functional organization of mammalian retina. *J. Neurophysiol.*, 1953, **16**, 37–68.

Livingston, R. B. Central control of receptors and central transmission systems. In *Handbook of Physiology, Neurophysiology* I, Washington, D. C.: American Physiological Society, 1959. Pp. 741–760.

Marshall, W. H. Observations on subcortical somatic sensory mechanisms of cats under nembutal anesthesia. *J. Neurophysiol.*, 1941, **4**, 25–43.

Marshall, W. H., C. N. Woolsey, and P. Bard. Observations on cortical sensory mechanisms of cat and monkey. *J. Neurophysiol.*, 1941, **4**, 1–24.

Mehler, W. R., M. E. Feferman, and W. J. Nauta. Ascending axon degeneration following anterolateral cordotomy. *Brain* (in press).

Mountcastle, V. B. Modality and topographic properties of single neurons of cat's somatic sensory cortex. *J. Neurophysiol.*, 1957, **20**, 408–434.

Mountcastle, V. B., P. W. Davies, and A. L. Berman. Response properties of neurons of cat's somatic sensory cortex to peripheral stimuli. *J. Neurophysiol.*, 1957, **20**, 374–407.

Mountcastle, V. B., and E. Henneman. Pattern of tactile representation in thalamus of cat. *J. Neurophysiol.*, 1949, **12**, 85–100.

Mountcastle, V. B., and E. Henneman. The representation of tactile sensibility in the thalamus of the monkey. *J. comp. Neurol.*, 1952, **97**, 409–440.

Mountcastle, V. B., and G. F. Poggio. Unpublished observations, Laboratory of Physiology, The Johns Hopkins University, 1958.

Mountcastle, V. B., and T. P. S. Powell. Central neural mechanisms subserving position sense and kinesthesis. *Bull. Johns Hopkins Hosp.*, 1959a, **105**, 173–200.

Mountcastle, V. B., and T. P. S. Powell. Neural mechanisms subserving cutaneous sensibility, with special reference to the role of afferent inhibition in sensory perception and discrimination. *Bull. Johns Hopkins Hosp.*, 1959b, **105**, 201–232.

Poggio, G. F., and V. B. Mountcastle. A study of the functional contributions of the lemniscal and spinothalamic systems to somatic sensibility. Central nervous mechanisms in pain. *Bull. Johns Hopkins Hosp.*, 1960, **106**, 266–316.

Powell, T. P. S., and V. B. Mountcastle. The cytoarchitecture of the postcentral gyrus of the monkey, Macaca mulatta. *Bull. Johns Hopkins Hosp.*, 1959a, **105**, 108–131.

Powell, T. P. S., and V. B. Mountcastle. Some aspects of the functional organization of the cortex of the postcentral gyrus of the monkey: A correlation of findings obtained in a single unit analysis with cytoarchitecture. *Bull. Johns Hopkins Hosp.*, 1959b, **105**, 133–162.

Puletti, F. Cutaneous tactile units in cat. *Physiologist*, 1959, **2**, 96.

Ratliff, F., and H. K. Hartline. The responses of *Limulus* optic nerve fibers to patterns of illumination on the receptor mosaic. *J. gen. Physiol.*, 1959, **42**, 1241–1255.

Renkin, B. Z. Response of cortical neurons in the somatic sensory area I of the cat in relation to position and frequency of electrical stimulation within cutaneous receptive fields. Doctoral dissertation, The Johns Hopkins University, 1959.

Rose, J. E., and V. B. Mountcastle. The thalamic tactile region in rabbit and cat. *J. comp. Neurol.*, 1952, **97**, 441–490.

Rose, J. E., and V. B. Mountcastle. Touch and kinesthesis. In *Handbook of Physiology. Neurophysiology* I. Washington, D. C.: American Physiological Society, 1959. Pp. 387–429.

Rose, J. E., and C. N. Woolsey. Cortical connections and functional organization of the thalamic auditory system of the cat. In H. F. Harlow and C. N. Woolsey (Editors), *Biological and Biochemical Bases of Behavior*. Madison: Univ. of Wisconsin Press, 1958. Pp. 127–150.

Tunturi, A. R. Further afferent connections to the acoustic cortex of the dog. *Amer. J. Physiol.*, 1945, **144**, 389–394.

Walker, A. E. *The Primate Thalamus*. Chicago: Univ. of Chicago Press, 1938.

Whitlock, D. G., and E. R. Perl. Cutaneous afferent projections through ventrolateral funiculi to rostral brainstem of cat. *J. Neurophysiol.*, 1959, **22**, 133–148.

Woolsey, C. N. "Second" somatic receiving areas in the cerebral cortex of cat, dog and monkey. *Fed. Proc.*, 1943, **2**, 55.

Woolsey, C. N. Patterns of localization in sensory and motor areas of the cerebral cortex. In *The Biology of Mental Health and Disease*. New York: Hober, 1952. Pp. 193–206.

Woolsey, C. N. Organization of somatic sensory and motor areas of the cerebral cortex. In H. Harlow and C. N. Woolsey (Editors), *Biological and Biochemical Bases of Behavior*. Madison: Univ. of Wisconsin Press, 1958. Pp. 63–82.

Woolsey, C. N., W. H. Marshall, and P. Bard. Representation of cutaneous tactile sensibility in the cerebral cortex of the monkey as indicated by evoked potentials. *Bull. Johns Hopkins Hosp.*, 1942, **70**, 399–441.

23

SVEN LANDGREN
Department of Physiology, Veterinärhögskolan, Stockholm

The Response of Thalamic and Cortical Neurons to Electrical and Physiological Stimulation of the Cat's Tongue

The response of the appropriate peripheral nerve fibers to tactile, thermal, and gustatory stimuli applied to the tongue has been extensively studied by previous workers (Zotterman, 1935, 1936; Pfaffmann, 1941; Zotterman, 1949; Hensel and Zotterman, 1951*a–e*; Dodt and Zotterman, 1952*a, b*; Dodt, 1952, 1953; Liljestrand and Zotterman, 1954; Cohen, Hagiwara, and Zotterman, 1955). The input to the central nervous system of afferent impulses evoked by certain defined tongue stimuli can be predicted, therefore, in great detail. The present study concerns the responses evoked by such stimuli in single units at higher levels of the path from the cat's tongue and compares these responses with those obtained from the afferent nerve fibers. The following features were investigated: (1) the localization of the thalamic and the cortical tongue projection areas; (2) the specificity of the response of thalamic and cortical neurons to different types of tongue stimuli; (3) the peripheral receptive field of thalamic and cortical neurons; (4) the pattern of impulse discharge evoked in thalamic and cortical neurons by certain defined stimuli applied to the tongue; and (5) the response of thalamic neurons to nociceptive cutaneous stimuli and to electrical stimulation of the mesencephalic reticular formation.

Methods

The experiments were carried out on preparations anesthetized with Nembutal, or on *encéphale isolé* preparations of cats anesthetized with thiopentone sodium, ether, or trichlorethylene during the operation. The spinal cord was blocked by local anesthesia at the level of C_2, and respiration was maintained artificially. Incisions and pressure points of the head holder were anesthetized locally.

Action potentials of thalamic and cortical units were recorded by glass micropipettes filled with 3 molar (M) KCl or with 4.5 M NaCl. The micromanipulator was of the type described by Eccles, et al. (1954), and the microelectrodes were stereotaxically guided to the thalamic tongue nucleus, as described by Appelberg and Landgren (1958).

Physiological stimuli—touch, pressure, warming, cooling, and flavored solutions, namely, 0.5 M NaCl, 0.01 M quinine HCl in Ringer's solution, 0.3 M acetic acid in Ringer's solution (pH 3), and distilled water—were applied to the tongue. Electrical stimulation was also applied to the surface of the tongue, the chorda tympani, the trigeminal component of the lingual nerve, and the dorsal ascending trigeminal tract. Further details concerning the technique are given by Cohen, et al. (1957) and by Appelberg, Kitchell, and Landgren (1959).

The positions of the recording and stimulating electrodes were verified on histological serial sections. For the sake of convenience, neurons responding to tactile stimulation are referred to as "touch cells" and those responding to cooling as "cold cells."

Results and Discussion

The localization of the thalamic tongue nucleus and the cortical tongue projection area

Mountcastle and Henneman (1949) and Rose and Mountcastle (1952) localized thalamic responses to tactile stimulation of the tongue to nucleus ventralis posteromedialis (VPM). This localization of the thalamic nucleus in the tactile path from the tongue was confirmed by Appelberg and Landgren (1958), who found the responsive region medially in the caudal half of the VPM (Horsley-Clarke coordinates A 8 to 10, L 3.5 to 4.5, H +1 to −1). The tongue nucleus does not reach the medial tip of the VPM, which is mainly occupied by small cells and extends to about 2 millimeters from the midline. In the dorsolateral direction the tongue area was bordered by neurons responding to tactile stimulation of the whiskers and the maxillary region of the face. Laterally face responses were obtained, and in the ventrolateral direction the tongue nucleus was bordered by neurons activated from the mandibular region and the contralateral forepaw. The tongue neurons were always grouped together along the same microelectrode track and were never observed among the face neurons. The findings, therefore, support the idea of a precise topographical organization within VPM.

Neurons that responded to cooling or warming of the tongue or to flavored solutions were found by Landgren (1959, 1960b) in the same region. The thalamic cells that responded to cooling of the tongue were found laterally in the medial region of VPM, which is mainly occupied by small neurons (Horsley-Clarke coordinates A 8 to 9, L 3.8 to 4.5, H 0 to −1).

The cortical tongue-projection area in the cat was localized by Patton and Amassian (1952) and by Cohen, et al. (1957). It is found dorsal to the rhinal fissure and the presylvian sulcus in the region where the ectosylvian and suprasylvian gyri join the sigmoid gyrus. Single neurons responding to tactile, thermal, or gustatory stimuli applied to the tongue were found close together within this area.

The specificity of response of thalamic and cortical neurons to different types of tongue stimuli

The majority of the thalamic and cortical neurons investigated were unimodal; that is to say, they responded to only one type of stimulus applied to the tongue. This is seen in Table 1, which summarizes results obtained from a population of 62 thalamic neurons and another population of 106 cortical neurons. All these neurons were fully tested with mechanical, thermal, and gustatory stimuli applied to the tongue, and all responded to at least one of these stimuli. (For further details, see Landgren, 1957b, 1959, 1960a, b.)

The unimodal group comprised 94 per cent of the thalamic neurons and 75 per cent of the cortical neurons. The number of thalamic neurons activated by more than one type of stimulus was very small and consisted of four cells responding to touch and to cooling of the tongue. In the cortical population the bimodal and multimodal neurons formed a considerably larger group (25 per cent), and the combinations of effective stimuli were more complex. In all neurons responding to more than one type of stimulus, one of the effective stimuli was of a mechanical nature (touch, stretch, or pressure).

Among the unimodal neurons, the touch cells formed a large group. These neurons were in fact the most easily obtained in both the thalamus and the cortex. The cold cells also formed a considerable group, but only one cortical and one thalamic neuron were activated by warming. The group of neurons responding to flavored solutions applied to the tongue was small. The thalamic neuron in this group was discharged by salt solution only and was not affected by bitter or acid solutions, distilled water, or thermal or mechanical stimuli. The unimodal cortical taste cells responded to more than one but not all qualities of taste (cf. Cohen, et al., 1957).

Table 1.   Response of Thalamic and Cortical Cells to Different Types of Tongue Stimuli (from Landgren, 1959)

Effective tongue stimuli	Number of responding	
	Thalamic cells	Cortical cells
Unimodal		
Touch	32	32
Stretch or pressure	4	29
Cooling	20	12
Warming	1	1
Taste	1	5
	58	79
Multimodal		
Touch and cooling	4	8
Stretch and cooling		9
Touch, cooling, and warming		3
Stretch, cooling, warming, and taste		3
Stretch, cooling, and taste		1
Stretch and taste		1
Touch and taste		2
	4	27
Total	62	106

The populations of thalamic and cortical neurons here presented are small and probably not representative of all types of neurons in the nuclei investigated, as the microelectrode technique is likely to select spikes that are easily recorded. Nevertheless certain conclusions may be drawn from the observations. Under the present experimental conditions, afferent impulses are obviously conducted to the thalamic and cortical levels along specific paths. At both levels specific neurons are in the majority among the type of cells selected by our technique. On the other hand, convergence between modalities seems to exist in both the thalamus and the cortex. The incidence of such convergence increases considerably at the cortical level.

The peripheral receptive field of thalamic
and cortical "touch cells"

The extension of the receptive field on the tongue was studied mainly in neurons responding to touch or pressure (cf. Cohen, et al., 1957; Landgren, 1960a). Considering the size of the receptive field, two different types of neurons were found in the thalamic and cortical cell populations. One of these had small well-defined receptive fields, 2 mm in diameter or smaller. About one-third of the thalamic neurons investigated belonged to this group, and some of the cortical neurons also showed this feature. The other type had larger receptive fields, and the majority of the thalamic and the cortical neurons belonged to this group. The fields of the thalamic neurons were, however, generally unilateral and limited to the tip, the edge, or the region of the dorsal papillae of the tongue. On the other hand, several cortical neurons showed large bilateral receptive fields that could include areas outside the tongue, for instance, on the nose. Thalamic neurons never showed such large fields.

It is interesting to compare the size of the receptive fields of the peripheral touch fibers with those of the thalamic and cortical touch cells. Maruhashi, Mizuguchi, and Tasaki (1952) describe the receptive field of the touch fibers from the skin of the cat as spotlike (1 to 2 mm²) and generally consisting of one, sometimes two, touch spots. These authors also investigated a small number of touch fibers in the lingual nerve of the cat and found that the fibers of the tongue did not differ from those of the skin. A group of thalamic and cortical neurons thus have receptive fields similar to those of the afferent nerve fibers, whereas most of the central neurons have considerably larger fields. The tactile path from the tongue to the cortex seems, therefore, to possess two different systems of neurons: one with point-to-point connections and a minimum of convergence, and another with an increasing degree of convergence at the consecutive levels of the path.

Neurons with ipsilateral receptive field were in the majority in the populations of thalamic and cortical cells investigated, but contralateral neurons were often found. They were sometimes recorded close to ipsilateral cells. As mentioned above, bilateral receptive fields occurred mainly in the cortical populations. The incidence of neurons with ipsilateral receptive field was highest in the population of thalamic neurons that responded to tactile stimulation of the tongue and mouth cavity, whereas contralateral receptive fields were in the majority among cells activated by touching the lips, nose, and whiskers (Landgren, 1960a).

The response of thalamic and cortical neurons to electrical and physiological stimulation

Neurons activated by tactile stimulation of the tongue showed the shortest latency of response seen in the thalamic as well as in the cortical populations: electrical stimulation of the tongue surface evoked a response in thalamic touch cells with a latency of 3 to 6 milliseconds and in cortical touch cells with a latency of 4 to 8 milliseconds. Thalamic and cortical neurons activated only by cooling of the tongue had a shortest latency that was two to three times longer than that of the corresponding touch cells.

The spikes of the thalamic and cortical touch cells were observed on the initial phase of the thalamic and cortical focal potentials. This

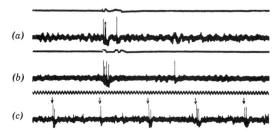

Fig. 1. Pattern of activity of a thalamic cell responding to light tactile stimulation of the cat's tongue (*a* and *b*) and to electrical stimulation (3 shocks per second) of the surface of the tongue (*c*). No delayed discharge; time marking, 50 cps; negativity signaled upward. (From Landgren, 1959.)

observation, as well as the short latency of their response, indicates that the touch cells studied were primary thalamic and cortical neurons. The difference in latency of 1 msec between the shortest thalamic and cortical responses observed suggests that these neurons are consecutive links in the specific tactile path from the tongue.

The typical response of a thalamic neuron to a brief touch of the tongue, or to a single electric shock applied to the afferent nerve or to the surface of the tongue, was a short burst of impulses (see Fig. 1). Similar responses to tactile stimulation of the skin have been described and carefully analyzed by Rose and Mountcastle (1954). We were interested mainly in the activity of the thalamic touch cells during the period immediately following the initial burst of spikes evoked by touch or electrical stimulation. After-discharge was looked for. In *encéphale isolé* preparations, however, an after-discharge was generally not evoked by these stimuli. The response pattern of the thalamic touch cells is thus similar to that of the peripheral touch fibers, which also respond phasically (cf. Zotterman, 1936; Maruhashi, Mizuguchi, and Tasaki, 1952). On the contrary, the cortical touch cells often showed a long series of after-discharge in response to tactile and electrical stimuli that evoked phasic responses in afferent fibers and in thalamic neurons. Figure 2 shows the response of a cortical neuron to tactile stimulation of the tongue, and Fig. 3 shows similar

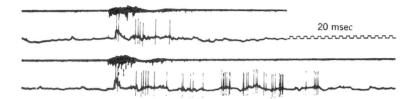

Fig. 2. Response pattern of a cortical cell discharged by touching the tongue (the lower of the two pairs of records). The afferent discharge is recorded from a cut branch of the contralateral lingual nerve to provide an indication of duration and form of the afferent volley from the tongue (upper records). Amplifier time constant, 50 msec; negativity signaled upward. (From Cohen, et al., 1957.)

responses to electrical stimulation of the afferent nerves. In both cases, the response consists of an initial group of spikes, a pause lasting 50 to 80 msec, and a series of after-discharges. The pause is due to inhibition of the cortical neuron, as shown by Fig. 3f, in which an injury discharge was blocked by the afferent volley. The lack of after-discharge in the response of the thalamic touch cells indicates that the excitatory inflow responsible for the after-discharge of the cortical neurons must be generated in the cortex or fed into the cortex via channels other than the specific afferent pathway.

The response of afferent fibers in the lingual nerve and chorda tympani to cooling and warming of the tongue was investigated quantitatively by Hensel and Zotterman (1951c) and by Dodt and Zotterman (1952a, b). The effect of changes from one tongue temperature to another was recorded. The temperature was changed by means of a thermode technique described by Hensel, Ström, and Zotterman (1951). A number of thalamic neurons that responded only to cooling of the tongue were tested by this technique (cf. Landgren, 1960b). The response is shown in Fig. 4. The temperature of the tongue was lowered suddenly from a constant level of 36° C to 24.5° C (Fig. 4b), held at this temperature for about 80 sec (Figs. 4b to e), and then raised again to 36° C (Fig. 4e).

Impulse-frequency histograms from this experiment are shown in Fig. 5. The frequency was counted in 1-sec intervals. A sudden fall in the temperature of the tongue thus caused an immediate increase in the discharge frequency of the thalamic cold cell. The maximum frequency was reached within 4 sec, and the curve then fell to an

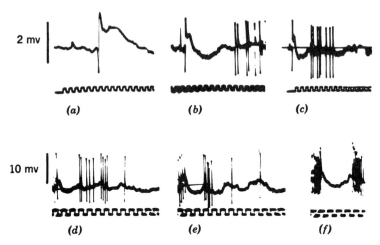

Fig. 3. Response pattern of cortical touch cells discharged by electrical stimulation of the afferent path. (*a* to *c*) Records from a cell discharged by a single electric shock to the ipsilateral trigeminal component of the lingual nerve. Note the single spike riding on a negative potential wave which was followed by a slow positive potential, during which the cell was silent, and then a burst of action potentials on an irregular negative wave of low amplitude. Time: (*a*) 1 msec, (*b*) 5 msec, (*c*) 10 msec. (*d* to *f*) The response of a cortical touch cell to a single electric shock to the ipsilateral chorda tympani. Note inhibition of high-frequency spontaneous activity in (*f*). Time, 20 msec; time constant for all records, 50 msec; negativity signaled upward. (From Cohen, et al., 1957.)

approximately constant level. In some experiments the temperature was kept constant for a period of up to 4 min, and the adapted frequency (steady discharge) of the thalamic neuron remained at a constant level throughout this period.

The initial maximum frequency, as well as the frequency of the steady discharge, increased with increasing intensity of the stimulus, that is to say, with larger drops in the temperature of the tongue, as can be seen by comparing the lower and upper histograms of Fig. 5. The present material does not, however, allow a quantitative analysis of the steady discharge at different temperatures.

The response of the thalamic cold cells to cooling of the tongue was thus very similar to that recorded by Hensel and Zotterman

(1951c) from the afferent fibers. The frequency of the steady discharge of the thalamic neurons was higher than that of the peripheral fibers. The steady discharge frequency of the fibers never exceeded 10 impulses per second, whereas values between 10 and 30 impulses per second were observed in thalamic neurons. The steady discharge frequency of the thalamic neurons was also less regular than that of the peripheral nerve fibers. Apparently random variations within a

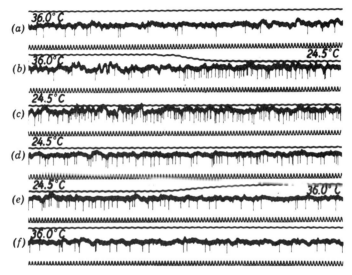

Fig. 4. The response of a thalamic neuron to cooling and warming of the tongue. The temperature of the thermode applied on the tongue surface is recorded in the top beam; 50 cps on lower beam. Records (a) to (c) and (d) to (f) are continuous. The time interval between records (c) and (d) was 70 sec; the temperature of the tongue was kept at 24.5°C during the interval. Negative spikes recorded downward. (From Landgren, 1960b.)

range of 10 impulses per second were generally observed in the thalamus. The gradually increasing initial frequency seen in the response to cooling in Figs. 4 and 5 was not observed in all cold cells. Generally the maximum frequency was reached within the first second after cooling. Although the above-described pattern of response to cooling of the tongue was seen in most thalamic cold cells, some neurons responded phasically to drops in temperature that would be expected to give a steady discharge in the afferent fibers. Such responses may indicate the existence of thalamic response patterns differing from those seen in the periphery.

Warming the tongue caused a decrease in the impulse frequency of

the thalamic cold cells.   This is shown in Figs. 4e and f, and in the
upper histogram of Fig. 5 and in Fig. 6a.   The postexcitatory depression

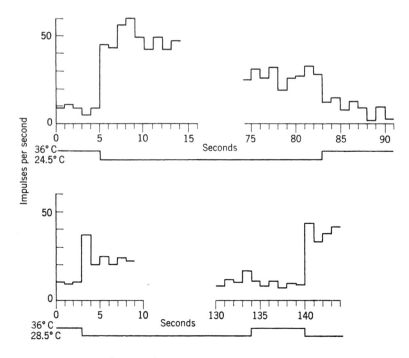

Fig. 5.   Histograms showing the impulse frequency of the neuron of Fig. 4 when
responding to changes of the tongue's temperature between constant levels.
(From Landgren, 1960b.)

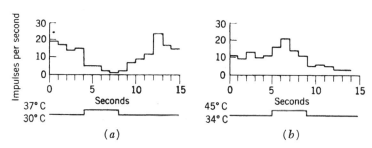

Fig. 6.   Impulse-frequency histograms showing the response of two thalamic
neurons to warming of the tongue.   (From Landgren, 1960b.)

typical for the corresponding responses in the peripheral nerve fibers
was less obvious in the records from thalamic cold cells.

As seen in the histogram of Fig. 6a, the activity of the cold cell was reduced to about one impulse per second when the tongue was flooded with Ringer's solution at 37° C. Similar effects were always observed when the tongue was warmed to body temperature. Flooding with water or Ringer's solution at 45 to 46° C caused an increase in the frequency of discharge in some of the thalamic cold cells, as

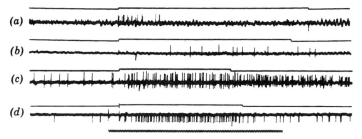

Fig. 7. Different types of impulse discharge recorded from cortical cells responding to cooling of the tongue only (*a* to *c*), and to cooling and touch (*d*). Time, 50 cps. The signal on the top beam marks the application of Ringer's solution at 10° C to the tongue. Negativity signaled upward. (From Landgren, 1957a.)

is shown by the histogram in Fig. 6*b*. The increase in activity at temperatures above 45° C was expected, since a similar effect was demonstrated by Dodt and Zotterman (1952*b*) in the cold fibers from the tongue.

Only one thalamic neuron was found that responded to warming of the tongue with an increase in impulse frequency. It was spontaneously active with a frequency of 3 to 5 impulses per second at a tongue temperature of about 32° C. Flooding the tongue with Ringer's solution at 42° C increased the frequency to about 20 impulses per second. Cooling, touch, pressure, spontaneous movements of the tongue, or flavored solutions applied to the tongue had no effect. This neuron was found close to two other cells that responded only to cooling of the tongue.

Cortical neurons that responded to cooling of the tongue have been described by Landgren (1957a). Phasic as well as persistent responses were observed, as is shown in Fig. 7. Unfortunately these neurons were not tested with the thermode technique used in the investigation of the afferent fibers and thalamic cold cells. Cooling was induced by flooding the tongue with water or Ringer's solution of known temperature. The temperature of the tongue surface was not recorded, and the response to constant tongue temperatures was not investigated. In spite of the less rigorous control of the stimulus, it was possible to observe that the latency, duration, and frequency of the cortical cold-cell discharge were dependent on the intensity of the cooling of the tongue. The more intense the cooling was, the shorter was the latency, the higher the frequency, and the longer the duration of the response. A brief cooling of the tongue (1 to 2 sec) often caused a persistent after-discharge consisting of rhythmic bursts. The bursts appeared with about 2-sec intervals of silence, and the rhythmic activity lasted for about half a minute. This type of activity was not observed in the thalamic cold cells.

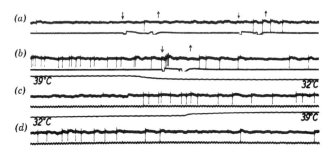

Fig. 8. Records from a thalamic neuron responding to touch (*a* and *b*) and to cooling (*c* and *d*) of the tongue. Records (*a*) and (*b*) show the beginning and end of a period of repetitive tactile stimulation of the tongue (1 per second). The interval between records (*a*) and (*b*) was 22 sec. Tactile stimulation signaled by deflections on lower beam and indicated by arrows. Record (*d*) is a continuation of (*c*). Thermode temperature signaled on top beam; time, 50 cps; negative spikes recorded downward. (From Landgren, 1960a.)

It may be of interest to describe the response of the bimodal neurons activated by mechanical stimuli and by cooling of the tongue, since these neurons showed an interesting effect from repetitive tactile stimulation that was not as obvious in the unimodal touch cells. The effect was observed in thalamic as well as cortical neurons of this bimodal type, and a typical response is shown in Fig. 8. The records shown in this figure were obtained from a thalamic neuron that responded to cooling of the tongue in a way similar to the unimodal thalamic cold cells previously described. The effect of a drop in the temperature of the tongue from 39 to 32° C and rewarming to 39° C is shown in Figs. 8c and d. The response of this neuron to tactile stimulation of the tongue is shown in Figs. 8a and b; its latency was long and variable at the beginning of repetitive stimulation (Fig. 8a). The number of spikes evoked by each touch was small, and the frequency in the burst low. The spontaneous background activity was about one impulse per second. Later in a series of repeated touching of the receptive field, the background activity increased, and each response appeared with a shorter latency and consisted of a burst of spikes of high frequency (Fig. 8b). This type of response differs obviously from that of the unimodal touch cells with their regular latency and small variation in number of spikes. Some of the latter neurons did, however, show an increase in background activity when stimulated repetitively, but this increase was relatively small.

The response of thalamic neurons to nociceptive and reticular stimulation

The effect of nociceptive cutaneous stimuli, and of electrical stimulation of the mesencephalic reticular formation, on the activity of single thalamic touch cells was investigated by Landgren (1959). The nociceptive stimuli used were pinching of the pinna or the nose or pricking the gingiva, and 17 thalamic neurons were tested. Of these neurons, 9 showed an increased and 4 a decreased rate of firing in response to pinching; 4 cells were not affected by such stimuli. All these neurons responded to tactile stimulation of the tongue within a well-defined receptive field. The effect of pinching, however, was not elicited from localized areas, and the thalamic cells did not respond to touch or moderate pressure within the pinched areas. The increase in impulse frequency caused by pinching developed gradually and after a relatively long latency. It lasted considerably longer than the period of stimulation (cf. Landgren, 1959, Fig. 4).

The long latency, gradual onset, and long duration of the nociceptive response, as well as the lack of localized receptive field, suggested that the nociceptive effects were mediated via the reticular system. This assumption was supported by the observation that thalamic touch cells were influenced by electrical stimulation of the mesencephalic reticular formation. Single electric shocks to the reticular formation evoked no response, but tetanic stimulation (6 volts, 250 per second for 4 to 6 sec) considerably increased the discharge frequency of the spontaneously active thalamic cells (see Fig. 9). Also in this case the effect developed gradually and lasted for about 60 sec after the end of the reticular stimulation.

The effect of reticular stimulation on the specific response of the thalamic touch cells was observed in a few cases. The latency of the response decreased, and the number of spikes in the bursts of response seemed to increase. However, the spontaneous background activity increased simultaneously, as seen in Fig. 9, and this effect swamped the specific response. It is thus not clear from these observations whether the specific response in the tactile path was facilitated or blocked by reticular stimulation. The fact that both increase and decrease in the activity were seen in thalamic neurons after nociceptive stimuli may indicate that excitation as well as inhibition of these neurons may be exerted via the reticular formation.

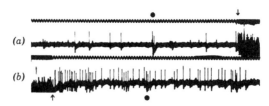

Fig. 9. Response of a thalamic touch cell to electrical stimulation of the tongue evoked (*a*) before and (*b*) after a period of electrical stimulation of the mesencephalic reticular formation (6 volts, 250 per second, for 6 sec). Note that the increased background activity in (*b*) obscures the specific response, which is indicated by dots. Arrows mark the beginning and end of the reticular stimulation. Time, 50 cps; time constant, 5 msec. (From Landgren, 1959.)

Summary

The reactions of single neurons at the thalamic and cortical level of the specific sensory path from the tongue of the cat were studied. The response of the afferent nerve fibers, the thalamic neurons, and the cortical neurons in this path were compared.

The extracellular spike potentials of the neurons were recorded with microelectrodes filled with KCl or NaCl, placed in the respective nuclei by means of a stereotaxic technique. Anesthetized, as well as *encéphale isolé*, preparations were used.

The thalamic nucleus in the tactile path from the tongue occupied a region about 1 mm in diameter and was localized medially and caudally in the nucleus ventralis posteromedialis. It did not reach the most medial part of this nucleus, and in the lateral direction it was bordered by neurons responding to tactile stimulation of the perioral region. The tongue neurons were always grouped together and never appeared mixed with face neurons.

The cortical projection area of the tactile path from the tongue was found on the orbital surface of the brain in the region where the ectosylvian and suprasylvian gyri join the sigmoid gyrus.

Neurons responding to cooling or warming of the tongue or to flavored solutions applied to the tongue were recorded in the same regions.

The majority of the thalamic and cortical neurons investigated were unimodal, that is, responded to only one type of physiological tongue stimulus. Bimodal and multimodal neurons were observed, however, and they occurred more frequently in the cortical than in the thalamic population of neurons (see Table 1).

The extension of the peripheral receptive fields of thalamic and cortical neurons responding to touch or pressure on the tongue was investigated. One group of these neurons showed small and well-defined receptive fields (1 to 2 mm in diameter) similar to those of the afferent touch and pressure fibers from the skin. Another group of neurons showed wider receptive fields, the cortical cells having larger, often bilateral, fields and the thalamic ones less extensive unilateral fields.

Thalamic touch cells responded to electrical stimulation of the tongue surface with a latency of 3 to 6 msec. The latency of the cortical touch cells was 4 to 8 msec. Thalamic and cortical neurons responding to cooling of the tongue showed a latency two to three times longer than that of the corresponding touch cells.

The response pattern of the thalamic touch cells, evoked by electrical stimulation or by a brief touch of the tongue, was similar to that of the peripheral touch fibers and consisted of a burst of spikes without after-discharge. Such stimuli evoked a more complex response, which consisted of a burst of spikes, a pause in the discharge, and a long-lasting series of after-discharge in the cortical touch cells.

The response of the thalamic cold cells to constant tongue temperatures and to changes between such temperatures was mainly similar to the response of the peripheral cold fibers, but there were certain differences. This was true also for some of the cortical cold cells. Phasic responses were observed in thalamic and cortical cold cells to stimuli that were expected to evoke a steady discharge in the peripheral nerve fibers.

A small number of thalamic and cortical neurons were activated by warming or application of flavored solutions on the tongue.

Unimodal thalamic touch cells were influenced by nonlocalized, nociceptive stimulation of the skin and also by electrical stimulation of the mesencephalic reticular formation. Generally an increase in the discharge frequency was observed. A decrease in the activity occurred in some of the neurons, however, and some were not at all affected by nociceptive and reticular stimuli.

References

Appelberg, B., R. L. Kitchell, and S. Landgren. Reticular influence upon thalamic and cortical potentials evoked by stimulation of the cat's tongue. Acta physiol. scand., 1959, **45**, 48–71.

Appelberg, B., and S. Landgren. The localization of the thalamic relay in the specific sensory path from the tongue of the cat. Acta physiol. scand., 1958, **42**, 342–357.

Cohen, M. J., S. Hagiwara, and Y. Zotterman. The response spectrum of taste fibres in the cat: A single fibre analysis. Acta physiol. scand., 1955, **33**, 316–332.

Cohen, M. J., S. Landgren, L. Ström, and Y. Zotterman. Cortical reception of touch and taste in the cat: A study of single cortical cells. Acta physiol. scand., 1957, **40**, Suppl. 135.

Dodt, E. The behaviour of thermoceptors at low and high temperature with special reference to Ebbecke's temperature phenomena. Acta physiol. scand., 1952, **27**, 295–314.

Dodt, E. Differential thermosensitivity of mammalian A-fibres. Acta physiol. scand., 1953, **29**, 91–108.

Dodt, E., and Y. Zotterman. Mode of action of warm receptors. Acta physiol. scand., 1952a, **26**, 345–357.

Dodt, E., and Y. Zotterman. The discharge of specific cold fibres at high temperatures (the paradoxical cold). Acta physiol. scand., 1952b, **26**, 358–365.

Eccles, J. C., P. Fatt, S. Landgren, and G. J. Winsbury. Spinal cord potential generated by volley in the large muscle afferents. J. Physiol., 1954, **125**, 590–606.

Hensel, H., L. Ström, and Y. Zotterman. Electrophysiological measurements of depth of thermoreceptors. J. Neurophysiol., 1951, **14**, 423–429.

Hensel, H., and Y. Zotterman. The response of the cold receptors to constant cooling. Acta physiol. scand., 1951a, **22**, 96–105.

Hensel, H., and Y. Zotterman. The persisting cold sensation. Acta physiol. scand., 1951b, **22**, 106–113.

Hensel, H., and Y. Zotterman. Quantitative Beziehungen zwischen der Entladung einzelner Kaltefasern und der Temperatur. Acta physiol. scand., 1951c, **23**, 291–319.

Hensel, H., and Y. Zotterman. Action potentials of cold fibres and intracutaneous temperature gradient. J. Neurophysiol., 1951d, **14**, 377–385.

Hensel, H., and Y. Zotterman. The response of mechanoreceptors to thermal stimulation. J. Physiol., 1951e, **115**, 16–24.

Landgren, S. Cortical reception of cold impulses from the tongue of the cat. Acta physiol. scand., 1957a, **40**, 202–209.

Landgren, S. Convergence of tactile, thermal and gustatory impulses on single cortical cells. Acta physiol. scand., 1957b, **40**, 210–221.

Landgren, S. The thalamic and cortical reception of afferent impulses from the tongue. In G. E. W. Wolstenholme and M. O'Connor (Editors), Pain and Itch. Ciba Study No. 1. London: Churchill, 1959. Pp. 69–83.

Landgren, S. Thalamic neurones responding to tactile stimulation of the cat's tongue. Acta physiol. scand., 1960a, **48**, 238–254.

Landgren, S.   Thalamic neurones responding to cooling of the cat's tongue. *Acta physiol. scand.*, 1960*b*, **48**, 255–267.

Liljestrand, G., and Y. Zotterman.   The water taste in mammals. *Acta physiol. scand.*, 1954, **32**, 291–303.

Maruhashi, J., K. Mizuguchi, and I. Tasaki.   Action currents in single afferent nerve fibres elicited by stimulation of the skin of the toad and the cat. *J. Physiol.*, 1952, **117**, 129–151.

Mountcastle, V. B., and E. Henneman.   Pattern of tactile representation in thalamus of cat.   *J. Neurophysiol.*, 1949, **12**, 85–100.

Patton, H. D., and V. E. Amassian.   Cortical projection zone of chorda tympani. *J. Neurophysiol.*, 1952, **15**, 243–250.

Pfaffmann, C.   Gustatory afferent impulses. *J. cell. comp. Physiol.*, 1941, **17**, 243–258.

Rose, J. E., and V. B. Mountcastle.   The thalamic tactile region in rabbit and cat.   *J. comp. Neurol.*, 1952, **97**, 441–490.

Rose, J. E., and V. B. Mountcastle.   Activity of single neurones in the tactile thalamic region of the cat in response to a transient peripheral stimulus. *Bull. Johns Hopkins Hosp.*, 1954, **94**, 238–282.

Zotterman, Y.   Action potentials in the glossopharyngeal nerve and in the chorda tympani. *Skand. Arch. Physiol.*, 1935, **72**, 73–77.

Zotterman, Y.   Specific action potentials in the lingual nerve of the cat. *Skand. Arch. Physiol.*, 1936, **75**, 105–119.

Zotterman, Y.   The response of the frog's taste fibres to the application of pure water. *Acta physiol. scand.*, 1949, **18**, 181–189.

24

CARL PFAFFMANN, R. P. ERICKSON, G. P. FROMMER, and
B. P. HALPERN
Brown University

Gustatory Discharges
in the Rat Medulla and Thalamus

As a sensory system, taste is relatively poor in the range of sensory qualities it elicits and the fineness of the discriminations it can mediate. Yet it is noteworthy in that certain taste stimuli can elicit strong acceptance or preference responses on the one hand, and strong aversion or avoidance responses on the other. The term, sensory communication, often implies information transmission with an emphasis on content and the richness of cognitive detail; yet we must not neglect the total behavior sequence that eventuates in response. In its simplest terms the fundamental response to sensory stimulation may be one of two alternatives, (1) a positive response of approach and acceptance, or *adience,* to use E. B. Holt's term (1931), or (2) a negative response, avoidance or rejection (*abience*). Sensory stimulation determines which one of these two classes of response will occur. The study and elucidation of this total sequence, and the chain of neurohumoral events leading thereto, are basic to the understanding of the physiology of behavior.

The sense of taste may be used as a model behavioral system in this enterprise, for taste stimuli elicit very definite positive and negative responses. Such behavior is relatively simple and stable and readily amenable to study, compared with the more complex *learned* avoidance and/or approach behaviors. Unfortunately our knowledge of the purely stimulus-response relations is better than our knowledge of their physiology, especially as regards the central nervous system. The studies to be described in this paper were designed to provide information on the primary receiving systems of taste in the central nervous system. These studies are basic to the further study of the physiological substrates of the behavior controlled by gustatory stimulation.

In contrast to the many electrophysiological studies of the peripheral taste nerves, there have been only a few studies of the electrical activity in the central nervous system elicited by gustatory stimuli. A number of years ago Ectors (1936) and Gerebtzoff (1939a) in Bremer's laboratory reported changes in the ongoing electroencephalogram of the rabbits' masticatory cortex (Bremer, 1923) following taste stimulation. A more recent report of localized potentials in the cerebral cortex has appeared (Inanaga, Nakao, and Fuchiwaki, 1953). Still more recently, workers in Zotterman's laboratory have recorded single spike activity in a small number of gustatory units in the tongue somatosensory area of the cerebral cortex of the cat (Cohen, et al., 1957; Landgren, 1957), and Appelberg and Landgren (1958) from the same laboratory have reported single-element activity in the tongue receiving area of the thalamus caused by taste stimuli. The present paper will present data on both multiunit and single-unit recordings from the medulla, and multiunit recordings from the thalamus, of the rat following the application of taste stimuli to the anterior tongue surface (cf. Erickson, 1958; Halpern, 1959).

Prior anatomical studies have pointed to the cephalic portion of the nucleus of the tractus solitarius as the medullary terminus for the primary gustatory afferents and as the site of origin of the second-order gustatory neurons (Allen, 1923; Gerebtzoff, 1939b; Torvik, 1956) although contrary claims have implicated the nucleus of Staderini (Krieg, 1942). Our original probings in search of the medullary taste area, therefore, were made in the general region of the rat's solitary nucleus. Probings for the thalamic area were likewise guided by the earlier clinical, anatomical, and behavioral ablation studies, which pointed to the more medial portions of the posterior ventromedial nucleus of the sensory thalamus (Adler, 1934; Gerebtzoff, 1939a, b; Patton, Ruch, and Walker, 1944; Anderson and Jewell, 1957). Appelberg and Landgren's report (1958) of taste recordings in this area appeared while our initial studies were in progress (Blum, Ruch, and Walker, 1943).

Method

All experiments were carried out on albino rats of the Sprague-Dawley strain under Nembutal anesthesia (pentobarbital sodium, Abbott, 60 milligrams per cubic centimeter). Recordings have now been made in over 60 rats weighing from 188 to 390 grams (males and females) with a number of different observations in mind. I shall report only on the locus and general nature of the neural discharge in this area. To date we have observed thalamic responses in some 19 rats of the same strain under the same conditions of anesthesia. Further experiments on both areas are still in progress.

The trachea was cannulated and the animal's head mounted firmly in a holder that clamped over the snout but left the buccal cavity relatively clear. In studying the medulla, the central portions of the occipital and interparietal bones were removed, and the cerebellum was ablated by aspiration to expose the floor of the fourth ventricle. A three-coordinate electrode manipulator positioned the recording electrode of enameled nickel chrome wire (Tophet C) which varied from 25 to 113 microns in diameter in different experiments. The ground and reference electrodes were placed on exposed bone or muscle surrounding the incision, which was kept warm and moist by a covering of cotton soaked in Ringer's solution. A Grass amplifier, cathode-ray oscillograph, audiomonitor plus integrator circuit, and Esterline-Angus recording milliammeter were used to obtain quantitative records of the multiunit activity. Single-unit activity in the medulla was recorded with fluid-filled (3 molar KCl) pipettes connected to the cathode follower of the amplifier. Photographic records were made with a cathode-ray oscillograph and Grass kymograph camera.

Electrode placements were made under direct visual control in the medulla, using the dorsal cochlear nucleus, obex, and midline as reference guides. When searching for the taste-reactive area, the recording electrode was moved dorsoventrally in 0.05- to 0.10-millimeter steps. The procedure for the thalamic recordings was similar except that the electrodes were placed under stereotaxic control through small burr holes in the skull. In this case the electrodes were driven through the cortex and underlying structures to the appropriate depth. Records of electrical activity in the thalamus so far have been made only with the wire electrodes.

Gustatory stimulation was carried out by means of a flow system and tongue chamber in which the anterior tongue could be enclosed to provide uniform stimulation. In some cases the flow chamber was fitted with a small aperture to permit tactile stimulation. Tap water at different temperatures in the flow system served as temperature stimuli and was monitored by a thermocouple in the tongue chamber. Taste solutions were made up in distilled water, but tap water was used for rinsing, all at room temperature.

Histological study of the electrode tracks was made for six medulla preparations and six thalamic preparations. Four of the medulla preparations contained only a single active electrode track resulting from a 113-micron wire. In these instances the electrode had been advanced to a depth that yielded a maximal response. Other non-taste electrode tracks in the same preparations were advanced to a uniform depth of 2.5 mm to aid in distinguishing them from the active placement. The thalamic preparations contained one or more active tracks. Animals were killed by perfusion through the heart with isotonic saline followed by 10 per cent formalin. The medulla preparations were embedded in celloidin and stained according to Weil's procedure. Sections were cut at 20 microns and stained at every fifth section except in the immediate vicinity of the electrode tracks, where every section was stained. The thalamic preparations were cut as frozen sections and stained with cresyl violet.

Results

Medulla

The wire electrodes yielded a typical multiunit response from a circumscribed area of the medulla at an average position 2.63 mm anterior to the obex and 1.6 mm lateral to the midline at a depth of 1.03 to 1.23 mm below the floor of the fourth ventricle. The active region is approximately 0.4 mm wide, 0.5 mm long, and 0.2 mm thick. We have since found that the area can be located reliably by searching just posterior to the dorsal cochlear nucleus along the rostral portion of a diagonal connecting the obex with a point 3 mm anterior to the obex and 2 mm lateral to the midline. Figure 1 is a diagrammatic sketch of the floor of the fourth ventricle, based on measurements with the three-coordinate manipulator. The dots indicate points where the recording electrodes entered the substance of the medulla in a dorsoventral direction. The encircled points indicate those from which responses to taste stimuli could be recorded at the appropriate depth. The preparation shown is of particular interest because all the active points were surrounded by electrode placements that were unresponsive to taste stimulation. The active area was found at an average depth of 1.03 mm in this preparation. Most taste areas were spontaneously active at a low level without stimulation. The beginning of water flow might be associated with a slight transient increase in activity, whereas the flow of taste solution over the tongue caused a marked discharge that continued as long as the solution remained on the tongue. The discharge returned to the resting level when the solution was rinsed off the tongue (see Fig. 2). Most taste points were also responsive to stimulation of the tongue by touch and temperature, particularly cooling, although occasional points were encountered that responded only to taste. These would be encountered

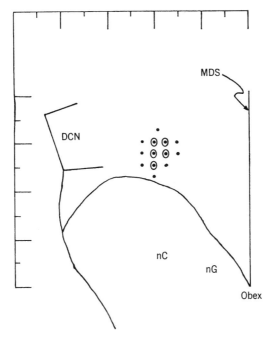

Fig. 1. Semidiagrammatic dorsal view of the left side of the floor of the fourth ventricle of a rat. The dots represent the point of entry of a 50-micron electrode moved dorsoventrally. Each encircled dot indicates an electrode track along which taste responses were recorded at an average depth of 1.03 mm. DCN, dorsal cochlear nucleus; MDS, median dorsal sulcus; nC, cuneate nucleus; nG, gracile nucleus.

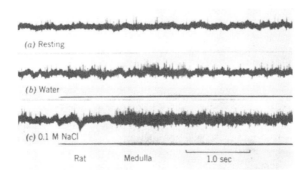

Fig. 2. Oscillograph records of electrical activity in the bulbar taste site following taste stimulation of the anterior tongue. (*a*) Resting activity; (*b*) water flow during period indicated by the signal line; (*c*) 0.1 M NaCl flow indicated by signal line. The apparently long latency of response in (*c*) is an artifact due to the length of tubing through which the solution flowed before reaching the tongue chamber. Electrode diameter, 50 microns.

in a track that might show taste, touch, and temperature sensitivity at points just above or below. Electrode tracks slightly more lateral to the taste-responsive positions were found to be responsive to tactile and pressure stimulation of the snout, chin, jaw, and teeth, especially of the incisors. Placements more medial to the taste points could not be activated by mechanical stimulation of the face or mouth region.

Fig. 3. Photomicrograph of a 20-micron section of the rat's brain stem (Weil stain). The marker points to the end of an electrode track in the central portion of the rostral solitary nucleus. The track had been made by a 113-micron electrode which was driven into the medulla until the gustatory response had just passed its maximum amplitude. The electrode was then withdrawn.

The responsive taste areas were ipsilateral to the area of the tongue being stimulated. In a couple of experiments, the external auditory meatus was irrigated with a 2 per cent procaine solution in order to block transmission in the chorda tympani as it crossed the tympanum and middle ear. This eliminated the taste discharge reversibly. The response returned after the procaine was removed and the meatus flushed with Ringer's solution.

The histological studies show that the active electrode sites were located in the rostral portion of the nucleus of the tractus solitarius (see Fig. 3). In only one experiment have we encountered taste activity at another locus. This was at an electrode depth of 2 mm, somewhat lateral to the usual placement. This might have resulted from activity in one of the fasciculi of the nervus intermedius as it traversed the brain stem. Presumably activity in both the afferent fibers to and efferent fibers from the nucleus could be recorded by the appropriate placement of electrodes. We believe that all our preparations, save that just mentioned, have been recorded from the nucleus.

The magnitude of electrical activity in the medulla in terms of integrator deflection as a function of stimulus concentration very closely resembles the relation found in the chorda tympani nerve itself (see Fig. 9). Most commonly the relation between stimulus concentration and magnitude of integrator deflection is a sigmoid function of the logarithm of the concentration. There are detectable responses to quinine, HCl, and NaCl at 0.001 M concentration. Sugar is first clearly detected at 0.01 M concentration. The electrolytes HCl and NaCl yield much larger values at maximal response than do the non-electrolytes, quinine and sucrose; sucrose reaches a value of about 15 to 30 per cent of the salt response on the average, but the relation for sugar may differ at different electrode depths. In every preparation where variation in response with depth was examined, the relative magnitude of the sugar response increased and the relative magnitude of the salt response decreased in the more ventral position. Such a change has been observed with a change in depth from 0.1 to 0.15 mm. Further work will be necessary in order to ascertain whether other regional differences exist within this area and how this may relate to the discrimination of taste "quality."

The similarity between the periphery and medulla is further apparent in records of single-unit activity. Earlier work had shown that, although the individual afferent nerve fibers for taste are differential in their sensitivity, they are not rigidly specific; thus many units will respond to more than one of the classical four basic tastes (Pfaffmann, 1941, 1955; Cohen, Hagiwara, and Zotterman, 1955; Kimura and Beidler, 1956; Fishman, 1957; Beidler, 1957). Figure 4 illustrates the discharge of a medullary unit that responds to each of the basic taste stimuli, and indeed to tactile as well as temperature changes. This unit has a very general sensitivity. Other units however, have been found that respond to taste stimuli and indeed to only some solutions as, for example, only salt and acid (see Fig. 5). Furthermore, the relative frequency of discharge of individual units varies with the particular taste stimulus. Figure 6 is a bar graph showing the relative frequency of discharge shown by some 14 different single elements during the first second of stimulation by the basic taste stimuli. In all, 53 single units have been studied with essentially the same dis-

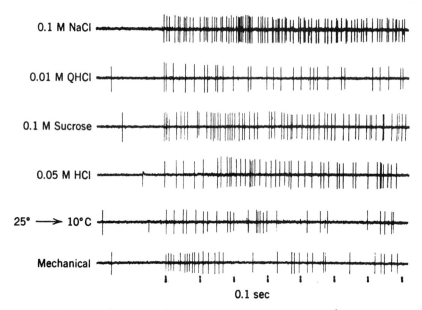

Fig. 4. Response of a unit with a broad range of sensitivity for taste and other modalities. Taste solutions were presented at room temperature. Abrupt cooling with water from 25° to 10° C caused a discharge of impulses, but warming did not. A tap with a glass rod provided mechanical stimulation. The records have been aligned with the beginning of the impulse discharge.

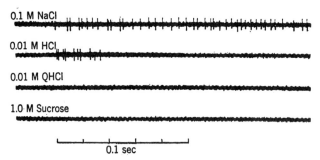

Fig. 5. Response of a unit responsive to NaCl and HCl but not to quinine hydrochloride or sucrose. Negativity up.

tribution of sensitivities as is shown in Fig. 6. Differential sensitivity of the taste units is reflected more in the frequencies of discharge to different stimuli than in an all-or-none specificity for one or another taste stimulus.   Such a pattern of differential activity closely resembles that found at the periphery.   The distribution of sensitivity appears to be largely a heterogeneous one; no major types or classes of receptor sensitivity are apparent from the single-unit studies of taste at the periphery or in the medulla of the rat.

The close similarity of the responses in the medulla and periphery raises the question of whether first-order or second-order units are under examination.   It is possible that both are being sampled, but

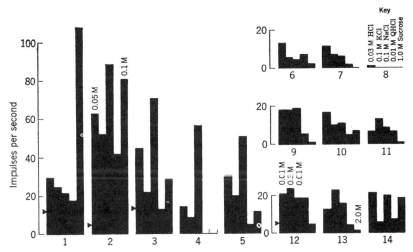

Fig. 6.   Bar graph showing the responsiveness of 14 of 53 units sampled in the solitary nucleus.   The ordinate shows the frequency of impulses per second during the first second of discharge to the five test solutions indicated in the key.   Other concentrations used in units 2, 12, and 13 are indicated in the appropriate columns.

it is assumed that the majority of the units observed are second-order elements, because the cells of the nucleus might be expected to yield larger potentials than would fibers.   In several instances we have observed a notch on the record of the rising phase of the single-unit impulse much like that described as a prespike potential in unit-cellular recording (Frank, 1959).

The broad range of sensitivity of some of the chemically sensitive units raises the possibility that they might be part of the reticular formation rather than part of the nucleus of a specific sensory pathway.   The former seems improbable for several reasons.   (1) The animals were in all cases under deep barbiturate anesthesia, which usually makes it difficult to influence reticular units by peripheral stimulation (Amassian and DeVito, 1954).   Here the chemically sensitive units were very responsive, and many could be observed in a very small area. (2) No responses could be evoked from these units (or in this area) other than by tongue stimulation. A large percentage of reticular units are sensitive to wide skin areas as well as to other sensory inputs (for example, Amassian and DeVito, 1954; Scheibel, et al., 1955). (3) The responses evoked by stimulation of the tongue were very sharply localized; indeed it was often difficult to locate the responsive area. As previously indicated, it was only about 0.5 mm long and 0.4 mm in diameter and was localized in the nucleus of the solitary tract. If the responses were reticular, they should have been observed throughout the medial medulla.

It is not possible to say whether the units recorded in the medulla show greater generality of response to taste stimuli than do the peripheral units, for the present data do not permit of such quantitative comparisons. Many medullary units show not only a generality of response to taste stimuli but also a cross-modality sensitivity. Of 43 chemically sensitive units tested with tongue cooling, 40 showed an increase in activity; 3 did not respond. An abrupt temperature drop as small as $3°$ C was effective in some cases, and an abrupt drop from 25 to $10°$ C was roughly equivalent to the presentation of 0.1 M NaCl in stimulating efficiency. Warming the tongue had less effect upon the medullary discharge. Of 22 units so tested, 15 showed a slight decrease in activity but the remaining 7 showed no observable change. Of 38 units tested with mechanical stimulation, 27 showed an increase and 11 no change. Similar cross-modal stimulation has been encountered at the cortex and thalamus of cat and has been attributed to "convergence" of neural inputs or pathways in the central nervous system (Landgren, 1957). Yet such cross-modal stimulation can take place at the periphery. We have seen taste units in the chorda tympani nerve that respond to mechanical stimulation of the tongue (Pfaffmann, 1941), and Fishman (1957) has shown that nearly all taste units in the same nerve of the rat will be stimulated by cooling to $10°$ C. Furthermore, histological evidence shows that the nerve filaments arising from the plexus at the base of the taste buds can be traced into the surrounding epithelium. These are the so-called extrageminal fibers as contrasted with the intrageminal fibers that end within the taste bud and terminate on the gustatory sense cells (Kolmer, 1927). It is conceivable that some cross-modal effects may result from such innervation. Cross-modal CNS effects may differ only quantitatively from those observed at the periphery.

Thalamus

Our observations on the thalamus to date have been made only with the gross wire electrodes. The responses here are remarkably similar to those found in the medulla except that in the thalamus there is more background activity which often has a rhythmic character. Figure 7 illustrates an integrator recording of the thalamic responses to taste stimulation. The region from which such activity may be

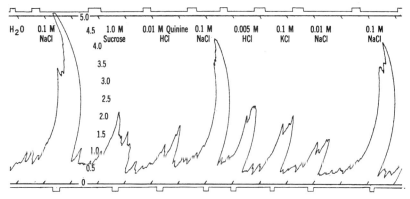

Fig. 7. Integrator responses from the thalamic taste area. Record reads from right to left; solutions marked above the responses. Top signal marker indicates the beginning and duration of taste stimulation, lines immediately above and below record indicate 1-minute intervals, bottom line indicates beginning and duration of rinse water flow.

recorded has been found, in 19 rats, to lie at an average position 1.5 mm lateral to the midline at a depth of 6 to 7 mm below the skull surface, in a plane corresponding to Craigie's plate 18 (1925), Gurdjian's Fig. 8 (1927), Krieg's frontal plane 54 (1946), and deGroot's plane A 3.8 (1959), all at the level of the thalamic nuclei. The area measures approximately 0.5 mm wide by 0.5 mm long by 0.3 mm deep. The taste-responsive area shows little response to tactile or temperature stimulation of the tongue. The latter sensitivities are found at a somewhat more lateral placement, but here there is little or no taste response. More lateral placements of the electrodes show a distribution of points for tactile sensitivity of the shoulders and body surface in the manner described for other species (Mountcastle and Henneman, 1949; Rose and Mountcastle, 1952). Histological study shows that the region from which taste activity can be recorded is in the most medial tip of the ventrobasal nuclear complex. This corresponds

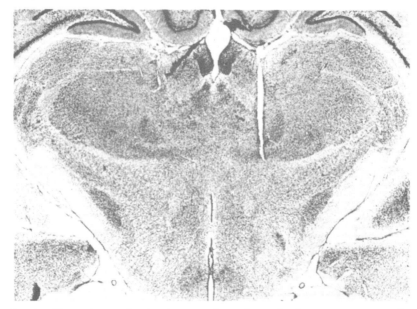

Fig. 8. Photomicrograph showing an electrode track that terminates at the ventral border of the ventrobasal nuclear complex. The movement of the electrode in the ventral direction was stopped as soon as the magnitude of the response to 0.1 M NaCl passed its maximum and began to decline. Frozen section stained with thionin.

to the nucleus commissura intraventralis of Gurdjian (1927) which seems equivalent to the nucleus ventralis posteromedialis (VPM) of other species.

Figure 8 shows a photomicrograph of an electrode track in an experiment in which only one electrode was inserted. The downward advance of the electrode was halted just as the taste response began to diminish so that the bottom of this track was still in the reactive area. Taste responses were recorded at a depth of 6.60 to 7.00 mm, so that only the bottom 0.40 mm of the track represents the active taste area. This corresponds to the ventral nuclear region in which Benjamin and Akert (1959) have recently found degeneration after ablation of the cortical taste area. The general location of the electrically active area agrees with their findings and with the expectations based on prior studies.

The response of this region to the basic taste stimuli and a series of NaCl concentrations in general resembles that found in the medulla, which in turn resembles that found in the chorda tympani nerve. To date only the anterior tongue has been stimulated and, as in the chorda tympani and medulla, the strongest response is to sodium chloride solutions.   Likewise the relative magnitudes of the response to 0.1 M KCl, 0.005 M HCl, 1.0 M sucrose, and 0.01 M quinine hydrochloride are similar to those found at the periphery and in the medulla.   It is clear that at this level we are recording at least from second-order and probably from third-order elements.   The same relative magnitudes of response seem to be preserved throughout.

Discussion

It is in connection with the quantitative similarities of the peripheral, medullary, and thalamic salt response that I wish to consider certain behavioral relations.   The behavioral effects of taste stimulation may be studied very conveniently by means of the two-bottle preference method which C. P. Richter (1942) used to such good advantage in his studies of dietary self-selection.   Typically the animal is provided with two or more drinking containers which are left in place on the living cage for ad libitum consumption.   In the two-bottle procedure, one contains water, the other a taste solution.   Preference (or aversion) is indicated by the volume of fluid taken from each container.   In the ideal case, the animal takes equally from both containers when each contains water or when one contains a subthreshold solution.   By a gradual increase in the solution concentration on successive days, a point is reached where the animal begins to take more solution than water, if it is a preferred solution.   The intake will increase with successive increases in concentration until some maximal concentration (peak preference) is reached.   Beyond this point, intake falls off as concentration is raised still further.   Ultimately, the higher concentrations may be avoided or taken in very small amount so that nearly all the fluid consumed in a 24-hour period will be water.   In other cases, such as quinine, only an aversion may be observed for the suprathreshold concentrations.

A typical preference-aversion curve taken from some of our earlier studies is plotted in Fig. 9 for sodium chloride solution presented to the rat.   The normal rat shows a definite and reproducible preference for mild salt solutions, even when it is on a completely adequate diet and no salt need is apparent.   The preference can, of course, be enhanced or depressed by increasing the salt need or feeding excess salt in the diet; yet the normal animal displays a strong and definite salt preference.   The figure also shows the typical chorda tympani response curve, together with the points from the medulla.   All points are relative to the value for 0.1 M NaCl. The points for the magnitude of response at each neural site are quite similar. Note that the behavior, the salt preference, lies within the operating characteristics of the sensory neural system. In the normal animal, preference does not become apparent until the afferent input has reached a significant

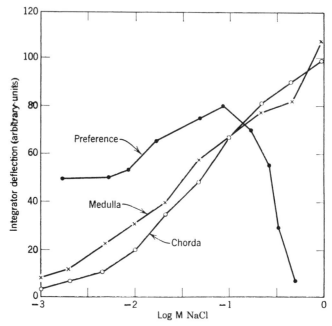

Fig. 9. A composite graph showing the magnitude of the integrator deflection in arbitrary units (maximum = 100) in the chorda tympani nerve and in the medullary taste area for different concentrations of NaCl. Preference behavior of the normal animal is indicated as percentage preference, that is, amount of salt solution ingested in a 48-hour period as a percentage of the total fluid (NaCl solution plus H_2O) ingested during the 48-hour test period.

level, some 20 to 30 per cent of maximal discharge. From this point on, behavior as percentage preference is an increasing function of the stimulus concentration up to the peak preference point. A behavioral inversion occurs at a concentration where the intensity of the neural input is still increasing. We have as yet been unable to find any change in the sensory neural activity, beyond that of magnitude, that might correspond to this point of inversion. Other studies show that the maximal point and the preference threshold, as well as degree of preference or aversion, can be manipulated. The enhanced intake of the adrenalectomized rat shows the effect on such behavior of increasing the physiological need, and increasing the amount of salt in the diet can be shown to depress the preference. An important area of investigation will be to search for changes in the central neural systems which may be related to such changes in behavioral response.

So far we have studied only the peripheral NaCl threshold for electrical activity in the chorda of the adrenalectomized rat and have found no difference between the normal and adrenalectomized animal. The receptor sensitivity does not seem to be markedly shifted by the changes in blood chemistry consequent to adrenalectomy (Pfaffmann and Bare, 1950). A similar conclusion is drawn from other studies using the conditioned-reflex method to determine taste thresholds (Carr, 1950; Harriman and MacLeod, 1953). In addition, we have studied the magnitude of chorda tympani discharge to sugar after lowering the blood sugar level by insulin injections (Pfaffmann and Hagstrom, 1955). No change in discharge from the peripheral receptors was noted. Insulin injection in the intact animal will cause an enhanced preference for sugar solutions (Richter, 1942; Soulairac, 1947).

Another behavioral observation is relevant in this connection, namely, that the salt preference can be reduced if the order of presenting the solutions is reversed, so that the strong and usually avoided concentrations are presented first. As the descending series is presented, the intake rises to reveal a preference function, but the whole response is depressed and the preference threshold itself may be elevated. In this reversed series, there is no reason to believe that the peripheral sensitivity has been significantly changed. Rather, the prior experience with the strong aversive salt solutions presumably generalizes to the taste of the weaker salt solution so that the response to them is now reduced. Such a change in sign or significance of the incoming sensory neural message can result, therefore, either from the effects of experience or from changes in physiological state, as described above. For the moment, however, we have been able to show only that there is a general correspondence between the range of stimulus concentrations that elicit preference or aversion behavior and those concentrations that initiate neural activity. The delineation of those changes in the sensory neural input that might be related to the observed behavioral effects must await further research.

Summary

This paper has described certain new observations on the taste relays in the central nervous system in the medulla and thalamus of rat. In the medulla electrical activity was recorded in the rostral portion of the nucleus of the solitary tract; in the thalamus activity was recorded from the medial tip of the ventrobasal complex. The discharges in the medulla and thalamus are very similar to the discharges recorded peripherally in the chorda tympani nerve. Single-unit analysis at the level of the medulla showed that the response patterns to chemicals resembled those previously described for the single fibers in the chorda tympani nerve. Previous studies emphasizing the patterned nature of the gustatory afferent input thus were extended to the level of the medulla. In addition, many chemically sensitive units were found to display cross-modality sensitivity, for they could be activated by mechanical and/or temperature stimulation. In the latter case, cooling to 10° C usually elicited a discharge, whereas warming reduced the level of activity. It was not possible, on the basis of available evidence, to determine the extent to which these cross-modality effects depend on central convergence, because similar cross-modal stimulation had been reported in prior studies of single fibers of the chorda tympani nerve.

Finally a correlation between behavior and certain quantitative features of the activity in the periphery and medulla was attempted for the responses to saline solutions. It was shown that behavior and neural discharge corresponded over the same general range of stimulus intensity, but at the same time, behavior did not follow neural activity directly. In normal animals, the preference threshold tended to occur at a value higher than that at which the first detectable electrophysiological response occurred. The preference increased with increasing stimulus concentration up to a maximal point beyond which the preference began to decline. No obvious change in the sensory input, save that of magnitude, has been observed at the inversion point in the response to saline solutions. The further delineation of the physiological substrate underlying the acceptance of some stimuli as contrasted with the avoidance of others was discussed as an important problem for further research.

Acknowledgments

The research described here was supported in part by a contract with the Office of Naval Research and a grant from the National Science Foundation. Dr. Halpern held a PHS Research Fellowship, 1958–59, and Mr. Frommer held a National Science Foundation Predoctoral Fellowship, 1957–59.

We wish to acknowledge the assistance of Mr. Ira Brown who, under the supervision of Dr. George Anderson, Director of Laboratories, Providence Lying In Hospital, embedded and cut the celloidin sections of the medulla. We wish also to acknowledge the assistance and guidance of Dr. R. M. Benjamin of the Department of Neurophysiology, University of Wisconsin, in preparing the histological sections of the thalamus.

References

Adler, A.   Zur Topik des Verlaufes der Geschmackssinnfasern und anderer afferenten Bahnen im Thalamus. *Z. ges. Neurol. Psychiat.*, 1934, **149**, 208–220.

Allen, W. F.   Origin and distribution of the tractus solitarius in the guinea pig. *J. comp. Neurol.*, 1923, **35**, 171–204.

Amassian, V. E., and R. V. DeVito.   Unit activity in reticular formation and nearby structures. *J. Neurophysiol.*, 1954, **17**, 575–603.

Anderson, B., and P. A. Jewell.   Studies on the thalamic relay for taste in the goat. *J. Physiol.*, 1957, **139**, 101–107.

Appelberg, B., and S. Landgren.   The localization of the thalamic relay in the specific sensory path from the tongue of the cat. *Acta physiol. scand.*, 1958, **42**, 342–357.

Beidler, L. M.   Facts and theory on the mechanism of taste and odor perception. In J. H. Mitchell, Jr., N. J. Leinen, E. M. Mrak, and S. D. Bailey (Editors), *Chemistry of Natural Food Flavors.*   Chicago: Quartermaster Food and Container Institute, 1957.

Benjamin, R. M., and K. Akert.   Cortical and thalamic areas involved in taste discrimination in the albino rat. *J. comp. Neurol.*, 1959, **111**, 231–260.

Blum, M., T. C. Ruch, and A. E. Walker.   Localization of taste in the thalamus of *Macaca mulatta. Yale J. Biol. Med.*, 1943, **16**, 175–191.

Bremer, F.   Physiologie nerveuse de la mastication chez le chat et le lapin. Centre cortical du goût. *Arch. int. Physiol.*, 1923, **21**, 308–353.

Carr, W. J.   The effect of adrenalectomy upon the NaCl taste threshold in rat. *J. comp. physiol. Psychol.*, 1950, **43**, 377–380.

Cohen, M. J., S. Hagiwara, and Y. Zotterman.   The response spectrum of taste fibres in the cat: A single fibre analysis. *Acta physiol. scand.*, 1955, **33**, 316–332.

Cohen, M. J., S. Landgren, L. Ström, and Y. Zotterman.   Cortical reception of touch and taste in the cat: A study of single cortical cells. *Acta physiol. scand.*, 1957, **40**, Suppl. 135.

Craigie, E.   *An Introduction to the Finer Anatomy of the Central Nervous System.* Philadelphia: Blakiston, 1925.

Ectors, L.   Etude de l'activité électrique du cortex cérébral chez le lapin non narcotisé ni curarisé. *Arch. int. Physiol.*, 1936, **43**, 267–298.

Erickson, R. P.   Responsiveness of single second order neurons in the rat to tongue stimulation. Doctoral dissertation, Brown University, 1958.

Fishman, I. Y. Single fiber gustatory impulses in rat and hamster. *J. cell. comp. Physiol.*, 1957, **49**, 319–334.

Frank, K. Identification and analysis of single unit activity in the central nervous system. In *Handbook of Physiology. Neurophysiology* I. Washington, D. C.: American Physiological Society, 1959. Pp. 261–277.

Gerebtzoff, M. A. Recherches oscillographiques et anatomophysiologiques sur les centres cortical et thalamique du goût. *Arch. int. Physiol.*, 1939a, **51**, 199–210.

Gerebtzoff, M. A. Les voies centrales de la sensibilité et du goût et leurs terminaisons thalamiques. *Cellule*, 1939b, **48**, 91–146.

deGroot, J. The rat forebrain in stereotaxic coordinates. USAF report, Contract No. A.F. 49(639)-384, OSR, ARDC, 1959.

Gurdjian, E. The diencephalon of the albino rat. *J. comp. Neurol.*, 1927, **43**, 1–114.

Halpern, B. P. Gustatory responses in the medulla oblongata of the rat. Doctoral dissertation, Brown University, 1959.

Harriman, A. E., and R. B. MacLeod. Discriminative thresholds of salt for normal and adrenalectomized rats. *Amer. J. Psychol.*, 1953, **66**, 465–471.

Holt, E. B. *Animal Drive and the Learning Process.* New York: Holt, 1931.

Inanaga, K., H. Nakao, and H. Fuchiwaki. Electrical excitation on the cerebral cortex by gustatory stimulation. II. Localized evocation type potentials of the cerebral cortex by gustatory stimulation. *Folia psychiat. neurol. jap.*, 1953, **7**, 7–10.

Kimura, K., and L. M. Beidler. Microelectrode study of taste bud of the rat. *Amer. J. Physiol.*, 1956, **187**, 610.

Kolmer, W. Geschmacksorgan. In W. v. Möllendorf (Editor), *Handbuch der mikroskopischen Anatomie des Menschen.* III. Berlin: Springer, 1927. Pp. 154–191.

Krieg, W. J. S. *Functional neuroanatomy.* New York: Blakiston, 1942.

Krieg, W. J. S. Accurate placements of minute lesions in the brain of the albino rat. *Quart. Bull. Northwestern Univ. Med. Sch.*, 1946, **20**, 199–209.

Landgren, S. Convergence of tactile, thermal and gustatory impulses on single cortical cells. *Acta physiol. scand.*, 1957, **40**, 210–221.

Mountcastle, V., and E. Henneman. Pattern of tactile representation in thalamus of cat. *J. Neurophysiol.*, 1949, **12**, 85–100.

Patton, H. D., T. C. Ruch, and A. E. Walker. Experimental hypogeusia from Horsley-Clarke lesions of the thalamus in *Macaca mulatta*. *J. Neurophysiol.*, 1944, **7**, 171–184.

Pfaffmann, C. Gustatory afferent impulses. *J. cell. comp. Physiol.*, 1941, **17**, 243–258.

Pfaffmann, C. Gustatory nerve impulses in rat, cat and rabbit. *J. Neurophysiol.*, 1955, **18**, 429–440.

Pfaffmann, C., and J. K. Bare. Gustatory nerve discharges in normal and adrenalectomized rats. *J. comp. physiol. Psychol.*, 1950, **43**, 320–324.

Pfaffmann, C., and E. C. Hagstrom. Factors influencing sensitivity to sugar. *Amer. J. Physiol.*, 1955, **183**, 651.

Richter, C. P. Total self-regulatory functions in animals and human beings. *Harvey Lect.*, 1942, **38**, 63–103.

Rose, J. E., and V. B. Mountcastle. The thalamic tactile region in rabbit and cat. *J. comp. Neurol.*, 1952, **97**, 441–490.

Scheibel, M., A. Scheibel, A. Mollica, and G. Moruzzi. Convergence and interaction of afferent impulses on single units of reticular formation. *J. Neurophysiol.*, 1955, **18**, 309–331.

Soulairac, A. La physiologie d'un comportement: l'appétit glucidique et sa régulation neuro-endocrinienne chez les rongeurs. *Bull. Biol. Paris*, 1947, **81**, 273–432.

Torvik, A. Afferent connections to the sensory trigeminal nuclei, the nuclei of the solitary tract and adjacent structures. *J. comp. Neurol.*, 1956, **106**, 51–141.

25

PATRICK D. WALL

Department of Biology and Center for Communication Sciences
Massachusetts Institute of Technology

Two Transmission Systems for Skin Sensations

It is our intention in this paper to study the role of the relay nuclei in a sensory transmission pathway. The term "relay nuclei" originated after anatomical studies and physiological studies that applied brief electrical stimuli to the input fibers. The arrival of a synchronized volley of impulses over the input fibers results in a brief burst of impulses leaving the cells in the nucleus. From results of this variety, it appears that the cells of the relay nuclei hand on the characteristics of the input, and that some amplification results from the repetitive discharge of impulses following the arrival of single impulses in the afferent fibers. The extent of this amplification effect can be modified by the action of other systems playing onto the relay nuclei. If this were the only function of these cells, the term "relay" would be apt, but the study of individual cells with the use of natural stimuli shows them to possess much subtler and more interesting properties. By virtue of the particular afferent fibers that converge onto these cells and by their interconnection with neighbors, the output of the cells reflects certain aspects of the peripheral stimulus and modifies the temporal pattern of the afferent impulses so that the output pattern does not necessarily represent a summation of the input patterns. Rather than acting as "relays," the cells act as "funnels and filters."

The cells examined here are those that receive afferent fibers from the skin. The peripheral nerve fibers of the skin enter the spinal cord in the dorsal roots. After entering, they divide into branches and terminate on nearby cells in the dorsal horn of the gray matter of the spinal cord. Some of the entering skin fibers send long running branches which travel the full distance up the dorsal columns of the spinal cord and terminate on cells in the caudal end of the medulla. The fibers from the leg end on cells in the nucleus gracilis. We shall compare the organization and input-output properties of two groups of cells that receive direct fibers from the skin, the first, the dorsal horn interneurons, and the second, the nucleus gracilis.

The afferent skin nerve fibers in the peripheral nerves of cat have been studied extensively by Adrian (1928), Zotterman (1939), Maruhashi, Mizuguchi, and Tasaki (1952), and others. The fibers vary in size with a monotonic distribution of diameters from the large myelinated fast-conducting A fibers to the small C fibers. Physiological knowledge is limited almost entirely to the larger fibers because until the recent work of Iggo (1959) no method was available for studying the properties of the fibers conducting below about 15 meters per second. Most or perhaps all of the fibers in this larger diameter group respond to distortions of the skin; the larger the fiber, the lower the threshold and the quicker the adaptation of the response to continued pressure stimulation. Some of the smaller pressure-sensitive fibers are also sensitive to temperature changes. Compounds that cause damage and itch on the skin result in a continuous barrage of impulses in the smaller afferent fibers. The reason for the high pressure sensitivity of the larger fibers can no longer be related to the presence of specialized transducing structures in the skin but is probably related to the anatomy of the clusters of the fine terminals in the skin. There may no longer be any justification for discussing the peripheral skin nerve fibers in terms of separate specialized groups, with each group selecting a particular variety of environmental change. Rather the fibers should be thought of as a single group within which there is a continuous variation of properties from the large low-threshold fibers to the small fibers with high threshold and low adaptation and temperature sensitivity. The justification for this simpler view has been presented elsewhere (Wall, 1960).

The effect of a volley arriving over these afferent fibers is to produce repetitive firing in the cells on which they end. Since we intend to show that the temporal pattern of the impulses leaving the cells may be significant, it will be necessary first to enquire into the mechanism of this repetitive firing. After presenting these results, we shall present, for the two groups of cells, their organization and the nature of their input and output.

The Primary Spinal Neurons Receiving Skin Afferents

Methods

Cats were used in all these experiments. Most of the experiments were carried out in unanesthetized preparations in which the spinal cord was sectioned at the first cervical segment, with the brain destroyed by arterial ligation. The lumbar region of the spinal cord was exposed and placed under oil. One of the main problems in microelectrode recording is the achievement of complete stability. It is relatively simple to fix the bony structures by impaling the pelvic iliac crests with pins and by clamping one of the lumbar vertebrae. However, some pulsations of the cord remain, largely because the ventral plexus of veins pulsates. Partial stabilization can be obtained by sewing the dura, which has been cut down the dorsal midline, to the surrounding muscles, so that the cord lies in a hammock of dura. The final stabilization is achieved by placing sutures into the four dentate ligaments that surround a segment. These four sutures are then gently raised to lift the cord about a half millimeter so that the blood-vessel pulsation is no longer transmitted to the cord. The last two maneuvers are carried out while the dorsal vessels of the cord are observed through a microscope, so that no embarrassment of the circulation occurs. Recordings are taken through intracellular microelectrodes filled with saturated potassium chloride.

Since in all these experiments only the time of occurrence of action potentials was to be observed, it was possible to eliminate grid-current problems by use of a capacity-coupled head stage. Results were checked with electrodes filled with sodium sulphate and, where possible, by extracellular recording, in order to eliminate the damaging effect of the recording electrodes. The position of the recording electrodes was determined by cutting them off, leaving them in position, and locating them by the simple technique of fixation, dehydration, and clearing (Wall, et al., 1956). Stimulation of the skin was by sudden application of pressure, heating by radiation, localized crushes, and cowhage, an itch-producing substance. Electrical stimulation was applied to dorsal roots and peripheral nerves that had been dissected free and placed on stimulating electrodes. In all experiments the animals were paralyzed by intravenous Flaxedil and were artifically respired.

The origin of repetitive discharge

When a dorsal root carrying the afferent fibers into the dorsal horn of the spinal cord is stimulated by a brief electric shock, a single nerve impulse is set off in each of the fibers. The volley of impulses sweeps into the synaptic region of the first central cells. These cells respond with a burst of impulses which considerably outlasts the time of arrival of the afferent barrage. Two different types of theory have been advanced to explain this prolongation of the effect of an afferent volley. The first, suggested by Barron and Matthews (1938), proposed that a long-lasting excitatory process is set up in the synaptic region by the arrival of the afferent volley. A prolonged disturbance occupies the ends of the fibers and gives rise to the dorsal root potential which indicates that the endings of the fibers remain in a hyperexcitable state for well over 100 milliseconds after the volley's arrival. This prolonged depolarization of the endings may give rise to a burst of impulses which travels backwards over the afferent skin nerve fibers and is called the dorsal root reflex. The generation of this dorsal root reflex is evidently closely related to the prolonged depolarization of the endings of the fibers in which the impulses run. When the membrane potential of the endings falls below a critical level, sufficient current flows to the depolarized region to generate nerve impulses in the nearby parts of the nerve fiber. The firing continues as long as the ends remain sufficiently depolarized. In a repetitive burst originating in this manner, the stimulating disturbance determines the beginning and end of the burst and possibly its frequency, but since the membrane is oscillating it does not directly determine the particular time of occurrence of an impulse. Thus the pattern of firing is determined by the general level of the stimulating disturbance, and the particular time of a given impulse is decided by the preceding impulse in the burst. In such a system, the interjection of an additional impulse into the middle of the burst will retime the subsequent impulses in the burst, since each fiber contains an oscillating "pacemaker" (Wall, 1959, 1960).

The second type of theory for the origin of a repetitive discharge was suggested by Lorente de Nó (1939) among others. According to this theory, the cells are driven to fire repetitively by neighboring cells that proceed to bombard each other with impulses after the arrival of the afferent volley. The "reverberating circuit" is one such arrangement of interconnecting cells that prolongs the effect of an arriving volley.

Dorsal horn interneurons were examined in order to try to differentiate between these two theories. Microelectrodes were placed in the axons of primary central cells. They were known to be primary central cells because of the time and pattern of their discharge, which clearly differentiated them from the afferent fibers. The microelectrode was intentionally placed in the axon in order to avoid, as far as possible, damage to the somadendrite region of the cell by the recording electrode. No use was to be made of the slow potentials that may be recorded from the cell body, since we believe that in this complex situation these potentials are at present uninterpretable. It was known that the microelectrode was within the axons, partly because of the absence of large slow potentials and partly by the technique of using the microelectrode as a stimulator in order to determine the threshold of the surrounding membrane. It was found that ten times more current was necessary to produce an action potential in the region of the cell body than in the axon.

The pattern of discharge of these neurons in response to electrical stimulation of the dorsal roots has been reported by Frank and Fuortes (1956). The discharges vary in frequency, duration, and regularity. In an unanesthetized preparation, almost all the cells are spontaneously active. It is immediately obvious that the effect of the stimulus is not simple, since the burst response is followed by a very prolonged suppression of the spontaneous firing, and a repetition of the stimulus during this silent period results in a shorter burst response. The burst responses may have a simple pattern of initially high frequency, gradually declining, but quite complex repeated patterns are also frequently observed. When successive bursts to repeated stimuli are examined, it is seen that the variation in latency of a particular component impulse within the burst increases as the latency of the impulse increases. In other words, the accuracy of the timing of the later components is considerably lower than that of the first responses. When the intensity of the stimulus is decreased, the length of the burst decreases, so that at threshold for the cell only one response is evoked. There is only a slight shift in latency—less than 1 millisecond—for the first impulse evoked by a maximal or threshold peripheral stimulus. A careful observation of the transition between maximal and threshold stimuli is most interesting. The shortening of the length of the discharge is accompanied by very little change in the frequency of the remaining impulses. There are no signs in these cells that the impulses appear at longer and longer intervals as the stimulus is weakened.

Observations of this nature may be suggestive of the mechanisms producing the burst, but they are clearly not adequate to decide the issue. In order to locate the pacemaker, the recording electrode within the cell was used as a stimulating electrode. It is interesting to note that a brief stimulus of a single cell never produced repetitive firing of the cell, even if several impulses were produced in the cell in rapid succession. This implies that more than one cell must be set in action if repetitive firing is to occur. The single impulse produced by the intracellular stimulation was made to interact with the repetitive firing of the cell produced by stimulation of a peripheral nerve. The intracellular stimulation was arranged at such a time that the impulse produced would sweep into the cell-body region and precede and abort one of the impulses in the repetitive discharge. This disturbance of the expected pattern produced a very clear-cut result (Wall, 1959). The interjected impulse disturbs the time of appearance of the impulses that immediately follow it, but the subsequent parts of the pattern occur at the expected time without interference. This implies that the pattern is determined by some mechanism outside the region invaded by the interjected impulse and therefore not within the cell under examination. It is clear that this external firing mechanism is not merely determining the frequency of firing, for, if that were the case, the occurrence of an interjected impulse within a cell would decide the particular time of occurrence of the next impulse. It is evident then that the particular time of occurrence of impulses in the pattern of discharge is determined outside the cell, probably by bombardment of the cell by impulses from neighbors. This mechanism contrasts with the internal-oscillator type responsible for the dorsal root reflex (Wall, 1959) and for the tonic rhythm of motor neurons (Morrell, et al., 1956).

Since after-discharges may be very prolonged, the problem of their explanation merges with the problem of explaining "spontaneous" activity. The same two groups of theories could be offered. We have attempted to analyze the theories in the same way. An impulse interjected into a single cell does not appear to influence the probable time of firing of the next expected impulse in the "spontaneous" firing. This suggests that here too the discharge is the consequence of the mutual bombardment of a group of cells. One might expect, if this were the case, that there would be preferred firing frequencies in the pattern of spontaneous firing, as a consequence of the particular anatomical connections of the neighboring cells. We have been unable, however, to find any significant deviations from the Poisson distribution of the intervals between successive impulses. The absence of preferred firing frequencies does not, of course, eliminate the possibility that the cells are discharging as a consequence of bombardment by the neighbors.

There are two consequences if this conclusion is accepted. The first is that the pattern of firing of a cell may be under relatively accurate control and that a mechanism exists for the production of complex patterns. This follows from the suggestion that the cells are responding to bombardment by a particular impulse, rather than to a synaptic disturbance with a long time course. In essence, it is suggested that each order of the control mechanism has a relatively short time constant. The second consequence is that the so-called "spontaneous" firing need not be regarded as a measure of the instability of the cell, or as noise through which the signal must be transmitted, but rather as a consequence of the activity of neighbors and of afferent bombardment, so that the "spontaneous" activity itself becomes a signal.

Organization of the interneurons

A horizontal lamina of cells lies across the top of the dorsal horn in the lumbar spinal cord, ventral to the substantia gelatinosa. We have shown (Wall and Cronly-Dillon, 1960) that the fast nerve fibers in a skin nerve end among these cells. Some of the cells send their axons into the dorsolateral column of the white matter, where they run in the position classically assigned to the dorsal spinocerebellar tract. The same cells are presumed to send axons onto deeper cells within the dorsal horn of the spinal cord. All these cells respond to light touch of the skin. Recordings were taken from single cells, usually by penetrating the axons running up in the dorsolateral funiculus. They are known to be primary central cells because fast fibers of the skin nerve end in this region and because they respond to stimulation of the dorsal root or peripheral nerve with a synaptic delay averaging 1.2 msec. The afferent fibers sweep into the spinal cord in the entrance zone of the dorsal root and travel medially over the cap of the dorsal horn. The fibers then hook laterally and penetrate the lamina of cells from its medial edge. A clear topographic map exists within the lamina. As would be expected, there is a rostral-caudal representation of the skin of the animal in such a way that the tail and perineal skin fibers project onto the caudal-sacral segment cells, whereas the cells in the third lumbar segment receive fibers from the flank and anterior thigh region. However, within each segment there is also a medial lateral organization in the lamina, so that the medial cells receive fibers from the distal region of the leg and paw and the lateral cells receive fibers from the more proximal parts of the leg.

Each cell subserves its own special area of skin. The shape of these areas is usually oval, with the long axis running along the axis of the leg. No cell was found to subserve a shattered or irregularly shaped peripheral area. The average area (in millimeters) subserved by a peripheral afferent fiber that responds to light touch was found to be 5×4, the smallest observed was 2×2, and the largest, 18×10. In contrast the primary cells had an average receptive field of 63×32 mm, the smallest being 30×15, and the largest, 150×35. These areas are shown in Fig. 1. It is immediately apparent that many peripheral fibers from a contiguous area of skin must converge onto each single central cell. These areas were remarkably stable during procedures designed to vary the excitability of the central cells, such as tetanization, strychninization, asphyxia, etc. This absence of signs of a subliminal fringe and the absence of a surrounding inhibitory region suggest strongly that the area subserved by the cell originates from a straightforward convergence of excitatory fibers onto the cell, and that the area is not formed by the accidental coincidence of extensively ramifying terminal arborizations, nor is it formed by the active processes implied by an "inhibitory surround." We are forced to conclude, therefore, that many fibers from an area of skin end on a particular cell, and that fibers from the surrounding skin do not.

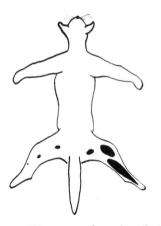

Fig. 1. Diagram to show the relative areas subserved by input fibers and by the dorsal horn cells on which the fibers end. The fibers shown on the left responded to light touch in an average area of 5 × 4 mm; the smallest found was 2 × 2 mm, the largest, 18 × 10. The cells shown on the right responded to light touch with an average receptive field of 63 × 32 mm, the smallest, 30 × 15, the largest, 150 × 35. The position of the areas on the leg is not significant. The results show that many fibers must converge onto each cell.

We have now to ask the anatomical or embryological question, by what route do these many fibers converge onto a particular cell? The question is made more difficult by the apparent chaos in the organization of the afferent fibers as they enter the spinal cord over the dorsal roots. The large dorsal roots such as L 7 or S 1 contain fibers derived from all parts of the leg. Even subdivisions of these large roots still appear to subserve very large areas of leg. If this were the actual state of the peripheral nerves, the embryologist would have to explain how nerve fibers from one area of skin could mix with fibers from many other areas and yet finally regroup together on a single cell. In order to solve this problem, recordings were made from a single cell, and its peripheral receptive field was determined. Small cuts were then made through the spread-out sheet of afferent fibers as they entered the spinal cord. It was discovered (Wall, 1960) that all the fibers ending on one of these cells were contained within a very small area of the afferent mass. It is postulated that the fibers converging onto a single cell are gathered together in a microbundle in their peripheral course.

The output of the interneurons

It has been shown in the previous section that many nerve fibers that respond to light touch converge onto a single cell, since the area subserved by the cell is considerably bigger than that of a single afferent fiber. We can now ask the question, do other types of fiber also converge onto this cell from the periphery? The cells of this group are monopolized by afferents from the skin and do not respond to muscle afferents. The cells on which proprioceptive afferents end lie deeper in the dorsal horn than do the cells that we have studied here.

When large afferent fibers that respond to light touch or hair movement are studied, we find that their discharge characteristics are marked by a rapid adaptation to continuous light pressure, and that their response does not increase if the pressure applied to the skin is increased beyond light pressure. The dorsal interneurons respond to a light touch on their receptive area with a rapid burst of impulses, which rapidly declines to the resting discharge of the cell even if the pressure is maintained. The discharge of the cell somewhat outlasts the afferent volley, as would be expected from the properties of repetitive discharge discussed above. However, the response of the cell to light touch only exaggerates the properties of the large afferent fibers. If the pressure on the skin is increased to a medium pressure, the initial burst of impulses is similar to that evoked by light touch, but adaptation is much less marked. This must mean that these cells are connected not only to the largest afferent fibers but also to some of smaller diameter. If the pressure is increased further to a crushing pressure, the response is exaggerated and lasts throughout the duration of the stimulus. These results are shown for a single cell in Fig. 2.

Further evidence for the convergence of many types of fibers onto these cells is shown in Fig. 3. The cells respond to heating of the skin. Their sensitivity to temperature is not high—a minimum skin temperature rise of 2° C is required—but the effect is quickly reversed so that the skin is not damaged. The effect is not due to piloerection. Recently we have found far more sensitive cells which also responded to touch in the trigeminal nuclei concerned with the skin of the nares. This low sensitivity to temperature change may be a consequence of the type of innervation in furred skin. The characteristics of pressure and temperature responses of these cells are shown in Figs. 2 and 3 by a shift in the frequency distribution of the pattern of impulses. It will be noted that the patterns of high-pressure and high-temperature stimulation are similar.

After damage to the receptive area by crushing or pin pricks, the cell does not return to its normal pattern of firing. Instead of the

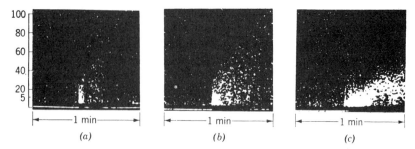

Fig. 2. Reaction of a dorsal horn cell to three pressure stimuli of increasing intensity. Recording by microelectrode within a single cell. Stimuli: (a) 2 gm suspended from hairs in middle of cell's receptive field; (b) 20 gm suspended from hair; (c) pinch applied directly to skin. On the left of each picture, the cell is firing spontaneously without disturbance of the skin; in the middle, the stimulus was applied and remained on for the rest of the recording. Each dot represents a nerve impulse; the height above the base line represents the interval, shown in milliseconds on the scale, between the recorded impulse and the preceding one. Dots on the bottom line represent impulses that occurred at intervals greater than 100 msec after the preceding one. The pictures show the rapid adaptation to the small stimulus and the decreasing adaptation as the stimulus intensity is increased.

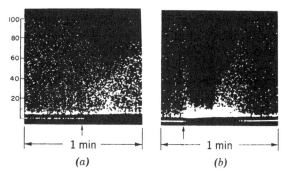

Fig. 3. Reaction of a dorsal horn cell that is sensitive to skin pressure and to changes in skin temperature. The method of recording is the same as that shown in Fig. 2. Heat was applied to the cell's receptive field by a radiant lamp for 15 sec after the arrow. In (a) the skin temperature rose a maximum of 4° C, and in (b) 12° C. The activity of the cell is shown before, during, and after the application of heat. The pattern of firing produced by this stimulus should be compared with that produced by pressure.

general distribution of intervals between impulses, brief bursts of impulses occur, not unlike those seen following electrical stimulation of a peripheral nerve. These brief high-frequency bursts of 4 to 6 impulses are followed by relative inactivity of the cell. The bursts occur several times per second at irregular intervals. The cell still responds to increased pressure on the receptive field. A similar pattern of activity is produced by applying cowhage, an itch-producing substance, to the receptive field.

Finally, a most interesting pattern of discharge of the cells is evoked by continuous light vibration of the skin. Pressing a plate lightly on the skin area 60 times per second, drives the large afferent fibers so that the cells are bombarded by one or more impulses from each large fiber every time the pressure on the skin is increased. The cells follow the varying pressure pattern with a remarkably regular pattern of firing, so that as the pressure is increased the cell responds with one, two, or three impulses, followed by a period of silence until the next increase in skin pressure. This contrasts with the prolonged burst produced by a single application of pressure, and it is apparent that the cells are partially inhibited during the silent period, since electrical stimulation of the afferent fibers produces a much shorter response in the cell than in the absence of vibration. It appears, therefore, that continuous firing of the large afferent fibers produces a mixed excitatory and inhibitory effect on the cell. We have already seen another example of this mixed effect in which an electrical stimulus to a dorsal root is followed first by a burst response, and then by a prolonged disappearance of the spontaneous activity. One mechanism would explain these two observations, namely, an inhibitory feedback onto the cell which tended to limit its firing and which was set off by the discharge of the cell. The effect of such a feedback would be to extend the dynamic range of the cell so that, although capable of responding to small afferent volleys, it would still be able to respond to larger and larger afferent barrages without saturation. Such a mechanism would be expected to respond at a particular level of continuous afferent barrage with an alternating output. The afferent barrage produced by damage to the skin or by itch-producing substances appears to be a continuous one, and yet, as has been described, the output of these cells alternates between bursts and silence. The postulated inhibitory feedback may be one way in which such an alternating output is achieved.

The Dorsal-Column–Medial-Lemniscus System

We have dealt with a system of cells within the spinal cord which receive afferent nerve fibers from the skin. Now we can turn to a second group of cells which also receive direct afferents from the periphery. The dorsal columns in the cat are made up of ascending branches of afferent nerve fibers that terminate on two groups of nuclei, the nucleus gracilis and cuneatus in the dorsal caudal end of the medulla. Fibers originating in the hindleg and entering the cord over the lumbar and sacral dorsal roots divide and pass up the cord to end on the more medial and caudal of the two nuclei, the nucleus gracilis. The axons of these cells cross to the opposite side of the brain and proceed on as the medial lemniscus, which terminates in thalamic nuclei. The questions we shall ask of this system are the same as those asked of the dorsal horn system: what is the nature of the input to the cells, what is their organization, and what is their output?

All fibers entering the dorsal part of the spinal cord run for a short distance within the dorsal column system, but it is not clear from the literature whether all types of fiber contribute to the dorsal columns up to the nucleus gracilis. It was established by Lloyd and McIntyre (1950) that the largest of the proprioceptive fibers ended on the cells of Clarke's column and did not proceed farther up the cord than the lower thoracic segments. Glees and Soler (1951) studied the number of fibers that penetrated to the medullary relay nuclei from the dorsal roots and concluded that only 25 per cent of the entering fibers ended in the nucleus gracilis. The method they used was to count those fibers that stain with osmic acid by the Marchi technique after section of the dorsal root. This method selects out the large myelinated fibers, and some of their results were to be expected from the demonstration that the large proprioceptive fibers do not contribute to the input of nucleus gracilis. However, the decrease in the size of the input is greater than could be explained in this way, and therefore one could presume that many of the myelinated skin fibers also were not traveling all the way to the medulla. We could expect then that the total number of the input fibers would be smaller, but we had to discover whether the loss was limited to particular types of fiber or whether the decrease was unselective. The response of the nucleus gracilis to afferent volleys evoked by electrical stimulation had been reported by Therman (1941) and Amassian and DeVito (1957). The issues of particular interest here were not found in the literature.

Methods

Cats were used in all these experiments. Most animals were under light Dial anesthesia, but all results were confirmed in a small number of decerebrate animals. Recordings from the dorsal columns were taken through glass intracellular electrodes or by platinum-tipped indium extracellular electrodes. Recordings within the nucleus were by means of extracellular platinum-tipped indium electrodes. The problem of stability is severe in this region and could not be solved by the rigging method used in the spinal cord. One method tried was immediately abandoned: the medulla was stabilized by running two thin needles transversely through the medulla just rostral to the gracile and cuneate nuclei, but in our hands this method always led to a prompt vascular collapse. The method best suited for investigating this region was adopted after consultation with Dr. Carreras, now at the Department of Physiology, Johns Hopkins Hospital. The region between the arch of the first cervical vertebra and the foramen magnum was exposed with the head slightly flexed. A lucite tube was sealed with dental-impression cement around the area, and the dura was cut. The pia over the nucleus was incised, and the electrode placed over the pial opening. The area and tube were then flooded to a depth of 1 cm with mineral oil, and embedding paraffin wax with a melting point of 45° C was then poured over the top of the oil. This reseals the cranial contents and prevents further pulsation of the brain stem. The paraffin wax is at a sufficient distance from the nerve tissue and the oil, a poor conductor of heat, so that the surface of the medulla does not suffer from heat damage. The cylindrical electrode may be driven with ease through the surrounding wax.

The nature of the input

The medial part of the dorsal columns, which contain the fibers of hindleg origin, was searched with microelectrodes in the first cervical segment, as had been done previously in the dorsal-root entrance zone. Large touch fibers were entered with peripheral stimulus properties identical to those observed in the dorsal root. It is apparent for two reasons that these fibers send collaterals into the lumbar dorsal horn and are not separate from those fibers that end in the dorsal horn. The first reason is that these fibers carry an equivalent of the dorsal root reflex, as has been described by Hursh (1940). We have discussed the origin of these impulses previously (Wall, 1959) and have shown that it originates from the terminal arborization of skin nerve fibers in the dorsal horn. These impulses flow antidromically out of the dorsal roots but, since there are other collaterals in the dorsal column, the volley also proceeds orthodromically up the dorsal column. This means that, unless special precautions are taken, each volley of afferent impulses, set off in the periphery and traveling up the dorsal columns, is followed by a second volley, which has echoed off the terminals of the same fibers in the dorsal horns. The second reason for believing that the fibers in the dorsal column also end in the dorsal horn is that antidromic stimulation of the dorsal columns in the cervical region results in ventral root reflexes similar to those observed when the large fibers of a peripheral skin nerve are stimulated.

In addition to the fibers that responded to light touch of the skin, fibers with a medium pressure threshold were observed, as well as some that responded to joint movement. However, we were quite unable to observe fibers that had a high threshold or that responded to temperature changes. Furthermore, the fibers easiest to observe in the dorsal roots, the tonic proprioceptors, were entirely absent.

In order to confirm these findings with another technique, the following experiment was set up. The cord was exposed in the lumbar, lower thoracic, and upper cervical regions. The lower lumbar and all sacral ventral roots were cut. The nerves to gastrocnemius and the sural nerve were sectioned peripherally in the peroneal fossa and placed on recording electrodes. Stimulating electrodes were placed on the dorsal columns at C 2, T 10, L 3, and L 7. The records shown in Fig. 4 are the responses produced at the ends of the two peripheral nerves by the arrival of antidromic impulses from the four stimulus points. It is apparent that almost no fibers from the muscle nerve penetrate into the thoracic and cervical dorsal columns, and it is probable that these few are of joint origin. In contrast, a considerable proportion of the fibers from the skin nerve proceed all the way up the dorsal column. An analysis of the front of the compound action potential shows that the conduction velocity of the fast fibers drops by a factor of one-half when the fibers enter the upper thoracic and cervical part of the dorsal columns. A similar finding was reported by Lloyd and McIntyre (1950) for the fibers of muscle origin ending on Clarke's column. This decrease in conduction velocity spreads out the compound action potential. A step-by-step analysis of the changes in the compound action potential recorded at the periphery was made as the stimulus point was moved up the cord. From these data it became evident that fibers in the slower half of the A group were failing to penetrate more than a few segments up the dorsal columns, and no signs of C fibers of peripheral origin could be detected. The dorsal root reflex was eliminated from these records, first by stimulation at 15 per second, since this reflex fails to follow high rates of stimulation, and second by the overheating of the cord to 39° C, which also suppresses the effect.

It is evident, therefore, that only the largest fibers from peripheral skin nerves and some joint fibers penetrate the cervical dorsal columns from the lumbar and sacral dorsal roots.

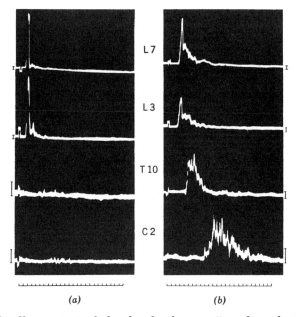

(a)   (b)

Fig. 4. The fiber content of the dorsal columns. Recording electrodes were placed (a) on the cut nerve of gastrocnemius and (b) on the sural nerve in the popliteal fossa. All relevant ventral roots were cut. Stimulating electrodes were placed on the dorsal columns at L 7, L 3, T 10, and C 2. The stimulus was maximal for the response lasting 0.4 msec and was applied at 20 per second in order to abolish the dorsal root reflex. The gain was changed for the various stimulus points; the vertical line marks 0.5 mv. The volley recorded consists of antidromic impulses originating under the stimulating electrodes and running out to the cut end of the nerve. It will be seen that few fibers from the muscle nerve penetrate to the upper cervical region, whereas many of those from a skin nerve reach this region.

The organization of the nucleus

As has been shown most recently by Glees and Soler (1951) there is an orderly arrangement of the afferent fibers. The medial caudal cells respond to skin stimulation of the tail and perineal region, and the lateral rostral cells to the skin innervated by the lumbar segments. A small number of cells can be detected, lateral to gracilis cells, that respond to stimulation of the flank, and lateral to these cells is the cuneate nucleus responding to foreleg stimulation. The dorsal column nuclei are, therefore, really a single nucleus with the skin of the whole body except the face represented. In addition to the medial lateral representation of the segments, however, we found that there was a most striking topographical map in the dorsal ventral direction.

The afferent fibers stream into the nucleus from its dorsal surface. The cells closest to the point of entry of the fibers responded to receptive areas at the periphery of the limb, whereas the deeper cells were stimulated from more proximal parts.

Thus the two dorsal column nuclei and the joining band of cells may be said to contain a topographic map of the skin of the cat on the same side with the tail medial and caudal, the neck rostral and lateral with the animal lying on its back so that the cells representing the feet lie dorsally in the most dorsal parts of the gracile and cuneate nuclei.

The nature of the output

We have already seen that the input fibers to this beautifully organized group of cells are quite restricted in their function, as compared with the full spectrum that makes up the input to the dorsal horn. One expects, therefore, to discover a much more restricted repertory of stimuli to which these cells will respond, in contrast with the wide range found in the group of dorsal horn cells. This was in fact the case; almost all the cells responded to light pressure of the skin within their receptive field and showed a slight increase of response if the pressure was increased to a medium pressure. The response to steady pressure rapidly adapted. Further increases of pressure failed to produce any further augmentation of the response. No cells were observed that responded to temperature, skin damage, or itch-producing substances. The only exceptions to these cells were a small proportion that responded only to joint movement and some that responded to both light and medium pressure on skin around a joint, and to movement of the joint. The receptive fields of the cells were smaller than those observed for the dorsal horn cells. No signs were seen of an inhibitory surround about these cells' receptive areas when the skin was tested by application of mild pressure, but, as in the case of the dorsal horn cells, massive heavy stimulation of the surroundings did reduce the number of impulses evoked from the excitatory area of skin.

Discussion

We have described two types of cells, one in the dorsal horn and the other in the dorsal column nuclei, both receiving direct afferents from the skin, but with extremely different properties. We suggest that the main reason for the difference in their properties is that only a small part of the available spectrum of peripheral nerve fibers ends on the dorsal column nuclei. Since only peripheral fibers of the largest diameter send branches up to the dorsal column nuclei, these cells can process only the type of information transmitted along the largest fibers. These fibers respond to small distortions of the skin and adapt rapidly. In contrast, the dorsal horn interneurons respond to all types of skin stimuli. The simplest explanation, since we know that microbundles of large fibers converge onto these cells, is that these microbundles contain samples of all other types of fiber from the skin area that the cell subserves. Although we have presented evidence that the large fibers end on these cells, we have no definite evidence that the smaller fibers end directly on them, and in fact it is possible that some of the smaller fibers project onto these cells by way of the substantia gelatinosa whose function remains a mystery. It is known that the dendrites of the cells we are considering extend up into the region of the substantia gelatinosa, and the anatomy of the axons of substantia gelatinosa cells is still largely unknown because of their extremely small diameter. We can, however, conclude that the primary difference between the two relay nuclei is decided by the anatomy and embryology of the afferent nerve fibers ending on them.

Turning now to a discussion of the function of these cells, there is little doubt that the dorsal column—medial lemnicus system is a true sensory pathway, although it is evident that it can handle only a restricted repertory of the sensory modalities. In contrast, the dorsal horn interneurons, which have been studied here, apparently respond to all types of skin stimuli to which the animal is known to respond. Examples of these modality-convergent central systems are now multiplying. There are two possible explanations for their function. In the first, one would assume that these cells are involved in a stimulus-response system in which modality is largely disregarded, such as the flexor reflex, the startle response, some autonomic responses, or a "nod is as good as a wink" system. The other explanation is that these are "common-sense" cells in which the modality of the stimulus, which has given rise to the over-all increase of activity, is preserved in the temporal pattern of the impulses. At present we are inclined to this latter view, partly because of our inability to find other primary central cells that preserve the specificity of modalities, which has been claimed for the peripheral nerve fibers. However, one must emphasize that the restrictions of present-day neurophysiological techniques do not allow an exhaustive search to be made of a region; consequently the function of all cells in the region cannot be described.

One reason for believing that cells of the dorsal-horn interneuron type might be involved in the sensory pathways of the skin of man comes from the comparison of results obtained on the cat neurons with sensation derived from a vibrated area of skin in man (Wall and Cronly-Dillon, 1960). We have shown above that, during continuous vibration of the skin, the response of these cells is partly inhibited, so that a much larger stimulus of any type is required to drive the cell into high-frequency discharge. If these cells were involved in a sensory pathway, one would expect that vibration of the skin would affect their ability to transmit information about other modalities. It is found in man that light skin vibration raises the threshold of the vibrated area for touch, pain, and temperature. An interaction of the various modalities undoubtedly takes place at all points in the nervous system from the periphery to the highest centers. However, the confirmation of the predictions derived from observations of the cat's primary central neurons adds some weight to the suggestion that some parts of our sensory pathways may contain common-sense cells.

If cells in a sensory pathway respond to all sensory modalities, and yet the behavior of the animal is modality-specific, there would have to be cells that responded to certain aspects of the discharge pattern in the common-sense cells and to the number of cells active. It is necessary in such a system to postulate cells that can "read" the code contained within the temporal pattern of impulses. It will be necessary to search for such cells, using natural stimuli or artificial stimuli designed to imitate certain aspects of the pattern. It is clear that a sudden electrical stimulus applied to a mixed nerve would give no useful information, since the massive synchronized volley would contain aspects of every possible stimulus pattern, with the result that any selective properties of the transmission system would be overwhelmed by such an onslaught. The results of experiments using electrical stimuli, or even such abrupt "natural" stimuli as clicks of sound or flashes of light, may be used to establish the presence of connections between parts of the nervous system. However, if properties of connections are sought rather than simple anatomy, it is evident that the system must be tested with stimuli that the system has been evolved to handle. Cells have been observed in the dorsal horn which responded to electrical stimuli to the largest peripheral nerve fibers and might therefore be assumed to be involved in the light-touch transmission pathway. However, touch to the skin failed to evoke responses in these cells, and only heavy distortion of the skin stimulated them. Presumably, the electrical stimulus had bombarded the cells with an adequate volley, owing to its synchronization, whereas the natural light pressure stimuli were not able to produce this adequate pattern.

The suggestion that modality of skin sensation may be transmitted by the pattern of the afferent barrage has been made by Weddell and his co-workers (1955). Their reasons for making these suggestions are quite indirect, owing largely to their inability to confirm that specialized endings must exist in skin sensitive to the various modalities of sensation. They also found signs of fibers sensitive to both pressure and temperature, as had Hensel (1951).

Since these reports, the specialization of the peripheral nerve fibers has been further reduced by the findings of Iggo (1959) that the C fiber group of small fibers includes some that respond to small distortions of the skin. However, even if there are no specialized endings, and even if some fibers respond to stimuli of more than one modality, it is somewhat premature to conclude that there are no specialized fibers. The alternative view expressed here is that the separation of nerve fibers of the peripheral skin into specialized groups may have originated from inadequate samples. The specialized groups may have been artificially separated out from a continuous distribution. It must be emphasized, however, that this is only an extension of the previously held views and in no way suggests a lack of specialization in peripheral nerve fibers. The point to be made here is that a group of central cells has been found which sums the properties of different types of peripheral nerve.

There are two consequences of suggesting a common-sense type of cell in a sensory pathway. First, the separate temporal patterns produced by the different modalities of the stimulus must remain separate, no matter how the stimulus is applied. In criticism of Weddell, Iggo (1959) has objected that sensation continues long after the application of the stimulus when adaptation may have erased the differences between patterns. In the instances shown here, this is obviously not the case, especially since the past history of a stimulus is perceived as well as the situation at the moment of sensation.

A large amount of information is contained in the available types of activity, that is to say, the temporal pattern of activity in single cells since the onset of the stimulus, the number of cells active, and the relative activity of cells in the two different systems. Where different stimuli produced a similar pattern, the cells onto which the common carrier cells project would be unable to differentiate between these similar patterns. Autonomic cells of the cord, for example, react to both vibration and heating of the skin. High temperatures above 45° C and high distorting pressures to the skin produce similar patterns of discharge in the spinal interneurons of cat and similarly painful sensations in man.   Thus in a sensory system depending on common-sense cells, we may expect to find interaction between the modalities, points of confusion, and particular stimuli producing illusions.   The second consequence of a common-sense system is that cells must exist which will react only to certain aspects of the pattern of bombarding impulses.   We know already of such cells in the motor system and of the classical concept of the "adequate stimulus."   Now they must be sought in the sensory transmission pathways.

The final point for discussion relates to the nature of sensation, and it is best discussed in relation to the findings on itch.   It has been shown that the dorsal horn cells may produce impulses in short bursts interspersed with inactivity.   After a fast burst of impulses the cells are inhibited, and it is suggested that these cells have an inhibitory mechanism playing back onto them as a consequence of their own activity.   If the afferent barrage is steadily increased, the cells will also increase their output, but a level of input may be reached at which the output alternates between following the input and being inhibited by the feedback.   If the input is further increased, the output will increase further because the inhibition is overridden.   This means that a particular section of the continuously variable steady input has been translated into a characteristic pattern.   This pattern, it is suggested, is the basis of the sensation of itch.   If this is true, it implies that the possibility of a separate sensation called itch is the consequence of the anatomy and physiology of the relay nucleus, rather than the presence of specialized peripheral receptors.   Filtering properties of this subtle order may exist in other sensory relay nuclei and clearly would determine the nature of what can be sensed.

Summary

Microelectrode studies have been made of two groups of cells, one in the dorsal horn of the lumbar region of the spinal cord, and the other, the nucleus gracilis.   Both these groups receive direct afferent fibers from the skin.   A comparison is made of these two types of cells involved in the transmission of information from the skin.

1. The cells act as amplifiers.   The origin of the amplification is discussed.

2. The cells act as points of convergence of fibers from different parts of the skin.   Each cell subserves a larger area of skin than any

one afferent fiber. The anatomy of this convergence is shown to depend on the presence of microbundles of fibers.

3. The cells act as points of convergence of fibers of different types: the cells of nucleus gracilis respond only to the largest fibers in the skin nerves; the dorsal horn cells respond as though all types of skin fibers terminated on them.

4. The temporal impulse pattern of activity of the cells on which many types of fiber converge differs, depending on the nature of the stimulus. It is suggested that subsequent cells can differentiate certain aspects of this pattern.

5. The output pattern of impulses is a consequence of two factors, the nature of the afferent fibers converging on the cells and the interconnections between the cells. These factors determine what may be sensed by subsequent stages in the nervous system.

Acknowledgments

This work is supported in part by the U. S. Public Health Service, the Teagle Foundation, and the Bell Telophone Laboratories, and in part by the U. S. Army (Signal Corps), the U. S. Air Force (OSR), and the U. S. Navy (ONR).

The author wishes to thank Drs. McCulloch, Lettvin, and Pitts for their help, advice, and encouragement.

References

Adrian, E. D. *The Basis of Sensation.* London: Christophers, 1928.

Amassian, V. E., and J. L. DeVito. La transmission dans le noyau de Burdach (Nucleus cuneatus). *Colloq. int. Centre nat. Rech. sci.*, 1957, No. 67, 353–393.

Barron, D. H., and B. H. C. Matthews. The interpretation of potential changes in the spinal cord. *J. Physiol.*, 1938, **92**, 276–321.

Frank, K., and M. G. F. Fuortes. Unitary activity in spinal interneurones of cats. *J. Physiol.*, 1956, **131**, 424–435.

Glees, P., and J. Soler. Fibre content of the posterior columns and synaptic connections of nucleus gracilis. *Z. Zellforsch.*, 1951, **36**, 381–400.

Hensel, H. The response of mechanoreceptors to thermal stimulation. *J. Physiol.*, 1951, **115**, 16–24.

Hursh, J. B. Relayed impulses in ascending branches of the dorsal root fibers. *J. Neurophysiol.*, 1940, **3**, 166–174.

Iggo, A. A single unit analysis of cutaneous C afferent fibres. In G. E. W. Wolstenholme and M. O'Connor (Editors), *Pain and Itch.* Ciba Study No. 1. Boston: Little, Brown, 1959.

Lloyd, D. P. C., and A. K. McIntyre. Dorsal column conduction of group I muscle afferent impulses and their relay through Clarke's column. *J. Neurophysiol.*, 1950, **13,** 39–54.

Lorente de Nó, R. Transmission of impulses through cranial nerve nuclei. *J. Neurophysiol.*, 1939, **2,** 402–464.

Maruhashi, J., K. Mizuguchi, and I. Tasaki. Action currents in single nerve fibres elicited by stimulation of the skin of the toad and cat. *J. Physiol.*, 1952, **117,** 129–151.

Morrell, R. M., K. Frank, M. G. F. Fuortes, and M. C. Becker. Site of origin of motoneurone rhythms. *XX Physiol. Congr. Abstr.*, 1956, 660–661.

Therman, P. O. Transmission of impulses through the Burdach nucleus. *J. Neurophysiol.*, 1941, **4,** 153–166.

Wall, P. D. Repetitive discharge of neurons. *J. Neurophysiol.*, 1959, **22,** 305–320.

Wall, P. D. Cord cells responding to touch, damage and temperature of the skin. *J. Neurophysiol.*, 1960, **23,** 197–210.

Wall, P. D., and J. Cronly-Dillon. Pain, itch and vibration. *AMA Arch. Neurol.*, 1960, **2,** 365–375.

Wall, P. D., W. S. McCulloch, J. Y. Lettvin, and W. H. Pitts. The terminal arborisation of the cat's pyramidal tract determined by a new technique. *Yale J. Biol. Med.*, 1956, **28,** 457–464.

Weddell, G. E., E. Palmer, and W. Pallie. The morphology of peripheral nerve terminations in the skin. *Biol. Rev.* 1955, **30,** 159–195.

Zotterman, Y. Touch, pain and tickling: an electrophysiological investigation of cutaneous sensory nerves. *J. Physiol.*, 1939, **95,** 1–28.

26

RAÚL HERNÁNDEZ-PEÓN

Brain Research Unit, Medical Center of Mexico City

Reticular Mechanisms
of Sensory Control

In contrast with the great amount of work that has been devoted in the past to the study of the control of motor outflow, our knowledge about the regulation of sensory input is a recent neurophysiological acquisition. The traditional concept, which assumed that all the impulses generated at the receptors are simply relayed to the cerebral cortex for their final integration, has been shaken by the discovery of centrifugal mechanisms that modify the transmission of sensory impulses at all the levels of the specific afferent pathways. From time to time, descending fibers coursing along with specific afferents have been pointed out by anatomists and neglected by physiologists. In addition to the specific afferent pathways containing ascending and descending fibers, a great deal of anatomical and physiological evidence has revealed within the brain the existence of "nonspecific" sensory systems where impulses of various sensory modalities converge. An important part of this recently recognized polysensory system lies in the central core of the brain stem, and it is usually referred to as the *reticular system*. The participation of this central region in the regulation of the activity of all sensory paths has become increasingly evident in the last few years and is the subject of the present paper. The data that I propose to review will be derived mainly from some of our studies relevant to the topic, including part of the most recent experiments.

Effects of Reticular Stimulation upon the Activity of Sensory Neurons

At the spinal cord

In 1954 Hagbarth and Kerr found in curarized cats that electrical stimulation of the brain-stem reticular formation blocked postsynaptic spinal afferent volleys from lumbar dorsal roots. This unexpected finding was the first demonstration that the reticular system is able to influence the transmission of sensory messages within the central nervous system as far down as the first synapse. The blocking interaction between centripetal and centrifugal impulses was interpreted by those authors as resulting from true inhibition of postsynaptic neurons. This interpretation is strongly supported by recent microelectrode studies of spinal sensory units. Indeed, Hagbarth and Fex (1959) have found that electrical stimulation of various central structures, such as the brain stem, the cerebellum, and the sensorimotor cortex, either inhibits or increases the firing of spinal postsynaptic sensory units.

Spinal visceral afferent responses evoked by single shocks applied to the splanchnic nerve have also been found to be abolished by stimulation of the bulbar reticular formation (Tolle, Feldman, and Clemente, 1959).

Since the spinal neurons tested represent only one group of second-order sensory neurons, it is of utmost interest to know whether other second-order neurons in the somatosplanchnic path are also influenced by reticular stimulation. This possibility has been explored independently at the following sensory nuclei: the spinal fifth, the gracilis, and the cuneatus.

At the spinal fifth sensory nucleus

By recording, from the spinal trigeminal sensory nucleus, the synchronized afferent volleys evoked by stimulating the infraorbital nerve, Hernández-Peón and Hagbarth (1955) were able to observe a definite reduction in the size of the nuclear potential during electrical stimulation of the brain-stem reticular formation or of the sensorimotor cortex. More recently, in cats with electrodes permanently implanted, in collaboration with Davidovich and Miranda, I have confirmed and extended the results. Electrical stimulation of the mesencephalic reticular formation diminished the postsynaptic potential recorded from the spinal fifth sensory nucleus and evoked by tactile or nociceptive stimuli delivered to the face, *pari passu* with the behavioral arousal reaction. In contrast with the modification of the postsynaptic wave, the presynaptic spike of the evoked potential remained unchanged (Fig. 1).

At the gracilis and cuneatus nuclei

In confirmation of the hypothesis that presumes that this descending influence is common to all the initial relays of the somatosplanchnic

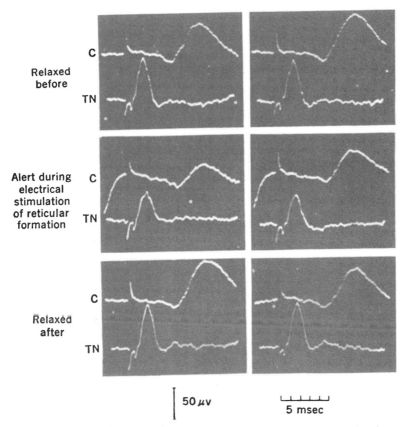

50 μv

5 msec

Fig. 1. Effects of electrical stimulation of the mesencephalic reticular formation upon the evoked potentials recorded simultaneously from the spinal fifth sensory nucleus (TN) and from the face area of the somatic sensory cortex (C). The potentials were evoked by single shocks of nociceptive intensity applied to the face. Both the cortical and bulbar nuclear potentials were clearly reduced during alertness induced by brief reticular stimulation (2 sec), whereas the presynaptic spike of the descending trigeminal root remained unchanged.

path, it was found that brief electrical stimulation of the mesencephalic tegmentum or sensorimotor cortex inhibited the postsynaptic activity of the gracilis nucleus evoked by single shocks applied directly to the dorsal columns of the spinal cord (Hernández-Peón, Scherrer, and Velasco, 1956). However, the primary spike representing the activity of first-order sensory neurons was unaffected. In experiments carried out by the writer in 1955 (Hernández-Peón, 1955), the splanchnic evoked potentials recorded from the gracilis nucleus were also diminished by stimulation of the mesencephalic reticular formation.

Dawson (personal communication), recording sensory evoked potentials from the cuneatus nucleus, confirmed the inhibitory effects of cortical and reticular stimulation.

From all the above-mentioned results, it seems warranted to conclude that activation of the brain-stem reticular formation inhibits sensory transmission at the first synapses of the somatosplanchnic path.

At the cochlear nucleus

In preliminary experiments cited by Hernández-Peón, Scherrer, and Jouvet (1956), Jouvet, Berkowitz, and Hernández-Peón observed a

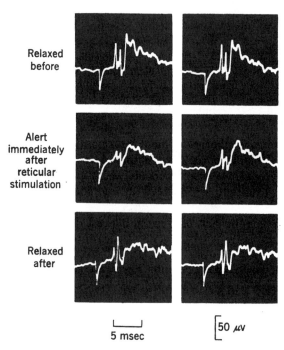

Fig. 2. Effects of electrical stimulation of the mesencephalic reticular formation upon the auditory potentials recorded from the dorsal cochlear nucleus in a cat with electrodes permanently implanted. The evoked potentials were reduced significantly during the alertness induced by brief reticular stimulation (2 sec).

definite reduction of auditory evoked potentials recorded from the dorsal cochlear nucleus during repetitive electrical stimulation of the mesencephalic tegmentum. Later, however, Jouvet and Desmedt

(1956) and Desmedt and Mechelse (1958) found inhibition of the cochlear-nucleus potentials only from stimulation of a region located laterally within the ascending auditory pathway. In contrast with these negative results, Killam and Killam (1958, 1959) reported that electrical stimulation of the brain-stem reticular formation inhibited the auditory potentials recorded from the cochlear nucleus, and that the inhibition was intensified by chlorpromazine. Under the action of this drug, the threshold of reticular stimulation was lowered, and the duration of the inhibitory effect at the cochlear nucleus was prolonged.

The reticular influence upon acoustic volleys recorded from the dorsal cochlear nucleus has more recently been confirmed in our laboratory (Bach-y-Rita et al., 1961). It was observed that, in cats with electrodes permanently implanted, brief electrical stimulation of the mesencephalic reticular formation elicited diminution of the auditory potentials, together with behavioral alertness not oriented to the acoustic stimulus (Fig. 2).

It seems as though the transmission of auditory impulses at the level of the cochlear nucleus is under the control of a complex descending system of fibers, and it is likely that a subtle functional organization will be found in the origin as well as in the termination of those descending fibers that end around the cells of the first auditory relay. Aside from differences in experimental techniques, the complexity of such an anatomical arrangement might explain the apparently contradictory results mentioned above.

At the retina

There is evidence that the spontaneous and the evoked retinal activity can be centrifugally altered by activity in the reticular system of the brain stem. In 1955 Granit found that electrical stimulation of the mesencephalic reticular substance elicited posttetanic potentiation or inhibition of single retinal ganglion cells. By recording photically evoked potentials from the optic tract and therefore from a population of axons of retinal ganglion cells, Hernández-Peón, Scherrer, and Velasco (1956) confirmed Granit's finding. These authors observed either potentiation or diminution of the retinal potentials following electrical stimulation of the mesencephalic or pontine reticular formation. It was not clear, from either of the studies mentioned above, which conditions were responsible for each of these opposite effects. As suggested by Hernández-Peón, Scherrer, and Velasco (1956) it is possible that the two different retinal effects elicited by reticular stimulation might be the result of predominant activation of specific inhibitory or facilitatory centrifugal fibers with different modes of termination at the retina.

At the olfactory bulb

The olfactory pathway is also under the control of the arousal system of the brain stem, as has been recently demonstrated. Although it was known that electrical stimulation of intralaminar thalamic nuclei (Arduini and Moruzzi, 1953) and of basal rhinencephalic structures (Kerr and Hagbarth, 1955) influences the electrical activity of the olfactory bulb, evidence regarding direct reticular control of

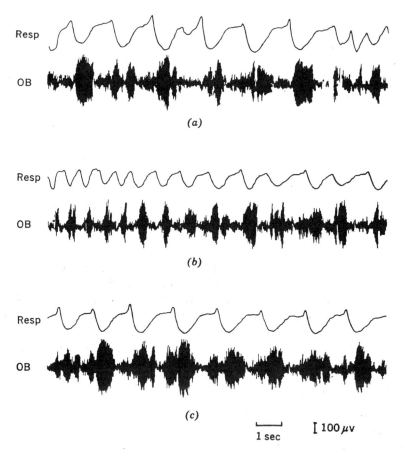

Fig. 3. Simultaneous recordings of the respiratory movements (Resp) and the electrical activity from the olfactory bulb (OB) in an alert cat with electrodes permanently implanted. The bursts of the rhythmic activity in the olfactory bulb diminished during sniffing (first part of tracing *b*).

the first olfactory synapse was lacking until the experiments of Lavín, Alcocer-Cuarón, and Hernández-Peón (1959). In freely moving cats, electrodes implanted in the olfactory bulb recorded bursts of rhythmic activity (34 to 48 per second) whenever the sleeping or relaxed awake animal was alerted by any kind of sensory stimulation. However, these "arousal discharges" decrease in amplitude during olfactory stimulation (Fig. 3). The same activating effect can be induced by direct electrical stimulation of the mesencephalic reticular formation *pari passu* with the well-known behavioral arousal response. Further evidence about the reticular participation in the centrifugal activation of the olfactory bulb has been provided by lesion experiments (Hernández-Peón, 1960; Hernández-Peón, Lavín, et al., 1960). In fact, lesions in the mesencephalic tegmentum or in the vicinity of the anterior commissure prevented the appearance of the olfactory "arousal discharges" otherwise elicited by sensory stimuli.

At the specific thalamic nuclei

The excitability of specific thalamic as well as second-order sensory neurons is also influenced by activity in the brain-stem reticular formation. Hernández-Peón, Scherrer, and Velasco (1956) recorded simultaneously photically evoked potentials from the lateral geniculate body and from the optic tract of cats immobilized by curare. They observed that, during and following electrical stimulation of the mesencephalic or pontine reticular formation, the thalamic responses were depressed, even though the retinal potentials were facilitated. In freely moving cats with electrodes permanently implanted in the lateral geniculate body and in the mesencephalic reticular formation, Hernández-Peón, Guzmán-Flores, et al. (1957) observed diminution of photic geniculate potentials during the behavioral arousal induced by brief electrical stimulation of the reticular formation. Depression of these potentials has been more recently confirmed by Bremer and Stoupel (1959) in the *encéphale isolé* preparation. In addition, our Belgian colleagues have shown that the same reticular stimulation elicited a remarkable facilitation of the geniculate volleys evoked by an electric shock applied to the optic nerve, and they interpreted both kinds of effects as resulting exclusively from excitatory interactions. However, recent microelectrode studies by Arden and Söderberg (1959) have provided evidence of inhibition of the lateral geniculate cells from extraretinal sources. Arden and Söderberg found that, in the animal with intact eyes, whistling and other noises that desynchronized the EEG increased the resting discharges of many neurons of the lateral geniculate nucleus. But when the optic nerve was blocked, the same acoustic stimuli, as well as nociceptive stimuli, brought about an abrupt decrease in the firing rate of lateral geniculate cells, associated with the appearance of fast activity in the EEG. These results indicate the existence of inhibitory afferents to the lateral geniculate cells which do not arise in the retina. It is quite probable that these inhibitory fibers originate in a central region of sensory convergence related to arousal. Indeed, the recent experiments of Söderberg reported at this symposium have demonstrated beautifully that the activity of lateral geniculate neurons is inhibited by direct electrical stimulation of the brain-stem reticular formation and is enhanced by transection of this structure.

The influence of reticular stimulation on transmission at the specific somatic thalamic nuclei was disclosed by King, Naquet, and Magoun (1957). They reported that the facilitatory interaction between two consecutive responses in the internal capsule evoked by stimulation of the medial lemniscus was depressed during the EEG arousal elicited by reticular stimulation. The same interaction was enhanced during EEG synchronization following brain-stem lesions. More recently, Appelberg, Kitchell, and Landgren (1958) have shown that reticular stimulation and sensory arousal reduced or abolished the negative component of the thalamic potentials evoked from the tongue or from the dorsal ascending trigeminal tract, and that often the negative component was replaced by a positive potential.

At the cortex

Since the original discovery of Moruzzi and Magoun (1949) of the desynchronizing action of the brain-stem reticular formation upon cortical electrical activity, it has been known that reticular stimulation abolishes the after-discharges of sensory cortical evoked potentials. In 1954, Bremer, using his *encéphale isolé* preparation, observed a reduction of the primary auditory evoked potentials during arousal. The blocking effect of arousal upon cortical evoked potentials has been widely confirmed and extended to various sensory pathways both in acute experiments and, under more physiological conditions, in animals with electrodes permanently implanted. In a recent study on the effects of reticular stimulation upon cortical evoked potentials, Bremer and Stoupel (1959) have disclosed two different effects depending on the testing stimulus. Whereas the potential evoked by a stimulus applied to the receptors (flash, click, or cutaneous shock) was always reduced or abolished, the cortical response elicited by an electric shock applied to the central pathway (optic nerve or thalamic nuclei) was remarkably facilitated. Dumont and Dell (1958) have also observed facilitation of the cortical potential evoked by a shock applied to the optic chiasma when the mesencephalic reticular formation was stimulated. Thus it appears that reticular stimulation can produce both facilitatory and blocking interactions at the cortical level, just as has been observed at lower stations of the sensory pathways.

Reticular Tonic Action

Tonic actions within the central nervous system are disclosed by release methods. An indication of tonic centrifugal inhibition and facilitation in sensory systems has been provided by the effects on the awake brain of central anesthetics and localized lesions.

Effects of central anesthetics

Ever since the classic paper of Derbyshire, et al. (1936), neurophysiologists have known that cortical evoked potentials are more prominent during deep than during light anesthesia. In line with traditional thinking, this paradoxical effect was misinterpreted for many years: the evoked signal was thought simply to stand out better against the decreased background activity. Actually, if one measures carefully the absolute amplitude of the cortical evoked potentials, there is no doubt that a real enhancement is induced by the anesthetics.

The exploration of subcortical levels of specific afferent pathways has revealed that the enhancement of sensory signals during anesthesia also occurs at thalamic stations and as far down as the first sensory synapse. Hagbarth and Kerr (1954) reported that the spinal afferent volleys from lumbar afferents are increased during pentobarbital or chloralose anesthesia. Hernández-Peón, Scherrer, and Velasco (1956) observed similar enhancement of evoked potentials during pentobarbital anesthesia at the spinal fifth sensory nucleus, the gracilis nucleus, and the lateral geniculate body. Hagbarth and Höjeberg (1957) have confirmed in human studies that the somatic evoked potentials recorded from the white substance are larger during deep than during light anesthesia. The amplitude and duration of the rhythmic activity induced in the olfactory bulb by a puff of air applied to the olfactory mucosa in awake cats is also increased during barbiturate anesthesia (Fig. 4).

All these observations indicate that the central anesthetics employed release inhibitory influences which act tonically during wake-fulness upon those lower sensory stations. But in addition to their releasing effects upon sensory inhibition, barbiturates eliminate centrifugal facilitation of certain sensory neurons, as was demonstrated in the olfactory bulb by Hernández-Peón, Lavín, et al. (1960) and in the spinal cord by Hagbarth and Fex (1959).

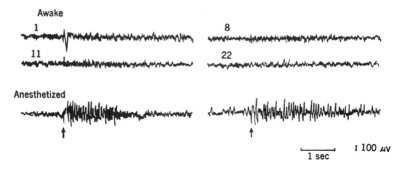

Fig. 4. Effects of barbiturate anesthesia upon the electrical activity of the olfactory bulb induced by olfactory stimulation (puffs of odorized air delivered through a catheter permanently implanted in the nasal cavity). In contrast to the high-voltage olfactory discharge induced in the anesthetized cat, the low-voltage activity induced in the awake cat progressively declined during 22 repetitions of the olfactory stimulus.

If there are tonic inhibitory and facilitatory influences upon sensory transmission, which can be impaired by pharmacological actions, the question arises concerning the origin and trajectory of those centrifugal influences. Although anatomical evidence is meager, some lesion experiments point out the importance of the participation of the mesencephalic reticular formation in the centrifugal mechanisms described for sensory control.

Effects of central lesions upon sensory activity at the spinal fifth sensory nucleus

The finding that certain central anesthetics released descending tonic inhibitory influences upon sensory synapses led naturally to the question, which are the central structures involved in the centrifugal activity? An initial approach to this question was undertaken by Hernández-Peón and Scherrer (1955a), who found that lesions in the mesencephalic tegmentum, which produced EEG patterns of sleep, enhanced the spinal trigeminal postsynaptic potential evoked by single shocks applied to the infraorbital nerve.

Effects of central lesions upon the electrical activity of the olfactory bulb

As has been mentioned before, the "arousal discharges" recorded from the olfactory bulb in the alert cat were abolished by lesions in the septal region or in the mesencephalic reticular formation which rendered the cat unconscious. On the other hand, the same lesions

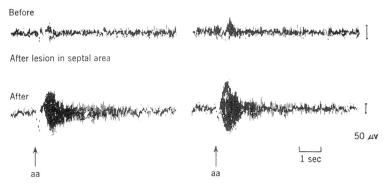

Before

After lesion in septal area

After

50 μv

1 sec

aa   aa

Fig. 5. Enhancement of the induced activity in the olfactory bulb by a lesion in the septal area. The activity was evoked by puffs of air odorized with amyl acetate (aa) and delivered through a polyethylene tube permanently implanted in the nasal cavity of an awake cat with permanently implanted electrodes in the olfactory bulb.

enhanced the electrical activity induced in this peripheral structure by olfactory stimulation (Fig. 5). These results indicate that the elements of the olfactory bulb involved in the arousal discharges are different from those directly activated by olfactory stimulation. It may also be concluded that the mesencephalic reticular formation exerts a double centrifugal influence (facilitatory and inhibitory) which constantly modifies the excitability of peripheral olfactory neurons.

Effects of spinal section upon spinal afferent volleys in decerebrate cats

Although the lesion experiments described here have provided un-questionable evidence about the essential role of the brain-stem reticular formation in the tonic centrifugal control of lower sensory synapses, they do not permit any conclusions to be drawn about the source of those influences. Indeed, it may be that the lesions have interfered not with the site of origin but with the path of the cen-trifugal influence. This question has been further clarified by successive elimination of the higher parts of the central nervous system which might be concerned with the descending mechanisms of sensory control. Hernández-Peón and Brust-Carmona (1961) have shown that a high spinal transection (at C 2) remarkably enhanced tactile evoked potentials recorded from the lateral column of the

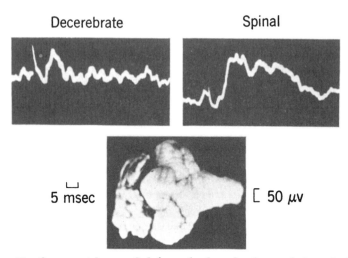

Fig. 6. Tactile potentials recorded from the lateral column of the spinal cord and evoked by single electric shocks of mild intensity applied to the skin. The left tracing was obtained in a cat with the brain stem transected in the rostral part of the mesencephalon, and the right tracing was recorded shortly after transection of the spinal cord in the upper cervical region.

spinal cord in decerebrate cats (Fig. 6). In line with these results, Holmqvist, Lundberg, and Oscarsson (1959) have also reported that the evoked discharges recorded from four different ascending spinal pathways in decerebrate cats are enhanced after section of the spinal cord. Their experiments have further revealed that the descending tracts responsible for the supraspinal tonic inhibition are located in the dorsal part of the lateral funiculus. Obviously, this tonic descending inhibition arises in the mesencephalo-rhombencephalic level of the brain.

Filtering of Sensory Impulses during Attention and Habituation

Sensory inhibition and facilitation in intact cats

The functional significance of the reticular mechanisms that control the sensory inflow to the brain has been disclosed by correlated behavioral and electrophysiological studies in animals with electrodes permanently implanted in the sensory paths.

It has been discovered that the amplitude of sensory evoked potentials, recorded along the specific afferent pathways from the first synapse up to the cortex, varies according to the degree of attention to, or distraction from, the tested sensory stimulus. Thus, auditory evoked potentials from the dorsal cochlear nucleus are reduced when the cat is attentive to visual, olfactory, or somatic stimuli (Hernández-Peón, Scherrer, and Jouvet, 1956). Retinal potentials evoked by flashes of light and recorded from the optic tract have also been observed to diminish when the cat seemed attentive to acoustic (Hernández-Peón, Guzmán-Flores, et al., 1957), olfactory, or somatic stimuli and when the cat focused its attention upon a rat (Palestini, Davidovich, and Hernández-Peón, 1959). By the same token, tactile evoked potentials recorded from the spinal fifth sensory nucleus (Hernández-Peón, 1959; Hernández-Peón, Lavín, et al., 1960) and from the lateral column of the spinal cord (Hernández-Peón and Brust-Carmona, 1961) are reduced or abolished when the cat is attentive to visual, acoustic, olfactory, or somatic stimuli (Fig. 7).

Another instance of inattention promoted by repetition of a given stimulus which thus becomes meaningless to the individual is known as *habituation*. There is a great deal of evidence that the development of this type of "negative learning" in the relaxed awake animal is associated with an oscillating but progressive decline of the corresponding afferent volleys recorded from second-order sensory neurons, as well as from thalamic and cortical levels (Hernández-Peón, 1960). I have proposed the term *afferent neuronal habituation* for the electrophysiological correlates of habituation observed in awake cats along the specific sensory pathways. There is evidence of habituation to auditory stimuli at the dorsal cochlear nucleus (Hernández-Peón and Scherrer, 1955b; Hernández-Peón, Jouvet, and Scherrer, 1957; Galambos, Sheatz, and Vernier, 1956); of photic habituation at the retina (Palestini, Davidovich, and Hernández-Peón, 1959) as well as at the lateral geniculate body and the visual cortex (Hernández-Peón, Guzmán-Flores, et al., 1956, 1958; Ricci, Doane, and Jasper, 1957; John and Killam, 1959; Cavaggioni, Giannelli, and Santibañez, 1959); of olfactory habituation at the olfactory bulb (Hernández-Peón, Alcocer-Cuarón, et al., 1957); and of tactile habituation at the spinal fifth sensory nucleus (Hernández-Peón, 1960; Hernandez-Peón, Lavín, et al., 1960) and at the lateral column of the spinal cord (Hernández-Peón and Brust-Carmona, 1961). A remarkable observation common to all these experiments is that afferent neuronal habituation is released and prevented by barbiturate anesthetics.

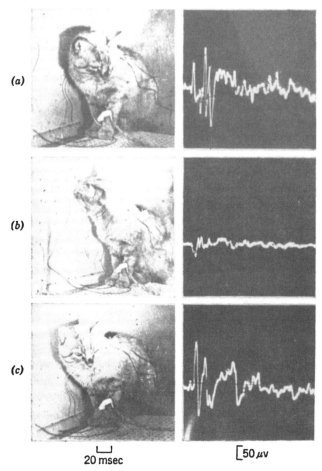

Fig. 7. This figure illustrates the blockade of tactile evoked potentials recorded from the lateral column of the spinal cord during attention elicited by and focused upon an olfactory stimulus. The records were taken (*a*) when the cat was relaxed; (*b*) when the cat was attentively sniffing fish odor delivered through a tube into the cage; and (*c*) when the cat was relaxed again shortly after the odor was removed.

Although the sensory impulses generated by stimuli out of the span of attention are blocked, the afferent messages elicited by the stimuli attended to are facilitated. Selective focusing of attention on a given stimulus can be induced experimentally by Pavlovian conditioning, and thus the changes in the potentials evoked by the conditional

stimulus have been studied.   Facilitation of sensory transmission in-
duced by conditioning has been observed at the cochlear nucleus
(Hernández-Peón, Jouvet, and Scherrer, 1957; Galambos, Sheatz, and
Vernier, 1956), at the retina (Palestini, Davidovich, and Hernández-
Peón, 1959), and at the olfactory bulb (Hernández-Peón, Alcocer-
Cuarón, et al., 1957), as well as in widespread subcortical and cortical
regions (Hernández-Peón, 1957).

Sensory inhibition in decorticate cats

Since cortical activation can produce inhibition of lower sensory
relays, just as direct reticular stimulation does, it is convenient to
assess which is the functional role of the corticifugal mechanisms
in the electrophysiological events observed during attention and
habituation.   This question has been approached experimentally by
Hernández-Peón and Brust-Carmona (1961), who have observed
habituation of tactile evoked potentials from the lateral column of
the spinal cord in cats with chronic bilateral removal of the neocortex
(Fig. 8).   Although the stability and selectivity of attention are impaired by decortication, these experiments have shown that the cortex
is not necessary for the establishment of afferent neuronal habituation,
which is one of the simplest types of learning.

Sensory inhibition after interruption of specific pathways

It is natural that, in describing the functional role of the descending fibers coursing along with the ascending afferents of the specific
sensory pathways, some authors suggest that those descending fibers
may be responsible for the sensory suppression observed during distraction and habituation.   However, this hypothesis is not supported
by recent experimental findings.   Bach-y-Rita et al. (1961) have
clearly shown that, in cats with bilateral interruption of the auditory
pathway at the mesencephalon, the auditory evoked potentials from
the dorsal cochlear nucleus are reduced by habituation and may be
reduced also when the cat is distracted by stimuli of various sensory
modalities (Fig. 9).

Sensory inhibition after reticular lesions

Since lateral lesions in the mesencephalon do not prevent the sensory
inhibition at the cochlear nucleus elicited by distraction and habituation, it seems warranted to conclude that these centrifugal effects
are most probably mediated by the central network of the brain stem.
Indeed, the original experiments of Hernández-Peón and Scherrer

(1955*b*) and Hernández-Peón, Jouvet, and Scherrer (1957) have shown that an extensive lesion of the mesencephalic reticular formation released and prevented acoustic habituation at the cochlear nucleus. By the same token, mesencephalic tegmental lesions have also been found to release and prevent habituation of olfactory in-

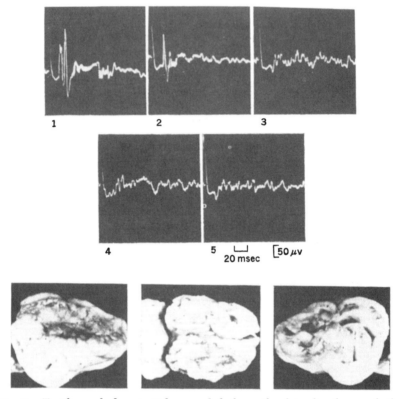

Fig. 8. Tactile evoked potentials recorded from the lateral column of the spinal cord in a cat with chronic bilateral removal of the neocortex. The mild cutaneous shocks were delivered at intervals of 10 sec.

duced activity in the olfactory bulb (Hernández-Peón, 1960). In these experiments, lesions involving the pontine tegmentum did not prevent habituation of those olfactory neurons. It must be pointed out that lesions in the same areas have an opposite effect upon vestibular habituation. Whereas pontine lesions prevent habituation of postrotatory nystagmus, mesencephalic lesions and complete transection of the brain stem at the mid-collicular level do not

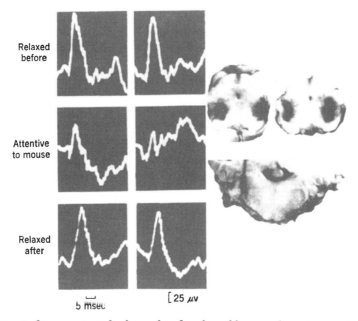

Relaxed
before

Attentive
to mouse

Relaxed
after

5 msec

[25 µv

Fig. 9. Auditory potentials from the dorsal cochlear nucleus in a cat with bilateral electrolytic lesion of the lateral part of the mesencephalic tegmentum. A clear reduction of the evoked potentials was observed when the cat was looking attentively at a mouse, just as happens in the intact cat.

prevent habituation to rotation (Hernández-Peón and Brust-Carmona, 1961).

The results mentioned above indicate that the different areas of the reticular formation do not participate to the same extent in the development of habituation in the different sensory pathways and suggest that the anatomical and functional organization of the brain-stem reticular formation extends to the centrifugal systems of sensory control.

Sensory inhibition in decerebrate cats

Further evidence concerning the unessential role of supramesencephalic structures for habituation of lower sensory stations has been recently obtained in my laboratory. Bach-y-Rita, et al. (1960) have observed habituation of auditory evoked potentials recorded from the dorsal cochlear nucleus in cats with the brain stem transected above the mesencephalon (Fig. 10). In similar mesencephalic preparations Hernández-Peón and Brust-Carmona (1961) recorded habituation of tactile evoked potentials from the lateral columns of the spinal cord

(Fig. 11). They further observed that a high spinal transection enhanced remarkably the habituated spinal potentials, which diminished
again upon repetition of the tactile stimuli. These results support

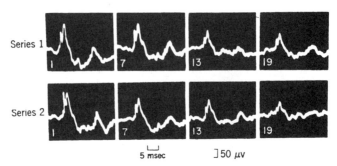

Fig. 10. Acoustic habituation at the dorsal cochlear nucleus in a cat in which
all of the brain rostral to the posterior hypothalamus was removed. A small dose
of Flaxedil was injected in order to prevent movements of the animal placed
in the stereotaxic instrument. The clicks were repeated at intervals of 1 sec.
Between the first and second series of successive clicks, there was an interval
of 5 min without acoustic stimulation.

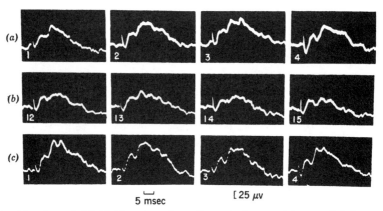

Fig. 11. (a and b) Habituation of tactile evoked potentials recorded from the
lateral column of the spinal cord in a decerebrate cat; (c) the enhancement
obtained immediately after transection of the spinal cord at C 2.

the idea of descending influences of brain-stem origin, acting upon
sensory neurons, and they also indicate that habituation can occur
in the isolated spinal cord.

A Hypothesis regarding Neuronal Circuits of Sensory Control

The assembled evidence just reviewed permits the formulation of
a working hypothesis that may aid in the investigation of the neuronal

circuits concerned with the regulation and integration of sensory signals. But before entering into speculative realms it is convenient to summarize certain established facts.

1. The brain-stem reticular system is a region where impulses of all sensory modalities converge. It is reached by impulses from the lower segments of the specific afferent paths as well as by those arising from the cortical receiving areas.

2. The same central region is able to decrease or increase the excitability of most sensory neurons, and thus to inhibit or facilitate sensory transmission at all the levels of the specific afferent paths.

3. The centrifugal control of sensory paths exerted by the reticular system is tonic and selective.

It follows that the core of the brain stem may be looked upon as a form of "high command" which constantly receives and controls all information from the external and internal environment, as well as from other parts of the brain itself. But at a given moment only a limited part of the information reaches this central area, and a large number of informing signals are excluded. The exclusion of afferent impulses from sensory receptors takes place just as they enter the central nervous system. Therefore, the first sensory synapse functions as a valve where sensory filtering occurs.

We might conceive the reticular mechanisms of sensory filtering as formed by a feedback loop with an ascending segment from second-order sensory neurons to the reticular formation and a descending limb carrying impulses in the opposite direction. It is not unlikely that both centripetal and centrifugal limbs of the loop contain specific facilitatory and inhibitory fibers with reciprocal neuronal connections, as illustrated in Fig. 12. Such an arrangement would prevent over-activation of sensory neurons and, therefore, an excessive bombardment of the brain by afferent impulses. In this way, the dynamic equilibrium operating at the entrance gates of the central nervous system would preserve the delicate and selective mechanisms of sensory integration.

Since the centrifugal filtering influences must necessarily depend upon the "decision-making system" involved in sensory integration, it seems reasonable to conceive that important integrative mechanisms lie at the brain-stem level. This hypothesis is in keeping with the chaotic derangement of behavior produced by brain-stem lesions, which is more disturbing than the impairment produced by extensive cortical damage. Indeed, Hernández-Peón and Brust-Carmona (1961) have observed in cats that restricted mesencephalic tegmental lesions that did not yield somnolence abolished alimentary and defensive conditioned responses, which were easily established in decorticate animals.

It is evident, therefore, that the integrity of the rostral portion of the reticular system is more necessary, at least for basic processes of sensory integration, than the higher parts of the specific sensory pathways.   This view does not disregard the important role of the specific

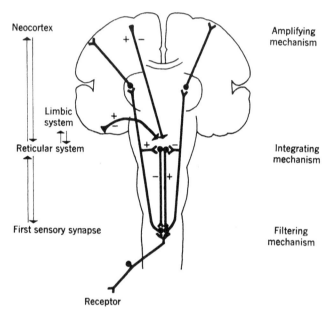

Neocortex

Amplifying
mechanism

Limbic
system

Reticular system

Integrating
mechanism

First sensory synapse

Filtering
mechanism

Receptor

Fig. 12.   Schematic representation of some of the main neuronal circuits involved in the transmission and integration of sensory impulses.

paths: by their divergent spatial organization, they amplify sensory information from localized groups of receptors and thus permit fine and complex discriminations.

Figure 12 illustrates a working hypothesis according to which three fundamental levels of the central nervous system participate in the amplification, integration, and filtering of sensory signals.   It must be emphasized that the oversimplified diagram by no means pretends to include all the neuronal systems involved in the regulation and integration of sensory impulses, nor is it claimed that the postulated functions are exclusive to the corresponding levels.

Summary

Experimental data are presented relevant to the influence of the brain-stem reticular system upon specific sensory neurons. Electrical stimulation of the reticular formation elicits inhibitory or facilitatory interactions at all the synaptic stations of the specific afferent pathways. These effects have been observed in the somatic, auditory, visual, and olfactory pathways. On the other hand, lesions involving the central core of the brain stem enhance synchronized afferent volleys. It can be concluded that during wakefulness there is a tonic inhibitory action of the supraspinal structures that regulates sensory messages. Recordings from unanesthetized cats with electrodes permanently implanted have shown that sensory evoked potentials from the lowest sensory synapses are reduced when the animal seems behaviorally inattentive to the test stimulus. For instance, the tactile evoked potentials recorded from the lateral column of the spinal cord diminish when the cat is attentively sniffing fish odor. Likewise, the spinal afferent potentials are reduced by monotonous repetition of the tactile stimuli. Both types of "sensory suppression" are still obtained in the dorsal cochlear nucleus after bilateral interruption of the paraspecific descending pathway coursing along the specific auditory pathway. Since tactile and acoustic habituation are also obtained in decerebrate cats, it is evident that the cortex is unnecessary for the manifestations of plastic inhibition at the lowest sensory synapses.

It is suggested that the reticular system of the brain stem resembles a "high command" in that it receives all kinds of information from the external and internal environment. In turn this region of polysensory convergence filters by feedback circuits all the sensory impulses at their entrance to the central nervous system. In this way a filtering mechanism is closely linked to those mechanisms that select the information to be amplified at higher levels of the brain.

References

Appelberg, B., R. L. Kitchell, and S. Landgren. Reticular influence upon thalamic and cortical potentials evoked by stimulation of the cat's tongue. *Acta physiol. scand.*, 1958, **45**, 48–71.

Arden, G. B., and U. Söderberg. The relationship of lateral geniculate activity to the electrocorticogram in the presence or absence of the optic tract input. *Experientia*, 1959, **15**, 163–169.

Arduini, A., and G. Moruzzi. Sensory and thalamic synchronization in the olfactory bulb. *EEG clin. Neurophysiol.*, 1953, **5**, 235–242.

Bach-y-Rita, G., H. Brust-Carmona, J. Peñaloza-Rojas, and R. Hernández-Peón. Papel funcional de las vías descendentes suprabulbares en la habituación neuronal aferente del núcleo coclear. *Resum. III Congr. mex. Cienc. fisiol.,* 1960, 29–30.

Bach-y-Rita, G., H. Brust-Carmona, J. Peñaloza-Rojas, and R. Hernández-Peón. Absence of para-auditory descending influences on the cochlear nucleus during distraction and habituation. *Acta neurol. latinoamer.,* 1961, **7,** 73–81.

Bremer, F. The neurophysiological problem of sleep. In J. F. Delafresnaye (Editor), *Brain Mechanisms and Consciousness.* Oxford: Blackwell, 1954.

Bremer, F., and N. Stoupel. Facilitation et inhibition des potentiels évoqués corticaux dans l'éveil cérébral. *Arch. int. Physiol. Biochem.,* 1959, **67,** 240–275.

Cavaggioni, A., G. Giannelli, and G. Santibañez. Effects of repetitive photic stimulation on responses evoked in the lateral geniculate body and the visual cortex. *Arch. ital. Biol.,* 1959, **97,** 266–275.

Derbyshire, A. J., B. Rempel, A. Forbes, and E. F. Lambert. Effects of anesthetics on action potentials in cerebral cortex of the cat. *Amer. J. Physiol.,* 1936, **116,** 577–596.

Desmedt, J. E., and K. Mechelse. Suppression of acoustic input by thalamic stimulation. *Proc. Soc. exp. Biol. Med.,* 1958, **99,** 772–775.

Dumont, S., and P. Dell. Facilitations spécifiques et non-spécifiques des réponses visuelles corticales. *J. Physiol.,* 1958, **58,** 261–264.

Galambos, R., G. Sheatz, and V. G. Vernier. Electrophysiological correlates of a conditioned response in cats. *Science,* 1956, **123,** 376–377.

Granit, R. Centrifugal and antidromic effects on ganglion cells of retina. *J. Neurophysiol.,* 1955, **18,** 388–411.

Hagbarth, K.-E., and J. Fex. Centrifugal influences on single unit activity in spinal sensory paths. *J. Neurophysiol.,* 1959, **22,** 329–338.

Hagbarth, K.-E., and S. Höjeberg. Evidence for subcortical regulation of the afferent discharge to the somatic sensory cortex in man. *Nature, Lond.,* 1957, **179,** 526–527.

Hagbarth, K.-E., and D. I. B. Kerr. Central influences on spinal afferent conduction. *J. Neurophysiol.,* 1954, **18,** 388–411.

Hernández-Peón, R. Central mechanisms controlling conduction along central sensory pathways. *Acta neurol. latinoamer.,* 1955, **1,** 256–264.

Hernández-Peón, R. Discussion on interpretation of conditioning on the basis of electroencephalographic data. In IV int. Congr. EEG clin. Neurophysiol., *Acta med. belg.,* 1957, 450–455.

Hernández-Peón, R. The centrifugal control of afferent inflow to the brain and sensory perception. *Acta neurol. latinoamer.,* 1959, **5,** 279–298.

Hernández-Peón, R. Neurophysiological correlates of habituation and other manifestations of plastic inhibition (internal inhibition). In Moscow Colloq. on EEG of Higher Nervous Activity. *EEG clin. Neurophysiol.,* 1960, **Suppl. 13,** 101–114.

Hernández-Peón, R., C. Alcocer-Cuarón, A. Lavín, and G. Santibañez. Regulación centrífuga de la actividad eléctrica del bulbo olfatorio. In *I Reun. Cient. Cienc. Fisiol. Montevideo,* 1957, 192–193.

Hernández-Peón, R., and H. Brust-Carmona. Functional role of subcortical structures in habituation and conditioning. In A. Fessard, R. W. Gerard, and J. Konorski (Editors), *Brain Mechanisms and Learning*. Springfield, Ill.: C. C Thomas, 1961. Pp. 393–408.

Hernández-Peón, R., A. Davidovich, and M. Miranda. Habituation to tactile stimuli in the spinal trigeminal sensory nucleus. *Acta neurol. Latinoamer.* (in press).

Hernández-Peón, R., C. Guzmán-Flores, M. Alcaraz, and A. Fernández-Guardiola. Photic potentials in the visual pathway during attention and photic habituation. *Fed. Proc.*, 1956, **15**, 91–92.

Hernández-Peón, R., C. Guzmán-Flores, M. Alcaraz, and A. Fernández-Guardiola. Sensory transmission in visual pathway during "attention" in unanesthetized cats. *Acta neurol. latinoamer.*, 1957, **3**, 1–8.

Hernández-Peón, R., C. Guzmán-Flores, M. Alcaraz, and A. Fernández-Guardiola. Habituation in the visual pathway. *Acta neurol. latinoamer.*, 1958, **4**, 121–129.

Hernández-Peón, R., and K.-E. Hagbarth. Interaction between afferent and cortically induced reticular responses. *J. Neurophysiol.*, 1955, **18**, 44–55.

Hernández-Peón, R., M. Jouvet, and H. Scherrer. Auditory potentials at the cochlear nucleus during acoustic habituation. *Acta neurol. latinoamer.*, 1957, **3**, 114–116.

Hernández-Peón, R., A. Lavín, C. Alcocer-Cuarón, and J. P. Marcelin. Electrical activity of the olfactory bulb during wakefulness and sleep. *EEG clin. Neurophysiol.*, 1960, **12**, 41–58.

Hernández-Peón, R., and H. Scherrer. Inhibitory influence of brain stem reticular formation upon synaptic transmission in trigeminal nucleus. *Fed. Proc.*, 1955a, **14**, 71.

Hernández-Peón, R., and H. Scherrer. "Habituation" to acoustic stimuli in cochlear nucleus. *Fed. Proc.*, 1955b, **14**, 71.

Hernández-Peón, R., H. Scherrer, and M. Jouvet. Modification of electric activity in cochlear nucleus during "attention" in unanesthetized cats. *Science*, 1956, **123**, 331–332.

Hernández-Peón, R., H. Scherrer, and M. Velasco. Central influences on afferent conduction in the somatic and visual pathways. *Acta neurol. latinoamer.*, 1956, **2**, 8–22.

Holmqvist, B., A. Lundberg, and O. Oscarsson. The relationship between the flexion reflex and certain ascending spinal pathways. *Experientia*, 1959, **15**, 1–4.

John, E. R., and K. F. Killam. Electrophysiological correlates of avoidance conditioning in the cat. *J. Pharmacol. exp. Therap.*, 1959, **125**, 252–274.

Jouvet, M., and J. E. Desmedt. Contrôle central des messages acoustiques afférentes. *C. R. Acad. Sci., Paris*, 1956, **243**, 1916–1917.

Kerr, D. I. B., and K.-E. Hagbarth. An investigation of olfactory centrifugal fiber system. *J. Neurophysiol.*, 1955, **18**, 362–374.

Killam, K. F., and E. K. Killam. Drug action on pathways involving the reticular formation. In H. H. Jasper, et al. (Editors), Henry Ford Hosp. int. Sympos., *Reticular Formation of the Brain*. Boston: Little, Brown, 1958. Pp. 111–122.

Killam, E. K., and K. F. Killam. Phenotiazine pharmacologic studies. *Res. Publ. Ass. nerv. ment. Dis.*, 1959, **37**, 245–265.

King, E. E., R. Naquet, and H. W. Magoun. Alterations in somatic afferent transmission through thalamus by central mechanisms and barbiturates. *J. Pharmacol. exp. Therap.*, 1957, **119,** 48–63.

Lavín, A., C. Alcocer-Cuarón, and R. Hernández-Peón. Centrifugal arousal in the olfactory bulb. *Science*, 1959, **129,** 332–333.

Moruzzi, G., and H. W. Magoun. Brain stem reticular formation and activation of the EEG. *EEG clin. Neurophysiol.*, 1949, **1,** 455–473.

Palestini, M., A. Davidovich, and R. Hernández-Peón. Functional significance of centrifugal influences upon the retina. *Acta neurol. latinoamer.*, 1959, **5,** 113–131.

Ricci, G., B. Doane, and H. Jasper. Microelectrode studies of conditioning. Technique and preliminary results. In IV int. Congr. EEG clin. Neurophysiol. *Acta med. belg.*, 1957, 401–415.

Tolle, A., S. Feldman, and C. D. Clemente. Effects of bulbar stimulation and decerebration on visceral afferent responses in the spinal cord. *Amer. J. Physiol.*, 1959, **196,** 674–680.

27

GEOFFREY B. ARDEN and ULF SÖDERBERG

Nobel Institute for Neurophysiology, Karolinska Institutet, Stockholm

The Transfer of Optic Information through the Lateral Geniculate Body of the Rabbit

Discharges are evoked in the cells of the lateral geniculate body when light falls on the eye, but, in addition, there is a resting activity that continues when the eye is not stimulated. Light-evoked responses have been studied by several authors (Bohm and Gernandt, 1950; Tasaki, Polley, and Orrego, 1954; De Valois, et al., 1958; Erulkar and Fillong, 1958; Arden and Liu, 1960a, b; Marriott, Morris, and Pirenne, 1959), but the resting activity has received less attention. Recent experiments have shown that, when gross electrodes are chronically implanted in the optic pathways of otherwise intact animals, the evoked potentials that can be recorded are modified if auditory and olfactory stimuli are presented together with the photic stimuli (Hernández-Peón, et al., 1956, 1957; Hernández-Peón in this volume), and the geniculate response varies with the animal's state of attention. In addition, studies on the lateral geniculate body, both with gross electrodes (Evarts and Hughes, 1957a, b; Schoolman and Evarts, 1959) and with single-cell recordings (Bishop, Burke, and Davis, 1959), have shown that the lateral geniculate synapses modify optic nerve impulses.

In the experiments described below, the resting activity of single lateral geniculate cells was monitored for relatively long periods, with the object of exploring its source, some ways in which it could be altered, and the ways in which such modification affected the responses evoked by retinal illumination. The plan of the experiments was to study the discharges of lateral geniculate cells under conditions in which the nervous input to the lateral geniculate body differed widely. The input could be reduced by section of parts of the brain, or by removal of retinal activity through ischemia or ocular evisceration, and it could be increased by photopic or auditory stimulation or electrical stimulation of the brain stem.

Methods

The experiments were performed on rabbits. These animals have very little binocular vision (Walls, 1942), and the complications that are caused by binocular interaction (Bishop and Davis, 1953; Cohn, 1956; Erulkar and Fillenz, 1958) are thereby largely avoided. The rabbits' lateral geniculate body is not distinctly laminated (Rose, 1935).

Operative procedures

The first type of experiment was performed on *encéphale isolé* preparations since it was desirable to avoid the effects of anesthetics during the experiment. Initially, however, the rabbit was anesthetized with ether in a stream of 94 per cent O_2 and 6 per cent CO_2. The trachea, femoral artery, and femoral vein were cannulated. The spinal cord was sectioned at the level of C_1, and the animal respired artificially. Blood pressure was maintained by a subcutaneous injection of isopropyl-noradrenaline (Isodrin). The head of the animal was mounted in a Horsley-Clarke headholder, and further operative procedures were carried out as detailed below. All pressure points and wounds were locally anesthetized with Lidocaine. After the preparation had been completed, the ether anesthesia was discontinued, and the animal was placed in a shielded box to dark adapt. Its temperature was maintained by hot-water bottles, and glucose saline was routinely given to replace fluid loss. In some experiments the animals were immobilized with Flaxedil.

Initially small holes were drilled through the cranial vault for the recording electrodes. In later experiments the skull was opened more widely and the lateral geniculate body exposed by sucking away the overlying cerebral cortex to such an extent that the optic radiation was interrupted. In a further series, the cortical surface was exposed as far forward as the bregma, and the whole of the hemispheres with the exception of the frontal cortex was removed by suction.

The *cerveau isolé* preparation was used for some experiments. The ether anesthetic was employed, the cranial vault exposed, opened widely, and the hemispheres removed, as already described. A midcollicular section was then made by suction, a procedure that also damaged the anterior colliculus. In this preparation the only possible neural connections of the lateral geniculate were to the retinas, the rostral part of the brain stem, the basal ganglia, the olfactory system, and the frontal cortex.

A modified *cerveau isolé* was also made by removing the skull over the cerebellum, which was then sucked away, and a midpontine section was made just rostral to the entry of the fifth nerve, as has recently been described by Batini, et al., 1959*a*, *b*, *c*. In this preparation the brain is active, unlike the classical *cerveau isolé* of Bremer (1935) but for these experiments the point was that more of the rostral position of the brain stem remained intact and in functional connection with the geniculate and the cortex than is the case in the classical preparation.

Preparation of the eye

The eyelids and corneas were locally anesthetized. The eyelids were held open by stay sutures, and the eyes were coated with liquid paraffin to prevent drying. A fine syringe needle was inserted into the anterior chamber and connected to a pressure system. Usually the pressure was held at 35 millimeters Hg, but it could be rapidly increased to pressures well above the arterial level. The ischemia thus produced in the eye stopped all retinal and optic-nerve activity within 3 to 6 minutes (Noell and Chinn, 1950; Noell, 1951; Arden and Greaves, 1956; Bornschein, 1958). Provided the ischemia does not continue for more than 10 to 20 min, the eye recovers completely once the circulation is readmitted.

Recording and display

The responses of single lateral geniculate cells were recorded. The electrodes were glass micropipettes, 4 molar NaCl filled, tip resistance 5 to 20 megohms. Positive spikes of 1 to 20 millivolts were obtained and amplified via cathode followers and amplifiers. The usual criteria for single-cell recordings were established.

The electrodes were either introduced through small holes in the cranial vault and lowered to the correct position, or, when the cortex had been removed, inserted under direct vision. In the latter case, a large brass ring was fastened with dental cement to the parts of the cranial vault bones that remained, and the electrode was inserted through the ring. The system was then filled with warm Ringer's solution and sealed with dental wax. Histological examination checked the electrode positions after the end of the experiments.

The electrocorticogram (EEG) was recorded with silver-silver chloride ball electrodes which were sealed with dental cement into drill holes made in the skull.

In some experiments the arterial blood pressure was measured with an ELEMA manometric bridge. A photocell monitored the light stimuli applied to the eye and could be used to trigger the time base of the cathode-ray oscilloscope.

Photocell, blood pressure, and EEG were recorded on an Offner dynograph. In addition, a continuous record of the spike discharges of the lateral geniculate cells was made by an EKCO rate meter. The rate meter was suitably biased to discriminate spikes from base-line noise. When its time constant was set as short as 0.1 sec, single spikes were visible as small deflections, while high-frequency bursts of spikes produced single large deflections, and the resulting records gave a graphic and semiquantitative representation of the pattern of discharge of a cell over long periods. The spikes were also displayed on a cathode-ray screen and photographed intermittently, as a check of the discrimination and to be sure that the same cell remained under the electrode. Quantitative work was carried out from the spike photographs.

Stimulation

A flickering light could be focused into either of the animal's eyes. The source was a tungsten filament lamp, run at constant voltage. An area of the retina of approximately 15 degrees of visual angle was evenly illuminated. The rabbit's optical system is very poor, and the actual area of retina illuminated was greater than that calculated on a geometric basis. Nevertheless, by no means all the retina was illuminated, and since there is a fairly precise anatomical projection of the retina on the geniculate body (Brouwer, Zeeman, and Houwer, 1923; Brouwer, 1927) it was not to be expected that all the cells in the geniculate body would be activated by the light stimulus. The microelectrode was moved through the geniculate body until a suitable area was found (see Arden and Liu, 1960a). The interposition of neutral filters enabled the light intensity to be varied over a wide range. The maximum intensity available was about 180,000 lux.

Auditory stimulation was carried out by whistling, hand clapping, and other noises. These rather crude stimuli were adequate for the purpose of the experiments, for it was apparent that though the discharge rate of lateral geniculate cells was often influenced by noise (and was indeed very sensitive to such stimuli) there was no specific response to sound, the alteration being related to the alteration in the activity of the brain as a whole.

Electrical stimuli were delivered from a square-wave stimulator which fed into an isolation transformer. The stimulating electrodes were bipolar concentric needles, whose uninsulated tips were placed in the mesencephalic reticular formation or in the thalamus. Usually the EEG showed patterns of sleep or drowsiness, and stimuli caused short periods of activation. The blood pressure, heart rate, and EEG were unchanged when the intraocular pressure (IOP) was altered.

Results

The results described below are based on the recordings made of the activity of more than 150 cells over periods lasting from several minutes to hours. A vast heterogeneous material has been collected but, unfortunately, so great is the diversity of the responses of the different cells of the lateral geniculate body that the series is not nearly large enough for a statistical evaluation. The description of results is therefore limited to certain topics, and no attempt is made at a comprehensive survey of the findings.

Nature of the resting activity

When an *encéphale isolé* preparation is allowed to remain quietly in the dark, the cells of the lateral geniculate body continue to fire. If several trials show no resting activity, the preparation is usually in poor condition. Unlike the peripheral sense organ or the retinal ganglion cell (Kuffler, Fitzhugh, and Barlow, 1957), the spikes are not discharged rhythmically. Usually high-frequency bursts of two, three, or more spikes are seen, and each burst is followed by a pause that may be prolonged. The frequency in the burst is about 300 per second, and 5 to 20 spikes may occur in each second. During a burst the spike height may decline and then recover, as in inactivation processes (Granit and Phillips, 1956). A histogram of intervals between successive spikes plotted against the frequency of occurrence of any interval shows that, in the great majority of cases, the shortest intervals occur with the greatest frequency (Fig. 1a). A similar type of resting activity has been found in the midbrain (Amassian and Waller, 1958), the thalamus (Katsuki, Watanabe, and Maruyama, 1959), and the cerebral cortex (Martin and Branch, 1958; Hubel, 1959; and others). This type of resting activity is not to be expected if the lateral geniculate cells passively relay the spikes that originate in retinal ganglion cells.

Now in the classical *cerveau isolé* it was found that many of the cells did not have a resting discharge, even though the preparation was in good condition, thus supporting the idea that the retina is not the sole source of the normal geniculate resting discharge. The relative importance of the retina in the maintenance of the geniculate resting discharge can be determined in another way, by abolishing retinal activity. When the intraocular pressure is increased and all retinal activity is prevented, any remaining discharge in the geniculate cells must be caused by activity in the brain itself.   In the

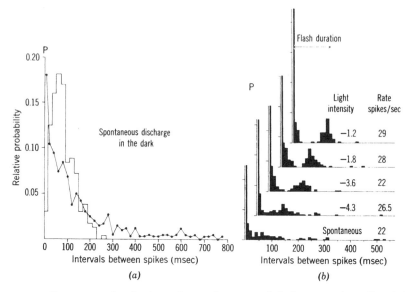

Fig. 1. Frequency distribution of impulses recorded from single cells. (*a*) Comparison of geniculate and retinal ganglion cell discharge.   Relative probability of obtaining any interval between successive spikes (ordinate) and duration of interval (abscissa). Histograms, retinal spikes; connected solid points, geniculate spikes.   Class intervals for the two observations the same. More than 200 spikes counted for each histogram.   The retinal discharge varies about a mean frequency (mean interval, 80 msec; mode, 60 msec).   The geniculate discharge is in bursts; the most frequent interval is the shortest (mean, 140 msec; mode, 80 msec).   (*b*) Alteration in the discharge of a geniculate cell produced by flickering lights of varying intensity.   Histograms of interspike intervals as in (*a*).   In the dark (lowest diagram) the cell fires irregularly in bursts.   During a flicker it fires at end of the on and off of the light.   The bursts are unchanged by the light but become more and more nearly synchronized with increasing stimulus intensity.   Flicker rate, 3.9 per second; light intensity in log units below maximum of 100,000 lux. (From Arden and Liu, 1960*b*.)

encéphale isolé the resting discharge continues when retinal activity ceases, and the number of spikes discharged per unit time may be unaffected or only slightly reduced.   In the *cerveau isolé*, on the other hand, all resting activity stops under these conditions.   No evidence for true "spontaneous" geniculate activity has been found.

The geniculate resting discharge is therefore to a great extent maintained by either the cortex or the brain stem, and the relative importance of the two could be determined by making a decorticate *encéphale isolé*. This preparation resembles the simple *encéphale isolé*, in that the geniculate cells' resting discharge continues (if present) after retinal activity has been halted by an increase in IOP. However, even in the best decorticate *encéphale isolé* preparations, it was possible to find cells that had little or no resting discharge. It is not clear whether this was due to the brain removal *per se* or to the animal's general condition, which was easily impaired by the drastic operative procedures.

It seems, therefore, that there is a sizable pathway to the geniculate body from the midbrain, and this "second input" is more effective in maintaining the geniculate resting discharge than is the retina. There may also be a corticogeniculate input, but it seems to be of less importance.

Long-term fluctuations in resting activity

In the *encéphale isolé*, the rate of the resting activity of geniculate cells often varies through the recording period. Some of these variations are artifacts, due to cell damage, or movement caused by respiratory and arterial pulsations, but others are certainly not. In particular, rhythmic fluctuations in the resting discharge (period 5 to 10 sec) are seen which may persist for hours (Fig. 2). These rhythms could be found in only a part of a group of neighboring cells. The rhythm could on occasions be stopped or rephased or its period decreased by afferent stimuli or during "arousals." It was easily halted by small doses of barbiturate.

Another much less rhythmic variation in the discharge rate could be correlated with the changes in the pattern of the EEG (Fig. 3). As the EEG changed from large-amplitude, slow-wave activity, to small-amplitude, fast activity ("arousal"), the firing rate changed with it. In animals with periodic spindles in the EEG, this correlation was most striking. In some lateral geniculate cells, the firing rate increased with arousal, and in others it decreased, but the pattern was constant for any one cell, unless the conditions were altered in other ways (see below). Such variations in discharge rate with changes in the EEG wave form have been described for cortical cells (Li and Jasper, 1953). It may be that the observed correlation is due to a common mechanism driving both cortical and geniculate cells, or that the cortical rhythms cause the alterations in the geniculate discharge. After almost complete decortication, correlation between the geniculate firing rate and the activity of the remaining strip of cortex could still sometimes be observed, so that the first alternative is to be preferred.   In several experiments, especially after a narrow exposure of the lateral genicu-

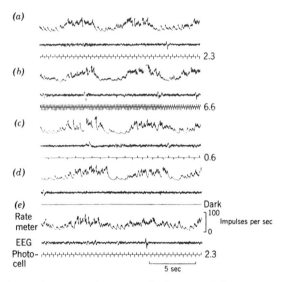

Fig. 2.   Rhythmic changes in the resting discharge and the interaction with light-evoked responses.   Sections of records of responses to a flickering light stimulus (frequency indicated at right).   Note that the responses to light are clearly visible only in the periods of low activity.   Photocell records on an a-c channel; downward deflection, light on, upward, light off.

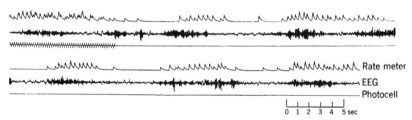

Fig. 3.   Two continuous strips of records of geniculate firing rate, EEG, and light signals.   Bursts of high voltage in the EEG (*encéphale isolé*) indicate good correlation to changes of activity of a lateral geniculate cell.   Note that light-evoked responses are also affected.

late body, local seizures developed in the remaining cortex.   Under these conditions, the responses to light in the lateral geniculate body sometimes continued unaffected.   This is additional evidence against the presence of a very important corticogeniculate pathway, and it is strengthened by the observation that convulsants have little effect on the geniculate body (Söderberg and Arden, 1961).

Auditory stimulation

When the recording electrode was placed in the *medial* geniculate body, cells were found in which prompt and distinctive responses were evoked by noises. In the *lateral* geniculate body also, the firing rates of many cells (*encéphale isolé*) were affected though not in a clear-

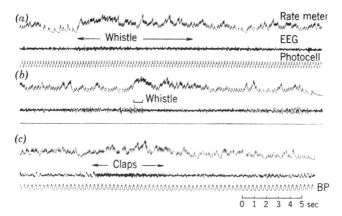

Fig. 4.   Paper records of responses to light and sound in an *encéphale isole* rabbit.   Rate-meter record from left geniculate body.   Light to the right eye. (*a*) A whistle causes EEG "arousal" and increases the discharge rate.   The cell begins to follow the flickering light stimulus.   (*b*) A short whistle which causes EEG arousal increases the resting discharge rate.   Note that the effects outlast the stimulus both in the cortex and in the geniculate body.   (*c*) A series of claps induces EEG arousal and increases geniculate firing.   The arousal is not accompanied by changes in the blood pressure (BP) or pulse rate.

cut manner.   Often a series of claps would produce a gradual prolonged change in the discharge rate (Figs. 4 and 5).   The alteration in firing rate was also correlated with EEG changes: a sound that activated the EEG altered the firing rate, whereas one that did not cause arousal had no effect.   On the other hand, if the EEG already showed low-voltage fast activity, sounds could still affect the firing rate.   In all the cells studied, arousing stimuli produced the same effect as "spontaneous arousal."   Both acceleration and slowing of the resting discharge were seen.

These experiments suggest that the "second input" is in fact a path-

(a)   (b)   (c)

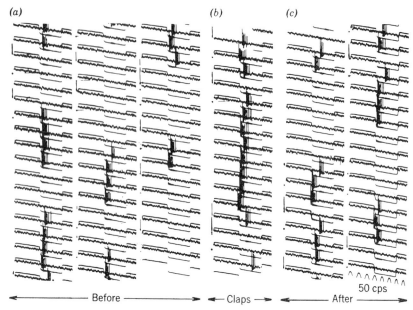

◄──────── Before ────────► ◄─ Claps ─► ◄──────── After ────────►   50 cps

Fig. 5. Spike records from *encéphale isolé* rabbits to show the effects of a series of claps on the response to light. Moving paper, sweep triggered by photocell, whose output (on = up) is shown on the second beam. (*a*) Unaroused condition. The cell fires bursts of spikes sporadically. The bursts are, however, synchronized to the light flashes. (*b*) During a period when the rabbit was aroused by a series of hand claps, the bursts appeared more frequently and were still well synchronized with the light stimulus. In addition, each burst contained an average of one more spike than during the periods (*a*) before and (*c*) after the clapping. Note also that bursts with short latency have more spikes than long-latency bursts. In fact, careful measurements on original records show that the arousing stimulus significantly decreased the latency.

way from the reticular activating system of Moruzzi and Magoun (1949).

The response to light

Recently some peculiarities in geniculate responses to light were described and analyzed by Arden and Liu (1960*b*). The same phenomena have been repeatedly observed in the present series of experiments. In the *encéphale isolé,* not all lateral geniculate cells respond to light, and those that did responded briefly with bursts very similar to those that are found in the resting activity. Repeated flashes produce responses that differ in latency and in the number of spikes evoked. If a slow flicker is used as a stimulus, the total number of spikes discharged may be no more than in a corresponding period of darkness, and each response consists of a single spike (Fig. 6) or a brief burst which might well be found in the resting activity. Under these circumstances, one can tell that a group of spikes is an evoked response only because it is synchronized with the stimulus.

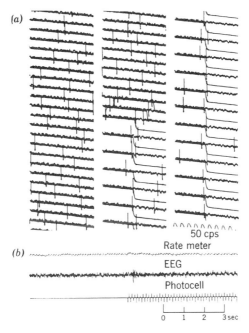

Fig. 6. Records of lateral geniculate activity show that average firing rate is the same in the dark and during flicker, but that the flicker synchronizes the spikes with the light stimulus. (*a*) Photographs of spike responses. Resting discharge (free-running sweep, moving paper) followed by responses to repetitive light flashes (sweep triggered by beginning of flash-photocell records, upper trace). Positive upward; read left to right; photocell record deflected upward during a period of light. (*b*) Same events on continuous paper record. Rate-meter tracing shows small deflections that correspond to the discharge of each spike. The discharge is unusual because there are no "bursts."

This holds even in cases where the resting activity is not in bursts (Fig. 6). An increase in stimulus intensity does not as a rule systematically increase the actual number of spikes, or the frequency of the spikes in the response, but the synchronization of the response with the stimulus is improved. In addition, many cells are seen which respond to light by a decrease in the discharge rate and which will stop discharging when a repeated "flickering" stimulus is applied.

Furthermore, the cells remain silent for the period of the intermittent illumination. It was previously suggested that this diminution of activity was due to inhibition, and the synchronization of response with stimulus, which is achieved without any increase in firing rate, was explained as a result of almost synchronous excitation and inhibition of lateral geniculate cells by retinal afferents of opposite potentialities (Arden and Liu, 1960b; Marriott, Morris, and Pirenne, 1959; Posternak, Fleming, and Evarts, 1959). Further support for this hypothesis has been obtained. In these cells where illumination causes a decrease in the number of spikes, the resting activity continues after abolition of retinal activity (cf. Fig. 10). The response to light must therefore be an active inhibition. In addition, this inhibition can be selectively removed during the first moments after the rise in intraocular pressure (see below).

In the *encéphale isolé* the responses of a cell to a series of identical light flashes may vary a good deal. This variation is often related to fluctuations in the resting activity. Thus, in Fig. 3, the amplitude of the evoked responses is greater during an EEG "spindle" than after. In Fig. 2, the slow rhythmic changes in resting discharge are superimposed on the light-evoked discharge, and during the periods of maximum activity the evoked responses are submerged.

Arousals

In the *encéphale isolé,* the state of activity of the brain is subject to wide, apparently "spontaneous," fluctuations, and it is often difficult to be sure how arousing stimuli affect the evoked responses. It is clear, nevertheless, that in both the presence and absence of adequate retinal illumination arousals can alter the discharge rate equally, and the alterations thus produced are as great as those produced by the light stimulus itself. In addition, a cell that did not respond to the flickering light stimulus could often be induced to do so by means of whistles or claps. Also, if a cell was responding regularly, such stimuli might evoke showers of "irrelevant" spikes that obscured the response to light.

The most marked difference between the *encéphale isolé* preparations (both decorticate and with cortex intact) and the *cerveau isolé* preparation is that, in the latter, almost all the cells respond to light and the responses are regular and reproducible (Fig. 7). Most cells increased their average discharge rate in response to light, but a few were silenced by this stimulus. This feature, combined with the low

resting activity and the uniformity of successive responses, makes a
striking contrast to the *encéphale isolé*, where there is a great divergence between different cells, and some are scarcely affected by
retinal illumination.   These differences must be due to removal of the
influence of the reticular activating system.   Since arousals can stop
geniculate discharge, it is clear that the second input must to some
extent inhibit the lateral geniculate.   But it was surprising to find so

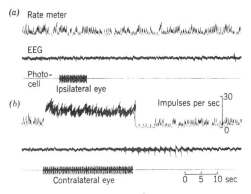

Fig. 7.   *Cerveau isolé.*   (*a*) Low resting discharge, absence of effect of stimulating
ipsilateral eye.   (*b*) Contralateral eye: great increase in rate of discharge.   After
end of illumination, resting discharge decreases and, transiently, bursts cease to
appear.   The events are recorded on very slow paper, a fact that explains the unusual shape of the EEG.

great a difference between the *encéphale* and *cerveau isolé* preparations.   It is concluded that some brain-stem mechanism is primarily
responsible for the failure of lateral geniculate cells to be influenced
by light stimuli.

The midpontine *cerveau isolé* is intermediate in its properties between the *encéphale isolé* and the classical *cerveau isolé*.   The resting
discharge is lower, and the response to light more regular than in the
encéphale isolé.   Moreover, with this preparation it is easy to activate, by electrical stimulation, that remainder of the rostral brain stem
that is in functional connection with the lateral geniculate body, and
to study the alteration of the light-evoked discharge.   The effect of
stimulation is prolonged and resembles the action of arousing stimuli
in the *encéphale isolé*, and the geniculate cells cannot be "driven"
by the electrical stimuli.

Even in this preparation, more than one type of response was observed after electrical stimulation.   In some cells the resting discharge
rate was greatly increased; in others it was decreased or unaffected.

Strikingly, the temporal distribution of the spikes was also affected: discharges in bursts were supplanted by rhythmic discharge.

When the electrical stimulus occurs during a series of responses to a flickering light stimulus, the number and the frequency of the evoked spikes may increase in some cells and decrease in others. Examples are shown in Figs. 8 and 9. Analysis of such results, however,

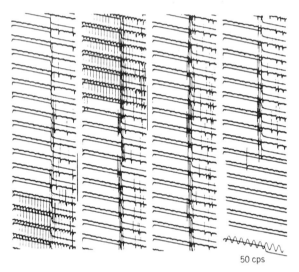

50 cps

Fig. 8. *Cerveau isolé,* midpontine section. Activity in the lateral geniculate body. The four columns of records are continuous. The responses to light are evoked with greater regularity than in an *encéphale isolé.* Reticular-formation stimulation, as indicated by the shock artifacts in the first two columns, increases the frequency and number of spikes evoked by the light. The effect appears slowly, is maximal at the cessation of the electrical stimulation, and outlasts it considerably.

shows that an increase in the number of spikes may act in two different ways on the actual response. In Fig. 5 the light-evoked spikes are increased in number and frequency; in other cases, when the total number of spikes was increased by reticular stimulation, many of the extra spikes seem to have been irrelevant since they fell in between the light-evoked spikes and obscured them.

Effects of reversible retinal ischemia

Soon after the intraocular pressure was increased, the resting-discharge rate of the lateral geniculate body began to increase. This occurred in almost every instance. In the *cerveau isolé* where geniculate cells had no resting discharge, this effect was most dramatic, but

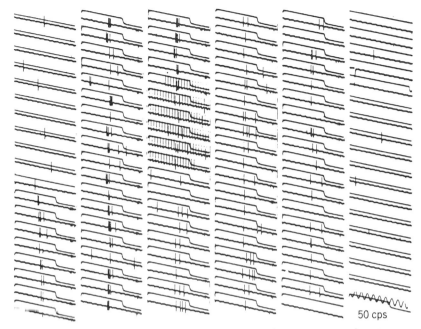

50 cps

Fig. 9. *Cerveau isolé,* midpontine section. Geniculate activity. The six consecutive columns are continuous records. At beginning and end there were periods of darkness. Electrical stimulation of the thalamic reticular formation (shock artifacts in third column) markedly reduces the ability of this geniculate cell to respond to flickering light.

accelerations of the resting-discharge rate were also seen in the *encéphale isolé.* The increased rate was not associated with alterations in blood pressure, heart rate, or EEG. In the *cerveau isolé,* there was no change in respiration. The alterations in the discharge rate are, therefore, unlikely to be related to any painful stimulation. At the same time, light-evoked discharges increased, and, in those cells that were previously inhibited, the light now caused an accelerated discharge (Fig. 10). In other cells the light response turned from excitation to inhibition. As the ischemia continued, all retinal activity dwindled, the geniculate resting discharge decreased once more, and light-evoked responses vanished. In the *encéphale isolé* and the midpontine *cerveau isolé,* resting activity still remained. When the circulation was readmitted, the geniculate resting discharge again increased. Light now increased firing rate; later the original state of affairs returned. In one case it was possible to go through this cycle three times while recording from the same geniculate cell. An explanation of these phenomena is deferred to the section on Discussion.   In several cases, it was observed that the effects of auditory stimulation were changed by an increase in intraocular pressure.   A cell whose discharge was normally accelerated by arousal changed its characteristics, and, after complete abolition of retinal activity, whistles and claps caused a slowing of the discharge.

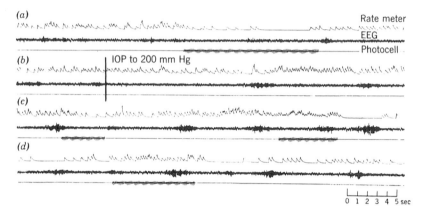

Fig. 10.   The initial effect of increasing the intraocular pressure (IOP) on lateral-geniculate firing rate and EEG to show the reversal of a light response. Four continuous records.   (*a*) Light silences the discharge of this cell.   (The light intensity is constant, but the signal is distorted by the chopper amplifier.) After IOP rise, as indicated in (*b*), the resting activity begins to increase.   Note the spindles in EEG.   This cell's activity shows no correlation with the EEG. (*c*) The inhibitory effect of light has diminished and, at end of line, has re-versed—the cell now follows the flicker! (*d*) The resting activity has decreased, and light clearly facilitates the resting discharge.   The resting disharge con-tinues at the level seen in the end of the row, but after another minute the light becomes ineffective in altering the geniculate discharge rate—retinal activity has been halted.

Discussion

The second input

The experiments reported above have shown that the lateral genicu-late body has an important input from the brain stem as well as from the retina.   Though this has been denied by some investigators on the basis of transsynaptic degeneration studies (O'Leary, 1940; see also Walls, 1953), other histologic studies have shown that there is a fiber tract running toward the lateral geniculate body (Cajal, 1955, p. 392; cf. Bürgi, 1957), and Scheibel and Scheibel (1958) have found reticular formation neurons that terminate in the geniculate bodies.

Additional evidence for the existence of a second input has been presented by Chavaz and Spiegel (1957) who recorded rhythmic potentials with electrodes implanted in the lateral geniculate body both before and after section of the optic nerve. In addition, the present experiments suggest that the second input is even more important than the optic nerve in the maintenance of the resting activity of the lateral geniculate cells, and that stimuli that cause arousal or supposed changes in "attention" can greatly influence the discharge rate. That this input comes from the brain stem is indicated by the fact that auditory stimuli and brain-stem stimulation are still effective in decorticate preparations. It is believed, therefore, that the second input is identical with the reticulogeniculate fibers described by the anatomists. These findings are in substantial agreement with the findings of Hernández-Peón, et al., and with recent observations by Hubel and Wiesel and by J. D. Green (personal communications). Some of the alterations in the light-evoked responses are so large that they suggest that the information that reaches the cortex depends very greatly upon whether the animal is aroused or not. There is, in fact, some evidence that the rat's ability to use visual information in the performance of motor acts is discontinuous and does depend on the presence of other stimuli (Lashley, 1942). The human correlate to these findings is largely unknown.

The responses described for the lateral geniculate body of the rabbit correspond in many ways to the extensive series of findings made by Jung and co-workers on the cortex of the cat (Baumgarten and Jung, 1952; Jung, 1958). In particular, the cells in Jung's classification that do not respond to single flashes but can be made to do so by stimulation of the intralaminar thalamic nuclei (type A) are very similar to lateral geniculate cells of the rabbit. It is possible that the cortical findings are in part secondary to lateral geniculate effects—though it is to be remembered that there may be a species difference, and the effects of arousal are more marked in the cat's cortex than in its lateral geniculate body (Bremer and Stoupel, 1959; Long, 1959).

It is possible that the brain-stem reticular formation affects the retina by way of the centrifugal retinal pathway described by Granit (1955) and Dodt (1956). The experiments reported here were originally planned to investigate this point. Since, when the retina is inactivated, the correlation between EEG and geniculate discharge rate may alter and even reverse, the brain-stem arousal system may in fact influence the retina. The alternative possibility remains that optic nerve fibers may pass into the midbrain and there affect the second input. The anatomical basis of such an assumption is the tractus peduncularis transversus of von Gudden (1881), later called "the posterior accessory optic tract" and described, for example, by Gillilan (1941). This tract has not been interrupted in any of the present experiments. It is, however, quite large as well as myelinated in the rabbit (v. Gudden, 1881), where its fibers (both crossed and uncrossed) run from the optic tract to the n. ped. transv. and to a number of midbrain nuclei and to the subthalamus.

Inhibitory retinal afferents

Recently, Arden and Liu (1960b) suggested that geniculate cells were inhibited by the activity of a fraction of the optic nerve fibers. Since complete abolition of the retinal resting activity almost always (in the *encéphale isolé*) leaves the geniculate cells with a resting discharge, whereas retinal illumination can cause complete silence, active retinal inhibition of the lateral geniculate must occur. This has also been inferred by Posternak, Fleming, and Evarts (1959) on the basis of evoked potentials. The changes that occur after an increase in IOP may be explained on this basis as an alternative to a centrifugal loop of control, if it is also supposed that many of these retinal cells that cause excitation of the lateral geniculate body are less sensitive to ischemia than the cells that cause inhibition. The inhibitory cells will then cease their discharge first, and the resting firing rate will increase in the geniculate cells. The alternative is that ischemia causes an increase in those retinal discharges that excite the geniculate cells. But Noell (1951), when he recorded optic nerve impulses during asphyxial block in mammalian retinas, never observed any increase in retinal discharges after an IOP increase (see also Bornschein, 1958). On the contrary, the activity diminished gradually and smoothly with time, though some cells appeared more resistant than others. It is likely, therefore, that the geniculate body is released from inhibition in the moments that follow an increase of IOP, and this release may be responsible for the sensation of "entoptic light" that results from pressure on the eye (Schubert, 1958).

The explanation given above implies that inhibitory geniculate responses contain a masked excitatory component. This can sometimes be demonstrated by altering the intensity of the light stimulus. Further, a cell that is inhibited by turning the stimulus light on and off can sometimes be made to discharge by moving the light across the retina (Söderberg and Arden, 1961). A similar finding has been reported by Hubel (1959) for cells in the optic cortex and also for the geniculate cells (Hubel, personal communication).

Significance of the burst discharge

If the burst discharge occurs naturally, it has great significance for the transmission of optic information. It might be thought to be an artifact, produced by irritation of the cell by the microelectrode tip, but this is almost certainly not the case. In some sites (for example, the retina) bursts do not appear even when the electrode is deliberately made to deform the cell (Kuffler, 1952), whereas many authors using different techniques have reported bursts from many parts of the cerebrum, cerebellum, and midbrain. Since in the present experiments the activity of one cell could be recorded for hours, and since during that period the cell continued to discharge in bursts, injury from the electrode tip seems unlikely. On the other hand, when a cell was damaged by the microelectrode tip, either an injury discharge began suddenly, or it was preceded by a period of accelerated discharge during which the burst disappeared. In *cerveau isolé* preparations there was frequently no resting discharge, which argues against the "irritation theory." Finally, under some circumstances (in the after-discharge that sometimes follows the end of a train of stimuli, or in arousal) the periodic bursts may change into a pattern in which spikes appear at random (Hubel, 1959). In the retina, the resting discharge has a skewed distribution about a mean frequency (Kuffler, Fitzhugh, and Barlow, 1957), and a stimulus is signaled by increase or decrease in the rate of spike production.

The lateral geniculate light-evoked responses consist of events that are common in the resting discharge, and this is puzzling enough; but, in addition, the evoked responses vary, depending upon the state of activity of the rest of the brain. The variations in response to identical stimuli are often greater than can be produced by alteration of the stimulus parameters. This makes it very difficult to see how a pulse-frequency code similar to the retinal one could work in the lateral geniculate body. The present results are not, however, unlike those of other workers who have shown that in other higher centers the responses to afferent stimuli are brief, and the relation of stimulus intensity to response is tenuous (Mountcastle, Davies, and Berman, 1957; Katsuki, Watanabe, and Maruyama, 1959). Analysis of the lateral-geniculate records suggests that the mechanism of intensity discrimination might be based upon the degree of synchronization of the responses of the various cells in a group (Arden and Liu, 1960b). This mechanism, which would have to be quite complex (Bullock, 1957), is at least possible, but there is still no experimental evidence for its existence.

When the brain-stem reticular formation is activated, the firing rate of cells in the lateral geniculate body may either increase or decrease. The attenuation or reinforcement of the evoked responses is independent of whether more or fewer spikes are discharged. This is difficult to reconcile with any pulse-frequency code but is compatible with the synchronization mechanism mentioned above, which operates on the signal-to-noise ratio. However, the arousing stimuli were presented under most unnatural conditions. It is likely that the brain stem operates in many different ways: the animal may be generally alerted or its attention may be directed toward a particular sensation. Depending on the condition, the evoked response may be quite different (Horn and Blundell, 1959), and experiments in acute preparations give no information on this point.

Summary and Conclusions

Geniculate responses recorded from rabbits (*encéphale isolé*, decorticate *encéphale isolé*, classical and midpontine *cerveau isolé*) show the following peculiarities:

1. The resting discharge is often irregularly distributed with most of the spikes appearing in bursts.

2. The response to illumination usually consists of similar bursts.

3. Responses to successive light stimuli vary. The average rate of discharge is largely independent of the light stimulus, particularly in the *encéphale isolé*. The bursts are synchronized with the stimulus. Synchronization varies with the intensity of the stimulus.

4. Both in the dark and in flickering light, the average firing rate and the distribution are affected by "arousing" stimuli. Both increases and decreases in discharge rate are seen. Spontaneous alterations in "arousal" are accompanied by similar effects. Electrical stimulation of the brain-stem reticular formation causes the same effect as natural "arousing" stimuli.

5. Stimuli that desynchronize slow-wave activity in the electroencephalogram (EEG) are often effective in changing the geniculate activity, even in a state of "arousal." The geniculate discharge thus responds over a wider range than the EEG.

6. Removal of the optic-tract input can modify or reverse the geniculate response to unspecific "arousing" stimuli. Sometimes these stimuli were effective only when the optic-tract input was intact.

7. When the optic nerve is blocked, there is a resting discharge

that is closely related to the state of activity of the brain stem, thus indicating a second input to the lateral geniculate body from this structure. This is a confirmation of the anatomical findings of Scheibel and Scheibel (1958) who demonstrated the presence of reticulogeniculate fibers.

8. As the optic-nerve conduction fails during a period of acute ocular ischemia, the lateral-geniculate discharge rate momentarily increases. This and other evidence suggest that geniculate cells are under inhibitory retinal influence.

The geniculate seems to act as an interpreter of retinal information, as a recoder, and as a mixer. We have attempted to investigate the new code, and to suggest lines on which reinterpretation could be based. However, the diversity of the responses of the geniculate cells is so great that the material collected is not nearly large enough for a statistical evaluation.

Acknowledgment

The research reported in this paper was in part sponsored by the Office of Scientific Research of the Air Research and Development Command, United States Air Force, through its European Office, under Contract AF61 (052)-119.

Geoffrey Arden held an Alexander Piggot Wernher Memorial Fellowship in Ophthalmology. His present address is Institute of Ophthalmology, Judd Street, London, England.

References

Amassian, V. E., and H. J. Waller. Spatiotemporal patterns of activity in individual reticular neurons. In H. H. Jasper, et al. (Editors), *Reticular Formation of the Brain.* Boston: Little, Brown, 1958. Pp. 69–108.

Arden, G. B., and D. P. Greaves. The reversible alterations of the electroretinogram of the rabbit after occlusion of the retinal circulation. *J. Physiol.,* 1956, **133,** 266–274.

Arden, G. B., and Y.-M. Liu. Some types of response of single cells in the rabbit lateral geniculate body to stimulation of the retina by light and to electrical stimulation of the optic nerve. *Acta physiol. scand.,* 1960a, **48,** 36–48.

Arden, G. B., and Y.-M. Liu. Some responses of the lateral geniculate body of the rabbit to flickering light stimuli. *Acta physiol. scand.,* 1960b, **48,** 49–62.

Batini, C., F. Magni, M. Palestini, G. F. Rossi, and A. Zanchetti. Neural mechanisms underlying the enduring EEG and behavioral activation in the midpontine pretrigeminal cat. *Arch. ital. Biol.*, 1959a, **97,** 13–25.

Batini, C., G. Moruzzi, M. Palestini, G. F. Rossi, and A. Zanchetti. Effects of complete pontine transsections on the sleep-wakefulness rhythm: the midpontine pretrigeminal preparation. *Arch. ital. Biol.*, 1959b, **97,** 1–12.

Batini, C., M. Palestini, G. F. Rossi, and A. Zanchetti. EEG activation patterns in the midpontine pretrigeminal cat following sensory deafferentation. *Arch. ital. Biol.*, 1959c, **97,** 26–32.

Baumgarten, R. von, and Jung, R. Microelectrode studies on visual cortex. *Rev. Neurol.*, 1952, **87,** 151–155.

Bishop, P. O., W. Burke, and R. Davis. Activation of single lateral geniculate cells by stimulation of either optic nerve. *Science*, 1959, **130,** 506–507.

Bishop, P. O., and R. Davis. Bilateral interaction in the lateral geniculate body. *Science*, 1953, **118,** 241–243.

Bohm, E., and B. Gernandt. Comparison of off/on-ratios in retina and geniculate body. *Acta physiol. scand.*, 1950, **21,** 187–194.

Bornschein, H. Spontan- und Belichtungsaktivität in Einzelfasern des N. Opticus der Katze. I. Der Einfluss kurzdauernder retinaler Ischämi. II. Der Einfluss akuter Jodazetatvergiftung. *Z. Biol.*, 1958, **110,** 210–231.

Bremer, F. Cerveau "isolé" et physiologie du sommeil. *C. R. Soc. Biol., Paris,* 1935, **118,** 1235–1241.

Bremer, F., and N. Stoupel. Facilitation et inhibition des potentials évoqués corticaux dans l'éveil cérébral. *Arch. int. Physiol. Biochim.*, 1959, **67,** 240–275.

Brouwer, B. *Anatomical, Phylogenetical and Clinical Studies on the Central Nervous System.* The Herter Lectures of the Johns Hopkins Univ., Vol. 17. Baltimore: Williams and Wilkins, 1927.

Brouwer, B., W. P. C. Zeeman, and A. W. Houwer. Projection of retina on primary centers. *Schweiz. Arch. Neurol. Psychiat.*, 1923, **13,** 118–137.

Bullock, T. H. Neuronal integrative mechanisms. In B. T. Scheer and T. H. Bullock (Editors), *Recent Advances in Invertebrate Physiology.* Eugene, Oregon: Univ. of Oregon Press, 1957.

Bürgi, S. Das Tectum Opticum, seine Verbindungen bei der Katze und seine Bedeutung beim Menschen. *Dtsch. Z. Nervenheilk.*, 1957, **176,** 701–729.

Cajal, S. R. *Histologie du Système Nerveux, II.* Madrid: Instituto Ramón y Cajal, 1955. Chapter 15.

Chavez, M., and E. A. Spiegel. The functional state of sensory nuclei following deafferentation. *Confin. Neurol.*, 1957, **17,** 144–152.

Cohn, R. Laminar electrical responses in lateral geniculate body of cat. *J. Neurophysiol.*, 1956, **19,** 317–324.

De Valois, R. L., C. J. Smith, S. T. Kitai, and A. J. Karely. Response of single cells in monkey lateral geniculate nucleus to monochromatic light. *Science*, 1958, **127,** 238–239.

Dodt, E. Centrifugal impulses in rabbit's retina. *J. Neurophysiol.*, 1956, **19,** 301–307.

Erulkar, S. D., and M. Fillenz. Patterns of discharge of single units of the lateral geniculate body of the cat in response to binocular stimulation. *J. Physiol.*, 1958, **140,** 6P–7P.

Evarts, E. V., and J. R. Hughes.   Relation of posttetanic potentiation to sub-normality of lateral geniculate potentials. *Amer. J. Physiol.,* 1957*a*, **188,** 238–244.

Evarts, E. V., and J. R. Hughes.   Effects of prolonged optic nerve tetanization on lateral geniculate potentials. *Amer. J. Physiol.,* 1957*b*, **188,** 245–248.

Gillilan, L. A.   The connections of the basal optic root (posterior accessory optic tract) and its nucleus in various mammals. *J. comp. Neurol.,* 1941, **74,** 367–408.

Granit, R.   Centrifugal and antidromic effects on ganglion cells of retina. *J. Neurophysiol.,* 1955, **18,** 388–411.

Granit, R., and C. G. Phillips.   Excitatory and inhibitory processes acting upon individual Purkinje cells of the cerebellum in cats. *J. Physiol.,* 1956, **133,** 520–547.

von Gudden, B.   Über den Tractus peduncularis transversus. *Arch. Psychiat.,* 1881, **11,** 415–423.

Hernández-Peón, R., C. Guzmán-Flores, M. Alcaraz, and A. Fernández-Guardiola.   Sensory transmission in visual pathway during "attention" in unanesthetized cats. *Acta neurol. Latinoamer.,* 1957, **3,** 1–8.

Hernández-Peón, R., H. Scherrer, and M. Velasco.   Central influences on afferent conduction in the somatic and visual pathways. *Acta neurol. Latinoamer.,* 1956, **2,** 8–22.

Horn, G., and J. Blundell.   Evoked potentials in visual cortex of the unanaesthetized cat. *Nature, Lond.,* 1959, **184,** 173–174.

Hubel, D. H.   Single unit activity in striate cortex of unrestrained cats. *J. Physiol.,* 1959, **147,** 226–238.

Jung, R.   Coordination of specific and nonspecific afferent impulses at single neurons of the visual cortex.   In H. H. Jasper, et al. (Editors), *Reticular Formation of the Brain.* Boston: Little, Brown, 1958. Pp. 423–434.

Katsuki, Y., T. Watanabe, and N. Maruyama.   Activity of auditory neurons in upper levels of brain of cat. *J. Neurophysiol.,* 1959, **22,** 343–359.

Kuffler, S. W.   Neurons in the retina: Organization, inhibition and excitation problems. *Cold Spring Harbor Sympos. quant. Biol.,* 1952, **17,** 281–292.

Kuffler, S. W., R. Fitzhugh, and H. B. Barlow.   Maintained activity in the cat's retina in light and darkness. *J. gen. Physiol.,* 1957, **40,** 683–702.

Lashley, K. S.   Mechanism of vision; autonomy of visual cortex. *J. genet. Psychol.,* 1942, **60,** 197–221.

Li, C.-L., and H. Jasper.   Microelectrode studies of electrical activity of cerebral cortex in the cat. *J. Physiol.,* 1953, **121,** 117–140.

Long, R. G.   Modification of sensory mechanisms by subcortical structures. *J. Neurophysiol.,* 1959, **22,** 412–427.

Marriott, E. H. C., V. B. Morris, and M. H. Pirenne.   The absolute visual threshold recorded from the lateral geniculate body of the cat. *J. Physiol.,* 1959, **146,** 179–184.

Martin, A. R., and C. L. Branch.   Spontaneous activity of Betz cells in cats with midbrain lesions. *J. Neurophysiol.,* 1958, **21,** 368–370.

Moruzzi, G., and H. W. Magoun.   Brain stem reticular formation and activation of the EEG. *EEG clin. Neurophysiol.,* 1949, **1,** 455–473.

Mountcastle, V. B., P. W. Davies, and A. L. Berman. Response properties of neurons of cat's somatic sensory cortex to peripheral stimuli. *J. Neurophysiol.*, 1957, **20**, 374–407.

Noell, W. K. Site of asphyxial block in mammalian retinae. *J. appl. Physiol.*, 1951, **3**, 489–500.

Noell, W. K., and H. I. Chinn. Failure of visual pathway during anoxia. *Amer. J. Physiol.*, 1950, **161**, 573–590.

O'Leary, J. Structural analysis of lateral geniculate nucleus of cat. *J. comp. Neurol.*, 1940, **73**, 405–430.

Posternak, J. M., T. C. Fleming, and E. V. Evarts. Effect of interruption of the visual pathway on the response to geniculate stimulation. *Science*, 1959, **129**, 39–40.

Rose, M. Das Zwischenhirn des Kaninchens. *Mém. Acad. Pol. Sci. Lettr.* (*Cl. Sci. math. nat.*), 1935, **B8**, 5–104.

Scheibel, M. E., and A. B. Scheibel. Structural substrates for integrative patterns in the brain stem reticular core. In H. H. Jasper, et al. (Editors), *Reticular Formation of the Brain*. Boston: Little, Brown, 1958. Pp. 31–55.

Schoolman, A., and E. V. Evarts. Responses to lateral geniculate radiation stimulation in cats with implanted electrodes. *J. Neurophysiol.*, 1959, **22**, 112–129.

Schubert, G. Ein entoptisches Hypoxie-Phenomen. *Z. Biol.*, 1958, **110**, 232–235.

Söderberg, U., and G. B. Arden. Single unit activity in the rabbit lateral geniculate body during experimental epilepsy. In R. Jung and H. Kornhuber (Editors), *The Visual System: Neurophysiology and Psychophysics*. Berlin, Göttingen, Heidelberg: Springer, 1961.

Tasaki, I., E. H. Polley, and F. Orrego. Action potentials from individual elements in cat geniculate and striate cortex. *J. Neurophysiol.*, 1954, **17**, 454–474.

Walls, G. L. *The Vertebrate Eye and its Adaptive Radiation*. Cranbrook Institute of Science, Bull. No. 19. Bloomfield Hills, Mich.: Cranbrook Press, 1942.

Walls, G. L. *The Lateral Geniculate Nucleus and Visual Histophysiology*. *Univ. Calif. Publ. Physiol.*, 1953, **9**, 1–100. Berkeley and Los Angeles: Univ. of Calif. Press, 1953.

28

KENNETH D. ROEDER
Tufts University

ASHER E. TREAT
The College of the City of New York

The Reception of Bat Cries by the Tympanic Organ of Noctuid Moths

Analysis of Neural Information in Multi-Unit Systems

Description of the behavior of an animal in terms of the individual contributions and interactions of its receptor cells, neurons, and effector units is the major objective of a considerable segment of biological research. This massive problem can hardly be touched until it is possible to describe what goes into the nervous system, that is, the form in which aspects of the external world are coded in the digital pattern of nerve impulses that appears to be the only form of rapid communication between one part of the nervous system and another.

Microelectrode techniques have revealed local graded events of unsuspected variety and complexity on the somatic and dendritic surfaces of receptor cells and neurons (see Bullock in this volume), but they have not shaken the classical belief that the interaction of neurons at distances greater than 0.5 millimeter or so depends upon the propagation of a pattern of digital nerve impulses. However complex the "domestic" affairs of a neuron or neuron group may be, each neuron reports at a distance in the simplest of symbols and in a code that is reasonably easy to interpret.

The electrical signals of intraneuronal events and unit fiber responses have been detected in most parts of the vertebrate nervous system. In most cases the value of these individual signals in assessing the sensory input that releases some adaptively significant behavior pattern is akin to that of a public opinion poll conducted on a population of unknown size by questioning one or two individuals who are able only to answer "yes." It is usually not possible to obtain even a reasonable simultaneous sample of the activity of the many thousands of receptors that in vertebrates run in parallel from the major sensory surfaces to the central nervous system. Since this is technically so difficult at present, another approach is to choose a sensory surface so simple that most or even all of the nerve-impulse traffic between it and the central nervous system can be decoded at one time.

The insects have much to offer as subjects in the solution of this general problem. Their behavior is often highly complex and at the same time stereotyped and predictable. In insects some axons are larger in diameter than those found in vertebrates, and, since the body size of insects is limited for other reasons, the total neuron population of an insect's central nervous system is of necessity much smaller. This implies an economy of neurons involved in any given behavior pattern, particularly when speed is an important response characteristic, as in the actions of prey and predator (Roeder, 1959). For instance, the startle pattern of a cockroach to a puff of air depends on three or four internuncial neurons, the jumping movement of a locust is evoked by a short burst of impulses in a single motor nerve fiber (Hoyle, 1955), and the total feeding pattern of a fly can be released by applying sugar solution to a single chemosensory cell (Dethier, 1955; Hodgson and Roeder, 1956). It is the object of this paper to illustrate these possibilities by describing the information-gathering capacity of the two-receptor system found in the ear of certain moths.

Bats and Moths

The brilliant and extensive studies by Griffin and others (reviewed by Griffin, 1958) have proved beyond doubt that bats flying in darkness are able to locate quite small objects by means of the echoes of a succession of vocal pulses. These pulses contain fundamental frequencies as high as 80 kilocycles per second, have durations as short as 1 millisecond, and may be repeated by the bat 5 to 100 times a second. Some of the prey of insectivorous bats, ranging from small midges to fair-sized moths, appear to be located on the wing from the echoes of such pulses.

Speculation that certain moths are able acoustically to detect the approach and thereby avoid the attack of predatory bats goes back to 1877 (White, 1877). An auditory function was assigned to the tympanic organ (Eggers, 1919, 1925), and it was shown that moths of several families react to ultrasonic stimuli by changes in flight pattern (Schaller and Timm, 1950; Treat, 1955). Haskell and Belton (1956) recorded tympanic nerve responses to stimuli from 8 to 20 kcps, a range that was extended by Roeder and Treat (1957) to over 100 kcps.

Ultrasonic reception by the tympanic organ has been demonstrated by electrophysiological methods in moths of the families Noctuidae, Notodontidae, Arctidae, Amatidae, Geometridae, and Cymatophoridae. A tympanic organ is absent in the Sphingidae, Saturnidae, and most of the Cossidae—families containing generally large-sized species. Many families of microlepidoptera also lack known auditory organs.

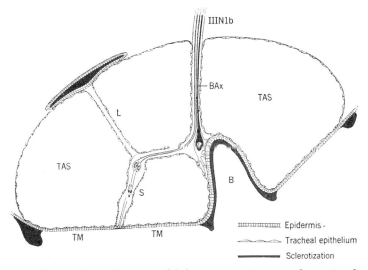

Fig. 1. Schematic frontal section of left tympanic air sac and associated structures, redrawn with modifications from Eggers (1919). B, *Bügel;* BAx, axon of B cell; L, ligament; S, scoloparium containing acoustic sense cells and scolopes; TAS, tympanic air sac; TM, tympanic membrane; IIIN1b, tympanic nerve.

The Tympanic Organ

The anatomy of the tympanic organ has been described in detail elsewhere (Eggers, 1919; Roeder and Treat, 1957; Treat, 1959) and only a brief account of its structure will be given here (Fig. 1). The tympanic membrane faces obliquely rearward into a concavity between the metathorax and the abdomen. The tympanic scoloparium is attached to the inner surface of the tympanic membrane and extends obliquely forward across the air-filled tympanic cavity. It is suspended loosely in the tympanic cavity by its attachment to the tympanic membrane, by a ligament, and by the tympanic nerve which arises from it. The scoloparium contains two receptor cells (A cells), each of which extends in the direction of the tympanic membrane as a fine process, terminating in a highly refractile object known as a scolops. The two scolopes usually lie at slightly different levels within the scoloparium and may differ slightly in size. They are thought to be the sound-sensing structures, but much remains to be learned about their ultrastructure and physiology.

The tympanic nerve, composed of two A axons, passes from the scoloparium to a cuticular support, the *Bügel*, where it is joined by the axon of a large stellate or pear-shaped cell, the B cell, lying in the tracheal epithelium that ensheathes the *Bügel*. The tympanic nerve, now containing two A axons and the B axon, leaves the tympanic cavity and passes forward and downward to join a major nerve trunk and eventually enter the dorsolateral surface of the pterothoracic ganglion. Most of the nerve activity described below was picked up by an electrode hooked under the tympanic nerve near its junction with the major nerve trunk. An indifferent electrode was inserted usually into the abdomen.

The Tympanic Nerve Response

Nerve impulses generated by the two acoustic receptor cells are recorded only from an electrode placed as described above. Destruction of the tympanic scoloparium abolishes all neural and behavioral responses to ultrasound. Therefore, the nerve spike pattern in the tympanic nerve or, more correctly, in both tympanic nerves, appears to contain all the impulse-coded information reaching the insect via this sense modality. An electrode under the tympanic nerve in unstimulated condition registers an irregular and apparently spontaneous discharge of spikes attributable to the A fibers, and a more-or-less regular discharge of larger spikes that originate in the B cell. The B spikes recur 5 to 20 times a second in most cases and are completely uninfluenced by acoustic stimulation or by the A fiber response, but they can be driven to higher frequencies by mechanical distortion of the sclerites enclosing the tympanic organ. The behavioral role of the B cell is unknown, but its physiological activity has been described in some detail (Treat and Roeder, 1959).

The acoustic response is illustrated in Fig. 2, which shows the effects of a continuous pure tone at different intensities. In (a) the intensity was just above threshold for one A unit. The tone was switched on abruptly and the switching transient generated a short burst of spikes followed by a ragged spike discharge as the tone continued. With an increase in sound intensity (b) the A discharge increased in frequency but adapted rapidly, dropping to about 50 per cent of the initial frequency in 0.1 sec. B fiber activity can be distinguished in records (a) and (b) as the larger, slowly recurring spike. In (c) a further increase in sound intensity pushed the A unit to higher fre-

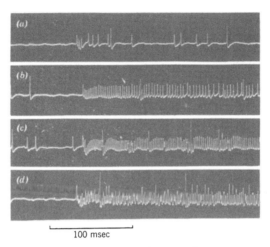

100 msec

Fig. 2. Tympanic nerve response of *Prodenia eridania* to a pure tone of 40 kcps. (a) Threshold response in most sensitive acoustic unit. Sound intensities: (b) 7 db, (c) 16 db, (d) 23 db above sound intensity of (a).

quencies of discharge, and the appearance of occasional double spikes indicates responses in the second, less sensitive A unit. In (d) both A units responded, the more sensitive at a steady high frequency and the less sensitive at some lower frequency. Adaptation of the less sensitive A unit is evident toward the end of the record, whereas the more sensitive A unit maintained a high discharge frequency, the stimulus being so far above its threshold. In this case the thresholds of the two A fibers to the 40-kcps tone differed by 15 to 20 db. A similar situation has been found in other noctuid species. There is no evidence that this threshold difference is dependent upon the frequency of the stimulating tone.

The response of the tympanic organ to sound pulses is illustrated in Fig. 3. Here again, the two-step nature of the response is evident. In (a) and (b) the intensity of the sound pulse was sufficient to excite only the more sensitive A unit, which responded with two and three spikes, respectively. In (c) the higher sound intensity elicited four spikes from the more sensitive A unit, and one spike (interposed between the first pair of spikes from the more sensitive unit) in the less sensitive A unit. In (d) both A units responded with several spikes.

These and other laboratory studies (Roeder and Treat, 1957) make it possible to summarize the characteristics of the A receptors in the noctuid tympanic organ.

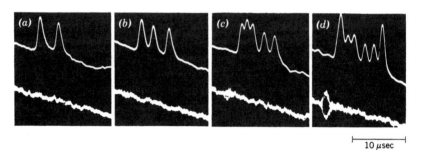

10 μsec

Fig. 3. Tympanic nerve response of *Prodenia eridania* to a brief shaped pulse of sound at 30 kcps. Upper trace, tympanic nerve response; lower trace, sound pulse recorded by condenser microphone about 1 foot farther from sound source than the tympanic preparation. Intensity of the pulse was increased from (a) through (d).

1. The intensity of an ultrasonic stimulus is neurally coded (a) by the two steps in sensitivity of the A units, (b) by the frequency of discharge in one or both A units, and (c) with short sound pulses, by the length of the after-discharge in one or both A units (see Fig. 3).

2. The frequency range of the sound to which the A units respond extends from about 3 kcps to over 100 and possibly as high as 240 kcps. Maximum sensitivity appears to lie between 15 and 60 kcps, although this measurement may have been limited by the frequency range of the available measuring instruments.

3. There is no evidence for any neural mechanism whereby the tympanic organ can discriminate between different sound frequencies within this range.

4. The rapid adaptation of the A units to a continuous tone and the tendency to continue to discharge spikes for some time after the cessation of a click or short sound pulse suggest that the tympanic organ is particularly fitted for the reception of pulsed sounds. After-discharges lasting 60 to 70 msec have been obtained to sound pulses with a duration of only 1 to 2 msec under approximately anechoic conditions in the field (see below).

Tympanic Nerve Responses to Bat Cries

These characteristics leave little doubt that the tympanic organ is adapted to respond to the cries of bats. Direct demonstration of this was obtained when a bat was released and flew in the laboratory (Roeder and Treat, 1957). (A simultaneous recording of bat cries and tympanic nerve responses will be shown in Fig. 5.) However, the laboratory experiments left unanswered several questions: the maximum range of the tympanic organ as a bat detector, its function as a trigger for evasive or other action on the part of the moth, the directional properties of binaural preparations, and the response to other sounds likely to be encountered by moths in flight.

Attempts to answer some of these questions were begun with field experiments in July 1958. A great deal remains to be done, particularly with respect to localization of the sound source, but some of the preliminary observations will be reported below. The first observations were made with a single channel of recording equipment at the summer home of one of us (Treat). This is situated in a hilly rural area near Tyringham, Massachusetts, and was chosen because the activities of bat and moth populations in the region had been under study for a number of years. Later experiments were carried out at Tyringham and Concord with two channels of recording equipment and a stereophonic tape recorder.

A table bearing a dissecting microscope, electrode manipulators, and preamplifiers (Grass P-8) was placed in an area over which feeding bats were likely to fly. Moths were captured at an ultraviolet fluorescent light and restrained by pinning on cork. The tympanic nerves were dissected with the insect either dorsal side or ventral side up instead of sagittally transected as in the earlier experiments (Roeder and Treat, 1957). Surrounding apparatus was arranged on the table so that both tympanic organs were exposed to a relatively unobstructed sound field above the horizontal plane. The preamplifiers were connected by 50-ft leads to a-c operated equipment so as to minimize electrical and acoustic artifacts.

The cries of bats

The same background of apparently random and spontaneous A spikes and the regularly recurrent B spike was obtained under these field conditions even when the night was windless and apparently quiet. The arrival of a bat in the area was unequivocally signaled by groups of spikes recurring 8 to 20 times a second (Fig. 4). The move-

ments of the bats were followed as far as possible by means of a floodlight, and it became evident that the number and frequency of the
spikes in each group increased as the bat approached (Fig. 4a).
In a loudspeaker the tympanic nerve response took the form of a series
of short tones that rose in pitch and increased in duration as the bat
approached. The reverse occurred as a bat passed over and away
from the preparation. Each group contained from 1 to 20 spikes,
recurring initially 700 to 800 times per second and dropping at the

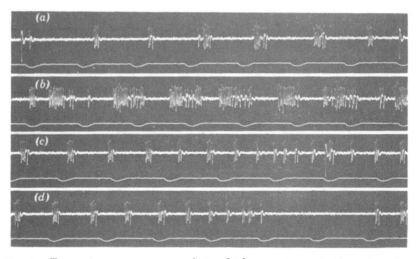

Fig. 4. Tympanic nerve response of *Graphiphora c-nigrum* to the cries of a
naturally feeding bat in the field. (*a*) An approaching bat emitting cruising
pulses; (*b*) pulses emitted when bat was close to the preparation; (*c*) and (*d*)
continuous record of an approach and two buzzes by a bat. Time line, 100-msec
intervals.

end of the group to an irregular 100 to 200 per second. In groups
containing less than 7 or 8 spikes, only one of the A fibers was active
(Fig. 4a); in longer groups, spikes belonging to both A fibers were
evident (Fig. 4b).

After a little experience it became quite easy to interpret some of
the maneuvers of the flying bats from the tympanic nerve response
rendered audible by a loudspeaker or visible by an oscilloscope. A
cruising bat elicited a fairly regular sequence of spike groups, usually
5 to 20 per second. The change in pitch and duration of individual
spike groups gave a rough indication of the approach, nearness, and
departure of the bat, as described above. A sudden increase in the
repetition frequency of the groups indicated an increase in the fre-

quency of the bat's cries—the "buzz" made by a bat when it detects and attacks a flying insect (Griffin, 1958, p. 186 ff.). Two buzzes appear in Fig. 4c and d and one in Fig. 5f. Although the bat was not observed during the buzz recorded in Fig. 4, the timing of the spike groups suggests that the bat made two attacks in rapid succession. The pause following the second buzz (Fig. 4d) perhaps indicates that

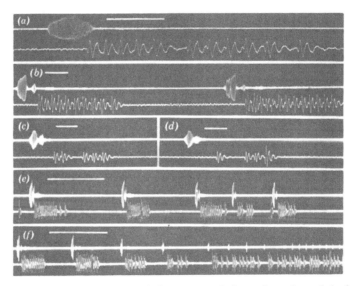

Fig. 5. Simultaneous recording of the cries made by a flying bat while feeding in the laboratory (upper traces), and the tympanic nerve responses of a moth (*Agroperina dubitans*) to the same cries (lower traces). The spike potentials have been distorted by filtering. (a) to (d) Responses to cries of varying intensity; (e) and (f) continuous sequence of cries ending in a buzz. Time line: 10 msec in (a) through (d), 100 msec in (e) and (f). For further details, see text.

the bat captured its prey on the second pass. Fig. 4b was obtained as a bat flew close to the table bearing the preparation. Both A fibers are active, and the after-discharge elicited by each cry lasts for 60 to 70 msec. The double nature of some of the spike groups in this sequence suggests that the tympanic organ was detecting both the original cry of the bat and its echo from a neighboring wall. Separation of the spikes into two and possibly three subgroups in Fig. 5 indicates the detection of echoes, some of which are also apparent in the acoustic record.

A more vivid picture of the relation between the physical aspects of the bat's cry and the form in which it is coded by the moth's ear is provided by Fig. 5. These records were obtained during an experiment carried out in collaboration with Frederick Webster of the Lincoln Laboratories, Massachusetts Institute of Technology, who kindly made available ultrasonic microphones and high-speed tape-recording equipment, as well as captive bats and a room especially constructed for studying their flight. A Granith microphone recorded the cries made by a bat (*Myotis*) flying in a restricted space. The same cries were detected by the tympanic organ of a moth (*Agroperina dubitans;* lower trace) on a table 8 to 10 feet away from the microphone. Both were recorded on tape at 120 inches per second, and subsequently played back at 3.75 inches per second, at which time the oscillogram was made. The bat was generally nearer to the microphone when the latter registered its cries, so that the latency of the nerve response reflects the difference in the lengths of the sound paths to the physical and to the biological detectors, as well as the sensory latency and the conduction time in the tympanic nerve. The relation between the duration of the bat cry and the tympanic nerve response is shown in Fig. 5a. Figures 5b, c, and d show variations in the number of spikes in the spike group as the loudness of the bat cry varies at the tympanic organ. Figures 5e and f show the details of a buzz. During the rapid succession of bat cries the tympanic nerve response becomes almost continuous.

Range

The sensitivity of the tympanic organ is such that a bat in flight is detected in all parts of even a very large room. In the field the tympanic organ responded to the cries of bats flying outside the range of vision with the available floodlight.

The range of the tympanic organ was estimated by observing bats (probably *Myotis*) as they left their roost in a barn during the early dusk. The table bearing the preparation was placed about 200 yards from the barn and in the path taken by most of the bats, many of which flew a fairly straight course. An observer listened to the tympanic nerve discharge through headphones as he moved "upstream" between the preparation and the roost. Since the bats were quite visible at this hour and generally flew singly, it was not difficult to find the point at which the characteristic nerve response began while a bat was flying directly over the observer. The ground distance was 100 feet from the preparation for a bat approaching at an estimated altitude of 30 to 40 feet. Retreating bats while flying at the same altitude were detected until they had passed 30 to 40 feet beyond the preparation. These measurements were obtained with a specimen of *Graphiphora c-nigrum,* one of the more sensitive noctuid species examined.

Experiments with artificial sound pulses and clicks have shown that the tympanic organ is capable of discriminating differences in sound intensity. This would seem to enable a moth to detect changes in the range of a flying bat. Changes in the range of a sound source can be determined from its relative loudness only if the source is relatively nondirectional, its signals are constant in intensity and, when the tympanic organ is the detector, of constant duration. A bat in natural flight may vary both the intensity and duration of its cries which are emitted in a somewhat directional pattern (Griffin, 1958); consequently, even though the tympanic organ is capable of serving as a good range detector, its effectiveness in this respect may be reduced by maneuvers of the bat. Changes in the number of spikes contained in each of the successive spike groups of Fig. 4*a* suggest an approaching bat. In other recordings the number of spikes in consecutive groups varied in an irregular manner even when the bat was known to have made a more-or-less straight approach. Presumably this was due to the causes listed above. The decline in the number of spikes in each group during the two buzzes in Figs. 4*c* and *d* does not indicate that the bat was at this time moving away, since the intensity of the cries is known to drop as their frequency increases during a buzz (Griffin, 1958).

Direction

Attempts are being made to find out whether the direction of a source of pulsed sound is encoded in the tympanic nerve response. Insufficient data are available, but the preliminary observations are of some interest.

The directional sensitivity of a single tympanic organ was mapped in the field so as to minimize the effects of echoes. The sound source was a plane-surfaced condenser loudspeaker (Roeder and Treat, 1957). Its radiating surface was directed vertically upward and at a spherical reflecting surface, so that the sound was radiated in a more-or-less nondirectional manner. Square pulses of 1-msec duration and constant intensity, fed into the loudspeaker, provided a series of clicks of constant intensity.

This sound source was placed on radii to the moth at 45-degree intervals and moved toward and away from the tympanic preparation until a tympanic nerve response was elicited of arbitrarily chosen magnitude (between one and two spikes in the most sensitive A fiber for a dozen or so clicks). The distance along the radius at which this arbitrary response occurred was then measured and entered on a polar plot for each of the eight radii. Measurements have not yet been made in planes other than with the moth inclined 30 degrees to the horizontal.

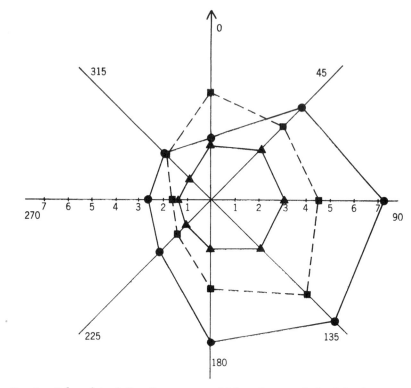

Fig. 6. Polar plot of the distances at which a 1-msec click of fixed intensity elicits the same response from the right tympanic organ when placed at various angles relative to the median axis of the moth (0 to 180 degrees). The moth was headed toward 0 degree and inclined upward at about 30 degrees to the plane of the measurements. Distances in meters from the moth are indicated on the 90-to-270 degree line. Triangles and solid line, *Acronycta;* squares and broken line, *Graphiphora;* circles and solid line, *Lucania.*

The method was sensitive and gave reproducible results for a given specimen. The three species tested (Fig. 6) showed very similar directional properties, although in this and other experiments both different species and individuals of species showed marked differences in sensitivity.

The somewhat directional sensitivity of a single tympanic organ suggests that the cries of a bat might produce a differential response in right and left tympanic organs which would locate the position of the bat relative to the moth. A number of binaural recordings were made from single specimens as bats maneuvered overhead.

The results of these experiments are unclear and require further analysis. Since the threshold response of the tympanic organ is all or none (the presence of a single A spike), a differential response was more evident when bats were flying at extreme range. The bats (probably *Myotis*) remained at some distance in this experiment, and the length of the after-discharge and the spike frequency in the two nerves gave some indication of the movements of the bats. This differential effect was evident when the tape was played stereophonically and the nerve activity was made audible.

In another experiment the relation between the moth preparation and the movements of the bats seemed to be ideal for binaural localization. Red bats (*Lasiurus borealis*) repeatedly approached and passed over the experimental table at an altitude of 6 to 10 feet and at right angles to the long axis of the moth. Under these circumstances noctuids of several species gave little evidence of a consistent change in the differential response in the two ears. In most cases both ears detected the bat simultaneously, and the number of spikes in the groups recorded from the right and left tympanic nerves seldom differed by more than one spike in groups of five or six. Occasionally one ear would completely miss a bat cry registered by the other, even though it registered preceding and following cries. Frequently the buzz was registered by one ear only, but on the whole it was not possible to determine from the binaural recordings whether a bat was approaching from the right or left side. Since these findings do not correspond with what would be expected from the directional studies made with a single tympanic organ, the possibility of localization by moths must await further study.

Conclusions

The main object of this paper has been to determine the informational content of the tympanic nerve response to the cries of bats. Detection, and in a sense amplification, of pulsed sounds against a continuous background is made possible by the complementary properties of rapid adaptation to a continuous tone and after-discharge to a brief sound pulse. The considerable range and low directionality to pulsed sounds are what would be expected of a warning device.

As a detector of changes in the range of a sound source the tympanic organ would seem to be quite effective. However, in the detection of the range of a bat by its cries, this property may be counteracted somewhat by variations in the loudness of the bat's cries. In spite of the directional properties of each ear, the ability of a binaural preparation to detect the direction from which a bat is approaching seems to be quite limited. The bat's direction can be determined from the differential spike response in both ears only when the intensity of the cries reaching the moth is quite low. In the field this occurs when the bat is at extreme range or when the cries are emitted at lower intensity during a buzz.

It must be emphasized that these conclusions are based entirely upon what can be read from the afferent spike potentials in the tympanic nerve. They cover the information available to the moth via its auditory sense, and, although they may set limits, they can in no way predict the reactions the moth might show to auditory stimuli. Under laboratory conditions, intact moths, whether restrained or in free flight, have rarely shown clear behavioral reactions to actual or recorded bat cries, even while responsive to sounds of other kinds such as those of the Galton whistle, the jingle of keys, and hissing sounds made orally by the observer (Treat, 1955; and unpublished observations by the present authors). Until the reason for this is known, the evasive behavior observed in the presence of hunting bats in the field cannot with certainty be attributed to acoustic stimuli alone, although it seems scarcely credible that so valuable an adaptation for the detection of an approaching predator could have gone unexploited.

Thus we are in the anomalous position of knowing more about the information-gathering properties of a sense organ from its afferent output than about the behavior with which it is concerned. It will be necessary to make a closer study of the behavior of free-flying moths in the presence of recorded bat cries under various conditions, as well as the much more difficult task of observing the conditions and results of actual encounters between bats and moths in the field. Electrophysiological studies of the synaptic interaction of the A fibers and B fiber with second-order neurons within the pterothoracic ganglion may also help to answer some of these questions.

Further field studies will be directed also to the question whether ultrasonic stimuli other than bat cries are behaviorally significant to moths. Tympanic responses have been reported to sounds associated with the flight movements of other moths at close range (Roeder and Treat, 1957), and in the present field experiments, tympanic nerve responses were obtained to the stridulations of several insects, some at a considerable distance. Regular tympanic responses were also noted to stimuli of unknown origin and nature.

Summary

1. Some of the difficulties in assessing the informational capacity of the multi-unit sensory surfaces found in vertebrates can be avoided by working with the relatively simple sensory systems often encountered in invertebrates.

2. The tympanic organ of noctuid moths contains two acoustic receptor cells and one nonacoustic sense cell of unknown function.

3. Electrophysiological study of activity in the tympanic nerve shows that the acoustic receptors differ only in threshold. One or both will respond, depending on the sound intensity, from 3 kcps to over 100 kcps. There is no evidence of pitch discrimination. Sound intensity is neurally coded by the two steps in sensitivity provided by the acoustic cells, by the frequency of the spike discharge in one or both, and in the case of brief sound pulses by the length of the after-discharge. Both acoustic cells adapt rapidly to a pure tone and appear to be specialized for the reception of brief sound pulses.

4. The response of the tympanic organ to the echo-locating cries made by flying bats was studied in the field. Bats could be detected at distances of 100 feet or more by means of the characteristic sequence of spike groups in the tympanic nerve. This pattern was readily distinguished from tympanic responses to other sounds, and the range and many of the maneuvers made by a flying bat could be interpreted from the nerve recording.

5. Each tympanic organ is somewhat although not markedly directional to the source of a sound pulse. In some cases the direction of a bat relative to the preparation could be interpreted from binaural recordings when its cries reached the moth at low intensity. In other cases the directional significance of the binaural responses was obscure.

Acknowledgment

This work was supported by grants from the National Science Foundation and the U. S. Public Health Service. Some of the equipment used was obtained under a previous contract between the U. S. Chemical Corps and Tufts University.

References

Dethier, V. G.   The physiology and histology of the contact chemoreceptors of the blowfly. *Quart. Rev. Biol.*, 1955, **30,** 348–371.

Eggers, F.   Das thoracale bitympanale Organ einer Gruppe der Lepidoptera Heterocera. *Zool. Jahrb.* (*Abt. Anat.*), 1919, **41,** 273–376.

Eggers, F.   Versuche über das Gehör der Noctuiden. *Z. vergl. Physiol.,* 1925, **2,** 297–314.

Griffin, D. R.   *Listening in the Dark.*   New Haven, Conn.: Yale University Press, 1958.

Haskell, P. T., and P. Belton.   Electrical responses of certain lepidopterous tympanal organs. *Nature, Lond.,* 1956, **177,** 139–140.

Hodgson, E. S., and K. D. Roeder.   Electrophysiological studies of arthropod chemoreception.   1. General properties of the labellar chemoreceptors of Diptera. *J. cell. comp. Physiol.,* 1956, **48,** 51–76.

Hoyle, G.   Neurophysiological mechanism of a locust skeletal muscle. *Proc. roy. Soc.,* 1955, **B143,** 343–367.

Roeder, K. D.   A physiological approach to the relation between prey and predator. *Smithson. misc. Coll.,* 1959, **137,** 287–306.

Roeder, K. D., and A. E. Treat.   Ultrasonic reception by the tympanic organ of noctuid moths. *J. exp. Zool.,* 1957, **134,** 127–158.

Schaller, F., and C. Timm.   Das Hörvermögen der Nachtschmetterlinge. *Z. vergl. Physiol.,* 1950, **32,** 468–481.

Treat, A. E.   The response to sound in certain lepidoptera. *Ann. ent. Soc. Amer.,* 1955, **48,** 272–284.

Treat, A. E.   The metathoracic musculature of *Crymodes devastator* (Brace) (Noctuidae) with special reference to the tympanic organ. *Smithson. misc. Coll.,* 1959, **137,** 365–377.

Treat, A. E., and K. D. Roeder.   A nervous element of unknown function in the tympanic organs of moths. *J. Insect Physiol.,* 1959, **3,** 262–270.

White, F. B.   (Untitled communication) *Nature, Lond.,* 1877, **15,** 293.

29

YASUJI KATSUKI

Tokyo Medical and Dental University

Neural Mechanism
of Auditory Sensation in Cats

Concerning the function of the cochlea of the higher animals, Helmholtz first established the hypothesis called the "resonance theory" and later the "place theory" in a strict sense. This theory says that the sound waves that have reached the cochlea are analyzed completely in the cochlea, and, from each cochlear partition, nerve discharges evoked by the component sounds, into which the original sound has been analyzed, are transmitted through nerve fibers to the auditory cortex. Thus the frequency and the intensity of sound are perceived at the cerebral cortex.

In contrast, there is another hypothesis called "the telephone theory," originated by Rutherford (1886). According to this hypothesis, the sound waves are not analyzed in the cochlea at all. Each nerve fiber sends signals to the auditory cortex, like a telephone, and the analysis of sound is made completely in the neural network in the brain. However this latter hypothesis had no concrete experimental evidences in Rutherford's time, and the earlier hypothesis was therefore more widely believed to be valid.

In 1943 Békésy made it possible, by an ingenious method, to see the actual vibratory movement of the cochlear partition of human ears, as well as those of certain mammals, and he demonstrated that the sound analysis in the cochlea is made in a specific way. According to his observation, a considerable region of the cochlear partition is forced into vibration by even a single, though not necessarily a strong, pure tone. Following this discovery, Tasaki, Davis, and Legouix (1952) also confirmed the above conclusion (the space-time pattern theory) by recording microphonic potentials from each turn of the cochlea of guinea pig, with pairs of thin wire electrodes which were introduced into both the scala vestibuli and the scala tympani through very tiny holes on the bony wall of the cochlea. Those experimental

561

results brought the site of complete sound analysis into obscurity again and reminded us once more that the mechanism of sound analysis may be explainable in part by the telephone theory.

With refinement of the electrophysiological techniques, Tasaki (1954) and Tasaki and Davis (1955) elucidated the peripheral auditory neural mechanism in guinea pig considerably by recording the electric responses of single auditory neurons to sound stimulation. Nevertheless, very few studies have been attempted at the higher centers of the brain, with the exception of the one by Erulkar, Rose, and Davies (1956). From the experimental results of Békésy (1943) and those of Tasaki, Davis, and Legouix (1952), we have come to expect that the higher auditory centers in the brain are concerned first with the completion of the sound analysis and then with the perception of sound.

Several authors, Galambos and Davis (1943), Galambos (1944), Thurlow, et al. (1951), Gross and Thurlow (1951), Galambos (1952); Galambos, et al. (1952); Rose and Galambos (1952), Hilali and Whitfield (1953), have already studied the nerve responses to sound stimulation with classical techniques at different levels of the auditory pathway in the brain, but their results are hardly comparable with each other because of the diversity of their experimental conditions. With the technique described below, Katsuki, et al. (1958) and Katsuki, Watanabe, and Maruyama (1959) have successfully recorded the electric responses of single auditory neurons to airborne sonic, and if necessary ultrasonic, stimulation by means of a superfine microelectrode from various levels at the auditory pathway in cat. Recordings have been made from the cochlear nerve, the dorsal and ventral cochlear nuclei, the trapezoid body, the inferior colliculus, the medial geniculate body, the cerebral primary and secondary auditory areas, and the ascending reticular system. This report will be concerned with the results obtained in those experiments.

Materials and Method

Experimental results described in this article were obtained from over 800 adult cats weighing 2 to 3 kilograms. The use of anesthetics for surgical operation affected the activities of neurons seriously, and therefore the operations were made with several different methods, and the results obtained were compared with each other.

Cannulation of the trachea was first performed in all cases. In most of the animals, the microphonic potentials were led from the round window of the tested ear by introducing a thin silver-wire electrode into the bulla through a small opening on the bullar bone, in order to determine whether or not the stimulus sound could reach the cochlear end organ.

In order to expose various regions of the brain where the responses of neurons were recorded, the skin and muscle underneath were cut, and then the skull was opened to expose the dura. Under the binocular microscope the dura was cut, and a microelectrode was introduced very slowly with a micromanipulator into the brain tissue through an intact pia.

For the recording of a single neuron response, a very thin glass micropipette filled with 3 molar KCl solution was used as a recording electrode. The diameter of the electrode tip was less than 0.2 micron, and its ohmic resistance ranged between 30 and 50 megohms. The amplifiers used were a conventional cathode-follower preamplifier and a high-gain resistance capacitance and, if necessary, d-c amplifiers.

Responses to stimuli were mostly photographed on running film by means of a specially designed cathode-ray oscilloscope with three channels, one for the nerve response, one for the stimulus (mostly for the microphonic), and one that indicated time. These three records were photographed simultaneously on a film.

The stimuli used were continuous pure tones at various frequencies and also brief tone bursts at 44 fixed frequencies between 30 and 100,000 cycles per second. The duration of each burst was about 100 milliseconds but could be altered when desired.

For the production of tone bursts, an apparatus was designed in such a way that the tone bursts were automatically produced in succession by means of a rotary switch driven by a motor. Details of the apparatus have been reported elsewhere by Katsuki, Watanabe, and Maruyama (1959). Three high-fidelity loudspeakers built by M. Kato were used in combination to produce sounds, and a specially designed condenser speaker (Shinagawa Musen Co.) produced the ultrasonic waves.

The intensities of the sounds produced, both continuous and transient brief bursts, were calibrated with the automatic recording apparatus of Brüel and Kjaer. The precision microphone took the place of the animal ear in the sound-proofed room where the experiments were performed, and the results confirmed satisfactorily that the frequency-response curve was almost flat within ±4 decibels in the range from 100 cps to 10 kcps, and above 10 kcps it deviated gradually. Actual measurements were not made above 20 kcps, but in the range below 20 kcps the two loudspeakers were so adjusted that the tweeter and the condenser speaker coincided with each other.

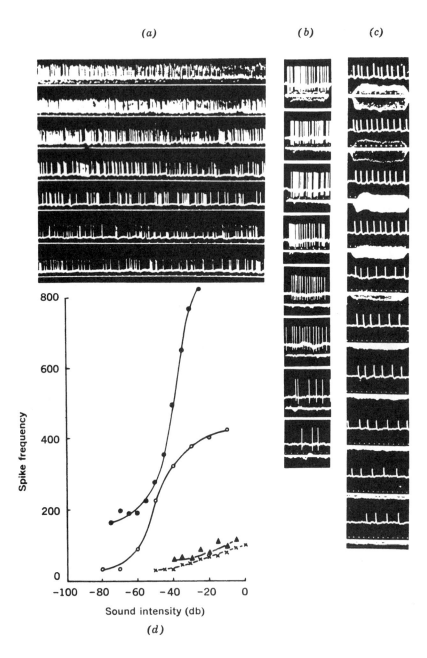

All these stimuli were delivered to the animal ear in a free field, and considerable attention was paid to the reflection of sound waves with very high frequencies.

The animals were isolated in the sound-proofed room, and, except for the microelectrodes, all control of the equipment took place outside the room.

The immobilization of the animal and, in addition, of the brain tissue was very critical, because single neuronal responses should be recorded over a long period, as in serial experiments. Details of the experimental techniques and also of the results will be found in Katsuki, et al. (1958) and Katsuki, Watanabe, and Maruyama (1959).

Results

Response pattern

Marked spontaneous discharges of neurons were usually found at the peripheral regions, whereas at the upper regions they were encountered much less frequently, and at the cortex most neurons were quiet. When the sound stimuli were presented to the ear, the frequency of discharge at the peripheral regions increased with an increase in the intensity of sound (Figs. 1a, b, c); the relation was sigmoid between the frequency of impulses per second and the intensity of sound in decibels (Fig. 1d). This relation was maintained up to the subcortical level. But the rate of increase of the impulse frequency in response to stronger sounds was different at each region. The higher the level, the lower in general became the rate, as shown in Fig. 1d.

In contrast to this, brief responses, that is, on, off, or on-off responses, began to appear at the inferior colliculus. Those responses were encountered more often at the higher levels, and finally at the super-

Fig. 1. Relation between the spike frequency of a single neuron and the sound intensity obtained from (a) the cochlear nerve, (b) the trapezoid body, and (c) the medial geniculate body. The intensity of sound decreases from top to bottom by 10 db in all cases. The stimulus sounds in (b) and (c) are tone bursts at 9000 and 6000 cps, respectively; the stimulus sound in (a) is a pure tone at 870 cps. The spontaneous discharge of the neuron in the cochlear nerve is shown at the bottom of (a). In the other two areas there was almost no spontaneous discharge.

Relations are shown in the curves at the lower left (d). The solid circles refer to the neuron at the cochlear nerve; the open circles to the trapezoid body; and the crosses to the medial geniculate. The curve with triangles was obtained from the deep layers at the cortex.

ficial layers of the cortex almost all neurons showed only this type of response. In these cases there was no clear sigmoid relation, as was seen at the lower levels, between the frequency of impulses and the intensity of sound (Fig. 1). This is one of the most conspicuous differences between the neurons at the periphery and those at the cortex.

Response area

Measurements were made of the response area of the neuron at different levels, that is, the cochlear nerve, the dorsal cochlear nucleus, the trapezoid body, the inferior colliculus, the medial geniculate body, and the cortex (Fig. 2). The stimuli used were continuous pure tones and tone bursts. As described above, all neurons at the periphery respond to both continuous and brief sounds, whereas many neurons at the upper brain respond only to brief sounds. Comparisons were made of the response areas measured with these two types of sounds at the peripheral regions. Areas responding to the two kinds of sounds practically coincided.

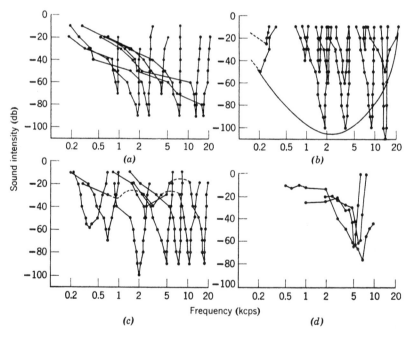

Fig. 2. Response areas of single neurons obtained from (a) cochlear nerve, (b) inferior colliculus, (c) trapezoid body, and (d) medial geniculate body.

As reported by Tasaki (1954) and by Galambos and Davis (1943), the frequency range in which the neuron was activated was very wide for sounds of sufficient intensity. From a low-frequency sound up to a certain high-frequency sound, each neuron responded continuously. When the intensity of the sound was reduced step by step, the neuron responded less and less to low tones, while the upper limit remained almost unchanged, and the response frequency range became narrower and narrower. The so-called characteristic frequency of the neuron was thus obtained finally.

As is well known, the eighth-nerve bundle in the cat contains many nerve cells that have migrated from the dorsal cochlear nucleus. Consequently, many responses of neurons to sound stimulation recorded from the nerve come from the secondary neurons. In order to get only the responses of the primary neuron, the microelectrode was inserted deep into the internal meatus after the cerebellum was removed by suction. Some of the responses thus obtained were found to be different from those we have reported earlier.

The response areas obtained from the primary neurons were of two types. One is a well-known type, but the other is not. The first type is asymmetrical against the axis of the characteristic frequency; that is to say, the upper limit is a sharp cut-off, whereas on the low-frequency side of the characteristic frequency the threshold shows a gradual rise toward the lower limit (Fig. 3a). The second type of area, however, is rather symmetrical against the axis of characteristic frequency: on the higher-frequency side the thresholds also show a gradual rise, as is shown in Fig. 3b. Most of the neurons that show symmetrical areas have the low characteristic frequencies. When the characteristic frequencies of the neurons were high, almost all of them had asymmetrical areas. These results reminded us of Lorente de Nó's description (1933) of the different innervation modus of the inner and outer hair cells in the organ of Corti. One or very few inner hair cells are innervated directly by a single cochlear nerve fiber, but the nerve fibers that innervate the outer hair cells turn at right angles below the outer hair cells and run for a third of a spiral or even farther towards the round window and innervate many hair cells.

In consequence of this peculiar mode of innervation of hair cells and of the characteristic vibratory movement of the cochlear partition discovered by Békésy (1943), the cochlear nerve fibers may be classified into two groups, those that innervate the inner and those that innervate the outer hair cells. And it is my opinion that the fiber that has an area of the symmetrical type may innervate the outer hair cell, whereas the fiber with an asymmetrical type may innervate the inner hair cell. But such a classification cannot be made distinctly when the characteristic frequency of the neuron is very high, because the difference in innervation modus between the inner and the outer hair cells is not very large.

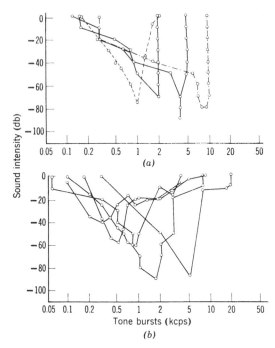

Fig. 3. Response areas obtained from primary neurons: (a) usual asymmetrical type; (b) symmetrical type. The ordinate is intensity of sound in decibels; the abscissa, the frequencies of tone bursts.

Davis (in Stevens and Davis, 1938) concluded from experimental results on the waltzing guinea pig that the thresholds for sound stimulation of the inner and outer hair cells are different: the threshold of the outer hair cell is lower than that of the inner hair cell. We made some comparisons of the thresholds of fibers for various sounds among the fibers, but we did not find obvious differences. Only statistically did the fibers with symmetrical areas show lower thresholds than those with asymmetrical areas. These two types of response area were found mostly at the primary neurons.

At the level of the dorsal cochlear nucleus, the response area was reduced on the low-frequency side as if it were partly cut down, while the upper-limit frequency was kept constant as a sharp cut-off.

At even higher levels, the response areas were much more reduced in size on the low-frequency side, and they were much more sharply limited than at the periphery. The narrowest area was thus obtained

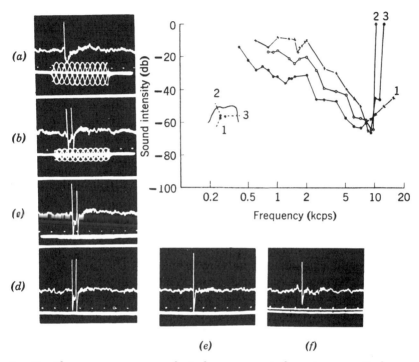

Fig. 4. Three response areas of single neurons at the primary cortical area (the ectosylvian gyrus). The inset shows the location of recorded neurons; their depths are: (1) 900 μ; (2) 100 μ; (3) 700 μ. Responses of a neuron to 8000-cps sounds with intensities decreasing by 10 db from (a) to (f). Upper beam, neural response; lower beam, microphonic; time marked in 10-msec intervals.

at the medial geniculate body, as shown in Fig. 2. In contrast, the response areas of cortical neurons were revealed to be unexpectedly wide, though the response pattern of impulses was different, as is shown in Fig. 4.

Interaction of responses to two sounds delivered simultaneously

Reports on the interaction of neuronal responses to two-sound stimulation have been published by Galambos (1944), Hilali and

Whitfield (1953), and others. In addition, Allanson and Whitfield (1956) discussed this problem from the point of view of information theory. It is indeed of great importance to know more details about the interaction of neurons.

When discharges of a single neuron were encountered at the cochlear nerve, the effect was examined of a second tone on the responses to tone bursts of different frequencies. As a second tone a continuous pure tone was generally used. Most of the responses to the first tone bursts were not disturbed at all when the intensity of the second tone was weak. But as the intensity of the second tone was increased step by step, responses to the first tone began to be suppressed, and a further increase in the second tone suppressed the firing of the neuron completely for as long as the tone bursts continued. When the intensity of the second tone was very strong, there was no obvious response to the first stimulus, and the responses to the second tone took the place of those to the first, provided the frequency of the second tone was so chosen as to be within the response frequency range of the neuron.

The inhibitory effect of the second tone on the response of the neuron to tone bursts depended somewhat on the frequency of the second tone but not much if the frequency of the second tone was within the response range, as shown in the upper part of Fig. 5. The inhibitory frequency range in which the impulses in response to successive tone bursts of a certain intensity were suppressed by a second tone was found in general to be at the high-frequency and low-frequency sides of the characteristic frequency of the neuron. By changing the intensity of the tone bursts it was possible to determine, for a certain sound, the response area of the neuron obtained by simultaneous presentation of two sounds, and also one or two inhibitory areas at either one or both sides of the characteristic frequency. These response areas, as well as the inhibitory areas, were of course closely related to the intensity of the second stimulus, as has been mentioned already.

The lower figure in Fig. 5 is an example obtained at the root of the cochlear nerve. The areas shown with broken and solid lines were obtained by successive tone bursts at different intensities with and without a second tone of 10 kcps at —17 db. The areas outlined with dotted lines are inhibitory areas. The points represented by solid and open circles were obtained from the film on which the responses were continuously recorded.

Sometimes spontaneous discharges of neurons, which were usually found at the lower neural levels, were inhibited by sound stimuli.

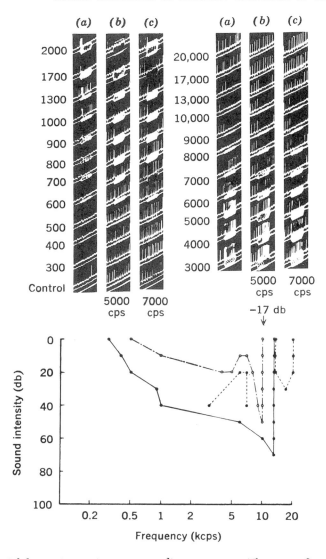

Fig. 5. Inhibitory interaction among auditory neurons. The upper figure shows the suppression of response discharge to tone bursts in the inferior colliculus. The frequencies of the tone bursts are given at the left. Column (*a*) shows the responses of a neuron to tone bursts alone; the suppression of spontaneous discharge is seen at the higher end of the response range. Columns (*b*) and (*c*) show the suppression of discharge in response to tone bursts with a tone in the background at 5000 and 7000 cps, respectively. The lower figure shows the response area of a neuron obtained with tone bursts alone (solid line), and the response area (broken line) and inhibitory areas (dotted lines) obtained by tone bursts with a background tone of 10 kcps at −17 db delivered simultaneously.

Such inhibitory phenomena were rarely observed when the recording was made of the responses from the distal portion of the cochlear nerve near the internal auditory meatus. But they were encountered on almost half the successful recordings from the proximal portion of the nerve in the vicinity of the cochlear nuclei. From the experimental results mentioned above, it is reasonable to conceive of this inhibitory effect of the second tone on the responses to the first tone as of synaptic origin, since many nerve cells of the secondary auditory neurons are found at the root of the cochlear nerve.

Although this inhibitory interaction between the responses of a neuron to two sounds delivered simultaneously was observed very distinctly in the peripheral region, similar phenomena have also been found at the upper levels, even at the level of the medial geniculate body or of the cortex. Such an inhibitory mechanism may certainly be the basis of the narrowness of the response area at the higher levels of the pathway. Accordingly, the inhibitory area on the low-frequency side of the characteristic frequency may play an important part in narrowing the response areas at the higher levels. And at the same time the inhibitory area on the high-frequency side of the characteristic frequency may play a similar part in reducing the response areas of the primary neurons which have a symmetrical response area.

The abolition of the spontaneous discharges of the neuron by single-sound stimulation must have some relation to the mechanism of reducing response areas, and the second tone delivered simultaneously with the first one may result in similar but stronger inhibitory effects on the activity of the neuron at the higher levels. The ability of the neuron to discriminate frequency may gradually improve as the nerve impulses ascend through synapses, one by one, from the periphery to the thalamic level.

As mentioned above, the responses to single-sound stimulation obtained from neurons at the superficial cortical layers were very brief, and most of them were on-type responses with very few impulses. Erulkar, Rose, and Davies (1956) have also reported similar results.

The response areas of cortical neurons were unexpectedly wide. With those neurons, the inhibitory interaction of the responses to simultaneous two-sound stimulation was not obvious, whereas an apparent facilitatory interaction was often observed. When the frequencies of two sounds were in a certain relation, the response was intensified, and the brief response changed to a continuous one. The frequency relations that accomplished this change were the following: (1) when the two sounds were harmonically related; that is, the ratio of the frequencies of the sounds was 1:1 or 1:2 or 1:3, · · · or (2) when the difference in frequency between two sounds was 100 or 200, or some such difference, that is, when the number of beats per second ranged from scores to hundreds.

In the harmonically related tones, the frequencies of the two sounds share some of the same upper partials. Therefore this case corresponds to intensifying the intensity of a single sound. However, as mentioned already, the number of impulses of a cortical neuron does not increase with an increase in intensity of a stimulus sound. Thus intensifying the response in this way cannot be explained simply as an increase in the intensity of the sound. Moreover it is very difficult to find perfect coincidence of the frequencies of two sounds produced by two separate sources, as in the present experiment. Small differences between the frequencies of two sounds were usually found by a careful examination of the microphonics.

With the beating tones, repetitive firing was very often observed when the frequency difference, that is, the number of beats, was between, say, 50 and 200 or more per second. In this range the number of impulses was highly correlated with the number of beats. At present we do not know whether the mechanism of the harmonically related or beating tones is essentially the same, but the number of beats produced by two sounds seems to be of great importance. Since a very small or a very large number of beats did not produce the repetitive impulses, it may be concluded that the beating sounds may activate the neuron repetitively like the repetition of a brief tone burst.

The responses at the peripheral neurons to beating sounds were in the form of grouped impulses, each group of which was correlated with each wave of the beat. When such a group of impulses reached the cortex repetitively, a cortical neuron might be able to be activated repetitively too.

However, as Fig. 6 shows, the amplitude of a beating sound also plays an important role in the repetition of discharges. When the amplitudes of two sounds are equal, the amplitude of the beating sound is maximum, and when the difference in their amplitudes is large the discharge is not repetitive.

Thus it becomes clear that the repetition of discharge in a cortical neuron depends upon both the amplitude and the frequency of the beating sounds, and the repetition may not be due to the synaptic facilitatory interaction among cortical neurons. As regards the beat frequency, measurements showed us that the repetitive discharge occurred rather easily when the beats were between 50 and 200 per second, and at around 100 per second the repetition was most promi-

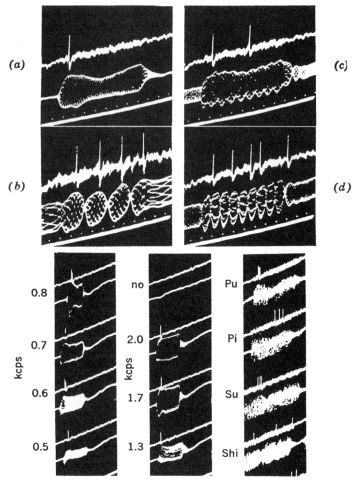

Fig. 6. The upper figures show (a) the response of a cortical neuron to a tone burst of 3000 cps; (b) responses to beats between 3000 and 3050 cps; (c) responses to beats between sounds at 3000 and 3100 cps, whose intensities are different; and (d) almost the same as (c). The lower figures show responses of a neuron to tone bursts (first two columns), whose frequencies are given at the left side of each column, and a human voice (third column) producing the sounds shown at the left of the column.

nent. When the beats were too few or too many, the repetition could not be observed.

Such a numerical relation between the beat frequency and the repetition of discharge may depend upon the adaptation and the rate of recovery of the cortical neuron.

At the start we expected more complicated neuronal interactions among cortical neurons. But the relation was found to be rather simple. We encountered the repetitive discharge with two-sound stimulation only in special cases.

In such cases, the beating sound does not always produce repetition of discharge. When the characteristic frequency of the neuron was high, two or three different beating sounds could activate the neuron. But a particular numerical relation of beating sounds to characteristic frequency has not so far been found. Though further studies may elucidate the complicated relations among cortical neurons, at the present stage of our knowledge nothing further can be said on the cortical neurons. Among psychological phenomena, harmony or consonance in musical sounds is very important. Since the time of Helmholtz (1862) it has been believed that beating sounds are closely related to musical consonance.

It may not be possible to compare directly such human psychological phenomena with the characteristics of the cat's cortical neurons, but the bases of our sensation may be concerned somewhat with such characteristics of neurons.

Slow potentials recorded from the peripheral region

When a microelectrode was inserted in the root of the eighth nerve bundle or in the dorsal cochlear nucleus at the medulla, in response to a tone burst whose duration is measured in scores of milliseconds, a negative sustained slow potential as long as the tone burst was obtained by the use of a high-gain d-c amplifier. The shape of this potential was quite similar to the rectified envelope of the microphonics evoked by the tone burst. With the advance of the electrode into the brain tissue, the size of the negative potential increased gradually. After reaching the maximum value, say 1 or 2 millivolts, the potential began to decrease gradually. Sometimes the polarity of the potential changed during its advance. But the positive potential is very unstable compared with the negative potential. Occasionally the positive spike discharges of a single neuron in response to a burst appeared superimposed on these slow potentials. As a result of a further advance of the electrode or of successive changes in the frequency of the sound, these discharges disappeared.

The biological origin of this potential was confirmed by several methods, and a definite conclusion was obtained from the fact that this potential completely disappeared when the eighth nerve was cut.

The sustained slow potentials evoked by a tone burst were indeed stable, but by changing the frequency of the burst the potential could

be abolished.   Therefore the origin of this potential was considered
to be a large cell group which was specific for each sound frequency.
With simultaneous two-sound stimulation, the slow potential in response to a tone burst appeared to be abolished by a background
pure tone.   However, careful examination of the response revealed
that a background tone caused a sustained deviation of the beam by
which the slow potential due to a burst was hidden.   Thus it was
found that the abolition of the slow potential is only apparent and is
not due to the synaptic inhibition.   Positive spike discharges often
appeared superimposed on a negative or positive slow potential.   They
were more unstable than the slow potential.   These spike potentials
may not be directly related to the slow potentials, just as the spike
discharges of the cortical neuron have no direct relation to the evoked
potential there.   Such slow potentials as those obtained at the dorsal
cochlear nucleus were obtained, though not often, at the higher
levels, for example, at the trapezoid body or the inferior colliculus.
At the higher levels, however, the potential was not so sustained, and
it gradually decreased in size and changed finally to the so-called
evoked potential obtained at the cortex.   This tendency agrees well
with the change in the adaptation of discharge in neurons from the
periphery to the cortex.   The origin of these slow potentials is still
not clear.   However the remarkable ramification of the dendrites of
a neuron at the synapses may perform an important role in producing
the slow potential.

Tonotopic localization

A beautiful map of the tonotopic localization obtained with surface
electrodes on the cortex of dog has been shown by Tunturi (1950,
1955b) and by Hind (1953).   In the present study an analysis of
single-neuron activity has resulted in a general tendency that agrees
with Tunturi's report, that is, the neurons situated at the anterior
part of the ectosylvian gyrus responded to high-frequency tones, those
at the posterior part to low tones, and those at the middle part to
medium high tones.   However, the arrangement of each neuron was
not so systematic, and the real situation must be far more complicated.

According to our experimental results, tonotopic localization at the
cortex may not be so important for discriminating sound frequency,
because these results have led us to conclude that the discriminations
of frequency and intensity are performed at the thalamic level.   At
the lower levels, attempts were made to find tonotopic localization
along the auditory tract.   Of course, there is a certain tendency, but

we could not find anywhere, except for the cochlea, such a beautiful map of the systematic arrangement of neurons as the one obtained at the cortex by Tunturi.

Responses of a neuron in the ascending reticular system to sound stimulation

Recordings of responses of single neurons were made at the reticular formation of the midbrain and at the anterior nucleus group of the thalamus. In recording in the reticular formation, an electrode was inserted deep into the midbrain just between the superior and inferior colliculus. In the thalamus, the electrode was inserted through the cortex toward the nucleus, making a certain angle to the cortical surface.

Recording of responses of a single neuron was, of course, made under an intravenous injection of Flaxedil with calm artificial oxygen inhalation, and no barbiturates were used.

In both cases the most important characteristics of neurons belonging to this system were found to be the following.

1. Most of the neurons responded to a very wide range of frequencies and responded to any sound in this range almost equally. In other words, they had no characteristic frequency to which they responded with particular sensitivity (Fig. 7).

2. The thresholds for sounds of neurons in the reticular system were in general much higher than the thresholds of neurons in the classical auditory pathway.

3. The adaptation of discharge was slow. Most neurons had spontaneous discharge, and the responses of many neurons to sound stimulation were continuous as long as the sound continued.

4. Among these neurons, many showed remarkable after-discharge. After the end of sound stimulation the discharges continued for quite a while.

5. Though many neurons showed the usual response pattern—which was a continuous repetitive discharge—many others showed a particular pattern, which started with a brief period of inhibition and was followed by noticeable discharges. Also in some cases inhibition of discharges appeared as long as the sound continued.

6. The nerve to the fore- and hindlegs was stimulated electrically. Many neurons in these regions responded to this kind of stimulation, but we have not so far seen neurons that respond to both acoustic and electric stimulation on the leg nerve.

7. At the reticular system of the midbrain the number of neurons responding to acoustic stimulation was relatively large, whereas at the

thalamus, n. ventralis anterior, only a small number of neurons responded to acoustic stimulation, but many neurons were responsive to the electrical stimulation of the legs.

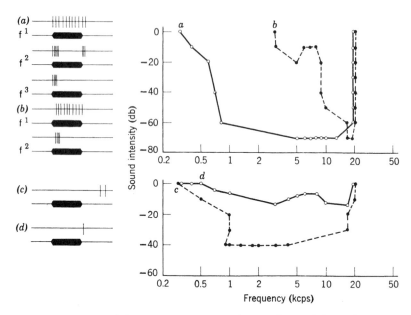

Fig. 7. Responses of four neurons to tone bursts obtained from the reticular formation in the midbrain: f^1, f^2, and f^3 show various response patterns of the neurons for sounds of different frequencies. Note the wide response areas where there is no specific frequency sensitivity and a relatively high threshold. The letters (a) to (d) identify the same neuron in the two parts of the figure.

Auditory area II

The secondary auditory area at the cortex was also studied similarly. Most neurons whose responses were recorded were situated deep in the cortex. In many ways the neurons encountered there were similar to those encountered at the reticular system, though their thresholds for sound stimulation were much higher than those obtained at the reticular system. Most neurons did not respond to brief tone bursts but only to strong sustained tones.

From only those results described above, it is not clear with what kind of functional activities the neurons in the secondary auditory area are concerned. But it is reasonable to consider that the secondary area may be concerned with much higher integrative function than the primary area.

Conclusion

From all the experimental results obtained, the following conclusion was drawn.

The analysis of complex sounds is made partly in the cochlea; additional analysis is considered to be made while the impulses of each auditory neuron, in response to sounds, ascend to the cortex; and the analysis is completed at the medial geniculate body because the narrowest response area of the neuron is found at this region. The wide response areas at the periphery are reduced step by step at the synapses of each ganglion on the auditory pathway.

The mechanism for the gradual narrowing of response areas at the higher levels of brain may be the inhibitory interaction of neurons. Recording the potentials from the auditory neuron could be done only extracellularly, and therefore the inhibitory interaction observed on the auditory system was not identified as the synaptic inhibition observed at the motoneuron in the spinal cord of cat. However, quite recently Wiesel (1959) succeeded in recording the intracellular potential from the retinal ganglion cells of cat. The mechanism of the inhibitory interaction of the auditory neuron may not be so different from that of the retina.

The sigmoid relation between the number of spikes per unit time and the intensity of sound could be followed up to the geniculate body but not at the cortex. The response pattern of the neuron was quite brief at the cortex and much different from the other regions described above.

From all these results it was concluded that the discriminations of the frequency and intensity of sounds are made at the medial geniculate body in the thalamus, and the integration of the component sounds analyzed may be made at the cortex where the beating sounds among these component sounds play an important role. Butler, Diamond, and Neff (1957) and Tunturi (1955a) also support in part this conclusion.

The responses of single neurons to sound stimulation could be recorded from the secondary cortical auditory area, but not so easily as from the primary area. They were mostly recorded from the deep layers. These responses were thought to be concerned not with the discrimination of the frequency and intensity of sound but with their integration.

Responses of single neurons were also recorded from the ascending reticular system, especially from the midbrain and the anterior thalamus. The response pattern of some neurons there was similar to that observed at the auditory tract. However, noticeable differences were also found: the thresholds of neurons were in general relatively high, and most neurons had no characteristic frequency; in other words, their response areas were flat for various sound frequencies, and they had no sharply tuned frequency. Moreover some of them showed conspicuous after-discharge after cessation of the stimulation. Suppression of spontaneous discharges and also suppression of responses to acoustic stimulation were seen. Though other authors, for example, Machne, Calma, and Magoun (1955) and Sheibel, et al. (1955), have reported it, convergence of afferent impulses on these neurons has not so far been observed.

From these experimental results it may be said that the ascending reticular system also is not concerned with sound analysis, but further details of the function of this system are still obscure.

Summary

1. Electrical responses of a single neuron to sound stimulation were recorded from several levels of the auditory tract in cats: from the cochlear nerve, the dorsal and ventral cochlear nuclei, the trapezoid body, the inferior colliculus, the medial geniculate body, and the auditory cortical area. The ascending reticular system was also studied.

2. The response patterns of impulses were continuous so long as the sound lasted, except at the cortical level. The number of impulses increased as the intensity of the sound was increased. The relation between the impulse frequency and the sound intensity in decibels was sigmoid. This relation held from the peripheral nerve up to the medial geniculate body. The pattern at the superficial layers of the cortex was brief, and the impulses adapted rapidly. The responses there were of the on, off, and on-off types, and the relation between the impulse frequency and the sound intensity was not sigmoid but irregular.

3. The response area of a single neuron was measured. At the periphery the area was quite wide, and the higher the level, the narrower the area became. The narrowest area was found at the medial geniculate body, whereas rather wide response areas were obtained from the cortical neuron.

4. From the primary neuron two types of areas were obtained. One is the well-known asymmetrical type. The other has, instead

of a sharp cut-off at the higher frequency side of the characteristic, a gradual rise in threshold for different sounds. The second type is, rather, a symmetrical type. The neurons with low characteristic frequencies had a response area of the symmetrical type, whereas almost all neurons with high characteristic frequencies had asymmetrical response areas. The neurons with symmetrical response areas may innervate the outer hair cells, and the asymmetrical ones the inner hair cells.

5. Interaction of responses was found at any level when two sounds were presented to the animal simultaneously. At the lower levels inhibitory interaction predominated. The inhibitory effect of a sound on the response of a neuron to brief tone bursts was found at either one or both sides of the characteristic frequency of the neuron. It may be owing to this mechanism that the response area apparently becomes narrower and narrower at the higher levels of the brain.

6. Most responses recorded from the superficial cortical layers were brief. But with simultaneous two-sound stimulation, those responses became repetitive when the two sounds produced a beating sound. With 50 to 200 beats per second the repetition of discharges occurs easily. It may be that the mechanism of the repetition of discharge is not due to the synaptic facilitation, but that the group of impulses resulting from the beating waves that come from the lower region activate the cortical neurons as repeated brief sound stimulation.

7. Slow potentials evoked by the tone bursts were found most often at the dorsal cochlear nucleus, less at the ventral nucleus, and rarely at the inferior colliculus. The polarity of this potential was usually negative but was sometimes positive. Impulse discharges due to the tone bursts were often superimposed on the slow potentials with which they might have no direct relation. The origin of this potential may be a large cell group that is specific for each frequency and makes a strong potential field. The change in the shape of this potential at different levels of the brain may depend on the adaptation of the neuron at that level.

8. Recordings of the responses of neurons were made from the secondary area at the cortex. They were not so easy to make, however, as those from the primary area. The thresholds of most of the neurons were generally high for various sounds, and no neurons were specific for sound frequencies.

9. The responses of neurons were recorded from the ascending reticular system, especially from the midbrain and the thalamus. The thresholds of those neurons for sound stimulation were relatively high, and most of the neurons had no specificity for a particular sound fre-

quency. They also showed the remarkable after-discharge that the neurons in the classical auditory pathway never showed. The inhibitory interactions of responses similar to those observed in the auditory tract were also seen among these neurons.

10. The bases of various psychological phenomena in human auditory sensation are discussed in relation to the present experimental results.

References

Allanson, J. T., and I. C. Whitfield. The cochlear nucleus and its relation to theories of hearing. In C. Cherry (Editor), *Information Theory; Papers Read at a Symposium.* New York: Academic Press, 1956. Pp. 269–286.

Békésy, G. v. Über die Resonanzkurve und die Abklingzeit der verschiedenen Stellen der Schneckentrennwand. *Akust. Z.,* 1943, **8,** 66–76.

Butler, R. A., I. T. Diamond, and W. D. Neff. Role of auditory cortex in discrimination of changes in frequency. *J. Neurophysiol.,* 1957, **20,** 108–120.

Erulkar, S. D., J. E. Rose, and P. W. Davies. Single unit activity in the auditory cortex of the cat. *Bull. Johns Hopkins Hosp.,* 1956, **99,** 55–86.

Galambos, R. Inhibition of activity in single auditory nerve fibers by acoustic stimulation. *J. Neurophysiol.,* 1944, **7,** 287–303.

Galambos, R. Microelectrode studies on medial geniculate body of cat. III. Response to pure tones. *J. Neurophysiol.,* 1952, **15,** 381–400.

Galambos, R., and H. Davis. The response of single auditory-nerve fibers to acoustic stimulation. *J. Neurophysiol.,* 1943, **6,** 39–58.

Galambos, R., J. E. Rose, R. B. Bromiley, and J. R. Hughes. Microelectrode studies on medial geniculate body of cat. II. Response to clicks. *J. Neurophysiol.,* 1952, **15,** 359–380.

Gross, N. B., and W. R. Thurlow. Microelectrode studies of neural auditory activity of cat. II. Medial geniculate body. *J. Neurophysiol.,* 1951, **14,** 409–422.

Helmholtz, H. v. *Die Lehre von den Tonempfindungen,* 1862.

Hilali, S., and I. C. Whitfield. Responses of the trapezoid body to acoustic stimulation with pure tones. *J. Physiol.,* 1953, **122,** 158–171.

Hind, J. E. An electrophysiological determination of tonotopic organization in auditory cortex of cat. *J. Neurophysiol.,* 1953, **16,** 475–489.

Katsuki, Y., T. Sumi, H. Uchiyama, and T. Watanabe. Electric responses of auditory neurons in cat to sound stimulation. *J. Neurophysiol.,* 1958, **21,** 569–588.

Katsuki, Y., T. Watanabe, and N. Maruyama. Activity of auditory neurons in upper levels of brain of cat. *J. Neurophysiol.,* 1959, **22,** 343–359.

Lorente de Nó, R. Anatomy of the eighth nerve. I. *Laryngoscope,* 1933, **43,** 1–38; II. *Laryngoscope,* 1933, **43,** 327–350.

Machne, X., I. Calma, and H. W. Magoun. Unit activity of central cephalic brain stem in EEG arousal. *J. Neurophysiol.,* 1955, **18,** 547–558.

Rose, J. E., and R. Galambos. Microelectrode studies on medial geniculate body of cat. I. Thalamic region activated by click stimuli. *J. Neurophysiol.,* 1952, **15,** 343–358.

Rutherford, W.   A new theory of hearing.   *J. Anat. Physiol.*, 1886, **21**, 166–168.

Scheibel, M., A. Scheibel, A. Mollica, and G. Moruzzi.   Convergence and interaction of afferent impulses on single units of reticular formation.   *J. Neurophysiol.*, 1955, **18**, 309–331.

Stevens, S. S., and H. Davis.   *Hearing: Its Psychology and Physiology*.   New York: Wiley, 1938.

Tasaki, I.   Nerve impulses in individual auditory nerve fibers of guinea pig.   *J. Neurophysiol.*, 1954, **17**, 97–122.

Tasaki, I., and H. Davis.   Electric responses of individual nerve elements in cochlear nucleus to sound stimulation.   *J. Neurophysiol.*, 1955, **18**, 151–158.

Tasaki, I., H. Davis, and J.-P. Legouix.   The space-time pattern of the cochlear microphonic, as recorded by differential electrodes.   *J. acoust. Soc. Amer.*, 1952, **24**, 502–519.

Thurlow, W. R., N. B. Gross, E. H. Kemp, and K. Lowy.   Microelectrode studies of neural auditory activity of cat.   I. Inferior colliculus.   *J. Neurophysiol.*, 1951, **14**, 289–304.

Tunturi, A. R.   Physiological determination of the arrangement of the afferent connections to the middle ectosylvian auditory area in the dog.   *Amer. J. Physiol.*, 1950, **162**, 489–502.

Tunturi, A. R.   Effect of lesions of the auditory and adjacent cortex conditioned reflexes.   *Amer. J. Physiol.*, 1955a, **181**, 225–229.

Tunturi, A. R.   Analysis of cortical auditory responses with the probability pulse.   *Amer. J. Physiol.*, 1955b, **181**, 630–638.

Wiesel, T. N.   Recording inhibition and excitation in the cat's retinal ganglion cells with intracellular electrodes.   *Nature, Lond.*, 1959, **183**, 264–265.

30 *A. FESSARD*

Collège de France, Paris

The Role of Neuronal Networks in Sensory Communications within the Brain

Sensory communications within the brain are most often understood as those corresponding to one single sense modality. Neurophysiologists who have worked in this field of brain research have been occupied mainly in exploring the specific structures (primary pathways and relay nuclei, specific projection areas of the cerebral cortex), recording evoked potentials, measuring velocities and latencies, and making explicit the input-output functions. It is quite clear, however, that the brain, even when studied from the restricted point of view of sensory communications, must not be considered simply as a juxtaposition of private lines, leading to a mosaic of independent cortical territories, one for each sense modality, with internal subdivisions corresponding to topical differentiations. It is now too well established that the primary ascending pathways send off abundant collaterals through which sensory signals of different origins can invade various subcortical structures of the brain stem, diencephalon, and basal nuclei, and also that the so-called nonspecific or associative areas of the cortex are responsive to peripheral stimulations of different kinds. More recently, even the motor cortex has been shown to behave in this way (see Buser and Imbert in this volume), to say nothing of the cerebellum. The chief consequence—from the point of view of sensory communications—of this extensive and multisensory invasion of so many regions of the brain is that afferent signals of all sorts are thus offered more than one opportunity to meet and interact, and this is assuredly one of the major functions the brain has to perform. As we have good reasons, based on electrophysiological as well as histological data, to believe that these regions of intersensory communications play their roles mainly by virtue of their polyneuronic, networklike structural features, it has seemed to us relevant to the general topic of this symposium to consider the properties of *neuronal networks* in relation to the problem of intersensory communications within the brain.

As is common practice, invasion of nonspecific brain structures by sensory signals can be traced through by recording the related electrical events—either *evoked potentials* with gross-electrode techniques, or *spikes* and/or *local slow waves* when microelectrodes are used. Here we should refer to the abundant literature that has appeared in the last ten years during which the names of Jasper, Magoun, Moruzzi, and their associates have stood out. These authors were concerned mainly with the reticular formation of the brain stem and its diencephalic extension. For our theme, it is certainly the best example, for it is well known that the reticular formation is at the same time a region of multisensory convergences and one of rich interconnectivity among its elements. But this is not the place for even a rapid survey of these epoch-making studies. The purpose here is rather to contribute to attenuating the impression, often left by the results obtained in these investigations, that the reticular formation might be, if not the sole, at least by far the most powerful neuronal pool for intersensory communications. For this purpose, some other examples borrowed from the recent work of my associates, D. Albe-Fessard, P. Buser, and their respective collaborators, will be given here. They have explored several subcortical structures as well as the so-called associative areas of the cerebral cortex, with the idea of differentiating the respective properties of these neuronal fields according to the way in which they receive and utilize sensory signals from various receptor categories or sites.

The most difficult task has been to detect evoked potentials from the nonspecific areas of the cerebral cortex in awake animals. Because these potentials are small and relatively labile, adequate methods must be found for extracting signal from noise. One method is to use chloralose: the *secondary evoked potentials* (SEP), as these potentials are often called, disappear under barbiturate anesthesia, but for unknown reasons chloralose has the effect of synchronizing and enhancing them.

From the example shown in Fig. 1—a cat's brain under light chloralose anesthesia—one can see that, compared to the primary evoked potentials, the cortical SEPs exhibit greater latencies (often around 15 to 20 milliseconds or more) and also much longer durations, undoubtedly owing to spread of the components, since both characteristics strongly suggest that these SEPs are the outcome of a polyneuronal propagation.

Whereas the example in Fig. 1 is concerned with somatic projections only—but without any somatotopic arrangement and with a bilateral representation—Fig. 2 illustrates the multisensory aspect. Systematic millimeter-by-millimeter exploration of the associative cortex of the unanesthetized cat by Buser and Borenstein (1959) has revealed an intricate "mosaic" distribution of widespread heterosen-

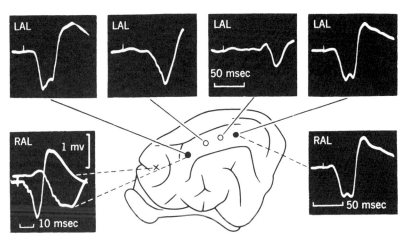

Fig. 1.   Cortical responses of a cat's brain under light chloralose anesthesia. Stimulation of one anterior leg, either left (LAL) or right (RAL).   Records signaled with dotted lines are contralateral responses; those with solid lines are ipsilateral ones.   Left lower record: primary evoked potential (cross, specific area) and secondary evoked potential (SEP) at fast time scale for comparison.   All SEPs are bilateral (an example is shown at right).   Open circles, places where responses are of longest latencies, are most labile, and disappear first when anesthesia becomes less light.   (After Albe-Fessard and Rougeul, 1958.)

sory projections.   This represents, of course, favorable conditions in which intersensory communications can take place by bringing specific messages nearer to one another; but it does not tell us anything about the way in which two or more heterogeneous signals actually communicate with each other.

To the modern neurophysiologist, it is clear that this question cannot be dealt with unless he speculates, and when possible experiments, at the neuronal level.   There he can make use of the vast body of evidence that has emerged mainly from a systematic experimentation on the spinal cord, because since Sherrington led the way there is no reason why processes similar to those resulting from the convergence of impulses onto single spinal neurons—*spatial summation* and *facilitation* or *inhibition, occlusion, after-effects during the recovery cycle*—should not take place within the brain's neuronal fields as well. As a matter of fact, these basic properties have been recognized there, either by common statistical criteria in macro-explorations or, more recently, by the more direct evidence given by microexplorations at the neuronal level. Figure 3 illustrates a case in which an inhibitory (3) and a facilitatory (4) effect (depending on

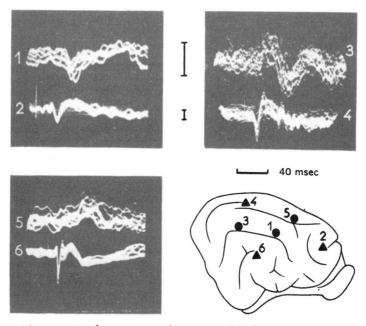

Fig. 2. A comparison between secondary (1, 3, 5) and primary (2, 4, 6) evoked potentials from three different sensory modalities: somesthetic (1, 2, bipolar derivations) and, in monopolar derivations, visual (flash, 3, 4) and auditory (click, 5, 6). Triangles, places within primary areas; circles, places within associative areas. Ten tracings are superimposed in each record. Unanesthetized cat; calibration in amplitudes, 100 μv. (After Buser and Borenstein, 1959.)

the timing) are exerted on a short discharge (in response to cutaneous stimulation) by the precession of a subliminal (local) response provoked in the same neuron by a cutaneous stimulation applied to another region of the body.

This same example of a unit in the caudate nucleus can also be utilized to illustrate the fact that the reticular formation is not the only subcortical structure devoid of sensory specificity;* this is shown

* Plurivalence of neurons in the reticular formation was first confirmed with microelectrode explorations by von Baumgarten, Mollica, and Moruzzi (1953), Amassian and DeVito (1954), and Scheibel, et al. (1955).

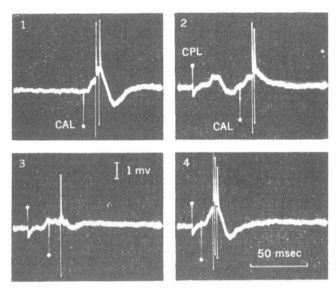

Fig. 3. Extracellular microelectrode recordings from a unit in the caudate nucleus (cat under chloralose). Responses to contralateral anterior leg stimulation (CAL) either alone (1) or preceded (in 2, 3, 4) by the response of the same cell to stimulation of the contralateral posterior leg (CPL). This latter response is a diphasic slow wave—a complex excitatory postsynaptic potential (EPSP)—followed by a complex inhibitory postsynaptic potential (IPSP) exerting a facilitatory (2, 4) or inhibitory (3) effect, depending on a delicate timing. (After Albe-Fessard, Oswaldo-Cruz, and Rocha-Miranda, 1958.)

in Fig. 4 (NC), where one can see how the same cell in the caudate nucleus fires in response to somatic, visual, or auditory stimuli. A similar plurivalence is—somewhat unexpectedly—exhibited by neurons in the red nucleus (NR, magnocellular part) but with a different response pattern (Massion and Albe-Fessard, 1959); whereas the centrum medianum of the thalamus (CM) responds mainly, though not exclusively, to somatic stimuli, and in a diffuse way, involving heterotopic convergences (Albe-Fessard and Gillett, unpublished). Multisensory plurivalence has also been proved to exist in nucleus centralis medialis and in zona incerta (Albe-Fessard, Oswaldo-Cruz, and Rocha-Miranda, 1960), whereas association of signals coming almost exclusively from one sense modality (somatic, visual, or auditory) has been observed by Buser, Borenstein, and Bruner (1959) in regions of the thalamus they describe as "specific associative systems." Several other examples could be given. In each particular structure, the special nature and organization of all converging afferences must be related to its own functional role. But our purpose here is not to dis-

cuss this comparative aspect of the question, and we should limit ourselves to the consideration of more general properties, namely, to
those related to the incorporation of the most fundamental dynamic
aspects of central activity into neuronal networks.

What are these fundamental aspects of neurodynamics that we must
use in order to get a better understanding of the intersensory communication mechanisms? Those that have been alluded to above

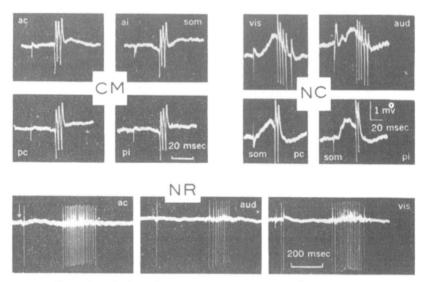

Fig. 4. Examples of plurivalent neurons in centrum medianum (CM), caudate
nucleus (NC), and red nucleus (NR). Stimulations are either visual (vis),
auditory (aud), or somatic (som). In CM, all the stimulations were somatic.
In ac, pc, ai, pi, the first letter means anterior or posterior, the second ipsi- or
contralateral. (After Albe-Fessard and various collaborators.)

(facilitation, inhibition, occlusion, phases of the recovery cycle) are
the classical processes that are observed when the ordinary double-
shock technique is used; and the basic results so obtained come
chiefly from the study of monosynaptic reflexes. It is clear that these
oversimplified and artificial conditions can provide us with only limited results and inspire theories that cannot go very far toward solving
most of our problems. Let us recall those properties that introduce
complications into the whole picture, as soon as we try to understand
the fate of sensory messages when they invade the nonspecific regions
of the brain.

1. Sensory messages appear generally in the form of *trains of impulses.*

2. Neurons onto which different sensory messages can converge are often, in awake animals, in a state of quasi-permanent *repetitive firing*, a consequence of sustained depolarization due to an incessant random bombardment by afferent impulses coming from everywhere; this is a property of major importance which has often been overlooked, for relatively few microexplorations of brain structures have as yet been made on nonanesthetized animals.

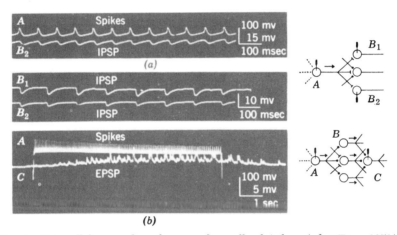

(a)

(b)

Fig. 5.   Intracellular recordings from ganglion cells of *Aplysia* (after Tauc, 1959), with two hypothetical diagrams explaining the kind of experiment performed. (*a*) Inhibitory postsynaptic potentials (IPSP) monosynaptically produced by a train of spikes delivered by an autoactive cell *A*. (*b*) Much slower time· scale. Cell *A* is artificially rendered autoactive by slight depolarization of its membrane (see sudden jump of the base line). Complex excitatory postsynaptic potentials (EPSP) are picked up from cell *C*. Observe delay of initiation and prolongation of activity.

3. Any change in the excitatory state of these self-firing neurons is expressed in *frequency variations*, so that synaptic transmission of elementary messages through active networks never occurs, save exceptionally, in the form of the one-to-one correspondence of impulses familiar to the macroelectrophysiologist.   An example of this property in a very simple case—ganglion cells of Aplysia (Tauc, 1959)—is given in the tracing of Fig. 5b, to be compared to those of Fig. 5a: there one can see how complex is the time pattern of nonpropagated activities (synaptic potentials) exhibited by a silent cell *C* (actually rendered silent by slight hyperpolarization) when it is indirectly activated by a regular train of impulses dispatched through a set of interneurons.

4. Complications also arise from the fact that any repetitive activity,

when made more (excitation) or less (inhibition) intense induces pre- and postsynaptic changes which may outlast their cause: these are the so-called *plastic changes,* which manifest themselves either in long persistences of frequency variations or in prolonged alterations of transmissive potentialities, which can be manifested in postsynaptic potentials when a test shock is applied, or in both. These important phenomena of central dynamics will not be discussed here; we have given them more attention in Fessard (1960); Fessard and Szabo (1959, in press); Fessard and Tauc (1958).

5. Alterations in the *space pattern* of a sensory message must be considered together with alterations in its time pattern, as soon as one considers the multifiber aspect of the information conveyed by this message: but whereas changes in the space pattern may be very slight in a simple transmission system composed of parallel chains of neurons, distortion such as occurs in a passive network, a fortiori in an active one, may become considerable.

We can now return to our central question, that of the role of neuronal networks in intersensory communications. The plurivalence of their neurons makes it clear—be it said once more—that they are the sites of election for these communications and ensuing interactions. But now we have a better idea of the price we have to pay in order for these performances to take place: first of all, that of a nonnegligible distortion, or sometimes complete transformation of the different incoming messages, even before they have encountered and started their integrative operations. But this is not all. It is easy to understand that if sets of parallel private lines—one set for each sense modality and submodality—had to supply *separately* plurivalent neurons, one at least for each possible combination, this would require a number of side-by-side pathways that would greatly exceed the capacity of the brain to contain them. As a matter of fact, there is no other economic solution to this topological problem than that of a *reticular* arrangement of neuronal connections. But this economy of routes is obtained at the expense of their privacy: said otherwise, segments of the routes utilized by messages to different cells are often held in common. This sharing of lines limits the capacity of the system to transmit simultaneous messages ("busy lines") as soon as there is some degree of spatial complexity. The corresponding psychological aspects of such limitations are evident. However, these effects of occlusion may be attenuated when sufficiently simple messages are involved, since a network may offer several alternative routes to the same point. This may be the main reason for the maintenance of function in most nervous structures, in spite of severe mutilations, although not beyond a certain degree of functional complexity.

Another important aspect of the question of interactions between converging signals is that in which long-lasting traces of activity enter into play. A priori, neuronal fields of intercommunications, as are networks, should be ideal places for the formation of new associative links, provided their neurons are endowed with some mechanism(s) for persistence of effects. We have every reason to believe that such mechanisms exist, as there is little doubt that subcortical reticular systems (reticular formation of the brain stem, nonspecific nuclei of the thalamus) play an important part in the formation of conditioned reflexes (see Fessard and Gastaut, 1958). As concerns the nature of the persisting traces, it is well known that the so-called "reverberating circuits" have been considered with great favor by many investigators, although satisfactory proof of their reality is still lacking (see, however, Verzeano and Negishi, 1959); and neurons in networklike structures may be interwoven in such a way that many closed loops of short-axon neurons are present and make possible a long persistence of recurring activities. But it is even more likely that a variety of plastic modifications takes place, owing to prolonged activity or inactivity (Eccles, 1953; Eccles and McIntyre, 1953), since there is every chance that cumulative changes occur within structures in which the common behavior of sensory messages includes repetitive firing and after-discharges, chainlike and circular transmissions, and ephaptic incremental interactions favored by close packing of neurons. As a consequence, one is led to believe that, among the possible plastic changes produced under these conditions, processes such as tetanic and posttetanic potentiations should acquire great importance. This aspect has been considered elsewhere by Fessard (1960). Let it only be pointed out here that, owing to common interneurons, such as are abundantly present in networks, potentiation created by increases in the natural spike frequency can, in spite of being a presynaptic phenomenon, perform transfers of facilitation from one kind of afferents to another. More generally, convergence, autogenic repetitive firing, and potentiation form a cooperative chain of dynamic processes that associate integration in neural space—due to convergence—with integration in time—due to accumulation of plastic increments of latent facilitation at axon endings.

A delicate point in problems of sensory communication is that of the moment when, and the place where, a sensory signal loses its identity after its association with others so as to form an integrated whole. This is a point from which problems of integration should replace those of simple communication, but it is clear that the distinction cannot be an absolute one and up to a certain point may appear somewhat conventional. The track of a single-unit message is doomed to be rapidly lost when one tries to follow it through a neuronal field endowed with networklike properties, within which the elementary message readily interacts with many others. On the contrary, a multiunit pattern may preserve its dominant features much further in spite of distortions of different kinds. When the multiunit pattern under consideration is composed of afferent signals belonging to different sense modalities, the problem may, or may not, still be called one of sensory communication. This is a matter of definition, and we may choose to extend the notion of "communication" to pattern-to-pattern input-output relationships. Unfortunately, we still lack principles that would help us describe and master such operations in which heterosensory communications are involved. These principles may gradually emerge in the future from an extensive use of multiple microelectrode recordings, together with a systematic treatment of data by modern electronic computers, so that pattern-to-pattern transformation matrices can be established and possibly generalized. The only special case that has given rise to some theoretical approach is that of association between two heterogeneous signals in the experimental situations offered by conditioning (Uttley, 1958).

For the time being, it seems that we should do better to try to clear up such principles as seem to govern the most general transformations—or transfer functions—of multiunit homogeneous messages during their progression through neuronal networks. Some knowledge of that sort should at any rate precede any speculation about the way two (or more) multiunit homogeneous messages of different modalities communicate with each other to give rise to an integrated pattern of neural activity—a method representing a more analytical and, up to now, an easier approach than the pattern-to-pattern transformation matrix just alluded to.

A certain amount of speculation has already been applied by different authors to the problem of progression and transformation of messages through networks. Obviously, it requires some knowledge of the "connectivities" within the network under consideration, that is to say, of the arrangement and relationships of its component neurons. Unfortunately, this is a point upon which our information is rather scarce, and we have too often to replace lacking data by insufficiently founded assumptions. There have been few skillful experts devoted to the laborious Golgi method since Cajal's work (see, for instance, Scheibel and Scheibel, 1957); and although a gradual increase of precise knowledge in these matters is undoubtedly to be expected in the future, the huge number of elements and the extreme complexity of their interconnections leave no hope that we can ever know to the last detail, or fully grasp as a significant whole, the distribution of activities within actual neuronal networks.

In order to get a correct representation of at least their main dynamic features, recourse to models—either mathematical or physical —offers a useful though sometimes misleading method. To begin with, simplified neural circuits of special designs have been hypothesized in which operations known to occur in nature can be performed artificially by the model. Such attempts have never been and could not be taken very seriously by the majority of neurophysiologists. It would not be sound to assume that we could safely infer close resemblances in internal organization and mechanisms from similarities in performance. A different approach, in diametrical opposition to the preceding one, assumes that the structural and dynamic parameters of neuronal networks are randomly distributed. Probabilistic theories have been proposed which we believe are of value in explaining the statistical properties of neural nets (Rapaport, 1950; Shimbel, 1950). An impressive model of that sort is the one by Beurle (1956), who started with reasonable initial assumptions regarding the neurophysiological processes involved. In Beurle's model, propagation and interference of excitation waves through large populations of randomly connected neurons are considered, and an appropriate mathematical treatment is applied which leads to suggestive deductions concerning the total behavior of the system.

However, as Gerard expressed it (1960), "a random population with random connections is a weak postulate; any additional structuring must give richer attributes." As a matter of fact, all neuronal assemblies are machinelike as well as stochastic in their total behavior. However, their intrinsically organized properties, far from being the result of a complicated and diversified circuitry, as in a radio set, are only the consequence, we believe, of the few types of elementary connections we know to exist between neurons. Beyond that, little more than mere repetition of the same elementary connection pattern determines the special structure or architectonics of a given network and endows it with its statistical properties. Some typical aspects of such linkages in neuronal fields are diagrammatically represented in Figs. 6a to d. They illustrate the most common arrangements of neurons in synaptic networks, where concurrent, "intercurrent," and recurrent connections contribute in different ways and proportions to the general patterning of each system.

With some ten to forty thousand neurons per cubic millimeter, such tridimensional systems of interconnections must be endowed with a richness of dynamic potentialities that is quite unimaginable and certainly defies full description, particularly since the coexistence in a single network of different elementary connection patterns is a reasonable assumption. Although it seems most likely that coexistence by itself should result in new properties that could not have been

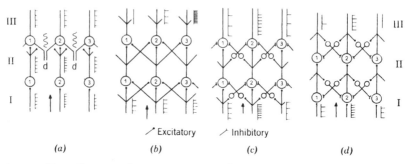

Fig. 6. Four diagrams showing some typical connection patterns that are supposed to exist, and most probably to coexist (together with others), in real networklike structures of the brain. Impulse discharges of conventional frequencies are also represented, except for step II in (*b*), (*c*), and (*d*). For more details, see text.

predicted from a separate study of each particular connection pattern, we believe that such a study must be undertaken if we are ever to get at some understanding of the role neuronal networks may play in sensory communications. Some familiar cases of simple connection patterns are considered below, with no other ambition than to recall in qualitative terms what general transmission properties these arrangements of neurons may be responsible for when they are invaded by incoming trains of impulses. We believe that some useful and unsophisticated evidence can thus be ascertained before any mathematical treatment be assayed.

Figure 6*a* is a schema corresponding to old views that ignored the existence of synaptic convergences. According to these views, transmission of sensory impulses obeyed the one-to-one rule and occurred along parallel chains of independent neurons (*a*I to *a*II). This was in agreement with the misleading feeling that sensory messages, in order to be reliable, should retain their strict individual characteristics, at least up to the territories where integrative interactions take place. As concerns the mechanism of these interactions, very little was then made explicit, save perhaps that electric field actions were often favored as possible agents of integration. Today the importance and even the existence of such "ephaptic" interactions are still a matter for discussion. Although definitive proof is still lacking, we think that such parallel couplings as those schematized in Fig. 6a at level III between dendritic trees (d) may play some role, if not the main one, as factors for synchronization of slow waves and for determining the frequency of the consequent trains of spikes. Synchronization is well known to be a common feature of the brain's electrical activities, one by which several foci of rhythmic discharges, the so-called *pacemakers,* are formed and obliged to compete with each other. This is one of the multiple aspects of the integration processes that do not concern us here, apart from their being the natural consequence of all sensory communications.

It would not be enough, however, to say that sensory communication processes result only in integration phenomena. Apart from cases of generalized hypersynchrony (slow waves of deep sleep, coma, epilepsy, all of which are accompanied by total loss of consciousness) synchronization factors, inasmuch as they operate within restricted neuronal pools, may involve the intervention of factors of *segregation,* which may give rise to contours or emphasize contrasts, and thus be held responsible for the organization of perceptive field. Now it seems that a major advance has occurred in the last few years, as regards experimental evidence that such factors exist. They actually manifest themselves through *sharpening* or *focusing effects,* which are revealed when distributions of evoked potentials, picked up from successive relay stations along ascending pathways, are compared. This appears particularly striking from the investigations of Sumi, Katsuki, and Uchiyama (1956) in audition, and of Mountcastle (1957), in somesthesia. But what may be the basic structures and elementary mechanisms responsible for such effects?

Linear connections like those in Fig. 6a are by their essence devoid of all organizing virtue, except perhaps by way of electric field interactions, which, apart from and in contradistinction with their synchronizing influence, could be thought of as being capable of localized inhibiting (anodal) effects; or, more ambitiously, they could be held responsible for the whole instantaneous organization of the perceptive field (Köhler, 1951). Such views sound most unlikely today, or at best they can be said to lack serious experimental support. The fact is that sharpening effects are already conspicuous in the outflow of impulses from some sensory relay nuclei of primary ascending pathways, but, rather than field actions, one is led to involve some property of overlapping fields of axon collaterals, as schematized in Fig. 6b, since it is now known that these nuclei are so constituted. They also contain a certain number of interneurons: briefly said, they are moderately but safely endowed with networklike properties, and it is difficult not to think that they might owe their sharpening power to this monotonous structural pattern rather than to a special and complicated circuitry. The same thing could obviously be said, with even more likelihood of being true, of polyneuronal fields of neurons like those present in nonspecific subcortical nuclei or cortical associative territories. There we can safely assume the presence of abundant interneurons, either excitatory or inhibitory, interwoven in three-dimensional patterns of all kinds. Figures 6c and d present highly simplified diagrams in which only intercurrent inhibitory neurons appear.

Figure 6b (or equivalent ones) shows the simplest type of network connectivity, one that has been taken as a prototype by several authors (for instance, Eccles, 1953). As has already been explained (Fessard, 1954), it is easy to see how such a simple structural pattern can give rise to a sharpening effect, provided one assumes a sigmoid shape for the transmission curve at each synaptic level (I, II, III, and so on). Such a shape is a common feature of monosynaptic transmissions. For simplicity, let us assume that the same curve can be utilized at each synaptic level (Fig. 7a). The point k, where the curve intersects the bisector is a critical point: if the density of impulses is such that the total excitation exceeds the value of its abscissa, the propagation through the network will be incremental; but if the message is too weak, as is supposed here in the figure, it will become less and less important at each step, until somewhere within the network it fails to be transmitted. The characteristic input-output curve for the whole network is thus changed into an abrupt or step function, as represented in Fig. 7b. A schematic application of this principle is represented in its simplest form in Fig. 6b. Along the chain of neurons 1, a message of low frequency is shown fading out; with neurons 2 the transmission is slightly decremental, whereas it is strongly incremental in 3. The net result is an accentuation of contrast between the impulse frequencies in 1 and those in 3.

With the introduction of inhibitory interneurons into our schematic networks, the sharpening effect appears to be a direct consequence of the existing cross-connections (see Figs. 6c and d), if we assume that these predominate over the uncrossed ones (not represented here). Two kinds of connections are distinguished here, although they are likely to coexist in fact. They simply represent a generalization of what is now known to exist in the spinal cord at the motoneuron level: diagram (c) generalizes to a large population of neural chains the cross-connection principle that operates in reciprocal inhibition; diagram (d) shows retroactive inhibitory cross-connections similar to those discovered in the spinal cord by Renshaw (1941) and studied in great detail by Eccles, Fatt, and Koketsu (1954).

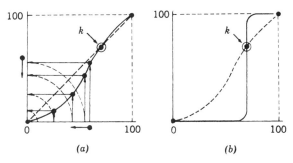

(a)   (b)

Fig. 7. (a) Theoretical S-shaped input-output curve relative to the transmission of groups of impulses through one barrier of monosynaptic junctions. The intersection with the bisector k is a critical point, as explained in the text. This figure is intended to show how transmission through several successive synaptic barriers, such as are present in networks, is necessarily decremental as shown here, if the representative point starts below k. (b) The resulting steplike transmission curve of a network containing a sufficient number of successive synaptic barriers.

In these two diagrams, the maxima of input activities have been placed in the center (neurons numbered 2) so that it is easy to realize how a discharging focus can be sharpened: it appears to be an obvious consequence of the fact that all neurons of the central column exert a more powerful inhibitory influence on their environment of less active units, forward in (c), backward in (d), than they themselves undergo from these. The result is a *relative* enhancement of the central output activities as compared to the marginal ones. (Our diagram is of course greatly simplified, and we have to think of many more neurons in series and in parallel.) Here the conventional absence of cross-interactions other than inhibitory ones has obliged us to show a general attenuation of the rate of firing from input to output; but in real networks excitatory interneurons are also present, together with collaterals like those in (b), and the total action could, at least partly, compensate for the frequency decrease and eventually accentuate the sharpening effect.

Other complications and other connection patterns could be considered, if sufficient justification from histological and/or microphysiological data can be brought out. For instance, a pattern similar to the one in Fig. 8 has to be considered in certain cases, because it has been found to characterize the connections in the reticular formation of the brain stem (Scheibel and Scheibel, 1957). It may act as an amplifying device, useful to counteract a general tendency to excessive inhibitory and degenerative effects when sharpening mechanisms are at work. But a clear understanding of how such connection patterns and others may work requires more precise knowledge of the extension, orientation, and distribution of the more-or-less profuse axon collaterals and of the ramifications displayed inside the dendritic field. The Scheibels' study appears as a model in this respect.

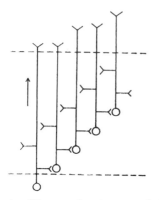

Fig. 8. Diagram showing one of the possible types of conduction circuits through the reticular core of the brain stem. (Slightly modified from Scheibel and Scheibel, 1957.)

In most of the tentative network diagrams that have been proposed, recurrent connections form a more-or-less important part of the general design and are generally assumed to have great functional significance. It has been a long time since Lorente de Nó (1933) drew attention to the structural reality of such closed loops in the cerebral cortex, and it was rapidly assumed that they should support the concept of self-re-exciting neuron chains, which was proposed by Alexander Forbes (1929) to explain spinal reflex afterdischarges. Reverberating circuits have, since then, often been held responsible for perpetuation of firing activities, as observed, for instance, in isolated portions of cerebral cortex, a question submitted to deep analytical studies by Burns (1951, 1958). Some authors have gone so far as to suggest a role for the reverberating circuits in the persistence of memory traces, a highly controversial question beyond the scope of the present theme. As concerns the problems of sensory communication, we thought it wiser here to refrain from reproducing one of the various diagrams illustrating the reverberation principle that have been proposed by several authors including ourselves (see Burns, 1951; Eccles, 1953; Burns, Grafstein, and Olszewski, 1957). As a matter of fact, we still lack sufficient knowledge about the topological arrangements of these neuronal circuits, and histological pictures say nothing about the nature—whether excitatory or inhibitory—of the recurrent neurons. Actually, these circuits seem to be there to secure more often a system of self-protective inhibitory action than one of regenerative feedback. As shown by the frequent occurrence of delayed hyperpolarization waves after the activation of neurons in different cerebral structures (see Fig. 3, but intracellular records are more convincing, as exemplified in Albe-Fessard, 1960), the spinal Renshaw cell system is likely to be a very common pattern there too. A great number of recurrent interneurons would then be involved in such systems of negative feedback.

Prolonged firings, however, have to be explained; it is often claimed that they have brought out full evidence of the reality of impulse circulations: for us, this statement is one more example of the undue discredit which still surrounds the idea that autogenic repetitive firing might explain at least a certain number of such facts. Periodic firing is no more than a formal appearance that may be produced by diverse causes, and autorhythmic activities are known to arise commonly when excitable elements undergo protracted depolarizing actions. It seems hardly appropriate that, at the moment when the generality and importance of potential gradients of long duration, such as occur within central structures, are discovered, one should go on denying the existence of and possible functional roles for autorhythmic modalities of cellular activity. Patterns of autorhythmic discharges, inasmuch as they would arise in consequence of some persisting electrotonic action immediately after the occurrence of a previous well-organized activity, would themselves appear to retain some features of spatiotemporal organization within the neuronal field considered. Contrary to that, the process of reverberation, when it operates through the multiple intricate microloops and zigzag recurrent pathways of a finely woven network, seems, as can easily be imagined, to be devoid of all ordering virtue. Relative to the problems of sensory communication, reverberation appears to be a factor of blur and random diffusion for sensory messages, in a word, a source of noise. This is not to say that we should disregard it entirely, but it is difficult indeed to consider it as an organizing factor for any operation of message modeling and patterning.

For us, the major principle of dynamic organization in simple sensory communication and heterosensory interference is the one that, in various kinds of interplay and rivalry, opposes *excitation* and *inhibition* within networklike structures. It is clear that this principle is nothing but a generalization of Sherrington's law of reciprocal innervation, in the various cases in which neuronal cross-connections have become so numerous and intricate that "network" is an appropriate term for the responsible structure. The principle applies to

vertebrate retinas (Kuffler, 1952) and a striking example of its validity
is also offered by the lateral eye of *Limulus* in which inhibition is
exerted mutually among the receptor units (ommatidia) as has been
shown by Hartline and his collaborators (Hartline, Wagner, and Ratliff, 1956; Hartline and Ratliff, 1957; see also Ratliff in this volume,
and Tomita, 1958). As concerns the cerebral structures, which we
already know to behave in a similar way, let us recall the above-
mentioned investigations by Mountcastle (1957), Katsuki, et al.
(1958), together with those presented at this symposium by Dr. Jung
(see also Jung and Baumgartner, 1955), and in Dr. Hubel's contribution to the discussion. There is little doubt that many other cases
will be discovered and analyzed in the near future.

According to classical views, the principle of antagonism, or cooperation, between excitation and inhibition, has for a long while been used
as a useful and illuminating way to describe and even to explain the
greatest number of operations that can be achieved by the central
nervous system; but only three levels of complexity are usually recognized, namely: (1) the level of behavior where we know, for instance,
what essential importance to give to cortical inhibitory processes, as
inferred from experiments on learning and conditioned reflexes, according to Pavlov's original conception, more-or-less revised and
completed by his followers; (2) the level of gross organization of interconnected centers, at which it appears that no one of the main operations performed by the CNS—be it of coordination, subordination, or
regulation—can be accomplished without the joint or successive intervention of excitatory and inhibitory processes; (3) the level of a
neuron surface, where the rivalry of excitation and inhibition is played
between synaptic loci and finally takes on the aspect of conflicts between chemically antagonistic transmitters and electromotive forces
of opposite direction.

To these three levels, it has seemed to us that between the last two,
an intermediate step should be introduced—or if already recognized
its importance emphasized—at which excitation and inhibition interact within a relatively homogeneous field of interconnected e and i
neurons, the networklike component of all nuclear, reticular, or laminar
structures to which sensory messages are conveyed and within which
they prepare to give rise to integration patterns.

Summary

1. A Symposium on Sensory Communication has to consider the
problem of how afferent messages from different origins can meet and
interact within the brain.

2. The reticular formation of the mesencephalon is now well recognized as a site of election for heterosensory convergences and interactions, but there are many other structures endowed with similar properties, as has been revealed by recent electrophysiological explorations. Among these, we can enumerate a number of nonspecific thalamic nuclei, zona incerta, caudate nucleus, red nucleus, cortical associative areas, and motor cortex. From these places, long-latency "secondary" evoked potentials can be picked up, as well as unitary discharges generated in neurons that respond to different sensory modalities.

3. We have called such neurons "plurivalent," and "network" has seemed to be an appropriate term for structures in which they are abundant, for such a widely distributed plurivalence cannot happen without a great wealth of interconnectivity among the constituent units.

4. Mechanisms of interaction between converging afferent impulses can be described only partly in terms of neuronal facilitation, inhibition, and occlusion, at least inasmuch as these processes are considered in their simple aspects under the artificial conditions of classical experimentation (double-shock technique). Most often, sensory communication actually requires the propagation and transmission of trains of impulses, these being distributed within the structures according to certain spatial patterns that may undergo complex changes at each of the relay stations encountered. There transmission of a mixture of excitatory and inhibitory impulses takes place, which results in the frequency modulation of a permanent autogenic firing, rather than in the familiar one-to-one correspondence between input and output volleys. Persistence of effects due to plastic changes, and perhaps also to closed-chain circulation of impulses, introduces further complications, in addition to endowing the system with new functional properties. These may play a part in the formation of conditioned reflexes.

5. When neuronal networks are studied, problems of sensory communication rapidly become those of sensory integration, but no sharp separation can be made, except conventionally, between the two categories of problems, especially as network properties are already present at the level of the lowest primary relay stations.

6. Probabilistic theories seem to be more adequate than machine-like models to describe the most general properties of neuronal networks, although both may be misleading as long as we lack precise knowledge about the plan of each network's internal connections.

7. Reasonable deductions from the consideration of some of the simplest neuronal connection patterns that we know to exist already

seem possible, however, and appear to fit in with the resultant properties exhibited by the whole neuronal assembly. Attention has been drawn particularly toward the property one can describe as a sharpening or focusing, one by which contrasts in the distribution of neuronal discharges are emphasized. This has been demonstrated to be the case by several workers, in the fields of somesthesia, audition, and vision.

8. Three types of theoretical structural patterns, in which "intercurrent" connections are regularly interwoven and repeated, have been considered particularly. The simplest type has its cross-connections provided only by axon collaterals, whereas, in the other two, cross-connections are due to interpolated inhibitory neurons, directed either forward or backward. In this way the principles that operate at the spinal-cord level in reciprocal or recurrent inhibition are generalized.

9. An explanation has been given of how the input-output transfer functions for the multifiber messages through these prototypes of many networks could take the form of sharpening or focusing effects.

10. The principle of reverberation or regenerative feedback, as it may operate within networks, appears to us, not as one capable of introducing order, but rather as one that would diminish the information content of sensory messages. For us, the major principle of dynamic organization in intersensory as well as in simple sensory communications is one that, in various kinds of interplay and rivalry, conjoins or opposes excitation and inhibition within networklike structures.

References

Albe-Fessard, D. Sur l'origine des ondes lentes observées en dérivation intracellulaire dans diverses structures cérébrales. *C. R. Soc. Biol.*, 1960, **154,** 11–16.

Albe-Fessard, D., and E. Gillett (unpublished results).

Albe-Fessard, D., and A. Rougeul. Activités d'origine somesthésique evoquées sur le cortex non-spécifique du chat anesthésié au chloralose: rôle du centre médian du thalamus. *EEG clin. Neurophysiol.*, 1958, **10,** 131–152.

Albe-Fessard, D., E. Oswaldo-Cruz, and C. E. Rocha-Miranda. Convergences vers le noyau caudé de signaux d'origines corticale et hétérosensorielle. Etude unitaire de leurs interactions. *J. Physiol.*, 1958, **50,** 105–108.

Albe-Fessard, D., E. Oswaldo-Cruz, and C. E. Rocha-Miranda. Activités evoquées dans le noyau caudé du chat en réponse à des types divers d'afférences. I. Etude macrophysiologique; II. Etude microphysiologique. *EEG clin. Neurophysiol.*, 1960, **12,** 405–420; 649–661.

Amassian, V. E., and R. V. DeVito. Unit activity in reticular formation and nearby structures. *J. Neurophysiol.*, 1954, **17**, 575–603.

von Baumgarten, R., A. Mollica, and G. Moruzzi. Influence of the motor cortex on the spike discharges of bulbo-reticular neurons. *EEG clin. Neurophysiol.*, 1953, Suppl. **3**, 68.

Beurle, R. L. Properties of a mass of cells capable of regenerating pulses. *Phil. Trans. roy. Soc.*, 1956, **B240**, 55–94.

Burns, B. D. Some properties of isolated cerebral cortex in the unanesthetized cat. *J. Physiol.*, 1951, **112**, 156–175.

Burns, B. D. *The Mammalian Cerebral Cortex.* London: Arnold, 1958.

Burns, B. D., B. Grafstein, and G. Olszewski. Identification of neurons giving burst response in isolated cerebral cortex. *J. Neurophysiol.*, 1957, **20**, 200–210.

Buser, P., and P. Borenstein. Réponses somesthésiques visuelles et auditives recueillies au niveau du cortex "associatif" suprasylvien chez le chat curarisé non anesthésié. *EEG clin. Neurophysiol.*, 1959, **11**, 285–304.

Buser, P., P. Borenstein, and J. Bruner. Etude des systèmes "associatifs" visuels et auditifs chez le chat anesthésié au chloralose. *EEG clin. Neurophysiol.*, 1959, **11**, 305–324.

Eccles, J. C. *The Neurophysiological Basis of Mind: The Principles of Neurophysiology.* Oxford: Clarendon Press, 1953.

Eccles, J. C., and A. K. McIntyre. The effects of disuse and of activity on mammalian spinal reflexes. *J. Physiol.*, 1953, **121**, 492–516.

Eccles, J. C., P. Fatt, and K. Koketsu. Cholinergic and inhibitory synapses in a pathway from motoraxon collaterals to motoneurones. *J. Physiol.*, 1954, **126**, 524–562.

Fessard, A. Mechanisms of nervous integration and conscious experience. In J. F. Delafresnaye (Editor), *Brain Mechanisms and Consciousness.* Oxford: Blackwell, 1954. Pp. 200–236.

Fessard, A. Le conditionnement considéré à l'échelle du neurone. In Moscow Colloq. on EEG of Higher Nervous Activity. *EEG clin. Neurophysiol.*, 1960, **Suppl. 13**, 157–183.

Fessard, A., and H. Gastaut. In *Le Conditionnement et l'Apprentissage.* Paris: Presses Universitaires de France, 1958.

Fessard, A., and T. Szabo. Possibilité d'un transfert de la facilitation posttétanique dans une chaîne disynaptique. *J. Physiol.*, 1959, **51**, 465–466.

Fessard, A., and T. Szabo. La facilitation de post-activation comme facteur de plasticité dans l'établissement de liaisons temporaires. In *Brain Mechanisms and Learning.* Oxford: Blackwell (in press).

Fessard, A., and L. Tauc. Effets de répétition sur l'amplitude des potentiels post-synaptiques d'un soma neuronique. *J. Physiol.*, 1958, **50**, 277–281.

Forbes, A. *The Foundations of Experimental Psychology.* Worcester: Clark University Press, 1929.

Gerard, R. W. Neurophysiology: An integration (molecules, neurones and behavior). In *Handbook of Physiology. Neurophysiology III.* Washington: American Physiological Society, 1960. Pp. 1919–1965.

Hartline, H. K., and F. Ratliff. Inhibitory interaction of receptor units in the eye of *Limulus. J. gen. Physiol.*, 1957, **40**, 357–376.

Hartline, H. K., H. G. Wagner, and F. Ratliff. Inhibition in the eye of *Limulus. J. gen. Physiol.*, 1956, **39**, 651–673.

Jung, R., and G. Baumgartner. Hemmungsmechanismen und bremsende Stabilisierung an einzelnen Neuronen des optischen Cortex: Ein Beitrag zur Koordination corticaler Erregungsvorgänge. *Pflügers Arch. ges. Physiol.*, 1955, **261**, 434–456.

Katsuki, Y., T. Sumi, H. Uchiyama, and T. Watanabe. Electric responses of auditory neurons in cat to sound stimulation. *J. Neurophysiol.*, 1958, **21**, 569–588.

Köhler, W. Relational determination in perception. In L. A. Jeffress (Editor), *Cerebral Mechanisms in Behavior.* The Hixon Symposium. New York: Wiley, 1951. Pp. 200–243.

Kuffler, S. W. Neurons in the retina: Organization, inhibition and excitation problems. *Cold Spring Harbor Sympos. Quant. Biol.*, 1952, **17**, 281–292.

Lorente de Nó, R. Studies on the structure of the cerebral cortex. I. The area entorhinalis. *J. Psychol. Neurol. Lpz.*, 1933, **45**, 381–438.

Massion, J., and D. Albe-Fessard. Caractéristiques différentielles des réponses aux stimulation sensorielles des deux parties du noyau rouge. *C. R. Acad. Sci.*, 1959, **248**, 3737–3749.

Mountcastle, V. B. Modality and topographic properties of single neurons of cat's somatic sensory cortex. *J. Neurophysiol.*, 1957, **20**, 408–434.

Rapaport, A. Contribution to the probabilistic theory of neural nets. *Bull. math. Biophysics*, 1950, **12**, 109–121.

Renshaw, B. Influence of discharge of motoneurons upon excitation of neighboring motoneurons. *J. Neurophysiol.*, 1941, **4**, 167–183.

Scheibel, M., and A. Scheibel. Structural substrates for integrative patterns in the brain stem reticular core. In H. H. Jasper, et al. (Editors), *Reticular Formation of the Brain.* Boston: Little, Brown, 1957. Pp. 31–55.

Scheibel, M., A. Scheibel, A. Mollica, and G. Moruzzi. Convergence and interaction of afferent impulses on single units of reticular formation. *J. Neurophysiol.*, 1955, **18**, 309–331.

Shimbel, A. Contributions to the mathematical biophysics of the central nervous system with special reference to learning. *Bull. math. Biophysics*, 1950, **12**, 241–275.

Sumi, T., Y. Katsuki, and H. Uchiyama. Electric responses of auditory neurons in cat to sound stimulation. *Proc. Jap. Acad.*, 1956, **32**, 67–71.

Tauc, L. Preuve expérimentale de l'existence de neurones intermédiaires dans le ganglion abdominal de l'Aplysie. *C. R. Acad. Sci.*, 1959, **248**, 853–856.

Tomita, T. Mechanism of lateral inhibition in eye of *Limulus*. *J. Neurophysiol.*, 1958, **21**, 419–429.

Uttley, A. M. A theory of the mechanism of learning based on the computation of conditional probabilities. In *I Congr. int. Cybernet.*, Namur. Paris: Gauthier-Villars, 1958. Pp. 830–856.

Verzeano, M., and K. Negishi. Neuronal activity and states of consciousness. In *Abstr. Commun. XXI int. Congr. physiol. Sci.*, Buenos-Aires, 1959, 287–288.

31

P. BUSER and M. IMBERT
Centre de Physiologie nerveuse du C.N.R.S. et Faculté des Sciences, Paris

Sensory Projections to the Motor Cortex in Cats: a Microelectrode Study

From studies performed on cats it now seems to be quite well established that, for a given peripheral stimulus (visual, auditory, or somatic), cortical evoked responses can be identified, not only on the corresponding primary projection field, but also on several other loci. Various experimental conditions permit these conclusions: studies under chloralose anesthesia, or on unanesthetized preparations immobilized with curare, as well as observations on chronic implanted animals. Sensory responses thus recorded outside the projection area differ from primary evoked potentials in several characteristics, such as longer duration, longer latency, higher susceptibility to depressive actions and also to barbiturate anesthesia, and so forth. We have suggested for these activities the name *secondary** or *irradiation* potentials.

In an attempt to classify these *nonprimary* responses, by using only the formal criterion of localization, one may distinguish three different categories.

A first group of responses comes from the associative cortex (suprasylvian and anterior lateral gyrus). On unanesthetized curarized preparations, the number of "active" points is rather high in those regions, as revealed by a systematic oscillographic exploration (Buser and Borenstein, 1959). Deep chloralose narcosis causes secondary potentials to increase in size and at the same time restricts their superficial extent to limited foci. Thus a characteristic topography emerges

---

* This term—meaning "nonprimary"—was chosen as the most "neutral" description of the mechanisms possibly involved, better than "associative" or "nonspecific." The term does not imply a correspondence or similarity between these responses and the phenomena obtained with deep barbiturate narcosis, and described as *secondary discharges*, by Derbyshire, et al. (1936) and Forbes and Morison (1939).

for each sensory modality: roughly, two suprasylvian foci may thus be identified for vision, two for audition (Buser, Borenstein, and Bruner, 1959),* and two for somatic sensitivity (Albe-Fessard and Rougeul, 1958); the anterior lateral gyrus also seems to possess some localized projections for all three modalities (Amassian, 1954).

A second group of projections, still incompletely worked out, involves the medial aspect of the neocortex. It appears that, here also, visual, auditory, and somatic foci may be identified independent of any other cortical projection system (Bruner, 1960).

Finally, there is a third group, localized in the "motor" cortex. Previous data from other investigators (Gastaut and Hunter, 1950; Wall, Remond, and Dobson, 1952; de Haas, Lombroso, and Merlis, 1953; Hunter and Ingvar, 1955; Feng, Liu, and Shen, 1956), as well as more recent observations from our laboratory (Buser and Borenstein, 1956; Ascher and Buser, 1958; Rougeul, 1958), have pointed out that responses to visual and auditory stimuli exist in addition to somesthetic responses in the motor cortex of cats under one or another of the experimental conditions previously mentioned—chloralose, curare, or "chronic" preparations. And recording from the pyramidal tract has revealed a close correspondence in time between these "motor irradiation potentials" and the pyramidal efferent discharges elicited by a given sensory stimulus. Topographically, the frontal areas activated in response to visual and auditory stimuli overlap almost completely, a situation that is not usual in the suprasylvian or the medial cortex. And finally, facilitation is observed at both levels, cortex and pyramids, when two stimuli from different modalities are separated by a short interval.

The investigations we are going to speak about here were suggested by these interaction effects between modalities. Hence, it seemed highly probable from macroelectrode recordings, although it had not been demonstrated, that the pathways for different sense modalities converge either upon the cortical cells themselves, or at least somewhere on the afferent route to the motor cortex.

A microelectrode study on the motor cortex was undertaken, therefore, which would give some additional information on this particular type of sensory irradiation at the cortical level.

Methods

Sixty cats were used, either deeply anesthetized with chloralose (9 centigrams per kilogram) or unanesthetized and immobilized with

---

* And also Thompson and Sindberg (1960).

curare; in the latter case, surgery was performed under ether, and the animal was put under artificial respiration at the time it was given curare (Flaxedil).

In most cases, pyramidal activity was surveyed; two bipolar concentric electrodes (tip distance for each electrode, 0.2 millimeter; distance between electrodes, 2 millimeters) were introduced stereotaxically into the pontine pyramidal tract, one on each side of the midsagittal plane. The final placement was made by physiological control: when we stimulated electrically one motor cortex and recorded from the ipsilateral deep electrode, a short-latency (0.5-millisecond) response suddenly appeared as the electrode reached the tractus. This short-latency response has been described by Patton and Amassian (1953) and others.

Recordings from single cortical units were made with microelectrodes, either tungsten wires electrolytically sharpened and further isolated with varnish except at the very tip, or glass micropipettes pulled with a microforge of the de Fonbrune type and filled with 3 molar KCl. The tip diameters measured 2 to 5 microns for tungsten, about 1 micron for glass.

After removal of the dura and sometimes of the pia, the electrode (metal or glass) was introduced very slowly by means of a micrometric screw. During the experiments, as much care as possible was taken to prevent the cortex from drying or cooling off.

A conventional amplifying and recording system was used (954 acorn-tube cathode follower, a-c amplifiers, Cossor double-beam cathode-ray oscillograph 1049, Recordine Alvar camera). As a rule, only extracellular fast activities (spikes) were recorded in these experiments, and no particular attention was paid to very slow variations or a fortiori to d-c shifts that might accompany cellular penetrations.

Sensory stimuli were delivered at low frequencies (0.5 to 0.1 per second); for somatic stimulation, brief electric shocks were delivered to each leg through two needles inserted under the skin (no further specification of the somesthetic submodalities was considered in this investigation); brief flashes (from a neon bulb) and short clicks were the visual and auditory stimuli.

Results

In the general organization of the "sensorimotor" cortex of the cat, a distinction is usually made between purely somatic areas (areas I and II) and the more rostral motor cortex. According to architectonic criteria (Campbell, 1905; Winkler and Potter, 1914; Langworthy, 1928), the somatic cortex covers three different fields: areas 1, 3, and 5, whereas the "motor" cortex consists of areas 4 and presumably 6; a small dimple, sometimes hardly visible, lying behind the sulcus cruciatus, may be considered as the separation between the two regions. Hence, the cortical zone containing giant pyramidal cells (Betz cells) lies not only on the anterior, but also on the rostral part of the posterior sigmoid gyrus.

Consequently, studying the "motor" area included exploration of at least this whole region—the anterior sigmoid and frontal part of the posterior sigmoid gyrus, and of the coronal gyrus. Furthermore, records were also taken, for comparison, from the somatic area I lying somewhat more caudally (close to anterior lateral and anterior suprasylvian gyrus).

As specified before, experiments were performed under two experimental conditions: (1) with the animal under chloralose anesthesia and (2) with the animal unanesthetized and immobilized with curare. Though the results obtained with the two procedures are comparable, it is convenient to consider them separately.

Experiments under chloralose anesthesia

Penetrating into the cortical layers brings up units, some of which are spontaneously active over the whole period of recording (5 to 30 minutes), whereas others are silent and fire only to peripheral stimuli. No particular localization of active or inactive cells was found, but chloralose unlike curare seems to provide somewhat unfavorable conditions for sustained spontaneous activity.

Neurons thus isolated were tested for the different sensory stimuli mentioned earlier: subcutaneous shocks at various points on the skin, flashes, and clicks, applied either singly or in combination.

With this method, two main categories of observations were made. A. Groups of units that differ in sensory responsiveness can be identified: (1) elements that respond to any of the stimuli applied, somatic, visual, or auditory, have been called *polysensory* units (group I); (2) neurons that respond only to *somatic* stimuli (group II) are of two kinds; some are polyvalent or "atopic," being activated from more than one peripheral locus (subgroup IIa); others respond with a spatial specificity, following the classical organization of the somesthetic pathways (subgroup IIb). B. The topography of these units is dependent upon their pattern of responsiveness. A majority of the units encountered within the motor area are polysensory. As the explored point is progressively shifted toward the more caudal somatic area, the number of polysensory elements decreases rapidly, making place for neurons that are exclusively somatic (group II). Subgroup IIa mainly occupies the frontal edge between somatic and motor cortex, whereas subgroup IIb is localized more to the posterior. These results are summarized in Table 1.

Table 1. Topographical Distribution (in Percentages) of Group I, IIa, and IIb Neurons in Experiments under Chloralose (200 Units)

Gyrus	Polysensory I	Somatic polyvalent IIa	Somatic somatotopic IIb
Anterior and oral posterior sigmoid	92	8	—
Posterior sigmoid (middle strip)	18	72	10
More caudal zones	—	37	63

Some additional details concerning these various types of cells will now be considered.

Polysensory units (group I)

Topography. Polysensory units may be found in the whole pericruciate area (Fig. 1), including the anterior sigmoid and the oral portion of the posterior sigmoid gyrus, as well as the upper part of the gyrus coronalis. Toward the frontal pole, exploration has not been carried out far enough to allow definite conclusions to be drawn about the anterior spread of this type of cell. As to the posterior limit, it coincides rather well with the end of area 4 as identified by anatomy (see Discussion).

We consider next the cortical depth of these recorded cells. Here it is interesting to note that, whereas single units may be found at variable depths during penetration, most of the elements actually responding to the different stimuli applied lie in deep layers (beyond 1000 microns). This is not the case in the somatic cortex, however, as will be described below.

Types of responses. For a given polysensory unit, both latency and number of spikes per stimulus depend on the modality and strength of the stimulus applied (Fig. 2). Some quantitative data for pericruciate elements studied from this point of view are given in Table 2.

The range of variability for latencies is rather high, depending not only (as expected) on the intensity of the stimulus applied but also

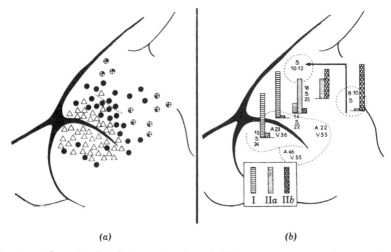

(a)   (b)

Fig. 1. Schematic distribution of units of different groups, according to their sensitivity to peripheral stimuli. Experiments under deep chloralose. (a) Over-all topography, each "point" representing at least five observed units. Triangles, polysensory elements (group I); filled circles, somatic nonsomatotopic (group IIa); quartered circles, somatotopic units (group IIb), responding to stimulation of contralateral hindleg (upper-right quarter filled), or foreleg (lower left-quarter filled). Broken line indicates theoretical posterior limit of area 4. (b) Histograms of distribution of group I, IIa and IIb units recorded in anterior sigmoid gyrus and different levels behind sulcus cruciatus. Total number of units, 201. Numbers give rough indications of range of latencies for somesthetic stimuli (S), and mean values for flashes (V) and clicks (A).

Table 2. Quantitative Data on a Single Sample of Perisigmoid Cortex, and Over-all Limits of Latencies Obtained on About 100 Units

	Single Sample		100 Units
	Latency (msec)	Most frequent number of spikes	Over-all limits of latencies (msec)
	Mean — Limits		
Somesthetic A_c*	22.5 — 18–28	1	12–22
A_i*	23.1 — 20–28	1	12–25
P_c*	25 — 23–28	2	12–25
P_i*	26.4 — 24–32	1	13–29
Auditory	25 — 22–28	2	25–45
Visual	50 — 40–52	2	45–90

* A, anterior; P, posterior; c, contralateral; i, ipsilateral.

on the rate of stimulation (rates above one per second increase latency very rapidly). In addition, the variability fluctuates for one unit as time goes by and differs from one unit to another. This is especially true for visual and auditory stimuli. Since in those areas

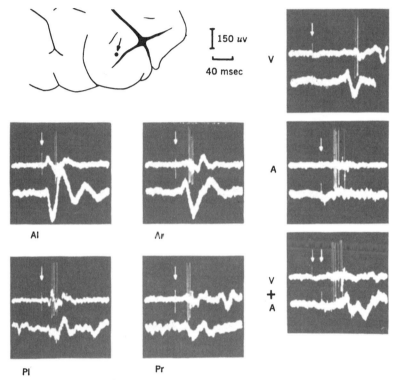

Fig. 2. Polysensory neuron of perisigmoid cortex. Right cortex under deep chloralose; no spontaneous firing; negative deflection upward for microelectrode recording. Typical responses to somesthetic stimulation of each limb, anterior or posterior, left or right (Al, Ar, Pl, Pr), to a flash (V), a click (A), and a closely spaced flash and click (V + A). Lower trace, activity from pontine pyramidal tract, recorded with a bipolar concentric electrode (tip distance, 0.5 mm) in right pyramidal tract.

we were dealing with neither a homogeneous population nor standard (experimental and physiological) conditions, we felt that any statistical treatment of the data might be misleading. Thus the values we have given for latencies to flashes or clicks are rather rough approximations of the range (Fig. 1b).

For somesthesia, the values obtained in spite of their variability show some interesting variations with both the place of the stimula-

tion and the cortical topography of the recorded unit.   The influence
of these parameters will be considered below.

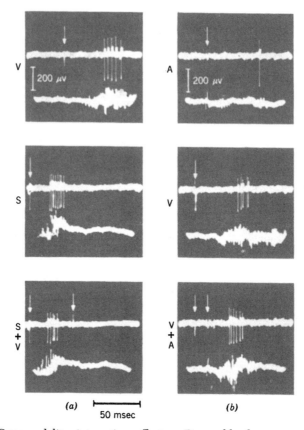

Fig. 3. Cross-modality interaction effects.   Deep chloralose; no spontaneous
firing; negative deflection upward.   Lower trace, recording from pontine pyramidal
tract.   (*a*) S +V, occlusion of a response to flash (V) by a preceding somesthetic
stimulus (S).   (*b*) V + A, facilitation following closely spaced flash (V) and
click (A).

Cross-modality interaction effects.   By combining two stimuli of
different modalities, each of which separately activates a given cell,
we may observe interactions between individual responses of cortical
cells.

At intervals from 20 to 50 msec, the second response is usually
occluded by the first.   This may result in the disappearance of, or
at least a reduction in, the number of spikes, and an increase in their
latency (Fig. 3*a*).

With shorter delays between stimuli, striking facilitation frequently occurs, especially in the combination of flash and click.   The modifications shown in Fig. 3*b* are quite typical: the final response appears earlier than the first of the two component responses and contains a greater number of spikes than a simple arithmetic summation would predict (Figs. 2 and 3*b*).

No general assumption can be made regarding the interval necessary for obtaining facilitation; actually, approximate coincidence in time of the individual responses seems to provide the best conditions.   With this in mind, it is obvious that the optimal interval depends entirely upon the latencies of the individual responses (that is, types and strength of stimuli, state of the preparation, and so on).

Such dynamic processes as occlusion and facilitation frequently parallel what is observed in the pyramidal tract at the pontine level (Fig. 3), as has been previously described (Buser and Ascher, 1960).

Somatic units (groups IIa and IIb)

In explorations of the caudal part of the posterior sigmoid gyrus (Fig. 1), the anterior lateral, the anterior suprasylvian, and the lateral portion of the coronal gyrus, a quite different pattern of sensory responsiveness has been found in the majority of the units encountered.   It is clear that these units respond to somatic stimulation only.

As has already been specified, the behavior of these neurons still depends on their location to an extent that suggests a subdivision into the two subgroups, II*a* and II*b* (Fig. 1).

Subgroup IIa. These units are spread over a rather small area, lying caudal to the polysensory zone.   They almost never respond to both light and sound, though some of them respond to sound alone. On the other hand, they are somatically *polyvalent* in that they respond to stimulation of more than one bodily part; in general they respond to stimulation of all four legs, sometimes to two forelegs, two hindlegs, or one side.   The precise pattern of response varies among the cells, and we have been unable to find any systematic correlation with the location of the recorded cell.

Subgroup IIb involves somatic area I in which the somatotopic arrangement is known to dictate the responsiveness of a given cell. Nothing need be added to previous, more detailed studies by others (see Discussion).

In the exploration of these more caudal zones, especially those of somatic area I, a rather striking difference appears in comparison with the motor-cortex responses: as soon as the cortex is penetrated, units are found, even in the upper layers, that respond to stimulation of

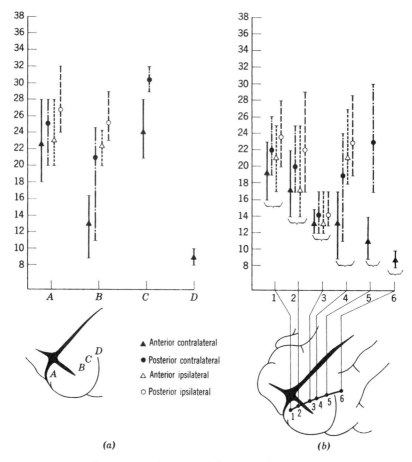

Fig. 4. Average latencies and range of latencies for responses to somesthetic stimuli (shocks to each leg). (*a*) Values from a single experiment at four cortical locations (the mean values are based on 30 trials). (*b*) Values from 200 different units (20 different experiments).

the body. This stands in contrast to the motor cortex where the responding units lie in the deep layers.

Variation in latency of somesthetic responses. An interesting fact is revealed when we consider the latencies of the somatic responses in relation to the position of the puncture along an anteroposterior axis (Figs. 1*b* and 4).

This spatial evolution is strikingly complex. Evidently latencies for all four legs do not decrease progressively from the motor cortex to the somatic cortex. A zone of minimal latency and minimal variability of latency of responses to both ipsilateral legs can be defined near the sulcus cruciatus. The changes are even more complex for the contralateral legs, since they depend upon the position of the explored axis with respect to the sagittal plane. The latency of response to stimulation of one contralateral leg goes through a minimum at the same time as the latency for the ipsilateral legs; for more caudal locations the latency for this contralateral response increases again, whereas the latency for the other contralateral leg decreases still more until it reaches a value equal to that of the specific point in somatic I (Fig. 4).

This complex variation of latency with location distribution suggests the existence of two separate projection systems for somesthesis, one to the motor region, and one to the somatosensory area, as has already been shown by macroelectrode recordings (Albe-Fessard and Rougeul, 1958; Buser and Ascher, 1960). Whereas the latter originates in the ventral posterior group of the thalamus, it is highly probable that the projection system of the motor region finds its way through thalamic nuclei lying more medially (see Discussion).

Experiments under curare

No fundamental differences characterize the results obtained on unanesthetized preparations from results observed with chloralose. Some quantitative variations are nevertheless to be expected; they will be considered now.

Spontaneous activity

First, the number of units spontaneously active is markedly higher in the absence of chloralose. Furthermore, the patterns of this repetitive firing may vary with time, for a given unit; they probably depend upon the characteristics of the general cortical rhythm; they sometimes consist in sustained discharges of relatively high frequency and at other times show a tendency toward a decrease in firing rate or toward a grouping of the spikes into bursts separated by silent periods. As will be seen, such modifications, already described by others (Li and Jasper, 1953; Ricci, Doane, and Jasper, 1957), may bear upon the responsiveness of the cell to sensory inputs.

Sensory responses

In our experiments, any response to a peripheral stimulus from a spontaneously active cell consisted, not in a single series of successive spikes, but in a more-or-less complex modification of the spontaneous discharge pattern. Sometimes there was a decrease in firing rate,

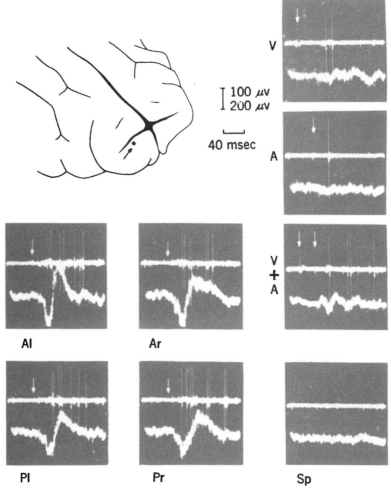

Fig. 5. Polysensory neuron of anterior sigmoid cortex. Unanesthetized preparation, immobilized under curare; right cortex. The same abbreviations are used as in Fig. 2. No spontaneous activity, as shown in the figure in the lower right-hand corner (Sp). Calibration, 100 μv (upper channel) and 200 μv (lower channel). Lower trace, pyramidal tract activity.

sometimes an increase, sometimes decreases and increases in alternation.

If we now consider the modes of responding of the cells to sensory stimulation and their topographical distribution, it appears that the main characteristics remain unchanged. Units of each of the three groups are found under curare as well as under chloralose: polysensory (group I), somatic polyvalent (Group IIa), and somatotopic (group IIb) neurons (Fig. 5).

Differences appear first when statistical data of their topographical distribution are considered: thus it appears that the polysensory units are no longer the only elements found in the anterior sigmoid gyrus. But this area now also contains a certain number of somatic polyvalent units (IIa) that do not clearly respond to either sound or light. On the contrary, the purely somatic distribution under curare (IIa versus IIb) hardly differs from that observed under chloralose, and IIa cells still lie in the perisigmoid cortex, whereas IIb cells are found in more caudal locations (Table 3).

Table 3. Topographical Distribution (in Percentages) of Group I, IIa and IIb Neurons in Experiments under Curare (100 Units)

Gyrus	Polysensory I	Somatic polyvalent IIa	Somatic somatotopic IIb
Anterior and oral posterior sigmoid	64	36	—
Posterior sigmoid (middle strip)	21	69	10
More caudal zones	—	30	70

Some modalities of responses may also be mentioned. The pattern of response of a given polysensory unit, spontaneously firing, is, in general, similar for all four limbs to the response obtained under chloralose (acceleration and/or inhibition); on the other hand, with curare, light and sound stimulation act similarly to somatic stimulation. Finally, interaction effects may be observed that are rather similar to those obtained under chloralose. One additional observation is that the responsiveness of neurons seems to depend upon the rate of spontaneous activity: during a period of sustained random firing—often corresponding to a desynchronized cortical global rhythm —sensory stimuli have little effect, whereas fully developed responses may be observed later on, at the time when spontaneous activity has decreased or shows a tendency toward a grouping of the spikes.

Discussion

Single-unit analyses have often been performed on the neocortex. The main purpose has been to clarify and make precise the pattern of neuronal organization within a given sensory channel and receptive area. Analyses have been carried out of the somesthetic systems (Amassian, 1953; Li, Cullen, and Jasper, 1956; Mountcastle, 1957; Mountcastle, Davies, and Berman, 1957); of the auditory system (Erulkar, Rose, and Davies, 1956; Katsuki, Watanabe, and Maruyama, 1959); and of the visual system (Jung, Creutzfeldt, and Baumgartner, 1956; Hubel, 1958). There have also been investigations concerned with detailed electrophysiological properties of cortical cells by Albe-Fessard and Buser, 1955; Phillips, 1956a, b; and others. Apparently but few studies have been undertaken with these same techniques and oriented toward the problems of convergence of different sensory inputs onto the neocortex. Amassian (1954) reports convergence for some optic, acoustic, and somatic stimuli for neuron pools in the anterior lateral gyrus of cats under chloralose. Ricci, Doane, and Jasper (1957) identify effects of visual stimuli on single units of the motor cortex of unanesthetized monkeys during conditioning experiments. Finally, Li's studies on activation or inhibition of the motor Betz cells by central (Li, 1956a, b) or tactile stimulation (Li, 1959) also deal with functional properties of these neurons. This scarcity of data contrasts with the growing number of results obtained from different subcortical levels in the same line. So far as the cat's brain is concerned, our results suggest that the motor cortex behaves as a converging neuropile. This is similar to what has previously been demonstrated in the reticular formation (literature cited in Rossi and Zanchetti, 1957), in the hippocampus (Green and Machne, 1955), in the amygdala (Machne and Segundo, 1956), in the lenticular nucleus (Segundo and Machne, 1956), in the caudate nucleus (Albe-Fessard, Oswaldo-Cruz, and Rocha-Miranda, 1958), and in the posterior lateral nucleus of the thalamus (Borenstein, Bruner, and Buser, 1959) and so forth.

It is far from certain from our results precisely which category of cortical neurons represents the "polysensory" elements in the motor cortex. Since we did not test the recorded units for responsiveness to antidromic stimulation of the medullary pyramids, as did Martin and Branch (1958) and Li (1959), the indications at hand are only indirect. It is nevertheless worth mentioning that although, during penetration, units are found at many cortical levels it is obvious that the great majority of neurons actually reacting to sensory stimulation lie deep in the cortex; in most cases, any evaluation of the real depth is somewhat illusory, owing to the complex folding of the frontal cortex. But sometimes, when the puncture was made perpendicular to an unfolded part, measurements pointed toward the fifth layer, which is thought to contain most of the large pyramidal cells. Another point is the rather close correspondence between single-unit events and the pattern of pyramidal discharges recorded at both levels, especially when facilitation or occlusion occurs. In those cases, it is reasonable to consider the observed neurons not necessarily as Betz cells but as elements having some relation to the pyramidal-tract activity. Nevertheless, it is worth mentioning that, in the cat, other efferent fibers lead from the motor cortex to subcortical nuclei and are presumably involved in "extrapyramidal" control of motor activity (for references, see Lindsley, 1952).

As has been pointed out earlier, in explorations of the more posterior regions of somatic area I, the number of neurons found to be responsive to sensory stimulation (an adequate somatic shock in that particular case), was much higher than in the motor cortex. That the responsive neurons are located in a particular deep layer, as was found in the motor cortex, does not seem to be the rule for the purely somatic area (a fact clearly shown by Mountcastle, 1957). This observation probably supports the assumption that the number of neurons that can be activated by the sensory projections is noticeably higher in the somatic than in the motor region.

One possible bearing of our results concerns the question of the specific characteristics of the "pyramidal motor cortex" in contrast to the somatic projection field. It seems clear that the cortical area lying anterior to somatic I possesses a system of polysensory and converging afferents much different in their organization from that of the specific somatic projections.

However, whether the criterion used here may be sufficient to define the "motor" cortex remains questionable. Neither the contour of origin of the pyramidal tract, as defined by anatomical (Gobbel and Liles, 1945; Chambers and Liu, 1957) or physiological methods (Woolsey and Chang, 1947; Lance and Manning, 1954), nor the area defined by its property of eliciting localized movements when stimulated (Garol, 1942), may be strictly confined to the perisigmoid cortex. The fact that at least one early component of the efferent-reflex pyramidal discharge has been shown to be correlated with the primary evoked potential of somatic area I to a contralateral tactile shock (Brookhart and Zanchetti, 1956; Buser and Ascher, 1960) also suggests that a pyramidal outflow may be directly determined from the somatic cortex.

Finally, it seems probable—even if it has not been thoroughly demonstrated—that the best anatomical correlate for the polysensory frontal area defined here could be deduced from architectonic studies:

its cortical extent appears in fact rather close to that of area 4 and possibly area 6.

The question now arises, what pathways are responsible for the projections to the motor cortex that we have considered here.

So far as the somatic modality is concerned, it should be recalled that a "nonsomatotopic" projection system toward the motor cortex has previously been shown to pass by nucleus centrum medianum of the thalamus. One of its cortical projection fields is fairly coincident with the perisigmoid cortex, especially its presigmoid division (Albe-Fessard and Rougeul, 1958).

From other investigations that deal mostly with visual "irradiation" projections toward the frontal cortex, apparently no definite proof has emerged about which pathways are involved, though it is highly probable that at least in part the primary projection area is by-passed (Wall, Remond, and Dobson, 1952; de Haas, Lombroso, and Merlis, 1953; Hunter and Ingvar, 1955; Buser and Ascher, 1960). More information should be obtained about the thalamic pathways that are involved in those frontal projections, for vision as well as for audition.

With these problems of pathways incompletely solved, the question is still partly open, whether the convergences observed occur at the cortical level or somewhere on the way to it. In fact, for the somatic system itself, convergences appear to occur at the thalamic level, or even below (Albe-Fessard and Rougeul, 1958). But it is far from certain that the other pathways—visual and auditory—simply follow the same secondary route, or a parallel one, and converge at the same subcortical level as do the somatic messages. This point should also be left open for further investigation.

In summary, this much can be said about the cat:

1. Anterior to the somatic area I there lies a "polysensory" area on which somatic as well as visual and auditory projections can be identified. This field involves area 4 and presumably area 6 and thus occupies the major part of the "pyramidal cortex," though some pyramidal efferents originate from different cortical zones (somatic II and possibly I).

2. Sensory projections from the visual, auditory, and somatic systems activate the same neuronal elements. These polysensory neurons lie deep in the cortex and may be concerned at least to some extent with the cells of origin of the pyramidal tract.

On the basis of this reasoning, it may be proposed that the motor cortex must be considered functionally as representing some sort of a common pool for sensorimotor integration. This property is almost certainly not shared by the specific primary areas and probably also not so regularly by the association areas.

Summary

By cortical explorations with extracellular microelectrodes, a study was undertaken, on cats, of the responsiveness of single neurons from the "sensorimotor" cortex to different kinds of sensory stimulations— somatic (four legs), visual, and auditory. Sixty animals were used, either deeply anesthetized with chloralose or unanesthetized and immobilized with curare. The region explored involved the following gyri: anterior and posterior sigmoid, coronal, anterior suprasylvian, and anterior lateral.

Two main sets of data were obtained under chloralose:

1. Several types of neurons can be distinguished in terms of their responsiveness to sensory stimuli: *polysensory* units, activated by any of the stimuli applied (group I); *somatic* units (group II) responding to somesthesis only, either indifferently to all four legs (group IIa) or to only a given part of the body, in accordance with the classical somatotopic organization of the somesthetic pathways (group IIb).

2. Elements thus characterized are not scattered throughout the whole sensorimotor cortex but are rather precisely grouped in space according to their responsiveness: polysensory elements lie around the sulcus cruciatus, with an extension fairly coincident with that of architectonic area 4 and 6 (conventional "motor cortex"); group IIb somatic units lie in the classical somatosensory zone (anterior lateral and anterior suprasylvian gyrus, and so on), as reported by other investigators; and group IIa lie between in a transition zone between "motor" and "somatic" areas.

Under curare, the results are fundamentally similar: differences appear in the relative number of units of group I and group IIa, with more elements in group IIa now identified in the motor cortex.

These results confirm our previous finding in macroelectrode investigations on cats, that the "motor" cortex is a polysensory area, having projections simultaneously from visual, auditory, and somesthetic pathways. That these different sensory systems may activate the same neuronal element is thus demonstrated by the present exploration on single units, which shows that facilitations or occlusions occur between responses from different origins.

Mention is made also of the level at which those interactions may occur, the possible identification of part of the recorded cells with Betz pyramidal neurons, and some functional implications of our results, such as differences between the "motor" and the "somatic" components of the sensorimotor cortex.

Acknowledgments

This work was supported in part by a grant from the Office of Scientific Research of the Air Research and Development Command, United States Air Force, through its European Office, under contract No. AF 61(052)103.

We are indebted to Mrs. Laplante, Miss Giacobini, and Miss Giraud for their technical assistance.

References

Albe-Fessard, D., and P. Buser.   Activités intracellulaires recueillies dans le cortex sigmoïde du Chat: participation des neurones pyramidaux au potentiel évoqué somesthésique. *J. Physiol.*, 1955, **47**, 67–69.

Albe-Fessard, D., E. Oswaldo-Cruz, and C. E. Rocha-Miranda.   Convergences vers le noyau caudé de signaux d'origine corticale et hétérosensorielle. Etude unitaire de leurs interactions. *J. Physiol.*, 1958, **50**, 105–108.

Albe-Fessard, D., and A. Rougeul.   Activités d'origine somesthésique, évoquées sur le cortex non-spécifique du Chat anesthésié au chloralose: rôle du centre médian du thalamus. *EEG clin. Neurophysiol.*, 1958, **10**, 131–152.

Amassian, V..E.   Evoked single cortical unit activity in the somatic sensory areas. *EEG clin. Neurophysiol.*, 1953, **5**, 415–438.

Amassian, V. E.   Studies on organization of a somesthetic association area, including a single unit analysis. *J. Neurophysiol.*, 1954, **17**, 39–58.

Ascher, P., and P. Buser.   Modalités de mise en jeu de la voie pyramidale chez le Chat anesthésié au chloralose. *J. Physiol.*, 1958, **50**, 129–132.

Borenstein, P., J. Bruner, and P. Buser.   Organisation neuronique et convergences hétérosensorielles dans le complexe latéral postérieur "associatif" du thalamus chez le Chat. *J. Physiol.*, 1959, **51**, 413–414.

Brookhart, J. M., and A. Zanchetti.   The relation between electrocortical waves and responsiveness of the cortico-spinal system. *EEG clin. Neurophysiol.*, 1956, **8**, 427–444.

Bruner, J.   Réponses visuelles et acoustiques au niveau de la face médiane antérieure du cortex chez le Chat sous chloralose. *J. Physiol.*, 1960, **52**, 36.

Buser, P., and P. Ascher.   Mise en jeu réflexe du système pyramidal chez le Chat. *Arch. ital. Biol.*, 1960, **98**, 123–164.

Buser, P., and P. Borenstein.   Réponses corticales "secondaires" à la stimulation sensorielle chez le Chat curarisé non-anesthésié. *EEG clin. Neurophysiol.*, 1956, Suppl. **6**, 89–108.

Buser, P., and P. Borenstein. Réponses somesthésiques, visuelles et auditives, recueillies au niveau du cortex "associatif" suprasylvien chez le Chat curarisé non-anesthésié. EEG clin. Neurophysiol., 1959, 11, 285–304.

Buser, P., P. Borenstein, and J. Bruner. Etude des systèmes "associatifs" visuels et auditifs chez le chat anesthésié au chloralose. EEG clin. Neurophysiol., 1959, 11, 305–324.

Campbell, A. W. Histological Studies on the Localization of Cerebral Function. Cambridge: University Press, 1905.

Chambers, W. W., and C. N. Liu. Cortico-spinal tract of the cat. J. comp. Neurol., 1957, 108, 23–55.

Derbyshire, A. J., B. Rempel, A. Forbes, and E. F. Lambert. The effect of anaesthetics on action potentials in the cerebral cortex of the cat. Amer. J. Physiol., 1936, 116, 577–596.

Erulkar, S. D., J. E. Rose, and P. W. Davies. Single unit activity in the auditory cortex of the cat. Bull. Johns Hopkins Hosp., 1956, 99, 55–86.

Feng, T. P., Y.-M. Liu, and E. Shen. Pathways mediating irradiation of auditory and visual impulses to the sensorimotor cortex. Congr. int. Physiol., Bruxelles, 1956, 997.

Forbes, A., and R. Morison. Cortical response to sensory stimulation under deep barbiturate narcosis. J. Neurophysiol., 1939, 2, 112–128.

Garol, H. W. The "motor" cortex of the cat. J. Neuropath. exp. Neurol., 1942, 1, 138–145.

Gastaut, H., and J. Hunter. An experimental study of the mechanism of photic activation in idiopathic epilepsy. EEG clin. Neurophysiol., 1950, 2, 263.

Gobbel, W. G., and G. W. Liles. Efferent fibers of the parietal lobe of the cat (Felis domesticus). J. Neurophysiol., 1945, 8, 257–266.

Green, J. D., and X. Machne. Unit activity of rabbit hippocampus. Amer. J. Physiol., 1955, 181, 219–224.

de Haas, A. M., C. Lombroso, and J. K. Merlis. Participation of the cortex in experimental reflex myoclonus. EEG clin. Neurophysiol., 1953, 5, 177–186.

Hubel, D. H. Cortical unit responses to visual stimuli in non-anesthetized cats. Amer. J. Ophthalmol., 1958, 46, 110–122.

Hunter, J., and D. H. Ingvar. Pathways mediating metrazol induced irradiation of visual impulses. EEG clin. Neurophysiol., 1955, 7, 39–60.

Jung, R., O. Creutzfeldt, and G. Baumgartner. Microphysiologie corticaler Neurone; Koordination und Hemmungsvorgänge im optischen und motorischen Cortex. Colloq. int. Microphysiol., Centre nat. Rech. sci., 1957, 67, 411–434.

Katsuki, Y., T. Watanabe, and N. Maruyama. Activation of auditory neurons in upper levels of brain of cat. J. Neurophysiol., 1959, 22, 343–359.

Lance, J. W., and R. L. Manning. Origin of the pyramidal tract in the cat. J. Physiol., 1954, 124, 385–399.

Langworthy, O. R. The area frontalis of the cerebral cortex of the cat, its minute structure and physiological evidence of its control of the postural reflex. Bull. Johns Hopkins Hosp., 1928, 42, 20–65.

Li, C.-L. The facilitatory effect of stimulation of an unspecific thalamic nucleus on cortical sensory neuronal responses. J. Physiol., 1956a, 131, 115–124.

Li, C.-L. The inhibitory effect of stimulation of a thalamic nucleus on neuronal activity in the motor cortex. J. Physiol., 1956b, 133, 40–53.

Li, C.-L. Some properties of pyramidal neurons in motor cortex with particular reference to sensory stimulation. J. Neurophysiol., 1959, 22, 385–394.

Li, C.-L., C. Cullen, and H. H. Jasper. Laminar microelectrode analysis of cortical unspecific recruiting responses and spontaneous rhythms. *J. Neurophysiol.*, 1956, **19,** 131–143.

Li, C.-L., and H. H. Jasper. Microelectrode studies of the electrical activity of the cerebral cortex of the cat. *J. Physiol.*, 1953, **121,** 117–140.

Lindsley, D. Brain stem influences on spinal motor activity. *Res. Publ. Ass. nerv. ment. Dis.*, 1952, **30,** 174–195.

Machne, X., and J. P. Segundo. Unitary responses to afferent volleys in amygdaloid complex. *J. Neurophysiol.*, 1956, **19,** 232–240.

Martin, A. R., and C. L. Branch. Spontaneous activity of Betz cells in cats with midbrain lesions. *J. Neurophysiol.*, 1958, **21,** 368–379.

Mountcastle, V. B. Modality and topographic properties of single neurons of cat's somatic sensory cortex. *J. Neurophysiol.*, 1957, **20,** 408–434.

Mountcastle, V. B., P. W. Davies, and A. L. Berman. Response properties of neurons of cat's somatic sensory cortex to peripheral stimuli. *J. Neurophysiol.*, 1957, **20,** 374–407.

Patton, H. D., and V. E. Amassian. Single and multiunit analysis of the cortical stage of pyramidal tract activation. *Congr. int. Physiol., Montreal,* 1953, 666–667.

Phillips, C. G. Intracellular records from Betz cells in the cat. *Quart. J. exp. Physiol.*, 1956*a*, **41,** 58–69.

Phillips, C. G. Cortical motor threshold and the thresholds and distribution of excited Betz cells in the cat. *Quart. J. exp. Physiol.*, 1956*b*, **41,** 70–84.

Ricci, G., B. Doane, and H. H. Jasper. Microelectrode studies of conditioning: technique and preliminary results. *Congr. int. Sci. neurol., Bruxelles,* 1957, 401–415.

Rossi, G. F., and A. Zanchetti. The brain stem reticular formation. Anatomy and physiology. *Arch. ital. Biol.*, 1957, **95,** 199–435.

Rougeul, A. Observations électrographiques au cours du conditionnement instrumental alimentaire chez le chat. *J. Physiol.*, 1958, **50,** 494–496.

Segundo, J. P., and X. Machne. Unitary responses to afferent volleys in lenticular nucleus and claustrum. *J. Neurophysiol.*, 1956, **19,** 325–339.

Thompson, R. F., and R. M. Sindberg. Auditory response fields in association and motor cortex of cat. *J. Neurophysiol.*, 1960, **23,** 87–105.

Wall, P. D., A. G. Remond, and R. L. Dobson. L'action des afférences visuelles sur l'excitabilité du système moteur. *Rev. Neurol.*, 1952, **87,** 164–166.

Winkler, C., and A. Potter. *An Anatomical Guide to Experimental Research on the Cat's Brain.* Amsterdam: Verluys, 1914.

Woolsey, C. N., and H. T. Chang. Activation of the cerebral cortex by antidromic volleys in the pyramidal tract. *Res. Publ. Ass. nerv. ment. Dis.*, 1947, **27,** 146–161.

32

RICHARD JUNG

Abteilung für Klinische Neurophysiologie, Universität Freiburg

Neuronal Integration in the Visual Cortex and Its Significance for Visual Information

In the following pages I shall treat two topics, one neurophysiological and the other psychophysiological: (1) the integration of information from various afferents by neurons in the visual cortex, resulting in convergence of specific retinal and vestibular receptors and of nonspecific reticulothalamic afferents; (2) the correlation of these neuronal mechanisms in the cat with subjective visual perception in man.

I shall survey first some of the work on extracellular recording of neuronal discharges in the primary visual area of the cat, done in our laboratory during the past eight years. Second I shall discuss more in detail the recent experiments on neuronal mechanisms of vision by Baumgartner, Grüsser, Grüsser-Cornehls, and their collaborators, some of which are not yet published.

It is the purpose of this paper to summarize the results of our experiments on the coordination of retinal and nonretinal afferents and on neuronal integration in the visual cortex. I hope to show in the discussion that the neuronal organization of visual mechanisms and some apparently contradictory neuronal responses to stimulation by diffuse light and patterned light can be explained by two mechanisms of inhibition in antagonistic and synergistic neurons, and that nearly all neuronal findings can be understood when a definite information value is assigned to the specific antagonistic neuronal response types: brightness information to the B neurons and darkness information to the reciprocally functioning D neurons of the visual cortex. Thus I wish to show the remarkable agreement between the results of neuronal recordings and psychophysiology. Subjective sensory information may be used as a guide for our neurophysiological experiments. The microphysiology of cortical neurons only corroborates with modern techniques the detailed visual observations of the classical sensory physiologists of the nineteenth century.

Methods

Glass capillaries of the Ling-Gerard type with tips 0.5 to 3 microns in diameter, filled by the method of Tasaki, et al. (1954), were used to record extracellularly single-neuron activity from the cortex of cats in Bremer's *encéphale isolé* preparation with artificial respiration.

For physiological stimulation of the visual cortex two methods were employed: (1) a diffuse white light, shed on a ground glass disc placed in front of the eyes with dilated pupils; continuous illumination at 30 to 500 lux, intermittent stimulation with light and dark phases of equal duration at 30 to 500 lux, and light flashes, 0.3 millisecond in duration at 3000 to 12,000 lux; and (2) pattern stimulation of the eye for simultaneous white-black contrast, provided by a grid of bright and dark stripes, illuminated from behind, that was exposed at different positions of the contrasting fields to the receptive area of the recorded cortical neuron (Baumgartner and Hakas, 1959, in preparation).

Electrical stimulation of the optic nerve, the reticulothalamic system, or the cortex was provided by a thyratron stimulator or by square-wave pulses 0.5 msec in duration.

Since, during vestibular stimulation, it was necessary to avoid activation of other receptors and movements of the animal, labyrinthine polarization was provided at the round window of the ear by weak galvanic currents of 0.1 to 0.3 milliampere, as described by Grüsser, Grüsser-Cornehls, and Saur (1959). These currents are known to excite vestibular receptors only, without acoustic manifestations. In all experiments with diffuse light the *encéphale isolé* cat showed spontaneous eye movements. For pattern stimulation and labyrinthine stimulation, eye movements were abolished by application of curare or by cutting the external eye muscles.

A six-beam cathode-ray oscillograph, built by J. F. Tönnies, was used to record simultaneously several neuronal spikes, electrocorticograms, and a photocell record of light stimulation.

In the following discussion we use the terms "neuronal responses" or "neurons" in interpreting the spike records from units in the visual cortex. No distinction is made here between axons and cell bodies, both of which are parts of the neuron.*

Results

Types of responses of cortical neurons

In the course of our experiments it became necessary to classify the various neuronal responses to visual and other stimuli. Since the response to light stimulation is obviously the most important for the visual system, we have adopted such a classification since our first papers (von Baumgarten and Jung, 1952; Jung, von Baumgarten, and Baumgartner, 1952). The letters A, B, C, D, and E are used for different response patterns of cortical neurons to "light-on" and "light-off," thus expanding the classical distinctions of retinal "on," "off," and "on-off" elements (Fig. 1). Further work has confirmed the reality of these neuronal groups, and it has been possible to relate them to certain visual information values and to demonstrate that these groups represent fixed patterns and inherent properties of these neuronal systems with their afferent connections from retinal receptors. The variations of the responses to light-on and off with patterned stimulation of the retina (Fig. 6) becomes understandable only when one assigns a distinct information value to the main neuronal types. It can be shown that these neuronal groups have a specific significance for visual messages coming from the retina. No significant correlations exist between these neuronal response types and those to thalamic stimulation (Akimoto and Creutzfeldt, 1957–1958; Creutzfeldt and Akimoto, 1957–1958) or vestibular stimulation (Grüsser and Grüsser-Cornehls, 1960; Grüsser, Grüsser-Cornehls, and Saur, 1959).

In the following pages the responses of cortical neurons are classified after binocular light stimulation, as illustrated in Fig. 1. Most cortical neurons show the same response to monocular stimulation of either the ipsilateral or contralateral eye but no response to stimulation to the other eye. Thus the combinations of the monocular types—B-A, C-A, D-A or A-B, A-C, and so forth—are by far the most frequent in Grüsser and Grüsser-Cornehls's results (1960; see also Grüsser, Grüsser-Cornehls, and Saur, 1959). Most neurons in the

* This distinction is made in some special studies on the lateral geniculate body in order to exclude presynaptic optic nerve fibers (Grüsser and Saur, 1960), and in some studies of receptive fields in the visual cortex, in order to exclude optic radiation fibers.

visual cortex respond only to *monocular* retinal afferents. Thus stimulation of a single eye with light would show many nonresponsive

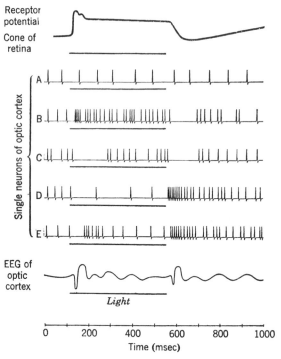

Fig. 1. Five types of neuronal responses of the visual cortex to light and dark stimulation and their relation to receptor excitation and to the EEG (from Jung, Creutzfeldt, and Grüsser, 1957). Topmost graph shows receptor potential recorded with microelectrode, intracellularly from outer plexiform layer of the retina. Bottom graph shows cortical potentials with on and off effect from gross electrode recording on cortical surface (macrorhythms). Graphs A–E, schematic representation of discharges of different neuronal types: A neuron, no reaction to light or dark; B neuron, activated by light, inhibited by dark with delayed after-activation (similar to on element of retina). C neuron, inhibitory break for both light and dark; D neuron, inhibited by light and activated by dark (reciprocal of B neuron, similar to off element of retina). E neuron, pre-excitatory inhibition precedes delayed activation by light, early activation by dark (similar to on-off elements of retina).

neurons, which would respond, however, to illumination of the other eye or to binocular light or dark stimuli.

A *neurons* are not responsive to diffuse light or dark stimuli but may respond to thalamic or vestibular stimulation. They seem to be a heterogeneous group. Although the number of A neurons was estimated in 1955 to be about half the neurons of the visual cortex (Baumgartner and Jung, 1955; Jung and Baumgartner, 1955), it is actually smaller. The more fully our methods of visual stimulation have developed (flicker, binocular and patterned stimuli), the fewer unresponsive A neurons we have found. Jung, Creutzfeldt, and Baumgartner (1957) found that some A neurons can be activated rhythmically by flicker, although they may not respond to single light flashes. Some inactive B, D, and E neurons, not to be confounded with A neurons, can be activated by thalamoreticular stimulation to respond to light stimuli (Akimoto and Creutzfeldt, 1957–58). Others may respond to moving stimuli, as described by Hubel (1958), and still others to patterned light at the border of light and dark contrast (Baumgartner and Hakas, 1962) or to optic-nerve stimulation. Even supramaximal optic-nerve stimulation cannot discharge all cortical neurons, however, and 28 per cent remain unresponsive (type 1a of Grützner, Grüsser, and Baumgartner, 1958). After optic-nerve stimulation combined with illumination to the other eye, 16 per cent of 211 neurons remained unresponsive. It is believed that A neurons are a stabilizing system to maintain a medium background of excitation in the cortex (Jung, 1958a) and may be a reserve system for special visual functions that cannot yet be defined exactly.

B neurons show on responses and off inhibition at diffuse illumination. They constitute the best-defined neuronal group in the visual cortex (Jung, 1953a; Jung, von Baumgarten, and Baumgartner, 1952; Jung, Creutzfeldt, and Grüsser, 1957) and represent about a fourth of the cortical neurons in area 17. They correspond to retinal on elements and seem to be correlated with brightness information.

C neurons show inhibition to both light-on and off. They are rare (3 to 5 per cent), and their information value cannot yet be determined. After stimulation of one eye an inhibitory C response may be obtained from some neurons that show on responses of the B type from the other eye (Grüsser and Grüsser-Cornehls, 1960).

D neurons show off responses and on inhibition at diffuse illumination (Jung, 1953a; Jung, Creutzfeldt, and Grüsser, 1957), corresponding to off elements in the retina. They show strong primary inhibition to brief light flashes, followed by late activation (Baumgartner, 1955). D neurons mediate darkness information.

E neurons respond with strong off responses and constant on inhibition followed by late activation to light-on, corresponding to on-off elements in the retina. They show a shorter pre-excitatory inhibition after brief light flashes than do D neurons (Baumgartner, 1955).

Responses of the E type were found frequently in about 18 per cent of cortical neurons during the first years of our experiments in which faint patterns during illumination were not rigorously excluded. The proportion of E neurons has decreased and of D neurons increased as the diffuseness of light stimuli has been brought under increasing control. E neurons may show transitions to D neurons (Jung, 1953a). Responses of the E type can also be obtained from D neurons near the border contrast of patterned stimuli. An abnormal on-off response similar to E neurons may occur in an asphyxiated cortex when respiration is insufficient. These "pseudo E neurons" can be restored to their original responses (B or D) if anoxic damage is avoided. From results on contrast stimulation (Baumgartner, 1961; Baumgartner and Hakas, 1962) it seems probable, though not yet certain, that the information value of E neurons corresponds to the darkness information of D neurons, and that E neurons may be only a subgroup of the D system.

After *nonspecific thalamic stimulation,* five different response types (I–V) were found in the neurons of area 17 (Akimoto and Creutzfeldt, 1957–58; Creutzfeldt and Akimoto, 1957–58). Type I shows no response, whereas the other types respond by a different pattern of activation or inhibition after relatively long latencies. Except in types I and III, all neurons show inhibitory silent periods before or after activation. The latencies are more variable and longer after thalamic stimulation than after light stimulation. For responses after *vestibular stimulation,* see Fig. 4.

After *optic-nerve stimulation,* four different response types have been described (Grüsser and Grützner, 1958b; Grützner, Grüsser, and Baumgartner, 1958).

These different responses to nonretinal stimuli show no statistical relation to the neuronal responses that follow light stimulation. Neither thalamoreticular nor optic-nerve stimulation alters the A, B, C, D, and E responses to light stimuli. These neuronal responses seem to be fixed patterns of the retinocortical chain, which can only be modified by contrasting light stimuli.

Neuronal responses to patterned light

Our earlier experiments from 1951 on were all done with diffuse light stimulation until in 1958 Baumgartner (1961; Baumgartner and Hakas, 1959, 1962) began to use patterned light with white-black contrast. Although the neuronal responses obtained in contrasting fields seem to differ from those in diffuse light, the results may be explained by the same laws of reciprocal inhibition of antagonistic neurons in the same field, by lateral inhibition of synergistic neurons in surrounding fields, and by the information value of antagonistic neuronal systems. The mechanism of lateral inhibition was elucidated by Hartline's work on *Limulus* (1949; Hartline and Ratliff, 1957) and Kuffler's experiments on spot stimulation of the cat's retina (1953; Barlow, FitzHugh, and Kuffler, 1957). Similar neuronal organizations to Kuffler's retinal receptive fields (on and off centers and surrounding zones of reciprocal responses) were also found in cortical neurons by Baumgartner and Hakas (1959, 1960, in preparation) and Hubel and Wiesel (1959).

In pattern vision, information is usually conveyed by contours and not single spots of light. For this reason Baumgartner has used contrast patterns instead of spots of light as stimuli to explore the neuronal responses to patterned light and to determine the receptive fields of cortical neurons. He has done the following experiments in my laboratory. A grid of vertical or horizontal stripes was illuminated from behind and exposed to the receptive fields. The grid was shifted in steps that differed about 41 degrees from left to right and from right to left. Baumgartner and Hakas (1959, 1960, in preparation) compared the neuronal responses to light-on and light-off in these various positions. They recorded from three different levels of the visual system and found essentially the same contrast responses in axons of retinal ganglion cells in the optic nerve, in geniculate neurons, and in cells of the visual cortex of the cat. An example from a geniculate neuron showing reversed responses at the border contrast was presented at this symposium, in the discussion following Ratliff's presentation.

In the bright stripe of the grid, the various neuronal types respond qualitatively to light-on and off in the same way as they do in diffuse light. In the dark field, however, the neurons show mainly a reversed response to light-on and off as if they were in the dark at light-on and faintly illuminated at light-off. The intensity of the responses varies with different positions of the grid and the relation to the zone of bright-dark contrast. The spike frequency seems dependent upon the projection of this border zone to the receptive-field center of the neuron.

Figure 6 (which will be shown on p. 645) illustrates these neuronal reactions in various positions of the contrasting grid of bright and dark stripes: strict reciprocity is maintained in the responses of the antagonistic B and D neurons to patterned contrast stimuli. In light-activated B neurons the on response is greatly enhanced at the bright border zone to the dark stripe. In the middle zone of a bright stripe of about 5 degrees aperture, the neuronal responses have about the same frequency as in diffuse light stimulation (see Fig. 6, left and center). The B neurons discharge increasingly as the dark stripe is approached from the middle of the bright stripe (with a relative weak on response) and show a maximum on response, when the receptive-field center projects to the bright zone immediately at the border of the dark stripe (Fig. 6, lateral peaks above white stripe). When the projection of the field center of a B neuron crosses the dark contour, the on discharge is suddenly diminished or inhibited, and eventually a reversal of the responses to light-on and off occurs: a B neuron projecting upon the dark stripe is largely inhibited by light-on and activated by light-off.

These alterations and paradoxical on inhibitions and off discharges of B neurons correlate exactly with our subjective experience in simultaneous contrast: at light-on, the dark zone in the contrasting field appears blacker and the white zone whiter than the indifferent *Eigengrau*, the subjective gray visual background without illumination. The B neurons also conserve their type of discharge with a high-frequency peak and brief silent period when they are activated by light-off in the black zone, and they do not show the more gradual decline of the off responses of D neurons.

D neurons, which respond by inhibition to light-on and by activation to light-off, reverse their response when they cross the border from the bright to the dark stripe. Projecting to the dark stripe, they respond to light-on by activation with a continuously diminishing discharge. A short pre-excitatory inhibition may precede this unusual on activation. Since the off activation is not always completely suppressed in the dark stripe, on-off responses of D neurons are frequently seen in the border zone.

E neurons, which respond to both light-on and off, show reactions to contrast patterns similar to those of D neurons when their field projection approaches the contrasting border in the bright stripe. They reverse to strong on activation when the border of the dark stripe is crossed. In the median region of the bright or dark stripe, an on-off response reappears, but off activation is less pronounced in the dark than in the bright stripe.

Receptive-field centers. The diameter of the receptive-field center can be measured by comparing neuronal responses in the different grid positions of the white-dark contrast stimulus. Baumgartner and Hakas (1960) have compared these central cores of the receptive fields in retinal, geniculate, and cortical neurons. The receptive-field center, determined by the responses of the neurons at the border of the bright and dark stripes, was computed from the distance of the contrast borders during the activation period of the neuron. From the distance of the eye to the contrast object, together with the optical constants of the eye, the diameter of the receptive-field center on the retina was measured in millimeters. This receptive-field center is only the central part of the whole receptive field that Hartline (1940*a*) and Kuffler (1953) and collaborators (Barlow, FitzHugh, and Kuffler, 1957) determined in retinal ganglion cells. The surrounding area of inhibition cannot be measured precisely by Baumgartner's method. Baumgartner's receptive-field centers correspond approximately to Kuffler's on and off centers.

The horizontal diameter of the receptive-field centers of cortical neurons was significantly smaller than the field centers of retinal and geniculate neurons. Those cortical neurons that could be distinguished from radiation fibers had an average diameter about half that of the receptive-field centers of retinal and geniculate neurons (Baumgartner and Hakas, 1960).

The considerable overlap that Hartline has found in his first studies on the receptive field of retinal ganglion cells (1940*a*, *b*) should become smaller in cortical neurons when their receptive-field center is measured without the surrounding zone of lateral inhibition.

Convergence of retinal, reticulothalamic, and vestibular afferent impulses at single neurons of the visual cortex

Specific and nonspecific coordination. Most neurons of the visual cortex were found to receive convergent impulses from specific retinogeniculate, nonspecific reticulothalamic (Akimoto and Creutzfeldt, 1957–58; Creutzfeldt and Akimoto, 1957–58) and vestibular afferents (Grüsser, Grüsser-Cornehls, and Saur, 1959; Grüsser and Grüsser-Cornehls, 1960). Although some indications of convergence of specific and nonspecific afferents in the visual cortex were obtained with macroelectrodes (Bremer and Stoupel, 1958, 1959; Dumont and Dell, 1958; and Jasper and Ajmone-Marsan, 1952), it was not clear whether these impulses converged at the same cortical neurons. Li, Cullen, and Jasper (1956) were unable to activate the same neuron in the somatosensory cortex by stimulation of both specific and nonspecific thalamic nuclei, although Li (1956) described some facilitation by the latter. However, Akimoto and Creutzfeldt (1957–58) in our laboratory demonstrated the convergence of specific retinal and nonspecific thalamic afferents on single neurons of the visual cortex. About two-thirds of the neurons of area 17 examined showed this convergence and dependence of retinal and thalamic impulses. No consistent correlations were found between the various types of neuronal response after retinal stimulation (A, B, C, D, and E) and after single nonspecific thalamic stimuli (I, II, III, IV, and V). Some A neurons ordinarily unresponsive to light showed different responses to thalamic stimulation during illumination and darkness (Jung, Creutzfeldt, and Baumgartner, 1957). This observation makes it probable that A neurons receive subliminal impulses from the retina at cortical or subcortical levels that cannot be discovered in the usual spike records after stimulation by light alone. It has also been possible to drive weakly responding B, D, and E neurons to respond strongly to the same light stimulus during and after repeated thalamic stimulation. However, thalamic stimulation is not able to change the response type to diffuse retinal stimulation: type A remained A, B, B, and so forth, even when these neurons were stimulated by thalamoreticular shocks (Creutzfeldt and Akimoto, 1957–58).

The functional significance of this convergence is particularly evident with flickering light: the critical flicker-fusion frequency (CFF) of cortical neurons is, on the whole, raised by nonspecific thalamic

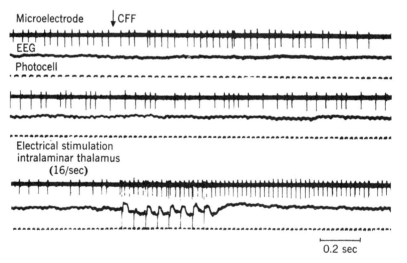

Fig. 2. Increase in flicker frequency and CFF of a B neuron in the visual cortex by thalamic stimulation (from Jung, Creutzfeldt, and Grüsser, 1957). Flicker at rising frequency gives regular neuronal discharges up to 18 per second. This is followed by breaks in firing as a sign of critical fusion frequency (CFF at 18 per second). At frequencies of 19 to 41 per second, the neuron can no longer respond to each flash and shows interruptions in firing. After a brief series of thyratron pulses (16 per second) in the intralaminar thalamus, the neuron follows the high flicker frequency of 41 per second and responds to each flash with a discharge. After 1 or 2 sec, interruptions in the response appear once again.

stimuli (Fig. 2; Creutzfeldt and Grüsser, 1959; Jung, Creutzfeldt, and Grüsser, 1957). Only in rare instances does inhibition occur and is CFF diminished. This lowering of the CFF is seen better after

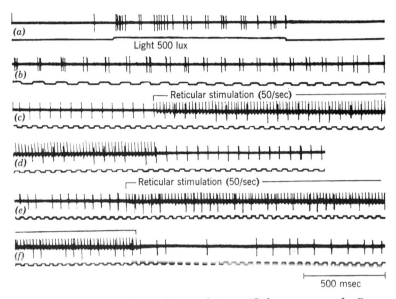

Fig. 3. Inhibitory effect of reticular stimulation on flicker responses of a B neuron in the visual cortex. (a) The B neuron shows little spontaneous discharge but typical activation during illumination with high-frequency initial discharge and periodic maintained discharge. (b) Responses to flickering light of 6 to 10 per second with gradual diminution of initial discharge until a stimulus-response rate of 1:1 is reached at the end. (c, d) Flicker at 16 per second gives a regular rate of 1:1 discharges. However, during reticular stimulation at 50 per second, the flicker responses are inhibited and become irregular with pauses and doublets. The inhibitory pauses disappear at cessation of stimulation, but a few doublets continue. (e, f) At flicker frequencies of 20 per second the limit of the critical-fusion frequency (neuronal CFF) is attained, and the neuronal response fails once. After reticular stimulation, multiple pauses appear again, and inhibitory pauses continue after cessation of stimulus.

The effects of stimulation by the same implanted electrode in the mesencephalic reticular formation in the same freely moving cat were observed behaviorally and filmed by Grüsser before these neuronal recordings were made: 50 stimuli per second of identical strength resulted in the arrest of spontaneous activity and searching movements of the head and eyes to the upper visual field and both sides.

reticular stimulation (Fig. 3). The facilitation and inhibition of cortical neurons and the alteration of neuronal CFF may correspond to the influence of attention and fatigue on visual perception and on the flicker-fusion frequency.

The long and variable latencies following stimulation of the non-specific thalamus are in striking contrast to the short and practically constant latencies of neurons following optic-nerve stimulation or high-intensity illumination. Creutzfeldt and Akimoto have offered the following hypothesis to explain this difference: specific optic-nerve afferents discharge cortical neurons directly, probably via axosomatic synapses over a disynaptic pathway from the geniculate body; but nonspecific afferents act on these neurons only as modulators, possibly by axodendritic synapses and over multisynaptic pathways, thus facilitating or inhibiting cortical neurons that receive convergence from various sources.

Vestibulovisual coordination. Experiments have shown (Grüsser and Grüsser-Cornehls, 1960; Grüsser, Grüsser-Cornehls, and Saur, 1959) that nearly all the neurons of the visual cortex that the authors recorded were activated by labyrinthine polarization. They found further that this activation altered the responses of cortical neurons to light stimulation. Their results can be summarized as follows. Four types of response to labyrinthine polarization were observed in cortical neurons (Fig. 4): α, no activation by labyrinthine polarization; β, activation by onset, not by cessation, of labyrinthine polarization; γ, no response to onset, but activation by cessation; δ, activation by onset and cessation. Type δ was most frequent, type α was rarely seen. When the same neuron was stimulated by both positive and negative labyrinthine polarization, nearly all combinations of responses ($\alpha - \alpha +$, $\delta - \alpha +$, and so forth) could be elicited; δ responses to both stimuli ($\delta - \delta +$) provided the most frequent combinations.

During labyrinthine polarization, the activated neurons showed slow adaptation within 3 to 10 sec. Similarly, following cessation of polarization, the frequency of the neurons decreased within 3 to 10 sec to the level of spontaneous activity when they were activated by termination of the stimulation. Definite inhibition of neuronal discharge could not be observed during and after labyrinthine stimulation. The latencies of neuronal activation following labyrinthine polarization were variable and measured 25 to 200 msec. The bursts of neuronal discharges following brief light flashes were increased in all light-responsive neurons (B, D, and E) by labyrinthine polarization; the duration of the silent periods sometimes decreased, sometimes increased. The critical flicker frequency was raised significantly in most of the cortical neurons during neuronal activation by polarization of the labyrinth (Grüsser and Grüsser-Cornehls, 1960). The activation of cortical neurons during or after labyrinthine polarization decreased slowly within several seconds. Constant rhythmic interruptions were not apparent, but some fluctuations of discharge rate resulted in bursts at higher frequencies. This was particularly evident when the labyrinth was polarized while neuronal discharges were induced by flickering light. In this case, some pauses in discharge, occurring at intervals of 200 to 800 msec, were recorded. However no clear relation between these pauses and nystagmic eye movements induced by labyrinthine polarization was found in those experiments in which the eye movements were not abolished by curare or other means.

The relation of neuronal activity in the visual cortex to vestibular nystagmus and to optokinetic nystagmus elicited by moving patterns needs further investigation.

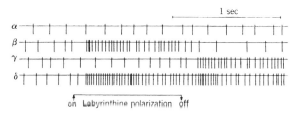

Fig. 4. Four types of responses of neurons in the visual cortex during and after labyrinthine polarization (schematic drawing, from the results of Grüsser and co-workers). α, no response (rare); β, activated by onset, no response from cessation of polarization; γ, no response from onset, activated by cessation of polarization; δ, activated by onset and by cessation of labyrinthine polarization (most frequent).

Binocular coordination. The visual cortex is usually considered as an area in which binocular integration occurs between the two halves of the visual fields. However single-neuron recordings of Grüsser-Cornehls and Grüsser (unpublished), which compare monocular and binocular diffuse light stimuli, have not been able to demonstrate binocular convergence for most neurons of area 17. These investigators collected data on neuronal responses following monocular and binocular stimulation in various combinations, and the results of their experiments with a special binocular stimulator have now been worked out and will be published in detail.

In the geniculate relay, optic-nerve fibers of the two eyes end separately in different layers, and accordingly Grüsser and Saur (1960) have found that geniculate neurons are activated by monocular stimuli only. A minority of the neurons showed less evoked activity to binocular stimulation than to monocular stimulation. Genuine binocular convergence in geniculate neurons, as described by Erulkar and Fillenz (1958), was not confirmed when stray light from the other side was rigorously avoided.

In the cat's visual cortex most neurons respond only to stimulation from one eye. Some modulation of monocular activation by stimulation of the other eye or binocular convergence may occur in about one-third of the neurons of area 17. Genuine binocular convergence at neurons that respond to monocular stimulation of either eye separately is rare: only about one-tenth of the cortical neurons in area 17 show clear binocular convergence. Binocular interaction manifests itself mainly by inhibition from one eye depressing the excitation from the other dominant eye, when identical stimuli are presented to both eyes. The type of response of these neurons to light stimulation usually differs from one eye to the other. (An example is shown in Fig. 2 of Grüsser, Grüsser-Cornehls, and Saur, 1959; this neuron was activated by ipsilateral light, like a B neuron, and inhibited by contralateral light-on and off, like a C neuron.) Most neurons classified as binocular B neurons also show on responses of the B type to stimulation from one eye but are, like A neurons, unresponsive to stimulation from the other eye. Summation of similar responses from the two eyes at single neurons was exceedingly rare, although recently Hubel and Wiesel (1959) described such binocular summation for corresponding parts of the visual field to stimulation with a spot of light. Grüsser and Grüsser-Cornehls obtained their results by diffuse light stimulation with pattern vision excluded. Contrasting stimuli may give some different results.

The predominance of monocularly influenced neurons in the visual cortex, found by Grüsser-Cornehls and Grüsser, indicates that binocular integration cannot be a simple convergence of both monocular afferents on the same neurons of area 17. If such neuronal convergence occurs at all, it needs higher visual centers above the primary receiving area, as the coordination for stereoscopic vision probably does also. The relation of neuronal discharges to binocular flicker is discussed below.

Thus binocular coordination is possible only on the basis of independent monocular projection in geniculate and cortical neurons. This monocular neuronal activation in the final paths of the cortex agrees well with psychophysiological experience, which has shown that each eye independently develops a complete visual image that is only secondarily modified by the mechanisms of binocular rivalry and stereoscopic vision.

Correlations of cortical neuronal activity in cats with subjective visual sensation in men

The psychophysiological correlations with our findings in the neuronal system of the visual cortex have been described in other papers (Jung, 1959). In summary, the following twelve parallels have been found between neuronal discharges in the primary visual cortex of the cat (area 17) and the psychophysiological results of sensory experiments with human subjects.

Diffuse illumination of light-adapted eyes

1. *Brightness of light sensation and Weber-Fechner law.* In the human eye, equal increments of subjective brightness correspond to a logarithmic function of the intensity of the stimulating light, according to the Weber-Fechner law. Correspondingly, the discharges of the B neurons of the visual cortex that are activated by light increase approximately with the logarithm of the stimulus intensity, if light adaptation is held constant. Simultaneous with the activation of B neurons, there is inhibition of the D neurons, which are activated by darkness. The brighter the light stimulus, the more rapidly the B neurons fire, and the stronger is the inhibition of the D neurons (Jung and Baumgartner, 1955).

2. *Flicker fusion.* The fusion of flickering light (subjective CFF) occurs at frequencies of about 50 per second (at a light intensity of 500 lux) in both man and cat (Kappauf, 1936). The critical flicker frequency of single neurons (neuronal CFF) is defined as the flicker rate at which individual neurons fail to respond to each flash. This neuronal CFF, up to which single neurons can follow flickering light, varies under similar conditions in different cortical neurons between 5 and 50 per second. The highest neuronal CFF of single cortical neurons, 50 per second (at 500 lux), corresponds to the maximal flicker frequency of the EEG and the subjective CFF in humans (Grüsser and Creutzfeldt, 1957). The CFF of retinal receptors (cones) is much higher (Grüsser, 1957, 1960).

3. *Porter's law.* Subjective CFF occurs at higher flicker frequencies as the light intensity is increased. Correspondingly, the neuronal CFF of cortical neurons is also higher at higher light intensities of the flicker.

4. *Brightness enhancement (Brücke-Bartley effect).* In subjective experiences described in 1864, Brücke found that maximal brightness at different flicker frequencies always occurs below the flicker fusion

(subjective CFF). Correspondingly, the maximal impulse frequency of all cortical neurons responding to flickering light (1:1 light-dark ratio) is below the flicker frequency of their neuronal CFF. In an average cell population of cortical B neurons the mean maximal impulse frequency occurs around flicker rates of 10 per second, which corresponds to Bartley's brightness enhancement (Bartley, 1959), although in individual neurons it may vary between 3 and 25 per second (Grüsser and Creutzfeldt, 1957). This maximum is also found in retinal neurons (Grüsser and Creutzfeldt, 1957) but can be explained only partly by properties of retinal receptors (Svaetichin, 1956b; Grüsser, 1957, 1960; Grüsser and Rabelo, 1958; Hartline, 1938b).

5. *Alterations of CFF by attention and thalamic stimulation.* Just as subjective CFF can be raised by attention and arousal, the neuronal CFF, that is, the maximal frequency up to which single cortical neurons can follow flickering light, may also be raised by stimulation of non-specific thalamic nuclei and the reticular formation (Jung, Creutzfeldt, and Grüsser, 1957; Creutzfeldt and Grüsser, 1959; Jung, 1958b).

6. *Local adaptation.* In subjective experience, a bright image fades out within seconds when the projection on the retina is held constant by a mirror system. Correspondingly, the discharge frequency of cortical neurons in response to continuous illumination—after the initial peak of discharge has waned—gradually decreases until an average rate of discharge is reached at around 10 per second, which is also present in the absence of a light stimulus. This neuronal behavior is probably due to a gradual decline in the receptor potential in the outer plexiform layer of the retina (Grüsser, 1960).

7. *After-images following brief light stimuli (successive contrast).* After-images show a rhythmic alternation of light and dark phases. Correspondingly, cortical neurons also show rhythmic activation and inhibition phases with similar time course (Jung, von Baumgarten, and Baumgartner, 1952; Jung and Baumgartner, 1955; Jung, Creutzfeldt, and Grüsser, 1957). After brief light flashes, the primary image corresponds to the primary activation of B neurons and the Purkinje second after-image to the secondary activation; the third after-image of Hess corresponds to the tertiary activation of B neurons following a single flash. The dark intervals between images and after-images correspond to the activation of D neurons and the inhibition of B neurons. This is shown in Fig. 5 (Grüsser and Grützner, 1958b).

8. *Subjective coordination of visual perception with vestibular stimuli.* The vestibular component in the regulation of the spatial stability of the visual world (*Raumkonstanz der Sehdinge*) corresponds

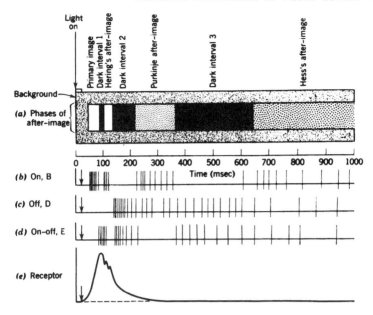

Fig. 5. Correlations of subjective after-images of man and neuronal discharges in retina and visual cortex of the cat, following a brief light flash (300 lux) (modified from Grüsser and Grützner, 1958b). (*a*) Scheme of successive phases of after-images (according to Fröhlich, 1929). Time in milliseconds. (*b* to *e*) Schematic responses of neurons and receptors: (*b*) retinal on neurons and cortical B neurons; (*c*) retinal off neurons and cortical D neurons; (*d*) retinal on-off neurons and cortical E neurons; (*e*) receptor response of the outer plexiform layer of the retina with intracellular recording.

The light flash is marked by a descending arrow. In (*b–e*) the arrows are shifted to the right because Fröhlich's subjective *Empfindungszeit* is 20 to 40 msec longer than the latency of B neurons in the visual cortex. The shaded area surrounding the after-images signifies the background of observation and the *Eigengrau* of the eye.

The scheme combines retinal and cortical responses, although cortical neurons show lower frequency and stronger periodicity. A pause in the primary activation of cortical B neurons is concurrent with the initial E discharge. During longer illumination both correspond to the *bande noire* of Charpentier (Jung, 1961) and after short flashes probably to dark interval 1. The scheme of Grüsser and Grützner (1958b) has been corrected appropriately, as they also described a pause in on-off-neurons between 200 and 450 msec.

to different types of modulation and activation of the majority of the cortical neurons of area 17 by polarization of the labyrinth, shown in Fig. 4 (Grüsser and Grüsser-Cornehls, 1960). This vestibular activation was not found at geniculate neurons (Grüsser, Grüsser-Cornehls, and Saur, 1959).

9. *Subjective brightness with monocular or binocular illumination and binocular rivalry.* In the human eye, there is very little subjective binocular brightness summation, and strong rivalry occurs between the corresponding areas of the two eyes when different stimuli are conveyed from their receptive fields (Hering's binocular rivalry). Correspondingly, 90 per cent of the cortical neurons of area 17 are activated or inhibited only by monocular stimulation and are not influenced by the other eye alone, although some modulation may occur in about 30 per cent of cortical neurons with binocular stimulation.

Only a small number of neurons can be influenced by each of the two eyes separately. Binocular convergent impulses manifest themselves chiefly as inhibition from one eye to light activation of the other eye, as one might expect from binocular rivalry (Grüsser and Grüsser-Cornehls, 1960, and unpublished work).

10. *Binocular flicker sensations.* Since Sherrington's experiments (1897, 1906) it is well known that binocular synchronous and alternate flicker of the two eyes has little mutual influence on visual experience, and that each monocular mechanism develops independently a complete visual image. Only at the limit of CFF does binocular flicker show a little higher frequency than monocular flicker. Correspondingly, there is little if any reinforcement by synchronous binocular flicker and little reduction by alternate binocular flicker in cortical neurons. The neuronal CFF of the majority of neurons in area 17 is determined by only one eye (Grüsser-Cornehls and Grüsser, unpublished experiments).

11. *Regulation of visual attention.* The variation in attention for visual stimuli may have its neuronal correlate in the convergence of specific retinal and nonspecific thalamic impulses upon cortical neurons. A subjective visual sensation in the central area of the visual field, evoked solely by an arousing stimulus without illumination (the *Schreckblitz* of Ebbecke, 1943; the *Weckblitz* of Ahlenstiel, 1949), may correspond to the arousal of cortical neurons of area 17. Nonspecific thalamic impulses modify the discharge pattern of most visual neurons after a relatively long latency of between 20 and 150 msec. The number of neurons responding to retinal afferents can be increased by stimulation of the nonspecific system, which increases the readiness of cortical neurons to respond to visual impulses (Akimoto and Creutzfeldt, 1957–58; Creutzfeldt and Akimoto, 1957–58; Jung, 1958b).

Patterned light stimulation (white-black contrast)

12. *Simultaneous contrast at the margin of white and black fields.* The subjective phenomena of border contrast have their objective

correlates in certain changes in neuronal responses that depend upon their relation to the border of the white and dark stripes (Fig. 6). The cortical B neurons show a maximum of discharges at light-on when their receptive field is stimulated by white light at the margin of a dark field. They show a minimum of discharges at light-on and

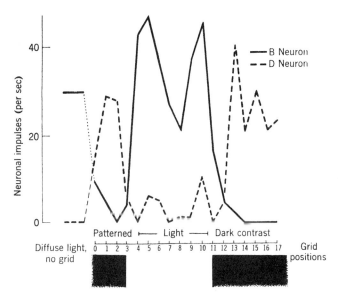

Fig. 6. Diagram of responses to contrast pattern and diffuse light in a cortical B and D neuron from the same experiment (from unpublished experiments of Baumgartner). The responses to light-on are plotted for spike frequencies per second on the ordinate (spikes counted in the first 500 msec following light-on) in relation to the projections of the receptive field to contrast pattern and diffuse light, on the abscissa.

The bright stripe has a visual angle of 5 degrees 41 minutes and is exposed in grid positions 3 to 11 by steps moving across the receptive fields of the neurons from left to right. Reciprocal activation and inhibition of the two antagonistic neurons and contrast enhancement of the discharge at the border of bright and dark stripes are clearly shown in the peaks and troughs of the upper graphs. This neuronal behavior may be explained satisfactorily by two factors: (1) *reciprocal inhibition* of antagonistic neurons in the same receptive field, and (2) *lateral inhibition* of synergic neurons in the surrounding field.

a reversal of their responses to light-on and off when their receptive field is situated in the dark stripe. A receptive field corresponding to the dark stripe reverses the discharge of B neurons to the D type of response: at light-on the B neuron is inhibited, and at light-off it is activated, although the total light shed on the retina is certainly increased at light-on and diminished at light-off. A reversal of the response may occur in D neurons; that is to say, they change to activation at light-on and inhibition at light-off when their receptive fields correspond to the dark stripe (Baumgartner and Hakas, 1959, 1960, 1962; Baumgartner, 1961; Fig. 6). Although D neurons then seem to respond like B neurons at light-on, they conserve their peculiar type of discharge at light-off and do not show the high-frequency peak of initial on response of the B neurons. In other words, D neurons exposed to the dark stripe respond to light-on as they usually do to light-off, after they have been exposed to diffuse light or to the bright stripe.

This neuronal behavior can be understood in physiological and psychological terms, if one accepts the following *four assumptions*, which are discussed later in detail:

1. Retinal on and off neurons as well as cortical B and D neurons discharge reciprocally. When the on or B neurons are activated, the corresponding off or D neurons are inhibited in the same retinal or cortical region.

2. Retinal on and off neurons as well as cortical B and D neurons show lateral inhibition of the same type of neurons in the surrounding retinal or cortical area during light adaptation. Lateral facilitation of neurons of the reciprocal type may be secondary to the lateral and reciprocal inhibition.

3. Activation of the on-center neurons in the retina and the B neurons in the cortex corresponds to the perception of brightness.

4. Activation of the off-center neurons in the retina and D and E neurons in the cortex corresponds to the perception of darkness.

Assumptions 1, 3, and 4 are supported by experiments with both diffuse and contrast light stimulation. Assumption 2 is derived from experiments with contrast light in cats (p. 632) and spots of light in *Limulus* (Hartline, 1949; Hartline and Ratliff, 1957) and in cats (Kuffler, 1953; Barlow, FitzHugh, and Kuffler, 1957). Although Kuffler's 1953 paper stresses the variability of response in the same retinal neuron, the later paper, with Barlow and FitzHugh (1957), discusses also the relation of his "on center" to a sensation of whiteness and of his "off center" to a sensation of blackness; in addition, it offers an explanation of simultaneous contrast by lateral inhibition in the retina.

Discussion

I propose to regard the neuronal mechanisms described above as special examples of some general principles of sensory information in the cortical neuronal system. The simplified conception that a primary receiving area for a given sense modality contains mosaic representation of the receptors cannot be maintained. Since we began to work on the neurons of the visual cortex seven years ago, this neuronal system has taught us many new things not only about visual mechanisms but also about neuronal integration in the cortex, the convergence of specific and nonspecific afferents, and the coordination of different receptor mechanisms. In addition to the striking correlations between subjective sensation and objective neuronal mechanisms, some of the intimate regulations of cortical functions can be investigated in the visual cortex. For these reasons the visual neuronal system may be a useful model for elucidating the coordination of information at the cortical level in general.

Although the gross electrophysiology of the visual cortex has been explored extensively with macroelectrodes for twenty-five years, since the pioneer work of Bishop, Bartley, and their associates (Bartley, 1936; Bartley and Bishop, 1933; Bishop, 1933; Bishop and Clare, 1952; Bishop and O'Leary, 1938), it has made only a limited contribution to the psychophysiology of vision. This is apparent in Bartley's recent review (1959), which appeared during this symposium. Microphysiological methods, first applied to retinal elements by Hartline (1938a), Granit (1946, 1950, 1955), and their collaborators, and to the visual cortex by our group in Freiburg (von Baumgarten and Jung, 1952; Jung, von Baumgarten, and Baumgartner, 1952) have brought more positive results in a relatively shorter time. Many parallels have been found between neuronal events and psychophysiological phenomena (summarized on the preceding pages, and in Jung, 1959).

The comparison of subjective visual sensation in man and neuronal discharges in cats has obvious limitations. Psychologists and philosophers may also object to such a correlation as a matter of principle. However, if we exclude color vision and foveal functions, and if we limit our investigations to light-dark sensations and corresponding neuronal activity, the cat may perhaps be regarded as a suitable experimental animal. The cat uses binocular vision similarly to the human, and it has precise optomotor and vestibulovisual coordination. Kappauf's behavioral experiments (1936) have shown that flicker fusion in the cat corresponds approximately to flicker fusion in man. If we are conscious of our limitations, the conclusion may not be too bold that comparable, although not identical, neurophysiological mechanisms are at work in the cortical visual systems of both man and cat.

The cat, as a predominantly nocturnal animal, is not, of course, appropriate for studies of color vision because it seems to be color-blind, although some kind of cone vision is present in the central retinal areas. In accordance with this, Grüsser (1957) was not able to find reversed membrane responses of receptor potentials in the cat's retina following complementary color stimulation, as Svaetichin (1956*a*, *b*) described in fishes.

Hubel's (1958) interesting findings of neuronal responses to moving stimuli in the freely moving cat show that some cortical neurons have special responses to directional movements of objects. Further experiments will probably clarify the neurophysiological basis of movement vision when they are combined with investigations of vestibular stimuli (Grüsser and Grüsser-Cornehls, 1960; Grüsser, Grüsser-Cornehls, and Saur, 1959) and of optokinetic nystagmus.

Correlations between subjective sensory phenomena and objective neurophysiological functions will be possible only for relatively simple kinds of optic stimuli and for rather primitive visual sensations. However, Baumgartner and Hakas's results (1959, 1960) on neuronal contrast phenomena show that researches on simultaneous contrast and moving stimuli promise to yield further psychophysiological correlations with the neurophysiological basis of pattern vision.

If we try to correlate neuronal responses with visual experience, it is not sufficient to describe the on, off, and on-off responses to certain visual stimuli in a population of neurons. We should attempt to distinguish the different response types according to their information value and to explain these neuronal responses by neurophysiological mechanisms that resemble those found elsewhere in the central nervous system. This kind of explanation is limited by the lack of good intracellular records from cortical neurons that record clearly membrane alterations of excitation and inhibition. But we hope to show that systematic application of extracellular recordings allows some explanations of neuronal organization by two inhibitory mechanisms. Kuffler (1953) has shown in the retina that the same neuron may respond differently to light-on and off when the same stimulus arrives at different locations of its receptive field. These variations in response do not mean that the information value of discharge in certain neuronal types cannot be defined and that these various response types would not represent exact neuronal communication mechanisms. Baumgartner and Hakas's experiments on simultaneous contrast (1959, 1960, in preparation) have shown clearly that these apparently contradictory responses to diffuse and patterned light can be understood only when the information value of discharge of these neurons is correlated with certain subjective sensations.

In this context I should like to explain our terminology for the different types of response of cortical neurons. I have often been asked why we call the cortical units that correspond to retinal on elements B neurons and the cortical equivalents of off elements D neurons. First, these letters are used to characterize cortical neurons in contrast to neurons at lower levels. Second, it would not make much sense to speak of "on neurons" when in some conditions they may show only off responses. When we speak of "on responses" or "off responses" these terms undoubtedly have a descriptive value, but they do not characterize the sensory information of these actions when the same neuron discharges in one situation at light-on, in another at light-off. When we call them B or D neurons, according to their type of response to binocular diffuse illumination (Fig. 1), this terminology has the advantage of indicating at the same time the information value of the neurons as brightness- or darkness-signaling units.

We leave it open whether a similar terminology might be applied to the neurons of lower retinal and geniculate levels. Here the time-honored terms on, off, and on-off elements have been traditional since the work of Hartline (1938a, b) and Granit (1946, 1950, 1955), although Kuffler's experiments (1950) have shown variability in these responses. Baumgartner has found that on elements of the optic nerve and geniculate body as well as cortical B neurons respond under certain contrast conditions by opposite reactions; they may reverse their on activation to inhibition by light-on and to activation by light-off when their receptive field projects to a black object.

Functional significance of convergence in the visual cortex

Our results demonstrate the convergence and coordination of various afferent impulses from the eye, the labyrinth, and the nonspecific thalamus at neurons of the visual cortex. This cortical area is generally considered only as a specialized receiving area for messages coming from the retina and as a binocular projection of both homonymous retinal fields. However, the function of its neuronal apparatus is apparently an integration of visual information from the retina with the results of vestibulo-optokinetic coordination and with the activity of the nonspecific system of the brain stem.

Many questions still remain about the functional significance and the mechanism of this integration and of its relation to other sense organs. The correlations of neuronal coordination in the visual cortex with psychophysiological experiences that have been described indicate that objective foundations for various subjective visual phenomena are not beyond the reach of our present experimental methods.

Neurophysiology, psychology, and technical science may each contribute to such a synthesis.

The relation of retinal and thalamoreticular convergence to sensory facilitation by attention and possible differences in the synaptic transmission of specific and nonspecific afferents have been discussed by Akimoto and Creutzfeldt (1957–58; Creutzfeldt and Akimoto, 1957–58). The significance of this convergence for special attentional mechanisms of the vestibulovisual coordination of ocular movements and for the stability of the visual world has been mentioned briefly in another paper (Jung, 1958b). The functional significance of specific and nonspecific coordination is evident from the alteration of the frequency up to which visual neurons are able to follow flickering light (neuronal CFF) by thalamic and reticular stimuli, described above and illustrated in Figs. 2 and 3.

The significance of nonspecific reticular afferents to the visual cortex is also demonstrated by the following experiments. Neuronal recordings from area 17 during retinal anoxia (described in Baumgartner, Creutzfeldt, and Jung, 1961) have shown a consistent difference after mesencephalic transection: intraocular ischemia knocks out spontaneous discharges in the *cerveau isolé* preparation but not in the *encéphale isolé*. These experiments permit the following conclusions: acute exclusion of specific afferent impulses depresses "spontaneous" neuronal activity in area 17 only when the connections with the lower brain stem are interrupted; thus activity of cortical neurons can be maintained by the ascending brain-stem connections (cf. Magoun, 1952)—including the reticular formation and the lower cranial nerve afferents—in the absence of specific afferents. One may conclude that both specific and nonspecific afferents are important to the maintenance of the neuronal activity of the visual cortex. The neurons of a primary receiving area of the cerebral cortex, although driven by specific afferent impulses, are also dependent upon nonspecific influx from the lower brain stem for the maintenance of normal "spontaneous" activity. Specific visual afferents and the nonspecific thalamic nuclei seem insufficient to maintain an active waking state without the lower brain stem, as suggested by Bremer's old experiments (1938) on the electroencephalogram in *encéphale* and *cerveau isolé* preparations. It appears that the corticothalamic visual system needs some afferent input from the periphery (either from the adequate specific receptors or from other afferents, for example, vestibular or trigeminal) over the reticular formation to drive the cortex and to maintain average neuronal activity during the waking state. The cortical neurons may maintain some activity if either the specific retinal or the nonspecific brain-stem influx is cut off. But neuronal activity in the visual cortex drops nearly to zero when both brain-stem reticular and eye afferents are blocked, although connections with the thalamus remain intact. However, we have induced only acute short depression of specific afferents by local retinal anoxia, which causes pain in *encéphale isolé*, and have not excluded them for a longer time by cutting both optic nerves. Such chronic specific deafferentation should be investigated by further experiments in *cerveau isolé* and *encéphale isolé* preparations after cutting of the trigeminal and optic nerves.

Although reticular functions are generally and loosely called nonspecific, Jung and Hassler (1960) have pointed out that "attention" contains many "specific" mechanisms of attentive behavior, which are special sensorimotor functions of the extrapyramidal and reticular systems. The visual motor functions and their coordination with vestibular afferents are mainly a performance of subcortical mechanisms, but precise optokinetic reactions are coordinated by the visual cortex.

What is the functional significance of vestibuloretinal convergence in the primary visual cortex? Is it only a nonspecific arousal or is it also a specific sensory regulation? We believe the latter to be the more important mechanism for sensory information, as suggested earlier (Jung, 1958b) and discussed in detail by Grüsser and his collaborators (Grüsser and Grüsser-Cornehls, 1960; Grüsser, Grüsser-Cornehls, and Saur, 1959).

The long latency and after-effect of vestibular activation of cortical neurons following labyrinthine polarization indicate a complex and polysynaptic mechanism. Because polysynaptic chains with vestibular afferents have been demonstrated anatomically in the reticular formation, one might assume that these labyrinthine effects would travel over the nonspecific reticular system and that vestibular activation would represent a nonspecific arousal. Against such an explanation the following arguments may be used. First, the *encéphale isolé* cat usually has an EEG of the arousal type, and further sensory arousal from trigeminal, auditory, or olfactory afferents usually has little or no effect on the neurons of the visual cortex. Second, sensory arousal also causes inhibitory phenomena (Jung, Creutzfeldt, and Grüsser, 1957; Whitlock, Arduini, and Moruzzi, 1953) and does not result in a constant activation to the double frequency of cortical neuronal discharge, as does specific vestibular stimulation in the absence of eye movements. Third, reticular stimulation often causes inhibition of cortical visual neurons (Fig. 4).

Psychophysiological observations favor the assumption that a specific vestibulovisual coordination exists in the reticular formation and projects to the visual cortex to regulate the stability of the visual world (*Konstanz der Sehdinge* of the old sensory physiologists). Such "specific" vestibular afferents from the reticular system might be able to influence the neurons of the visual cortex of the *encéphale isolé* preparations in a more differentiated manner than a nonspecific arousal. Moruzzi (1954) and his co-workers have shown that, besides convergence of sensory stimuli, the reticular formation contains many specific elements activated only by special sensory stimuli. The complex organization of vestibular, optomotor, and visual coordination needs special mechanisms in the so-called nonspecific reticular system in order to regulate localized visual attention by head and ocular movements. Some neuronal mechanisms of these specific and nonspecific functions are demonstrated by the work of Duensing and Schaefer (1957a, b, 1958): they found two types of neurons in the brain-stem reticular formation during nystagmus, one having exact time relations to the nystagmic phases and the other showing only diffuse activation.

The long latencies of the cortical neuronal responses to labyrinthine stimulation and their purely activating effects, described by Grüsser and Grüsser-Cornehls (1960) and Grüsser, Grüsser-Cornehls, and Saur (1959), present some difficulties to the interpretation of specific visual functions of the vestibulocortical connections. One would expect from a specific interaction of vestibuloreticular and retinogeniculate afferents a complex interaction of excitatory and inhibitory effects, such as are exhibited by all other cortical afferent mechanisms, including direct electrical stimulation of the cortex (Creutzfeldt, Baumgartner, and Schoen, 1956). One might explain the lack of inhibition following vestibular stimuli by assuming that interaction of inhibition and excitation occurred at lower levels of the brain-stem vestibulovisual system and that only positive activating effects of this interaction were conducted from the brain stem to the higher cortical level. According to von Holst's principles of *Reafferenz* (von Holst, 1951; von Holst and Mittelstaedt, 1950) only those afferent messages that are different from the afferent and efferent impulses would be conducted from the lower centers of optovestibular regulation to the higher cortical centers (von Holst's *Exafferenz,* 1951).

Long latencies after labyrinthine polarization, similar to those found in cortical neuronal responses, were also observed in eye movements. Thus the 30- to 50-msec latency of vestibular activation in visual neurons may not be too long for visual coordination. Also the qualitatively different types of response (α, β, γ, δ, described by Grüsser, Grüsser-Cornehls, and Saur, 1959, and shown in Fig. 4) indicate a mechanism more complex than diffuse nonspecific activation. In order to elucidate this vestibulovisual coordination, further experiments should investigate optokinetic stimulation, compare responses of the visual cortex to labyrinthine stimulation with and without eye movements, and study the relation of these phenomena to the vestibular projection areas of the cortex (Mickle and Ades, 1954; Walzl and Mountcastle, 1949).

Two antagonistic neuronal systems for brightness and darkness information

1. *B neurons and on elements correlated with brightness perception; D neurons and off elements correlated with darkness perception.* Experiments using diffuse illumination have shown good correlation between the discharge frequency of B neurons and brightness perception, in accordance with the Weber-Fechner logarithmic relation. The discharge frequency of D neurons at light-off is similarly dependent upon the intensity of the preceding light and successive contrast. These correlations with brightness and darkness information are shown more clearly by the responses to patterned stimulation, described by Baumgartner and Hakas (1959, 1960, 1962).

The information value of the visual neuronal types can be defined for the B and D neurons according to the psychophysiological correlations 1, 3, 4, 6, and 10, set forth above. All neuronal phenomena in the B and D system can easily be understood when we describe their functions in psychophysiological terms. Then we may say: the discharge frequencies of B neurons correspond to brightness information, the discharge frequencies of D neurons to darkness information in a given receptive field of the retina.

Simplifying these correlations, we may call the B neurons "brightness neurons" and the D neurons "darkness neurons," if we can permit such loose psychophysiological analogies. The significance of these neuronal discharges for brightness and darkness information is as valid for diffuse light stimulation (correlations 1, 3, 4, and 6) as for pattern stimulation with light-dark contrast (correlation 10). However, these relations are valid only when eye movements are excluded. With moving eyes or with vestibular stimulation, the conditions for the stability of the visual world become more complex and cannot yet be defined in clear physiological terms. For this we need more experiments on vestibulovisual relations and optokinetic nystagmus.

Baumgartner (in press) has shown that neuronal discharges at the border of contrasting fields vary with the broadness of the stripes, in agreement with subjective border contrast in human perception. Corresponding results were obtained from threshold measurements of human vision by Harms and Aulhorn (1955). Thus, neuronal discharges of the visual cortex in the cat show good psychophysiological correlations with human visual perception, not only for diffuse light stimulation but also for pattern stimulation.

The B and D neurons represent two clearly antagonistic groups with reciprocal responses to stimulation by diffuse light and patterned light. In earlier papers (Jung, 1958a; Jung and Baumgartner, 1955) we have described the C, D, and E neurons, which show primary inhibition after light-on, as a neuronal group opposed to the B neurons, which show primary activation after light-on. We have demonstrated that these neuronal groups maintain a balance of reciprocal activation and inhibition in the visual cortex, but we have not defined the information value of these neurons. Now since Baumgartner's and Hakas's (1959) findings of neuronal correlations with pattern vision (see also Baumgartner, in press; Jung, 1959) and the close correspondence of neuronal responses with after-images (Grüsser and Grützner, 1958a; Jung, Creutzfeldt, and Grüsser, 1957), we can say more about their significance for visual experience: we correlate the B neurons with brightness information and the D neurons with darkness information. When we disregard the rare C neurons, which are inhibited by both light-on and light-off, as elements probably correlated with inhibitory processes, only the information value of the E neurons remains to be discussed.

The question arises whether the E neurons, which correspond to the on-off elements of the retina, are only signaling changes of illumination, without regard to brightness or darkness information, and whether they are particularly important for the discrimination of moving stimuli. However, the perception of movement is based on rather complex mechanisms involving optomotor and vestibular regulation (pp. 651–652, 661). Therefore movement perception cannot simply be correlated with the on-off system of neurons. The late response of the E neurons to light-on also makes it unlikely that they constitute an arousing system that activates other types of neurons upon any change in degree or location of illumination.

2. *Are E neurons an independent on-off system or a variant of the D and off system?* It remains to be decided whether the retinogeniculate on-off system and the E neurons in the cortex constitute an independent system that signals any change in illumination in their recep-

tive fields or detects moving dark or light stimuli, independent of their lightness or darkness information. The preponderance of on-off elements in the peripheral retinal field, where pattern vision is rather poor and movement responses prevail, seems to favor an independent system *for movement perception.* The behavior of E neurons during successive contrast of after-images seems also to indicate a function different from that of the D system. However, many other observations make it more probable that the retinogeniculate on-off elements and the cortical E neurons contribute to the D system of darkness information. The constancy of pre-excitatory inhibition by onset of light in on-off elements and E neurons, the weakness and variability of the late on discharge as well as the predominance of inhibition during long diffuse light stimulation and during pattern stimulation of a bright field, and finally the evident preponderance of off activation over the late and variable on response—all these combine to make E neurons similar to off elements and D neurons. Some E neurons show also spontaneous suppression of their on responses for a certain time and then are indistinguishable from, or show transitions to, D neurons, as has been discussed earlier (Jung, 1953a). Thus most E neurons may be different reaction types of D neurons. These observations all suggest a similar information value for darkness in the on-off and E systems. The D and E neurons together would be the antagonists of the cortical B neurons. This would fit well the conception of a relative balance of excitation and inhibition in cortical neurons (Jung and Baumgartner, 1955; Jung, 1958a). Such a balance would be shifted from the medium level of spontaneous discharge of both types of antagonistic neurons to the preponderance of an "agonistic" and suppression of an antagonistic neuronal system during illumination (B preponderance) or vice versa during darkness (D-E preponderance). Thus there would be only two antagonistic neuronal apparatuses: one, the *B system with predominant on excitation* and *brightness information,* the other, the *D-E system with predominant off excitation* and *darkness information.* When we disregard the A neurons and the very rare C neurons, the number of these two groups—on or B, off or D-E— would be approximately the same in retina, geniculate body, and cortex. This equal number would be better able to maintain relative equilibrium at a medium level of activity than would the combination of a predominant B system with many neurons and a weak D system with only a few neurons.

Off elements and D neurons may mediate shadow reactions, which are of high biological significance, especially in lower forms (Hartline, 1938b). On-off elements and E neurons participate in these shadow reactions because their off responses have shorter latencies. It seems probable that the late on response in on-off elements and E neurons would be similar to a postinhibitory rebound following the on inhibition, and is not a genuine on excitation of long latency.

All this favors a dualistic theory of two antagonistic neuronal systems, the B or *brightness system* showing on activation in diffuse light and the D or *darkness system* (including the E neurons) showing on inhibition and off activation after diffuse illumination.

In spite of the probable functional similarity of the D and E groups, it may be advisable to describe the responses of the two groups separately and to await further experiments with moving stimuli before putting them definitely into the same box.

Reciprocal inhibition and lateral inhibition as basic mechanisms of neuronal organization

Two relatively simple inhibitory mechanisms are basic to neuronal responses to stimulation with diffuse and patterned light and may be used to explain the coordination of neuronal organization in the visual system: (*a*) *reciprocal inhibition* of antagonistic neurons in the same region; and (*b*) *lateral inhibition* of synergistic neurons in neighboring regions.

We prefer also to explain some secondary activation phenomena in these neuronal systems by inhibition and consequent disinhibition: examples of such phenomena are after-activation of the same neurons in successive contrast following postexcitatory inhibition (Fig. 5) and simultaneous activation of antagonistic neurons in neighboring fields. The reciprocal organization of this neuronal system, with its tendency to balance both activation and inhibition in antagonism, will necessarily result in a preponderance of excitation of antagonistic neurons when the agonistic group is inhibited, as in Hartline's disinhibition in the *Limulus* eye. During light stimulation, excitation is probably prevalent in the on system, first in the retina, because it is directly activated by the receptors, and second in the relay-stations—retina, geniculate body, and cortex—because only excitatory spikes can be transmitted by nerve fibers and not inhibition as such. Inhibition may result only from synaptic action by a special transmitter (Eccles, 1957). However, the silence of inhibitory pauses may also cause a relative excitation in a reciprocally organized system.

Several other visual phenomena, which have been found by psychophysiological experiments and which were discussed earlier, may be explained by these two principles of neuronal inhibition: for example, the Weber-Fechner logarithmic relation of stimulus and response (by interaction of receptive fields and increasing lateral inhibition with increasing light intensity), successive contrast (by antagonistic periodicity of reciprocal inhibition), and the pattern effects of simultaneous contrast (by lateral inhibition). The mechanisms of lateral inhibition in the organization of receptive fields have been discussed for the retina by Barlow (1953), Barlow, FitzHugh, and Kuffler (1957), Kuffler (1953), and Kuffler, FitzHugh, and Barlow (1957), and for the cortex by Baumgartner (in press).

A simple device to demonstrate the reciprocal function of the two systems (the on and off neurons in the retina and the B and D neurons in the cortex) and to prove the prevalence of the on system over the off system is stimulation with brief flashes of light, consisting of nearly simultaneous on and off stimuli. These responses of neurons to flash were first investigated by Baumgartner (1955) in the visual cortex and later by Grüsser and Rabelo (1958) in the retina. The results showed that the neurons of the on and off systems are not discharged simultaneously following a light flash. The on system with terminal activation of cortical B neurons predominates first, showing primary discharge simultaneously with a brief inhibition of the on-off neurons (E neurons in the cortex) and a longer inhibition of the off neurons (D neurons in the cortex). The following periodic and reciprocally alternating after-discharges of both systems can be explained by successive interaction of excitatory and inhibitory processes and reciprocal action of the on and off systems (Grüsser and Rabelo, 1958). These mechanisms seem to be similar to reciprocal innervation and post-inhibitory rebound in the spinal cord. The alternating phases of reciprocal activation and inhibition of the "brightness system" (B neurons) and the "darkness system" (D and E neurons) offer a neurophysiological explanation of the after-images following light flashes as summarized in Fig. 5. In psychophysiological investigations, Ebbecke (1920) and Fröhlich (1929) have compared these subjective phases of the after-image with Sherrington's principles of reciprocal innervation and successive induction in the spinal cord. Sherrington himself had applied these principles to flicker vision in 1897, although objective data on neuronal discharges in the visual and spinal system were not yet available at that time. This is another example of mutual stimulation between neurophysiology and psychophysiology.

Significance of simultaneous contrast, receptive field centers, and lateral inhibition in the cortex

The physiological significance of simultaneous-contrast mechanisms in the retina and cortex is evidently the functional compensation of physical and physiological irradiation of light effects, as Hering demonstrated (1878, 1920, 1931). The normal imperfections of our dioptric apparatus, denounced by Helmholtz (1896b), can be compensated effectively by these physiological mechanisms of lateral inhibition and reciprocal field organization of neurons. Contrast mechanisms are present even in entirely differently organized eyes of lower forms, as has been shown by Hartline and collaborators in *Limulus* (1949; Hartline and Ratliff, 1957). But in cat and man simultaneous-contrast mechanisms are not confined to the retina; they are at work as well in the cortex. Contrast phenomena at higher levels were first demonstrated psychophysiologically by Hering's binocular contrast (1920, 1931). Monocular cortical contrast mechanisms are apparently responsible for the narrowing of the receptive fields of cortical neurons, found by Baumgartner and Hakas (1960) and described above.

These contrast mechanisms showing lateral inhibition of excited neurons are not specific to the visual system. It seems to be a general principle of neuronal organization that local excitation causes a surrounding field of neuronal inhibition. Jung and Tönnies (1950), using coarse cortical recordings of cortical potentials, and Creutzfeldt, Baumgartner, and Schoen (1956), using microelectrode data on single neurons in the sensorimotor cortex, have described constant inhibitory responses at the fringe of single electrical stimuli of the normal cortex. Excitatory irradiation prevails only in abnormal conditions of a preconvulsive state following repetitive stimulation (Jung and Tönnies, 1950). Kuffler's findings (1953) of the organization of on and off responses in the center and margin of retinal receptive fields show the same principles of lateral inhibition. The experiments of Barlow, FitzHugh, and Kuffler (1957) demonstrate further that lateral inhibition is not constant and can be diminished and reversed to lateral excitation and irradiation during dark adaptation. A similar sheet of inhibition surrounding a core of excitation was found in the somatosensory system and its cortex by Mountcastle (1957; Mountcastle, Davies, and Berman, 1957). These authors also describe reciprocal inhibition and different types of response in cortical neurons. Simultaneous inhibitory and excitatory reactions have also been seen in neuronal populations of other regions of the cortex even under such complex conditions as the conditioned reflexes (Ricci, Doane, and Jasper, 1957). Continuously discharging neurons similar to A neurons of the visual cortex were found in the motor cortex by Ricci, Doane, and Jasper (1957), in the sensory cortex by Cohen, et al. (1957), and in the auditory cortex by Erulkar, Rose, and Davies (1956).

The apparent discrepancies between Baumgartner's results and those of Hubel and Wiesel (1959) on the diameter of receptive fields in cortical neurons may be due to their methods and to the particular field shapes and the surrounding zone of lateral inhibition at the cortical level. Hubel and Wiesel with their spotlight method found relatively large receptive fields, whereas Baumgartner with his simultaneous-contrast method found smaller field centers in the cortex than in the retina and the geniculate body. Baumgartner and Hakas's method measured only the core of activation (called the receptive-field centers) mostly in horizontal directions. Hubel and Wiesel (1959) determined the whole receptive field with the surrounding area of lateral inhibition and found differently shaped field centers, elongated and elliptoid, mainly with the long axis vertical. These differences in the shape of receptive fields explain some effects specific to the direction of movement, first described by Hubel (1958) in cortical neurons. Baumgartner's investigations and other experiments on moving contrast stimuli, which are now in progress in our laboratories and which were briefly mentioned by Grüsser, Grüsser-Cornehls, and Saur (1959), have shown that the irregularities that sensitize the neurons to moving stimuli are predominantly found in D neurons, activated by light-off and dark contrast, or in E neurons, which may belong to an enlarged darkness system. Baumgartner and Hakas's findings (1960, in preparation) of smaller receptive-field centers in cortical neurons may be explained by a pronounced effect of lateral inhibition in the cortex. From psychophysiological experience, it seems probable that the mechanisms of lateral inhibition are not confined to the retina but are active also in supraretinal relay stations of the visual system. Thus central contrast mechanisms may also be able to improve the dim picture of scotopic vision, even when lateral inhibition is suppressed in the retina during dark adaptation, as was shown by Barlow, FitzHugh, and Kuffler (1957): the receptive-field center is narrowed down in the cortex by progressive action of lateral inhibition and simultaneous contrast. We believe that only the central core of the whole receptive field yields positive information. The surrounding halo, which responds reciprocally, varies in different conditions and may represent only a regulatory phenomenon of contrast and lateral inhibition. The area about which the neuron provides information, therefore, seems to be the receptive-field center, as determined by Baumgartner's contrast method, and not the whole receptive area from which illumination influences the firing of units, as defined by Hartline (1940a), Kuffler (1953), and Hubel and Wiesel (1959).

Consideration of the receptive-field center as the essential area of neuronal visual information suggests the apparent paradox: cortical neurons see more sharply than retinal neurons, owing to central mechanisms of lateral inhibition and simultaneous contrast. Lateral inhibition is certainly not confined to visual mechanisms, as is shown by Mountcastle's findings (1957) in somatosensory neuronal systems. A narrowing of the response area of single neurons to tone frequencies was found by Katsuki, et al. (1958) between the cochlear nerve and the colliculus, but not further up in the cortex, apparently caused by similar neuronal contrast mechanisms in lower acoustic relay stations. It seems probable that lateral inhibition is a general principle of afferent communication in various modalities, designed to increase sensory discrimination through contrast.

The synaptic mechanism of reciprocal and lateral inhibition is not yet clear and should be investigated by intracellular recordings similar to Eccles's work on the spinal cord (1957; Brock, Coombs, and Eccles, 1952; Phillips, 1956). Some indications for an electrotonic mechanism probably located on dendrites may be derived from Granit's work on the alteration of on and off responses by retinal polarization (1946; Gernandt and Granit, 1947).

Local adaptation and eye movements as essential factors of visual perception

Under normal conditions, immobile objects in the surroundings change their projection on the retina continuously during normal eye and body movements. Images artificially stabilized by mirror systems fade out quickly (Riggs, et al., 1953; Ditchburn and Ginsborg, 1953). This fading caused by local adaptation can be prevented by normal eye movements and is diminished by small tremor or flicker movements but not by the slow-drift component of fixation. Maintenance of vision is possible only when eye or head movements occur. Consequently, constant illumination on the retina is not a physiological stimulus, since the response fades as a result of local adaptation. Alternating light-dark stimuli, such as flickering light, come closer to the normal stimulation of our photoreceptors. Autrum (1957) has stressed this point and has discussed some parallels in the eyes of insects and lower animals. Of course insect and *Limulus* eyes are very different from the vertebrate eye and contain special features for movement vision. However, the general principles of visual physiology were discovered by Hartline in the eyes of *Limulus* (1940b, 1949).

Hubel (1958, 1959) has described special cortical neurons, which are activated only by directional movements but not by diffuse illumination. Grüsser and Kornhuber in unpublished experiments found that most neurons of the visual cortex are activated by moving bright and dark stripes in both opposite directions and various planes. Mainly in D and E neurons, different responses to opposite movements may prevail for certain directions. It seems probable that these direction-specific differences are due to an asymmetrical extension of the inhibitory surroundings of the neuronal receptive fields. But it is not yet certain whether those field asymmetries are constant features of the receptive field or functional fluctuations, varying in time. Baumgartner and Hakas's contrast investigations (1959, 1960, 1962) lean toward the latter explanation.

The apparent stability of the visual world during eye and body movements (Hering's *Konstanz der Sehdinge*) suggests that the central retinal projection cannot be a fixed retinocortical relation but it must depend upon continuously varying regulations. The central projection of the retinal image is regulated by very precise optovestibular coordination with eye and body movements, according to von Holst's principle of *Reafferenz* (1951; von Holst and Mittelstaedt, 1950). Although the retinal image is shifted over a considerable distance during active eye movements, our perception of the external world remains stable. Only passive eye movements (finger pressure in otherwise normal conditions) (Helmholtz, 1896*b*) or the intention to gaze when eye movements are hindered, say, by fixed eyeball or paralyzed eye muscles (Kornmüller, 1931) result in a sensation of movement of the surroundings. Conversely, moving surroundings cause optokinetic nystagmus, which tends to hold the retinal image at the same location, although it may be seen as moving in the right direction. During the quick phase of nystagmus the movement in the opposite direction that one might expect is suppressed. When the slow phase lags behind, during optokinetic nystagmus in reduced attention (Jung, 1953*b*), there is an illusion of more rapid apparent movement. The mechanism of nystagmus, which contributes to some extent to movement perception, is coordinated in the reticular formation of the brain stem from convergence of retinal, vestibular, and cortical impulses. Thus the visual perception of movement is the result of a complex regulation of retinal, vestibular, and proprioceptive messages with attentiveness and is primarily coordinated in the brain stem and elaborated in the cortex. It is not due simply to moving images on the retina, except in movements of discrete objects that are not followed by fixation.

In 1951 von Holst explained the complex vestibulovisual *"reafferent"* and voluntary regulation of eye movements and visual perception by means of rather simple schematic models. It is still not possible to express these models as precise synaptic connections or special neuronal mechanisms of the various structures of the visual, reticular, and vestibular systems.

Limitations of mechanical models of neuronal coordination

A comparison of mechanical communication devices with the information mechanisms of the living nervous system may bring to light some parallels, but the fundamental differences between the two cannot be overlooked.

We have seen some admirable examples of complicated computing machines in the Laboratory of Communications Biophysics at MIT and have witnessed their effectiveness in extracting information data from complex physiological processes. In their proper and very special functions these machines surpass the capabilities of their creators, the human brains, in speed and accuracy. However, to a neurophysiologist the differences in functional elements between these machines and the brain are impressive.

We should not concede too quickly that our brain uses machinelike patterns of functional interconnections for extracting useful information from the sense organs. The comparison of these machines with sensory signal-coding mechanisms of the living nervous system is severely handicapped by the immense number of neuronal units and the variety and richness of neuronal coordination and convergence in the brain.

Computers are very different from the mammalian nervous system. In fact, their characteristics may even be quite opposite to those of the brain. In common with neuronal systems, digital computers have a binary, all-or-none principle and homeostatic feedback, but their "neurons" have not the thousands of parallel lines of nerve fibers, the rich convergence and integrative faculty of CNS synapses,* nor the adaptive and learning functions of brain centers. The limited and rigid capacity of their performance might better be compared with insect ganglions, which have very few nerve cells and very precise and fixed patterns of machinelike "instinctive" function, than with the

---

* Estimations of the number of optic nerve fibers in cats and higher mammals vary between 100,000 and 1,000,000. This number is then multiplied in the next neuronal relay stations of geniculate body and cortex. These enormous numbers of parallel channels and multiple connections are characteristic of the nervous system but would be quite unusual in computing machines. Thus quantitative as well as qualitative differences between nervous and mechanical apparatus must be considered.

immense number and variety of neurons with their interconnections found in mammalian brains. Machines may be better at selecting certain signals from noise, but the brain is very much better at distinguishing meaningful signals from very similar ones that are meaningless.

Another difference between neuronal and computer systems is the significance of background activity. The background discharge of the neuron is essentially different from the noise or cross talk in electronic machines. In computing devices, noise is kept at a minimum level and normally does not interfere with actual integrating functions. In contrast, the background activity of the nervous system is of about the same order as the signals operating on and within it. It maintains a medium level of activity. Thus the background activity of normal neuronal "noise" with its spontaneous discharges has a positive function. It might be compared to the heating currents of an electronic apparatus. In the visual system, spontaneous neuronal "noise" corresponds subjectively to the *Eigengrau* (the gray background of visual perception in the dark). Although Adrian (1957–58) has assumed that neuronal "noise" has a disturbing action in the olfactory system, background activity in the visual system seems far from masking sensory effects. On the contrary, spontaneous activity of neuronal "noise" is a physiological necessity for effective perception of brightness and darkness. Spontaneous neuronal discharge (*Ruheaktivität der Neurone*) maintains a basic level of activity from which the two reciprocal systems, the B and the D neurons, deviate in opposite directions through activation or inhibition (Jung, 1958*a*; Jung and Baumgartner, 1955). In contrast to a silent system that can be activated in only one direction, these changes in the two directions avoid fixed thresholds and assure the adaptability and flexibility of the visual system as well as of the central nervous system in general.

Our nervous system has so many neurons to fire and so many nerve channels to use that it can afford some redundancy in its information. The CNS does not need to be too economical in controlling and reducing afferent information in the periphery, if it has a central device to select the information of particular significance. It can afford the extravagance of receiving much information at once or the same information twice or a thousand times with only slight variation. These repeated stimuli do not result in disturbing function but rather in the positive effect of adaptation or learning. Small variations in the information may make the message more adequate to the external conditions or more interesting to the central receiving apparatus. Thus some central barrier may be passed because a certain critical detail fits some previous experience. If the detailed variation is un-important, it does not hinder habituation to the repeated sensory stimulus, which then falls into the black box of neglect and oblivion in which so much uninteresting information ends every day. Dr. Hernández-Peón has, during this symposium, given us some good examples of habituation to sensory stimuli and of interrelations with other sense modalities (see also Hernández-Peón, et al., 1957).

In fact, the brain nearly always receives varied information about the same thing from a large number of receptors and from different sense organs. Multisensory information results if these various sensory messages are allowed to synthesize by means of additional atten-tional mechanisms. Such multisensory and synesthetic perceptions are common in our psychological experience. They are also most effective in influencing the behavior of animals. It is well known that the absence of the associated flavor from one sense organ (for example, olfactory) makes an otherwise interesting and vivid sequence of visual and acoustic perceptions quite uninteresting to an animal; thus a cat does not pay attention to performance on a sound film or to images in a mirror.

For experimental conditions in neurophysiology we have to use simpler stimuli confined to one or two sense modalities. But our ex-perimental results on single neurons show that multisensory conver-gence is a prominent feature in cortical sensory mechanisms. Even in a so-called primary cortical receiving area, as in area 17 of the visual system, several sense modalities are coordinated. Landgren has demonstrated at this symposium similar convergence in neurons of the sensory cortical area. Moruzzi (1954) and his co-workers have shown that convergence of different sense modalities may also occur at neurons of the reticular formation. Therefore we may assume that the so-called nonspecific afferent impulses that reach the cortex from the thalamoreticular system may also contain multisensory com-ponents in their integrated messages.

It would be very helpful for the neurophysiologist if information engineers would pay more attention to special neuronal mechanisms of the cortex which can be so beautifully demonstrated in the visual system. As neurophysiologists we have to accept and to investigate the nervous system as it is. We cannot change its mechanisms except by lesions. On the other hand, communication engineers can build and alter their models at will to approximate biological realities. In other words, we should be glad if our colleagues in the fields of technical communication would not only construct their fine computing machines to facilitate physiological analysis, but would also build neuronal models that resembled brain mechanisms. Then perhaps instead of telling us how the neurons of the nervous system should work according to their theories, they would be able to base their models and theories on neurophysiological findings in actual neuronal networks, such as the visual system. Then we might expect useful results from the cooperation of neurophysiologists, psychologists, and engineers—which is the main purpose of this symposium.

Appreciation of nineteenth century sensory physiology

Nearly all the correlations that we have found between neuronal discharges in cats and visual sensations in man originate in the basic findings of the great pioneers in subjective visual physiology of the last century, mainly by Purkinje, Helmholtz, Hering, and von Kries. Consequently, a discussion of these correlations seems to me appropriate. Subsequent to Purkinje's findings (1819), Aubert's observations (1865) on vestibular influence on visual verticality, Fechner's psychophysics (1860), and Brücke's findings· on flicker fusion and brightness enhancement (1864), Helmholtz's work (1896b), and Hering's observations and theories (1878, 1920, 1931)—all important discoveries in the physiology of vision were made and tested on man by precise measurement of subjective sensations. These pioneers discovered the laws of successive contrast in the train of after-images and the various phenomena of simultaneous contrast, visual-vestibular coordination, binocular rivalry, and so forth. The general principles of regulated stability of the visual world were developed clearly by Hering (1920, 1931), although the special neuronal mechanisms assuring this stability remained unknown. Our observations made with microelectrodes are no more than objective demonstrations at the neuronal level of these laws and principles, formulated by the classical psychophysiologists.

We are made humble when we compare our elaborate microelectrode techniques and our rather sophisticated interpretations with the insights that the great sensory physiologists of the nineteenth century had through precise observations of purely subjective experience. The now fashionable concepts of information and communication theories were not unknown to these scientists. In 1868 Helmholtz (1896a) foresaw the essentials of our modern communication theories when he contrasted the limited number of 26 letters of the alphabet through which the rich variety of written communication is achieved with the enormous number of visual receptors and neurons that produce the innumerable visual sensations. He also pointed out some differences between mechanical models having only single units for special functions and brain mechanisms working with millions of parallel units.

Psychophysiology and neurophysiology have too long dwelt apart, ignoring the many correlations between their two ways of exploring the same object of sensory communication. Coordination of these two lines of research will be possible when scientists remain aware of the differences in method and approach between subjective perceptual psychophysiology and objective neuronal electrophysiology. Then we may avoid muddling through the problems of sensory physiology by equivocations and succeed in achieving true convergence and mutual stimulation of the two scientific approaches.

Conclusions

I hope that our results will refute the reproach of atomistic research that we sometimes hear from psychologists who denounce investigations at the neuronal level. Although single-neuron recording picks out only a very few from the millions of cortical nerve cells, it does not result in a meaningless sample of brain functions if the sample is carefully selected and analyzed. An investigation of the neuronal activity in a regulated complex system like the cortex may reveal basic mechanisms if the system is examined with reasonable experiments under well-planned conditions. The theoretical bases for physiological experiments on sensory communication are supplied by subjective experience. Psychophysiology of the sense organs has preceded neurophysiological analysis of sensory mechanisms for more than 100 years, and it still provides the searchlights on this route of research.

The coordination of psychophysiological and neurophysiological experiments will lead us further than either of these approaches alone. The combination of the two may indicate a *via regia* to the exploration of human sensory information. The unilateral pursuit of only one method without regard to the other risks either blind neurophysiological recording or fanciful psychological hypotheses, and either of them may lead to minor sidetracks and end in a jungle of barren facts or luxuriantly growing speculations. With the help of the highly developed engineering techniques that facilitate our neurophysiological and psychophysiological research, a further advance on this path should not be too difficult and may elucidate some of the many unknown mechanisms of sensory communication.

Summary

Recent work from our laboratory on extracellular recordings of neuronal activity in the visual cortex of cats is described, and the results are compared with subjective visual perception in man. Neurons of the visual cortex integrate afferent impulses from the eye, from vestibular receptors, and from the nonspecific reticulo-thalamic system. These various afferents converge mostly on the same cortical neuron.

Specific retinal afferents and reticulothalamic stimuli result in certain response patterns of excitation and inhibition, which have been classified into five special neuronal types (A to E) for binocular stimuli by diffuse light and five types (I to V) for thalamic stimuli.

The types of neuronal response to light stimulation show all combinations with the types of response to thalamic stimulation. The majority of visual neurons are influenced by both specific and nonspecific afferents. The A neurons, which are unresponsive to light or darkness, respond mainly to thalamic stimuli. Nonspecific reticular or thalamic stimulation may drive many neurons to increased responses and may alter their ability to follow flickering light (CFF), although their fixed patterns of visual response may remain the same.

Vestibular stimulation by labyrinthine polarization activates most of the neurons in the visual cortex after a long latency of 25 to 250 msec. In contrast to retinal and nonspecific stimulation, inhibitory effects were rarely found after labyrinthine polarization.

The majority of the neurons in area 17 receive monocular specific impulses from one eye only, together with nonspecific and vestibular afferents. Some modulation of monocular activation or binocular convergence may occur in about one-third of the neurons of area 17. True binocular convergence from both eyes, acting chiefly in the form of inhibition from one eye depressing the excitation from the other eye, is found in only about one-tenth of cortical visual neurons.

Patterned light stimulation (bright-dark stripes) alters the neuronal responses to light-on and off, depending upon the projection of the neuronal receptive field to the border of bright and dark stripes. The responses correspond closely to the brightness sensation in simultaneous border contrast: B neurons show a maximum of discharges at light-on when stimulated by the bright field at the margin of the contrasting dark stripe. A minimum or reversed inhibitory response occurs at light-on when the receptive field of B neurons corresponds to the margin of the dark stripe. Vice versa, the on inhibition of

D neurons is changed to activation when their receptive field corresponds to the contrasting dark stripe, and the off responses of D neurons is maximal when it corresponds to the white stripe. The receptive-field centers of cortical neurons show significantly smaller diameters than the field centers of retinal and geniculate neurons. This seeming paradox—that cortical neurons "see" more sharply than retinal neurons—is explained by the action of lateral inhibition of corresponding cortical neurons in neighboring fields.

The organization of visual neurons in the cortex may be explained by two principles of inhibition, which were first described in the retina: (a) reciprocal inhibition of antagonistic neurons in the same region; and (b) lateral inhibition of synergistic neurons in neighboring regions.

The information value of the various neuronal response types to light and dark stimulation of the retina is discussed. Stimulation effects by diffuse light and by patterned light with bright-dark contrast indicate that B neurons are correlated with brightness information and D neurons with darkness information. It seems probable that E neurons, responding to light-on and off, are also correlated with darkness information.

The question is discussed, whether a dualistic conception of two antagonistic neuronal systems (on and B neurons versus off, on-off and D-E neurons) in retina and cortex may provide a better explanation of the neuronal coordination and relative balance of the visual neuronal system than the classical triple system of on, off, and on-off elements. In on-off E neurons, pre-excitatory inhibition to light-on is a constant feature, and the following late on discharge may be interpreted as an early postinhibitory rebound.

Twelve psychophysiological correlations between neuronal discharges in cats and visual sensation in man are described for brightness sensation, flicker fusion, Porter's law, brightness enhancement, alterations of CFF by attention and thalamic stimulation, local adaptation, after-images (successive contrast), coordination of visual perception with vestibular stimuli, binocular rivalry and brightness summation, binocular flicker sensations, regulation of visual attention, and simultaneous contrast.

The role of spontaneous activity of visual neurons is seen in the maintenance of a base level of activity (corresponding to the background of visual *Eigengrau*) from which both reciprocal systems, the B and the D neurons, may deviate in opposite directions as a result of activation and inhibition.

Limitations of mechanical models of neuronal coordination and differences between computers and brains are discussed. In computers the noise level is kept minimal in relation to signals, whereas in the brain the background discharge of "neuronal noise" is not very different in its magnitude from afferent signals and from their integration mechanisms. Machines are better at selecting signals from noise, but the brain is much better at distinguishing meaningful from similar meaningless signals.

The mechanisms of sensory communication may be studied best by comparison of neuronal action in sensory areas with psychophysiological experience.

References

Adrian, E. D. The control of the nervous system by the sense organs. *Arch. Psychiat. Nervenkr.*, 1957–58, **196**, 482–493.

Ahlenstiel, H. Der Weckblitz als hypnagoge Vision. *Nervenarzt*, 1949, **20**, 124–127.

Akimoto, H., and O. Creutzfeldt. Reaktionen von Neuronen des optischen Cortex nach elektrischer Reizung unspezifischer Thalamuskerne. *Arch. Psychiat. Nervenkr.*, 1957–50, **196**, 494–519.

Aubert, H. *Physiologie der Netzhaut.* Breslau: E. Morgenstern, 1865.

Autrum, H. Das Sehen der Insekten. *Studium Generale*, 1957, **10**, 211–214.

Barlow, H. B. Summation and inhibition in the frog's retina. *J. Physiol.*, 1953, **119**, 69–88.

Barlow, H. B., R. FitzHugh, and S. W. Kuffler. Change of organization in the receptive fields of the cat's retina during dark adaptation. *J. Physiol.*, 1957, **137**, 338–354.

Bartley, S. H. Temporal and spatial summation of extrinsic impulses with the intrinsic activity of the cortex. *J. cell. comp. Physiol.*, 1936, **8**, 41–61.

Bartley, S. H. Central mechanisms of vision. In *Handbook of Physiology. Neurophysiology* I. Washington, D. C.: American Physiological Society, 1959. Pp. 713–740.

Bartley, S. H., and G. H. Bishop. The cortical response to stimulation of the optic nerve in the rabbit. *Amer. J. Physiol.*, 1933, **103**, 159–172.

von Baumgarten, R., and R. Jung. Microelectrode studies on the visual cortex. *Rev. Neurol.*, 1952, **87**, 151–155.

Baumgartner, G. Reaktionen einzelner Neurone im optischen Cortex der Katze nach Lichtblitzen. *Pflügers Arch. ges. Physiol.*, 1955, **261**, 457–469.

Baumgartner, G. Die Reaktionen der Neurone des zentralen visuellen Systems der Katze im simultanen Helligkeitskontrast. In R. Jung and H. Kornhuber (Editors), *Neurophysiologie und Psychophysik des visuellen Systems*, Symposion Freiburg. Berlin, Göttingen, Heidelberg: Springer, 1961. Pp. 296–311.

Baumgartner, G., O. Creutzfeldt, and R. Jung. Microphysiology of cortical neurons in acute anoxia and in retinal ischemia. In T. S. Meyer and H. Gastaut (Editors), *Cerebral Anoxia and the Electroencephalogram.* Springfield, Ill.: C. C Thomas, 1961. Pp. 5–33.

Baumgartner, G., and P. Hakas. Reaktionen einzelner Opticusneurone und corticaler Nervenzellen der Katze im Hell-Dunkel-Grenzfeld (Simultankontrast). *Pflügers Arch. ges. Physiol.*, 1959, **270**, 29.

Baumgartner, G., and P. Hakas. Vergleich der receptiven Felder einzelner on-Neurone des N. opticus, des Corpus geniculatum laterale und des optischen Cortex der Katze. *Zbl. ges. Neurol. Psychiat.*, 1960, **155**, 243–244.

Baumgartner, G., and P. Hakas. Die Neurophysiologie des simultanen Helligkeitskontrastes: Reziproke Reaktionen antagonistischer Neuronengruppen des visuellen Systems. *Pflügers Arch. ges. Physiol.*, 1962, **274**, 489–510.

Baumgartner, G., and R. Jung. Hemmungsphänomene an einzelnen corticalen Neuronen und ihre Bedeutung für die Bremsung convulsiver Entladungen. *Arch. Sci. biol.*, 1955, **39**, 474–486.

Bishop, G. H. Cyclic changes in excitability of the optic pathway of the rabbit. *Amer. J. Physiol.*, 1933, **103**, 213–224.

Bishop, G. H., and M. H. Clare. Sites of origin of electrical potentials in striate cortex. *J. Neurophysiol.*, 1952, **15**, 201–220.

Bishop, G. H., and J. O'Leary. Potential records from the optic cortex of the cat. *J. Neurophysiol.*, 1938, **1**, 391–404.

Bremer, F. *L'activité électrique de l'écorce cérébrale.* Paris: Hermann, 1938.

Bremer, F., and N. Stoupel. De la modification des réponses sensorielles corticales dans l'éveil réticulaire. *Acta neurol. psychiat. belg.*, 1958, **58**, 401–403.

Bremer, F., and N. Stoupel. Facilitation et inhibition des potentiels évoqués corticaux dans l'éveil cérébral. *Arch. int. Physiol.*, 1959, **67**, 240–275.

Brock, C. G., J. S. Coombs, and J. C. Eccles. The recording of potentials from motoneurones with an intracellular electrode. *J. Physiol.*, 1952, **117**, 431–460.

Brücke, E. Über den Nutzeffekt intermittierender Netzhautreizungen. *Sitzber. Akad., Wiss. Wien (Math.-Nat. Kl.)*, 1864, **49 (II)**, 128–153.

Cohen, M. J., S. Landgren, L. Ström, and Y. Zotterman. Cortical reception of touch and taste in the cat: A study of single cortical cells. *Acta physiol. scand.*, 1957, **40**, Suppl. 135.

Creutzfeldt, O., and H. Akimoto. Konvergenz und gegenseitige Beeinflussung von Impulsen aus der Retina und den unspezifischen Thalamuskernen an einzelnen Neuronen des optischen Cortex. *Arch. Psychiat. Nervenkr.*, 1957–58, **196**, 520–548.

Creutzfeldt, O., G. Baumgartner, and L. Schoen. Reaktionen einzelner Neurone des sensomotorischen Cortex nach elektrischen Reizen. *Arch. Psychiat. Nervenkr.*, 1956, **194**, 597–619.

Creutzfeldt, O., and O.-J. Grüsser. Beeinflussung der Flimmerreaktion einzelner corticaler Neurone durch elektrische Reize unspezifischer Thalamuskerne. In *Proc. 1st int. Congr. neurol. Sci., Brussels*, Vol. III. *EEG, Clinical Neurophysiology and Epilepsy.* London: Pergamon, 1959. Pp. 349–355.

Ditchburn, R. W., and B. L. Ginsborg. Involuntary eye movements during fixation. *J. Physiol.*, 1953, **119**, 1–17.

Duensing, F., and K.-P. Schaefer. Die Neuronenaktivität in der Formatio reticularis des Rhombencephalons beim vestibulären Nystagmus. *Arch. Psychiat. Nervenkr.*, 1957a, **196**, 265–290.

Duensing, F., and K.-P. Schaefer. Die "locker gekoppelten" Neurone der Formatio reticularis des Rhombencephalons beim vestibulären Nystagmus. *Arch. Psychiat. Nervenkr.*, 1957b, **196**, 402–420.

Duensing, F., and K.-P. Schaefer. Die Aktivität einzelner Neurone im Bereich der Vestibulariskerne bei Horizontalbeschleunigungen unterbesonderer Berücksichtigung des vestibulären Nystagmus. *Arch. Psychiat. Nervenkr.*, 1958, **198**, 225–252.

Dumont, S., and P. Dell. Facilitations spécifiques et non spécifiques des réponses visuelles corticales. *J. Physiol.*, 1958, **50**, 261–264.

Ebbecke, U. Über zentrale Hemmung und die Wechselwirkung der Sehfeldstellen. *Pflügers Arch. ges. Physiol.*, 1920, **186**, 200–219.

Ebbecke, U. Über ein entoptisches Phänomen bei Schreck. *Klin. Mbl. Augenhlk.*, 1943, **109**, 190–193.

Eccles, J. C. *The Physiology of Nerve Cells.* Baltimore: Johns Hopkins Press, 1957.

Erulkar, S. D., and M. Fillenz. Pattern of discharge of single units of the lateral geniculate body of the cat in response to binocular stimulation. *J. Physiol.*, 1958, **140**, 6P–7P.

Erulkar, S. D., J. E. Rose, and P. W. Davies. Single unit activity in the auditory cortex of the cat. *Bull. Johns Hopkins Hosp.*, 1956, **99**, 55–86.

Fechner, G. T. *Elemente der Psychophysik,* Parts 1 and 2. Leipzig: Breitkopf and Härtel, 1860.

Fröhlich, F. W. *Die Empfindungszeit.* Jena: Gustav Fischer, 1929.

Gernandt, B., and R. Granit. Single fibre analysis of inhibition and the polarity of the retinal elements. *J. Neurophysiol.*, 1947, **10**, 295–302.

Granit, R. The distribution of excitation and inhibition in single fibre responses from a polarized retina. *J. Physiol.*, 1946, **105**, 45–53.

Granit, R. The organization of the vertebrate retinal elements. *Ergebn. Physiol.*, 1950, **46**, 31–70.

Granit, R. *Receptors and Sensory Perception.* New Haven: Yale University Press, 1955.

Grüsser, O. J. Receptorpotentiale einzelner Zapfen der Katze. *Naturwissenschaften*, 1957, **44**, 522.

Grüsser, O.-J. Rezeptorabhängige Potentiale der Katzenretina und ihre Reaktionen auf Flimmerlicht. *Pflügers Arch. ges. Physiol.*, 1960, **271**, 511–525.

Grüsser, O.-J., and U. Cornehls. Reaktionen einzelner Neurone im optischen Cortex der Katze nach elektrischer Labyrinthpolarisation. *Pflügers Arch. ges. Physiol.*, 1959, **270**, 31.

Grüsser, O.-J., and O. Creutzfeldt. Eine neurophysiologische Grundlage des Brücke-Bartley-Effektes: Maxima der Impulsfrequenz retinaler und corticaler Neurone bei Flimmerlicht mittlerer Frequenzen. *Pflügers Arch. ges. Physiol.*, 1957, **263**, 668–681.

Grüsser, O.-J., and U. Grüsser-Cornehls. Mikroelektrodenuntersuchungen zur Konvergenz vestibulärer und retinaler Afferenzen an einzelnen Neuronen des optischen Cortex der Katze. *Pflügers Arch. ges. Physiol.*, 1960, **270**, 227–238.

Grüsser, O.-J., U. Grüsser-Cornehls, and G. Saur. Reaktionen einzelner Neurone im optischen Cortex der Katze nach elektrischer Polarisation des Labyrinths. *Pflügers Arch. ges. Physiol.*, 1959, **269**, 593–612.

Grüsser, O.-J., and A. Grützner. Neurophysiologische Grundlagen der periodischen Nachbildphasen nach kurzen Lichtreizen. *Graefes Arch. Ophthalmol.*, 1958a, **160**, 65–93.

Grüsser, O.-J., and A. Grützner. Reaktionen einzelner Neurone des optischen Cortex der Katze nach elektrischen Reizserien des Nervus opticus. *Arch. Psychiat. Nervenkr.*, 1958b, **197**, 405–432.

Grüsser, O.-J., and C. Rabelo. Reaktionen einzelner retinaler Neurone auf Lichtblitze. I: Einzelblitze und Blitzreize wechselnder Frequenz. *Pflügers Arch. ges. Physiol.*, 1958, **265**, 501–525.

Grüsser, O.-J., and G. Saur. Monoculare und binoculare Lichtreizung einzelner Neurone im Geniculatum laterale der Katze. *Pflügers Arch. ges. Physiol.,* 1960, **271,** 595–612.

Grützner, A., O.-J. Grüsser, and G. Baumgartner. Reaktionen einzelner Neurone im optischen Cortex der Katze nach elektrischer Reizung des Nervus opticus. *Arch. Psychiat. Nervenkr.,* 1958, **197,** 377–404.

Harms, H., and E. Aulhorn. Studien über den Grenzkontrast. *Graefes Arch. Ophthalmol.,* 1955, **157,** 3–23.

Hartline, H. K. The response of single optic nerve fibers of the vertebrate eye to illumination of the retina. *Amer. J. Physiol.,* 1938*a,* **121,** 400–415.

Hartline, H. K. The discharge of impulses in the optic nerve of pecten in response to illumination of the eye. *J. cell. comp. Physiol.,* 1938*b,* **11,** 465–477.

Hartline, H. K. The receptive fields of optic nerve fibers. *Amer. J. Physiol.,* 1940*a,* **130,** 690–699.

Hartline, H. K: The effects of spatial summation in the retina on the excitation of the fibers of the optic nerve. *Amer. J. Physiol.,* 1940*b,* **130,** 700–711.

Hartline, H. K. Inhibition of activity of visual receptors by illuminating nearby retinal areas in the Limulus eye. *Fed. Proc.,* 1949, **8,** 69.

Hartline, H. K., and F. Ratliff. Inhibitory interaction of receptor units in the eye of Limulus. *J. gen. Physiol.,* 1957, **40,** 357–376.

Helmholtz, H. von. Die neueren Fortschritte in der Theorie des Sehens (1868). In *Vorträge und Reden,* Vol. I. 4 ed. Braunschweig: Vieweg, 1896*a.* Pp. 265–365.

Helmholtz, H. von. *Handbuch der physiologischen Optik.* 2 ed. Hamburg and Leipzig: G. Voss, 1896*b.*

Hering, E. *Zur Lehre vom Lichtsinne.* Vienna: Gerold, 1878.

Hering, E. *Grundzüge der Lehre vom Lichtsinn.* Berlin: Springer, 1920.

Hering, E. *Wissenschaftliche Abhandlungen.* Leipzig: Thieme, 1931.

Hernández-Peón, R., C. Guzmán-Flores, M. Alcaraz, and Y. Fernández-Guardiola. Sensory transmission in visual pathway during "attention" in unanesthetized cats. *Acta neurol. latino-amer.,* 1957, **3,** 1–7.

von Holst, E. Zentralnervensystem und Peripherie in ihrem gegenseitigen Verhältnis. *Klin. Wschr.,* 1951, **29,** 97–105.

von Holst, E., and H. Mittelstaedt. Das Reafferenzprinzip. *Naturwissenschaften,* 1950, **37,** 256–272.

Hubel, D. H. Cortical unit responses to visual stimuli in nonanesthetized cats. *Amer. J. Ophthalmol.,* 1958, **46,** 110–122.

Hubel, D. H. Single unit activity in striate cortex of unrestrained cats. *J. Physiol.,* 1959, **147,** 226–238.

Hubel, D. H., and T. N. Wiesel. Receptive fields of single neurones in the cat's striate cortex. *J. Physiol.,* 1959, **148,** 574.

Jasper, H. H., and C. Ajmone-Marsan. Thalamocortical integrating mechanisms. *Res. Publ. Ass. nerv. ment. Dis.,* 1952, **30,** 493–512.

Jung, R. Neuronal discharge. *EEG clin. Neurophysiol.,* 1953*a,* **Suppl. 4,** 57–71.

Jung, R. Nystagmographie: Zur Physiologie und Pathologie des optisch-vestibulären Systems beim Menschen. In *Handb. inner. Med.,* Vol. V. Berlin: Springer, 1953*b.* Pp. 1325–1379.

Jung, R. Excitation, inhibition and coordination of cortical neurons. *Exp. Cell Res.,* 1958*a,* **Suppl. 5,** 262–271.

Jung, R.   Coordination of specific and nonspecific afferent impulses ,at single neurons of the visual cortex.   In H. H. Jasper, et al.   (Editors), *Reticular Formation of the Brain*.   Boston: Little, Brown, 1958*b*.   Pp. 423–434.

Jung, R.   Microphysiology of cortical neurons and its significance for psychophysiology.   In *Festschrift Prof. C. Estable. An. Facult. Med. Montevideo*, 1959, **44**, 323–332.

Jung, R.   Korrelationen von Neuronentätigkeit und Sehen.   In R. Jung and H. Kornhuber (Editors), *Neurophysiologie und Psychophysik des visuellen Systems*, Symposion Freiburg.   Berlin, Göttingen, Heidelberg: Springer, 1961.   Pp. 410–434.

Jung, R., R. von Baumgarten, and G. Baumgartner.   Mikroableitungen von einzelnen Nervenzellen im optischen Cortex: Die lichtaktivierten B-Neurone.   *Arch. Psychiat. Nervenkr.*, 1952, **189**, 521–539.

Jung, R., and G. Baumgartner.   Hemmungsmechanismen und bremsende Stabilisierung an einzelnen Neuronen des optischen Cortex: Ein Beitrag zur Koordination corticaler Erregungsvorgänge.   *Pflügers Arch. ges. Physiol.*, 1955, **261**, 434–456.

Jung, R., O. Creutzfeldt, and G. Baumgartner.   Microphysiologie des neurones corticaux: processus de coordination et d'inhibition du cortex optique et moteur.   *Colloq. int. Microphysiol., Centre nat. Rech. sci.*, 1957, No. 67, 411–434.

Jung, R., O. Creutzfeldt, and O.-J. Grüsser.   Die Mikrophysiologie kortikaler Neurone und Ihre Bedeutung für die Sinnes- und Hirnfunktionen.   *Dtsch. med. Wschr.*, 1957, **82**, 1050–1059.

Jung, R., and R. Hassler.   The extrapyramidal motor system.   In *Handbook of Physiology. Neurophysiology* II.   Washington, D. C.: American Physiological Society, 1960.   Pp. 863–927.

Jung, R., and J. F. Tönnies.   .Hirnelektrische Untersuchungen über Entstehung und Erhaltung von Krampfentladungen: Die Vorgänge am Reizort und die Bremsfähigkeit des Gehirns.   *Arch. Psychiat. Nervenkr.*, 1950, **185**, 701–735.

Kappauf, W. E.   Flicker discrimination in the cat.   *Psychol. Bull.*, 1936, **33**, 597–598.

Katsuki, Y., J. Sumi, H. Uchiyama, and T. Watanabe.   Electric responses of auditory neurons in cat to sound stimulation.   *J. Neurophysiol.*, 1958, **21**, 569–588.

Kornmüller, A. E.   Eine experimentelle Anästhesie der äusseren Augenmuskeln am Menschen und ihre Auswirkungen.   *J. Psychol. Neurol.*, 1931, **41**, 354–366.

von Kries, J.   Die Gesichtsempfindungen.   In *Nagels Handbuch der Physiologie des Menschen*.   Braunschweig: Vieweg, 1905.   Pp. 109–282.

Kuffler, S. W.   Discharge patterns and functional organization of mammalian retina.   *J. Neurophysiol.*, 1953, **16**, 37–68.

Kuffler, S. W., R. FitzHugh, and H. B. Barlow.   Maintained activity in the cat's retina in light and darkness.   *J. gen. Physiol.*, 1957, **40**, 683–702.

Li, C.-L.   The facilitation of cortical sensorineuronal responses following stimulation of an unspecific thalamic nucleus.   *J. Physiol.*, 1956, **131**, 115–124.

Li, C.-L., C. Cullen, and H. Jasper.   Laminar microelectrode studies of specific somatosensory cortical potentials.   *J. Neurophysiol.*, 1956, **19**, 111–143.

Magoun, H. W.   An ascending reticular activating system in the brain stem.   *Arch. Neurol. Psychiat.*, 1952, **67**, 145–154.

Mickle, W. A., and H. W. Ades. Rostral projection pathway of vestibular system. *Amer. J. Physiol.*, 1954, **176**, 243–246.

Moruzzi, G. The physiological properties of the brain stem reticular system. In J. F. Delafresnaye (Editor), *Brain Mechanisms and Consciousness.* Oxford: Blackwell, 1954. Pp. 21–48.

Mountcastle, V. B. Modality and topographic properties of single neurons of cat's somatic sensory cortex. *J. Neurophysiol.*, 1957, **20**, 408–434.

Mountcastle, V. B., P. W. Davies, and A. L. Berman. Response properties of neurons of cat's somatic sensory cortex to peripheral stimuli. *J. Neurophysiol.*, 1957, **20**, 374–407.

Phillips, C. G. Intracellular records from Betz cells in the cat. *Quart. J. exp. Physiol.*, 1956, **41**, 58–69.

Purkinje, J. *Beiträge zur Kenntnis des Sehens in subjektiver Hinsicht.* Prague: J. G. Calve, 1819.

Ricci, G., B. Doane, and H. Jasper. Microelectrode studies of conditioning: technique and preliminary results. IV Congr. int. EEG clin. Neurophysiol. *Acta med. Belg.*, 1957, 401–415.

Riggs, L. A., F. Ratliff, J. C. Cornsweet, and T. N. Cornsweet. The disappearance of steadily fixed visual test objects. *J. opt. Soc. Amer.*, 1953, **43**, 495–501.

Sherrington, C. On reciprocal action in the retina as studied by means of some rotating discs. *J. Physiol.*, 1897, **21**, 33–54.

Sherrington, C. *The Integrative Action of the Nervous System.* London: Constable, 1906.

Svaetichin, G. Spectral response curves from single cones. *Acta physiol. scand.*, 1956*a*, **39**, Suppl. 134, 17–46.

Svaetichin, G. Receptor mechanisms for flicker and fusion. *Acta physiol. scand.*, 1956*b*, **39**, Suppl. 134, 47–54.

Tasaki, I., E. H. Polley, and F. Orrego. Action potentials from individual elements in cat geniculate and striate cortex. *J. Neurophysiol.*, 1954, **17**, 454–474.

Walzl, E. M., and V. Mountcastle. Projection of vestibular nerve to cerebral cortex of the cat. *Amer. J. Physiol.*, 1949, **159**, 595–603.

Whitlock, D. G., A. Arduini, and G. Moruzzi. Microelectrode analysis of pyramidal system during transition from sleep to wakefulness. *J. Neurophysiol.*, 1953, **16**, 414–429.

33 F. BREMER

Laboratoire de Pathologie Générale, Université de Bruxelles

Neurogenic Factors Influencing the Evoked Potentials of the Cerebral Cortex

Cortical reactivity has different connotations in neurophysiological literature, and it can be tested in various ways.

For the neurologist, the term alludes to the controllable lability of the human alpha rhythm, and especially to the ease with which this resting rhythm is diffusely disrupted by photic stimulation. A similar electrocorticographic "blocking reaction" characterizes, in the unanesthetized animal, the change from a condition of relaxation or drowsiness to a state of alertness or arousal. This change we know, since Moruzzi and Magoun's (1949) discovery, has as its immediate causal antecedent an increase in the continuous inflow of corticipetal impulses emitted by the mesencephalothalamic ascending reticular system. The psychophysiological importance of this arousal reaction has been emphasized recently by the demonstration that it participates in the neurophysiological events initiating Pavlovian conditioning— the so-called orienting reaction—and also, possibly, in the later phases of the processes that are the basis of temporary associations.

Another aspect of cortical reactivity concerns the experimental variations of the evoked potentials of receiving and association areas that respond to a synchronized volley of afferent impulses. Picked up by gross electrodes, these evoked potentials, which represent the integrated response of myriads of cortical units recorded in a volume conductor, involve various layers of the neocortical mantle that start to respond successively, as has been shown by depth recordings (Amassian, 1953; Amassian, et al., 1959; Tasaki, Polley, and Orego, 1954; Albe-Fessard and Buser, 1955; Perl and Whitlock, 1955; Li, Cullen, and Jasper, 1956a, b; Li, 1956a; Roitback, 1956; Mountcastle, 1957; Mountcastle, Davies, and Berman, 1957; see also Albe-Fessard,

1957; Bremer, 1958a; Rosenblith, 1959b; Purpura, 1959). The amplitude of these evoked potentials is a function of the number and
degree of synchronization of the activated cortical units. The "unit
responses" or unitary contributions are relatively slow potentials.
Their duration and shape, as well as the fact that they are conditionally
associated with a spike discharge, designates them as postsynaptic
(axosomatic and axodendritic?) excitatory potentials.

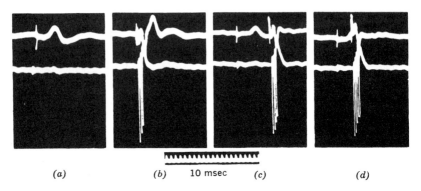

(a)   (b)   10 msec   (c)   (d)

Fig. 1. Simultaneous recording of evoked potential and unit spike discharge in the
somatosensory area. Cat, *cerveau isolé* preparation; monopolar record from the
surface (upper traces) and extracellular microelectrode record at a depth of 0.84
mm under the surface lead. (a) Response to shock on n. centromedianus; (b)
response to shock on n. ventralis posterolateralis; (c and d) facilitation of spike
discharge by preceding unspecific thalamic volley at intervals of 70 and 30 msec,
contrasting with depression of evoked averaging potential recorded from the
surface. (After Li, 1956a.)

It has often been observed that the response of a cortical cell,
recorded by a microelectrode in depth, may show a facilitation of its
axonal discharge, whereas the corresponding surface potential is
reduced or even completely obliterated (Fig. 1).

The deliberate simplicity of the stimulation that provokes the
cortical evoked potentials—a brief natural or electrical stimulus exciting
peripheral receptors or the corresponding afferent pathways—recalls
the use of the reflex twitch or of the ventral-root evoked potential
for the testing of spinal responsiveness. In both cases the primary
response may be followed by an after-discharge; this after-discharge
reveals the participation of secondary reactions of varying complexity,
which involve multisynaptic delayed transmissions and repetitive
responses of interneuronal aggregates, organized in reverberating circuits or endowed with true autorhythmicity. The evoked potential
of receiving and association areas thus provides information on the

first stages of the cortical processes that lead ultimately to perception. For the sake of completeness, a third aspect of cortical responsiveness should be mentioned here: the so-called direct responses of the superficial layers of the neo- and paleocortices to a brief electrical stimulus. Their great theoretical interest is attested by many studies following the first description of the phenomena by Adrian in 1937. However, the discussion of these direct responses is omitted in the present report. The reason for this omission is the uncertainty that still exists concerning the exact mechanism of their production and the poverty of information at hand on the dynamic factors of their experimental variations. For similar reasons the corticifugal discharges that accompany these potentials (see, for instance, Brookhart and Zanchetti, 1956; Zanchetti and Brookhart, 1958; Purpura, 1959) will also not be considered, although such discharges, in both their early and late components, constitute obviously valuable expressions of cortical reactivity.

My analysis will thus be concerned with the neurogenic factors that influence the evoked potentials of the mammalian cortex. Even with such a limitation, the scope of the topic remains a very broad one, and one that is obviously impossible to cover completely within the limits of this report.

It has long been known that, when recorded in an unanesthetized animal with intact encephalon, and even in an anesthetized animal, the evoked responses of a receiving area to an invariant testing stimulus may show wide variations in amplitude and waveform, although the conditions of the external and internal milieu are carefully controlled.

As Rosenblith (1959b) has remarked, this variability of evoked responses to identical stimuli is the more surprising in that the gross electrode, which records the activity of reasonably large neural populations, is already performing an averaging process. These threshold fluctuations of excitability, affecting simultaneously a great number of central units, have been described as correlated changes by Hunt (1955) and Frishkopf and Rosenblith (1958). These are distinct from local random changes, such as those related to the thermal fluctuations of ionic concentration (cf. Fatt and Katz, 1952; Fatt, 1954). The factors that have proved important in the correlated changes of cortical reactivity can be listed under the following headings: (1) postreactional modifications of excitability associated with the recovery cycle of thalamic and cortical neurons; (2) posttetanic potentiation; (3) inhibitory processes; (4) temporal summation of homosynaptic subliminal impulses; (5) spatial (heterosynaptic) summation of homologous subliminal impulses and occlusive convergence of supraliminal impulses; (6) heterosynaptic convergence of heterologous impulses.

Recovery Cycle

In the unanesthetized animal (intact cat or *"cerveau isolé"* preparation) the recovery cycle of the response of the visual area, studied by applying a pair of shocks to the radiations (Fig. 2*a*) shows an early phase, of about 25 msec duration, characterized by a progressive accentuation of the postreactional depression (Clare and Bishop, 1952; Schoolman and Evarts, 1959). Obviously, during this phase, a facilitatory recruitment counteracts cortically the refractoriness of the neurons that have been activated by the synchronized first afferent volley. As in the spinal cord, this early facilitation has the duration of a postsynaptic potential and may be explained by the summation of two membrane depolarizations. It may be an overt phenomenon, especially when the pair of stimuli follows immediately after a tetanic stimulation. As shown by the curves of Fig. 2*b*, this early facilitation is selectively sensitive to barbiturate anesthesia. The different components of the evoked potential do not generally recover simultaneously, since the surface-negative potential (which is conventionally interpreted as the response of the superficial dendritic feltwork) is more vulnerable than the surface-positive potential, which is thought to represent the axosomatic phase of the response. A late peak of facilitation may be observed in the region between 200 and 300 msec. Its significance is still uncertain (cf. also Marshall, 1949; Clare and Bishop, 1956). Obviously, it does not correspond to the supernormal phase that, in nerve and muscle fibers, is associated with a negative after-potential. No clear evidence of this supernormal phase of excitability in neurons of the central nervous system has been available since the paper of Lorente de Nó and Graham (1938) asserting its absence in the rabbit's oculomotor neurons. Besides, the intracellular recordings made by a number of authors failed to show a large long-lasting negative after-potential (see, however, recent observations by Kandel, et al., 1959, on hippocampal neurons). The condition for a brief phase of supernormal excitability could, however, appear when a neuron is made to conduct a train of impulses and recovers by developing a large positive after-potential. The depression of excitability accompanying the latter could be followed by a phase of supernormal excitability, as has been shown by Lorente de Nó and Laporte (1950) for the cervical ganglion of the turtle (cf. also Bremer, 1960).

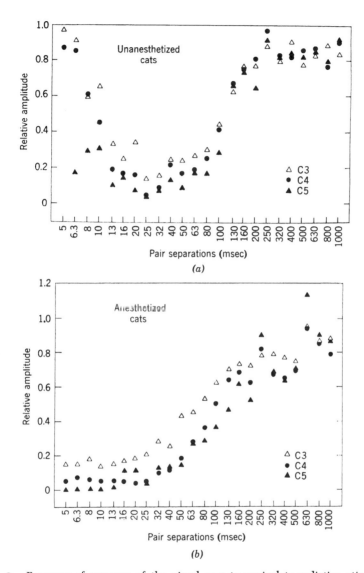

Fig. 2. Recovery of response of the visual area to geniculate radiation stimulation in the cat. Ordinate, relative amplitude (ratio of amplitude of test response to that of control response) of various components (C3, C4, C5) of the response of the visual cortex; abscissa, interval between conditioning and test stimuli plotted in milliseconds (log scale); means of results obtained on nine cats, (a) unanesthetized, (b) anesthetized with pentobarbital (30 mg/kg). (After Schoolman and Evarts, 1959.)

Posttetanic Potentiation

The phenomenon of posttetanic potentiation in cerebral responses has only recently been described. Its importance seems to be less than at the spinal level. In the response of the lateral geniculate body to optic-nerve impulses, studied by Hughes, Evarts, and Marshall (1956), it appears to be related to an increase in the so-called subliminal fringe (cf. also Marshall, 1949). The transient potentiation of the presynaptic-tract spike was not the determining factor of the phenomenon. As is the case for the posttetanic potentiation of spinal reflexes and of the neuromuscular transmission (see Eccles, 1957; Hughes, 1958), the mechanism of the phenomenon is still uncertain. A plausible assumption is that it involves a transitory modification of the process of chemical transmission at the synapse. Hughes and Rosenblith (1957) have concluded from experiments on cats that auditory sensitization after low-tone exposure may be a posttetanic potentiation occurring at the junction of the hair cells and the auditory nerve.

Inhibition

Surprisingly, evidence for the existence of active inhibition at the cortical level is still rather scanty. As pointed out by Mountcastle in this symposium, the distinction has not always been made with sufficient clearness between the cortical and the subcortical processes responsible for suppressions of cortical activities. The best indications for the action of an inhibitory process at the cortical level come from microphysiological studies in which the complication of a subcortical factor has been eliminated. The work of Tasaki, Polley, and Orrego (1954), Creutzfeldt, Baumgartner, and Schoen (1956), Li (1956b), Branch and Martin (1958), Phillips (1959), and Asanuma and Okamoto (1959) may be quoted in this context. On rare occasions (Tasaki, Polley, and Orrego, 1954; Asanuma and Okamoto, 1959), the indication has been obtained that the suppression of repetitive discharge of a cortical neuron was associated with an increase in the resting membrane potential of the cell. This recalls the description by Eccles and his associates (see Eccles, 1957) of the postsynaptic inhibitory potential of the spinal motoneuron. The evidence on the influence of inhibitory processes on evoked potentials at the pial surface is based either on pharmacological data (Purpura and Grundfest, 1956, 1957; Purpura, 1959); or on the study of the recovery cycle of the cortical response (Fleming, Huttenlocher, and Evarts, 1959; Posternak, Fleming, and Evarts, 1959). The evidence is inferential and does not yet admit of firm conclusions (cf. Brazier, 1954; Whitlock, Arduini, and Moruzzi, 1953).

A significant finding in the microphysiological experiments of Branch and Martin (1958) was the fact that the inhibitory suppression of the spontaneous discharge of a Betz cell in the motor area of the cat was not accompanied by a reduction in the responsiveness of the same cell to a volley of antidromic impulses. It was as if the inhibitory process, whatever its mechanism, had affected cortical inter-neurons, whose activity was necessary for the maintenance of the "spontaneous" discharge.

On the other hand, recent experiments on the cat by Phillips (1959), Brooks (1959), and Morrell (1959) have led these authors to the conclusion that repetitive antidromic volleys in the cat's pyramidal tract inhibit actively the discharges of pyramidal cells, which in Brooks's observations were evoked by electrical stimulation of the ipsilateral cortex or of the contralateral foot. It has been suggested that this inhibition is mediated by recurrent axon collaterals of pyramidal cells, and the phenomenon has been compared with spinal recurrent inhibition (the Renshaw phenomenon). It should be mentioned, however, that in observations by Patton and Towe (quoted by Patton and Amassian, 1960), made on cats anesthetized with chloralose, "in no instance was antidromic invasion followed by depressed excitability beyond that expected from refractoriness."

The study of the activity pattern of single hippocampal pyramidal cells has disclosed a type of inhibition in this cortex, which has already been described by Granit and Phillips (1956) in the cerebellar cortex: the "inactivation process." It is characterized by considerable de-polarization of the nerve cell, leading to initial excitation, followed by inactivation of spike generation, and finally by a hyperpolarization (von Euler and Green, 1960a, b). The physiological significance of the phenomenon is still uncertain. However, it seems to be involved in the mechanism of the "theta" hippocampal and diencephalic rhythm, interpreted as a relaxation-oscillation phenomenon.

Temporal and Spatial Summation of Functionally Homologous Impulses

The physiological significance of temporal summation (see Fig. 3) and of facilitatory convergence of homologous impulses needs no long comment. The "spatial" summation is probably the more important, for reasons which have been given by Eccles (1953).

A particularly interesting case of the phenomenon had seemed to be represented by the tonic facilitation exerted on receiving cortical areas, especially the visual area, by the constant inflow of subliminal afferent impulses. This facilitation, which had already been noticed by Marshall, Talbot, and Ades (1943), has been the subject of a special study by Chang (1952), whose observations have been confirmed by Malis and Kruger (1956), Bremer and Stoupel (1956), and Schoolman

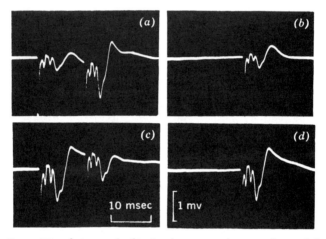

Fig. 3. Competition between facilitation by temporal summation and facilitation by reticular stimulation. Cat, *encéphale isolé;* monopolar surface recordings: responses to one or a pair of geniculate shocks at an interval of 11 msec. (*a* and *b*) Facilitation of the response to the second shock by temporal summation; (*c* and *d*) facilitation of the response by reticular arousal and simultaneous disappearance of facilitation by temporal summation, indicating that the neurons recruited by the two facilitatory processes belonged to the same subliminal fringe.

and Evarts (1959). The facilitatory process takes place in both the lateral geniculate body and the cortical networks, with the latter location probably the more important (Schoolman and Evarts, 1959). However, the exact mechanism of the phenomenon is still uncertain. Recent work by Arduini and Hirao (1960) tends to interpret it as being, at least partly, a release phenomenon, resulting from the suppression of the "dark discharge" of the retina.

The potentiation of the response develops progressively to a maximum in about 5 seconds and is sustained at a high level as long as the illumination is maintained. The onset and the cessation of retinal illumination are both followed immediately by a periodic variation of excitability which lasts for about a second and yields a curve of gradually damped oscillations. A similar cyclical sequence of events has been described by Gastaut, et al. (1951), Fleming and Evarts (1959), and Marshall (1958) in experiments in which a brief and intense photic flash was followed by test shocks at various intervals. An explanation of these cyclical fluctuations of the cortical response is suggested by the fact that, in the cat's auditory area, a similar periodic variation of cortical excitability often follows the response to a click or to a brief direct electrical stimulus (Chang, 1950, 1951; Bremer and Bonnet, 1950; Rosenzweig and Rosenblith, 1953). These papers show that the phases of increased cortical responsiveness were temporally— and probably causally—related to the thalamocortical rhythmical volleys and corresponding cortical waves that followed the initial stimulus. Similar rhythmical corticipetal volleys, which exert a facilitatory action on cortical responsiveness but remain subliminal, might explain the periodic fluctuation of cortical excitability in the visual area.

The potentiation (see Bremer, 1958b) exerted by callosal impulses on the responses of neocortical receiving areas to a volley of thalamo-cortical impulses can be considered as belonging also to the category of facilitatory convergence of functionally homologous impulses, on account of the general homotopy of the connections established by the neocortical commissure between left and right hemi-spheres.

An interesting aspect of callosal facilitation is that it does not always affect the over-all amplitude of the positive-negative evoked potential (as was the case in Figs. 4a and b) but may equally well change its waveform drastically. The modification (Figs. 4e and f) is always in the direction of a relatively greater increase in the surface-negative phase, sometimes as here at the expense of the positive phase. This recalls the cortical effect of stimulating drugs. As an explanation for this selective action of callosal facilitation on the negative phase of the sensory response, it has been suggested (Bremer, 1958b) that the callosal afferents, by their distribution to all cortical layers, including the superficial ones, could exert a special sensitizing effect on the dendritic component of the evoked potential.

The mutual facilitation mediated by commissural fibers between homologous receiving areas should be one of the accessory factors that contribute to interhemispheric bio-electrical synergy (Bremer and Stoupel, 1957).

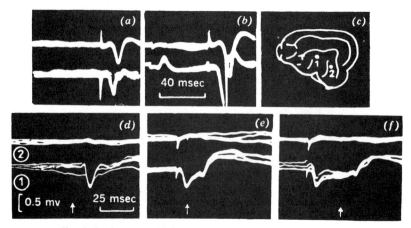

Fig. 4. Callosal facilitation of click responses. Cats, *encéphale isolé*. Monopolar surface leads on auditory area I (lower trace) and posterior ectosylvian, epi (upper traces). In each figure five successive sweeps are superimposed. (*a*) Click response alone; (*b*) click response conditioned by an electric shock to point on right auditory area I symmetrical to point 1 of (*c*); (*d*) another cat, click response (time of click stimulation indicated by arrow); (*e* and *f*) conditioning of response by a callosal volley, as in (*b*), at two different intervals. Note the homotopy of the callosal response; the potentiation in (*b*) of the response of epi, which is relayed by auditory area I; the striking modification of the shape of the facilitated click response in (*e*) and (*f*). (After Bremer, 1958*b*.)

Reticular Facilitation and Depression of Evoked Potentials

The importance of the convergence of heterologous impulses at thalamic and cortical levels, in cerebral dynamics, has been demonstrated in recent experimental work. This work stems from Moruzzi and Magoun's (1949) work on the reticular arousal mechanism. Their original contribution, as well as the numerous studies which followed it (see Magoun, 1958), had left unresolved a puzzling paradox: the fact that cortical primary and secondary responses to various sensory stimuli (photic flash, click, brief tactile or electrical stimulation of the skin) are generally reduced and sometimes markedly so (Figs. 5*a* and *b*; see also Figs. 9*e* to *i*) in reticulocortical arousal. In this arousal the spontaneous cortical activity reveals unquestionable activation, which is demonstrated by both surface recordings and microelectrode depth recordings, as in the experiments performed by Akimoto and Creutzfeldt (1958) in Jung's laboratory.

This was the situation when Gauthier, Parma, and Zanchetti (1956) made the interesting observation that the evoked potentials of the cat's somatosensory area are not depressed in reticular arousal when they are obtained in response to an electric shock applied on the thalamic relay nucleus. However, potentials of similar shape and amplitude evoked in the same area by brief skin stimulations show the usual inhibition.

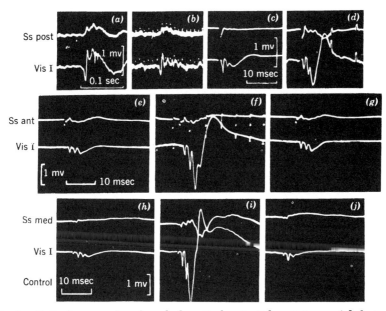

Fig. 5. Reticular arousal and evoked cortical potentials. Cats, *encéphale isolé* preparation; monopolar surface recordings, negativity upward. (*a*) Responses of the visual area and suprasylvian gyrus to a photic flash (pupil dilated); (*b*) response to the same stimulus during a reticular arousal produced by the stimulation (pulses at 100 per second, 0.3 msec duration; 1 volt) of the mesencephalic reticular formation; (*c* and *d*) same animal and same reticular stimulation, but test stimulus is now delivered to lateral geniculate body; note faster time scale. (*e* and *g*) Control responses of the visual and suprasylvian gyrus to a geniculate shock; (*f*) response recorded during a brief repetitive stimulation of n. centromedianus. (*h*) Responses of the visual area and suprasylvian gyrus to a geniculate shock; (*i* and *j*) responses to the same shock conditioned by a brief stimulus (5 shocks at 500 per second) to the mesencephalic reticular formation; (*i*) 60 msec and (*j*) 400 msec after end of reticular stimulation. Note appearance of a response of the association area during reticular arousal. (After Bremer and Stoupel, 1959*a*.)

Recent researches by Dumont and Dell (1958) and in our laboratory (Bremer and Stoupel, 1958, 1959*a*, *b*) conducted independently, have revealed a much more striking difference between the effects of cerebral arousal on the evoked potentials produced by "peripheral" and those produced by "central" stimulations. These experiments were performed on *encéphale isolé* cats. This preparation provides various advantages for such studies. One can mention the absence of systemic circulatory modifications during the arousal reactions (see Ingvar and Soderberg, 1958) and the functional stability of the brain, maintained over many hours.

Dumont and Dell recorded the response of the visual area to electrical pulses applied on the optic chiasma. In our experience the effects of reticular arousal have been studied comparatively on the responses of the visual, auditory, somatosensory, and association areas to brief stimuli applied either on the peripheral receptors or on cerebral structures (shock on the optic nerve or on the appropriate thalamic relay nucleus). Both mesencephalic-reticular and thalamic-reticular stimulation resulted in a striking facilitation of the evoked potentials produced by the "central" stimulus (Figs. 5c to g; see also Figs. 9a to c). This potentiation contrasted with the depression of the response in the same receiving area when it was evoked in the same experiment by a "peripheral" testing stimulus (Figs. 5a and b) and also with the disappearance of the late phases of the response. The surface-positive and surface-negative components of the evoked potential are equally increased in monopolar recordings. A brief train of shocks, and even a single shock, provided it precedes the test shock at an appropriate interval, may be as effective as the fast repetitive stimulation of the mesencephalic or thalamic-reticular location (Figs. 5h to j). As can be seen on the tracings of Figs. 5d and i, the so-called secondary responses recorded from association areas, especially the suprasylvian gyrus in the cat (see Buser, 1957), are also potentiated, and sometimes markedly so, in reticular arousal. Another noticeable characteristic of the phenomenon is that a reticular stimulus that is just supraliminal, judged on the bases of its electrocorticograph and ocular effects, exerts a facilitatory effect on all receiving areas of the neocortical convexity (see Bremer and Stoupel, 1959a, Fig. 10).

The potentiation of the responses coincides in the *encéphale isolé* cat, with the behavioral (ocular) signs of arousal. Besides, it characterizes similarly, in the same preparation, sensory awakening and arousal resulting from an injection of amphetamine (Bremer and Stoupel, 1959b). Barbiturate anesthesia abolishes it, but it can be observed in the condition of lasting sleep produced by a mesencephalic transection (stimulation of n. centromedianus).

The seat of the neural recruitment responsible for the increase in the size of the response is situated in both the thalamic relay nucleus and the cortical networks, as shown by simultaneous recordings of lateral-geniculate and cortical potentials in response to shock to the optic nerve (Figs. 6a to e). The thalamic contribution to the phenomenon seems, however, to be less important than the cortical one. Evidence of a thalamic factor may even be entirely lacking (Figs. 6f and g). Reticular facilitation still increases the surface-positive (axosomatic?) component of the evoked potential after its surface-negative

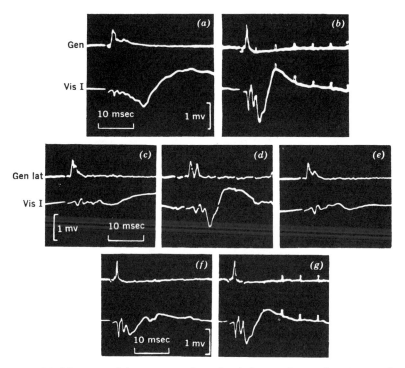

Fig. 6. Modifications of the responses from the thalamic relay nucleus in reticular arousal. Cat, *encéphale isolé*, three experiments; upper traces, bipolar records from lateral geniculate body; lower traces, monopolar records from visual area I, responses to shocks to contralateral optic nerve. (*b, d, g*) Responses facilitated by mesencephalic stimulation.

(axodendritic?) phase has been completely abolished as a consequence of the local application of a depressant. The true amplitude and duration of this positive phase is then revealed in the facilitated response (Fig. 7).

Although facilitation of the evoked potentials is normally associated with the desynchronization of the spontaneous electrical activity of the cortex, this other manifestation of arousal cannot be considered to be the condition of the potentiation. This is indicated by various

experimental data.   Potentiation remains quite obvious and may even
be increased in conditions, like chloralose anesthesia (Dell, 1960)
and atropine intoxication, where the electrocorticogram is markedly
and irreversibly slowed.   It can still be observed in complete absence

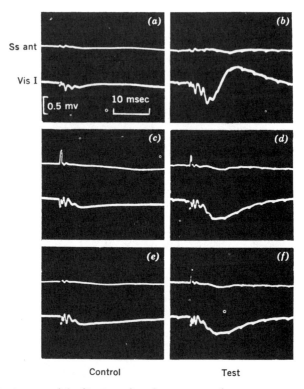

Fig. 7.   Persistence of facilitation of surface-positive slow component of cortical
response (Vis I) after elimination of surface-negative component. Cat, *encéphale
isolé;* test stimulus to geniculate body; arousal by repetitive stimulation of
mesencephalic reticular formation.   (*a* and *b*) Initial control and facilitated
responses of visual area (lower traces); (*c* to *f*) control and facilitated responses
during abolition of their surface-negative component after local application of
Nembutal. (Bremer and Stoupel, 1959*a*.)

of cortical spontaneous activity.   One even has the impression that
electrocortical desynchronization tends to reduce it.

A phenomenon of such regularity and magnitude deserves consideration.   It raises several more-or-less related questions.   The first concerns the neurophysiological mechanism of the potentiation.   Its basis
should be heterosynaptic spatial summation of unspecific—reticulo-
and thalamocortical—and specific—mainly sensory—impulses.

A study by Hunt (1955) of the facilitation of a monosynaptic spinal reflex in the cat may perhaps be of help in understanding the striking efficiency of a heterosynaptic convergence. Hunt has shown that a volley of afferent impulses from the nerve to lateral gastrocnemius soleus reaching the medial gastrocnemius motoneurons has by itself a very feeble transmitter potentiality. However, it exerts a powerful facilitation on the response of these motoneurons to a homonymous (medial gastrocnemius) volley. The author suggests that the remarkable efficiency of a heteronymous volley conditioning a homonymous one may find its explanation in the interspersed distribution of the synaptic knobs activated by the two kinds of afferent impulses that converge on the synaptic membrane of the tested motoneurons (cf. Lorente de Nó, 1938). If the same explanation should hold for the powerful facilitation of thalamic and cortical specific responses by ascending reticular impulses, the phenomenon would represent a special, but particularly important, case of the utilization by nature of a general principle of synaptic organization.

Whatever the explanation, reticular facilitation of the cortical evoked potentials is undoubtedly the expression of a neuronal recruitment from the subliminal fringe This can be shown by experiments in which reticular arousal is combined with other mechanisms of dynamogenesis, such as photic potentiation or temporal summation of two volleys of specific impulses. Such experiments (see Fig. 3) indicate that these three mechanisms of facilitation involve the same group of available elements in the population of cortical neurons. In the same context, reticular stimulation may magnify strikingly— and reversibly—the effect of a local application of strychnine at the receiving area (Fig. 8).

Psychophysiological Implication of Reticulocortical Facilitation

From a functional (psychophysiological) point of view, one may ask first with Rosenblith (1959a, b) whether the results of experiments made on an *encéphale isolé* or curarized preparation, and making use of highly artificial stimuli, have a bearing on the response of the intact animal to natural sensory stimuli. Before discussing this point, however, I must consider the serious difficulty that is raised by the difference in the effect of the same reticular arousal on the cortical responses evoked by what may be called "peripheral" and "central" stimulation. An important point should be mentioned at the outset. The reduction of the visual response to photic stimuli and the auditory

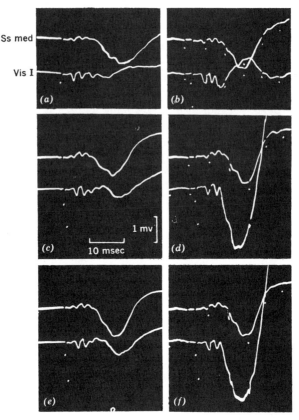

Fig. 8. Additive effects of local strychnine and reticular facilitation. Cat, *encéphale isolé*, responses of the middle suprasylvian gyrus and adjacent lateral gyrus (visual area) to shocks to the contralateral optic nerve. (*a* and *b*) Before, and (*c* to *f*) after the application of 0.25-per-cent solution of strychnine sulphate to the visual area; note that the effect of strychnine on the response of the visual cortex, which is barely visible in (*c*) and (*e*), becomes spectacular during each stimulation of the mesencephalic reticular formation; also that reticular facilitation represents here essentially a cortical process, as shown by the very slight increase in the radiation spike (spike 1) in traces (*b*), (*d*), and (*f*). (After Bremer and Stoupel, 1959a.)

response to clicks during reticular arousal cannot be accounted for by a subcortical blockade of the afferent specific impulses. The marked reduction of the cortical evoked potential in our experiments on clicks (which confirm Desmedt and La Grutta, 1957) is in contrast with the insignificant, or even complete absence of, modification of the response of the thalamic relay nucleus, recorded at the same time. This fact seems to leave only two possible explanations of the reduction, during reticular arousal, of the responses of the receiving areas to afferent impulses when these responses are evoked by a brief stimulation of peripheral receptors: either an active (true) inhibition of intracortical synaptic transmission, or the blocking of the specific impulses by occlusive convergence on common cortical neurons.

The first hypothesis encounters the a priori difficulty that one does not discern the functional significance that a diffuse inhibition of cortical operations could have in cerebral arousal. The second explanation assumes that the volley of corticipetal impulses evoked by the peripheral stimulus is unable, on account of its insufficient density, to overcome the relative refractoriness of cortical interneurons that are supraliminally activated by the reticular impulses. The highly synchronized volley evoked by the stimulus applied to either the optic nerve or the thalamic relay nucleus should be capable of overcoming the refractoriness of these units. To this would be added the response of the neurons recruited from the subliminal fringe. The ultimate result of the reticular conditioning would thus depend on the relative proportion of these three groups of cells in the cortical-neuron population. Some experimental data can be quoted in favor of this hypothesis.

When the response potential of the auditory area evoked by a click is reduced in reticular arousal, a modified waveform often indicates that the cortical reactional processes are accelerated in a way that recalls the modification characteristic of the facilitation of the response to a medial-geniculate shock. (Compare Figs. 9d to i with a to c.) Furthermore, when, as is the case in atropine intoxication, the electrocortical activation effect of the reticular stimulation has been abolished or greatly reduced, a potentiation of the response of the auditory area to a click can be observed in arousal. This observation recalls the experiments of Gellhorn, Koella, and Ballin (1954, 1955) who found that, in the lightly anesthetized but otherwise intact cat, painful peripheral or hypothalamic stimulation results in a potentiation of the cortical responses. Other evidences of reticular dynamogenesis of cortical potentials evoked by physiological sensory stimulation has been given by experiments in which a photic stimulus is repeated at frequencies exceeding 5 per second. This potentiation, which had been already demonstrated by the microphysiological observations of Creutzfeldt and Grüsser (1959; cf. also Creutzfeldt and Akimoto, 1958) has recently been described for the evoked potentials of the visual area (Steriade and Demetresco, 1960). In our own confirmatory observations, also made on *encéphale isolé* cats, the reversal of the reticular arousal effect on the response to the flickering light was clearly related to the greater synchrony of thalamocortical volleys (evidenced by the shape and amplitude of the geniculate potential), resulting from the rhythmic repetition of the stimulus. As it is for

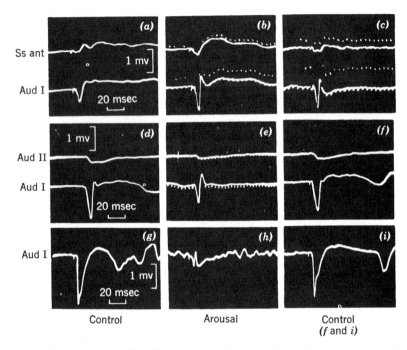

Fig. 9. Comparison of the effects of reticular arousal on the responses of the auditory area to geniculate shock and click. Cats, *encéphale isolé.* (*a*) Response to a medial geniculate shock; (*b* and *c*) responses to the same stimulus during repetitive stimulation of n. ventralis anterior for two different voltages; note the smaller facilitatory effect of the higher voltage; (*d* and *f*) controls, and (*e*) depression of click response during stimulation of the mesencephalic reticular formation; (*g* to *i*) similar depression (in *h*) of the click response during the arousal produced by a voice call.

the potentiation of the response to a geniculate stimulus, the seat of the facilitatory effect is mainly at the cortical level (cf. Brookhart, et al., 1957).

From the evidence at hand it seems that the fully waking brain is the seat of two physiological processes of opposite result. The one tends to reduce the number of cortical neurons that react in the receiving area to a peripheral stimulus, as a consequence of the desynchronization-activation of the cortical networks by the ascending reticular impulses. This depressive effect can be overcome in artificial experimental conditions in which the testing stimulus is applied more centrally and thus evokes a highly synchronized volley of afferent impulses. Such experiments permit the demonstration of a tendency toward a powerful facilitation of cortical operations in arousal. But even in responses produced by stimuli applied to peripheral receptors, the presence of this facilitatory process can be disclosed by the waveform of the evoked potentials as well as by the shortening of the refractory phase in the response of the visual cortex to a pair of flashes (Lindsley, 1958). Fuster's (1958) elegant psychophysiological experiments demonstrate a speeding up of visual recognition by the intact monkey during reticular arousal (see also Lindsley, 1958). Simpler psychophysiological experiments, on the other hand, show that auditory thresholds are better in the relaxed (almost drowsy) subject than in the aroused one (Rosenblith, unpublished experiments).

The selective suppression of the activation of nonrelevant cortical esthesioneurons, together with the promotion of the neurons tuned to the particular quality of the sensory stimulus or to its local sign, could be an important mechanism by which generalized thalamocortical awakening, or less diffuse cerebral arousals alerting to attention, would permit recognition of the stimulus pattern and allow adaptive behavioral responses.

Summary

The aim of this report was to survey some of the neurogenic factors that influence the electrical responses evoked in the neocortical gray matter by a brief volley of afferent impulses. Because of the importance and immediacy of the question, special emphasis has been given to heterosynaptic convergences of specific sensory and unspecific reticular impulses at the thalamic and cortical levels.

Like the mono- and polysynaptic reflex responses of spinal motoneurons to single electrical stimuli, the evoked potentials of the cerebral cortex are obviously the artificial product of experiments devised for analytical purposes. Yet the study of the factors of variation of these integrated cortical responses throws light on the physiological properties of the neuronal networks of the brain. By doing so, it represents an approach to the knowledge of the primary neural events, which, initiating an immensely complicated chain of neurophysiological processes, underlie perceptual integration.

References

Adrian, E. D.   Spread of activity in cerebral cortex.   *J. Physiol.*, 1937, **88**, 127–161.

Akimoto, H., and O. Creutzfeldt.   Reaktionen von Neuronen des optischen Cortex nach elektrischer Reizung unspezifischer Thalamuskerne.   *Arch. Psychiat. Nervenkr.*, 1958, **196**, 494–519.

Albe-Fessard, D.   Activités de projection et d'association du néocortex cérébral des mammifères. Les projections primaires.   *J. Physiol.*, 1957, **49**, 521–588.

Albe-Fessard, D., and P. Buser.   Activités intracellulaires recueillies dans le cortex sigmoïde du chat: participation des neurones pyramidaux au potentiel évoqué somesthésique.   *J. Physiol.*, 1955, **47**, 67–69.

Amassian, V. E.   Evoked single cortical unit activity in somatic sensory areas.   *EEG clin. Neurophysiol.*, 1953, **5**, 415–438.

Amassian, V. E., E. Berlin, J. Macy, and H. J. Waller.   Simultaneous recording of the activities of several individual cortical neurons.   *Trans. N. Y. Acad. Sci.*, 1959, **21**, 395–405.

Arduini, A., and Hirao, T.   Enhancement of evoked responses in the visual system during reversible retinal activation.   *Arch. ital. Biol.*, 1960, **98**, 182–205.

Asanuma, H., and K. Okamoto.   Unitary study on evoked activity of callosal neurons and its effect on pyramidal tract cells.   *Jap. J. Physiol.*, 1959, **9**, 473–483.

Branch, C. L., and A. R. Martin.   Inhibition of Betz cell activity by thalamic and cortical stimulation.   *J. Neurophysiol.*, 1958, **21**, 380–390.

Brazier, M. A. B.   The action of anesthetics on the nervous system.   In J. F. Delafresnaye (Editor), *Brain Mechanisms and Consciousness*. Oxford: Blackwell, 1954. Pp. 162–199.

Bremer, F.   Cerebral and cerebellar potentials.   *Physiol. Rev.*, 1958*a*, **38**, 357–388.

Bremer, F.   Physiology of the corpus callosum.   *Res. Publ. Ass. nerv. ment. Dis.*, 1958*b*, **36**, 424–448.

Bremer, F.   Mécanismes de facilitation des transmissions nerveuses centrales. *Actualités Neurophysiologiques*, 2 ser. Paris: Masson, 1960.

Bremer, F., and V. Bonnet.   Interprétation des réactions rythmiques prolongées des aires sensorielles de l'écorce cérébrale.   *EEG clin. Neurophysiol.*, 1950, **2**, 389–400.

Bremer, F., and N. Stoupel.   Interprétation de la réponse de l'aire visuelle à une volée d'influx sensoriels.   *Arch. int. Physiol.*, 1956, **64**, 234–248.

Bremer, F., and N. Stoupel.   Etude des mécanismes de la synergie bioélectrique des hémisphères cérébraux.   *Acta physiol. pharmacol. néerl.*, 1957, **6**, 487–496.

Bremer, F., and N. Stoupel.   De la modification des réponses sensorielles corticales dans l'éveil réticulaire.   *Acta neurol. psychiat. belg.*, 1958, **58**, 401–403.

Bremer, F., and N. Stoupel.   Facilitation et inhibition des potentiels évoqués corticaux dans l'éveil cérébral.   *Arch. int. Physiol.*, 1959*a*, **67**, 240–275.

Bremer, F., and N. Stoupel.   Etude pharmacologique de la facilitation des réponses corticales dans l'éveil réticulaire.   *Arch. int. Pharmacodyn.*, 1959*b*, **122**, 234–238.

Brookhart, J. M., and A. Zanchetti. The relation between electrocortical waves and responsiveness of the corticospinal system. *EEG clin. Neurophysiol.,* 1956, **8,** 427–444.

Brookhart, J. M., A. Arduini, M. Mancia, and G. Moruzzi. Mutual facilitation of cortical responses to thalamic stimulation. *Arch. ital. Biol.,* 1957, **95,** 139–146.

Brooks, V. B. Recurrent inhibition. *EEG clin. Neurophysiol.,* 1959, **11,** 614.

Buser, P. Activités de projection et d'association du néocortex cérébral des mammifères. I. Activités d'association et d'élaboration. Projections non spécifiques. *J. Physiol.,* 1957, **40,** 589–596.

Chang, H. T. The repetitive discharge of cortico-thalamic reverberating circuit. *J. Neurophysiol.,* 1950, **13,** 235–258.

Chang, H. T. Changes in excitability of cerebral cortex following a single electrical shock applied to the cortical surface. *J. Neurophysiol.,* 1951, **14,** 95–112.

Chang, H. T. Functional organization of central visual pathways. *Res. Publ. Ass. nerv. ment. Dis.,* 1952, **30,** 430–453.

Clare, M. H., and G. H. Bishop. The intracortical excitability cycle following stimulation of the optic pathway in the cat. *EEG clin. Neurophysiol.,* 1952, **4,** 311–320.

Clare, M. H., and G. H. Bishop. Potential wave mechanisms in cat cortex. *EEG clin. Neurophysiol.,* 1956, **8,** 582–602.

Creutzfeldt, O., and H. Akimoto. Konvergenz und gegenseitige Beeinflussung von Impulsen aus der Retina und den unspezifischen Thalamuskernen an einzelnen Neuronen des optischen Cortex. *Arch. Psychiat. Nervenkr.,* 1958, **196,** 520–538.

Creutzfeldt, O., G. Baumgartner, and L. Schoen. Reaktionen einzelner Neurone des sensomotorischen Cortex nach elektrischen Reizen. I. Hemmung und Erregung nach direkten und kontralateralen Einzelreizen. *Arch. Psychiat. Nervenkr.,* 1956, **194,** 597–619.

Creutzfeldt, O., and O.-J. Grüsser. Beeinflussung der Flimmerreaktion einzelner corticaler Neurone durch elektrische Reize unspezifischer Thalamuskerne. In *Proc. 1st. int. Congr. ncurol. Sci., Brussels, Vol. III. EEG, Clinical Neurophysiology and Epilepsy.* London: Pergamon, 1959. Pp. 349–355.

Dell, P. Discussion of the communication of Bremer. In Moscow Colloq. on EEG of Higher Nervous Activity. *EEG clin. Neurophysiol.,* 1960, **Suppl. 13,** 134–136.

Desmedt, J., and G. La Grutta. The effect of selective inhibition of pseudocholinesterase on the spontaneous and evoked activity of the cat's cerebral cortex. *J. Physiol.,* 1957, **136,** 20–40.

Dumont, S., and P. Dell. Facilitations spécifiques et non spécifiques des réponses visuelles corticales. *J. Physiol.,* 1958, **50,** 261–264.

Eccles, J. C. *The Neurophysiological Basis of Mind.* Oxford: Clarendon Press, 1953.

Eccles, J. C. *The Physiology of Nerve Cells.* Baltimore: Johns Hopkins Press, 1957.

von Euler, C., and J. D. Green. Activity of single hippocampal pyramids. *Acta physiol. scand.,* 1960*a,* **48,** 95–109.

von Euler, C., and J. D. Green. Excitation, inhibition and rhythmical activity in hippocampal pyramidal cells in rabbits. *Acta physiol. scand.,* 1960*b,* **48,** 110–125.

Fatt, P. Biophysics of junctional transmission. *Physiol. Rev.*, 1954, **34**, 674–710.

Fatt, P., and B. Katz. Spontaneous subthreshold activity at motor nerve endings. *J. Physiol.*, 1952, **117**, 109.

Fleming, T. C., and E. V. Evarts. Multiple response to photic stimulation in cats. *Amer. J. Physiol.*, 1959, **197**, 1233–1236.

Fleming, T. C., P. B. Huttenlocher, and E. V. Evarts. Effect of sleep and arousal on the cortical response to lateral geniculate stimulation. *Fed. Proc.*, 1959, **18**, 46.

Frishkopf, L. S., and W. A. Rosenblith. Fluctuations in neural thresholds. In *Symposium on Information Theory in Biology*. London: Pergamon Press, 1958. Pp. 153–168.

Fuster, J. M. Effects of stimulation of brain stem on tachistoscopic perception. *Science*, 1958, **127**, 150.

Gastaut, H., Y. Gastaut, A. Roger, J. Corriol, and R. Naquet. Etude électro-encéphalographique du cycle d'excitabilité corticale. *EEG clin. Neurophysiol.*, 1951, **3**, 401–428.

Gauthier, C., M. Parma, and A. Zanchetti. Effect of electrocortical arousal upon development and configuration of specific evoked potentials. *EEG clin. Neurophysiol.*, 1956, **8**, 237–243.

Gellhorn, E., W. P. Koella, and H. M. Ballin. Interaction on cerebral cortex of acoustic or optic with nociceptive impulses. *J. Neurophysiol.*, 1954, **17**, 14–21.

Gellhorn, E., W. P. Koella, and H. M. Ballin. The influence of hypothalamic stimulation on evoked cortical potentials. *J. Psychol.*, 1955, **39**, 77–88.

Granit, R., and C. G. Phillips. Excitatory and inhibitory processes acting upon individual Purkinje cells of the cerebellum in cats. *J. Physiol.*, 1956, **133**, 520–547.

Hughes, J. R. Post-tetanic potentiation. *Physiol. Rev.*, 1958, **38**, 91–113.

Hughes, J. R., E. V. Evarts, and H. H. Marshall. Post-tetanic potentiation in the visual system of cats. *Amer. J. Physiol.*, 1956, **186**, 483–487.

Hughes, J. R., and W. A. Rosenblith. Electrophysiological evidence for auditory sensitization. *J. acoust. Soc. Amer.*, 1957, **29**, 275–280.

Hunt, C. C. Monosynaptic response of spinal motoneurones to graded afferent stimulation. *J. gen. Physiol.*, 1955, **38**, 813–852.

Ingvar, D. H., and U. Söderberg. Cortical blood flow related to EEG patterns evoked by stimulation of the brain stem. *Acta physiol. scand.*, 1958, **42**, 130–143.

Kandel, E. R., W. A. Spencer, F. J. Brinley, and W. H. Marshall. Properties of hippocampal neurones as studied by intracellular recordings. *Fed. Proc.*, 1959, **18**, 305.

Li, C.-L. The facilitatory effect of stimulation of a thalamic nucleus on cortical sensory neuronal responses. *J. Physiol.*, 1956a, **131**, 115–124.

Li, C. L. The inhibitory effect of stimulation of thalamic nucleus on neuronal activity in the motor cortex. *J. Physiol.*, 1956b, **133**, 40–53.

Li, C.-L., C. Cullen, and H. H. Jasper. Laminar microelectrode studies of specific somatosensory cortical potentials. *J. Neurophysiol.*, 1956a, **19**, 113–130.

Li, C.-L., C. Cullen, and H. H. Jasper. Laminar microelectrode analysis of cortical unspecific responses and spontaneous rhythms. *J. Neurophysiol.*, 1956b, **19**, 131–143.

Lindsley, D. B. The reticular system and perceptual discrimination. In H. H. Jasper, et al. (Editors), *Reticular Formation of the Brain.* Boston: Little, Brown, 1958. Pp. 513–534.

Lorente de Nó, R. Analysis of the activity of the chain of internuncial neurones. *J. Neurophysiol.,* 1938, **1,** 207.

Lorente de Nó, R., and H. T. Graham. Recovery cycle of motoneurons. *Amer. J. Physiol.,* 1938, **123,** 388.

Lorente de Nó, R., and Y. Laporte. Synaptic transmission in a sympathetic ganglion. *J. cell. comp. Physiol.,* 1950, **35,** Suppl. 241, 41.

Magoun, H. W. *The Waking Brain.* Springfield, Ill., Charles C Thomas, 1958.

Malis, L. I., and L. Kruger. Multiple response and excitability of cat's visual cortex. *J. Neurophysiol.,* 1956, **19,** 172–186.

Marshall, W. H. Excitability cycle and interaction on geniculate striate system of cat. *J. Neurophysiol.,* 1949, **4,** 277–288.

Marshall, W. H. Temporal periodicities in the primary projection system. *Amer J. Ophthalmol.,* 1958, **46,** 99–106.

Marshall, W. H., S. H. Talbot, and H. W. Ades. Cortical response of the unanesthetized cat to photic and electrical afferent stimulation. *J. Neurophysiol.,* 1943, **6,** 1–15.

Morrell, R. M. Recurrent inhibition in cerebral cortex. *Nature, Lond.,* 1959, **183,** 979–980.

Moruzzi, G., and H. W. Magoun. Brain stem reticular formation and activation of the electroencephalogram. *EEG clin. Neurophysiol.,* 1949, **1,** 455–473.

Mountcastle, V. B. Modality and topography properties of single neurons of cat's somatic sensory cortex. *J. Neurophysiol.,* 1957, **20,** 408–434.

Mountcastle, V. B., P. W. Davies, and A. L. Berman. Response properties of neurons of cat's somatic sensory cortex to peripheral stimuli. *J. Neurophysiol.,* 1957, **20,** 374–407.

Patton, H. D., and V. E. Amassian. The pyramidal tract: its excitation and functions. In *Handbook of Physiology. Neurophysiology.* II. Washington, D. C.: American Physiological Society, 1960. Pp. 837–861.

Perl, R. E., and D. G. Whitlock. Potentials evoked in the cerebral somatosensory region. *J. Neurophysiol.,* 1955, **18,** 486–501.

Phillips, C. G. Actions of antidromic pyramidal volleys on single Betz cells in the cat. *Quart. J. exp. Physiol.,* 1959, **44,** 1–25.

Posternak, J. M., T. C. Fleming, and E. V. Evarts. Effect of interruption of the visual pathway on the response to geniculate stimulation. *Science,* 1959, **129,** 39–40.

Purpura, D. P. Nature of electrocortical potentials and synaptic organization in cerebral and cerebellar cortex. *Int. Rev. Neurobiol.,* 1959, **1,** 47–163.

Purpura, D. P., and H. Grundfest. Nature of dendritic potentials and synaptic mechanisms in cerebral cortex of cat. *J. Neurophysiol.,* 1956, **19,** 573–595.

Purpura, D. P., and H. Grundfest. Physiological and pharmacological consequences of afferent synaptic organizations on cerebral and cerebellar cortex. *J. Neurophysiol.,* 1957, **20,** 494–522.

Roitback, A. Metamorphoses of primary responses. In *Problems of the Modern Physiology of Nervous and Muscle Systems.* Acad. Sci. Georgian S. S. R., 1956. Pp. 243–256.

Rosenblith, W. A.   Sensory performance of organisms. *Rev. mod. Physics,* 1959*a*, **31,** 485–491; also in J. L. Oncley (Editor), *Biophysical Science, A Study Program.* New York: Wiley, 1959*a.* Pp. 485–491.

Rosenblith, W. A.   Some quantifiable aspects of the electrical activity of the nervous system (with emphasis upon responses to sensory stimuli). *Rev. mod. Physics,* 1959*b*, **31,** 532–545; also in J. L. Oncley (Editor), *Biophysical Science, A Study Program.* New York: Wiley, 1959*b.* Pp. 532–545.

Rosenzweig, M. R., and W. A. Rosenblith.   Responses to pairs of acoustic clicks at the round window and at the auditory cortex. *Psychol. Monogr.,* 1953, **67,** Whole No. 363.

Schoolman, A., and E. V. Evarts.   Response to lateral geniculate radiation stimulation in cats with implanted electrodes. *J. Neurophysiol.,* 1959, **22,** 112–119.

Steriade, M., and M. Demetresco.   Phénomènes de dynamogenèse aux différents niveaux des voies optiques pendant la stimulation lumineuse intermittente. *J. Physiol.,* 1960, **52,** 224–225.

Tasaki, I., E. H. Polley, and F. Orrego.   Action potentials from individual elements in cat geniculate and striate cortex. *J. Neurophysiol.,* 1954, **17,** 454–474.

Whitlock, D. G., A. Arduini, and G. Moruzzi.   Microelectrode analysis of pyramidal system during transition from sleep to wakefulness. *J. Neurophysiol.,* 1953, **16,** 414–429.

Zanchetti, A., and J. M. Brookhart.   Corticospinal responsiveness during EEG arousal in the cat. *Amer. J. Physiol.,* 1958, **195,** 262–266.

34

MARY A. B. BRAZIER, KEITH F. KILLAM,[1] and A. JAMES HANCE

Neurophysiological Laboratory,[2] Massachusetts General Hospital,
Research Laboratory of Electronics,[3] Massachusetts Institute of Technology, and
Department of Anatomy,[4] University of California, Los Angeles

The Reactivity of the Nervous System in the Light of the Past History of the Organism

Those physiologists whose field of investigation lies within the brain have long been restive with black-box models that strive to relate all the characteristics of the output to the extracorporeal parameters of the input. Such a line of argument derives from the past—from an earlier era of laboratory procedure in which the experimental animal was anesthetized to a level at which the brain no longer met the incoming stimuli with its own modifying excitatory and inhibitory reactions.

Modern physiology tells us that in such experiments centrifugal control of inflow, ascending reticular and midline thalamic influences, cerebellar modification and limbic system influence, as well as more complex factors introduced by the circulatory and humoral systems, have all been distorted by the experimenter, and the stimulus-response data that he collects are information, not about the functioning brain, but about a laboratory artifact that he has, in part, constructed.

In the normal animal the extracorporeal stimulus impinges on a nervous system already under stimulation from its proprioceptors, its interoceptors, and its *milieu intérieur*, and we are not justified in regarding it as being in a steady state.

[1] U.S.P.H. Senior Research Fellow. Now at Stanford University, California.

[2] Aided by grants from the U. S. Air Force [Office of Scientific Research, AF-49-(639)-98], the U. S. National Institute of Neurological Diseases and Blindness (B 369 Physiology), and the U. S. Navy (Office of Naval Research Nr101-445).

[3] Supported in part by the U. S. Army (Signal Corps), U. S. Air Force (Office of Scientific Research, Air Research and Development Command), and the U. S. Navy (Office of Naval Research).

[4] Aided by the Ford Foundation and by U.S.P.H. Grant B-611.

As techniques have developed that enable the investigator to study the electrophysiology of the brain in unanesthetized, freely behaving animals, evidence has grown that the changing degrees of interplay between the three great systems of the brain are paramount in determining the response to a given stimulus.

Of these three major functional systems of the brain we have heard in these sessions much about the specific afferent systems and something about the midline nonspecific system, but unfortunately nothing about the limbic system, for all three are involved in the subject that gives its title to this symposium. The purpose of this preamble is to emphasize the background of variables within the functioning brain against which the experiments that are to be reported have been designed, for these introduce yet another consideration, namely, variation with serial time. The results are being presented to emphasize that the response of the brain to a stimulus whose parameters are held constant is modified by the past experience of the organism. In other words, the output of the black box depends on its past history.

With the slow realization by Western scientists that the conditional reflex technique can give them a method for imposing a controlled experience on the animal has come an increasing wealth of information on the changing interaction of the various brain systems in the serial stages of training in the learning of a task.

The initial recognition of these serial changes came essentially from electroencephalographers skilled in the recognition of pattern in the records of both animal and man, and at first they had to meet considerable skepticism from other neurophysiologists who had not trained themselves in pattern recognition. Clearly some method or methods of automatic analysis, which would exclude or reduce the factor of human judgment, were needed to define these changes more exactly, and to convince the skeptic.

Three sets of experiments designed to demonstrate the effect of a controlled past experience will be briefly reported. The first example of changing response to identical stimuli is not a behavioral one but an electrocortical response.

Dr. Frank Morrell (1959), working in the Department of Neurology at the University of Minnesota, and following up some earlier observations he had made in collaboration with Dr. Jasper (Morrell and Jasper, 1956), observed that, if he exposed a rabbit to a noncyclic series of repetitive trains of flashes, each train always being preceded at a constant interval by a tone signal, there was a more-or-less orderly sequence of changes observable at specific sites on the cortex where he had implanted electrodes. From his EEG recordings he was able to recognize three stages through which the potentials of the rabbit's brain passed during exposure to trains of repetitive flash preceded by a tone signal. In summary these were:

Stage I: a long-lasting stage during which the acoustic stimulus elicited a generalized desynchronization.

Stage II: an extremely brief stage, often present for only ten trials (or even less), in which a rhythmic wave appeared in the recording from the visual cortex as soon as the tone was sounded and before the flash stimulation began.

Stage III: disappearance of the rhythmic response and appearance of a desynchronization localized to the visual cortex. This local desynchronization was considered by Morrell to represent the final form of electrographic conditioning. With the exception of stage II, this work will not be described in more detail, as Morrell's presentation of it to the American EEG Society's Symposium on "EEG and Conditioned Behavior" is in print (Morrell, Barlow, and Brazier, 1960).

Somewhat similar observations have been made by several workers, and the suggestion has been made by some that the rhythmic wave that appears so evanescently in stage II may reflect an anticipation by the brain of the flicker frequency heralded by the conditioning stimulus. In order to determine this point the Average Response Computer* in the Laboratory of Communications Biophysics, Massachusetts Institute of Technology, was called in as consultant.

We were fortunate in having Dr. Morrell come to work for a brief period in the laboratory at the Massachusetts General Hospital, where he and Dr. John Barlow ran a series of rabbits through a similar experience, recording the electroencephalograms on a 7-channel magnetic tape recorder, and submitting the recordings later to analysis by the computer at M.I.T.

Figure 1, reproduced from Morrell, Barlow, and Brazier, shows the result of averaging the records from one of the rabbits in the series. These are all averages of the electrocortical activity in the period between the tone signal that triggers the sweep and the onset of flicker. Averages of 10 trials, each 2 seconds in length, are shown in each trace, their serial incidence indicated at the right. The beginning deflection at the extreme right of each trace is caused by the first flash of the train.

The rhythmic activity that Dr. Morrell had perceived in his ink-

* This computer, designed and constructed at the Lincoln Laboratories of M.I.T., has been described by its designer (Clark, 1956).

written traces in stage II is now more clearly seen. In this rabbit it appeared most prominently in the averages of trials 60 to 70. What is this rhythm? It may bear no direct harmonic relationship to the frequency of the flash that the brain is being trained to expect, and in our experience it is a phenomenon that can be unrelated to the specific flash frequency. Rather we would propose that this may

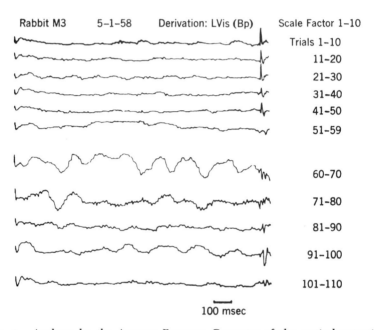

Fig. 1. Analyses by the Average Response Computer of the cortical potentials occurring in the visual cortex of a rabbit between a conditioning stimulus (tone) and the onset of light flickering at a rate of 10 per second. The sequence of trials averaged in each trace is indicated at the right. (From Morrell, Barlow, and Brazier, 1960.)

be a reflection of a transient dominance of limbic system arousal at this certain stage of training. In the rabbit the hippocampus lies little more than 2 millimeters below the visual cortex (from which these recordings were made). We know from the work of Green and Arduini (1954) that activation of the reticular formation, either by direct electrical stimulation or by normal stimulation of peripheral receptor organs, evokes a theta rhythm of high voltage, high enough, it is being suggested, to be recordable about 2 mm distant in a volume conductor.

The second series of experiments, illustrating the changing interplay between systems within the brain as the result of repetition of constant stimuli, concerns situations that develop during the establishment of a conditional reflex.

A paper has been published by John and Killam (1959) in which are reported many and subtle changes in the electrical recordings from multiple sites within the brain, as well as from its cortex, during conditioned avoidance training. To summarize the great amount of data presented, one may say that, during conditioning, stages were found where potentials arose in structures previously unresponsive to the stimulus, and modification was seen in the response of structures initially responsive.

One of the many interesting findings reported by John and Killam was a change at different stages of the training period in the potentials of the lateral geniculate nucleus to the repetitive flash that they were using as the conditioning stimulus. This was surprising, for of all the structures in the brain one would perhaps expect the classical relay nucleus of the visual system to be the most conservative in its response.

Fortunately it was possible for one of these authors (KFK) to come to work in the laboratory of the Massachusetts General Hospital for a period, so that some of the recordings could be put on a tape recorder and played back into the computer at M.I.T. The experiment was in many ways similar to those reported by John and Killam from work at Los Angeles, but in the experiments about to be reported the conditioned response was not avoidance of a shock but lever pressing, rewarded by milk. Electrodes were implanted in several sites within the brains of the cats in this series, but the present comments will be restricted to the events recorded at the lateral geniculate nucleus. These recordings were made some weeks after the implantation when the animals were unanesthetized and moving freely.

The signal was a 10-per-second flash, and the average response curves shown in Figs. 2a, b, and c are each the average of 432 responses of the lateral geniculate. This computation was processed in the Laboratory of Communications Biophysics at M.I.T., by the same computer that was used for the experiments of Morrell, just described. The computing procedure increases the signal-to-noise ratio so that detail previously hidden in background activity can now be studied.

In Fig. 2a is shown the average response of the lateral geniculate nucleus to 432 flashes during an early stage of learning when the cat was beginning to master the first step in responding appropriately to the 10-per-second flash. At this stage no limit of time was forced on the animal, the train of flashes being continued until he pressed the lever.

Averages of this kind are obtained by cross-correlating the electrical trace (recorded on magnetic tape) with a pulse timed by the flash. Thus the curves shown represent the average of only those potential

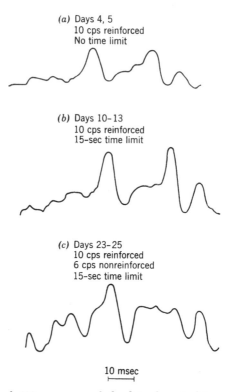

(a) Days 4, 5
10 cps reinforced
No time limit

(b) Days 10-13
10 cps reinforced
15-sec time limit

(c) Days 23-25
10 cps reinforced
6 cps nonreinforced
15-sec time limit

10 msec

Fig. 2. Average of 432 responses of the lateral geniculate nucleus to 10-per-second flash used as the signal that lever pressing will yield milk: (*a*) early stage of training when the cat is not yet pressed to respond within any given time limit; (*b*) the same cat has perfected the reflex response, including a time limit, for reward by milk, of response within 15 sec of onset of flash; (*c*) the same cat, having reached 100-per-cent-correct responses to 10-per-second flash, is now required to discriminate between this frequency and a 6-per-second flash that brings no reward. For interpretation see text. Averages computed by the Average Response Computer at the Laboratory of Communications Biophysics, M.I.T. (Unpublished records of K. F. Killam and J. S. Barlow.)

changes whose phases are consistently locked in time to the occurrence of the flash. Randomly occurring EEG potentials unrelated to the flash are averaged out.

The sweep starts at a constant interval before the incidence of each flash, and the flash artifacts are not included in the record. Since the signal is a 10-per-second flash, there is a 100-msec interval between the primary responses, and hence the first 25 msec of each of these curves represents the average residual effect in the lateral geniculate nucleus of each previous flash. In Fig. 2a this is negligible, that is to say, the effect of each flash has died out in the geniculate response by the time 75 msec have elapsed from the previous flash.

One notices in Fig. 2a that after the initial response there is another event, triple in form and building up to an amplitude as great as the initial response. If the primary response represents the initial arrival of the signal, this subsequent complex may be described as an electrical concomitant of the processing of that signal.

Figure 2b represents the average response from the same electrodes when the cat had fully learned the procedure and had become an automaton, responding correctly every time with a fully elaborated reflex. As these are also averages of 432 responses, the differences in the curves are significant. One notes an augmentation of the primary response, but an even greater augmentation of the complex that is here suggested for consideration as the sign of the processing that follows the receipt of the signal. Again one sees from the early part of the sweep that the average disturbance created in the geniculate nucleus by the flash does not last for the full 100-msec interval between flashes.

Figure 2c gives the curve obtained when this same fully trained animal was confronted with the necessity to discriminate between a 10-per-second and a 6-per-second flash. Only with the 10-per-second flash was lever pressing rewarded by milk. The brain now had to match the frequency it was experiencing against some stored information of a previously experienced frequency. The changes in the average response from the lateral geniculate nucleus apparently reflect this demand for a new form of processing. The primary response has become distorted, but more striking is the change in the complex that follows it and that presumably reflects a part of the elaboration procedure. Note also that the average effect of a flash is now very long lasting, occupying the whole of the 100-msec interval and even invading the rise of the curve of the newly entering subsequent flash. The stimuli are the same, but the reactivity has changed.

The last set of experiments to be reported here also have bearing on the effect that the past experience of the brain has on the response that a given stimulus will evoke from it. These experiments were begun in Los Angeles on a visit to Dr. Magoun's laboratories, and they have been continued at the Massachusetts General Hospital.

It has been shown by Livanov (1940) and since by several other workers on intersensory conditioning that, if one exposes an animal to trains of synchronous flashes and clicks, after a given time the same flash rate, without any click, will produce a repetitive response in the auditory cortex. This result can be readily confirmed in cortical recordings, but it seemed to us of interest to know whether this intermixing of the sense modalities could be detected in the deeper structures of the brain.

If an unanesthetized cat with implanted electrodes is exposed to a standard flash rate (without sound) many times, an ink-writing oscillograph indicates that the number of deep structures that respond may increase, but provided the flash is silent (for these experiments a silent fluorescent tube was used), the specific auditory pathway will remain apparently unresponsive except to the "on" of the train.

From experience with an automatic frequency analyzer in experiments that have no place in this report, it would appear that the various structures, if they react at all to repetitive stimulation with a repetitive response, do not necessarily adopt the stimulus frequency. Each structure responds within its preferred frequency band, which is usually rather narrow. The reticular formation, for example, responds in a frequency band grouped around a lower center frequency than the response of the specific relay nuclei. Undoubtedly this is directly related to the longer recovery time of reticular neurons (King, Naquet, and Magoun, 1957).

One cannot determine this characteristic from unanalyzed records such as those in Fig. 3a, but these are introduced to show that, although the reticular formation gradually assumes a rhythmic oscillation, no trace of rhythm can be detected in the auditory system in these ink-written records from a cat that has never experienced a click paired with the flash.

Quite a different result is obtained from the cat when he has experienced a click coincident with every flash of the train (Fig. 3b). With a past history of this kind, the auditory system reacts to flash alone with clearly marked responses, and even in this unanalyzed record there is a strong suggestion that the change is not restricted to the auditory cortex alone, for the medial geniculate nucleus is now clearly disturbed by the flash.

We were interested in studying this interaction of sense modalities with singly occurring stimuli, administered at random in order to avoid the influence of so-called "driving" with its implications of resonance.

Figure 4a shows superimposed oscillographic traces derived from the auditory system. The cat was in a sound-reduced compartment

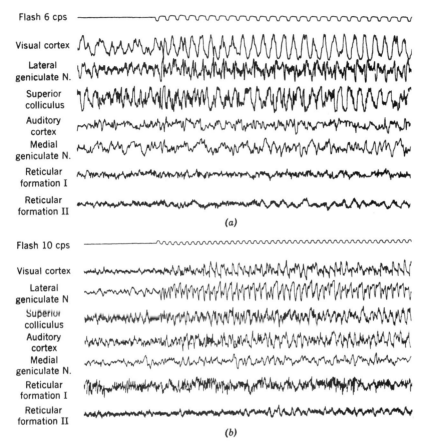

Flash 6 cps

Visual cortex

Lateral
geniculate N.

Superior
colliculus

Auditory
cortex

Medial
geniculate N.

Reticular
formation I

Reticular
formation II

(a)

Flash 10 cps

Visual cortex

Lateral
geniculate N

Superior
colliculus

Auditory
cortex

Medial
geniculate N.

Reticular
formation I

Reticular
formation II

(b)

Fig. 3. (a) Responses following the 6-per-second flash rate in the visual system, but not in the auditory system of an unanesthetized cat that has not experienced clicks paired with flicker. (b) The response of the same cat to 10-per-second flash alone after experiencing clicks paired with the flash at this rate. Note following by the auditory system in the absence of clicks. (Killam, Hance, and Brazier, unpublished records.)

in a steady silent fluorescent light. A rise in the signal line marks the moment when the light went off and a drop when it came on again. In the top record there is no response in the auditory system either to the "off" or the "on" of the light.

In the center trace a click has been introduced as the light went off. An initial response in the auditory cortex and medial geniculate nucleus is followed by two slow waves in both these structures. The fourth beam is recording from a position in the reticular formation where large responses to clicks were always present. Histology showed the tip of the electrode to be in a site analogous to the one which in the marsupial phalanger was found by Adey, Dunlop, and Sunderland (1958) to give large evoked responses to clicks. This was a part of the subparafascicular (SPF) system.

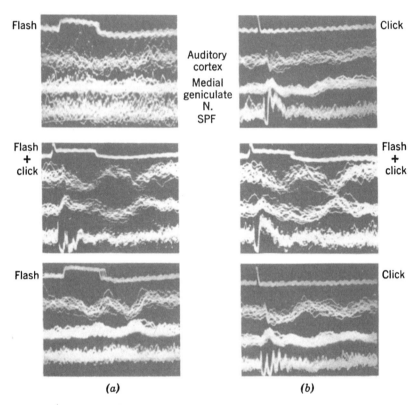

Flash

Click

Auditory
cortex
Medial
geniculate
N.
SPF

Flash
+
click

Flash
+
click

Flash

Click

(a)   (b)

Fig. 4. Superimposed traces from three sites in the brain of an unanesthetized cat showing: (a) response to flash before, during, and after pairing a click with the flash; (b) response to click before, during, and after pairing a flash with the click. (Killam, Hance, and Brazier, unpublished records.)

In the lowest record, the stimulus was identical with that in the top record, but the experience of the brain had changed: a flash alone now evoked the two slow waves in both auditory cortex and thalamus.

Not only does the auditory system give this mixed response to flash, but it also gives a mixed response to click, once it has had the experience of the pairing. This can be seen in the records of Fig. 4b. These are all recordings from the same sites as (a). The top record is the response to clicks of the animal who has experienced no pairing.

In the center record the click has been timed to coincide with the "off" of the flash. The slow waves seen in the previous experiment are again present. The lowest record is for the cat with click alone after experiencing the paired modalities. The slow wave brought in by the flash in the past is reproduced by the brain, although there is no flash stimulus in the present.

On return to the laboratory at the Massachusetts General Hospital, a new series of experiments was begun along these lines, with the object of processing the recordings through the computer in order to facilitate the detection of responses hidden from us in the single trace, or even in the superimposed traces for, of course, these latter do not in any way reduce the amplitude of the background activity.

In this series of experiments on unanesthetized cats with implanted electrodes, control runs were made to record responses to click alone and to flash alone before the animal was allowed to experience any pairing of stimuli. The stimuli were in all cases delivered at a rate of 1 per second. The cat was in a sound-reducing box, actually an old refrigerator cabinet lying on its side with one end replaced by a transparent plastic window. The stroboscope used was built into a sound-reducing box to muffle its click, and a second glass was placed in front of its light, so that an air space lay between the two glass fronts. The stroboscope was placed outside the cabinet containing the cat. In all the experiments a masking white noise was delivered, the intensity of which was regulated for each experiment by calibration of the input voltage to the loudspeaker. Clicks when used were also calibrated to a standard intensity in each experiment. The duration of each click was 100 microseconds.

Computer-averaged responses to flash after the cat had experienced pairing are shown for the reticular formation and centre median in Fig. 5. Neither of these sites is unresponsive to flash. The response of the reticular formation was of such high amplitude that it is reproduced at half the scale used for the other positions. The response of the reticular formation is, of course, merely confirmatory of the responses first demonstrated by French, von Amerongen, and Magoun (1952).

In Fig. 6 the response of the auditory cortex to flash is shown for contrast with response of the same electrode site to click. The response resembles spread from the visual cortex. The same scale has been used for both records.

When the control recordings were analyzed by the computer, it was found, as we had begun to suspect, that all these sites responded to

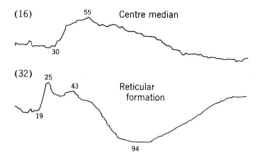

Fig. 5. Computer-averaged responses to 1-per-second flash alone in an unanesthetized cat who had experienced four trains of 300 flashes paired with clicks at 1 per second. The figures on the left are inversely proportional to the scale at which the traces are displayed; that is, the response in the reticular formation was of such high amplitude that it is reproduced at half the scale used for the centre median. Length of trace, 175 msec; unipolar recordings; the small numbers indicate latency times from the flash for various points on the curves.

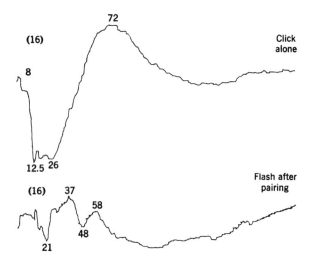

Fig. 6. Computer-averaged responses from the auditory cortex of an unanesthetized cat to 1-per-second click alone before pairing and to 1-per-second flash alone after experiencing a click paired with a flash 300 times. The response resembles spread from the visual cortex. Length of trace, 250 msec; unipolar recordings from middle ectosylvian gyrus; the numbers in parentheses indicate the scale factors used for the computation, and the other numbers the latency times from the stimulus.

both click alone and light alone, even before the animal had experienced these stimuli paired (Fig. 7). Although it may be possible to
assign some of the components of these curves to volume conduction
(for example, the response at 44 msec in the reticular formation and
superior colliculus), this explanation cannot be made to fit all the
data, especially when amplitude as well as phase is taken into account.

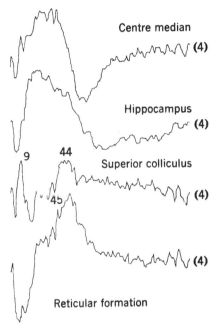

Centre median

(4)

Hippocampus

(4)

9 44

Superior colliculus

(4)

45

(4)

Reticular formation

Fig. 7. Computer-averaged responses to 1-per-second click from four deep
structures in an unanesthetized cat that had not experienced flash paired with the
click. All were recorded at the same amplifier gains and are displayed at the same
scale factor (4). Length of traces, 150 msec; unipolar recordings.

The usual effect of pairing was to increase the amplitude and introduce or emphasize a late wave component (such as had been detected
in the superimposed traces recorded by cathode-ray oscilloscope in
the UCLA series). For example, the reticular formation as shown in
Fig. 8 gives a greater response on the eleventh day of pairing click
with flash than on the first, and a click alone after this experience gives
a response of nearly twice the amplitude it evoked before the brain
had experienced pairing. (In all the figures the numbers to the left
of each trace are inversely proportional to the scale used in computing
the result.)

The late slow-wave response detected in the superimposed traces of the oscilloscopic records can be seen in Fig. 9 and was present in deeper structures as well as in the cortex.

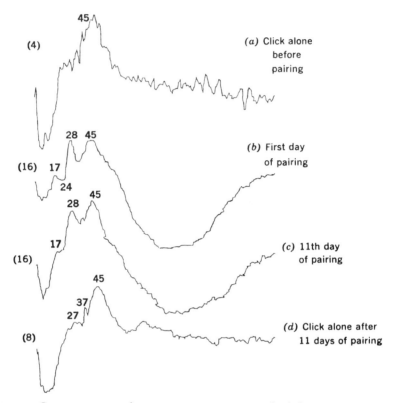

Fig. 8. Computer-averaged responses to 1-per-second click recorded in the reticular formation of an unanesthetized cat: (*a*) before pairing; (*b*) during pairing on the first day of intersensory conditioning; (*c*) on the eleventh day; and (*d*) response to click alone after this experience. The responses to paired stimuli are of such high amplitude that they are reproduced at one-quarter the scale of the top record. Note increase in response of reticular formation in trace (*d*) for which the write-out is at half the scale used for record (*a*) from the "naive" cat. Length of traces, 185 msec; unipolar recordings. Each test day consisted of four trains of 300 paired stimuli.

One additional comment should be made at this stage. The size of the response, at whatever stage it is evoked and whether by single or multiple sensory stimuli, will always depend on the state of the animal. In working with the fully functioning brain we are faced with a system that does not for long tolerate the isolation of a single variable. Difficult as it is to accomplish, the experimenter must attempt to keep his animal at the same level of general alertness in all the tests if his results are to be comparable. Although hard to

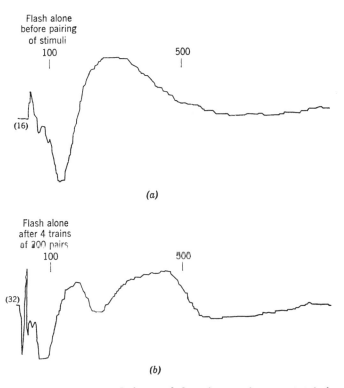

(a)

(b)

Fig. 9. Average responses to flash recorded in the visual cortex (*a*) before and (*b*) after pairing the flash with the click. Note development of late wave. These records have been plotted by the computer at one-fifth the time base used in previous illustrations in order to demonstrate the late wave. With this time base, the resolution of the early events is lost. Length of each trace, 950 msec. To aid the eye, the 100-msec and 500-msec intervals from the flash are marked. The small figures in parentheses at the beginning of the traces indicate the scale used for the computer write-out (the smaller the number, the greater the amplification). Note that the responses from the visual cortex increased in size after pairing (and hence are reproduced at half the scale used in the upper record).

achieve, this is easy to monitor, and for this reason simultaneous EEG records have been run in all the experiments reported here.

Space limits the display of changes found in all the brain centers from which responses were obtained in these experiments, but perhaps enough have been presented to explain why these records suggest to us that the intersensory response after pairing is an augmentation of one already there, and that the major changes are probably in the late slow fluctuations, such as were noticed in the oscilloscopic records. It would appear probable that all the parts of the brain take their individual parts in the subject of this symposium, namely, sensory communication.

Much further work is needed before a more definite statement can be made, but the possibility may be proposed that the so-called "temporary connections" set up by intersensory conditioning are not, in fact, temporary connections but already existing ones. It is interesting in this context to recall the proposals made by Gastaut (1958a, b), Fessard and Gastaut (1956), and others since, that conditioning may be explained on the basis of a center of convergence in the reticular formation.

The suggestion is made here that before conditioning the electrical signs of activity in these connections often lie below the noise level of the brain's ongoing activity, and that the influence of the pairing is to increase their amplitude to a level at which they can be seen.

The connection may be only one of very many somewhat diffuse pathways, but its repetitive use may result in a facilitated transmission—perhaps a membrane change—that brings it into dominance.

The possibility may even be entertained that further experiment may reveal similar electrophysiological information about other forms of conditional reflexes and that the concept of temporary connections may be superfluous. Should this eventually prove to be so, we would be relieved from speculations about protoplasmic outgrowths and the formation of new anatomical pathways as necessary mechanisms for the development of conditional reflexes.

Summary

The position is taken that the response of the brain to sensory stimulation depends more importantly on events within the brain than on the extracorporeal parameters of the stimulus. Not only do current activities in brain mechanisms influence the response, but past experience also has its effect.

Three series of experiments illustrating the influence of past experience on the reactivity of the brain to standard stimuli are reported. The first of these was carried out with Dr. Frank Morrell and Dr. John Barlow, the second with Dr. Keith Killam and Dr. Barlow, and the third in collaboration with Dr. Keith Killam and Dr. James Hance in the Department of Anatomy at the University of California at Los Angeles, continuing at the Massachusetts General Hospital with Miss Ruth Carpenter and Miss Margaret Magavern.

As a general conclusion the proposal is made that the so-called temporary connections set up by intersensory conditioning are not in fact temporary, but already existing connections whose responses are augmented by use. These responses often lie hidden below the noise level of the usual recording but can be revealed if processed through a computer that increases the signal-to-noise ratio.

Acknowledgment

It will be clear to the reader that there is one person to whom we are very greatly indebted, Prof. Walter Rosenblith. It is a pleasure to have an opportunity to express our appreciation to him and to his group, among whom we would especially like to mention Frank Nardo, who has been responsible for operating the computer in most of these studies.

References

Adey, W. R., C. W. Dunlop, and S. Sunderland. A survey of rhinencephalic interconnections with the brain stem. *J. comp. Neurol.*, 1958, **110**, 173–203.

Clark, W. A. Average response computer (ARC-1). Quarterly Progress Report, Research Laboratory of Electronics, MIT, 1956.

Fessard, A., and H. Gastaut. *Rapport sur les Bases Neurophysiologiques du Conditionnement*. Strasbourg: Société Française de Psychologie Scientifique, 1956.

French, J. D., F. K. von Amerongen, and H. W. Magoun. An activating system in the brain stem of the monkey. *Arch. Neurol. Psychiat.*, 1952, **68**, 577–590.

Gastaut, H. The role of the reticular formation in establishing conditioned reactions. In H. H. Jasper, et al. (Editors), *Reticular Formation of the Brain*. Boston: Little, Brown, 1958. Pp. 561–571.

Gastaut, H. Some aspects of the neurophysiological basis of conditioned reflexes and behaviour. In Ciba Foundation Symposium, *The Neurological Basis of Behaviour*. London: Churchill, 1958. Pp. 255–272.

Green, J. D., and A. Arduini. Hippocampal electrical activity in arousal. *J. Neurophysiol.*, 1954, **17**, 533–557.

John, E. R., and K. F. Killam. Electrophysiological correlates of avoidance conditioning in the cat. *J. Pharmacol. exp. Therap.*, 1959, **125**, 252–274.

King, E. E., R. Naquet, and H. W. Magoun. Alterations in somatic afferent transmission through the thalamus by central mechanisms. *J. Pharmacol. exp. Therap.*, 1957, **119**, 48–63.

Livanov, M. N. On the rhythmical stimulation and interrelations of the fields of the cerebral cortex. *Fisiol. Zh. USSR*, 1940, **28**, 172–194.

Morrell, F. Electroencephalographic studies of conditioned learning. In M. A. B. Brazier (Editor), Transactions of First Macy Conference, *The Central Nervous System and Behavior*. New York: Josiah Macy, Jr. Foundation, 1959. Pp. 307–374.

Morrell, F., J. S. Barlow, and M. A. B. Brazier. Analysis of conditioned repetitive response by means of the average response computer. In *Recent Advances in Biological Psychiatry*. New York: Grune and Stratton, 1960. Pp. 123–137.

Morrell, F., and H. H. Jasper. Electrographic studies on the formation of temporary connections in the brain. *EEG clin. Neurophysiol.*, 1956, **8**, 201–215.

35 THEODORE H. BULLOCK
University of California, Los Angeles

The Problem of Recognition
in an Analyzer Made of Neurons

Several previous speakers have referred to the destination of sensory input, the mechanism that finally analyzes the converging coded messages and compares them with criteria previously laid down for initiating appropriate motor action. But they have been too wise to say more than that such an operation is performed and must be provided for in any schema. The presumptuous effort in this communication will be to formulate the general problem and to consider some possibilities in neuronal terms.

To set out the problem more explicitly, let us consider those examples of animal behavior in which a predetermined stimulus configuration (innate or learned) evokes a characteristic either-or response, that is to say, a response or no response. This formulation embraces a large class of behavior, including the "take-off" of a fly, the onset of instinctive acts, and the identification of symbols, faces, and voices. The normal stimuli will impinge on several or many receptor units. Only certain patterns of activity of afferent nerve fiber, that is to say, certain combinations of labeled lines and certain temporal sequences, will evoke the characteristic response. The formula that will do so is built into the central nervous analyzer, either genetically or through learning. The analyzer may be said to perform an operation of recognition or decision on whether its criterion has been satisfied and its response is to be triggered.

The triggering formulas are in general complex. First, there is usually a set of permissive steady-state inputs—time of day, time of year, state of nutrition and hormonal balance, temperature, humidity, and so on. Then there are the timing and directing stimuli, which determine the moment of action and often its direction. These stimuli are, in familiar cases, visual or auditory and are therefore distinguished from other stimuli to the same sense organs on complex criteria. Processes of transformation or abstracting must be carried out to yield relatively simple outputs as a function of the relevant aspects of the stimulus pattern. These are not discussed here. Instead we take up the question of the final stage of processing when all preceding integrations are summed to yield a simple yes or no answer.

If the integration is performed in the nervous system, then there must be, before the command enters the effectors, a final determination by some competent functional unit of whether the criterion has been met. All relevant input and central predisposition must be read and finally evaluated by a single integrator. Whatever its composition, the integrator is a unit because it is necessary and sufficient for the performance of this act.

If averaging of many penultimate integrators is involved, something must perform the averaging and deliver an unequivocal answer—or else the action is not centrally determined. Quite possibly some actions are integrated finally only by mechanical averaging of the activity of motor units. But it seems likely that many actions are integrated centrally, such as the initiation of chirping by a cricket or of flight by a bird, or the recognition of faces or voices. For each of these events or, more precisely, for each pair or set of alternative events (are they "decisions"? I will try to avoid using the word), there must be a single final functional unit (or a number of equivalent units) upon which all preceding lines converge.

Since the final functional unit is, like a military general, competent to render an answer, it may be called a "decision" unit or a recognition-of-criterion unit or a high-level trigger unit. Not only are there many such units—one for each pair or alternative set of recognitions that the organism has acquired by birth and training—but there is physiological evidence to support the theoretical expectation that many successive levels of recognition on subcriteria occur, which gradually converge on the narrowest bottleneck in the stimulus-response sequence. (Obviously, conscious cognition is not invoked or required in this discussion.) The analogy to an army suggests the hierarchy of convergences and integrations of lines of incoming information, the encoding, decoding, and re-encoding, the predetermined differences in the value and meaning of the same message, depending on what line it comes on, the processing of data from lower levels in order to make interpretation possible, and other similarities. All these are features of our present picture of the nervous system.

Characteristics of Decision Making

In considering further the characteristics of decision making in a nervous system, it is reasonable to expect in different situations greater or less redundancy, that is to say, duplicate and equivalent recognition-of-criterion units with the same input. This concept would explain tolerance of injury and would provide for greater uniformity of threshold. Duplicate units would each have to receive from all inputs and lower integrative levels involved and command the same output paths—though they would not necessarily be identical in their answers and they may be influenced by each other. An uncritical number of units must agree closely in threshold to give the sharp onset characteristic of all-or-none or either-or actions, such as those we chose for illustration. If a certain threshold number or proportion of units were required, the structure at the next level, which would have to measure this number or proportion, would be the interesting and relevant unit in this context. The only requirement for the recognition-of-criterion units is that each of them incorporate in its answer all the preceding lines and stages of integration, so that it is competent to determine the next stage, whether it be an effector or a higher level. Each action, recognition, or sequential integrative level must finally be triggered by a single unit or an uncritical number of units with the same input requirement, within normal play. It is permitted and probable that in the nervous system some of the same input that influences the trigger unit influences units downstream from it, predisposing them or setting the stage for the trigger message. The downstream units are not equivalent to the trigger unit insofar as they do not receive information from all the relevant sources. Stimulation of a trigger unit alone may not precipitate the normal behavior because the stage is not set. There are a number of cases, however, where focal stimulation has released characteristic either-or complex acts, such as chirping in a cricket, and vivid recall in man.

In other words, there must be a nervous unit—but there may also be other units receiving from the same or overlapping input—entirely competent, on the basis of converging afferent lines, to decide whether a threshold signal of a predetermined pattern has risen above the background noise. The answer of this neural unit will bear a label, such as "left statolith appreciably displaced medially," and will encode the strength, rate of development, and duration. Then there must be a unit—though there may also be others—entirely competent to integrate this and other lines from right statocyst, leg nerves, and eyes to determine an output labeled perhaps "body detectably displaced in transverse plane." Among the various places to which this recognition goes is a unit—or many equivalent units—capable of integrating with lines from motor command centers to decide whether the tilt is imposed or expected from the last motor commands, and hence whether compensatory movements are to be initiated. This latter integration takes place beyond the afferent path, but the same principles apply: at each stage of comparison and integration of different preceding lines, a final common unit, or many parallel or nearly equivalent final common units, must perform the function. Parallel units at such a level must be very nearly equivalent, for if there were a meaningful difference in the significance of their activities there would have to be another level on which they converge, in order to be read and integrated.

Recognition-of-criterion units can be significantly labile in their judgments. As a result of their internal state, of the time elapsed since some previous action, of bias due to prevailing input or feedback (experience), or of "spontaneous" central initiative, a meaningful change may take place in the weight assigned to certain of their inputs or in their over-all threshold.

These units may be said to depend on maintained activity because they must be able to signal both increase and decrease in frequency; otherwise they would be only unidirectional, and a final balance between opposing messages would be deferred, as may be the case in symmetrical so-called "half-centers" (Gesell, Brassfield, and Lillie, 1954; Retzlaff, 1957). The final integration may then be between opposing muscles.

Such units are not required where the final integration of messages in different lines occurs mechanically in the joints. The output of trigger units must not be continuously graded in the cases of either-or behavior. Rather, we may reasonably anticipate that the output will consist of an abrupt change from one range of frequency of firing to a significantly different range.

Identity of Recognition-of-Criterion Units

What are these units which, like miniature sentient beings, receive all that goes before, act adaptively upon it, and pass on the result to the next level? They must be numerous, for many integrative levels in the path from receptor to effector may have a decision-making character, simply by having a sharply discontinuous relation between input and output. And there are many discrete aspects of behavior in which choices are made. Both lower and higher animals must possess such elements, since, in both, either-or behavior is a prominent part of their repertoire. It is not necessary, of course, that all the units be everywhere the same in composition or mechanism.

1. One possibility is that certain single neurons are high-level trigger units. We know of no smaller element than the neuron that can act in so unitary a way on the basis of converging nerve impulses. Every neuron is a decision-making element when it changes from a silent to an impulse-firing state. In this context only those neurons are interesting that integrate a complex input or control a large output. The capacity of a single neuron to integrate a large number of incoming impulses and to give a single-impulse answer, for example, the mediation of escape jerks by giant neurons like Mauthner's, demonstrates its suitability for a "decision" unit. This is true at least of those neurons whose response is go, no-go, but it is not at all limited to them. The single neuron's possibilities for complex evaluation and interaction of inputs, and for abrupt change from low- to high-frequency firing, fit it for higher-level and responsible recognition of criterion. In fact, we know experimentally of no larger *element* that can compare, evaluate, add, and multiply inputs in the central nervous system.

2. A theoretical alternative is that a mass of randomly connected neurons constitutes a trigger unit. Beurle (1956) has pointed out that activity with a sharp threshold can be propagated in a mass of a large number of cells—each assumed to have only some of the properties known for actual neurons—connected without specification, except that the probability of interconnection falls with distance. As an amplitude-sensitive switch, the mass of neurons can integrate by permitting subthreshold excitatory and inhibitory input to influence its critical level. Several loops leaving and re-entering the mass, also uncritical as to connections, can stabilize the amplitude of the propagated waves of excitation. The waves can terminate on one cell or in a specific link. In order to be useful, the mass of neurons must connect specifically at both input and output. If use were to produce a change in the connections, the size, or the constancy of some of the cells, then conditioning, trial-and-error learning, and recall could be explained. By postulating many local masses of this sort, each performing a limited integration, we may visualize something resem-

bling the brain. Storage of species characteristics and of individually learned information is in many cells, each a part of the storage of many items.

This concept of a mass of randomly connected neurons can meet the requirements under consideration, and it calls for only a modest degree of specific organization. There is a rather sharp distinction between the class of connections that is specific and critical, for example, inputs and outputs, and the class that is random within the given functionally defined mass. A useful nervous system would consist of a large number of unit masses.

The redundancy and the freedom from a critical number of neurons resemble features ascribed to the single-neuron alternative, but only in part. In Beurle's model the proportion of active neurons is critical, though the mass is the decision unit that measures this proportion, not a single neuron, or a number of equivalent independent neurons. The redundancy in Beurle's model, therefore, is of a different kind from that of the single neuron. The absolute amount of organization is the same: it consists of the connections into, out of, and between unit masses of single cells. But the redundancy required in Beurle's model is one of unspecifically connected cells, not duplication of the specific pathways. The physical basis of stored information in the two models is not different at the neuronal level. Neither model has been formally developed to a point where it is possible to compare the complexity of achievements it permits or to evaluate its relative advantages.

3. The third alternative seems a priori more attractive for the general case, in the light of current ideas about central nervous organization. This is a multiple-input, metastable feedback loop consisting of a definitely specified meshwork of mutually interacting neurons. In order that the meshwork should possess a threshold, a regenerative chain reaction is assumed to take place by positive feedback. The principal difference between the feedback loop and the single-cell model is that the constellation of inputs converging to provide the criterion of recognition or triggering does not need to converge on one cell: some of the necessary inputs may arrive at one cell of the network and some at another, up to any degree of complexity. Each cell then contributes an output to the others. But only if all of a certain predetermined set receives input in a predetermined amount and temporal pattern will the network constants provide the regeneration for self re-excitation. The difference between the multiple input feedback loop and the Beurle model is simply that the loop assumes a set of specified connections and functions and therefore presumably permits more complex input requirements for a given number of neurons.

The loop system resembles a flip-flop or a multivibrator circuit with two (or more) stable states. It may be self-restoring, or a specific input may be required to return it to one of its stable states. The output may be from one cell or from several or many of the cells; hence several output fibers may go to different fractions of the whole effector mechanism. In the simple form of this mechanism based on component cells having one axon, the cells in the efferent pathway next downstream will feel the impulses that are a part of the loop feedback before a loop threshold has been reached. In this case, since there must be a clear difference between outputs before and after threshold, we must suppose that a significant jump in frequency of impulses is the normal signal of triggering. Otherwise different forms of output must be invoked, for example, dendritic potentials for the interactions within the loop. The impulse output may now change from zero to a finite value or, as before, from one frequency to a considerably higher frequency. If the output frequency is continuously graded, then some later functional unit must read this and provide the sharp threshold required in a trigger unit.

This multiple-input, metastable feedback loop, suggested by MacKay (in discussion at this symposium), points to no one cell but to the whole loop as responsible for the recognition of input criterion and provision of a threshold. It can at the same time be the mechanism for many alternative actions by having more than two stable states. This feature might provide a mechanism for recognizing individuals or symbols, and to that extent it is possibly more important in higher vertebrates. More complexity and specification of connections are necessary than in the other alternatives. Like the alternative systems, it can, of course, employ redundant cells.

Another class of decisions is considered by Sholl and Uttley (1953), who advocate pattern recognition by a statistical process among many units, which would provide a continuous measure of pattern difference and thus an analog rather than a digital answer. Instead of "A differs from B" or "A does not differ from B," the continuous measure has the form "item A differs from ensemble C by more than does item B." Such a process is not at all precluded by the foregoing argument, but it can apply only to decisions other than those leading to recognition of familiar faces or to the take-off of a fly. Analog events are fundamental to both classes of decisions, but in the class of decisions considered here they eventuate in either-or answers.

Any or all three of the decision mechanisms "discussed" is consistent with present-day neurophysiology. The single-cell type is known to exist, for example, in giant fiber startle-response systems. Randomly connected masses are suggested by many neuropiles. The specific network with positive feedback is the most appealing, although inherently it does not offer any greater flexibility or complexity of achievement, economy of cells, or economy of specific or predetermined connections and transfer functions. All three mechanisms can provide the same features of performance—stability, redundancy, tolerance of injury, and modifiability.

Summary

The problem of recognition of a predetermined criterion by an analyzer of neurons is formulated. The class of behavior in which either-or actions, or recognitions, takes place is considered. The criteria of input from permissive, directive, timing, and other sensory signals are generally complex. Integration, processing, and decision making with respect to subcriteria take place sequentially in progressively more embracing stages. The final triggering must be made by a neurological unit upon which all input relevant to that choice converges and one that has a threshold. Three possible neuronal mechanisms are discussed: the single cell, the randomly connected mass, and the specifically connected, metastable feedback loop. All three are adequate. The first is known to occur, the third seems most likely for higher-level decisions.

References

Beurle, R. L. Properties of a mass of cells capable of regenerating pulses. *Phil. Trans. roy. Soc.*, 1956, **B240**, 55–94.

Gesell, R., C. R. Brassfield, and R. H. Lillie. Implementation of electrical energy by paired half-centers as revealed by structure and function. *J. comp. Neurol.*, 1954, **101**, 331–406.

Retzlaff, E. A mechanism for excitation and inhibition of the Mauthner's cells in teleost: a histological and neurophysiological study. *J. comp. Neurol.*, 1957, **107**, 209–226.

Sholl, D. A., and A. M. Uttley. Pattern determination and the visual cortex. *Nature, Lond.*, 1953, **171**, 387–388.

36

BURTON S. ROSNER

West Haven Veterans Administration Hospital and
Yale University School of Medicine

Neural Factors Limiting Cutaneous Spatiotemporal Discriminations

When two stimuli activate a sensory system in rapid succession, various psychophysical and physiological interactions occur. Psychologically, the stimuli may fuse into one continuous event, which sometimes seems "ragged"; or one stimulus may appear to be attenuated or even to be missing altogether. Corresponding neurophysiological experiments on "refractoriness" or "excitability cycles" demonstrate how a conditioning stimulus modifies electrophysiological responses to a succeeding test stimulus. This paper will consider some interrelations between psychophysical and physiological studies of responses to successive stimuli.

The following material concerns primarily the cutaneous system, and three major classes of experiments are distinguished. In the first experiments to be discussed, two temporally dispersed stimuli occupy noticeably different spatial positions. These studies emphasize the role of central neural events in temporal interactions, since the stimuli activate different subsets of receptors and peripheral axons. The second class comprises experiments in which two successive stimuli occur at the same locus, thereby bringing into play peripheral as well as central factors. The third class considers n-stimulus generalizations of the preceding two classes of experiments

Successive Stimulation of Two Different Loci

Psychophysical studies

Békésy (1957, 1959a) has described interactions between two transient stimuli that successively activate different cutaneous loci. His results show that two such equally intense stimuli fuse into a single event at temporal separations below about 1.5 milliseconds. This fused event seems to lie between the two sites of stimulation and closer to the earlier stimulus. At temporal separations beyond the critical value, subjects report two spatially separated sensations.

Dr. Ethel Matin (1959) in our laboratory has studied a closely related situation with electrocutaneous stimuli. Her experiments concern the suppression of a weaker stimulus by a nearby stronger

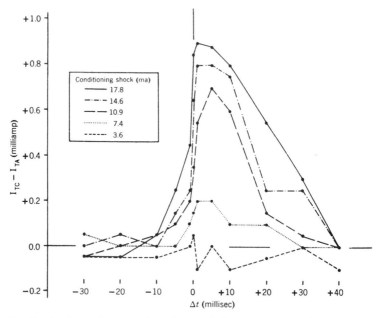

Fig. 1. Psychophysical suppression of a weak by a strong stimulus in two-point cutaneous stimulation. Abscissa, time difference between conditioning and test shocks. Ordinate, conditioned threshold minus absolute threshold. See text for fuller explanation.

one. In Matin's studies, an absolute threshold (I_{TA}) for a 100-microsecond electrical square wave was determined on the subject's middle or ring finger; these fingers were the "test" loci. Then the threshold was redetermined in the presence of a conditioning shock applied to the ipsilateral index finger. This second threshold is called a "conditioned threshold" I_{TC}. The intensity of the conditioning shock and the time between conditioning and test stimuli were varied systematically. The stimuli were delivered through surface electrodes from two specially designed constant-current isolation units, which prevented interactions through a common ground. Figure 1 shows some of the results of Matin's work; the middle finger was the test site in this experiment. The abscissa shows time between conditioning and test stimuli. Negative values signify that the test stimulus preceded the conditioning stimulus; positive values signify the reverse order. The ordinate is the difference between the conditioned threshold I_{TC} and the absolute threshold I_{TA}; the parameter is intensity of the conditioning stimulus. The data show that an intense, nearly simultaneous conditioning stimulus raises the absolute threshold at the test finger. This inhibitory effect occurs for interstimulus intervals between about -5 and $+35$ msec for the largest conditioning shock and diminishes as the conditioning shock becomes weaker. At the lowest conditioning intensity, there is some suggestion of facilitation at the test site. Control experiments showed, however, that intrasession variability of I_{TA} could occasionally reach 0.1 milliampere. Thus the possible facilitation with the weakest conditioning shock requires further confirmation.

The largest inhibitory effect for any conditioning intensity clearly occurs when the conditioning stimulus slightly precedes the test stimulus and not when Δt is equal to zero. The apparently retroactive inhibition at negative values of Δt may largely reflect the operation of the intensity-latency law: the latency of the response of a nerve fiber diminishes as the applied current is increased. When we allow for this intensity-latency effect, the positive values of Δt at which inhibition is maximal probably underestimate the true extent of the temporal lag necessary for the greatest inhibition. This temporal lag provides one possible mechanism for suppressing any psychophysical effects of traveling waves on the skin (Keidel, 1956). These waves, which arise from vibration of the skin, travel out from the point of stimulation. The greatest deformation of the skin at loci beyond the site of stimulation occurs slightly later than the maximal deformation at the site itself. Furthermore, traveling deformations are physically smaller than those at the source of the wave. Since Matin's data show that a stronger stimulus is most effective in suppressing a slightly later and weaker stimulus, a traveling wave would be less apparent subjectively.

Some time ago, von Frey and his students (Brueckner, 1901; Cook and von Frey, 1911; von Frey and Pauli, 1913) noticed an illusion of spatial displacement during two-point cutaneous stimulation with transient stimuli of unequal intensity. The weaker stimulus seemed displaced toward the stronger. This effect occurred at times in Matin's experiments; the test stimulus seemed shifted toward the conditioning stimulus. This apparent spatial displacement has an interesting parallel in audition. Both temporal and intensive differences provide cues for auditory localization. Békésy has pointed out the parallel between successive two-point cutaneous stimulations and successive dichotic stimulations. The illusory cutaneous displacement observed by von Frey provides a second parallel to auditory localization. In somesthesis and audition, the weaker stimulus is "pulled toward" the stronger.

Physiological studies

The properties of psychophysical inhibitory phenomena in somesthesis apparently depend upon processes in the central nervous system. Several workers have reported interference or inhibitory effects in the somatosensory system with two-point cutaneous stimulation (Amassian, 1952, 1953, 1954; Mountcastle, 1957, and in this volume; Malcolm and Smith, 1958). Probably the earliest report came from Marshall, Woolsey, and Bard (1937), who found that simultaneous stimulation of two cutaneous loci produced interference effects at the cerebral somatosensory cortex. Figure 2 shows a similar experiment from our laboratory on a cat anesthetized with pentobarbital sodium. In this experiment, the median and ulnar nerves of the left foreleg were resected; records were obtained simultaneously from the left radial nerve and the right cerebral somatosensory area I upon electrocutaneous stimulation of the first and second digits of the left foreleg. Nerve responses were recorded through a hypodermic needle which was insulated except for a small portion of its barrel; this electrode was slipped through the skin and manipulated until a maximally detectable response appeared. An indifferent electrode lay at a distance of about 2 centimeters, lateral to the recording electrode. Cortical responses were recorded from one point at the pial surface of the brain with a platinum wire macroelectrode. In Fig. 2, responses from nerve are in the left-hand column and those from the cortex are in the right-hand column. From top to bottom are responses to stimulation of the first digit alone, to stimulation of the second digit alone, and to stimulation of both digits together. The stimulus to the second digit came 0.5 msec earlier than the stimulus to the first, in order to make the negative peaks of the nerve responses coincide. The initial positivity in the peripheral records is somewhat larger than what we usually see with our recording method. The nerve responses summate perfectly when both digits are stimulated, whereas the cortical potentials show interference. The notch on the early portion of the negative phase of the nerve response to stimulation of the second digit marks the break between $A\beta$ and $A\gamma$ fibers, according to other experiments

Digit	Nerve	Cortex

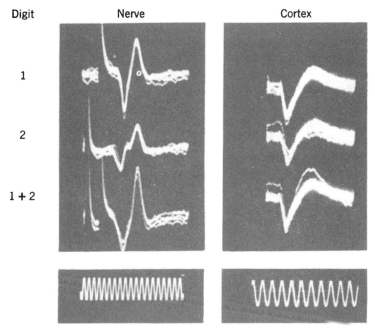

Fig. 2. Responses from radial nerve and contralateral cerebral somatic area I to electrocutaneous stimulation of (1) first digit of left foreleg alone, (2) second digit alone, and (1 + 2) both digits together. Nerve responses summate; cortical responses interfere. Calibrations, 5 kcps and 10 μvolt peak to peak (p-p) for nerve, 100 cps and 50 μvolt p-p for cortex.

of ours. Both fiber groups contribute to the cortical response, but the Aβ group accounts for a disproportionately high 60 to 70 per cent of the cortical potential (Rosner, et al., 1959).

Figure 3 shows that the cortical interference is reciprocal. The response marked C-2 was evoked by stimulation of the second digit; C-1 came from stimulation of the first digit. The recovery series in the remainder of the figure were obtained with a photographic overlap method. The series 1-2 shows the effects of prior stimulation of the first digit upon responses to stimulation of the second digit. The response at the far left comes from stimulation of the first digit, and the other potentials are cortical responses to stimulation of the second digit at various later times. Similarly, series 2-1 shows how stimulation of the second digit blocks succeeding responses to stimulation of the first digit. Whichever digit is stimulated first, it suppresses for some time responses to stimulation of the other digit. At 400 msec, which is the longest interstimulus interval shown, recovery is not quite complete in either series 1-2 or 2-1. The surface-positive portions of the blocked responses seem to recover faster than the surface-negative portions. Stimulation of the first digit, which yields a larger peripheral response than does stimulation of the second digit, also has a more severe blocking effect. Finally, the series 2-2 shows the effects of an earlier stimulus to the second digit upon re-

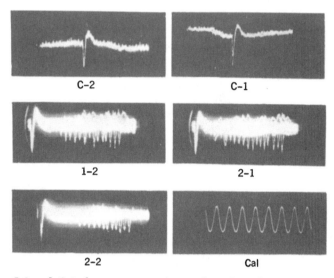

Fig. 3. C-2 and C-1 show response of contralateral cerebral somatic area I to stimulation of second and first digits of left foreleg. 1-2, Blockade of responses to stimulation of second digit at various times after a prior stimulus to first digit. 2-1, Blockade of responses to stimulation of first digit by prior stimulus to second digit. 2-2, Blockade when two successive stimuli activate second digit. Calibration, 20 cps at 50 μvolt p-p.

sponses to later stimulation of the same digit. Here the blocking is even more severe than in series 1-2, where the earlier stimulus activated the first digit.

Electrophysiological studies thus point to central processes which could play some role in psychophysical inhibitory effects on the skin. The time course of the electrophysiological effects is sufficiently long; in fact, with anesthetized preparations, the time course usually exceeds that of psychophysical functions. There are other similarities between the physiological and psychophysical results. Electrophysiological interference effects are reciprocal; they increase with the intensity of the blocking stimulus, and they decline as the distance between the stimuli increases.

There are certain discrepancies, however, between electrophysiological and psychophysical findings. For example, the effects shown in Fig. 3 occur only where the cortical projections of the two cutaneous loci overlap. Thus, the blocking shown in series 1-2 and 2-1 represents an excellent mechanism for increasing two-point acuity; but it cannot yet account for psychophysical spatial displacement effects, which represent a loss in two-point acuity. Furthermore, we have not found any obvious evidences of summation in our electrophysiological work, although summation occurs in Békésy's results and may have been present in Matin's studies. Finally, the evidence currently available does not imply that cerebral cortical processes are necessarily the sole or primary determinants of psychophysical inhibitory effects. Marshall's (1941) work on recovery cycles in somatosensory pathways suggests that thalamic centers, which we have not yet explored, may play a role in these psychophysical phenomena. The anatomical loci of the neural events underlying the psychophysical results remain an open question.

Successive Stimulation at One Locus

Psychophysical observations

This class of experiments with successive stimuli clearly involves both peripheral and central factors in the cutaneous system. When two brief, equally intense shocks stimulate a single cutaneous locus, they must be separated by 15 to 40 msec before the subject feels two temporally discrete events. The exact separation necessary for temporal resolution varies considerably among different observers. When the two shocks are felt as separate, the second feels less intense than the first for separations well beyond 40 msec. A similar situation obtains in audition, where Rosenzweig and Rosenblith (1950) report that two successive monotic clicks fuse into a continuous event for separations below 10 msec. Beyond this separation, the second click may seem softer than the first. Comparison of somesthesis and audition indicates, therefore, that the skin is somewhat more sluggish than the ear in its recovery from transient stimuli. In both systems, the time course of recovery depends partly on the intensities of the stimuli.

In contrast to these temporal blocking effects, Piéron and Segal (1939) have reported summation when two brief subliminal shocks are applied to the skin at separations between 40 and even 400 msec. They report that the threshold is lowered under these conditions and, furthermore, that the subject perceives the two shocks at intensities that are below the threshold for a single shock. We have consistently failed to repeat this finding in quite a few attempts with our constant-current square-wave generators. When the interval between the shocks is made a fraction of a millisecond, however, summation occurs at intensities that are below the threshold for a single shock. At these intervals the subject feels only a single pulse. Here, summation probably depends in part on summation of local electrotonic processes set up in peripheral nerve fibers by weak stimuli.

Electrophysiological observations

Peripheral limitations on temporal acuity in somesthesis may be studied directly in man by surface recording of nerve action potentials. We used a modified form of the technique described by Dawson and

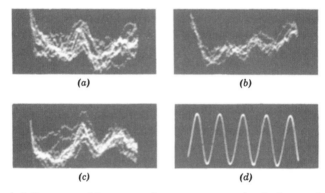

(a) (b)

(c) (d)

Fig. 4. (*a*) Response of human median nerve to single shock at index finger; (*b*) same when preceded by identical shock 2 msec earlier; (*c*) same when preceded by identical shock 4 msec earlier; (*d*) 1 kcps and 10 μvolt p-p.

Scott (1949) in order to obtain the results shown in Fig. 4. For this experiment, paired electrocutaneous stimuli were delivered to a single locus on the palmar surface of the index finger. A superficial recording electrode was placed over the median nerve on the skin of the palmar surface of the wrist; the nerve comes sufficiently near the surface here for recording. An indifferent electrode lay about 15 mm medial to the recording electrode, and a large, diffuse ground on the palm of the hand helped decouple the stimulating and recording systems. Each stimulus was 500 μsec long and about 1 ma in intensity. Figure 4*a* shows a control response to a single stimulus, and 4*b* and 4*c* show responses to a second stimulus when a prior stimulus occurred 2 msec and 4 msec earlier. At the 4-msec separation used in Fig. 4*c*, the nerve has substantially recovered. This experiment demonstrates again that the relative refractory period of mammalian A fibers ends in a few milliseconds. At the 4-msec delay, however, the subject still felt one sustained temporal event. For this and larger temporal separations, central factors must play the exclusive role in limiting temporal acuity for electrocutaneous pulses. For separations below 4 msec, both peripheral and central processes may circumscribe temporal resolution of the two successive stimuli.

Electrophysiological studies of the auditory system show that the cochlear microphonic in response to a second click is unaffected by an immediately preceding click. Action potentials from the eighth nerve, however, are reduced by a preceding click for intervals up to 100 msec, depending on intensity of stimulation (Rosenzweig and Rosenblith, 1950). Thus peripheral factors may play a very large role in limiting auditory temporal acuity. Direct comparison of these data with the results for skin shown in Fig. 4, however, must be made with caution. Acoustic stimulation with clicks activates specialized receptors, which in turn generate impulses in nerve fibers. Electrocutaneous stimulation probably activates peripheral nerve fibers directly, since specialized receptors apparently resemble mammalian postsynaptic membrane and therefore may not be electrically excitable (Grundfest, 1957). Thus results obtained with electrocutaneous stimulation may by-pass any bottlenecks to temporal acuity at the earliest stages of the cutaneous system. So-called "natural" cutaneous stimuli might show that the region of the receptor-nerve junction in somesthesis also imposes major limits on temporal acuity.

Our results with electrocutaneous stimulation nevertheless imply that central processes can restrict temporal resolving power in somesthesis. These central processes, exemplified in series 2-2 of Fig. 3, also can play a role in the apparent attenuation of a second stimulus by a first. This latter psychophysical effect, incidentally, probably constitutes part of the phenomenon of positive time error, where subjects underestimate the second of two equally intense stimuli for intervals up to about 500 msec. Certain features of psychophysical data, however, do not match current electrophysiological observations. For example, Uttal (1960) reports that subjects judge several electrocutaneous stimuli delivered a few milliseconds apart more intense than a single stimulus. Such temporal summation has not occurred in electrophysiological studies. Similarly, electrophysiological studies of central structures provide no parallel to the finding that two clicks less than 10 msec apart sound like a single continuous event with two "humps."

n-Stimulus Generalizations

Stimulation of a single locus

At low intensities, repeated stimulation of a single cutaneous locus with a train of electrical pulses evokes a sensation of vibration. This experiment shows that subjects can feel vibration at stimulus rates well over 100 pulses per second. Since subjects cannot resolve two pulses separated by less than 15 msec, vibratory sensations apparently would not be felt for pulse trains at rates greater than about 70 pps. Two hypotheses might help to account for this apparent contradiction between two-pulse and n-pulse findings. One would be a volley principle like that invoked by Wever (1949) for acoustic phenomena. It may be that all cutaneous neural elements do not fire in response to a repeated stimulus; one group may fire to one stimulus, a second group to the next stimulus, and so on, until the first group had recovered sufficiently for reactivation. A second possibility lies in the discrepancy that apparently exists between the physical rate of stimulation and the perceived rate of vibration. Békésy's (1959b) finding that the perceived rate of vibration drops almost three octaves for a 60-db increase in intensity shows that perceived rates of vibration may be considerably below the stimulus rate. This observation also speaks against any simple formulation of the volley principle.

Studies of evoked potentials at cerebral somatic area I (Rosner, 1956) show a continued decline to a steady plateau in the amplitudes of responses evoked by trains of stimuli. This decline becomes steeper as the pulse rate increases. Similar effects occur in peripheral nerves at even higher pulse rates (Gasser and Grundfest, 1936). Thus electrophysiological observations imply that repeated stimulation should yield poorer and not better temporal acuity. Systematic psychophysical studies of the growth of the feeling of vibration as a function of the number of pulses in a train may suggest an answer. Nor are these problems unique to somesthesis; Lindsley has reported at this symposium that in vision also temporal acuity is better when measured with flicker than when measured with only two successive flashes.

Two-point stimulation

Trains of pulses to a single locus represent the n-stimulus generalization of two pulses applied to a single locus. Similarly, repeated alternate stimulation of two cutaneous loci forms an n-stimulus generalization of the experiments reviewed earlier in this paper. When the interstimulus interval is adjusted to a fraction of a second, apparent movement is perceived. The stimulus seems to shuttle back and forth across the skin between the two sites of stimulation. This effect, which has been known for many years, is analogous to the better-publicized phenomenon of apparent visual movement. If apparent movement is viewed as a generalized form of two-point stimulation, an interesting theoretical possibility emerges. Apparent movement may be another reflection of the inhibitory and summative processes that have appeared in the experiments discussed above.

Summary

Psychophysical and electrophysiological studies of responses to successive stimuli bring out certain parallels between behavioral and physiological processes. Subjectively, stimulation of two cutaneous loci in rapid succession produces partial or complete suppression of the later stimulus, particularly when its intensity is low. This inhibitory effect diminishes as the time or the distance between the two stimuli increases. Electrophysiological studies of the central nervous system show similar interference effects between responses to nearly simultaneous stimulation of two cutaneous loci. These electrophysiological phenomena decline also as the distance or time between the two stimuli increases. But certain psychophysical observations have uncovered summation and spatial displacement during two-point cutaneous stimulation which are not matched by current electrophysiological findings.

Successive stimulation of a single cutaneous locus with brief electric shocks also produces inhibition. Subjects cannot resolve two shocks separated by less than 15 msec; at longer separations, the second pulse feels less intense. Recordings from human peripheral nerves demonstrate that peripheral factors cannot account for these observations for separations beyond 4 msec; central processes, exemplified by prolonged recovery cycles at thalamus and cortex, must also play a role in this situation. Direct comparison with data on the psychophysical effects of two successive monaural clicks indicates that the somesthetic system is somewhat more sluggish in its recovery than is the acoustic system. In both systems, however, some type of subjective summation occurs when two stimuli are presented too quickly for temporal resolution; these phenomena have no parallel in electrophysiological studies of central pathways. Finally, cutaneous vibration and cutaneous apparent movement may be viewed as n-stimulus generalizations of the preceding two-stimulus experiments.

Acknowledgment

This work was supported in part by grant M-1530 from the National Institute of Mental Health. I gratefully acknowledge the collaboration of Truett Allison in many of the experiments reported here.

References

Amassian, V. Interaction in the somatovisceral projection system. *Res. Publ. Assoc. nerv. ment. Dis.*, 1952, **30**, 371–402.

Amassian, V. Evoked single unit activity in the somatic sensory areas. *EEG clin. Neurophysiol.*, 1953, **5**, 415–438.

Amassian, V. Studies on organization of a somesthetic association area, including a single unit analysis. *J. Neurophysiol.*, 1954, **17**, 39–58.

Békésy, G. v. Sensations on the skin similar to directional hearing, beats, and harmonics of the ear. *J. acoust. Soc. Amer.*, 1957, **29**, 489–501.

Békésy, G. v. Similarities between hearing and skin sensations. *Psychol. Rev.*, 1959*a*, **66**, 1–22.

Békésy, G. v. Synchronism of neural discharges and their demultiplication in pitch perception on the skin and in hearing. *J. acoust. Soc. Amer.*, 1959*b*, **31**, 338–349.

Brueckner, A. Die Raumschwelle bei Simultanreizung. *Z. Psychol. Physiol. Sinnesorgane*, 1901, **26**, 33–60.

Cook, H., and M. von Frey. Die Einfluss der Reizstarke auf der Wert der simultanen Raumschwelle der Haut. *Z. Biol.*, 1911, **56**, 537–573.

Dawson, G. D., and J. W. Scott. The recording of nerve action potentials through skin in man. *J. Neurol. Neurosurg. Psychiat.*, 1949, **12**, 259–267.

von Frey, M., and R. Pauli. Die Starke und Deutlichkeit einer Druckempfindung unter der Wirkung eines begleitenden Reizes. *Z. Biol.*, 1913, **59**, 497–515.

Gasser, H. S., and H. Grundfest. Action and excitability in mammalian A fibres. *Amer. J. Physiol.*, 1936, **117**, 113–133.

Grundfest, H. Electrical inexcitability of synapses and some consequences in the central nervous system. *Physiol. Rev.*, 1957, **37**, 337–361.

Keidel, W. D. *Vibrationsreceptoren. Der Erschütterungssinn des Menschen.* Erlangen: Universitätsbund, 1956.

Malcolm, J. L., and I. D. Smith. Convergence within the pathway to cat's somatic sensory cortex activated by mechanical stimulation of the skin. *J. Physiol.*, 1958, **144**, 257–270.

Marshall, W. H. Observations on subcortical somatic sensory mechanisms of cats under Nembutal anesthesia. *J. Neurophysiol.*, 1941, **4**, 24–43.

Marshall, W. H., C. N. Woolsey, and P. Bard. Cortical representation of tactile sensibility as indicated by cortical potentials. *Science*, 1937, **85**, 388–390.

Matin, Ethel. Temporal aspects of cutaneous interaction with two-point electrical stimulation. Doctoral dissertation, Columbia University, 1959.

Mountcastle, V. Modality and topographic properties of single neurons of cat's somatic sensory cortex. *J. Neurophysiol.*, 1957, **20**, 408–434.

Piéron, H., and J. Segal. Sur un phénomène de facilitation rétroactive dans l'excitation électrique de branches nerveuses cutanées (sensibilité tactile). *J. Neurophysiol.*, 1939, **2**, 178–191.

Rosenzweig, M. R., and W. A. Rosenblith. Some electrophysiological correlates of the perception of successive clicks. *J. acoust. Soc. Amer.*, 1950, **22**, 878–880.

Rosner, B. S. Effects of repetitive peripheral stimuli on evoked potentials of somatosensory cortex. *Amer. J. Physiol.*, 1956, **187**, 175–179.

Rosner, B. S., E. Schmid, S. Novak, and J. T. Allison. Responses at cerebral somatosensory I and peripheral nerve evoked by graded electrocutaneous stimulation. *Amer. J. Physiol.*, 1959, **196**, 1083–1087.

Uttal, W. R. The three-stimulus problem: A further comparison of neural and psychophysical responses in the somesthetic system. *J. comp. physiol. Psychol.*, 1960, **53**, 42–46.

Wever, E. G. *Theory of Hearing.* New York: Wiley, 1949.

37   *ROBERT M. BOYNTON*
University of Rochester

Some Temporal Factors in Vision

Reciprocity

Probably the most fundamental relation in vision involving time is the reciprocity between time and intensity, known as Bloch's law, which is said to hold for short light flashes. Although there is some difference of opinion about the generality of the reciprocity law,* let us assume that it holds for some conditions at least and consider its significance.

The explanation of Bloch's law is certainly not simple. To dismiss it by stating that under certain conditions the eye responds to energy alone (without regard to its distribution in time) is not to offer an explanation but merely to restate the law. When a relation holds even as precisely as Bloch's law, one is tempted to suppose that the responsible mechanism must be located at or near the beginning of the chain of events between stimulus and response. By this reasoning, one might hope to show that the photochemical changes produced by flashes of equal energy but differing time distributions were identical. This position, however attractive, is blatantly impossible. The reasons are at the least the following:

1. Experiments by Wulff et al. (1958*a*, *b*) show that the largest part of the density changes of rhodopsin in solution that follow the onset of a very brief (20-microsecond) illuminating flash occurs within 0.5 millisecond of the flash. This suggests that if successive flashes were administered to the same receptor, within the critical duration of Bloch's law, the speed of the reaction would be sufficient to allow two discrete photochemical changes to occur. A suitable mechanism in the eye could theoretically respond to these as separate events and transmit a double signal to the brain.

---

* Whereas, for example, LeGrand (1957) states that Bloch's law holds exactly at all levels, Piéron (1952) insists that "the technical difficulties of experimentation and of obtaining measurements endowed with the necessary extreme precision have not enabled the question to be given a definite answer."

2. At threshold, the probability that two or more quanta will be absorbed in the same receptor is negligible; at threshold, then, summation* cannot be photochemical.

3. Much work has been done by Bouman (1955a, b) and his co-workers presenting double flashes separated in space as well as time. They found that complete summation occurs if two flashes are within a certain critical small distance and short time; it is perhaps even clearer in these experiments that separate receptors are involved and that summation cannot take place at a photochemical level alone.

Reciprocity, then, seems generally to involve a neural summation, and the fact that the relationship breaks down after a certain duration is reached (in the case of a single flash), or a time interval exceeded (in the case of two very brief flashes), may mean that the limit of temporal neural summation has been reached for the conditions being explored. This has been the usual view, and at threshold it is probably substantially correct.

Above threshold, however, certain interesting effects occur which do not seem to admit of such a simple interpretation. I would like now to describe an experiment by Katz (1959) which relates to this point.

Intensity-Time Interactions above Threshold

The subject is presented with a haploscopic view of two semicircular fields which appear juxtaposed in space. One of these fields, the comparison field, is always presented to the left eye. The other, the test field, is seen by the right eye. For a given experiment, the luminance of the comparison field is set at a standard value, and its duration is set at 200 msec. Successive presentations of the comparison field—and of the test field as well—are spaced in time to eliminate the adaptive effects of previous flash pairs. The experiments consist of an adjustment by the subject of the luminance of the test flash—whose duration is equal to or shorter than that of the comparison flash—until an equal brightness sensation is elicited. Thus in the usual case, the match involves an equal-brightness match of two

---

* The term "summation" is used since, within the critical duration in which intensity-time reciprocity holds, the system summates (or integrates) the light output without regard to its distribution in time. Neural summation is not necessarily implied.

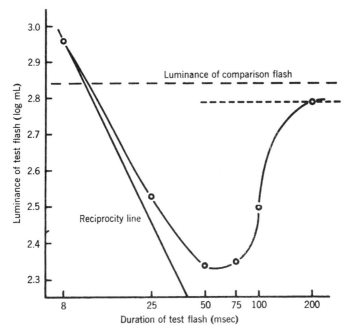

Fig. 1. Combinations of luminance and duration of test flashes required to match a 200-msec flash of 2.84 log mL in the opposite eye. The graph shows that, between 50 and 75 msec, the luminance of the test flash need be only about one-third that of the comparison flash in order for the two to appear equally bright. The reciprocity line has been drawn from the point at 8 msec to show that the efficiency of the eye is maximal for very short flashes despite the so-called "enhancement" effect (data from Katz, 1959; field size, 5 degrees; white light).

flashes of light of unequal duration.* Since the 200-msec flash of standard luminance is always delivered to the left eye, with the luminance of each shorter right-eye flash adjusted in turn to match, one may conclude that the luminances so obtained for the shorter flashes are those required to produce flashes of equal brightness; this involves only the assumption that flashes equal in brightness to a standard flash would themselves seem equally bright if compared with one another in turn.

A typical result from Katz is shown in Fig. 1. This figure is for a

* The results are little altered by varying in time the two extreme positions of the test flash relative to the comparison flash. When the two begin together under monocular presentation, the longer comparison flash produces a noticeable inhibition near the border of the short test flash; though only a part of the test flash is visible, the match is about the same as when the flashes terminate together. The condition of co-termination was used in most of the experiments.

single subject, field size, and luminance level, and though it represents only about 3 per cent of the data collected by Katz the results are typical for moderately high luminances, any area (from 0.5 to 5 degrees), and either a haploscopic or monocular comparison. We should note in Fig. 1 that reciprocity fails slightly between 8 and 25 msec. (Over the experiment as a whole, about 10 per cent more energy is required for the longer of these two short flashes. Unfortunately, 25 msec is on the border of the neural-summation range, so that the meaning of this is not unequivocal.) Further increases in time cause a still greater departure from reciprocity (much as occurs for threshold), but still further increases cause a given flash to grow dimmer, as revealed by the need to increase the luminance of the test flash to maintain equal brightness. Adding more light thus reduces brightness.

The most plausible explanation would seem to have something to do with the on-discharge that is characteristic of many of the neural units carrying the message about the flash to the brain. If the brightness sensation were related to the average number of impulses per unit time received during the flash, the obtained result would be expected. It should be pointed out in this regard that if time were quantized somehow in the input, and if brightness were related to the average input per time quantum, the same result would be expected.

Some readers will recognize this as a variation of the classical Broca-Sulzer (1902, 1903) experiment, which it is. We feel that the method is superior to those used previously because of the haploscopic presentation (used also by Alpern, 1950, 1953) and the use of a constant reference. stimulus. Haploscopic presentation rules out monocular retinal interaction effects, whereas the use of a constant reference stimulus negates the importance of sensitivity differences between the two eyes and permits unambiguous comparison between these data and the classical reciprocity data. Broca and Sulzer and their followers (for example, Stainton, 1928) had the subject adjust the luminance of a steady light for brightness, and there was therefore no constant reference stimulus.

The usual view that the Broca-Sulzer effect is an "enhancement" phenomenon is shown by Katz's study to be incorrect, in that very short flashes require the least energy for their perception.

Apparatus Required for Binocular Experiments

The apparatus required for an experiment like Katz's is illustrated in Fig. 2, which is taken from Kandel's (1958) dissertation. In general, it consists of a two-channel optical system, each channel of which has a source image at the plane of the shutters and another image in the pupil of the subject. By means of beam-splitting cubes, the Maxwellian view thus provided by each system may be brought into the

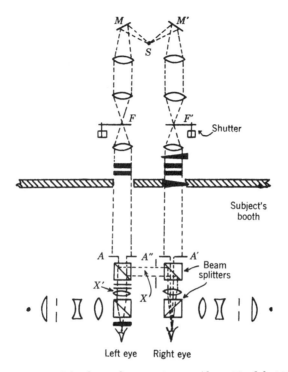

Fig. 2. Apparatus used for binocular experiments (from Kandel, 1958). Light from source lamp S is delivered by mirrors M and M' to two separate optical channels. Light is collimated and focused at F and F', where shutters are located. Light is recollimated and passed through neutral filters and wedges into subject's booth. With an opaque shield at X (between the two anterior beam splitters), the output of the two systems is delivered to the separate eyes. With an opaque shield at X', the outputs are superimposed and delivered only to the right eye. Auxiliary systems, projecting to left and right at bottom of diagram, provide adapting fields and/or fixation targets for the separate eyes.

separate eyes or superimposed in a right-eye view. The perceived areas of the fields are determined by apertures A, A', A'' at appropriate planes. Wedges and filters shown in solid black are used to control the luminance of the stimuli, and precautions are taken to monitor carefully the output of the stimulus lamp. Since the shutters cut each beam at a focal point, adequate (about 3 to 5 msec) rise-time characteristics are obtained. Katz used a large rotating motor-driven disc with auxiliary electromagnetic shutters for his work. The experiments of Kandel, to be described, involved electromagnetic shutters alone. The proper alignment of such systems is extremely difficult, and it is usual, therefore, to finish an experiment on one subject before starting to collect data on the next.

Transitional Thresholds

Until the late 1940's, most studies of light and dark adaptation ignored events occurring within the first few seconds following a sudden change in illumination. Two exceptions are the "instantaneous threshold" study of Blanchard (1918), in which an attempt was made to measure thresholds just at the offset of an adapting light, and the "alpha adaptation" reported by Schouten and Ornstein (1939). The latter study involved a binocular matching technique and indicated very rapid changes in "apparent brightness" (as evaluated by such a match) at the onset of an adapting stimulus. In 1947 Crawford reported that very abrupt threshold changes occurred at the onset and offset of a "conditioning" (adapting) stimulus. Later Baker (1953) investigated this more thoroughly for the offset condition—the beginning of dark adaptation—and although we at Rochester did one preliminary study along this line (Boynton and Bush, 1953) we have concentrated our attention upon the onset. In New York Battersby and Wagman (1959) have been working along similar lines, and in Holland Bouman (1955b) has conducted relevant experiments. It is gratifying to report that, since the agreement among laboratories is good, it will not be misleading if I confine my remarks to a description of some of the important aspects of our work.

The experiments consist of delivering to the eye (or eyes) a "conditioning-stimulus—test-flash (CS-TF) sequence," as illustrated in Fig. 3. The experiment begins with the subject adapted to some prevailing level (usually total darkness). A sequence is then delivered to the subject. A sample condition might involve a conditioning stimulus of a visual angle of 5 degrees, fixated centrally, with a duration of 0.5 sec, with a superimposed 1/50-sec test flash appearing, say, 0.05 sec following the onset of the conditioning stimulus. By manipulating the temporal position of the superimposed test flash, the threshold changes that occur as the eye goes from a dark-adapted state toward the light-adapted state can be investigated. A given threshold may be determined by any desired psychophysical method;

we have most often used a descending method of limits while exploring one conditioning interval* at a time. Enough time must be allowed between each sequence to permit the eye to return to its original state of adaptation.

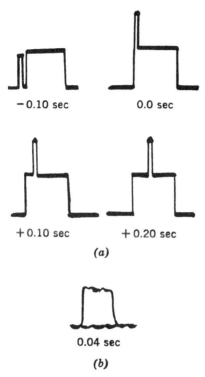

−0.10 sec   0.0 sec

+0.10 sec   +0.20 sec

(a)

0.04 sec

(b)

Fig. 3. Stimulus sequence used for transitional thresholds. (a) Four conditioning intervals; these are tracings of records produced by placing a phototube in the position of the eye and photographing the result on a cathode-ray oscillograph. (b) The test-flash characteristics on an expanded scale. (From Bush, 1955.)

Results are plotted as the log increment threshold versus conditioning interval. Some representative data are shown in Fig. 4 from Kandel (1958).

1. Threshold elevations occur at negative conditioning intervals up to at least 100 msec. This means that the threshold of a test flash may be elevated by a subsequent conditioning stimulus. Although I

* "Conditioning interval" is defined as the time between the onset of the conditioning stimulus and the onset of the test flash.

incline (Boynton and Triedman, 1953; Boynton, Bush, and Enoch, 1954) toward the view originally expressed by Crawford (1947) that differences in transmission time allow interference to occur between the coded representation of the two stimuli, this could be true only if the latency of response to the test flash were longer than that to the

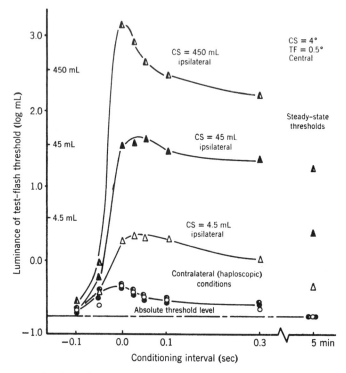

Fig. 4. Sample data from Kandel (1958) showing test-flash threshold as a function of conditioning interval (time between onset of conditioning stimulus and onset of test flash) for three luminances of conditioning stimulus and for ipsilateral and contralateral (haploscopic) conditions. In the ipsilateral case, the conditioning stimulus and test flash were delivered to the same eye, in the contralateral case to the separate eyes. Note that the effect under the contralateral condition is largely independent of the luminance of the conditioning stimulus.

conditioning stimulus. Ordinarily the test flash at threshold is both smaller and of lower luminance than the conditioning stimulus, and the explanation is plausible. However, Kandel (1958) has recently found conditions under which a leading test flash must be of higher luminance than the following conditioning stimulus in order to be visible at threshold. The explanation in terms of transmission time is definitely ruled out in this case and may indeed be of secondary importance in some of the more typical cases. If time is handled by the nervous system in discrete packages rather than as a continuous variable, any two stimuli that occur wholly or in part within the same "time frame" or information-processing period would be expected to interfere with one another as separately perceptible events. This has been called "quantization" of time by Stroud (1956).

2. The test-flash threshold just after the onset of the conditioning stimulus is as much as 100 times higher than it becomes after adaptation to the conditioning stimulus is complete. During the first few minutes following the onset of the conditioning stimulus, the threshold required for test-flash visibility drops. It seems likely that the transient elevation of the threshold is caused by some kind of masking effect produced by the burst of on-activity associated with the onset of the conditioning stimulus. As this on-activity subsides, the threshold drops, despite the decrease in receptor photosensitivity that must be occurring at the same time. The latter, if operating alone, would tend to drive the threshold in the opposite direction.

One consequence of this kind of thinking was the prediction that preadaptation of the eye, despite its known photodesensitization effect, could *reduce* the threshold of a test flash, provided the test flash in each case followed the onset of a constant conditioning stimulus. In an experiment to test this point (Boynton and Kandel, 1957; Boynton, 1958), it was found, for a conditioning stimulus of about 40 millilamberts, that increases in the level of preadaptation of the eye up to about 100 millilamberts resulted in a lowering of the test-flash threshold. The prediction, thus verified, involves the idea that light-adapting the eye will reduce the burst of on-activity to a given conditioning stimulus and thus the amount of masking.

3. The question of where, in the visual system, this transient masking takes place has been investigated. An early experiment (Boynton and Triedman, 1953) that at least indicated the plausibility of a peripheral locus showed that if the electroretinogram to CS-TF sequences is recorded, there is an interference between the response to each which makes it difficult to discriminate, in the graphic record, whether the test flash was present or not. This argued in favor of a peripheral locus. Recent experiments by Kandel (1958) involving ipsilateral and contralateral (haploscopic) presentation of the test and conditioning stimuli have clarified the issue to at least this extent: for the conditions that we have most often used, where the test flash is smaller than the conditioning stimulus, only a small interaction occurs between the visual systems associated with the separate eyes.

When the conditioning stimulus and test flash are delivered to contralateral eyes, no more than a two- or threefold rise in test-flash threshold is producible, *regardless of the intensity of the conditioning stimulus.* In experiments where the eye is completely adapted to the conditioning stimulus, no bilateral interaction occurs, which suggests a total independence of left- and right-eyed visual nervous systems. That we do find a significant effect is of importance because it shows that the two systems are not independent during the transient phase: the sensitivity of one can be momentarily altered by stimulation of the other. On the other hand, the small magnitude of the effect compared to the ipsilateral case indicates that the two systems are indeed largely independent so far as sensitivity is concerned. Furthermore, although the locus of the ipsilateral effect is largely peripheral, the contralateral experiments show that masking can also take place higher in the visual system. Thus a fraction of the ipsilateral interference may occur at higher levels—most likely at the visual cortex.

In order for the entire ipsilateral effect to be cortical, a much larger signal-to-noise ratio would seem to be required for discriminability at the cortex than at the periphery. This implies that, although a "discriminable" signal is sent from the eye to the brain, the visual brain is not able to make use of that signal. It appears most unlikely that the visual system, which seems peculiarly adapted to preserve and enhance small differences in contrast, could operate in such a way.

4. The relative size of the conditioning stimulus compared to the test flash is very important. A small test flash upon a large conditioning field allows discrimination to be made between the brightness of the test spot and the surround, even if the two are flashed simultaneously. However, if the conditioning stimulus is of the same size and retinal position as the test spot, the discrimination can be made only by successive comparisons in time. In experiments Kandel (1958) has found that thresholds become higher and more variable. The experimental differences between ipsilateral and contralateral conditions become smaller and may disappear altogether for stimulation of the peripheral retina when the onsets are near one another in time.

The implications of these results are as follows: although the ability of the higher visual centers (presumably the visual cortex) to mediate spatial discriminations based on intensive differences is very good, these centers are relatively poor at mediating discriminations based upon time alone.

An important and very significant exception is the contralateral equal-area condition in which one eye is completely adapted to the conditioning stimulus. In this case, the test-flash threshold is no different from the threshold for complete dark adaptation of the two eyes, regardless of the areas involved. One interpretation is that under steady-state conditions of adaptation, the prevailing rate of activity in the cortex produced by the conditioning stimulus is so low as to exert a negligible effect upon the perceptibility of the contralateral test flash.

Some of the results of the transient adaptation studies are summarized in Table 1.

Table 1. Factor by Which Threshold to Test Flash is Elevated above Threshold for the Dark-Adapted and Otherwise Unstimulated Eye

Data are from the experiment of Kandel (1958) for his eye and are for a conditioning stimulus having a diameter subtending 4 degrees at a luminance of 4.5 mL. The transient condition refers to thresholds obtained 0.05 sec following onset of conditioning stimulus; thresholds for the steady condition were taken after at least 5-min exposure to the conditioning stimulus. For the ipsilateral condition, the conditioning stimulus and test flash were each delivered to the right eye; for the contralateral condition, the conditioning stimulus was delivered to the left eye and the test flash to the right. Threshold contrast* is also given.

Condition of laterality	Diameter of test flash (degrees)	Temporal condition	Factor of elevation	Contrast at threshold (%)*
Ipsilateral	0.5	Transient	12	45
Contralateral	0.5	Transient	1.8	6.7
Ipsilateral	0.5	Steady	2.2	8.5
Contralateral	0.5	Steady	0.9†	3.7
Ipsilateral	4	Transient	1100	91
Contralateral	4	Transient	28	1.7
Ipsilateral	4	Steady	75	6.2
Contralateral	4	Steady	1	0.1

* Threshold expressed as a percentage of 4.5 mL. The concept *contrast* does not apply strictly to the contralateral condition.
† Not significantly different from 1.0.

Quantization of Time

It has been thrice suggested in the preceding discussion that the notion of an input quantization of time in vision is compatible with the results. The basic idea is that the visual input may be packaged in successive time frames, and that, therefore, any two events that occur within a given time frame and that depend upon a temporal discrimination alone for their perception cannot be discriminable. This idea has been mentioned by previous speakers on this program (perhaps most elegantly by Dr. Rushton) and has been expounded in some detail by Stroud (1956). It accords with the familiar fact that the visual presentation of sixteen still pictures each second, the standard rate for "motion" picture projection, provides a rather satisfactory illusion of continuous time-flow. Such time packages also appear consistent with what might be expected, from the standpoint of communication theory, of an efficient detection-transmission system. Numerous questions are raised by this theory, such as whether and how the time frames are triggered, and to what extent they are synchronized among the various components of the visual system handling information from different parts of the visual field. There would seem to be many important psychophysical and electrophysiological experiments that might be addressed to this question.

Flicker

One of the most intensively investigated areas in vision has been flicker. Of the thousands of studies done on this time-related subject, most have involved critical flicker frequency (cff) as the dependent variable. Perhaps the most fundamental relations involving flicker are the following: (1) the increase in cff associated with an increase in stimulus intensity; (2) variations in cff as a function of various complex temporal arrangements of a periodic stimulus in time. The latter range from the classical, often-investigated light-time fraction to very recent studies involving the interlacing of two frequencies within a single, repetitive pattern (Forsyth and Brown, 1959). The cff-versus-intensity function is so basic that one is surprised to find that no really satisfactory theoretical account exists, though several have been suggested. So far as the second class of phenomena is concerned, a plausible explanation exists in the theory of de Lange Dzn (1954), based on his own experiments and some early ideas of Ives and Kingsbury (1914). According to de Lange Dzn, the visual system performs a Fourier analysis on the stimulus pattern and then pays attention only to the fundamental frequency, ignoring the harmonics. In addition, the system is assumed to have a "filtering action," which causes the amplitude of the output, for a given amplitude of input, to decrease with increasing frequency. The behavior of a variety of complex stimulus patterns at a given intensity level may be accounted for in this manner. Also, the cff-intensity relation may be explained as a special case of intensity discrimination. Although why intensity discrimination itself improves with increasing luminance is still the subject of controversy, the principle of de Lange Dzn at least reduces the problem of flicker discrimination to a familiar domain.

There is, however, little hint in the theory of de Lange Dzn concerning how the alleged Fourier analysis is performed by the visual mechanisms. We have recently begun to turn our attention to this and other related problems, in a new kind of experiment that grew out of the conditioning-stimulus–test-flash type of study already described.

Our technique is to superimpose upon a flickering stimulus, a small test flash once each second. The test flash may fall in any part of either the light or dark phase of the cycle. It should be noted that the time intervals involved are of a different order of magnitude than those previously studied, and for this reason extremely brief test flashes must be used. A threshold is determined by the up-and-down method using a complex apparatus which records automatically and permits responses to be made to each flash (once per second). These measurements are repeated for other points in the cycle of the flickering stimulus. As many as thirty thresholds of the test flash, each based upon more than fifty yes-no judgments, may be determined in this way during a one-hour session. The apparatus, which took two years to build, has just recently been completed, and we are in the early stages of data collection.

A sample result is shown in Fig. 5. This is a record of the waxing and waning of the threshold of the test flash in response to a 30-cps flickering light. In this example, two flashes have been removed from each thirty to produce, once per second, a dark period five times longer than usual. The record shows a substantial recovery of sensitivity during this extra-long dark period, followed by an unusually large "on-response" to the first flash following the period of darkness. Then one sees a near obliteration of the "response" to the second flash, after which the system begins to return to normal.

It would seem that we have succeeded, by this technique, in providing an indirect picture of the visual *response* to a flickering light in the human subject. In time, we should be able to analyze such records to show the interrelations among the factors of photosensitivity and neural masking that seem clearly to be involved.

That we are able to obtain significant differences in the thresholds for test flashes separated by only two or three milliseconds bears discussion in terms of the time quantization idea previously mentioned. This result indicates that the sensitivity of the visual system is very precisely specifiable at a given moment. Variability of thresholds obtained under these transient conditions is not unduly large. These results would not be expected if the input were being quantized at the receptoral level. The results are, however, consistent with the idea of time quantization at some higher level of the visual system.

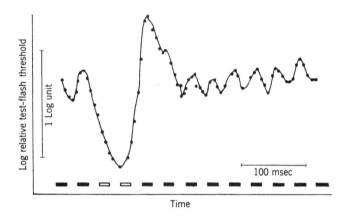

Fig. 5. Variations in the threshold of a test flash superimposed upon a 30-cps flickering light at various precisely timed points in the flickering-light cycle. Solid bars indicate the light phase of the flickering light; unfilled bars show the position of two flashes in each thirty that are omitted to cause a longer-than-usual dark period.

It should be noted in this connection that an observer will report a flicker of 30 cps as appearing substantially slower; further, if asked to count the number of flashes perceived in response to a finite number of flashes at this rate, he will report a substantially smaller number than actually presented (Cheatham and White, 1952; White and Cheatham, 1959; Forsyth and Chapanis, 1958). Our results also appear to confirm those of electroretinographic studies (Armington, White, and Cheatham, 1953) which indicate that the peripheral visual system is capable of following a flickering stimulus at a faster rate than can the central visual nervous system, although the peripheral mechanism does attenuate the amplitude of response to a progressively higher degree as flicker rate increases.

Summary

Three psychophysical experiments involving temporal relations in vision have been reported. The first shows that, even with improved experimental technique, high-intensity short flashes can indeed look brighter than longer flashes of the same intensity, thus indicating that the classical Broca-Sulzer effect is not an artifact. The second experiment was a study of the transitional fraction of a second between dark adaptation and the beginning of light adaptation, using the conditioning-stimulus–test-flash technique. From this second experiment it is clear that the nature of the discrimination involved may vary, depending on whether the eye has just been subjected to an increased level of stimulation or whether it has been allowed to adapt to it. The third experiment concerns the successful measurement of test-flash thresholds in the presence of a 30-cps flickering stimulus. Each of these experiments raises a number of questions regarding the neurophysiological organization, function, and interactions of the visual systems serving the two eyes. A common unifying principle, apparently consistent with the data, is the idea of temporal quantization of the visual input by the higher visual nervous system, so that the input is "packaged" into discrete time frames within which a purely temporal discrimination is not possible.

Addendum

If I may be permitted a few additional remarks, I should like to speak as a psychophysicist addressing those of you who are electrophysiologists.

We who are interested in psychophysics from the standpoint of a curiosity about underlying mechanisms are interested, I think, in the relation between conscious experience and neural events—and please note that I frankly say experience and not behavior. To a large degree, many of you seem to be interested in the same thing. Both of us are bothered, no doubt, by the realization that conscious experience is, by its private nature, outside the bounds of science; yet this stops neither of us from doing what we are doing, though it may prevent us from achieving ultimate answers.

But how can we best help one another toward a better understanding? The most precise information available about experience has been gained by using the observer as a null instrument—by finding combinations of physical variables that combine to produce the same response, be it threshold, brightness match, or what have you. There is no richness of experience in such results. In fact they tell us nothing about the subject's experience—directly. Nevertheless, psychophysical data obtained by treating the human observer as a null instrument have provided a framework within which many sensory physiologists operate, and have in themselves occupied a substantial portion of many textbooks on sensory physiology. The very indirection of the null-instrument approach is what gives it its scientific validity and makes it useful.

We use this technique in psychophysics because, until recently at least, it seemed the only possible one to use on the nonlinear, drifting, and uncalibrated device called the eye. The psychological scaling techniques described by Stevens also look promising, but at the moment the null-instrument results are much more precise. In attempting to relate our data to yours, however, we are often frustrated by the following two related factors:

1. A relatively casual attention to stimulus control. After the achievement of a truly remarkable state of the art as regards surgical and recording technique, it seems a shame to use whistles, hand claps, casual jabs, and unspecified lights as physical stimuli. It is those experiments in which wave length, energy, spatial distribution, and time are carefully specified that are likely to be most useful to us.

2. The very rare use of the null-instrument technique in your work. Take the matter of the interaction of intensity and time, for example, which has been worked out, so far as I know, only in *Limulus*, though I stand to be corrected. With your electrodes buried in the visual system, it would seem a relatively simple thing to determine where, and under what conditions, and with what response measures such a relation holds. The chances are good that flashes of equal time *times* intensity look the same to many animals, as well as to man. By following a program such as I suggest, you could at least establish the neural correlates of equivalent experiences. Surely this kind of work should antedate any serious efforts to find neural correlates for such things as form recognition, since the latter probably involves the comparison of equivalents also—between the coded representation of input, and that of past experience. These equivalents must be considerably more complex, however.

I offer these suggestions with a feeling of humility and a real appreciation for the magnificent work that you are doing. For our part, those of us in psychophysics who share your goals must do our best to keep abreast of your concepts, and to restrict our serious speculations to those that are not contrary to the facts that you have so far discovered.

References

Alpern, M. A study of some aspects of metacontrast. Doctoral dissertation, Ohio State University, 1950.

Alpern, M. Metacontrast. *J. opt. Soc. Amer.*, 1953, **43,** 648–657.

Armington, J. C., C. T. White, and P. G. Cheatham. Evidence for central factors influencing perceived number. *J. exp. Psychol.*, 1953, **46,** 283–287.

Baker, H. D. The instantaneous threshold and early dark adaptation. *J. opt. Soc. Amer.*, 1953, **43,** 798–803.

Battersby, W. S., and I. W. Wagman. Neural limitations of visual excitability, I: The time course of monocular light adaptation. *J. opt. Soc. Amer.*, 1959, **49,** 752–759.

Blanchard, J. The brightness sensibility of the retina. *Phys. Rev.*, 1918, **11,** 81–99.

Bouman, M. A. The absolute threshold conditions for visual perception. *J. opt. Soc. Amer.*, 1955a, **45,** 36–43.

Bouman, M. A. On foveal and peripheral interaction in binocular vision. *Optica Acta*, 1955b, **1,** 177–183.

Boynton, R. M. On-responses in the visual system as inferred from psychophysical studies of rapid-adaptation. *Arch. Ophthalmol.*, 1958, **60,** 800–810.

Boynton, R. M., and W. R. Bush. Dark adaptation and the instantaneous threshold. *Amer. Psychologist*, 1953, **8,** 324 [Abstract].

Boynton, R. M., W. R. Bush, and J. M. Enoch. Rapid changes in foveal sensitivity resulting from direct and indirect adapting stimuli. *J. opt. Soc. Amer.*, 1954, **44,** 56–60.

Boynton, R. M., and G. L. Kandel. On-responses in the human visual system as a function of adaptation level. *J. opt. Soc. Amer.*, 1957, **47,** 275–286.

Boynton, R. M., and M. H. Triedman. A psychophysical and electrophysiological study of light adaptation. *J. exp. Psychol.*, 1953, **46,** 125–134.

Broca, A., and D. Sulzer. La sensation lumineuse en fonction des temps. *J. Physiol.*, 1902, **4,** 632–640.

Broca, A., and D. Sulzer. La sensation lumineuse en fonction des temps. *C. R. Acad. Sci.*, 1903, **137,** 944–946; 977–979; 1046–1049.

Bush, W. R. Foveal light adaptation as affected by the spectral composition of the test and adapting stimuli. *J. opt. Soc. Amer.*, 1955, **45,** 1047–1057.

Cheatham, P. G., and C. T. White. Perceived number as a function of flash number and rate. *J. exp. Psychol.*, 1952, **44,** 447–451.

Crawford, B. H. Visual adaptation to brief conditioning stimuli. *Proc. roy. Soc.*, 1947, **B134,** 283–302.

Forsyth, D. M., and C. R. Brown. Flicker contours for intermittent photic stimuli of alternating duration. *J. opt. Soc. Amer.*, 1959, **49,** 760–763.

Forsyth, D. M., and A. Chapanis. Counting repeated light flashes as a function of their number, their rate of presentation, and retinal location stimulated. *J. exp. Psychol.*, 1958, **56,** 385–391.

Ives, H. E., and E. F. Kingsbury. The theory of the flicker photometer. *Phil. Mag.*, 1914, **28,** 708–728.

Kandel, G. L. A psychophysical study of some monocular and binocular factors in early adaptation. Doctoral dissertation, University of Rochester, 1958.

Katz, M. S. The perceived brightness of light flashes. Doctoral dissertation, University of Rochester, 1959.

de Lange Dzn, H. Relationship between critical flicker-frequency and a set of low-frequency characteristics of the eye. *J. opt. Soc. Amer.*, 1954, **44,** 380–389.

LeGrand, Y. *Light, Colour, and Vision.* New York: Wiley, 1957.

Piéron, H. *The Sensations.* New Haven: Yale University Press, 1952.

Schouten, J. F., and L. S. Ornstein. Measurements on direct and indirect adaptation by means of binocular vision. *J. opt. Soc. Amer.*, 1939, **29,** 168–192.

Stainton, W. H. The phenomenon of Broca and Sulzer in foveal vision. *J. opt. Soc. Amer.*, 1928, **16,** 26–37.

Stroud, J. M. The fine structure of psychological time. In H. Quastler (Editor), *Information Theory in Psychology.* Glencoe, Ill.: Free Press, 1956. Pp. 174–207.

White, C. T., and P. G. Cheatham. Temporal numerosity, IV: A comparison of the major senses. *J. exp. Psychol.*, 1959, **58,** 441–444.

Wulff, V. J., R. G. Adams, H. Linschitz, and E. W. Abrahamson. Effect of flash illumination on rhodopsin in solution. *Ann. N. Y. Acad. Sci.*, 1958a, **74,** 281–290.

Wulff, V. J., R. G. Adams, H. Linschitz, and D. Kennedy. The behavior of flash-illuminated rhodopsin in solution. *Arch. Ophthalmol.*, 1958b, **4,** 695–701.

38

J. Y. LETTVIN, H. R. MATURANA, W. H. PITTS, and W. S. McCULLOCH

Research Laboratory of Electronics, Massachusetts Institute of Technology

Two Remarks on the Visual System of the Frog

Part I. Form-Function Relations in the Retina

In two earlier papers (Lettvin, et al., 1959, and Maturana, et al., 1960) we described the operations on the visual image that were found in the fibers of the optic nerve in Rana pipiens. We remarked then that these operations seemed related to the anatomy of retinal cells. This note will show what kind of form-function relations we think hold. It is provisional, a program rather than a solution.

If we restrict ourselves to form perception such as applies to silhouettes, then we can say what qualities of the visual image are displayed in the output of the frog's retina. Neither color nor shading enters into the account, which is thereby incomplete. Nevertheless under that restriction we can account for almost all the fibers in the optic nerve and show that there are four operations on the image, each carried by a different population of axons. Each operation is invariant under change of average intensity of light by which a given silhouette-background combination is seen. Each operation is also uniformly distributed over the retina. There is, in addition, one group of fibers that respond to average illumination only, and we can say little about their distribution.

Now the anatomy of the frog's retina shows, according to Ramon y Cajal (1894), five types of ganglion cells whose axons make up the optic nerve. These types differ from each other by the shapes of their dendritic arbors—in the same way that species of trees differ when seen in outline. Furthermore the frog's retina has the same transverse structure from place to place along itself, there being no fovea, and the uniform distribution of the operations is matched by the symmetrical anatomy of the retina. Because the number of functions' we discriminate in the optic nerve matches the number of ganglion-cell types, it is natural to ask whether one can assign particular shapes to particular functions.

The material being combined arises from our own experiments and from Ramon y Cajal's 1894 treatise on the retina (which has been amplified by one of us, HRM).

Synopsis of anatomy

Rods and cones (Figs. 1, 2, 3). There are two sorts of rod and one kind of cone in the frog's retina. The red rods are most numerous, closely seconded by the cones. The green rods are relatively scarce. All the photoreceptors are uniformly distributed over the retina. The cell bodies of the rods and cones lie in the outer granular layer internal to the outer limiting membrane. Their "feet" issue into the outer plexiform layer. The feet of the red rods end in the outer zone of that layer, those of the cones in the middle zone, and those of the green rods in the inner zone.

Bipolar cells (Figs. 1, 2, 3). There are essentially three sorts of bipolar cells; their cell bodies lie in the middle granular layer. They are to be distinguished not so much on the basis of size as of connectivity, although the first two kinds are large and the third small. Type I has a dendrite that spreads out in the middle zone of the outer plexiform layer and connects widely and exclusively with the cones. Its axon bushes out widely in the outer zone of the inner plexiform layer. Type II has a dendrite that also branches in the middle zone of the outer plexiform layer, but in a very constricted way, and sends processes into the outer zone. (Ramon y Cajal thinks they connect exclusively with the red rods but is not sure.) Its axon arborizes at several levels of the inner plexiform layer. Type III has a dendrite that connects also over a constricted field, but seemingly with different types of photoreceptors, and in addition sends a process, the Landolt club, up between the somata of the photoreceptors to bulge out at the level of the outer limiting membrane. Its axon penetrates deepest in the inner plexiform layer but emits planar arbors at several of the outer levels.

Horizontal and amacrine cells (Figs. 1, 2). There are two sorts of lateral connectives in the retina. The horizontal cells have their bodies at the outer margin of the middle granular layer, and both dendrites and axons arborize in the outer plexiform layer. The amacrine cells have their bodies at the inner margin of the middle granular layer and emit wide-spreading processes into the inner plexiform layer. Different cells spread out in different levels, and there are

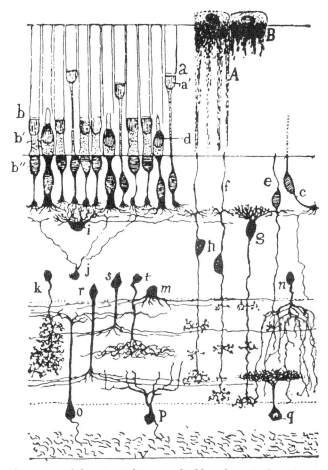

Fig. 1. The retina of frog in Golgi stain—highly schematic but showing spatial arrangement: g and h are bipolar cells; i is a horizontal cell; k, r, s, t, and m are amacrines; o, p, and q are ganglion cells. This is the bad off-hand diagram used for illustration in Ramon y Cajal's *Histologie du Système Nerveux* (Paris: Maloine, 1909–1911). The actual picture of frog's retina in exploded diagram is shown in Fig. 2.

several kinds of arbor—thin- and thick-branched, dense and rarefied. Amacrine cells do not have axons in the anatomical sense. No more is known about them, but they are commonly accepted as nervous elements.

Ganglion cells (Figs. 2, 4). The axons of these cells make up the optic nerve. There are half a million of them in the frog, body to body in a single sheet, the inner granular layer. The count tallies

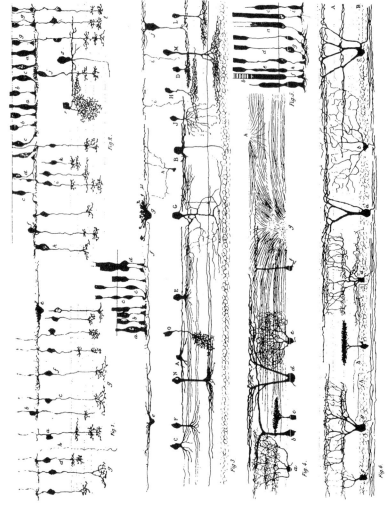

Fig. 2. Ramon y Cajal's detailed illustration of the frog retina from the 1894 study. It consists of six subfigures, as follows: (*Fig. 1*) Various bipolar cells; one horizontal cell is labeled *e*. (*Fig. 2*) Bipolars and rods and cones. (*Fig. 3*) *a, b, c, d* are rods and cones, *g* is a horizontal cell, *i* is type I bipolar in our classification; the cells with capital letters are amacrines. (*Fig. 4*) Some ganglion cells. (*Fig. 5*) Rods and cones. (*Fig. 6*) Ganglion cells: *e* shows the single-layer constricted dendritic spread; *a* is the single-layer extended spread; *c* is the many-level H distribution of dendrites; *a* and *g* are the many-level E distributions; *b* has a diffuse dendritic tree.

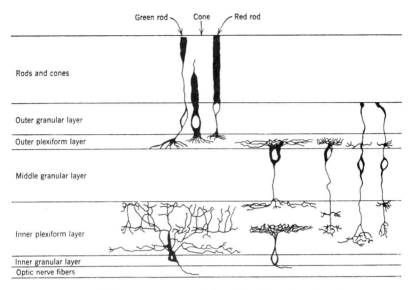

Fig. 3. Diagram of the retina excerpted from Fig. 2 to show the schema we are using in this paper.

with the number of fibers in the nerve (Maturana, 1959). Many of these cells have been erroneously thought to be glial because of the great inaccuracy of former estimates of the number of fibers in an optic nerve (Maturana, 1959). One can distinguish five types on the basis of shape and size of the dendritic trees, as revealed by Golgi stains (Fig. 4).

1. One-level constricted field (Fig. 4; *6e* in Fig. 2). These are the smallest of the ganglion cells; their axons are difficult to stain or see. The major dendrites extend only to the inner levels of the inner plexiform layer and there spread out in dense and constricted planar bush.

2. One-level broad field (Fig. 4; *6a* in Fig. 2). These are the largest of the ganglion cells. The major dendrites extend to the outer levels of the inner plexiform layer and there branch widely over a considerable area.

3. Many-level H distribution (Fig. 4; *6c* in Fig. 2). These are next to the largest of the ganglion cells. The major dendrites extend to the outer levels of the inner plexiform layer, as do those described in (2) above. However, they emit two widely spread arbors, one in the inner levels and one in the outer levels.

4. Many-level E distribution (Fig. 4; *6d* and *6g* in Fig. 2). These are next to the smallest ganglion cells. The major dendrites extend

only to the inner levels of the inner plexiform layer and there branch out in planar fashion. However, each branch emits twigs all along its course, and some extend into the outer levels of the inner plexiform layer, whereas others remain in the inner levels.

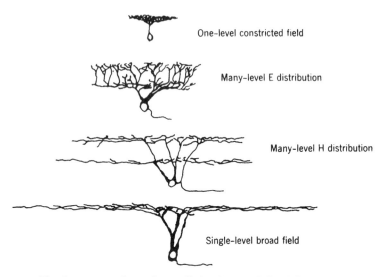

One-level constricted field

Many-level E distribution

Many-level H distribution

Single-level broad field

Fig. 4. The four types of ganglion cell (exclusive of the diffuse dendritic tree) compared for shape and relative size on the same scale.

5. Diffuse trees (Fig. 3; *6b* in Fig. 2). There are several sizes of these cells. The dendrites branch helter-skelter all over the inner plexiform layer and show no planar arrangement such as occurs in the other four kinds of cell.

You will notice that the dendritic arrangements shown in the figure suggest that there are two major subdivisions of the inner plexiform layer, although undoubtedly each subdivision can be divided into several levels.

Synopsis of physiology

As is apparent from the anatomy, any ganglion cell in the frog is connected via the bipolar cells at least to several hundred, or several thousand, photoreceptors. This connectivity is reflected in the observation by Hartline (1938) that any optic nerve fiber responds to visual events over an area of several degrees in diameter on the retina. That area is usually centered over the cell body; it is called the *receptive field*. We found four distinct types of receptive field for fibers in the optic nerve and one kind whose boundaries we could not measure. The types of receptive field are given below, together with the properties of the axons to which they belong.

Perhaps we had better say a word about our experimental procedure, although a full account appears in our other two papers. Our stimuli consisted of silhouettes of different size and shape, cut out of matte white, gray, or black paper and seen against matte white, gray, or black backgrounds. The targets were mounted on soft iron washers and moved against the background by means of magnets. They were viewed by reflected light, and the intensity of that light could be varied. Response of a nerve fiber was measured roughly by frequency and duration of firing. We took no records except those needed to illustrate our papers. Our question was not how great the response was to one or another manipulation, but rather which visual events produced greatest response and which produced least, and what aspect of the image could be varied without changing the response. We dealt with our own listening to the patterns of nerve spikes as the measure of extremes, just as one does with an a-c bridge. (Early in our work we had found that taking records hindered rather than helped this kind of research by leading to premature standardizing of method.)

The groups we discovered are as described in the following pages.

Unmyelinated fibers

These conduct at velocities of 20 to 50 centimeters per second. They make up 97 per cent of the population in the optic nerve (Maturana, 1959).

Group I, the boundary detectors (Fig. 5). These fibers have receptive fields, 2 to 4 degrees in diameter. They respond to any boundary between two grays in the receptive field, provided it is sharp. Sharpness of boundary rather than degree of contrast seems to be what is measured. The response is enhanced if the boundary is moved and is unchanged if the illumination of the particular contrast is altered over a very wide range. If no boundary exists in the field, no response can be got from change of lighting, however sharp that change. If a boundary is brought into the receptive field in total darkness and the light is switched on, a continuing response occurs after a short initial delay. Such axons are the "on" fibers of Hartline—a small well-focused spot of light is defined by a sharp boundary.

Group II, the movement-gated, dark convex boundary detectors (Fig. 6). These fibers have receptive fields of 3 to 5 degrees. They too respond only to sharp boundaries between two grays, but only if

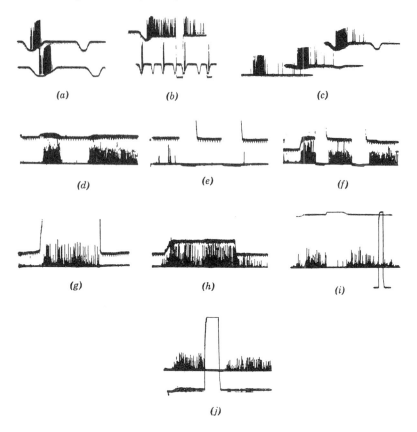

(a) (b) (c)

(d) (e) (f)

(g) (h) (i)

(j)

Fig. 5. The records were all taken directly with a Polaroid camera. The spikes are clipped at the lower end, just above the noise, and brightened on the screen. Occasional spikes have been intensified by hand for purposes of reproduction. The resolution is not good, but we think the responses are not ambiguous. Our alternate method of recording is by means of a device that displays the logarithm of the pulse interval of the signals through a pulse-height pick-off. However, such records would take too much explanation and would not add much to the substance of the present paper.

Operation 1. Contrast detectors. (*a*) This record is from a single fiber in the optic nerve. The base line is the output of a photocell watching a somewhat larger area than the receptive field of the fiber. Darkening is given by downward deflection. The response is seen with the noise clipped off. The fiber discharges to movement of the edge of a 7-degree black disc passed in one direction but not in the reverse movement. Time marks, 20 per second. (*b*) The same fiber continues to respond when the edge of the disc stops in its receptive field. The response disappears when the illumination is turned off and reappears when it is turned on. The lower line shows again the asymmetry of the response to a faster movement. Time marks, 20 per second. (*c*) The same fiber is stimulated to show an asymmetrical response to the 7-degree black object moved in one that boundary is curved, the darker area being convex, and is moved or has moved. That is to say, this fiber does not respond to changes of lighting, however sharp, if no boundary exists in the field, and it also does not respond to a sharp, straight boundary, moving or stationary, in the receptive field. It responds to a net curvature of the boundary if the darker gray is convex and the movement centripetal, or if the lighter gray is convex but is sufficiently relieved against the background to have a sharp shadow at its edge. It does not seem to matter how much darker the target is than the background beyond a certain very small contrast. When the curvature is so great that the target is a small area, about half the diameter of the receptive field, then, if the target is moved centripetally and stopped before the center of the receptive field, it sets up a long-lasting discharge. This discharge, however, is erased by a transient dimming of the light or by a shadow passed over the receptive field, and it will not recur until the boundary is moved again even very slightly. Here, also, small movements of the boundary enhance the prolonged discharge transiently. Again, the responses are invariant over a wide range of illumination, roughly that between dim twilight and bright noon. There are other remarkable aspects of this class of fibers, and they are discussed in our other two papers. Suffice it to say that they have not been seen before because they do not respond to spots of light whose boundaries are convex with respect to the darker gray and which are not moved.

---

direction and then in the reverse direction, and the stimuli are repeated under a little less than a 3-decade range of illumination in two steps. The bottom record is in extremely dim light, the top in very bright light. Time marks, 20 per second. (d) In the lower line, a group of endings from similar optic nerve fibers is shown recorded from the first layer in the tectum. A black disc 1 degree in diameter is moved first through the field and then into the field and stopped. In the upper line the receptive field is watched by a photomultiplier (see text), and darkening is given by upward deflection. Time marks, 5 per second for all tectal records. (e) Turning off and on of general illumination has no effect on these fibers. (f) A 3-degree black disc is moved into the field and stopped. The response continues until the lights are turned off but reappears when the lights are turned on. These fibers are nonerasable. (g) A large black square is moved into the field and stopped. The response to the edge continues so long as the edge is in the field. (h) The 3-degree disc is again moved into the field and stopped. When it leaves, there is a slight after-discharge. (i) A 1-degree object is moved into the field and stopped; the light is turned off, then on, and the response comes back. The light is approximately 300 times dimmer than in the next frame. Full on and off are given in the rectangular calibration at the right. (j) The same procedure as in (i) under very bright light. The return of response after reintroduction of the light seems more prolonged, but only because in (i) the edge was not stopped in optimal position.

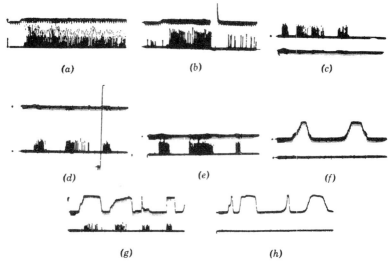

Fig. 6. *Operation* 2. Convexity detectors. The photomultiplier is used, and darkening is an upward deflection. (*a*) These records are all from the second layer of endings in the tectum. In the first picture, a 1-degree black disc is imported into the receptive field and left there. (*b*) The same event occurs as in (*a*), but now the light is turned off and then on again. The response is much diminished and in the longer record vanishes. These fibers are erasable. (*c*) The 1-degree disc is passed through the field, first somewhat rapidly, then slowly, and then rapidly. The light is very bright. (*d*) The same procedure occurs as in (*c*), but now the light has been dimmed about 300 times. The vertical line shows the range of the photomultiplier, which has been adjusted for about 3½ decades of logarithmic response. (*e*) A 1-degree black disc is passed through the field at three speeds. (*f*) A 15-degree wide black strip is passed through the field at two speeds, edge leading. (*g*) A 15-degree wide black strip is passed through the field in various ways, corner leading. (*h*) The same strip as in (*g*) is passed through the field, edge leading.

Both groups I and II can detect targets as small as 3 minutes of arc in diameter and are affected by very faint blurring of the edge. Their resolution is thus quite high.

Myelinated fibers (Fig. 7)

These fibers form about 3 per cent of the optic-nerve population. There are three peaks of conduction velocity, one between 1 and 5 meters per second, one at 10 meters per second, and one at about 20 meters per second. Group III is responsible for the first peak, group IV for the second, and we think the efferents to the retina give the third.

Group III, the moving or changing contrast detectors (Fig. 7).

These fibers have receptive fields, 7 to 11 degrees in diameter. They are, in effect, Hartline's "on-off" fibers with the differencing behavior

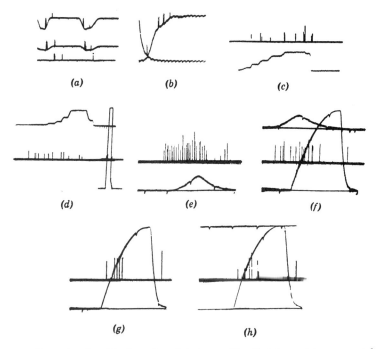

(a)   (b)   (c)

(d)   (e)   (f)

(g)   (h)

Fig. 7. *Operation* 3. Moving-edge detectors. The first two pictures are taken from a single fiber in the optic nerve. (*a*) A 7-degree black disc moving through the receptive field (the photocell was not in registration with the field). There is a response to the front and back of the disc independent of illumination. There is about a 300-to-1 shift in illumination between top and bottom record. Darkening is a downward deflection with the photocell record. Time marks, 5 per second. (*b*) General lighting turned off and on. Time marks, 50 per second. Note double responses and spacing. (*c*) This and succeeding records are in the third layer of endings in the tectum. Several endings are recorded but not resolved. Darkening is an upward deflection of the photomultiplier record. The response is shown to the edge of a 15-degree square moved into and out of the field by jerks in bright light. (*d*) The same procedure occurs as in (*c*) but in dim light. Calibration figure is at the right. (*e*) The response is shown to a 7-degree black disc passed through the receptive fields under bright light. The sweep is faster, but the time marks are the same. (*f*) The same procedure as for (*e*) but under dim light. The off and on of the photomultiplier record were superimposed for calibration. (*g*) Off and on response with about half a second between on and off. (*h*) Same as (*g*) but with 2 sec between off and on.

reported by Barlow (1953). However, they respond invariantly under wide changes of illumination to the same silhouette moved at the

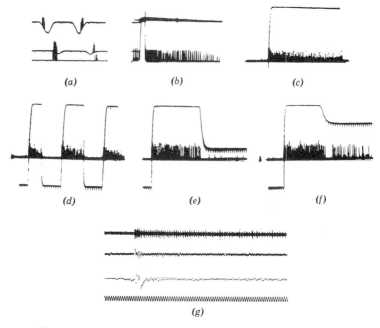

(a)   (b)   (c)

(d)   (e)   (f)

(g)

Fig. 8. *Operation* 4. Dimming detectors. (*a*) This and the next frame are taken from a single fiber in the optic nerve. Here we see the response to a 7-degree black disc passing through the receptive field. The three records are taken at three illumination levels over a 300-to-1 range. In the phototube record, darkening is a downward deflection. Time marks, 5 per second. (*b*) Light turned off and on—off shortly after one sweep began, on a little earlier on the next sweep. The fiber is silenced completely by the on of the light. Time marks, 5 per second. (*c*) In this and the next three frames, we are recording from the fourth layer of endings in the tectum. This frame shows the response to turning off the general illumination. (*d*) Light off and on at regular intervals. (*e*) Light off and on at a lower brightness. (*f*) Light off and on at a still lower brightness. (*g*) The synchrony of the dimming detectors. At the top are three or four fibers recorded together in the optic nerve when the light is suddenly turned off. The fibers come from diverse areas on the retina. In the second line are the oscillations recorded from the freshly cut retinal stump of the optic nerve when the light is suddenly turned off. In the third line are the oscillations recorded on the surface of the tectum, the visual brain, before the nerve was cut. Again the light is suddenly turned off. The last line is 20 cps. These records of synchrony were obviously not all made at the same time, so that comparing them in detail is not profitable.

same speed across the same background. They have no enduring response but fire only if the contrast is changing or moving. The response is better (higher in frequency) when the boundary is sharp or moving fast than when it is blurred or moving slow. Targets as small as 1 degree in diameter are effective throughout the whole receptive field. Sharp changes of illumination of a uniform field produce one or two on and off discharges. Local on or off in the center of the field produces more discharges over a longer period. *Group IV, the dimming detectors* (Fig. 8). These detectors are Hartline's "off" fibers. They respond to any dimming in the whole receptive field weighted by distance from the center of that field. Boundaries play no role in the response. Thus resolution is poor. The same percentage of dimming produces the same response, more or less independent of the level of lighting at the beginning. Turning the light off produces an enduring response characterized by marked rhythmicity. The on intensity needed to interrupt the discharge completely and promptly decays with length of time after the off.

Unclassified

Group V is rare. We cannot even say whether it has a receptive field in the usual sense. It fires at a frequency that is inversely related to the intensity of the average illumination. When the lighting is changed, that frequency slowly changes to its new level. It may be significant that group V fibers end in the same layer of the tectal neuropile as do those of group III.

Argument

If we should assume that the size of the dendritic field to some extent determines the size of the receptive field, and that the size of the cell body is also related to the diameter of the axon, as seems to be the case elsewhere in the nervous system, we would emerge with a fairly definite correspondence between cell types and operations. The one-level constricted field would be associated with the *boundary* detectors, the one-level extended field with *dimming* detectors, the many-level H-shaped with the *contrast movement and change* detectors, the many-level E-shaped with the *movement-gated, dark convex boundary* detectors. The diffuse type, which is rare, we associate, more or less by default, with the average light-level measuring group. Our initial assumption depends for its validity mainly on the idea that the amacrine cells do not act in such a way as to invert the dendritic-field/receptive-field relation, and this seems to be the case for several reasons. First of all, the diameters of the dendritic fields match well the angular diameters of the receptive fields. Second, the cell bodies are distributed in size in the same way as the dendritic fields, and, if the axon diameters reflect soma size, then the largest axons ought to have the largest receptive fields and the smallest axons the smallest fields; and this seems to be the case. Finally, receptive fields often appear to be not circular but elliptical or cardioid. In Maturana's pictures of the ganglion cells, both elliptical and cardioid dendritic fields appear.

This identification of the physiological operation with the type of dendritic tree purely on the basis of dimensions leads us to further but very speculative deciphering. We have taken the inner plexiform layer as consisting of two major strata, not denying that further subdivision may be important. However, suppose that the output of a ganglion cell reflects the sort of information it collects. There are two types of one-level cell. The first, which is concerned with boundaries but not with changes in average intensity, arborizes at the inner stratum; and the second, which is concerned with changes in level of lighting but is indifferent to boundaries, arborizes at the outer stratum.

Can the two many-layered types be considered as different combinations of information about boundary and information about dimming? For example, let the H type take some function of dimming in all its receptive field and combine it with some function of boundary in all that field. Let the E type take some function, over all the small areas of its receptive field, of how much dimming there is in any one small area combined with how much boundary there is in that area. One would expect, from such a view, that the E type could resolve more complex geometrical properties of the image in its receptive field than could the H type. For the E type can be thought combinatorial of many small dendritic processes of the H type. Thus we consider the H type as an event detector, responding to any change in distribution of light and boundaries on its receptive field. The E type, however, is more constrained, in that a change in distribution of light must be coherent with change in sharp boundaries. Therefore, the H type matches with the changing contrast detectors and the E type with the dark convex boundary detectors, and this match is consistent with relative dimensions of receptive and dendritic fields.

A much more specific guess, however, concerns the nature of the bipolars and their relations to the photoreceptors. The explicitness of Ramon y Cajal almost ensures that the information coming to the outer levels of the inner plexiform layer is concerned mainly with the activity of the cones (those bipolars that connect only to cones and have broad dendritic spreads in the outer plexiform layer), whereas the information to the inner levels is combinatory of the different photoreceptors (those bipolars that connect to rods and cones and have very constricted dendritic fields in the outer plexiform layer). Let us suppose for the moment that there are but two sorts of receptor, the red rods and the cones. Suppose again that the receptors somehow or other have an effect on the bipolars that is logarithmically related to the amount of light incident—after all, this can be argued from the dynamic range of the retina and the Weber-Fechner law. If the effects of the rods and cones on the bipolars (connected to both sorts of receptors) are opposite, then those bipolars connected only to cones ought to measure some function of total illumination or its change, whereas those connected to rods and cones ought to measure some function of local differences—in a word, something about contrast and boundaries—and that measure should be independent of average illumination. That this may in fact be the case is suggested by the work of Svaetichin and McNichol (1958) who, impaling an element somewhere between the outer and middle granular layers of the fish, found that as light intensity increased they got either continuous increases of depolarization or overpolarization of the element, depending on the color of the light.

It is very tempting to go further and construct a wonderfully complex model accounting for three photoreceptors, their relative densities, and so forth. But it would also be silly, for there are infinitely many that one may make (we have made some). The general notion that one sort of bipolar cell may "see" different kinds of photoreceptors as opposed in action, while another sort of bipolar sees one kind of photoreceptor only, is really the nucleus of the second guess and seems to be worth exploring. This notion is suggested by the anatomy; it is convenient to our deciphering of the ganglion cell function; it is not contradicted by any facts we know; and in fact it is supported to some extent by work on retinas of fish.

Part II. Collicular Cell Activity and Efferents to the Retina

In our other two papers we noted that the optic nerve fibers, having reached the colliculus, separate into four separate concentric sheets of terminals in the outer neuropile; that each stratum displays a continuous map of the retina in terms of one of the first four operations; that all four maps are in registration; and that, from the surface in, they occur in the order given, boundary detectors first, dimming detectors deepest. We also examined the cells in the tectum which receive from these terminals (Fig. 9). The great majority of them lie in the granular layer subjacent to the external neuropile and send their dendrites up to the very surface of the colliculus, as in Pedro Ramon's drawing. Some cells, but very few, lie in the external neuropile and look like horizontal connectives; these we shall not discuss. The cells in the granular layer do not emit their axons there; rather

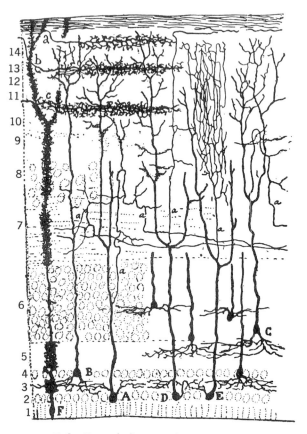

Fig. 9. Pedro Ramon's drawing of optic tectum in the frog.

the axons arise from the large ascending dendrites and travel laterally in what we call the "palisade layer" which separates the dimming detector endings from the granular layer.

There seem to be several kinds of these cells, as one might expect. We have not been able to define the subgroups at all well, but there are two major populations of polar types which we have named "newness" neurons and "sameness" neurons. The former have axons which can be seen physiologically in the optic nerve, so they are part of the efferent system. Their cell bodies lie in the outer zone of the granular layer beneath the palisade layer. The somata of the "sameness" neurons lie deep in the granular layer, very close to the ependyma.

Our description of the operations is not definitive—we are far from satisfied. Nevertheless it gives the flavor of the responses we have seen.

"Newness" neurons

These cells have receptive fields about 30 degrees in diameter, twice that of the dimming detectors. They are distributed so as to map continuously the visual field with much overlap. Such a neuron responds a little to sharp changes in illumination. If an object moves across the receptive field, there is a response whose frequency depends on the jerkiness, velocity, and direction of the movement, as well as on the size of the object. There is never an enduring response.

Different cells have different optimum directions for movement and different preferred sizes of objects. But what is remarkable about the responses of such a unit is a kind of adaptation which is best described thus: if we move an object along any particular diameter of the receptive field in one direction and get a response, then repeating that movement over the same path in less than 5 to 10 sec brings no response. Even with a 20-sec delay to the second stimulus, the second response is much reduced below the first. But the adaptation is only to that path. If we move the object for the stimulus along another diameter, at a 90-degree angle to the first, then the first stimulus does not seriously interfere with the second. Both paths can be adapted within a second, whereupon a third stimulus running over any diameter will not be very effective in under 10 sec. In some of these units the adaptation can be erased by a moment of complete darkness. One variant we have seen is the response to centrifugal movement reported by Andrew (1955) several years ago on our laboratory. A good fraction of such cells have transient responses to turning on or off the general illumination.

"Sameness" neurons

Each of these cells sees all or almost all of the visual field but has a "null region" that is different for different cells. We cannot say whether the null patches are so distributed as to map the retina in this layer. It is a bit embarrassing to present the following description so batrachomorphically, but at least it reflects what we have found so far. Every such cell, in fact, acts so complexly that we can hardly describe its response save in terms ordinarily reserved for animal behavior.

Let us begin with an empty gray hemisphere for the visual field. There is usually no response of the cell to turning on and off the illumination. It is silent. We bring in a small dark object, say 1 to 2 degrees in diameter, and at a certain point in its travel, almost anywhere in the field, the cell suddenly "notices" it. Thereafter, wherever that object is moved it is tracked by the cell. Every time it moves, with even the faintest jerk, there is a burst of impulses that dies down to a mutter that continues as long as the object is visible. If the object is kept moving, the bursts signal discontinuities in the movement, such as the turning of corners, reversals, and so forth, and these bursts occur against a continuous background mutter that tells us the object is visible to the cell.

When the target is removed, the discharge dies down. If the target is kept absolutely stationary for about two minutes, the mutter also disappears. Then one can sneak the target around a bit, slowly, and produce no response, until the cell "notices" it again and locks on. Thereafter, no small or slow movement remains unsignaled. There is also a place in the visual field, different for different cells, that is a sort of Coventry to which a target can retire and escape notice except for sharp movements. This Coventry, or null patch, is difficult to map. The memory that a cell has for a stationary target that has been brought to its attention by movement can be abolished by a transient darkness. These cells prefer small targets, that is, they respond best to targets of about 3 degrees.

There is also (we put this matter very hesitantly) an odd discrimination in these cells, which, though we would not be surprised to find it in the whole animal, is somewhat startling in single units so early behind the retina. Not all "sameness" cells have this property. Suppose we have two similar targets. We bring in target A and move it back and forth along a fixed path in a regular way. The cell sees it and responds, signaling the reversals of movement by bursts. Now we bring in target B and move it about erratically. After a short while, we hear bursts from the cell signaling the corners, reversals, and other discontinuities in the travel of B. Now we stop B. The cell goes into its mutter, indicating that what it has been attending to has stopped. It does not signal the reversals of target A, which is moving back and forth very regularly all the time, until after a reasonable time, several seconds. It seems to attend one or the other, A or B; its output is not a simple combination of the responses to both.

These descriptions are provisional and may be too naturalistic in character. However, we have examined well over a hundred cells and suspect that what they do will not seem any simpler or less startling with further study. There are several types, of which the two mentioned are extremes. Of course if one were to perform the standard gestures, such as flashing a light at the eye, probably the cells could be classified and described more easily. However, it seems a shame for such sophisticated units to be handled that way—roughly the equivalent of classifying people's intelligence by the startle response.

Some axons of the "newness" neurons have been seen in the optic nerve. To a degree we suspect their action to be reflected (?) in the response of the movement-gated, dark convex boundary detectors—or perhaps it is the other way round. At any rate, in a convexity detector, if we move a small dark object in rapidly and stop, we can get a continuing response after the stop unless the object has been moved in along the same path immediately before, within, say, a few seconds.

All these comments apply only to the frog. That must be emphasized. Neither the anatomy nor the receptive-field operations are the same in mammals or even in other amphibia.

Summary

Part I. We propose that a deciphering of retinal anatomy is possible. We find a correspondence in number between operational groups at the output of the retina and anatomical groups. The relative size of dendritic field to receptive field is such that the anatomical and functional groups can be matched. There are significant differences in the shapes of dendritic fields of different size and there are equally great differences in the operations done by fibers with receptive fields of different sizes. In attempting to decipher the anatomical differences by means of the functional differences, we come to the hypothesis that (with respect to the discrimination of silhouettes) the inner levels of the inner plexiform layer are concerned with boundaries, whereas the outer levels are concerned with average (or changes in the average) illumination. This hypothesis can be extended to account for the fact that bipolar cells that end exclusively in the outer levels of the inner plexiform layer are connected only to cones, whereas those that end deepest in the inner layers are connected to both sorts of rods as well as the cones. Our supposition is that those bipolar cells that receive information only from cones report average, or changes in average, illumination, whereas those that receive from rods and cones report the two types of photoreceptors in opposition to each other. If the photoreceptors affect the bipolars logarithmically with respect to the exciting light, the latter sort of bipolar cell will report some function of boundaries invariant under change of illumination. The different operations seen at the output of the retina can be considered combinatory of the sort of information we suppose comes to the inner plexiform layer via the bipolar cells. The different combinations we think are got by different ways of distributing dendrites.

Part II. We discuss two sorts of cells in the colliculus that receive information from the optic nerve fibers. One of them is concerned with detection of novelty in visual events, and the other with continuity in time of interesting objects in the field of vision.

Acknowledgments

This work was supported in part by the U. S. Army (Signal Corps), the U. S. Air Force (Office of Scientific Research, Air Research and Development Command), and the U. S. Navy (Office of Naval Research), and in part by Bell Telephone Laboratories.

Dr. Maturana is on leave from the University of Chile, Santiago, Chile.

References

Andrew, A. M. Report on frog colliculus. Quart. Prog. Rep., Res. Lab. Electronics, Mass. Inst. Technology, 1955, 77–78.
Barlow, H. B. Summation and inhibition in the frog's retina. *J. Physiol.*, 1953, **119,** 69–88.
Cajal, S. Ramon y. *Die Retina der Wirbelthiere.* Wiesbaden: J. F. Bergmann, 1894.
Hartline, H. K. The response of single optic nerve fibers of the vertebrate eye to illumination of the retina. *Amer. J. Physiol.*, 1938, **121,** 400–415.
Lettvin, J. Y., H. R. Maturana, W. S. McCulloch, and W. H. Pitts. What the frog's eye tells the frog's brain. *Proc. Inst. Radio Engr.*, 1959, **47,** 1940–1951.
Maturana, H. R. Number of fibres in the optic nerve and the number of ganglion cells in the retina of Anurans. *Nature, Lond.*, 1959, **183,** 1406.
Maturana, H. R., J. Y. Lettvin, W. H. Pitts, and W. S. McCulloch. Physiology and anatomy of vision in the frog. *J. gen. Physiol.*, 1960, **43 Suppl.,** 129–175.
Svaetichin, G., and E. F. McNichol, Jr. Retinal mechanisms for chromatic and achromatic vision. *Ann. N. Y. Acad. Sci.*, 1958, **74,** 385–404.

39

Comments

Fred Attneave*

In Defense of Homunculi

Perceptual phenomena are typically discussed in terms that seem to imply, or to take for granted, the existence of a perceiver—a homunculus who has some awareness of the outside world as a result of sensory data received and directs the activities of the organism accordingly. Yet most psychologists, including those who use such terms, will maintain stoutly that a "little man inside the head" is a gross absurdity, and will resort to elaborate circumlocutions rather than admit a homunculus to any systematic status. This taboo—for it virtually amounts to that—seems to have two major bases. The first is a morbid fear of ghosts, that is, a fear of admitting into one's thinking anything that might possibly be suspected of immateriality. The mention of a "perceiver" is likely to conjure up visions of a fluffy cloud of nonmatter hovering in the pineal region (perhaps connected with that body by a slender strand of ectoplasm), resistant to extirpation, insensitive to microelectrode stimulation, and generally quite beyond the pale of scientific investigation. The second objection, mentioned by Professor Boring in his opening address to this symposium, has to do with the supposedly regressive nature of the concept: if all the responsibility for perception and action is attributed to a homunculus, explaining his behavior poses exactly the same problem as explaining that of the whole organism, and we have got nowhere.

Now, it ought to be more obvious than it apparently is that both of these objections are directed at a little man of straw, and that one may perfectly well postulate a homunculus to which neither applies.

* University of Oregon.

In the first place, we may suppose—in the absence of overwhelming evidence to the contrary—that if a homunculus exists it must certainly be composed of neurons. In the second place it should be noted that we fall into a regress *only* if we try to make the homunculus do every-thing. The moment we specify certain processes that occur *outside* the homunculus, we are merely classifying or partitioning psycho-neural functions; the classification may be crude, but it is not in itself regressive. Indeed, one might even hypothesize a series of concentric or nested homunculi without falling into a regress, provided each con-tained an outer layer of functions not contained in the next smaller one

The homunculus concept of Professor Bullock is certainly neither ghostly nor regressive. He suggests that, for any given behavior, there must be at least one neuron that "decides," on the basis of ac-tivity in receptors and other neurons, whether to initiate that behavior or not. There may be as many "homunculi" of this sort as there are coherent behavior patterns. Although I have no disagreement what-ever with this argument, I wish to suggest a psychoneural system in which the homunculus has more nearly the character of a complex organism-within-an-organism and is still neither ghostly nor regressive.

What I shall present has very nearly the methodological status of a descriptive language: it is intended not as a theory, but rather as a framework within which a certain class of theories might be developed.

In Fig. 1, *P* stands for the perceptual machinery of the nervous system, *M* for the motor machinery, *A* for affective centers, and *H* for the homunculus.

The homunculus is conceived to have the following characteristics: (1) it is the region of convergence and interaction of objective infor-mation and affective or evaluative information; (2) its output (whether to *M* or to *P*) constitutes *voluntary behavior,* whereas, in contrast, reflexes are shown by-passing *H*, and highly automatized skills receive only gross (on-off or supervisory) voluntary control; (3) whatever is *conscious* is represented by activity within *H*; (4) whatever informa-tion is stored in permanent memory must first be represented by activity in *H* (though learning may occur much more generally; see below).

It is equally important, as suggested above, that we specify what the homunculus does *not* do. Pursuing the metaphor: let us com-pare his receptor surface with that of the whole organism. His dis-crimination of different states on a single sensory continuum like pitch or brightness is surprisingly crude, as if he possessed only about half a dozen receptors for any such continuum, though the correspondence

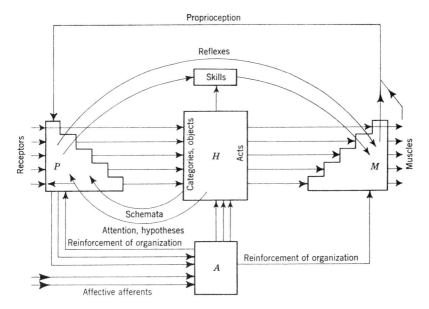

Fig. 1. Block diagram of a possible nervous system.

between these and the much greater number of external receptors seems highly flexible. The evidence that I am interpreting in this manner has been ably summarized for this symposium by Pollack, and I refer the reader to his paper in this volume.

More generally, it is supposed that each receptor at the surface of *H* is sensitive to a class of objects or events; for example, that the homunculus has a receptor for dogs, another for pianos, another for automobiles, and so on. This is to say that the over-all function of *P* is to classify receptor states of the organism in ways corresponding to stable and recurrent features of the external world (see Boring, 1952; Bruner, 1957). With respect to *H*, objects thus have the status of unitary stimuli. (For years, stimulus-response psychologists have got by with referring to objects as "stimuli," despite the bewildering variety of receptor states to which a single physical object may give rise.) What constitutes an object is variable, however. A line on paper may be an object; so may a letter composed of several lines; so may a word composed of several letters. Considerations of this sort suggest that *P* is organized in a hierarchical manner, and that *H* is in contact with various levels of the hierarchy, not merely the highest one. In some cases, through connections from the organism's receptors to *H* may exist, such that direct awareness of activity in these receptors is possible; in other cases H may receive only the output of one or more stages of analysis in P: what is or may be accessible to consciousness is an empirical issue, not to be settled by abstract arguments over the validity of the method of introspection or phenomenology.

Only about six or eight objects—or perceptual units, or "stimuli" in the molar sense, or "chunks" of information (Miller, 1956)—may be held in immediate memory at a given time. It is immaterial whether the items carry much or little information (as measured in bits) or whether they represent the activity of many or few of the organism's receptors. This bottleneck must be located at or near the receptor surface of the homunculus, since the competitive items are outputs of the perceptual machinery.

It is proposed that feedback loops, perhaps of several different kinds, exist between H and P. Attention may be voluntary, and we know from the recent work of Galambos, Hernández-Peón, and others that the facilitating and inhibiting effects of attention may extend quite far into the periphery. Perceptual hypotheses may involve selective facilitation or inhibition at higher levels of P. A perceptual schema is conceived as a pattern of activity in P that is modal or normative with respect to a given classification or output of P and that, when evoked, provides the basis for further and finer classifications (Attneave, 1957). It is suggested that schemata are evoked by activity at or near the receptor surface of H.

The motor system functions in a manner somewhat symmetrical with the perceptual. Whereas the function of P is to reduce a complex input to a unitary output, that of M is to receive a unitary order from H (or from P in the case of involuntary behavior) and elaborate it into a spatiotemporal pattern of muscular activity (cf. the "tree" or "pyramid" circuits employed in computers). This view is consistent with the subjective character of voluntary action. Moreover, it has been recognized since the 1890's, when Bryan and Harter (1897, 1899) studied the learning of Morse code by telegraph operators, that the acquisition of a predominantly perceptual skill, like receiving code or reading, and the acquisition of a predominantly motor skill, like sending code or typing, are remarkably similar processes. On either side learning is extremely slow, and typically shows several "plateaus," or periods of no progress. Bryan and Harter believed the resumption of progress following a plateau to be dependent upon the organization of material into larger units, either perceptual or motor. Subsequent research has supported this interpretation. The fact that the organizational systems eventually achieved tend to be valuable or adaptive rather than otherwise suggests the presence of paths from A to both P and M (Fig. 1). Alternatively, it might be supposed that this control is exercised via H.

In a discussion as incautious as this, a speculation on the neuro-anatomical correlates of the boxes in Fig. 1 can do little further harm. It seems likely that most of the operations attributed to P and M are accomplished in the cerebral cortex; A is perhaps roughly identifiable with the limbic system. For H I would seek a subcortical location. Penfield (1938, 1947)—whose work has exercised a major influence on my own thinking in this area—has insisted for a number of years that the "seat of consciousness" or "highest level of neural integration" must be located in the old brain. I shall not attempt to recapitulate his arguments, but I recommend them to the reader. Recent studies involving the reticular system make it a plausible choice for the homuncular role. If my understanding of Hernández-Peón's paper (in this volume) is correct, he is assigning the reticular formation almost exactly such a role.

Now the system blocked out above is not one that I propose to defend for the rest of my life. It is undoubtedly incomplete, even as a block diagram (for example, thinking and planning and also memory storage probably involve structures that should be shown as differentiated from H), and it may in fact be quite wrong. Worst of all, it bears a distressing resemblance to the common-sense conceptions of psychological structure held by many persons untutored in psychology.

Nevertheless, I would argue strongly that nothing in a system of this sort is inherently unscientific. The charge that an opponent is "assuming a little man inside the head" has been accepted as an argument stopper for too long. It is time that we reconsidered the scientific respectability of the homunculus.

References

Attneave, F. Transfer of experience with a class-schema to identification-learning of patterns and shapes. *J. exp. Psychol.*, 1957, **54**, 81–88.

Boring, E. G. Visual perception as invariance. *Psychol. Rev.*, 1952, **59**, 141–148.

Bruner, J. S. On perceptual readiness. *Psychol. Rev.*, 1957, **64**, 123–152.

Bryan, W. L., and N. Harter. Studies in the physiology and psychology of the telegraphic language. *Psychol. Rev.*, 1897, **4**, 27–53.

Bryan, W. L., and N. Harter. Studies in the telegraphic language. The acquisition of a hierarchy of habits. *Psychol. Rev.*, 1899, **6**, 345–375.

Miller, G. A. The magical number seven, plus or minus two: some limits on our capacity for processing information. *Psychol. Rev.*, 1956, **63**, 81–97.

Penfield, W. The cerebral cortex in man. I. The cerebral cortex and consciousness. *Arch. Neurol. Psychiat.*, 1938, **40**, 417–442.

Penfield, W.   Some observations on the cerebral cortex of man.   *Proc. roy. Soc.,* 1947, **B134,** 329–347.

H. B. Barlow

Three Points about Lateral Inhibition

1. Ratliff has described very elegant experiments that tell us a good deal about *what* the lateral inhibitory mechanism does and something about *how* it does it, but there remains a third question to ask. The fact that this mechanism has evolved independently in a wide variety of sensory relays suggests that it must have considerable survival value: *why* is this so?

The suggested answer is that it enables almost the same amount of information to be transmitted with a smaller expenditure of impulses. It is thus an example of a redundancy-reducing code and confers the advantages that Attneave (1954) and I have argued for. Figure 1 is intended to show the kind of transformation lateral inhibition imposes on a picture and to illustrate that this does reduce redundancy. Figure 1*a* is a typical scene at Endicott House; Fig. 1*b* shows the result of transforming the scene by a photographic process that mimics lateral inhibition. The method is to make a contact print through a negative of the original scene; the negative, instead of being illuminated with uniform light, is illuminated by an out-of-focus positive of the original scene. This gives an intensity of illumination in a small region p on the negative proportional to the average illumination of a larger region round p in the original. If the original is divided into small discrete regions m, n, and so on, illuminated at intensities I_m, and so on, this average on a line through p and extending s discrete regions on each side would be

$$\sum_{p-s}^{p+s} I_m/(2s + 1)$$

For a square or circular region around p the expression would be similar, but the limits of the summation would be complicated. Let us write

$$\sum I_m/b$$

where b is the number of regions averaged, the particular ones involved being understood without specification. The transmission of the negative at p will be proportional to $1/I_p$ if the gamma is unity:

(a)

(b)

Fig. 1. (a) F. Attneave and F. A. Geldard tackling a nonredundant stimulus. (b) The same after transformation by a process imitating lateral inhibition. Little information has been lost, but the redundancy has been considerably reduced.

so the light reaching the exposed paper will be

$$\sum I_m / I_p b$$

and after development its reflectance, R_p, will be

$$R_p = I_p \, b / \sum I_m$$

that is, it will be proportional to the intensity at that point in the original divided by the mean intensity over a region surrounding the point. Now if there were as many values of R_p in the transform as there are values of I_p in the original, the set of simultaneous equations would be soluble: actually, the terms in the summation for a point on the border of the transform must include points in the original for which there is no corresponding point in the transform, so the

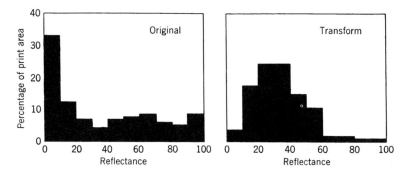

Fig. 2. Distribution of reflectance in Figs. 1a and b. The original (a) has mean reflectance 37 per cent, rms deviation from mean 32 per cent. The transform (b) has mean 34 per cent, rms deviation 17 per cent, so it would need less power in a signal-indicating modulation of a scanning beam.

equations are not soluble unless we already know something about the original (for example, that the intensity outside the border of the picture is zero). These edge effects mean that there is a loss of information, but it will be small if the area of the picture is large compared with the area over which the averaging is done. Figure 2 suggests that, associated with this small loss of information, there is a considerable reduction in the channel capacity required to transmit the picture. It shows the fraction of picture area that has a specified reflectance in the two cases, and it is clear that the spread of the distribution is considerably greater in the original than it is in the transform. In point of fact the variance is almost four times as great, or the rms deviation from the mean almost double. If the picture was scanned, the power of the signal required to modulate the beam would be almost halved; hence the channel capacity and the redundancy would be reduced. This argument is suggestive, not conclusive, for a number of reasons, including the serious inaccuracies of photographic measurements of this sort and the fact that we have not considered the effect of the transformation on the values of the discrete intensity levels employed, nor the effects of noise if this is present.

It will be noticed that in the above case the intensity of a signal at a point has been compared with the average of its value taken over surrounding regions. In the example of recoding given in my paper (p. 230) I suggested comparing it with a temporal average, but in other respects the processes are similar.

2. Both Katsuki and Mountcastle suggest in their papers that lateral inhibition occurs at successive relays in sensory pathways. It is interesting therefore to visualize what would happen if the process described above were repeated, and it is fairly easy to see that something qualitatively different emerges. Consider a simple white-black border, such as that at the edge of Geldard's shirt in the original. In the first transform this goes to gray-white-black-gray. In the second transform it would go to gray-black-white-black-gray. So the fringes multiply and spread out. Of course it may well be that a single stage of neural lateral inhibition does not go nearly as far as the photographic process described, but all the same it is clear that one should be cautious in suggesting that several stages of lateral inhibition do no more than can be done in a single stage.

3. The process described above reduces the redundancy associated with the fact that the luminance of a point in a visual scene is (usually) correlated with the local average luminance. If we look at a transformed picture such as Fig. 1b it might be possible to spot redundant features that still remain, and thus get clues about what is coded out at following relay points. An obvious feature is the presence of parallel white and black lines corresponding to edges of uniformly illuminated areas in the original. If one takes a high reflectance region of Fig. 1b there is a higher than average probability that a neighboring region will also be white, and that another neighboring region, at right angles, will be black. How could advantage be taken of this redundant feature to economize impulses?

The principle is to arrange that the feature that occurs more frequently than chance expectation shall be represented after coding by a smaller aggregate of impulses than were required before coding. So one would expect to find units that detected these features, and these units should inhibit activity in other units connected to the same input fibers. As far as I know, inhibition by a feature that occurs quite frequently, but has a low chance expectation of occurring, has not been described; but Hubel (1959) has actually found, recording from cells in the conscious cat's visual cortex, units that seem ideally suited to detect parallel adjacent white and black lines, which is the type of redundant feature described above. Unfortunately it would be disingenuous to claim that this confirmed the argument, and hence the hypothesis, because I knew of Hubel's result before presenting the argument. However Attneave was arguing along rather similar lines in 1954, and this clearly antedates the experimental result. Hubel's results also confirm another prediction of the hypothesis—namely, that units must be arranged to detect, not just movement, but movement in a specific direction, at a fairly early stage of analysis of the visual image. This prediction (Barlow, in press) was made before the experiment, but it must be admitted that it would probably follow from almost any other sensible idea about the analysis of the visual image!

References

Attneave, F. Informational aspects of visual perception. *Psychol. Rev.*, 1954, **61,** 183–193.

Barlow, H. B. The coding of sensory messages. In W. H. Thorpe and O. L. Zangwill (Editors), *Current Problems in Animal Behaviour.·* Cambridge: University Press (in press).

Hubel, D. H. Single unit activity in striate cortex of unrestrained cats. *J. Physiol.*, 1959, **147,** 226–238. Also personal communication.

Comment on Neural Quanta

In S. S. Stevens' figure pertaining to the neural quantum (p. 809), we have the percentage of occasions on which an increment of a given size was detected. In each successful detection the "threshold" must have been below the value of the increment used, so merely by differentiating the response curve we can obtain the frequency-distribution curve for the instantaneous values of the threshold itself. If Stevens' straight lines are correct, these distributions will be rectangles with absolutely vertical edges, the breadth being equal to the distance of the left-hand edge from zero. I have two comments to make, one about the shape, the other about the breadth, of this distribution.

1. An almost rectangular frequency distribution is not impossible, but it is sufficiently unusual to require rather careful substantiation before it is accepted. I agree that Stevens' results are suggestive,

but they do not quite convince me. Each point is subject to a sampling error because of the finite number of trials used to estimate the percentage heard. With 100 trials, this leads to a standard error of about ±5 per cent for the middle points, and about ±3 per cent for values near 10 per cent and 90 per cent, with a very asymmetric distribution of errors in the latter case. It would be interesting to see the results plotted *without* a thick straight line through them, and *with* the sampling error of each point indicated (because of the asymmetric distributions, fiducial or confidence limits would be still better). These errors make it difficult to assess "by eye" whether the rectangular distribution has been proved to exist or not, but there is a method of statistical analysis that should, with only minor modifications, enable the problem to be solved. This is the well-developed technique of probit analysis (Finney, 1947), which has been extensively used to investigate dosage-response relations, though its origins apparently go back to Fechner himself.

If the distributions do suffer from pronounced negative kurtosis, as Stevens suggests, then one would be prompted to take another look at the experimental method to see if there is a factor in the design of the experiment that would tend to discourage big deviations of the threshold from its mean. I believe that the technique of presenting blocks of stimuli of the same intensity will do just this. If the subject has not heard 19 out of 20 stimuli, he is unlikely to pay much attention to the twentieth, and will not hear that one either; likewise if he has heard 19, he will be more ready to hear the last one. This factor would operate less in the middle range, where he hears some weakly and is doubtful about others. So before accepting the rectangular distribution of thresholds I feel that one needs to obtain intensity response curves with random sequences of stimulus intensities, and to analyze them with an appropriate statistical technique.

2. Stevens claims that the breadth of the distribution is equal to the distance of the left-hand edge from zero. The evidence that the breadth is *proportional* to its distance from zero seems to be better than the evidence that it is *equal* to it, and this will be considered first.

For the reasons given above it will be assumed that the apparently rectilinear functions are approximations to, or distortions of, ogival psychometric functions. The proportionality of breadth and separation from zero then implies that the slope of the ogive is inversely proportional to the threshold intensity. If this is a valid experimental fact, does it provide unequivocal support for the neural quantum hypothesis, or can it be explained otherwise?

It turns out that these facts could very easily be explained otherwise,

for this inverse relation between slope and threshold is a direct prediction of another current hypothesis about the nature of sensory
thresholds. This hypothesis states that thresholds are internal decision
criteria that are automatically adjusted to such a value that false
positive responses occur at a constant, low rate. The hypothesis is a
very general one and requires only that there should be some kind of
random disturbance or noise, and that the threshold should be
adjusted to a level such that false positive responses would rarely be
caused by it. As far as the present prediction is concerned, it does
not matter what the source of noise is; it could be extrinsic or intrinsic,
in the apparatus, in the sense organ, or in the central nervous system.

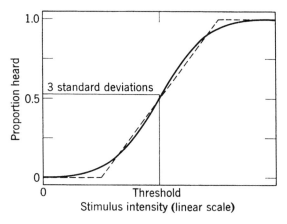

Fig. 1. Diagram showing dependence of the threshold intensity of stimulus and
the slope of the psychometric function upon the scatter of instantaneous values of
threshold and the false positive rate. The ogive drops to 0.001, the postulated
false positive rate, at zero stimulus intensity. The midpoint (threshold) then
lies 3 standard deviations to the right, and the dotted line approximating to the
ogive has a slope of 1/Threshold.

Figure 1 illustrates the argument. The continuous curve is the
normal ogive, along which a dotted straight line has been drawn as
an approximation. First travel down the ogive to zero stimulus
intensity. The ordinate here is very nearly zero, and it represents
the small probability that the instantaneous value of the threshold
will sink to or below zero stimulus intensity. On the rare occasions
when this does occur, something will be "heard" when no stimulus is
present; thus this probability represents the frequency of false positive responses. For the moment let us assume that this probability
is 1/1000. From the integral of the normal error function we find
that the ogive drops to 1/1000 at nearly 3 standard deviations from

the mean. Now the mean is the 0.5 point on the ogive, which is also the threshold; it follows that the threshold intensity of stimulus is equal to nearly three times the standard deviation of instantaneous values of threshold, if the false response rate is 1/1000. For other rates, the threshold will be a different multiple of the standard deviation, but two cautions are needed at this point. First, we are here dealing with the tail of a distribution, and, even among common types of distribution, tails vary a great deal. Second, things like guessing or carelessness on the part of the subject may have a relatively big effect on the tail, even when their effect elsewhere is small. These difficulties do not, however, alter the fact that, if the false positive rate is constant, the threshold should be a constant multiple of the standard deviation of the instantaneous values of threshold.

Now consider the slope of the ogive. I have drawn a dotted straight line through the 0.15 and 0.85 points. This does not depart from the ogive by more than ± 0.03 anywhere between the 0.10 and 0.90 levels, which incidentally shows what a tricky business it is going to be to prove rectilinearity experimentally. Now the slope of a line joining two fixed points on the ogive depends upon the standard deviation (σ) of the distribution from which it is derived, and in this case it happens to be exactly $1/3\sigma$. But in the previous paragraph it was shown that the threshold was equal to 3σ. Hence the slope will be equal to 1/threshold, or, allowing for different false positive rates and departures from normality, the slope will be inversely proportional to threshold.

There remains the third feature of psychometric functions that Stevens explained by the neural quantum hypothesis, namely, that the upper intercept is twice the lower. The straight line approximating to the ogive in Fig. 1 is not steep enough for this to be true, and to make it steep enough we should have to assume an improbably low false positive rate or a departure from normality. But in my experience visual, psychometric functions are not usually as steep as NQ theory demands, and there is the possibility that those factors in the design of the experiments which cause straightening of the ogive will also steepen it.

Obviously experiments are required to answer the following questions: (1) Do competent observers yield rectilinear functions when the experimental design is such as to avoid spurious *straightening?* (2) Is the slope 1.5/threshold (as predicted by NQ) if spurious *steepening* is avoided? (3) Does the slope vary with the false positive rate, as predicted by the alternative theory?

Meanwhile, it may give casual onlookers some wry amusement to observe that a psychologist is backing a physiological explanation, whereas a physiologist prefers a psychological explanation.

Reference

Finney, D. J. *Probit Analysis.* Cambridge: University Press, 1947.

T. H. Bullock

Four Notes for Discussion

Following presentation by Stevens

It strikes one outside this field quite forcefully that it is in a fair way to building up a science based on one species.

The kinds of questions being asked and the results so far are certainly of very great interest. They are highly relevant to a general science of behavior. But the concepts and principles would be greatly enriched if, instead of only special cases for man, we exploited some of the opportunities in lower forms.

There are many ways to measure how much a stimulus of a certain strength means to an animal, even though we cannot equate this with conscious sensation. The probability of a certain all-or-none response, the height of a jump, the angle of swimming down, the force developed against an opposing force, and others suggest themselves. More than ten years ago, von Holst was measuring the tilted posture of angel fish under the influence of different intensities of gravity (in a whirling chamber with running water) and of illumination from the side. Not only could he get the equivalences of these at different strengths but he could change the fish's weighting of light by putting a pinch of food juice in the water. The angel fish being a species that hunts by sight, it now tilted farther toward the light. Schöne has more recently found that two different measures of the response of shrimp to rotating stripes in the visual field show different behavior in relation to stimulus intensity—one linear, the other logarithmic.

Not only would further attention to lower forms contribute to the general value of the science of psychophysics but, in view of the need that has been strongly exposed here in the discussions for electrophysiological correlates, they might offer real advantages. Having nervous systems with fewer parts may significantly reduce the complex problem of where to put the electrodes! And, echoing a previous speaker, if you combine ablation, to which lower forms are more tolerant, that problem becomes simpler still. It will of course be difficult even so, and interpretation will be hard put to distinguish whether the correlate is still on the afferent path or already in the efferent path, instead of at the magic point just between! But if the effort is worth while, as I believe enthusiastically it is, the great variety of animals and responses offers an inexhaustible reservoir.

Following presentation by Woolsey

Referring to the large number of "auditory" areas or areas of cortex yielding evoked potentials to acoustic stimuli, I think a general point is not out of place, although it does not detract in the least from the findings that have been described. It would be well if we learned what our forms of stimulation mean to our animal subjects in terms of their normal behavior, or if we better designed our stimuli to resemble those the animal normally responds to. We could then interpret more easily the response we see.

Some animals have powerful built-in responses. Chickens react in opposite ways to a high-pitched "cheep" and a low-pitched "cluck." I can imagine an "auditory" area in which the high- and low-frequency zones would in fact be the regions of organization of these opposite commands.

Following presentations by Mountcastle, Landgren, and Buser

Mountcastle, Landgren, and Buser have cells that respond to stimulation of widely separated areas of skin and in forms calculated to involve several different modalities. What I want to call attention to in particular is the opportunity and desirability of using stimuli calculated to excite only thin afferent fibers. In each of several modalities there are both thick and thin fibers serving the same receptors, as has been pointed out especially by Katsuki. It is possible to stimulate each of these alone and also to block differentially. Zotterman among others has implicated the thin fibers in some nerves as more potent than the thick in determining reflexes. It would be well worth while to know their relative influence on single cells in the cortex and thalamus.

Following presentation by Jung

It is curious that we think nothing of stimulating the eye with a large sudden "on" but would not do the equivalent to the vestibular system—a blow to the head or a shock to the whole vestibular nerve. Dr. Jung very appropriately used a simulated normal stimulus to the latter receptors, calculated to excite some and inhibit others as natural stimuli do. But apart from lightning, animals probably never receive a sudden large illumination without a component of movement in the visual field. We know that light-on excites some receptors and inhibits others and that the system normally receives such events only with movement or in a small area contrasting with background. Since at the first stage of descriptive analysis all we want is a simple form of stimulus that is not too abnormal, would a small light spot or a pattern of spots perhaps moved in a simple way not be preferable to a bolt of lightning?

Commentary on Symposium

Hallowell Davis

It may be helpful to the experimental psychologists and to the communication experts of the symposium if I, as a neurophysiologist, repeat for emphasis a few points that have been made by one or more of our speakers. They are all points that I personally endorse most heartily. Some of them have emerged only recently as a growing convergence of opinion among neurophysiologists.

The first point is the complexity, and the nature of the complexity, of neural interaction in the gray matter of the central nervous system. These "centers" are more than *relay stations*. The nodes of Ranvier take care of the necessary power boosting to transmit impulses along axons, and the word "relay" should not be applied to gray matter. It implies something altogether too simple. There is always an interaction or an integration or a recording or a selective control. In the gray matter there are small neurons with short axons—axons so short that they may never develop full-blown all-or-none responses but may operate purely by electrotonic conduction or by graded local responses or by decremental conduction or by chemical neurohumors. Dr. Rushton emphasized the chemical aspects of neural action in these areas. *Only in the axons is the neural code a digital code.* I prefer to call the areas of interaction the *analog areas* in contrast to the *digital lines* that carry messages between them. Figure 1 sketches these relations.

Professor Fessard gave us a superb synthetic review of the concept of analog areas and also of parallel interacting systems. His paper best expressed the central core of neurophysiology of the symposium, and I endorse all the concepts he expressed. He called attention to the phylogenetically old reticular formation and some of the characteristics of its activity, including its lack of somatotopy. Dr. Mountcastle gave a dramatic illustration of this difference between the old, diffuse, slow but highly integrated, system and the new fast system that is adapted for discrimination and mensuration. I pointed out that G. H. Bishop notes that the new system is composed of large cells with large heavily myelinated fibers in contrast to the smaller fibers of the old system. It is worth noting that most of psychophysics, as well as most of neurophysiology, has been done on the new large-fibered systems. The recognition of the relation of the reticular formation, the limbic lobe, and so forth, to sensory systems and the cerebral cortex is a relatively recent development.

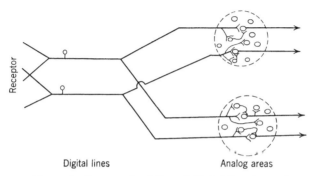

Digital lines   Analog areas

Fig. 1. Schematic diagram of relation of digital lines to analog areas.

Dr. Woolsey disorganized our previous concepts of the auditory cortex. It is now a series of areas, centering around the original "primary projection area" and perhaps also his new "suprasylvian area" as a center. Two or three neighboring areas show tonotopic localization, whereas others, more peripheral, show longer latencies, no tonotopy, and apparently an increasing degree of polysensory responses. It will be interesting to see just what relation these outlying auditory areas in Woolsey's animals bear to the areas of the human brain designated by Penfield as "speech areas," "interpretive cortex," and the areas that can evoke "memory patterns." Dr. Neff told us of certain rather complicated auditory discriminations that could not be relearned by cats after excision of all auditory areas. It may be significant that a common feature of these tasks is that they all require a short-term auditory memory.

The place of the cerebral cortex in the general scheme of organization of the central nervous system is not clear. Many of us hesitate to regard it as necessarily the "highest" level, but prefer to stress the close relation of each cortical area to the thalamic area with which it is most closely connected. Emphasis is now given to this vertical principle of organization and rather little to horizontal "cortical association pathways." This principle was not clearly stated at our symposium, but it is relevant to many of our problems.

The homunculus who lives in the head and "does the perceiving and the discriminating and makes the decisions" came in for some lively discussion, both in our formal sessions and in the dining room and the gun room. I am personally in sympathy with Dr. Bullock's arguments, particularly as elaborated in informal discussion. Dr. MacKay pointed out that an interconnected mass of neurons could function by analog principles as well as Dr. Bullock's single neuron or set of parallel neurons. The "decision center," I believe (with Penfield) must be relatively small, and should be located near the anterior end of the old part of the brain stem, not in the cortex.

Peter Elias*

A Note on the Misuse of "Digital" in Neurophysiology

Neurophysiologists at this meeting and elsewhere have distinguished the continuously graded electrical activity in the nervous system from the "digital" activity, meaning by the latter the spike potentials that obey the all-or-none law. On the side of the theory of automata, people engaged in proving theorems about what machines can do have parenthetically remarked that the nervous system itself is a computer, part analog and part digital. Putting these two statements together, it is natural to reach the conclusion that the pulse-operated part of the nervous system is the digital part, and the graded potentials represent the only analog activity present. But this is in fact a grave confusion of terminology. In its customary usage in the field of computers and data-processing devices, "digital" means much more than "pulse-operated." Although spike potentials in the nervous system may obey the all-or-none law and thus may not be able to represent continuous input variables by continuous variation in their amplitudes, the intervals between pulses are not so restricted. In fact the typical situation at the periphery seems to be a mapping from a continuous range of stimulus intensity to a continuous range of pulse frequency. This is an analog system in engineering terminology—a time-sampled one, but still analog, since the output pulse frequency varies continuously with variation in the input.

Even when discrete elements are present in the response—as in the

* Department of Electrical Engineering, Massachusetts Institute of Technology.

results of Rose and Mountcastle (1954), where as stimulus intensity increases the response jumps rather abruptly from one pulse per cycle to two and then at a later point to three, and so forth—there is still ample evidence of continuous parameters present in the waveform, for example, the interval at the output between pulses in a train and the duration of the period of the waveform. And these parameters vary continuously in the range between the discrete jumps in response. There is also evidence that fixed stimulus intensities at values near those that cause a change in number of pulses per period produce, in fact, output periods that sometimes have one and sometimes the other number of pulses. Consequently an averaging circuit later in the system could, by examining output for a sufficiently long time, deduce fine shades of variation in stimulus intensity merely from the count of pulses per period without making use of the other analog variables also present.

This is not to deny that digital operations take place in the nervous system. The fact that people can count and do logic shows that digital operation does in fact occur. But there is at least a serious question whether any digital activity takes place at the level of single nerve fibers, and it is certainly true that the great bulk of the activity that is cited by the neurophysiologist as digital is not so in fact.

Of course it is possible that, although continuous variation of a parameter like a pulse interval occurs in a neural signal, it is not meaningful. The system in its later stages may respond only to some digital aspect of a set of signals, such as the sequential order of arrival of spikes from different sources, independent of the exact time intervals that separate them. But this can never be deduced from observations of the signal alone. It can only follow from joint observation of the signal and all the responses it elicits at later stages in the system. Thus to call a signal "digital" is to make a very strong assertion as to its effects on all later stages, and some more neutral phrase, such as "pulse," is much more appropriate if the object is to describe the signal, and not to make assertions about all its possible influences in the nervous system. For the same reason, a subsystem in the nervous system should be described as "pulse-operated" rather than "digital" unless a strong assertion about function and mode of operation is intended.

Reference

Rose, J. E., and V. B. Mountcastle. Activity of single neurons in the tactile thalamic region of the cat in response to a transient peripheral stimulus Bull. Johns Hopkins Hosp., 1954, **94**, 238–282.

R. Jung

Note following Presentation by Ratliff: Baumgartner's Results on Lateral Inhibition and Reversal of Neuronal Responses as a Neurophysiological Basis of Border Contrast

The importance of Ratliff's findings for simultaneous contrast seems quite evident, and he has shown very clearly the parallels with human subjective experience. For those who think that the jump from *Limulus* to human experience might be too big, I should like to show some records that Dr. Baumgartner in my laboratory has obtained from the cat. They demonstrate that similar neuronal mechanisms are at work in various levels of the mammalian visual system, forming the basis of pattern vision.

Baumgartner has collected a great deal of material from all levels of the visual system, but to date only a preliminary note with Hakas (1959) without figures has been published. Therefore I show here an example of his work on simultaneous contrast in a geniculate neuron similar to the effects on cortical neurons, treated in my paper.

Baumgartner investigated the effects of contrasting stimuli on the neurons of the retina, the geniculate body, and the visual cortex in cats, using Bremer's *encéphale isolé* preparation and recording neuronal responses at light-on and off when a grid of black and white stripes was exposed at different positions in their receptive fields. The results are summarized in Fig. 1 for a geniculate neuron, showing reversal of the response with projection to a black field. Essentially the same is true for retinal neurons, as is indicated by Kuffler's (1953) and Barlow's work (with FitzHugh and Kuffler, 1957) and also for cortical neurons, as shown in other papers (Jung, 1959a, b). One interesting and surprising difference seems to be that the receptive field for cortical neurons is smaller than for the lower-level neurons in the optic nerve and the geniculate body.

The figure shows that the pattern and frequency of discharge of the neuron are dependent upon the relation of its receptive field to the contrast and are changed completely when the contrasting stimulus is moved over the receptive field: in the middle of the white field the discharge corresponds approximately to diffuse illumination, but when the receptive field is somewhat nearer to the contrasting black field at the margin of the white, the discharge is increased. With further movement into the black zone, the discharge diminishes again. Finally when the receptive-field center projects to the black zone, there occurs reversal of the discharge: inhibition during light-on and sometimes activation following light-off. As this neuron responds with an on discharge to diffuse light stimulation, let me call it an on neuron. But these on neurons, and similarly all our cortical B neurons, respond in a reverse manner by inhibition, similar to an off neuron, when they are stimulated at light-on by the black strip,

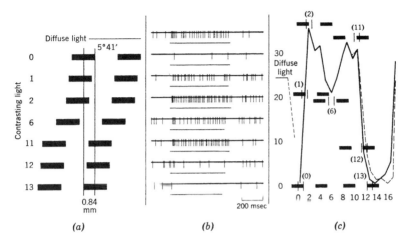

Fig. 1. Variation of neuronal discharge of an on neuron in the lateral geniculate body responding to different positions of the bright and dark zones in its receptive field. (*a*) Grid position in relation to the receptive-field center of the neuron: situation of the contrasting white and black stripes of the grid is marked in relation to the activated neuron. (*b*) Corresponding neuronal discharges to light (marked by horizontal bar). Diffuse light causes only weak responses of few spikes. In contrast, activation responses are maximal when the neuron is exposed to the white-black contrast at grid positions 2 and 11. Vice versa, inhibition of discharge occurs when the neuron is exposed to the black field; inhibition is maximal at the border of the white field in grid position 13. (*c*) Frequency diagram of the number of discharges in 0.5 sec following light-on at different grid positions. The dotted line shows a control of another spike count.

although the total light falling on the retina at light-on is markedly increased.

If you will allow me to use psychological terms to describe this behavior of the neurons more clearly, I would say the following: the on neurons of the retina or the geniculate body and the B neurons of the cortex respond to contrasting objects with a graded activation at light-on, corresponding to the increased subjective brightness of the receptive fields at the border of black stripes, and vice versa. Inhibition occurs when the receptor field lies in the black field: these

B neurons then respond to light-on with cessation of discharge when their field would appear dark subjectively at light-on. The reverse effect may be observed in off and D neurons: although they show activation at light-off in diffuse light or projecting on the white stripe, the discharge is inhibited by light-off when their receptive field has corresponded previously to the black stripe. Thus all these neuronal responses correspond to the subjective sensation in contrasting fields, when we assume that on or B neurons carry information of brightness and off or D neurons carry information of darkness from the retina to the visual cortex.

These observations of Baumgartner's show an elaborate mechanism of contrasting fields controlling pattern vision at three different levels of the visual system, similar to lateral inhibition in the retina of lower forms. They demonstrate a neurophysiological basis for contrast vision in the cat and confirm the parallels with human vision that Ratliff has drawn from *Limulus* eye.

References

Barlow, H. B., R. FitzHugh, and S. W. Kuffler. Change of organization in the receptive fields of the cat's retina during dark adaptation. *J. Physiol.*, 1957, **137**, 338–354.

Baumgartner, G., and P. Hakas. Reaktionen einzelner Opticusneurone und corticaler Nervenzellen der Katze im Hell-Dunkel-Grenzfeld (Simultankontrast). *Pflügers Arch.*, 1959, **270**, 29.

Jung, R. Microphysiology of cortical neurons and its significance for psychophysiology. In *Festschrift Prof. Cl. Estable, Anal. Facult. Med. Montevideo*, 1959a, **44**, 323–332.

Jung, R. Mikrophysiologie des optischen Cortex: Koordination der Neuronenentladungen nach optischen, vestibulären und unspezifischen Afferenzen und ihre Bedeutung für die Sinnesphysiologie. In *Medicine of Japan 1959*, 15. Gen. Ass. Japan Med. Congr. Tokyo, 1959b, **5**, 693–698.

Kuffler, S. W. Discharge patterns and functional organization of mammalian retina. *J. Neurophysiol.*, 1953, **16**, 37–68.

P. L. Latour* and M. A. Bouman

A Nonanalog Time Component in Visual Pursuit Movements

In reaching toward the ambitious aim of describing eye movements quantitatively in terms of a complicated closed-loop servomechanism model, time components such as latency times and reaction times are among the facts that must first of all be considered.

* Institute for Perception RVO-TNO, Soesterberg, The Netherlands.

Reaction time to sensory stimuli is easily accessible to accurate measurements with relatively simple technical arrangements when motor outputs are involved from the hands, the arms, the head, and other not very delicate parts of the human body. For detection of the exact moment at which a bulbus movement starts, less-simple techniques have to be and are being developed. These methods are well known and have been reviewed and discussed on their merits by Riggs, Armington, and Ratliff (1954). All such investigations start with the change in direction of reflected light, either directly incident on the cornea or via a mirror attached to the bulbus, or the change in electrical potential due to bulbus rotation picked up with electrodes on the skin around the eye. For our first studies we chose the latter method, and for laboratory use we developed, with the usual troubles, conventional apparatus—d-c amplifiers with high gain based on the chopper principle, direct recording of the output of these amplifiers, an electronic decade time counter, and so forth. The bandwidth covered was from 0 to about 30 per second; and the output noise corresponded to about 1 microvolt at the input with an input impedance of 200 kilohms, a drift not larger than 5 microvolts per half hour.

We started with very simple problems like reaction-time measurements in fixation movements in order to get acquainted with this type of work. It was expected that for a particular subject for a particular fixation movement some simple frequency distribution would be demonstrated by the corresponding reaction time around some most frequent value of about 0.18 sec. The truth proved to be different.

In a dark, electrically shielded room the dark-adapted test subject was situated opposite a television screen, 20 inches in diameter, at a distance of about 10 inches. On the screen a small bright test spot made step movements at irregular intervals, from a position 30 degrees to the right to a position 30 degrees to the left, and so on. Reaction times for these horizontal movements were measured with the decade counter, which was started by the stimulus and stopped by the change in electrical potential from the electrodes. An accuracy of about 5 msec was obtained with this arrangement.

It proved that in short sessions the frequency distribution of the reaction time for movements both to the left and to the right demonstrated values of preference that were equally spaced on the time scale, $t = \bar{t} \pm k\alpha$, in which $\bar{t}$ is the most frequent t value, $k = 0, 1, 2, \cdots$, $\alpha = 20$ to 40 msec (Figs. 1 and 2). In longer sessions this effect faded out. Some of the curves of the type shown in Fig. 1 were obtained with series of about 600 individual reaction times. With these large series we assured ourselves that it was not an artifact produced by statistical fluctuations in number of occurrences of the different values.

It was thought that the fading out in longer sessions was due to a slight continuous change in $\bar{t}$. This would mean that, when differ-

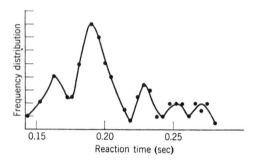

Fig. 1. The frequency distribution of 60 reaction times in a short run of about 25 sec.

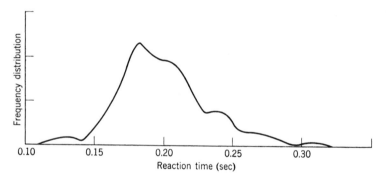

Fig. 2. The frequency distribution of 600 reaction times in a long run lasting about 30 min.

ences in succeeding reaction times for the same direction of movements were plotted, the dips in the distribution must return (Fig. 3). Indeed they do. In almost all cases we again found $\Delta t = \bar{\tau} \pm k\alpha$, in which Δt is the difference in reaction time in succeeding movements $\Delta t = t_n - t_{n+1}$. The maximum peak in Fig. 3 occurred at some positive time value $\bar{\tau}$. This points to the fact that there is indeed a slight continuous decrease in $\bar{t}$ in most cases. It is clear that $\bar{t}$ cannot continue this decrease systematically for a long time. When after some time a reverse in $\bar{t}$ occurs, it must be a fast change; otherwise for these long runs $\bar{\tau}$ would be at the point $t = 0$ (Fig. 4).

We next looked at the distribution in differences between succeeding differences, and the maximum peak was again not at the zero point. Hence the rise and fall of $\bar{t}$ in Fig. 4 contains t^2 components. As far as could be traced back in these experiments, applying this

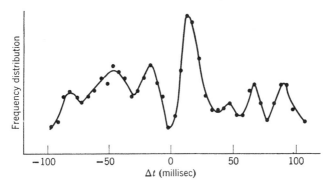

Fig 3. The frequency distribution of the difference in successive reaction times from Fig. 2.

procedure for a third time pointed to the existence of even t^3 components. We think $\bar{t}$ demonstrates in time a saw-tooth course with a period of about 20 sec: series of reaction times deduced from runs lasting less than 20 sec did not demonstrate the fading-out effect that made Fig. 1 change to Fig. 2.

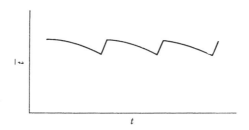

Fig. 4. The saw-tooth representation of $t = l_1 + l_2$.

When the reaction times for movements to the right are combined with those to the left, the effect disappears. In the same session on the same subject, $\bar{t}$, $\bar{\tau}$ and α are different for left and right movements. There seem to be different channels through which the two types of movements are handled.

The time component $k\alpha$ makes apparent some aspect of periodicity.

We found α values between 20 and 40 msec; these are frequencies between 50 and 25 per second. The mechanism underlying this periodicity in reaction times for fixation movements may be active as well when a simpler task is aimed at, such as fixation of a steady target.

There may be some relation with the frequency distribution in involuntary eye movements during steady fixation, about which Riggs, Armington, and Ratliff (1954) reported. Tremor components in the 20- to 50-per-second range occurred frequently in their work.

Of course it is premature to build a model based on these very first studies, but Fig. 5 presents a model with the same characteristics as

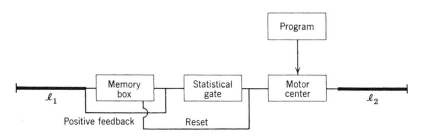

Fig. 5. The model.

our measurements have demonstrated. The stimulus arrives after a time l, at the input of the memory box. The output of this box is fed back positively to the input in such a way that it will start oscillating. By this mechanism, once a signal is applied to this unit, the unit itself repeats the signal periodically with increasing amplitude. The repetition rate is thought to be about 25 to 50 times per second, and the amplitude increases according to t^2. After about three periods the amplitude is saturated and remains essentially constant. In Fig. 6 the amplitude of this mechanism is represented. The output of the memory box is connected to a selecting device from which the threshold continuously changes statistically. Once a signal of the memory box succeeds in passing this statistical gate, the box is damped in order to clean the memory, by feedback from the gate to reset the box.

In the motor center the program decides on the adequate response, and this response acts on the eye muscles.

The time taken for the l_1 and l_2 range together is $\bar{t}$, α is the frequency with which the pulse pattern repeats in the memory box.

The saw-tooth behavior of Fig. 3 is attributed to $\bar{t}$ and may be influenced by training, motivation, and attention.

In summary, reaction-time measurements in fixation movements of the eye were expected to show a simple frequency distribution around some most-frequent value of about 0.18 sec. In contrast to this, it

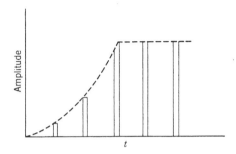

Fig. 6. The amplitude of the signal at the output of the oscillating amplifier as a function of time.

proved that for horizontal movements, which were the only ones studied, values of preference exist that are equally spaced, $t = \bar{t} + k\alpha$, in which $\bar{t}$ is the most frequent t value, $k = 0, 1, 2, \cdots$, and $\alpha = 20$ to 40 msec. In long sessions this behavior is hidden by a continuous decrease in $\bar{t}$, followed by a relatively fast increase in periods of about 20 sec.

Reference

Riggs, L. A., J. C. Armington, and F. Ratliff. Motions of the retinal image during fixation. *J. opt. Soc. Amer.*, 1954, **44**, 315–321.

W. D. Neff

On the Relationships between Physical, Physiological, and Psychological Variables

During this conference, evidence from physiological experiments in which response of peripheral nerves has been recorded when the external physical stimulus was varied has been compared with evidence from psychophysical experiments in which the behavioral response of human organisms has been measured when the external physical stimulus was varied. The latter, psychophysical data, have in some instances appeared to be explainable in terms of the former, physiological data. It may be satisfying to the investigator when the two sets of data seem to agree; it may also be misleading. A necessary intervening relationship is being ignored.

Every physiologist knows that any change in a dimension of the physical stimulus acting on a sensory receptor may result in more than one kind of change in the response of the nerve fibers excited by the sensory receptor. For example, when the sound pressure level of a pure tone is increased, more nerve fibers in the auditory nerve are excited, and an increase in the number of impulses per second occurs in some nerve fibers. And every psychophysicist knows that any change in a physical stimulus acting upon a sensory receptor may result in change in more than one attribute of sensation. For example, when the sound pressure level of a pure tone is increased, there is a change not only in loudness but also in other attributes of hearing, such as pitch and volume. In explaining behavioral response in terms of neurophysiological events, one needs to discover then what aspect of the neural change elicited by the physical stimulus is the basis of the behavioral response. For example, loudness may depend upon the number of nerve fibers excited, the frequency of firing in certain nerve fibers, the total flow of nerve impulses, or some higher-center change that depends upon one or more of these peripheral events.

It may be that in some cases the agreement between findings from electrophysiological and psychophysical experiments is not misleading but real. Final evidence is lacking, but change in the intensity of sensation, say, loudness, may be a function of change in total flow of nerve impulses, and the usual method of recording by means of a gross electrode on a nerve trunk may give a reasonably accurate measure of the total flow of impulses.

This discussion might be interpreted as an argument for more and better psychophysiological experiments. It is. Those of us who call ourselves psychophysiologists or physiological psychologists must take most of the blame for any confusion that may exist. In both amount of experimentation and development of our techniques of performing experiments, physiological psychology has lagged behind sensory physiology and psychophysics.

The chief method of physiological psychology in the investigation of sensory processes has been that in which behavioral responses to sensory stimuli were examined before and after ablation of neural centers or transection of neural pathways. This method has yielded valuable data, but data that have not always been easy to interpret in the attempt to fill the gap between electrophysiology and psychophysics. The use of electrical recording and electrical stimulation techniques via implanted electrodes in unanesthetized, behaving animals should supply new and useful information. In the experiments reported by Dr. Galambos, the value of the electrical recording technique has been demonstrated. I would like to describe very briefly an experiment in which electrical stimulation has been used to investigate the physiological events underlying sensory discrimination (Neff, Nieder, and Oesterreich, 1959).

Cats with electrodes implanted in auditory pathways or centers were trained to make a conditioned avoidance response (hindleg flexion) to an auditory stimulus (a train of clicks sounded over a loudspeaker). After a stable conditioned response had been established, immediate transfer of the conditioned response was observed in several cases when electrical stimulation of the auditory system was substituted for the external acoustic stimulus. A train of shock impulses to the central nervous system served as a stimulus equivalent to the sounds produced by the loudspeaker. In other cases the procedure was reversed. Animals were trained to respond to electrical stimulation of auditory centers, and responses were found to occur, without further training, when external sounds were substituted for the electrical stimulation. These results require more thorough investigation; there are obvious opportunities for errors that may lead to misinterpretations. Nevertheless, the method is one that holds promise for unraveling the coding and decoding mechanisms of the sensory nervous systems.

In conclusion, lest I be accused of overlooking the past, may I point out that the ideas expressed in the first part of my comment are not new. Fechner, the father of psychophysics, although necessarily limited in his knowledge of the neurophysiology of sensory systems, made the important distinction between inner and outer psychophysics, with inner psychophysics referring to the relation between sensation and neural events, outer psychophysics referring to the relation between sensation and the external physical stimulus (Boring, 1942, p. 37). A third relation that has been mentioned here is that between the physical stimulus and neural events.

References

Boring, E. G. *Sensation and Perception in the History of Experimental Psychology.* New York: Appleton-Century-Crofts, 1942.

Neff, W. D., P. N. Nieder, and R. E. Oesterreich. Learned responses elicited by electrical stimulation of auditory pathways. *Fed. Proc.*, 1959, **18**, 112. [Abstract.]

S. S. Stevens

Is There a Quantal Threshold?

The interesting review by Dr. Pollack of some of the accomplishments stemming from "detection theory" led to a day of lively discussion. Although no *post hoc* recounting can capture the full spirit of the interchange, the following pages try to display one of the points of view presented.

The existence of thresholds, both absolute and differential, has stood so long as a tenet in psychophysics that its outright denial is sure to attract our attention. But, as sometimes happens in such circumstances, this iconoclastic proposition turns out to be less of a shocker once its meaning is explained. (For a clear and informative account of this issue the reader should turn to Licklider, 1959.) As I understand it, the idea is simply that, when a human observer undertakes to detect a signal immersed in noise, his behavior has much in common with statistical decision theory. Being confronted with two statistical distributions, that of the noise and that of the noise plus the signal, the observer seems to behave as though he were testing a statistical hypothesis: given a sample, he tries to decide which of the two populations it came from. Depending on the "pay-off matrix," the observer may be timid or bold in his willingness to commit errors of one kind or another, and the degree of his daring helps to determine a boundary criterion (a cut-off) for the categories of his response. Since the parameters of the experiment can move the cut-off up or down the "decision axis," there is said to be no unique threshold in the sense of an all-or-none process.

It is interesting to note that complaints against the notion of a unique threshold have been voiced with persistent regularity. For example, American psychology's first Ph.D., Joseph Jastrow, produced a "theoretical refutation of the threshold theory" for Volume 1 of the American Journal of Psychology, some seventy years ago. It is often not clear exactly whose theory is being refuted in these polemics, for the adversary may turn out to be nothing more substantial than the "common notion," or the imprecise statements of the elementary text. In any case, Jastrow argues that the threshold is an arbitrary point on a distribution of responses, a distribution determined "without any break in the process, i.e., without any threshold."

Detection theory adopts a similar point of view, but brings interesting new tools to bear upon the problem. There appears to be suggestive evidence that the "detection-theory" model fits the discriminatory judgments that a person makes in certain kinds of situations, particularly when he is trying to dig a signal out of a noise. The model itself is a rich and interesting mathematical development. Although it was initially invented in another context, its potential isomorphism with a certain range of perceptual behavior opens up exciting possibilities for a mathematically "sophisticated" kind of psychophysics. The danger, of course, is that the formal model may steal the show. To many people the model may prove to be more interesting than any substantive, empirical outcome. As a matter of fact, some further investigations of the mathematical model have led Mathews (1960) to question the relevance of the mathematical theory to human perception.

Detection theory presumably applies, if at all, only to those situations in which the experimenter can decide whether the observer, when he categorizes a signal as present or absent, is right or wrong. Many experiments on absolute sensitivity are of this sort. The experimentor either turns on a signal or he does not; the observer says yes or he says no. Under other procedures the experimenter may present a signal in a certain location in time or space and force the observer to name or to "guess" the location. This procedure of "forced location," as I should like to call it, is particularly congenial to the detection experiment.

On the other hand, neither the model nor the procedure is readily applicable to many of the "threshold" problems that interest us. The *transition* threshold from cold to warm is one of many obvious instances. When different temperatures are applied to the skin, the observer can categorize them as warm or cold (or perhaps neutral), but the experimenter cannot score him right or wrong. Furthermore, the observer always knows when a stimulus has been presented, so that "catch trials" become impossible. Nevertheless, in the ordinary meaning of the term, there is certainly a threshold here, that is, a point or region of transition from warm to not warm, or cold to not cold. Incidentally, this categorizing of thermal stimuli illustrates the sense in which threshold determination may be regarded as nominal scaling (see Stevens, 1958).

Granted that transition thresholds are not grist for the mill of detection theory, what can we say about other threshold problems? In those experiments to which the detection paradigm is applicable, have we done away with the threshold? There is a manifest conflict here between those who say yes and those who say no, but this divergence may be more apparent than real. It may reduce to the simple proposition that when sufficient noise is present, either internally generated or externally added, a no-threshold model fits the behavior; but when noise is absent, the threshold emerges as an all-or-none step function of some kind. Noise, in other words, does not abolish the threshold; it merely obscures it.

This possibility suggests that our experimental strategy should include two lines of action. We should continue to explore the fertile and heuristic domain of detection theory (because signals often do in fact occur in noise), and we should study methods for reducing the noise in our experiments on differential sensitivity in order to see how the nervous system operates on pure signals, unobscured by noise. A complete suppression of noise may not be possible, of course, but a sufficient reduction may be achieved to allow a quantal step function to manifest itself in the action of the sensory system.

There is a suggestive analogy between this problem and the problem of the nature of electricity. When, after years of effort, R. A. Millikan finally succeeded in suppressing enough sources of noise in his oil-drop experiment, he was able to show that the charge on the electron is not normally distributed, as some evidence had suggested, but has a fixed, all-or-none value.

The threshold step function

How to reduce the noise in a discrimination process to a level low enough to reveal the "grain" in the sensory "continuum" is not easy to specify in advance. Those experiments that have been successful have seemed to involve a carefully contrived arrangement designed to make the observer's task as easy as possible, plus a fortunate selection of observers capable of maintaining an unwavering attention over extended periods of time. It is idle to expect that all observers will prove equal to the demands of this task.

Examples of the kinds of psychometric functions that can be obtained when competent observers are subjected to what appears to be an adequate method are shown in Fig. 1 (sample curves from Stevens, Morgan, and Volkmann, 1941). These six observers had each practiced for an hour or two before these data were recorded. After a couple of hours' work with two other observers the trials were discontinued, because these two observers failed to "settle down" to the point of giving consistent results. The general procedure was as follows.

The observer, comfortably seated in a quiet room, was asked to press a key whenever the steady, continuous 1000-cycle tone in an earphone appeared to change in frequency. Every 3 seconds an increment lasting 0.3 sec was added to the frequency of the tone. The increments were presented in series of 25, all the same size. The observer then rested for a few moments before another series of 25 increments was presented. Ordinarily about 200 judgments were obtained at each sitting.

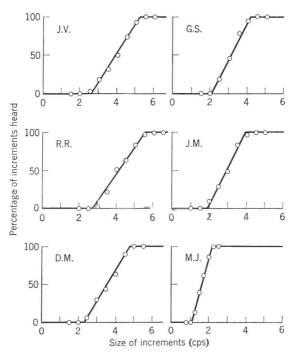

Fig. 1. Functions relating percentage of increments heard to the size of the increment in cycles per second for six different observers. To a steady, continuous reference stimulus (1000 cps), an increment lasting 0.3 sec was added every 3 sec. Each point is the average of 100 judgments. The lines were drawn to fit the points and at the same time to satisfy the slope requirement that the intercepts at 100 and 0 per cent should stand in the ratio of 2 to 1. The size of the NQ, measured in stimulus terms, is the intercept at 0 per cent.

Some observers found it easier to keep oriented, and to know precisely when to listen, with the aid of a light that flashed on midway between successive increments. Others found the light distracting, and for them it was turned off. (For further details, see Stevens, Morgan, and Volkmann, 1941.)

The psychometric functions in Fig. 1 have three characteristic features: (1) between 0 and 100 per cent they approximate straight lines, (2) the slope is inversely proportional to the intercept, and (3) the smallest increment that elicits a response 100 per cent of the time is twice as large as the largest increment that is never heard.

Granted that functions exhibiting these three features can be and have been obtained, what kind of a model can provide a reasonable blueprint of the mechanisms involved?   One attractive possibility rests on an assumption that discrimination is based on functional neural units, each of which operates in an all-or-none fashion, once its "threshold" is crossed.   Instead of calling this unit a quantum, perhaps we can avoid certain confusions with other usage by calling it an NQ, for "neural quantum."   This NQ is probably *not* a peripheral nerve fiber, but more than that we cannot yet specify.   How the NQ model works is as follows.

A steady stimulus of a given magnitude excites, at a particular instant, a certain number of the NQ units, and it does so with a little to spare; that is, there is a small *surplus* excitation p, insufficient to cross the threshold of the unit next in line, but available to be combined with the stimulus increment $\Delta\phi$.   When the increment $\Delta\phi$ is presented, it and the surplus p add together, and if their sum is large enough they excite one or more additional NQ.   Owing to an over-all fluctuation in the sensitivity of the organism—a fluctuation that is large relative to the size of the NQ and slow relative to the time taken for a stimulus increment to be added and removed—one value of the surplus p is as likely as any other.   Now if we measure the size of the NQ in terms of the stimulus increment Q that will just succeed always in exciting it, the value of $\Delta\phi$ that is just sufficient to complement p and thereby excite one additional NQ is given by

$$\Delta\phi = Q - p \tag{1}$$

A given $\Delta\phi$ will excite an additional NQ whenever $\Delta\phi \geqq Q - p$. Since p is distributed uniformly over the interval $0 \leqq p \leqq Q$, the additional NQ is excited a proportion of the time r_1 given by

$$r_1 = \Delta\phi/Q \tag{2}$$

Under the conditions of the experiment that led to the results in Fig. 1, the observer is not able to report when a single additional NQ is brought into play.   The reason is presumably because the over-all fluctuation in his sensitivity causes the "steady" tone to wax and wane—a process that would involve randomly occurring increments and decrements, each equal in size to a single NQ.   Against this fluctuating, steplike background, the observer adopts for his report the criterion of a double quantal jump.   Since the single quantal jump is indistinguishable from the background, the observer ignores it.

If two added NQ are thus needed to produce a response, Eq. 2 becomes

$$r_2 = \left(\frac{\Delta\phi - Q}{Q}\right) = \frac{\Delta\phi}{Q} - 1 \tag{3}$$

Or in terms of the percentage R of increments that an observer should be able to detect, we have

$$R = 100 \left(\frac{\Delta\phi}{Q} - 1\right) \tag{4}$$

This is the equation that gives a good account of the data in Fig. 1.

Applicability of the NQ model

The foregoing brief account, although it leaves many interesting questions unanswered, should be sufficient to make it clear that there is no necessary conflict between the concept of a quantal threshold step and the behavior of an observer confronted with a signal immersed in noise. Noises of many kinds and from many sources can obscure the quantal function, and in many experiments these sources of variability have turned curves like those in Fig. 1 into sigmoid forms resembling normal ogives. In practice, the sigmoid function is the more typical outcome; the rectilinear function seems to require very special procedures and the services of competent, well-motivated observers. The prediction of Eq. 4 stands as the limiting case—to be approached closer and closer as more and more sources of noise are neutralized.

The simple, well-known procedure of "constant stimuli" in which two stimuli, a "standard" and a "variable," are presented for judgment with a time interval between them is not adequate to produce the NQ functions. The time interval upsets the proverbial applecart by allowing room for the observer's sensitivity to change between standard and variable. Another procedure that ensures against rectilinear functions is the averaging of data from different observers, or sometimes even from different experimental sessions. This is because the average of two different rectilinear NQ functions will have a sigmoid form.

Miller's study (1947) with white noise demonstrates the important fact that the disrupting noise need not be in the nature of a background but may be the signal itself. In an earlier study with pure tones (Miller and Garner, 1944), impressive rectilinear functions consistent with the NQ hypothesis had been obtained, but when the signal was changed to a white noise the functions became ogives. It is important to note, however, that the "slopes" of Miller's ogives behaved as the NQ hypothesis predicts.

It should be noted that, since the NQ model assumes a continuous fluctuation in the over-all sensitivity of the organism, rectilinear psychometric functions should not result from measurements of the absolute threshold. There we should find sigmoid functions, reflecting such factors as the time distribution of the over-all sensitivity, the fluctuations inherent in the stimulus (for example, light quanta), or some combination of these.

In view of the demands laid down by the rigorous requirements for obtaining straight, clean NQ functions, it is not surprising that some authors have failed to find them. (For a review of the negative evidence, see Corso, 1956.) The surprising thing, perhaps, is that all in all so many satisfactory NQ functions have been recorded, beginning with those by Békésy in 1930 (see Stevens and Davis, 1938, pp. 145–147, and Békésy, 1960).

Although most of the work on the NQ has utilized acoustic stimuli, a study by Mueller (1951) gave strong evidence of NQ functions in vision. Mueller was not concerned with this issue, but his procedure happened to duplicate the principal requirements of the method of quantal increments. The result was that he obtained some 45 psychometric functions, most of which can be well described by straight lines. The slopes of these lines hover closely around the slope predicted by the NQ hypothesis. It is important to note that Mueller presented the various sizes of increments in a random order, not in groups of 25 identical increments. Although randomization of increments did not seem to work for Miller and Garner (1944), it may be possible under some circumstances to use this procedure.

Remaining problems

The questioning of the threshold concept by some of the workers in detection theory has served to revive a lively concern for the issue of whether an all-or-none step function can be found in the process of discrimination. This is all to the good, but interest in a problem is not a solution, and the question remains: how can we make the case for the NQ as compelling as the case for the unitary, all-or-none charge on the electron? One suspects that statistical tests and other data-threshing procedures will not eliminate the chaff in a manner convincing enough to compel agreement. Maybe the analogy with the story of the electron needs to be pushed even further and the solution placed on public view, much as Millikan exhibited his successful experiments. Anyone who doubted could watch the oil droplet accelerate as it picked up a unit charge representing a single electron. The principal difficulty in devising for the NQ a public demonstration that could be run off on demand would presumably lie with the problem of providing a suitable array of observers. The difference between a human observer and an electron is that the human observer is human. At any moment he may not keep his mind on his work—and thereby spoil the experiment.

Nevertheless, given a well-designed experiment and an observer capable of producing NQ functions with reasonable consistency, there are many tests that suggest themselves. Neisser (1957) addressed himself to several of these questions (sequential effects, length of runs, effects of catch trials, and so on), but many more parameters of the procedure deserve investigation. The steplike threshold process —the NQ as we have called it—is of sufficient concern in psychophysics and neurophysiology to justify a thorough search into the conditions of its realization.

References

Békésy, G. v. *Experiments in Hearing.* New York. McGraw-Hill, 1960.

Corso, J. F. The neural quantum theory of sensory discrimination. *Psychol. Bull.,* 1956, **53**, 371–393.

Jastrow, J. A critique of psycho-physic methods. *Amer. J. Psychol.,* 1888, **1**, 271–309.

Licklider, J. C. R. Three auditory theories. In S. Koch (Editor), *Psychology: A Study of Science. Study I: Conceptual and Systematic. Vol. II: Sensory Perceptual, and Physiological Formulations.* New York: McGraw-Hill, 1959.

Mathews, M. V. Mathematical detectability limits and signal perception. *J. acoust. Soc. Amer.,* 1960, **32**, 931 [Abstract].

Miller, G. A. Sensitivity to changes in the intensity of white noise and its relation to masking and loudness. *J. acoust. Soc. Amer.,* 1947, **19**, 609–619.

Miller, G. A., and W. R. Garner. Effect of random presentation on the psychometric function: Implications for a quantal theory of discrimination. *Amer. J. Psychol.,* 1944, **57**, 451–467.

Mueller, C. G. Frequency of seeing functions for intensity discrimination at various levels of adapting intensity. *J. gen. Physiol.,* 1951, **34**, 463–474.

Neisser, U. Response-sequences and the hypothesis of the neural quantum. *Amer. J. Psychol.,* 1957, **70**, 512–527.

Stevens, S. S. Problems and methods of psychophysics. *Psychol. Bull.,* 1958, **54**, 177–196.

Stevens, S. S., and H. Davis. *Hearing: Its Psychology and Physiology.* New York: Wiley, 1938.

Stevens, S. S., C. T. Morgan, and J. Volkmann. Theory of the neural quantum in the discrimination of loudness and pitch. *Amer. J. Psychol.,* 1941, **54**, 315–335.

40

WALTER A. ROSENBLITH

Center for Communication Sciences, Research Laboratory of Electronics
Massachusetts Institute of Technology

Editor's Comment

The chairman of the organizing committee of a symposium who is also the editor of the symposium volume has already had his say. If at this stage he prefers not to introduce new data from his own laboratory, he may feel tempted to summarize or explain the contents of the chapters, to pick trivial arguments with the contributors, or to pontificate on the great issues. The discussion that follows may partake of all these failings, no matter how great the effort to avoid them.

This comment owes much to the oral and written contributions of the participants. It hardly seems practical, however, to acknowledge in every instance the intellectual debt of the editor, especially since this debt may be compounded by partial misunderstandings or even distortions. Thus the absence here of footnotes, citations, or scholarly qualifiers should not be taken as a pretense to originality.

Throughout the animal kingdom, communication, whether within a species or between an organism and its external and internal environment, involves a multiplicity of sensory systems. These systems vary widely in sensitivity, resolving power, specificity, complexity, and adaptability.

Though it may seem trivial to state that every species is characterized by sensory communication problems of its own, such a view provides a useful corrective against too anthropocentric a bias, especially now when there is so much emphasis on the communication problems of man and his machines. In the course of evolution some species hit upon rather specialized solutions to their sensory problems. Though the study of these specialized sensory systems is by no means a study of curiosa, research into these most "seeing" alleys of evolution, where perception is marvelously acute, does not of itself lead to understanding sensory communication in its most general and symbolic aspects. Let me admit to a prejudice: in my view the arrow of evolution points toward both more highly developed nervous systems and increasingly complex tasks of communication. The properties that make language and other symbol systems effective tools of communication are not likely to be discovered in the nerve nets of coelenterates.

The search for principles of sensory communication seems to call for a mixed strategy, which in turn accommodates itself to a host of tactical solutions, as is apparent from the foregoing chapters. Thus there is ample freedom of choice regarding the aspects of sensory communication that an investigator may elect to work on. He may concern himself with the unique properties of sense organs, the neural encoding of sensory events, or the representation of properties of sensory stimuli in neural structures; on the other hand, he may choose to study interactions between sensory modalities, or to emphasize selective filtering operations, such as attending; he may search for relations between the detectability of a stimulus now present and the response to it that paid off in the past; he may select for study the behavior of a single cell, a particular organism, or a single species. But no matter how he partitions the sensory communication space, the investigator must sooner or later come to grips with the issue of biological veridicality. He then faces the question: How natural should his paradigms be?

Students of sensory processes have traditionally belonged to the larger tribe of students of input-output or stimulus-response relations. In the physical sciences, and often perhaps in engineering, enough is known about even the blackest of boxes so that reasonably critical experiments can be designed. In psychology and physiology, the situation is not so simple. Human observers, especially when they are laboratory inmates, usually accept with docility instruction in what responses they are to emit and what aspects of the stimulus they are to heed. Though they may acknowledge that instructional "set" affects the outcome of their experiments, some psychophysicists proceed as if they were somehow able to dispose of all distracting, contextual, and accessory influences, in the belief that the laws of pure sensation will thus be distilled. In their search for these laws, psychophysicists have amassed a mountain of useful data. The better their apparatus and the more standardized their methods, the more reliable are the S-R relations established within the confines of their laboratories. Once an experimental procedure is specified, thresholds can be measured for some senses with an accuracy that approaches that obtainable in some of the physical sciences. More recently, scaling procedures have yielded a psychophysical law whose generality is impressive within and across sense modalities. But such data, no matter how general or precise, deal only with isolated aspects of the organism's total sensory performance. When the attempt was made to predict how man's resolving power might be related to his ability to make absolute judgments, it became painfully obvious that we had no comprehensive theory to guide us. Puzzled experimenters had to run a whole gamut of experiments before they were able even to formulate the nature of the theoretical gap.

Physiologists, neuroanatomists, and neurophysiologists have been less stimulus-conscious than their psychophysical confrères. The difficulties encountered in attempting to get reliable responses from their physiological preparations scarcely warranted great refinement in the measurement of stimulus parameters. The biological training of these scientists had stressed the importance of studying sensory functions in a way that would approximate performance under natural conditions. Hence, snapping the fingers, whistling, or pinching the animal seemed to constitute both natural and adequate stimuli. But the search for stable responses soon pointed up the advantages of both anesthesia and electrical stimulation, and much knowledge of sensory pathways was uncovered by these means. Still, a comparison of maps of sensory areas that were current a scant fifteen years ago with contemporary maps from "more physiological" preparations proves highly instructive: the "wiring diagrams" have become progressively more complex, but, despite many beautiful experiments, the biological significance of the increasing multiplicity of representation remains yet to be established.

The foregoing remarks raise a query. We now know how to specify stimuli accurately, how to obtain stable responses, and how to establish a variety of psychophysical input-output relations. This know-how we deem necessary; but is it sufficient to understand the capability of an organism to partake in sensory communication?

Shannon's now classical model of a communication system (consisting of a source, an encoder, a channel, a decoder, and a receiver) deals with messages that flow over noisy channels between a source and a user. The source and the user obviously need to have a code in common. A code common to members of a given species is a critical element of the biological communication process. In one sense, each species represents a variation with a rather "particularized need-to-know." Some species are more thoroughly precoded than others; their message set is genetically determined, though certain releaser mechanisms are activated only during maturation. Other species, though also limited in their biological potentialities, are more moldable by learning and even by verbal instruction. Thus each species has a characteristic set of communication problems to solve, which relate to the search for food, the choice of a mate, the care of its young, or even the realm in which to manifest its exploratory behavior. To separate the study of sensory mechanisms from the substrate of communication problems to which they are biologically beholden may lead to severe distortions, a point that comparative physiologists and ethologists have repeatedly made.

Measures of naturalness are notoriously hard to come by, as is attested by those who have tried to assess the performance of high-fidelity equipment. Looking for principles of sensory communication is like steering between Scylla and Charybdis. In this instance, Scylla symbolizes the traditional view of the uniqueness of each species and each organism. Charybdis is then the current doctrinaire simplification by which experiments with repeatable outcomes, no matter what their biological veridicality, are adequate to reveal the organization of multistage and multipurpose systems by means of which organisms transact their sensory commerce.

We can hardly expect that a theory of sensory communication encompassing the several sense modalities will be the fortuitous by-product of the discovery of common features in the behavior of neuronal elements in sensory systems. The over-all performance of the entire organism in a biological context needs to be examined conjointly with the detailed mechanisms.

When biologically significant stimuli impinge upon reactive organisms in sequences patterned in both space and time, appropriate transducers assist in coding these patterns into spatiotemporally organized sequences of action potentials. The way in which an organism's nervous system processes these neural signals depends upon the stimulus pattern, other events in the organism's environment, and the organism's "needs," as well as the response behavior that it expects to emit. In other words, at least in higher organisms, physically identical stimuli do not give rise to invariant mosaic images on a cortical mirror. Instead, the processing of the incoming data consists in a variety of interdependent computational subroutines that are carried out both in parallel and in series. Such data processing is in no sense stereotyped: the availability of particular neural structures, as well as their relevance to the organism's task, will largely determine the role they will play in a given operation.

Perhaps we could grasp the functional neuroanatomy of sensory performance better if we were able to relate more meaningfully two broad classes of elements in neural wiring diagrams: those primarily concerned with furnishing coded representations of stimulus properties and those that are largely task-oriented. The extent to which given subsets of elements are redundant, or are capable of switching allegiance, needs to be elaborated before a model can be achieved that is both structurally and functionally adequate. At present we can only surmise that these two classes of elements or subsystems control each other in a flexible manner (a loud enough bang will divert attention from even the most captivating vista). Whatever hierarchical organization exists may be described in terms of feedback loops. At a level of evolution at which there exists significant encephalization, intensive sensory processing seems to take place as if the brain specified to the ear the acoustic events on which it would like a report, while at the same time the other sense departments were being advised, "Now is hardly the time. . . ."

However, those inputs whose information handling has temporarily been given a low priority contribute indirectly to the very sensory performance that they might be expected to hinder. In order for the organism to remain vigilant, the nervous system would seem to require moderate traffic over its sensory channels. Whether we attribute to this traffic a tonic or a stabilizing influence, it provides the circumstances in which the system can operate in a task-oriented and reasonably illusion-free manner. (Perhaps these considerations relate to the willingness of animals, including those young and human, to emit behavior for the sake of sensory rewards.)

One more point may be relevant. Recent studies of eye motion and eye-hand coordination underline the role played by these active interactions with the environment in the handling of sensory information. This *Abtastung* of the environment, which, incidentally, provides a mechanism of selective emphasis, is an instance of von Holst's *Reafferenz* principle. It enables the organism to process sensory data realistically by matching its expectations against the available input, and to check the validity of a given analysis of sense impressions by means of a complementary synthesis procedure.

An adequate theory of sensory function implies an adequate theory of brain function. And an adequate theory of brain function in its turn requires that the nervous system's behavioral repertory be predictably related to the behavior of the elements that compose it.

Kinetic gas theory is the classical example of an economical model in which a few postulates and statistical distribution functions bridge the gap between the molecules and the over-all properties that characterize the gas. Hardly anyone has seriously attempted to treat the entire nervous system as if it were a neuron gas. Instead, there has been an understandable reluctance to deal theoretically, or even experimentally, with the properties of populations or assemblies of neurons.

A passage by Adrian, written in the early 1930's is still the credo of the majority of neurophysiologists: "The nervous system is built up out of specialized cells whose reactions do not differ fundamentally from one another or from the reactions of other kinds of excitable cell. They have a fairly simple mechanism when we treat them as individuals: their behavior in the mass may be quite another story, but this is for future work to decide."

Today even the mechanism of the individual neuron has ceased to be fairly simple. Actually, computers have taught us that unless we know at least in outline the structural connectivity and program, it is hazardous to predict the behavior of a system from the properties of its individual components. We cannot limit ourselves to networks that are either wholly deterministic or wholly random, and whose elements are equally extreme. The specificity and plasticity of assemblies of neurons are based on both hierarchical and stochastic properties. No estimate of the information-handling capacities of such assemblies is meaningful unless we can enumerate the ensemble of possible states. As a matter of fact, we must do more: we must know which states are "forbidden," or impossible, for these highly ordered structures, and which states are "degenerate," i.e., not discriminably different. (The question of how a many-component system monitors its own state or even the state of one of its subsystems is, of course, without answer at this time.) At present we have neither instrumentalities nor concepts to assess the combined patterns of all-or-none and graded activities in a nervous system, or even in one of its subdomains.

It is in this respect that computers promise to contribute most effectively to rational model-making. Computers can simulate the behavior of networks or assemblies of neuronlike elements, or neuroids. Plausible properties can be postulated for these neuroids, and they can then be interconnected in prescribed ways. We can examine the behavior of such neuroidal networks as their geometrical and neuroidal properties are varied and assess the interaction of these properties with state and stimulus variables. From such experiments there may emerge reasonable measures of performance for these networks, as well as concepts for group properties that are commensurable with concepts such as excitability, inhibition, and refractoriness, all of which are really defined for single units only. Experiments of this kind may provide us with catalogues of interaction models, whose relevance we should be able to test in actual nervous systems. It may soon become possible to compare these models with observations on aggregates of neural elements *in vitro*, kept under strict physicochemical control.

From these varied approaches a mathematically oriented "systems neurophysiology" should emerge, whose concepts will attempt to make processes at various levels of biological organization reducible to common measure. In order to achieve this commensurability, we must expand the areas in which we make quantifiable observations in experiments that are biologically meaningful.

It is hardly possible to discuss the handling of sensory information without emphasizing temporal aspects of these operations. The preceding chapters contain numerous "time markers" along the route that leads from a discrete sensory stimulus to a discrete behavioral response. But we are far from achieving a complete or even coherent account of the temporal dimension of sensory performance.

We possess a wealth of data on reaction times to discrete stimuli (even to closely spaced pairs of stimuli); we have some knowledge of visual tracking performance; and an appreciable amount of data has been collected on temporal resolving power, among which we may want to include data on binaural localization. We are thoroughly familiar with the characteristic latencies of the "early" events that our gross electrodes record along the afferent pathways; these events, which are moderately sensitive to stimulus intensity, turn out to be tightly locked in time to the delivery of effective stimuli that are themselves most often transients. We are just beginning to study systematically the "later" physiological events, that is, events that occur along the time dimension in closer proximity to the behavioral response than to the stimulus. We are still uncertain whether these later events, some of which may be related to motor activity, can be assumed to be time-locked to either stimulus or response. Nor do we know how to interpret the time patterns of the activity evoked in single units; we are puzzled particularly by the "late responders" that we sometimes encounter among the single units, even at early stations of the afferent pathway. We know that physiological on and off responses outlast the transient stimuli that evoke them, and we know also that the ensuing after-effects of such responses are, almost anywhere in the central nervous system, too complex to be subsumed under the heading of refractoriness.

On the behavioral side, we lack data on response times for comparable sensory tasks in different species or in "modified"* organisms. We need to know more about how the time required to make a given sensory discrimination varies at successive stages of learning. We must gain a clearer understanding of how the set of possible stimulus-response transformations modifies the distribution of response times that goes with a given level of performance. We need specifically designed experiments to investigate how reaction time is affected by both the size and the nature of stimulus and response ensembles, and what role discriminability and recoding play in this context. We are beginning to acquire systematic data on man's ability to order sensory events within a given modality and across modalities; hopefully, these data will provide some clues about how the nervous system organizes and distinguishes sequences of sensory events and is able to switch between such sequences.

On the physiological side, we need more information on the timing of inhibitory phenomena, as well as the timing of events in efferent and descending pathways. We have as yet no adequate picture of the temporal relations that exist between events in those structures that furnish coded representations of stimulus properties and those that are task-oriented. We need to examine whether there are changes in the relative timing of certain neural aggregates that depend on the importance of the sensory message that is to be handled. We know already that, in monkeys at least, both the level and the speed of sensory performance can be influenced via the reticular formation, and there are data extant that relate reaction time to certain aspects of the human EEG. Here again, converging electrophysiological and behavioral evidence collected from the same organism may facilitate the discovery of critical interrelations, particularly when an organism's state is in some measure controlled by involving it in a biologically meaningful task.

In the future, studies of some temporal aspects of sensory communication are likely to make contact with studies of short-term memory. Some signposts along this path are already in existence. For are organisms not capable of "integrating" (albeit imperfectly) stimulus energy at absolute threshold over intervals of up to several seconds? Do not difference limens, for instance, deteriorate as the separation between reference and test stimuli increases? Does not reaction time depend significantly upon the length and variability of the foreperiod?

---

* The term "modified" is here a shorthand notation for split-brain preparations, animals in which parts of the brain have been ablated, human patients who have suffered neurological damage, and so forth.

A theoretical model of the mechanisms that underlie sensory communication must accommodate itself to a variety of boundary conditions, but no model can qualify as satisfactory unless it proves capable of predicting the real-time aspects of the communication process.

Let us be a trifle bold and quite foolish: let us try to imagine what a symposium on Principles of Sensory Communication might be like ten or fifteen years hence. It is hopeless to attempt to foresee the most exciting discoveries, and almost as useless to try to predict the topics that will dominate the discussion. There seems to be only one category of even halfway safe guesses: it may not be unreasonable to anticipate the new fields and approaches from which contributions to such a symposium are likely to come. The problems of neurophysiology and psychophysics will certainly continue to command overwhelming interest. But these problems may find themselves illuminated chiefly from the point of view of molecular biology and mathematics, and by techniques that will have been imported from solid state physics, biochemistry, pharmacology, computer and communications engineering, and the behavioral technology of that period.

By then we shall have learned much about how neural organization and varieties of coding reflect different ways of processing sensory information. We should also possess a clearer view of how events in the autonomic nervous system and other body systems interact with and regulate events in the central nervous system; the role played by sensory receptors imbedded in the central nervous system should prove of particular interest in this respect.

On the psychological side we are likely to find an empirically founded set of mathematical models dealing with a broad spectrum of sensory performance: covering, for instance, threshold phenomena, psychological scales for uni- and multidimensional continua, choice behavior, and even certain aspects of cognitive function. The developmental and geriatric aspects of information processing will have provided a lively area of research.

We shall certainly know a good deal more neurochemistry; the mathematics of complex systems will have made substantial progress; and there will be much new knowledge of those specific molecular mechanisms that underlie the organization of biological materials. Electrodes will have evolved through developments in molecular engineering. Computers will aid both experimenters and theoreticians in a variety of ways, some of which are already perceptible: for example, the processing of electrophysiological data, the simulation of sensory performance, and the simulation of the performance of assemblies of neuroids.

We may in ten or fifteen years have derived some useful generalizations from the widespread experience with man-machine (including man-computer) systems and from exposure to novel sensory environments. We shall certainly be able to telemeter converging data (electrophysiological, neurochemical, behavioral) from organisms while they interact with their environment. It should also be possible to sharpen the formulation of issues that are critical to our understanding of sensory function by incorporating the experimental subject into a "closed loop"—an arrangement wherein stimulus sequences are made contingent upon the physiological and psychophysical responses that the subject emits.

So long as these crude guesses are not interpreted as science fiction or otherwise taken too seriously, they may serve a purpose: to bring into focus the limited nature of our present armamentarium of theoretical and experimental tools.

What then do these 800-odd pages add up to? They constitute a record, rich in subscripts of this day, of a many-pronged search for principles of sensory communication. Nobody would maintain that this search has been entirely successful or, given our present knowledge of the functioning of the brain, that it could have been.

Philipp Frank, the contemporary philosopher of science, once remarked that few problems in nature come "labeled"—that is, attached either to the province of the sciences or that of the humanities. A similar condition can be assumed to hold within the sciences: temporary labeling of problems is necessary for purposes of analysis, but it is perilous to ignore the artificiality of such a procedure. Sensory communication is a textbook example of a problem complex that resists labeling: it pervades evolution, it accompanies the development and the decay of individual organisms, it is inextricably involved with learning, thinking, and consciousness, whatever meaning one may choose to give these terms. Its very nature belies the thought that we can choose a priori those scientific disciplines that are relevant. Thus we have little hope that the future of sensory communication will prove less interdisciplinary than its past. Such an outlook has serious consequences for research and for the education of the contributors to that symposium a decade hence. As one of the participants in the Endicott House Symposium, a distinguished neurophysiologist, put it, "We have . . . reached a point where we must realize that in order to go further in pursuing electrophysiological research we have to face a new organization of our laboratories as well as of our basic training."

INDEXES

Name Index

Ades, H. W., 236, 238, 239, 243, 244, 248, 253, 265, 272, 653, 682
Adey, W. R., 708
Adler, A., 456
Adrian, E. D., 431, 476, 663, 677, 820
Aguilar, M., 385
Ahlander, C., 212
Ahlenstiel, H., 644
Ajmone-Marsan, C., 635
Akert, K., 268, 466
Akimoto, H., 629, 631, 632, 635, 636, 638, 644, 650, 684, 691
Albe-Fessard, D., 586, 587, 589, 601, 608, 617, 620, 622, 678
Alcocer-Cuarón, C., 503, 509, 511, 512
Allanson, J. T., 570
Allen, W. F., 456
Allison, E. G., 208
Alpern, M., 742
Amassian, V. E., 413, 439, 463, 464, 486, 525, 588, 608, 609, 620, 675, 681, 728
von Amerongen, F. K., 709
Anderson, B., 456
Anderson, C. D., 91
Andersson, B., 205, 211
Andersson, S. A. A., 432
Andrew, A. M., 773
Appelberg, B., 438, 456, 504
Arden, G. B., 503, 521, 523, 524, 526, 529, 530, 532, 538–539
Arduini, A., 502, 651, 681, 682, 702
Armington, J. C., 752, 799, 802
Arteta, J. L., 238
Asanuma, H., 680
Ascher, P., 608, 615, 617, 621, 622
Attneave, F., 223, 780, 782, 786
Aubert, H., 665
Aulhorn, E., 654
Autrum, H., 660

Bach-y-Rita, G., 497, 511, 513
Baker, H. D., 744
Ballin, H. M., 691
Bard, P., 407, 414, 728
Bare, J. K., 469
Barlow, H. B., 173, 200, 220, 227, 381, 387, 391, 396, 525, 539, 633, 635, 646, 657–659, 767, 786, 796
Barlow, J. S., 701, 702, 704, 714
Barnes, R. B., 378
Barron, D. H., 478
Bartley, S. H., 642, 647
Baruch, J. J., 53
Batini, C., 523
Battersby, W. S., 744
Battin, R. H., 308
Baumgardt, E., 384, 397
von Baumgarten, R., 537, 588, 629, 631, 642, 647

Baumgartner, G., 200, 602, 620, 627–629, 631–636, 641, 642, 645, 646, 648–650, 652–655, 657–659, 661, 663, 680, 796, 798
Beebe-Center, J. G., 30
Beecher, H. K., 53, 56
Beidler, L. M., 147, 156, 461
Békésy, G. v., 7, 75, 79, 122, 123, 125, 128, 130, 320, 321, 561, 562, 567, 725, 728, 731, 734, 812
Belton, P., 547
Benjamin, R. M., 246, 466
Berglund, A.-M., 44
Berkowitz, E., 500
Berman, A. L., 406, 413, 432, 539, 620, 658, 675
Bertholf, L. M., 358, 371
Beurle, R. L., 595, 721, 722
Bice, R. C., 76, 84
Biddulph, R., 78
Birdsall, T. G., 96
Bishop, G. H., 126, 127, 139, 647, 678, 793
Bishop, P. O., 521, 522
Björkman, M., 43, 44
Black, R., 208
Blanchard, J., 744
Blix, M., 205
Bloom, G., 152
Blundell, J., 540
Bocca, E., 271
Bohm, E., 521
Bonnet, V., 238, 683
Borenstein, P., 251, 255, 587–589, 607, 608, 620
Boring, E. G., 777, 779, 805
Bornschein, H., 523, 538
Bouman, M. A., 382, 384–388, 390, 393–396, 398, 740, 744
Boynton, R. M., 744, 746, 747
Branch, C. L., 525, 620, 680, 681
Brassfield, C. R., 720
Brazier, M. A. B., 681, 701, 702, 707, 708
Bremer, F., 238, 243, 244, 248, 253, 432, 456, 503, 504, 523, 537, 628, 635, 650, 676, 678, 682–686, 688, 690
Brentano, F., 3
van den Brink, G., 384, 385, 388, 390, 393
Broadbent, D. E., 93
Broca, A., 742
Brock, C. G., 660
Brodmann, K., 404
Bromiley, R. B., 237, 248
Brookhart, J. M., 621, 677, 692
Brooks, V. B., 199, 681
Brouwer, B., 524
Brown, C. R., 750
Brown, P. K., 370, 372
Brown, P. S., 370
Brücke, E., 665
Brueckner, A., 727

827

Subject Index